SOCIETÀ ITALIANA DI FISICA

RENDICONTI

DELLA

SCUOLA INTERNAZIONALE DI FISICA

"ENRICO FERMI"

CXLVIII Corso

a cura di F. De Martini e C. Monroe

Direttori del Corso

VARENNA SUL LAGO DI COMO

VILLA MONASTERO

17 – 27 Luglio 2001

Computazione e informazione quantistica

2002

SOCIETÀ ITALIANA DI FISICA

BOLOGNA-ITALY

ITALIAN PHYSICAL SOCIETY

PROCEEDINGS

OF THE

INTERNATIONAL SCHOOL OF PHYSICS

"ENRICO FERMI"

Course CXLVIII

edited by F. De Martini and C. Monroe

Directors of the Course

VARENNA ON COMO LAKE

VILLA MONASTERO

17 – 27 July 2001

Experimental Quantum Computation and Information

2002

AMSTERDAM, OXFORD, TOKYO, WASHINGTON DC

Copyright © 2002 by Società Italiana di Fisica

ISBN 1 58603 270 4 (IOS Press)
ISBN 4 274 90536 5 C3042 (Ohmsha)
Library of Congress Catalog Card Number: 2002110640

Production Manager
A. OLEANDRI

Copy Editor
M. MISSIROLI

Publisher
IOS PRESS
Nieuwe Hemweg 6B
1013 BG Amsterdam
The Netherlands
fax: +31 20 688 33 55
e-mail: order@iopress.nl

Distributor in the UK and Ireland
IOS Press/Lavis Marketing
73 Lime Walk
Headington
Oxford OX3 7AD
England
fax: +44 1865 75 0079

Distributor in the USA and Canada
IOS Press, Inc.
5795-G Burke Center Parkway
Burke, VA 22015
USA
fax: +1 703 323 3668
e-mail: iosbooks@iospress.com

Distributor in Germany, Austria and Switzerland
IOS Press/LSL.de
Gerichtsweg 28
D-04103 Leipzig
Germany
fax: +49 341 995 4255

Distributor in Japan
Ohmsha, Ltd.
3-1 Kanda Nishiki-cho
Chiyoda-ku, Tokio 101
Japan
fax: +81 3 3233 2426

Supported by the European Commission, Research DG, Human Potential Programme,
High-Level Scientific Conferences HPCF-CT-2000-00164

Supported by Consiglio Nazionale delle Ricerche (CNR)

Supported by Istituto Nazionale di Fisica Nucleare (INFN)

Supported by Istituto Nazionale per la Fisica della Materia (INFM)

INDICE

F. De Martini and C. Monroe – Introduction . . . . pag. XVII

Gruppo fotografico dei partecipanti al Corso . . . . » XX

ATOMS AND CAVITY QED

S. Haroche, J. M. Raimond and M. Brune – Entanglement, complementarity and decoherence in cavity QED experiments . . . . » 3

  1. Foreword . . . . » 3
  2. Introduction . . . . » 4
  3. The cavity QED entangling machine . . . . » 8
    3˙1. The circular Rydberg atoms . . . . » 8
    3˙2. The superconducting cavity . . . . » 9
    3˙3. The atomic Ramsey interferometer . . . . » 10
  4. The quantum Rabi oscillation . . . . » 12
    4˙1. Vacuum Rabi oscillation . . . . » 12
    4˙2. Useful Rabi pulses . . . . » 13
    4˙3. Quantum Rabi oscillation in an applied field . . . . » 15
  5. Creation of an EPR pair . . . . » 16
  6. The quantum phase gate . . . . » 19
  7. Absorption-free detection of a single photon . . . . » 22
  8. Engineered entanglement of three quantum systems . . . . » 25
  9. Non-resonant entanglement . . . . » 28
    9˙1. Single-atom index effect and "Schrödinger cat" states . . . . » 28
    9˙2. A complementarity experiment . . . . » 29
    9˙3. Decoherence caught in the act . . . . » 30
  10. Conclusions and perspectives . . . . » 32

A. Kuhn and G. Rempe – Optical cavity QED: Fundamentals and application as a single-photon light source . . . . » 37

  1. Introduction . . . . » 37
  2. A short review of cavity quantum electrodynamics . . . . » 38

2˙1. Optical high-finesse cavities .... pag. 38
2˙2. Field quantization .... » 40
2˙3. Two-level atom coupled to a laser .... » 40
2˙4. Two-level atom coupled to a cavity .... » 41
2˙5. Normal-mode splitting .... » 43
2˙6. Enhanced spontaneous emission .... » 45
3. Three-level atoms .... » 47
4. Adiabatic passage .... » 49
4˙1. Population transfer and Fock-state preparation .... » 52
4˙2. Electromagnetically induced transparency .... » 53
4˙3. Emission of photons .... » 54
5. Single-photon sources .... » 54
5˙1. State of the art .... » 54
5˙2. Cavity-enhanced spontaneous emission .... » 54
5˙3. Vacuum-stimulated Raman scattering .... » 56
5˙4. Loss mechanisms .... » 57
5˙4.1. Non-adiabatic evolution .... » 58
5˙4.2. Spontaneous emission .... » 59
5˙5. Numerical simulation .... » 61
5˙6. Experimental realization .... » 61
6. Summary and outlook .... » 64

PH. H. BUCKSBAUM – Rydberg wave packets in quantum information science » 67

1. Introduction .... » 67
2. Basic properties of Rydberg atoms .... » 68
3. Forming superposition states .... » 73
4. Measuring sculpted Rydberg wave packets .... » 76
4˙1. Amplitude measurement .... » 77
4˙2. Phase measurement .... » 78
5. Quantum information applications .... » 79
5˙1. Data storage and retrieval .... » 79
5˙2. Data base searches with strong THz pulses .... » 81
5˙3. Entanglement in Rydberg atoms .... » 83
6. Conclusion .... » 85

H. WALTHER – Cavity QED—experiments with atoms and ions .... » 87

1. Introduction .... » 87
2. Conclusion .... » 99

S. L. ROLSTON – Quantum information processing with neutral atoms: Experimental challenges .... » 101

1. Introduction .... » 101
2. Neutral atom qubits .... » 102
3. Initialization .... » 109
4. Gates .... » 111

5. Coherence .... pag. 114
6. Measurement .... » 116
7. Summary .... » 116

BEC

E. A. CORNELL – Bose-Einstein condensation: The cure for decoherence arising from inhomogeneity .... » 121

E. ARIMONDO – Laser ionization of a Rb Bose-Einstein condensate towards quantum computation .... » 133

1. Introduction .... » 133
2. Ions within a BEC.. » 134
  2˙1. Ion motion .... » 134
  2˙2. "Pauli blockade .... » 136
3. Experimental setup .... » 136
4. Experimental results .... » 138
5. Toward quantum gates .... » 143

S. BURGER, F. S. CATALIOTTI, F. FERLAINO, C. FORT, P. MADDALONI, F. MINARDI and M. INGUSCIO – Bose-Einstein condensates in a 1D optical lattice .... » 147

1. Introduction .... » 147
2. Experimental setup .... » 148
3. Bose-Einstein phase transition in lower dimensions .... » 150
4. Expansion of a BEC from the combined trap................................... » 152
5. Superfluid dynamics .... » 155
6. Observation of a Josephson current in an array of coupled BECs .... » 157
7. Conclusions .... » 159

IONS

D. J. WINELAND – Trapped ions and quantum information processing .... » 165

1. Introduction .... » 165
2. Trapology .... » 166
  2˙1. RF micromotion .... » 167
3. Ion entanglement .... » 169
4. Motion decoherence.. » 172
  4˙1. High-temperature amplitude reservoir.................................... » 174
  4˙2. "Engineered" zero-temperature amplitude reservoir .... » 177

4˙3.   Motion decoherence in practice . . . .   pag.   179
5.   Entanglement-enhanced quantum measurement: "spin squeezing" . . . .   »   180
6.   Towards a quantum computer . . . .   »   184
    6˙1.   Improving motional coherence . . . .   »   184
    6˙2.   Phase coherence . . . .   »   184
    6˙3.   Internal-state amplitude coherence . . . .   »   185
    6˙4.   Multiplexing . . . .   »   188
7.   Summary . . . .   »   191

F. SCHMIDT-KALER, D. LEIBFRIED, J. ESCHNER and R. BLATT – Towards
quantum computation with trapped Ca$^+$ ions . . . .   »   197

1.   Introduction . . . .   »   197
2.   Scheme of an ion trap quantum computer . . . .   »   198
3.   Linear ion traps . . . .   »   200
4.   Ca$^+$ for quantum computation . . . .   »   202
5.   Laser cooling of trapped ions . . . .   »   204
    5˙1.   Doppler cooling . . . .   »   204
    5˙2.   Sideband cooling . . . .   »   205
    5˙3.   EIT cooling . . . .   »   206
    5˙4.   EIT-cooling of linear ion strings . . . .   »   208
6.   Addressing of individual ions . . . .   »   209
7.   Manipulation of the quantum information . . . .   »   210
8.   Conclusions and outlook . . . .   »   211

PHOTONS

H. ZBINDEN, N. GISIN, G. RIBORDY, D. STUCKI and W. TITTEL – Exper-
imental quantum communication . . . .   »   217

1.   Introduction . . . .   »   217
2.   Tools . . . .   »   217
    2˙1.   Photon counters . . . .   »   217
    2˙2.   Entangled photon pair sources . . . .   »   218
3.   Bell experiments . . . .   »   219
    3˙1.   Energy-time entanglement . . . .   »   220
    3˙2.   A long-distance Bell experiment . . . .   »   223
4.   QKD . . . .   »   225
    4˙1.   Introduction . . . .   »   225
    4˙2.   How QKD works . . . .   »   226
    4˙3.   QKD with faint laser pulses . . . .   »   228
    4˙4.   QKD with entangled photon pairs . . . .   »   228
    4˙5.   Faint laser pulses *vs.* entangled photons . . . .   »   230
5.   Conclusions . . . .   »   231

F. Sciarrino, E. Lombardi and F. De Martini – Delayed-choice entanglement— Swapping with vacuum one-photon quantum states . . . . pag. 233

P. Mataloni, G. Giorgi and F. De Martini – Frequency hopping in quantum interferometry: Efficient up-down conversion for qubits and ebits . » 241

QUANTUM COMPUTATION – QUANTUM MEASUREMENT

D. P. DiVincenzo – Quantum logic gates with semiconductor quantum dots » 251

1.   Introduction . . . . » 251
2.   Quantum dots for quantum computing . . . . » 252
     2'1.   Exchange coupling . . . . » 254
     2'2.   Quantum measurement . . . . » 256
     2'3.   Single-spin rotations . . . . » 258
     2'4.   Exchange-only quantum computing . . . . » 259
     2'5.   Complications and prospects . . . . » 260

J. I. Cirac, Luming Duan and P. Zoller – Quantum optical implementation of quantum information processing . . . . » 263

1.   Introduction . . . . » 263
2.   Basic concepts in quantum information theory . . . . » 265
     2'1.   Introduction . . . . » 265
     2'2.   Entanglement . . . . » 266
            2'2.1.   Entanglement of pure states . . . . » 266
            2'2.2.   Entanglement of mixed states . . . . » 270
     2'3.   Purification . . . . » 272
   · 2'4.   Quantum computing . . . . » 274
            2'4.1.   What is a quantum computer? . . . . » 274
            2'4.2.   Requirements . . . . » 275
            2'4.3.   Error correction . . . . » 278
     2'5.   Quantum communication . . . . » 281
            2'5.1.   Teleportation . . . . » 281
     2'6.   Quantum repeaters . . . . » 283
3.   Quantum information processing with single atoms and photons . . . . » 284
     3'1.   Introduction . . . . » 284
     3'2.   Trapped ions . . . . » 285
     3'3.   The model . . . . » 285
            3'3.1.   Single trapped ion . . . . » 285
            3'3.2.   Ions in a linear trap . . . . » 289
            3'3.3.   Ion trap quantum computer '95 . . . . » 290
            3'3.4.   Ion trap quantum computer 2000 . . . . » 293
     3'4.   Cavity QED . . . . » 295
            3'4.1.   Optical interconnects . . . . » 296

3˙5.   Collisional interactions for neutral atoms .... pag. 299

    3˙5.1.   Entanglement via coherent ground-state collisions .... » 299

    3˙5.2.   Rydberg atoms .... » 302

4.   Quantum information processing with atomic ensembles .... » 306

   4˙1.   Introduction .... » 306

   4˙2.   Interaction of light with atomic ensembles and collective enhancement of the signal-to-noise ratio .... » 307

    4˙2.1.   ΛI-level configuration .... » 308

    4˙2.2.   ΛII-level configuration .... » 310

    4˙2.3.   Four-level configuration .... » 313

   4˙3.   Scalable long-distance quantum communication .... » 317

    4˙3.1.   Entanglement generation .... » 318

    4˙3.2.   Entanglement connection through swapping .... » 320

    4˙3.3.   Entanglement-based communication schemes .... » 322

    4˙3.4.   Noise and built-in entanglement purification .... » 324

    4˙3.5.   Scaling of the communication efficiency .... » 326

   4˙4.   Other applications: Quantum light memory and single-photon source » 327

    4˙4.1.   Quantum light memory .... » 328

    4˙4.2.   Single-photon source .... » 331

   4˙5.   Applications in continuous variable quantum information processing » 332

5.   Summary .... » 336

C. MACCHIAVELLO – Basic concepts in quantum error correction .... » 341

1.   Introduction .... » 341

2.   Reducing errors via symmetrisation .... » 342

3.   Basics of classical error correction .... » 344

4.   Quantum error-correcting codes............................ » 347

   4˙1.   The three-qubit code............................ » 348

   4˙2.   Conditions for quantum error correction.................... » 349

   4˙3.   Quantum bounds.............................. » 352

   4˙4.   Quasi-classical codes .... » 353

5.   Conclusion .... » 355

G. FALCI, R. FAZIO and G. M. PALMA – Geometric quantum computation with Josephson qubits .... » 357

1.   Introduction .... » 357

2.   Geometric phases.................................... » 358

3.   Geometric quantum computation .... » 359

4.   Geometric phases in Josephson nanocircuits .... » 362

   4˙1.   Single-qubit geometric phase .... » 366

   4˙2.   Controlled geometric phase .... » 366

H. MACK, M. BIENERT, F. HAUG, F. S. STRAUB, M. FREYBERGER and W. P. SCHLEICH – Wave packet dynamics and factorization of numbers ... » 369

1.   Talbot effect, revivals and factorization .... » 369

2.   Model of Talbot effect................................. » 371

3. Mathematics of the Talbot effect . . . . pag. 372
    3`1. Free time evolution of a periodic structure . . . . » 372
    3`2. Quadratic phase factors . . . . » 373
    3`3. Integers and half integers of Talbot time . . . . » 374
4. Particle in a box . . . . » 374
    4`1. Propagation . . . . » 374
    4`2. Relation to the Talbot propagator . . . . » 376
5. Time evolution and autocorrelation function . . . . » 376
    5`1. Definition . . . . » 377
    5`2. Quadratic expansion of the energy spectrum . . . . » 377
6. Fractional revivals . . . . » 378
7. Factorization using wavepackets . . . . » 380
    7`1. General principle . . . . » 380
    7`2. Simulation . . . . » 381
8. Gauss sums and factorization . . . . » 382
9. Conclusions and outlook . . . . » 383

G. M. D'Ariano – Tomographic methods for universal estimation in quantum optics . . . . » 385

1. Introduction . . . . » 385
2. Definition of the problem . . . . » 388
3. Homodyne tomography, *i.e.* tomography for *continuous variables* . . . . » 388
    3`1. The balanced homodyne detector . . . . » 388
    3`2. Homodyne tomography . . . . » 389
    3`3. Why the name "tomography"? . . . . » 390
    3`4. Limitations of the Radon transform method . . . . » 390
    3`5. The exact method . . . . » 391
    3`6. Unbiasing noise from nonunit quantum efficiency . . . . » 393
    3`7. Multimode homodyne tomography . . . . » 396
    3`8. Improving statistical errors . . . . » 397
    3`9. Estimating ensemble averages of unbounded operators . . . . » 399
4. Pauli tomography, *i.e.* tomography for *qubits* . . . . » 400
5. Some experimental results . . . . » 401
6. Tomography of a quantum device . . . . » 402
7. Conclusions and future perspectives . . . . » 404

NMR, SQUID, STOPPING OF LIGHT

R. Laflamme, E. Knill, C. Negrevergne, R. Martinez, S. Sinha and D. G. Cory – Introduction to NMR quantum information processing . . . . . » 411

1. Liquid state NMR . . . . » 413
    1`1. NMR basics . . . . » 413
    1`2. A brief survey of NMR quantum information . . . . » 414
2. Principles of liquid state NMR QIP . . . . » 415
    2`1. Realizing qubits . . . . » 415

2˙2. One-qubit gates .... pag. 417

2˙3. Two-qubit gates ....   »   418

2˙4. Turning off the coupling ....   »   419

2˙5. Measurement ....   »   420

2˙6. The initial state ....   »   424

2˙7. Gradient fields ....   »   425

3. Examples of quantum algorithms for NMR ....   »   427

3˙1. The controlled-not ....   »   427

3˙2. Creating a labeled pseudo-pure state ....   »   430

3˙3. Quantum error correction for phase errors ....   »   433

4. Discussion...   »   435

4˙1. Overview of contributions to QIP ....   »   435

4˙2. Capabilities of liquid state NMR ....   »   437

4˙3. Prospects for NMR QIP ....   »   437

Y. YAMAMOTO, T. D. LADD, J. R. GOLDMAN and F. YAMAGUCHI – Solid-state crystal lattice NMR quantum computation ....   »   441

1. Introduction ....   »   441

2. Nuclear interactions and decoherence ....   »   443

2˙1. Nuclear interactions ....   »   443

2˙2. Decoherence ....   »   444

3. Materials for crystal lattice quantum computation ....   »   447

3˙1. Fluorapatite ....   »   447

3˙2. CeP ....   »   448

3˙3. Silicon ....   »   454

4. Gradient design ....   »   457

4˙1. Design considerations ....   »   457

4˙2. Magnetic-field calculation ....   »   458

4˙3. One-dimensional quantum computer ....   »   459

4˙4. Two-dimensional quantum computer ....   »   462

5. Decoupling and recoupling pulse sequences ....   »   463

5˙1. Heteronuclear decoupling concepts ....   »   463

5˙2. Hadamard decoupling and recoupling scheme ....   »   465

5˙3. Efficiency ....   »   466

5˙4. Homonuclear decoupling ....   »   467

6. MRFM readout ....   »   467

7. Scalability...   »   469

7˙1. Number of qubits..   »   469

7˙2. Number of logic gates ....   »   471

8. Conclusion ....   »   471

D. VION, A. AASSIME, A. COTTET, P. JOYEZ, H. POTHIER, C. URBINA, D. ESTEVE and M. H. DEVORET – Superconducting quantum bit based on the Cooper pair box..   »   475

1. Introduction: why superconducting tunnel junction circuits?............   »   475

2. The single Cooper pair box ....   »   479

3. Cooper pair box with orthogonal write and read ports ....   »   483

M. G. Castellano and F. Chiarello – Superconducting devices for quantum logic gates using magnetic flux states .... pag. 493

  1. Introduction .... » 493
  2. The building blocks: superconducting phase, magnetic flux quantization, Josephson effects .... » 494
      2·1. Superconducting phase .... » 495
      2·2. Magnetic flux quantization .... » 495
      2·3. Josephson effects .... » 495
  3. Basic Josephson devices .... » 496
      3·1. Josephson junctions .... » 496
      3·2. The superconducting interferometer .... » 498
      3·3. The rf-SQUID .... » 500
      3·4. The double SQUID .... » 500
  4. Qubit implementation .... » 502
  5. Qubit readout .... » 503
  6. Logic quantum gates .... » 505
  7. Decoherence .... » 507

M. Fleischhauer and C. Mewes – "Stopping" of light and quantum memories for photons .... » 511

  1. Introduction .... » 511
  2. Single-atom cavity QED .... » 512
  3. Temporary storage of photons in many-atom systems: electromagnetically induced transparency and slow light .... » 514
  4. Light stopping by adiabatic rotation of dark-state polaritons .... » 518
      4·1. Time-dependent group velocity .... » 518
      4·2. Polariton picture of EIT and "stopping" of light .... » 519
      4·3. "Stopping" of light .... » 520
      4·4. Adiabatic condition and non-adiabatic corrections .... » 522
  5. Quantum memory and decoherence .... » 524
  6. Summary .... » 528

Elenco dei partecipanti .... » 531

Introduction

Quantum mechanics is heralded as the most successful and revolutionary scientific theory in the 20th century, despite its bizarre features that defy our everyday experience. This dichotomy has fascinated students of physics over the last 70 years. The recent advent of quantum information science has once again brought quantum mechanical foundations to the forefront. We are now in a position to revisit some of the fundamental principles of quantum mechanics in the laboratory, as impressive advances in experimental techniques are making yesterday's gedanken experiment today's reality.

This Fermi Summer School of Physics on "Experimental Quantum Computation and Information" represents a primer on one of the most intriguing and rapidly expanding new areas of physics. In part, the interest in quantum information (QI) science is due to the discovery that a computer operating on quantum mechanical principles can solve certain important computational problems exponentially faster than any conceivable classical computer. But this interest is also due to the interdisciplinary nature of the field: the rapid growth is attributable, in part, to the stimulating confluence of researchers and ideas from physics, chemistry, mathematics, information theory, and computer science.

Physics plays a paramount role in QI science, as we realize that computing is itself a physical process subject to physical laws. The incredible growth of classical computers and information processors in the 20th century stems from Turing's notion that a computer is independent of the physical device actually being used; be they relays, vacuum tubes, or semiconductor transistors. As we strive to build useful quantum information processors into the 21st century, we thus look for any physical system that obeys the laws of quantum mechanics, from single photons and atoms to quantum superconducting devices. These Fermi lectures take us on a journey through these and other promising current experimental candidates for QI processing, spanning quantum optics and laser physics, atomic and molecular physics, physical chemistry, and condensed-matter physics. While this broad coverage of experimental physics represents a challenge to the student, such an appreciation of these fields will be critical in the future success of quantum technology. Indeed, the most exciting feature of QI science is that the technology ultimately leading to a quantum processor is likely presently unknown.

In all, twelve lecturers, thirteen seminar speakers, and forty-nine students from eleven countries participated in the School. We benefited from an idyllic location, a cooperative atmosphere, and a friendly interchange of scientific ideas between lecturers and students. It is our hope that this event will remain in the memories of its participants for many years to come, perhaps as an educational bookmark, or even as a moment when one has learned something new while enjoying the stunning scenery of Lago di Como and the surrounding mountains.

The organization of any school, and particularly one covering so many different fields in physics, cannot be accomplished without significant and generous assistance from individuals and agencies. We are especially indebted to the Italian Physical Society President Prof. Franco Bassani for his oversight of the school, and conference administrator Barbara Alzani for her continuous effort in making the Summer School run smoothly. We are also grateful to Ramona Brigatti, Francesca Fusina, and Matteo Pozzi for their efficient organization of activities in and near Varenna.

F. DE MARTINI and C. MONROE

Società Italiana di Fisica

SCUOLA INTERNAZIONALE DI FISICA «E. FERMI»

CXLVIII CORSO - VARENNA SUL LAGO DI COMO

VILLA MONASTERO 17 - 27 Luglio 2001

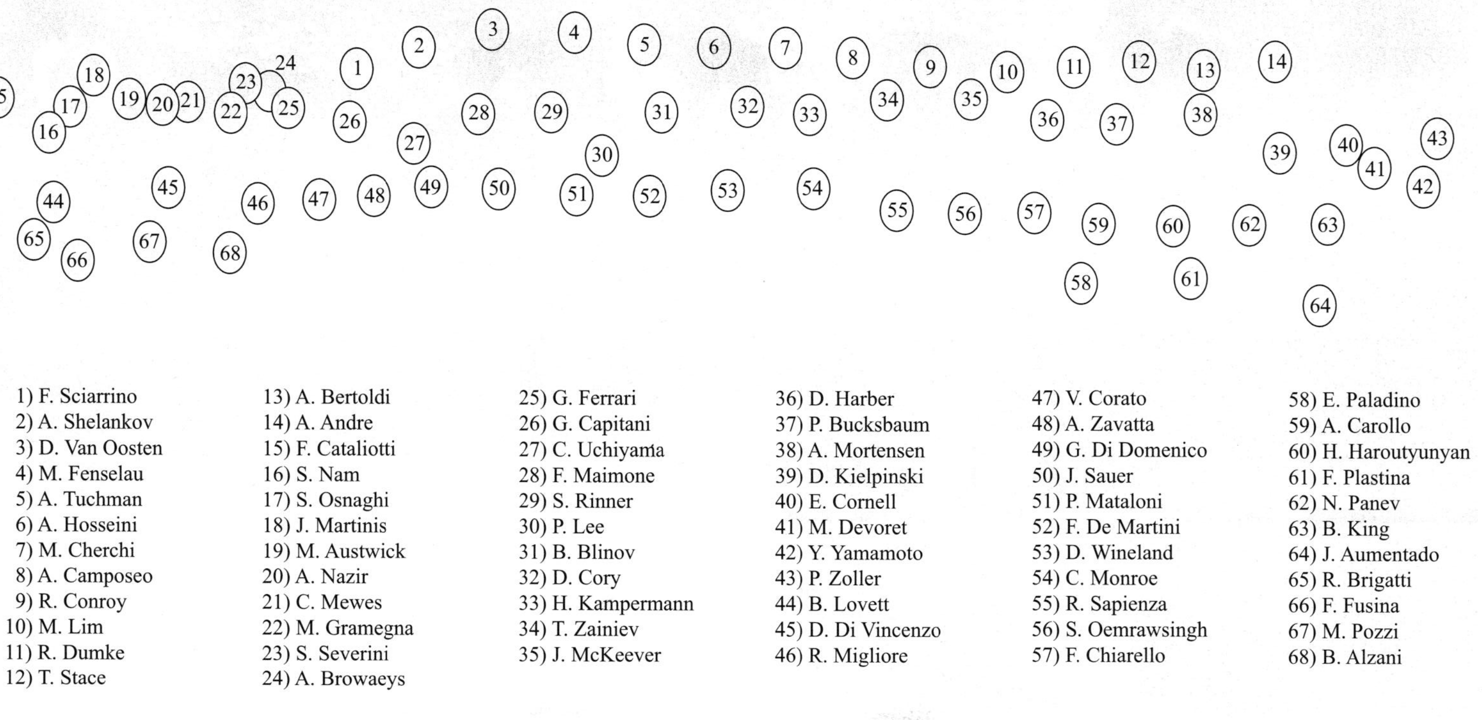

1) F. Sciarrino	13) A. Bertoldi	25) G. Ferrari	36) D. Harber	47) V. Corato
2) A. Shelankov	14) A. Andre	26) G. Capitani	37) P. Bucksbaum	48) A. Zavatta
3) D. Van Oosten	15) F. Cataliotti	27) C. Uchiyama	38) A. Mortensen	49) G. Di Domenico
4) M. Fenselau	16) S. Nam	28) F. Maimone	39) D. Kielpinski	50) J. Sauer
5) A. Tuchman	17) S. Osnaghi	29) S. Rinner	40) E. Cornell	51) P. Mataloni
6) A. Hosseini	18) J. Martinis	30) P. Lee	41) M. Devoret	52) F. De Martini
7) M. Cherchi	19) M. Austwick	31) B. Blinov	42) Y. Yamamoto	53) D. Wineland
8) A. Camposeo	20) A. Nazir	32) D. Cory	43) P. Zoller	54) C. Monroe
9) R. Conroy	21) C. Mewes	33) H. Kampermann	44) B. Lovett	55) R. Sapienza
10) M. Lim	22) M. Gramegna	34) T. Zainiev	45) D. Di Vincenzo	56) S. Oemrawsingh
11) R. Dumke	23) S. Severini	35) J. McKeever	46) R. Migliore	57) F. Chiarello
12) T. Stace	24) A. Browaeys			

58) E. Paladino
59) A. Carollo
60) H. Haroutyunyan
61) F. Plastina
62) N. Panev
63) B. King
64) J. Aumentado
65) R. Brigatti
66) F. Fusina
67) M. Pozzi
68) B. Alzani

ATOMS AND CAVITY QED

Entanglement, complementarity and decoherence
in cavity QED experiments

S. HAROCHE

Collège de France - 11 Place M. Berthelot, F-75231 Paris Cedex 05, France

J. M. RAIMOND and M. BRUNE

Laboratoire Kastler Brossel, Ecole Normale Supérieure
24 rue Lhomond, F-75231 Paris Cedex 05, France

1. – Foreword

Entanglement is an essential concept in quantum physics, which makes the microscopic world quite different from the classical macroscopic world of our daily experience. Entanglement leads to non-locality, an intriguing quantum property. It is also deeply related to the concept of complementarity and wave-particle dualism. Entanglement between a microscopic system and a macroscopic measuring device is also central in the theory of quantum measurement, leading to the famous Schrödinger cat paradox [1]. All the manifestations of entanglement seem strange because they are at odds with our intuition based on our observation of macroscopic objects made of a huge number of quantum particles. The appearance of classical behaviour when the size of objects increases is due to the phenomenon of decoherence [2], whose understanding again invokes the notion of entanglement between the system under study and its macroscopic environment. Up to recently, the manipulation of entangled systems of particles largely belonged to the realm of "gedanken experiments". Owing to the development of new techniques enabling us to isolate and study single quantum particles, it has now become possible to perform many of these thought experiments in the laboratory, realizing in this way textbook illustrations of basic quantum concepts.

Beyond their pedagogical quality, the experiments about entanglement are interesting because they demonstrate the possibilities of actually using entanglement to perform practical tasks in quantum information. By imprinting binary information in simple two-level quantum systems (so-called qubits) [3,4] and manipulating these systems, one can indeed perform some information processing tasks in ways quite different from what is possible according to classical physics. Quantum cryptography [5], teleportation [6] and quantum computation [7] become in principle possible. In practice, this last application is limited by decoherence, which tends to destroy coherence in the large systems required for useful implementation of quantum calculation algorithms. The study in the laboratory of simple systems illustrating the basic steps of entanglement engineering and the exploration of the mechanisms of decoherence in these systems are thus crucial steps for the physics of quantum information.

Various systems in quantum optics have been used to manipulate entanglement. Twin-photon beams and ions in traps have led to many beautiful demonstration experiments. In the field of cavity quantum electrodynamics, we are working at ENS with atoms interacting one by one with a quantum field made of a few photons stored in a microwave cavity. We have used this system to demonstrate various fundamental effects about entanglement, complementarity and decoherence. We reproduce in this lecture a detailed description of these experiments. The following text is an edited version of a colloquium article recently published in *Reviews of Modern Physics*. A copy of the transparencies used in the lectures of the Varenna school can also be found on our group web site [8].

2. – Introduction

The superposition principle is at the heart of the most intriguing features of the microscopic world. A quantum system may exist in a linear superposition of different eigenstates of an observable. It is then, in some way, "suspended" between different classical realities: a particle can be at two positions at the same time, a spin may point simultaneously towards two different directions, etc. When a measurement is performed, only one of these possibilities is actualized and the system is projected onto the corresponding eigenstate (wave-function collapse). It is impossible to get a classical intuitive representation of these superpositions. Their oddity becomes evident when one tries to transpose them to the macroscopic scale, as in the famous "Schrödinger cat" metaphor [1], describing a cat suspended between life and death.

When the superposition principle is applied to composite systems, it leads to the essential concept of entanglement. After two classical systems have interacted, each must be in a well-defined individual state, corresponding to definite results for any experiment. After two quantum particles have interacted, however, they can no longer be described independently of each other. Their "entangled" state is not a tensor product of eigenstates of observables pertaining to the two particles, which would describe independent systems with well-defined properties. It is instead a superposition of such products. The state of one particle is determined by a measurement performed on the other. Moreover, the same entangled state can be written in different forms, corresponding to different sets of non-commuting observables for the two particles. The state of one particle thus also depends upon the *nature* of the observable one has decided to measure on the other particle, even if the choice is made after the particles have separated. These quantum correlations are independent of the particle spatial separation and introduce a fundamental non-local aspect in the quantum world.

The non-classical properties of entangled states are clearly illustrated by the Einstein, Podolsky and Rosen [9] situation. Following Bohm's [10] analysis of the EPR problem, let us consider two spin-$(1/2)$ systems in the combined state:

$$(1) \qquad |\Psi_{\text{EPR}}\rangle = \frac{1}{\sqrt{2}}(|+_1, -_2\rangle - |-_1, +_2\rangle)$$

$$(2) \qquad = \frac{1}{\sqrt{2}}(|+_{\mathbf{u},1}, -_{\mathbf{u},2}\rangle - |-_{\mathbf{u},1}, +_{\mathbf{u},2}\rangle),$$

where $|\pm\rangle$ are the eigenstates of the spins along the Oz-axis ($|\pm_{\mathbf{u}}\rangle$ the states along an arbitrary direction $\mathbf{u}$) and the indices distinguish the two spins. $|\Psi_{\text{EPR}}\rangle$ being the rotation-invariant spin-singlet state, it takes the same form, as shown above, for any orientation of the quantization axis. All features of entangled states are clearly apparent in these expressions. Before any measurement, neither 1 nor 2 are described by a well-defined state. After 1 has been measured along axis $\mathbf{u}$, 2 points in the opposite direction along the same axis.

These basis-independent correlations cannot be understood in classical terms. Moreover, the statistical predictions of quantum mechanics contradict the results of any "local" theory, as they are expressed by the famous Bell inequalities [11, 12]. The experimental violation of these inequalities (for reviews, see [13, 14]) has vindicated quantum theory. More complex entangled states lead also to striking violations of locality. Greenberger, Horne and Zeilinger [15] proposed to use triplets of spin particles in the entangled state:

$$(3) \qquad |\Psi_{\text{GHZ}}\rangle = \frac{1}{\sqrt{2}}(|+_1, +_2, +_3\rangle + |-_1, -_2, -_3\rangle).$$

A single ideal experiment provides, in this state, opposite results for quantum mechanics and local theories.

Entanglement is also at the heart of quantum measurement. When two systems are in an entangled state, each of them can reveal information about the other, behaving as a measuring device. In a realistic measurement of a microscopic system, however, the meter is macroscopic. The situation is again reminiscent of the "Schrödinger cat" metaphor, with a meter evolving into a superposition of states with different classical properties. Such superpositions are extremely sensitive to the dissipative coupling between the meter and its environment. Entangled states involving macroscopic meters are rapidly transformed into statistical mixtures of product states, each of which describes the meter in a well-defined configuration correlated to the microscopic system in a corresponding eigenstate. This fast relaxation process is called "decoherence" [2, 16-19]. In fact, decoherence itself involves entanglement. The meter gets entangled with its environment. As the information leaks into the environment, the meter's state is obtained by tracing over the environment variables, leading to the final statistical mixture. This analysis is fully consistent with the Copenhagen description of a measurement.

Beyond these fundamental aspects, entangled states might have important applications for information transmission or processing. Elements of binary information can be coded in two-state quantum systems called qubits [3,4]. Contrary to ordinary bits, qubits can be in a quantum superposition of different logical values and combined into entangled states. EPR correlations between two qubits can be used to perform cryptographic key distribution [5]. By sharing EPR pairs, two operators can communicate in absolute secrecy. Quantum state teleportation [6] also uses the non-local features of the EPR pair to transmit the quantum state of a particle from one place to another at light velocity([1]).

More sophisticated entanglement manipulations could be used to perform interesting tasks, such as entanglement purification (*i.e.* the extraction of a subset of pure EPR states from a larger ensemble of particle pairs in a statistical mixture [20]). Quantum error correction codes [21] can be used to improve the quality of quantum transmissions [22]. Very complex entangled states could even be used to perform calculations out of the reach of ordinary computers. Factorization of integers [7,4], random searches [23], quantum systems simulations [24] could be performed by a "quantum computer" manipulating large sets of entangled qubits and operating faster than a classical computer. The qubits entanglement would be realized by small "entangling machines", called "quantum gates" [3,4]. They couple two qubits through a well-controlled conditional dynamics operation. The evolution of the state of one qubit (the "target") depends upon the state of the other (the "control"), which usually remains unaltered. A deterministic combination of gate operations leads to the final state, in which the "result" can be measured. These macroscopic qubits systems are however utterly sensitive to decoherence [25], which appears thus as a very severe bottleneck for quantum computing.

Fundamental tests as well as potential applications have triggered a considerable interest for experiments on basic quantum mechanics. The manipulation of complex entangled states puts very severe constraints on the experimental techniques. The individual systems should be prepared in a well-defined initial quantum state. They should be very well isolated from the environment and interact strongly with each other, as required for the realization of quantum gates. Their state should be accurately detected, with a high efficiency. Furthermore, individual qubit addressing is required for the engineering of the most general entangled state, scalable to arbitrary numbers of qubits, as well

---

([1]) Note that in all these schemes, as well as in the original EPR situation, no information can be transmitted faster than light even though the quantum collapse can be viewed as instantaneous. In the Bell's inequalities experiments, for example, each operator measures a random sequence on which the other has no deterministic control. All the information is contained in the correlations between the sequences measured by the observers. Classical information exchange is required to check these correlations. In teleportation, no state is received before a classical signal has been transmitted. There is never contradiction between non-locality and relativistic causality.

as for performing fundamental tests of quantum measurement theory. Many propositions have been made to implement these requirements. Solid-state devices (mesoscopic conductors [26], squids [27, 28], single-impurity spins, etc.) are under active consideration. Up to now, however, entanglement remains to be demonstrated in these systems. Nuclear magnetic resonance in liquid samples provides long relaxation times, spin-spin exchange interactions needed for quantum gates and sophisticated techniques developed for chemical analysis [29, 30]. Complex manipulations have been realized. However, they rely on very small deviations from thermal equilibrium and no clear-cut entanglement is involved [31, 32]. Moreover, measurements on individual spins are not feasible and only quantum averages can be detected.

Entanglement has so far been observed only in quantum optics. Photons in entangled states are spontaneously produced in atomic cascades or parametric down-conversion processes converting an incoming UV photon into two entangled visible ones. Polarization states can be easily manipulated, the photons propagating over long distances and being finally detected with a high efficiency. Entanglement between photons with different energies and times of arrival can also be realized [33]. Recently, these experiments have greatly benefited from the progress in optical fiber communication technology. Tests of Bell inequalities [12, 13], creation of GHZ triplets and non-locality tests [34], teleportation of quantum states [35-37] have been performed. Sophisticated quantum cryptographic systems, on the verge of industrial development, use correlated photons [33, 38-40]. In these experiments, entanglement results from the basic conservation laws in the initial spontaneous process and cannot be easily manipulated afterwards, due to the lack of an efficient photon-photon quantum gate.

Entanglement of slow or trapped matter particles offers other perspectives. Remarkable achievements have been obtained with trapped ions. Laser cooling techniques prepare a few ions in the vibrational ground state inside the trap. The long-lived internal structures can be entangled with the collective motion, providing quantum gate operation. Finally, laser-induced fluorescence techniques provide a selective detection of the ionic state with a near unity efficiency. Quantum gate operation [41], two-ion [42] and four-ion [43] entangled state have been demonstrated. Up to now, the condition of strong ion-ion coupling requires a very small spatial separation between the ions, making it difficult to address them individually.

Cavity Quantum Electrodynamics (CQED) [44] studies mixed atom/photon systems and offers other interesting tools for entanglement. A single two-level atom crossing a cavity mode gets entangled with the field, provided the coherent coupling overwhelms the dissipative processes ("strong-coupling regime"). Ordinary optical atomic transitions and very high-finesse cavities make it possible to meet the strong-coupling conditions [45] and to observe interesting quantum effects [46, 47]. The very fast time evolution of these optical systems did not make it possible so far to investigate entanglement directly.

Microwave CQED, with Rydberg atoms crossing superconducting cavities one by one [48, 49], offers an almost ideal system for entanglement studies. Relaxation rates are small and well understood. The atoms and the cavity can be prepared in pure states and the strong-coupling conditions are readily fulfilled. The atoms can be detected in a selective and sensitive way by field ionization. The time constants involved (millisecond range) are long enough to realize controlled and complex sequences. Finally, the quantum systems are separated by centimeter-scale distances and can be individually addressed.

This paper reviews Rydberg atom microwave CQED experiments on entanglement performed at ENS in Paris. We will focus on the main physical ideas, without insisting on experimental details, which can be found in review articles [50-52] or on our web site [8]. Section **3** presents a rapid overview of the experimental techniques. Section **4** describes the resonant atom-field interaction. The quantum Rabi oscillation of the atom in the cavity [53] provides all the ingredients to manipulate atom-cavity and atom-atom entanglement. These features are illustrated by the generation of an EPR atomic pair [54] in sect. **5**. We describe in sect. **6** the realization of a quantum gate [55]. The next two sections describe applications of this gate. In sect. **7**, we analyze a non-destructive measurement of a single photon [56]. In sect. **8**, we apply the quantum gate to the generation of an entangled triplet of the GHZ type [57]. Finally, sect. **9** reviews atom-cavity entanglement obtained through non-resonant dispersive interaction. We describe the generation of mesoscopic "Schrödinger cat" states of the field [58] and the study of their decoherence dynamics. We conclude (sect. **10**) by presenting some perspectives opened by these experiments.

3. – The cavity QED entangling machine

The experimental set-up is sketched in fig. 1(^{2}). Circular Rydberg atoms are produced in zone B by the excitation of a velocity-selected rubidium atomic beam effusing from oven O. They cross one by one a high-quality superconducting cavity C and are finally detected by the field ionization detector D. The whole set-up is cooled to around 1 K to minimize thermal field noise.

3`1. *The circular Rydberg atoms*. – Circular Rydberg states [61] correspond to large principal and maximum orbital and magnetic quantum numbers. The valence electron orbital is a thin torus centered on the atom's core, revealing quantum position fluctuations around the classical Bohr orbit. According to the correspondence principle, the properties of these states can be understood in classical terms. The transitions between neighboring circular states fall in the millimeter-wave domain for principal quantum numbers of the order of 50. Our experiments involve the three circular levels with principal quantum numbers 51, 50 and 49, called e, g and i, respectively (see bottom inset in fig. 1). The $e \Leftrightarrow g$ and $g \Leftrightarrow i$ transitions are at 51.1 GHz and 54.3 GHz, respectively. The degeneracy of the Rydberg manifold is lifted by a small electric field isolating the e, g and i states from the other non-circular levels. This field stabilizes the highly anisotropic orbit [62] and can also be used to tune the $e \Leftrightarrow g$ and $g \Leftrightarrow i$ transition frequencies using the quadratic Stark effect. The dipole matrix elements of these transitions, proportional to the radius of the circular orbit, are very large (1250 atomic units for the $e \Leftrightarrow g$ transition).

(^{2}) More details can be found in [53-60].

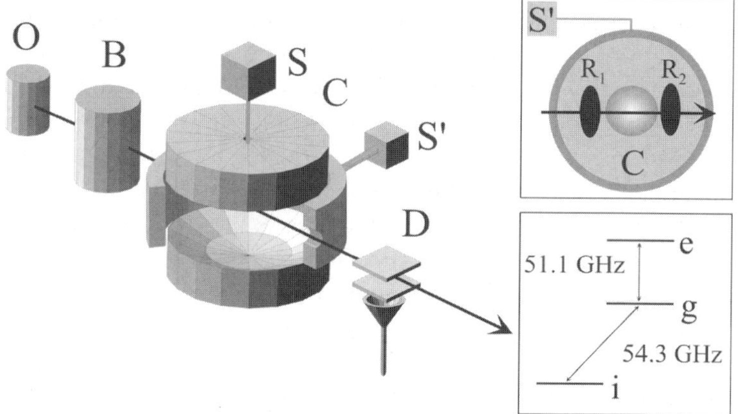

Fig. 1. – Experimental apparatus. Top inset: top-view of the cavity mode and of the two Ramsey field zones. Bottom inset: relevant circular levels.

The radiative lifetimes of e, g and i —of the order of $T_{\mathrm{at}} = 30\,\mathrm{ms}$— are much longer than those for non-circular Rydberg states. In free space, the atoms would propagate a few meters at thermal velocity before decaying. Radiative decay is thus negligible along the 20 cm path inside the apparatus. The circular levels are ionized in a moderate electric field (128 V/cm for e) inside the detector D. The resulting electrons are accelerated and counted with a 40 (± 10)% detection efficiency. Since e, g and i ionize in different fields, D is state-selective. All our information about the system's state is provided by this field-ionization detection.

The preparation of the circular levels in B [63] combines diode laser excitation and radiofrequency transitions. The process is pulsed and the atomic preparation time is known within a $2\,\mu s$ interval. The circular state purity is $\geq 98\%$. Atomic velocity selection is an essential ingredient for the control of experimental sequences on each atom. It is performed by Doppler-selective optical pumping techniques [54, 56] and results in a $\pm 2\,\mathrm{m/s}$ velocity class width. The position of each atom inside the apparatus is thus known with a $\pm 1\,\mathrm{mm}$ precision, an essential condition for individual atomic control.

The circular states excitation process prepares, on the average, 0.2 atoms per pulse, with Poisson statistics. Most pulses do not produce any atom and are rejected by the data acquisition software. In the remaining events, the probability for having two atoms in the same sample is about 20%. Half of these two-atom events are detected as such and rejected. Single-atom events are thus selected with a 90% probability.

3˙2. *The superconducting cavity*. – The cavity is an open Fabry-Perot resonator, made of two carefully polished spherical niobium mirrors facing each other (with each mirror having a diameter of 50 mm and a radius of curvature of 40 mm and with a distance between mirrors equal to 27 mm). The cavity sustains a Gaussian TEM_{900} mode at frequency ω, with a $w = 6\,\text{mm}$ waist, resonant or nearly resonant with the $e \Leftrightarrow g$ transition. This Fabry-Perot geometry is compatible with the application of a static electric field along the cavity axis, which is essential for the manipulation of circular states. Two small holes pierced in the center of the mirrors allow us to couple microwaves in and out of the cavity. The resonance frequency, tuned by mechanical translation of the mirrors, and the quality factor Q can thus be easily determined by cavity transmission experiments. A classical source S can be used to inject a small coherent field in C [64].

Our best cavity so far has a photon storage time $T_r = 1\,\text{ms}$ (corresponding to $Q = 3 \cdot 10^8$). This time is much longer than the atom-cavity interaction time (a few tens of μs) allowing for atom-cavity entanglement to be produced before relaxation processes set in. To obtain this T_r value, we use an aluminum ring enclosing the open space between the mirrors (shown open in fig. 1). It reflects the photons scattered by the mirror imperfections back into the mode and increases T_r by an order of magnitude. Small access and exit holes ($3\,\text{mm}$ diameter) are used for atomic beam access. Inhomogeneous stray electric fields in these holes randomly shift the atomic transition frequencies. Atomic coherences are spoiled, while the levels populations are not affected. All entangled state manipulations must thus be realized inside the cavity-ring structure.

At thermal equilibrium, the cavity mode contains about 0.7 thermal photons on the average, originating from thermal field leaks. This field is removed at the beginning of each experimental sequence. We send across C a few atomic pulses, each containing a few atoms prepared in the lower level g of the transition resonant with C. These atoms efficiently absorb the thermal photons and reduce the effective field temperature [56]. At the end of this "cooling sequence", the mean photon number is reduced down to 0.1. The experimental sequence then lasts a time no longer than $0.3\,T_r$ to limit thermal field build-up.

3`3. *The atomic Ramsey interferometer*. – The atoms can be prepared in a superposition of energy states before the interaction with C and mixed again after this interaction, making it possible to prepare and analyze complex entangled states. For this purpose, we use two pulses of classical microwave radiation, applied in the zones called R_1 and R_2 in the top inset of fig. 1. These pulses are generated by a second microwave source S' and injected in the cavity structure through a small hole in the ring. This auxiliary microwave field produces a standing-wave pattern inside the ring-mirror structure. Antinodes of this pattern, sandwiching the central cavity mode, are used to perform the pulses at the time an atom is crossing them. Different sequences of pulses can be applied on successive atoms by commuting the field source S'. Note that the fields applied in R_1 and R_2 have a very short relaxation time (in the nanosecond range), much smaller than the decay time T_r in the cavity mode C. This explains why these pulses can be described classically and do not produce any entanglement between atom and radiation [65]. They can be considered as classical tools to manipulate the atomic state superpositions.

Applying to an atom two successive $\pi/2$ pulses, R_1 and R_2, at a frequency ω_r that is nearly resonant with an atomic transition, for instance the $g \Leftrightarrow i$ transition at frequency ω_{gi}, one has a Ramsey separated field interferometer [66] (see fig. 2). The

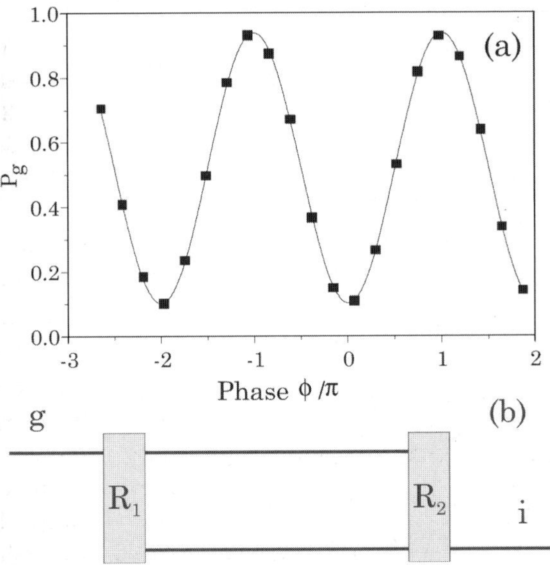

Fig. 2. – Ramsey fringes. (a) Signal observed for the $g \Leftrightarrow i$ transition. Probability P_g for detecting the atom in g *versus* the phase difference ϕ between the two $\pi/2$ pulses R_1 and R_2 expressed in units of π. The points are experimental, the line is a sine fit. (b) Diagram depicting the two paths followed by the atom in the Ramsey interferometer.

pulses mixing Rydberg states act as atomic internal state "beam splitters". To be specific, the $\pi/2$ R_1 pulse can perform the transformations $|g\rangle \rightarrow (|g\rangle + |i\rangle)/\sqrt{2}$ and $|i\rangle \rightarrow (-|g\rangle + |i\rangle)/\sqrt{2}$. After a time delay T, the second microwave pulse produces the transformations $|g\rangle \rightarrow (|g\rangle + \exp[i\phi]|i\rangle)/\sqrt{2}$ and $|i\rangle \rightarrow (-\exp[-i\phi]|g\rangle + |i\rangle)/\sqrt{2}$ with $\phi = (\omega_r - \omega_{gi})T$ describing the phase difference accumulated between the microwave source S' and the atomic coherence during time T.

Given that an atom is prepared in level g before R_1, the probability P_i for detecting it in level i after R_2 is the squared sum of two amplitudes corresponding to two atomic paths inside the interferometer (sketched in fig. 2(b)). The relative phase of these two amplitudes is ϕ. By repeating the experiment many times, we reconstruct the $g \rightarrow i$ transition probability, which oscillates *versus* ϕ, exhibiting the well-known pattern of Ramsey fringes. Ideally, $P_g = 1 - P_i = (1 - \cos\phi)/2$. Experimentally, our fringes have an 85% contrast (fig. 2(a)). The interaction of the atoms with the quantum field in C, between R_1 and R_2, modifies the phase and the amplitude of the fringes, revealing useful information about the atom-field coupling.

In some experiments, R_1 and R_2 can be used separately to prepare or analyse an atomic state superposition. It is useful to describe the atom undergoing the $g \rightarrow i$ (or $g \rightarrow e$) transition as a pseudo-spin 1/2 represented by a vector whose tip lies on a "Bloch sphere" of unit radius. The energy eigenstates correspond to the spin pointing in the "vertical Oz" direction. The classical microwave field produces a rotation of this pseudo-spin on the Bloch sphere. A $\pi/2$ pulse, for instance, acting on an energy eigenstate results in a spin lying in the "horizontal xOy" plane. Zone R_1 can thus be used to inject in C a spin pointing in a selected direction. Zone R_2, which rotates the spin before its detection along the Oz-direction in D, allows us to detect the spin along an arbitrary direction on the Bloch sphere. In the language of quantum information, the two-level systems are qubits, coding superpositions of $|0\rangle$ and $|1\rangle$ states. In this context, R_1 and R_2 allow us to perform the most general single qubits operations. For instance, the $\pi/2$ pulses are known as Hadamard transforms [4].

4. – The quantum Rabi oscillation

4'1. *Vacuum Rabi oscillation.* – The entangling mechanism in our experiments is based on the atom-cavity-field interaction. The simplest situation corresponds to an atom in level e entering the cavity, initially in its vacuum state $|0\rangle$. The cavity mode frequency ω is equal to the $e \Leftrightarrow g$ transition frequency ω_{eg}. The initial atom-cavity state, $|e, 0\rangle$, is coupled by dipole emission to $|g, 1\rangle$, describing an atom in the lower state g and a cavity containing one photon. Quantum oscillations are expected between these two states [50]. These "vacuum Rabi oscillations" [53,67] correspond to the oscillatory regime of spontaneous emission in a high-Q cavity.

Quantitatively, the situation is described by the well-known Jaynes and Cummings [68] Hamiltonian

$$(4) \qquad H = \hbar\omega_{eg}\sigma_z + \hbar\omega\left(a^\dagger a + \frac{1}{2}\right) - i\frac{\hbar\Omega}{2}f(x)(\sigma_+ a - \sigma_- a^\dagger),$$

where a and $a^\dagger$ are the photon annihilation and creation operators in the cavity mode, σ_z, σ_+ and σ_- the Pauli matrices of the atomic pseudo-spin. The atom-field coupling constant $\Omega/2$ is equal to $d\mathcal{E}_0/\hbar$, where $\mathcal{E}_0 = 1.5\,\mathrm{mV/m}$ is the r.m.s. vacuum field at the cavity center and d is the dipole matrix element for the $e \Leftrightarrow g$ transition ($\Omega/2\pi = 47\,\mathrm{kHz}$). The cavity mode Gaussian structure is described by the real function $f(x) = \exp[-x^2/w^2]$ ($x \propto vt$ is the atomic position along the beam, $x = 0$ on the cavity axis).

Let us consider first an atom at cavity center $x = 0$ with $v = 0$. If the system starts at time $t = 0$ from $|e, 0\rangle$, its state at time t is

$$(5) \qquad |\Psi_e(t)\rangle = \cos\left(\frac{\Omega t}{2}\right)|e, 0\rangle + \sin\left(\frac{\Omega t}{2}\right)|g, 1\rangle.$$

If the system starts in state $|g, 1\rangle$ instead, its state becomes at time t

$$(6) \qquad |\Psi_g(t)\rangle = \cos\left(\frac{\Omega t}{2}\right)|g, 1\rangle - \sin\left(\frac{\Omega t}{2}\right)|e, 0\rangle$$

(we use here the interaction representation). In general, these expressions describe a time-varying entanglement between the atomic and cavity systems. When the atom

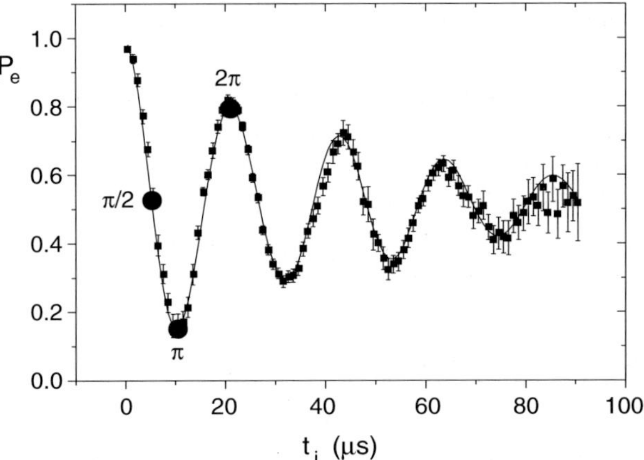

Fig. 3. – Vacuum Rabi oscillations. The atom in state e enters an empty resonant cavity. P_e denotes the probability for detecting the atom in e as a function of the effective interaction time t_i. Three important interaction times (corresponding to the $\pi/2$, π and 2π Rabi rotations) are indicated.

moves across the cavity, these expressions apply provided we replace t by an effective interaction time t_i taking into account the spatial variation of the coupling. When the atom interacts resonantly with the cavity across the full mode structure, $t_i = \sqrt{\pi}w/v$. Note that the atom-field entanglement is "frozen" when the atom leaves the cavity mode. The entanglement becomes then non-local.

We have observed (fig. 3) the quantum Rabi oscillation in vacuum by measuring the probability $P_e(t_i)$ that the atom remains in level e at time t_i [53]. Ideally, this probability oscillates at frequency $\Omega/2\pi$ (vacuum Rabi frequency):

$$(7) \qquad P_e = \frac{1 + \cos \Omega t_i}{2}.$$

The time t_i is determined by the atomic velocity. The damping of the experimental signal is mainly due to technical imperfections. During the first Rabi oscillation, a wide variety of atom-cavity entangled states can be obtained with high fidelity by properly choosing t_i.

4'2. *Useful Rabi pulses*. – When $\Omega t_i = \pi/2$ ("$\pi/2$ Rabi rotation"), the final atom-field state reads

$$(8) \qquad |\Psi_{\pi/2}\rangle = \frac{1}{\sqrt{2}}(|e,0\rangle + |g,1\rangle).$$

With an appropriate pseudo-spin states definition, it is the EPR state of eq. (1). The entanglement, created in a time of the order of $5\,\mu$s, lasts as long as the photon in the cavity (1 ms).

When $\Omega t_i = \pi$, the atom-cavity system, initially in $|e, 0\rangle$, ends up in the non-entangled state $|g, 1\rangle$. If the system starts from $|g, 1\rangle$ instead, it ends up in $-|e, 0\rangle$. This "π Rabi rotation" swaps the atom and cavity excitations. More generally, if the atom is initially in a superposition of e and g and the cavity in vacuum, the atom ends up in g, leaving in the cavity a superposition of the zero- and one-photon Fock states:

$$(9) \qquad (c_e|e\rangle + c_g|g\rangle)|0\rangle \longmapsto |g\rangle(c_e|1\rangle + c_g|0\rangle).$$

The reverse transformation, obtained for an atom initially in g interacting with a field in a coherent superposition of 0 and 1 photon Fock states, is

$$(10) \qquad (c_1|1\rangle + c_0|0\rangle)|g\rangle \longmapsto |0\rangle(-c_1|e\rangle + c_0|g\rangle).$$

The "π Rabi rotation" thus maps the state of one system onto the other. This mapping can be used to prepare or to detect the cavity field in an arbitrary 0 and 1 photon states superposition.

When $\Omega t_i = 2\pi$, the atom-field system evolves according to

$$(11) \qquad |e, 0\rangle \longmapsto -|e, 0\rangle, \qquad |g, 1\rangle \longmapsto -|g, 1\rangle.$$

After a full cycle of Rabi oscillation, the atom-cavity system experiences a global quantum phase shift π. Similar phase shifts are obtained for a spin-$(1/2)$ system undergoing a 2π rotation in ordinary space [69, 70]. Note that this 2π rotation also plays a central role in the micromaser trapping states [71]. Since $|g, 0\rangle$ is not affected by the atom-field coupling, the phase shift experienced by an atom entering C in g is conditioned to the presence of a photon inside the cavity. This conditional dynamics is, as we will see, essential for realizing a quantum phase gate.

The $\pi/2$, π and 2π Rabi pulses combined with classical Ramsey ones provide the basic ingredients of our entanglement "recipe". In general, the atomic velocity v is chosen to produce a 2π Rabi rotation for a full crossing of the cavity mode ($v = 503\,\mathrm{m/s}$). By applying an electric field across the cavity mirrors at appropriate times, the $e \Leftrightarrow g$ transition can be abruptly tuned out of resonance, freezing from then on the atom-cavity evolution. In this way, we can adjust t_i to the values corresponding to π and $\pi/2$ Rabi pulses. We can then combine, on a sequence of atoms crossing the cavity with the same velocity v, entanglement production by a $\pi/2$ rotation, state transfer by a π rotation and conditional dynamics by a 2π rotation, engineering in this way complex entangled states.

These manipulations rely on an accurate timing of operations on different atoms. These are possible only because the various relevant times of the experiment are properly ordered, as indicated in fig. 4. We have plotted there, on a logarithmic scale, the typical circular level lifetime T_{at}, the cavity damping time T_r, the Rabi period $T_\Omega = 2\pi/\Omega$, the typical Ramsey pulse duration $T_p = 1\,\mu\mathrm{s}$. The effective atom-cavity interaction time t_i can be varied between the time resolution of our timing unit (100 ns) and about $100\,\mu\mathrm{s}$

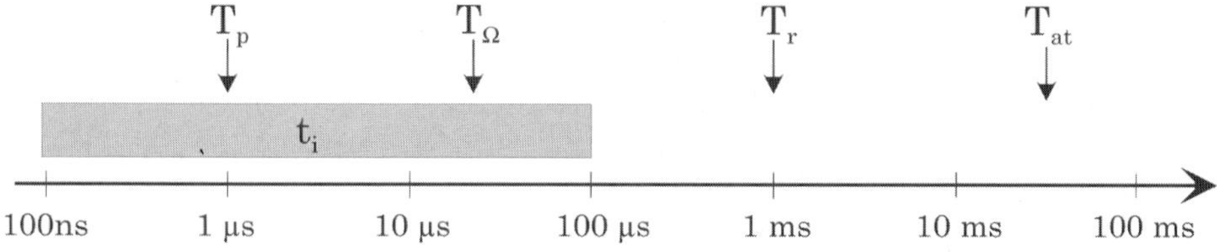

Fig. 4. – Relevant times of a CQED experiment, plotted on a logarithmic scale. T_{at}, T_r, T_Ω, t_i and T_p are defined in text. t_i can be varied in the range depicted by the gray bar.

(atoms with a velocity of $100\,\text{m/s}$ interacting resonantly with the mode during its full cavity crossing time). It is essential for these experiments that the relaxation times T_{at} and T_r are much longer than T_Ω (strong-coupling regime). It is also important that the time step resolution is much smaller than T_Ω.

4`3. *Quantum Rabi oscillation in an applied field.* – Atom-field entanglement is also produced when the atom interacts with a cavity field in the n-photon Fock state. For example, the system starting from the initial state $|e, n\rangle$ evolves into

$$(12) \qquad |\Psi_{e,n}(t_i)\rangle = \cos\left(\frac{\Omega\sqrt{n+1}t_i}{2}\right)|e, n\rangle + \sin\left(\frac{\Omega\sqrt{n+1}t_i}{2}\right)|g, n+1\rangle.$$

The maximally entangled EPR state is produced in a time $\pi/2\Omega\sqrt{n+1}$, faster than in the zero-photon case. Note that Rabi oscillations induced by pure Fock states of the field up to $n = 2$ have been recently observed [72].

Fock states are generally difficult to generate, but one can easily observe the evolution of an atom interacting with a small coherent field $|\alpha\rangle = \sum_n c_n|n\rangle$, where $c_n = \exp[-|\alpha|^2/2]\alpha^n/\sqrt{n!}$. The quantum Rabi oscillation signal appears now as a sum of sinusoidal terms at frequencies $\Omega\sqrt{n+1}$, weighted by the probabilities $|c_n|^2$ for finding the corresponding photon number in the coherent field:

$$(13) \qquad P_e = \sum_n |c_n|^2 \frac{1 + \cos\Omega\sqrt{n+1}t_i}{2}.$$

The corresponding experimental signal is shown in fig. 5(a) for an initial coherent field containing 0.85 photons on the average. We observe an oscillation collapse due to the dispersion of the Rabi frequencies and a subsequent "revival" [73] when the terms oscillating at different discrete frequencies come back into phase. The signal is damped by various imperfections. This signal provides direct evidence of field energy quantization. The field amplitude distribution is revealed by the frequency spectrum of the Rabi signal. This spectrum, obtained by a Fourier transform, is plotted in fig. 5(b). It exhibits well-separated discrete frequency components scaling as the square roots of the successive integers.

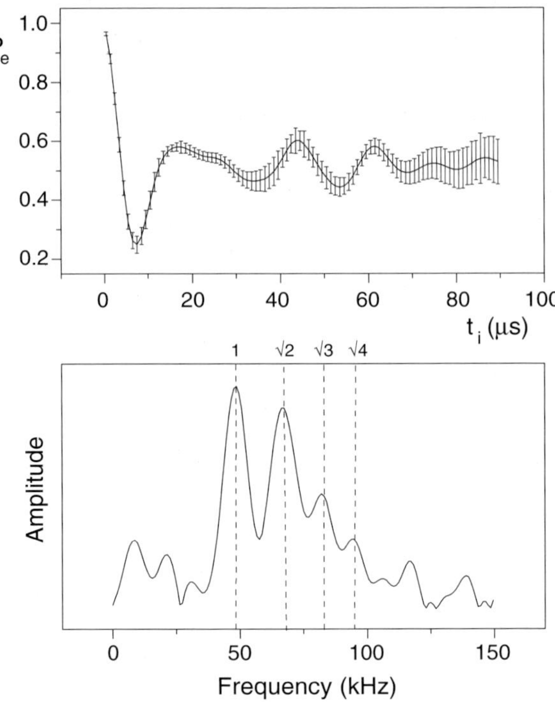

Fig. 5. – Quantum Rabi oscillation in a coherent field. The cavity contains initially a coherent field with 0.85 photons on the average. (a) Probability P_e for finding the atom in the initial state e as a function of the effective interaction time t_i. (b) Fourier transform of the signal in (a) revealing the discrete Rabi frequencies, occurring at the successive square roots of the integers.

5. – Creation of an EPR pair

In the simplest entanglement experiment [54], we let an atom A_1 undergo a $\pi/2$ Rabi rotation in an initially empty cavity. The resulting atom-cavity state is the EPR pair described by eq. (8). This entanglement persists after A_1 has left C. Detecting at some distance downstream the atom in a given quantum state instantaneously collapses the cavity mode in the correlated field state. For example, detecting the atom in level e (respectively, g) amounts to preparing a zero- (one-)photon Fock state in the cavity [60]. One can also, by detecting linear superpositions of e and g states (using a R_2^{eg} mixing pulse on the $e \Leftrightarrow g$ transition in front of the detector D), project the field onto corresponding superpositions of 0 and 1 Fock states. Contrary to photon number states, such superpositions have a non-zero electric-field expectation value, *i.e.* a non-uniform phase distribution. The phase information contained in the classical R_2^{eg} field has been transferred to the quantum cavity mode, via the atomic measurement and the non-local quantum correlations.

In order to read out the field state, we need a second atom A_2, prepared in g and undergoing a π Rabi rotation in a single-photon field. This atom carries away the cavity state

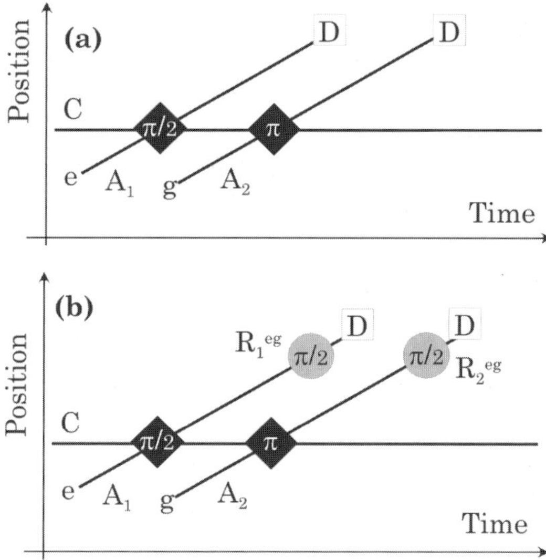

Fig. 6. – Temporal sequence of the EPR pair preparation. Space lines of A₁, A₂ and C in a position-*versus*-time diagram. The black diamonds indicate a resonant Rabi rotation (with the angle as a label). The open squares represent the detection events. The gray circles represent classical Ramsey pulses. (a) Timing of the EPR state preparation with a direct measurement of energy anticorrelations. (b) Timing of the transverse correlation experiment.

including its entanglement with A_1. In other words, detecting the A_1-C entanglement is equivalent to performing a A_1-A_2 two-atom EPR experiment. After A_2 has crossed C, the cavity is empty and disentangles from the atomic pair, which ends up in state

$$(14) \qquad |\Psi_{\mathrm{Pair}}\rangle = \frac{1}{\sqrt{2}}(|e_1, g_2\rangle - |g_1, e_2\rangle),$$

where the subscripts refer to the atom number. Since e and g represent, respectively, the $|+\rangle$ and $|-\rangle$ pseudo-spin states, $|\Psi_{\mathrm{Pair}}\rangle$ appears as the rotation-invariant spin-singlet state. The temporal sequence of this EPR atomic pair preparation is schematized in fig. 6(a). It exhibits the space lines of the two atoms and of the cavity mode in a qualitative position-*versus*-time diagram. The Rabi rotations and detection events are indicated by symbols (see figure caption).

In order to check the atomic entanglement, we have shown that the measurements of the two spins along an arbitrary direction are anticorrelated. In a first experiment, we directly check the anticorrelations of the atomic energies, corresponding to a measurement along the Oz-axis (see fig. 6(a)). We obtain, for the four possible detection channels, the probabilities $P_{eg} = 0.44$, $P_{ge} = 0.27$, $P_{gg} = 0.23$, $P_{ee} = 0.06$, with statistical errors of the order of 0.03. For a pure EPR pair, these probabilities should be $P_{eg} = P_{ge} = 1/2$, $P_{ee} = P_{gg} = 0$. Several experimental imperfections discussed in [54] account for the differences.

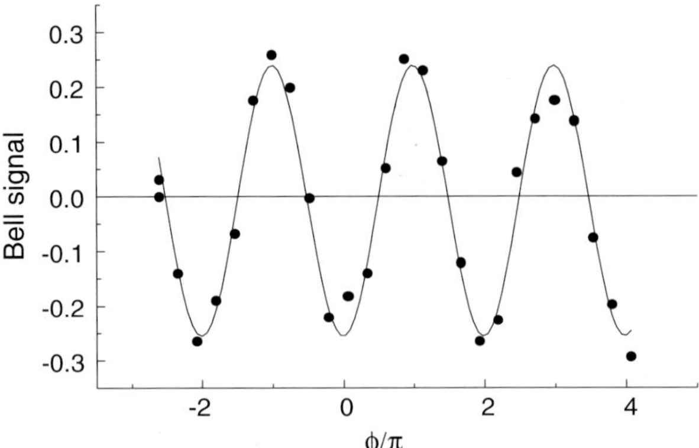

Fig. 7. – "Bell signal" plotted *versus* the relative phase ϕ (in units of π) of pulses R_1^{eg} and R_2^{eg}. The line is a sine fit.

We have also checked "transverse" spin component correlations, exhibiting the phase information transfer between the two atoms mediated by the non-local correlations. We make use of Ramsey $\pi/2$ pulses R_1^{eg} and R_2^{eg} applied, respectively, on atoms A_1 and A_2 on the $e \Leftrightarrow g$ transition after they have left C. The timing of this experiment is depicted in fig. 6(b) (the gray circles represent these classical pulses). The correlations between the two atomic detections are exhibited by plotting, *versus* the relative phase ϕ between the two Ramsey pulses, the "Bell signal"$(^3)$ $\langle \sigma_{x,1} \sigma_{\phi,2} \rangle = P_{g_1,g_2} + P_{e_1,e_2} - P_{g_1,e_2} - P_{e_1,g_2}$, where the σ's are Pauli matrices associated to the pseudo-spins and P_{a_1,b_2} is the probability for detecting A_1 in a and A_2 in b ($\{a,b\} = \{g,e\}$). This correlation should be ideally equal to -1 $(+1)$ for $\phi = 0$ ($\phi = \pi$) (perfect anticorrelation for a detection of the two particles along the same direction).

Figure 7 presents the corresponding experimental data. The contrast reduction is due both to the purity of the EPR state and to imperfections of the Ramsey pulses. Note that we are performing here a kind of Ramsey interferometry. We apply however the two pulses on different atoms. The observation of fringes in this case shows that the pair of atoms behave as a single system sharing quantum coherences.

The Bell signal contrast in this experiment is not high enough to observe a violation of the Bell inequalities [11]. In comparison with the experiments with photons [12, 13], the quality of our Ramsey spin "analyzers" is too low. Improvements of the set-up, briefly described in sect. **10**, will hopefully allow us to improve the situation, opening the way to a new-type Bell inequality experiment with massive particles. A similar experiment has recently been performed with entangled ions [74].

$(^3)$ This atomic correlation signal is similar to the photon polarization correlations used for experimental tests of Bell inequalities.

6. – The quantum phase gate

The conditional dynamics induced by the 2π Rabi rotation (eq. (11)) realizes an elementary quantum logic gate which can be used to entangle qubits in a "programmed" process. Let us consider an atom crossing the cavity, containing zero or one photon. The atom is either in the resonant state g or in the "spectator" state i (far off-resonance and thus not affected by the cavity interaction). If C is empty or if the atom is in i, the global state is unchanged. However, if the atom is in g and the cavity in $|1\rangle$, the global state undergoes the transformation $|g,1\rangle \mapsto \exp[i\Phi]|g,1\rangle$ with $\Phi = \pi$ (see eq. (11)). This is the conditional dynamics of a "quantum phase gate" [41]. This gate does not modify the photon number in the cavity.

If the atom enters in a coherent superposition of g and i (amplitudes c_g and c_i), the gate performs the transformations

$$(15) \qquad (c_g|g\rangle + c_i|i\rangle)|0\rangle \longmapsto (c_g|g\rangle + c_i|i\rangle)|0\rangle,$$

$$(c_g|g\rangle + c_i|i\rangle)|1\rangle \longmapsto (e^{i\Phi}c_g|g\rangle + c_i|i\rangle)|1\rangle.$$

The superposition is unchanged if the cavity is empty, while the phase of the atomic coherence is shifted by $\Phi = \pi$ when the cavity contains one photon. If the field is now in a superposition of zero and one photon states (amplitudes c_0 and c_1), we have

$$(16) \qquad |i\rangle(c_0|0\rangle + c_1|1\rangle) \longmapsto |i\rangle(c_0|0\rangle + c_1|1\rangle),$$

$$|g\rangle(c_0|0\rangle + c_1|1\rangle) \longmapsto |g\rangle(c_0|0\rangle + e^{i\Phi}c_1|1\rangle).$$

The phase of the field coherence is shifted by Φ by an atom in g, while it is unchanged by an atom in i. Finally, it is easy to see that, when both qubits are in state superpositions, the gate generates atom-cavity entanglement.

We have proved the coherent gate operation by performing two complementary experiments checking these transformations. To study the transformation described by eqs. (15), we either leave initially the cavity in the zero photon state or prepare it in the one-photon state by using the π Rabi pulse operation on a first "source" atom A_1 initially prepared in e (see sect. **5**). We then send a second atom A_2, initially in g, prepared in a superposition with equal weights of g and i ($c_g = c_i = 1/\sqrt{2}$) by a Ramsey pulse R_1^{gi}. This atom undergoes a full 2π Rabi rotation in C. We probe the atomic coherence by applying another $\pi/2$ Ramsey pulse R_2^{gi} before the detection in D. The timing of the experiment corresponding to the one-photon case is schematized in fig. 8(a). In the zero-photon case, atom A_1 is removed.

The Ramsey fringes observed on atom A_2 submitted to the R_1^{gi} and R_2^{gi} pulses are shown in fig. 9, as open diamonds (when the zero-photon state is initially prepared in C) and by black squares (when the cavity initially contains one photon). We observe clearly the π phase shift of the atomic coherence induced by the presence of one photon.

The complementary experiment testing eq. (16) involves a coherent superposition of 0- and 1-photon states. It is prepared by injecting in C a small coherent field $|\alpha\rangle = c_0|0\rangle +$

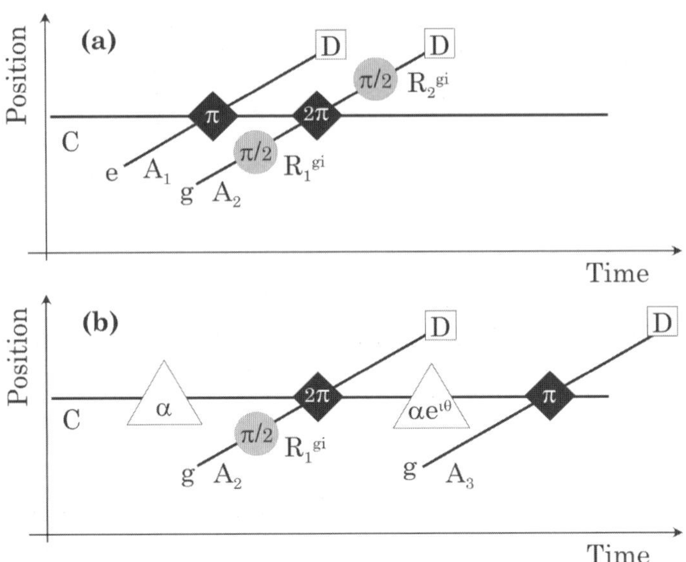

Fig. 8. – Timing of the quantum phase gate experiment. Same conventions as in fig. 6. (a) Check of the atomic coherence phase shift. A single-photon state is prepared in C by atom A_1. (b) Check of the field phase shift. The triangles represent coherent field injection in the cavity mode (amplitudes α and $\alpha \exp[i\theta]$).

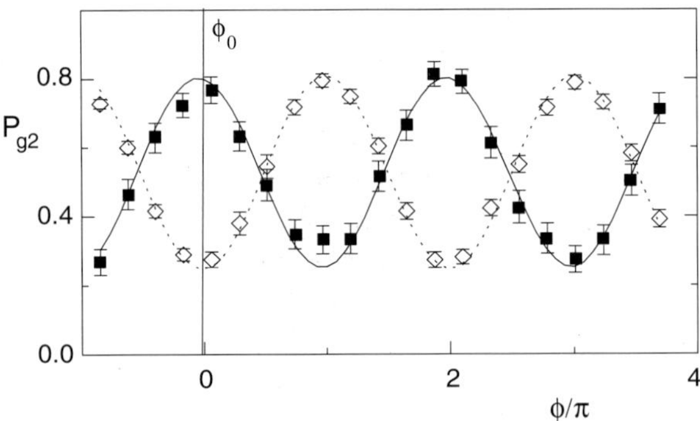

Fig. 9. – Test of the atomic coherence phase shift. Probability P_{g_2} for detecting atom A_2 in state g *versus* the phase ϕ of the Ramsey interferometer (units of π). Open diamonds: empty cavity. Solid squares: cavity containing one photon. The error bars reflect the statistical variance. The lines are sine fits. For phase $\phi_0 = 0$, indicated by a vertical line, the atomic state is directly correlated to the photon number (i for zero photon, g for one photon).

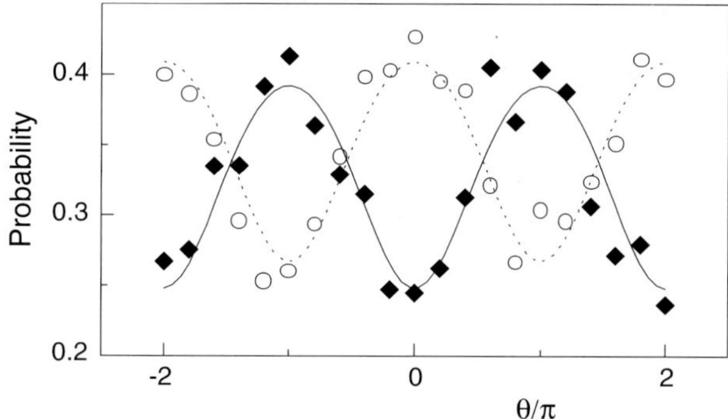

Fig. 10. – Test of the field phase shift. Conditional probabilities $P(e_3/i_2)$ (circles) and $P(e_3/g_2)$ (diamonds) *versus* θ (units of π) for detecting A_3 in e if A_2 has crossed C in i or g. Points are experimental and lines are fits based on a simple model.

$c_1|1\rangle + \sum_{n>1} c_n|n\rangle$ [64]. When the average photon number $|\alpha|^2$ is low (about 0.1), the c_n coefficients are negligible for $n > 1$ and, to a good approximation, $|\alpha\rangle \simeq c_0|0\rangle + c_1|1\rangle$, with $c_1 \simeq \alpha$. We then send atom A_2, prepared in a coherent superposition of g and i by a Ramsey $\pi/2$ pulse R_1^{gi}, and submit this atom to a 2π Rabi rotation in C (note that the R_2^{gi} pulse and atom A_1 are not used in this experiment). When A_2 is detected in i, the field phase should be unchanged. When A_2 is detected in g, the field state in C becomes $c_0|0\rangle + c_1 \exp[i\Phi]|1\rangle \simeq |\exp[i\Phi]\alpha\rangle = |-\alpha\rangle$. The classical phase of the coherent field is thus π-shifted.

To detect this shift, we analyze the field phase by a "homodyning" method: we inject, after A_2 has left C, a field with the amplitude $\alpha \exp[i\theta]$. It adds coherently to the field already present in C. The phase $\theta = T\Delta$ depends upon the detuning Δ between S and C ($T = 100\,\mu s$ is the delay between the two field injections). The amplitude of the resulting field ideally varies between 0 and 2α as a function of θ. The final field is probed by sending an atom A_3, initially in g, across C. It undergoes a π-Rabi pulse in the field of 1 photon. Hence, the probability P_{e_3} for detecting A_3 in e is ideally equal to the probability for finding a single photon in C. This probability is approximately equal to the average photon number. The timing of this experiment is depicted in fig. 8(b). The two field injections in the cavity mode are schematized by triangles on the cavity space-time line.

Figure 10 shows, *versus* θ, the conditional probabilities $P(e_3/i_2)$ and $P(e_3/g_2)$ for finding the probe A_3 in e provided A_2 was found in i (circles) or in g (diamonds). The lines are obtained by a simple model which accounts for the experimental imperfections by adjustable contrasts and offsets. The modulation in these signals reflects the interference of the two field pulses. When A_2 is detected in i, the probe absorption is minimum for $\theta = \pm\pi$ since the amplitudes of the two injected fields then add with opposite phases. When, instead, A_2 is detected in g, the probe absorption becomes minimum for $\theta = 0$, which exhibits clearly the π phase shift produced by atom A_2 in this case.

The data of figs. 9 and 10 demonstrate clearly the two complementary aspects of the quantum phase gate, expressed by eqs. (15) and (16). Note that the phase Φ can be tuned over the complete 0–2π range by adjusting the atom-cavity detuning [55].

7. – Absorption-free detection of a single photon

The quantum phase gate, sandwiched between two $\pi/2$ Ramsey pulses, realizes a "Control-Not" gate[4] which can be used to detect a single photon without absorbing it. A simple inspection of the signal in fig. 9 shows that, at a fringe extremum (phase $\phi_0 = 0$ depicted by a vertical line), the final atomic state is correlated to the photon number. In an ideal experiment, the atom exits in i if the cavity is empty, in g if C contains one photon. The atom appears as a microscopic measuring device, carrying away information about the photon number. This field intensity measurement does not change the photon number. Note that the question whether the photon left in C is the same as or a copy of the original one has no meaning in quantum mechanics, since photons in a field mode are indiscernible. The only pertinent notion is the cavity quantum state, which remains unchanged if it is in a 1-photon Fock state. This measurement is an absorption-free or "Quantum Non-Demolition" one, provided the field is in the $\{|0\rangle, |1\rangle\}$ subspace. Out of this subspace, the 2π Rabi rotation condition is not fulfilled for all photon numbers n (because of the $\sqrt{n+1}$ dependence of the Rabi frequency —see sect. 4). This method thus amounts to a single-photon QND detection (SP-QND) [56].

Quantum Non-Demolition measurements have been proposed in the '70s to improve the sensitivity of position or velocity measurements in gravitational wave detectors [75, 76]. A QND measurement realizes basically the ideal quantum measurement presented in quantum mechanics textbooks [77]. It gives as a result one of the eigenvalues of the measured observable and projects the measured system on the corresponding eigenstate. Provided it is also an eigenstate of the free Hamiltonian, there is no evolution after the measurement. Subsequent QND measurements give the same result again and again. Any difference between two consecutive measurements provides thus a very sensitive probe of a perturbation acting between them. Most realistic particle measurements in physics (including photon counting) are far from realizing these ideal conditions. Usually in photodetection, for instance, the photon is absorbed by a photosensitive surface. A QND detector for light must be perfectly transparent. Although this may appear as a strange property, it is, as this experiment shows, compatible with quantum laws.

In fact, QND measurements have been widely studied in quantum optics [78]. Most experiments so far have involved the interaction of a "signal" beam with a weak "meter" beam. Both beams travel together in a non-linear transparent medium (non-linear crystal [79], atomic vapor close to a resonance line [80], optical fiber [81]). The meter experiences an index of refraction proportional to the signal light beam intensity. This

---

(4) In this gate, one of the qubits, the "control", conditions, depending on its state, the evolution of the other, the "target". The latter switches state when the control is in state $|1\rangle$ and remains unchanged when the control is in state $|0\rangle$.

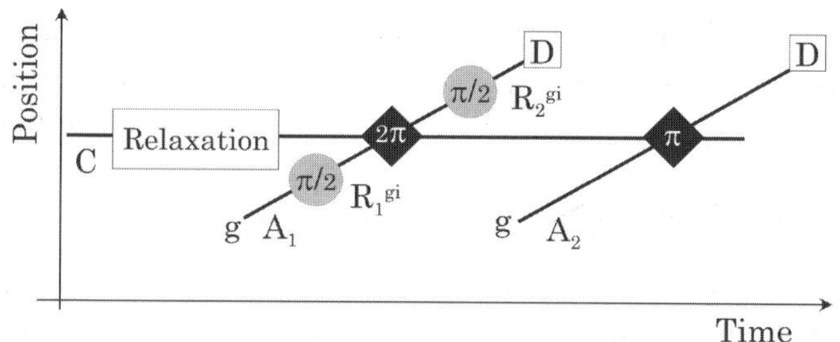

Fig. 11. – Timing of repeated measurement of a single photon: QND measurement of a thermal field followed by an absorptive one. Cavity relaxation after the cooling procedure builds up an initial 0.25 photon field.

index reflects directly on the meter beam phase at the exit of the medium, which is read out by comparison with a phase reference in an interferometric arrangement. At the output of the interferometer, the meter's intensity fluctuations directly reveal the signal ones. The signal intensity is not modified. The phase of the signal beam, however, is affected by the index changes due to the quantum fluctuations of the meter beam. In an ideal QND determination of the photon number, the signal phase is completely blurred.

In a variety of experimental conditions, the quantum correlations between the signal and meter have been demonstrated and improved over the years [78]. Repeated measurements have also been performed, exhibiting the essential properties of QND schemes [82, 83]. All of these experiments, relying on non-linear optical effects, operate only on macroscopic propagating laser beams, involving a very large number of photons.

Our SP-QND scheme bears strong analogies with these experiments. The signal is the single photon stored in the cavity. The meter is the circular atom, whose output phase is read by the Ramsey interferometer. The non-linear interaction is the 2π Rabi rotation. The extremely strong coupling of circular Rydberg atoms with millimeter-wave radiation makes it possible to push the QND principle down to the single-photon level [56]. Note that, as in other QND schemes, the SP-QND detection destroys phase information. If the field is initially in a superposition of zero- and one-photon states, the measurement reduces it to a Fock state, in which the phase information is completely lost. Note also that the average field energy is changed in an individual measurement. This is not surprising, since this measurement pins down the energy of a system initially presenting energy fluctuations.

In order to demonstrate the non-destructive feature of the SP-QND scheme, we have also realized repeated photon detections, either two successive QND detections, or one QND detection followed by an absorptive measurement. We will only describe the latter experiment, whose timing is depicted in fig. 11. We have performed the QND measurement on a small 0.25 photon thermal field, in which the probability for having more than one photon is small. We send atom A_1 to implement the SP-QND measurement. We then re-measure the cavity state by an absorptive measurement. A probe atom A_2 is sent in state g and undergoes a π pulse in a single-photon field. It thus exits in level e if the cavity contained one photon after the first measurement.

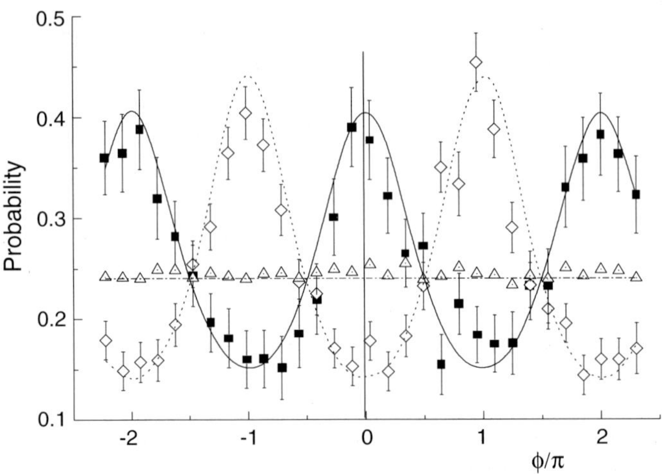

Fig. 12. – Repeated thermal field measurement. Square and diamond experimental points show the conditional probabilities $P(e_2/g_1)$ and $P(e_2/i_1)$ for finding the second atom in e, provided the first one has been measured in g or i, *versus* the Ramsey interferometer phase ϕ. The lines are obtained by numerical simulations. The triangles correspond to the probability $P(e_2)$ for finding A$_2$ in e when no A$_1$ atom is sent. The phase $\phi_0 = 0$ is indicated by a vertical line.

The results of this experiment are shown in fig. 12. Repeating the double measurement sequence many times, we have plotted *versus* the phase ϕ of the Ramsey interferometer the conditional probabilities $P(e_2/g_1)$ (squares) and $P(e_2/i_1)$ (diamonds) for detecting A$_2$ in e provided the SP-QND atom A$_1$ has been detected in i or g. We have also plotted, as a reference, the probability $P(e_2)$ for finding A$_2$ in e when no SP-QND atom A$_1$ is sent (triangles), *i.e.*, the results of a simple absorptive measurement of the initial thermal field.

The modulations of the absorption rates $P(e_2/g_1)$ and $P(e_2/i_1)$ reveal the field modifications due to the SP-QND measurement performed by A$_1$. Let us focus, for example, on the particular case $\phi = 0$. Atom A$_1$ indicates one photon when detected in g, zero photon when detected in i. In an ideal experiment, the field is then projected onto the corresponding state. A$_2$ should thus absorb a photon and be detected in e if A$_1$ is detected in g. We effectively observe that the probability $P(e_2/g_1)$ is larger than P_{e_2} for this phase setting. The *a posteriori* probability for having one photon after the SP-QND measurement is higher than the *a priori* one if A$_1$ is detected in g. This reveals the non-absorptive nature of the measurement and exhibits its repeatability. When the SP-QND atom A$_1$ is detected in i, the conclusions are reversed and the probability for getting A$_2$ in e is lower than the *a priori* one. Note also that, for phase $\phi = \pi/2$, the three probabilities $P(e_2)$, $P(e_2/g_1)$ and $P(e_2/i_1)$ are the same. For this phase, A$_1$ is always detected with a 50% probability in i or g. It does not convey any information on the field intensity and its detection does not modify the field state. Note that the sum of the $P(e_2/g_1)$ and $P(e_2/i_1)$ signals weighted by the probabilities for finding A$_1$ in g or i, respectively, which represents the mean photon number in the thermal field, is equal to $P(e_2)$, as required by ideal measurement theory.

The observed conditional probabilities do not reach the ideal zero and one values due to various experimental imperfections [56]. By performing a careful analysis of such signals, we have made a quantitative determination of the efficiency of our QND scheme. It allows to tell, with an 80% success rate, whether there is zero or one photon in C immediately after the passage of the meter atom. The probability that this atom has absorbed the photon, due to the imperfection of the 2π Rabi pulse, is about 10%.

8. – Engineered entanglement of three quantum systems

The quantum phase gate, which does not affect the photon number in C, can, in principle, be used to entangle an arbitrary number of atoms with the field. We made a first step in this direction by preparing an entangled state of three particles [57]. We prepare a maximally entangled state of the GHZ type by operating a quantum phase gate on an atom-cavity EPR pair. The sequence of operations could be modified to produce an arbitrary entangled state.

The entangled state preparation timing is shown in fig. 13(a). We send across C, initially empty, a first atom A_1 initially in e. It undergoes a $\pi/2$ Rabi rotation in C. The A_1-C EPR state is then given by eq. (8). A second atom A_2, initially in g, is prepared, before C, in $(|g\rangle + |i\rangle)/\sqrt{2}$ by a Ramsey pulse P_2^{gi}. This atom undergoes a 2π Rabi rotation in C, and thus performs a quantum phase gate operation with the cavity field as the control qubit. If C contains one photon, the A_2 coherence is phase-shifted by π. It stays unchanged if C is empty. The resulting A_1-A_2-C quantum state is

$$(17) \qquad |\Psi_{\text{triplet}}\rangle = \frac{1}{2}[|e_1\rangle(|i_2\rangle + |g_2\rangle)|0\rangle + |g_1\rangle(|i_2\rangle - |g_2\rangle)|1\rangle].$$

This three-particle entangled state can be rewritten as

$$(18) \qquad |\Psi_{\text{triplet}}\rangle = \frac{1}{2}\big[|i_2\rangle(|e_1,0\rangle + |g_1,1\rangle) + |g_2\rangle(|e_1,0\rangle - |g_1,1\rangle)\big],$$

describing an atom A_2 entangled with an A_1-C EPR pair, whose phase is conditioned to the A_2 state. Defining the pseudo-spins states $|+_i\rangle$ ($|-_i\rangle$) (with $i = 1, 2$) as $|+_1\rangle = |e_1\rangle$ ($|-_1\rangle = |g_1\rangle$), $|\pm_2\rangle = (|g_2\rangle \pm |i_2\rangle)/\sqrt{2}$ and $|+_C\rangle = |0\rangle$ ($|-_C\rangle = |1\rangle$), $|\Psi_{\text{triplet}}\rangle$ takes the form of the GHZ three-spin state:

$$(19) \qquad |\Psi_{\text{triplet}}\rangle = \frac{1}{\sqrt{2}}(|+_1,+_2,+_C\rangle - |-_1,-_2,-_C\rangle).$$

In order to check the A_1-A_2-C entanglement, we copy the state of C onto a third atom A_3, prepared in g and undergoing a π Rabi rotation in a single-photon field. Within a phase, A_3 maps exactly the state of C and the two atoms plus cavity correlations are directly reflected on the three-atom correlations.

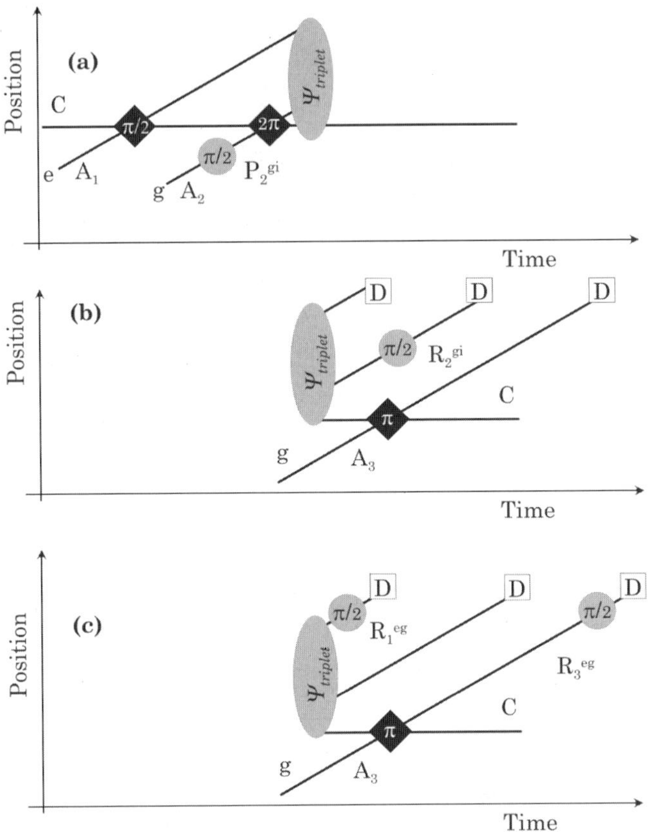

Fig. 13. – Timing of the three particle entanglement. (a) State preparation. (b) Test of "longitudinal" correlations. (c) Test of "transverse" correlations. Reprinted with permission from Rauschenbeutel *et al.*, *Science*, **288**, 2026. Copyright 2000, American Association for the Advancement of Science.

As in the analysis of the EPR pair (see sect. **5**), two sets of measurements in two orthogonal bases are required in order to characterize the three-particle entanglement. In a first experiment (experiment I, see timing in fig. 13(b)), we test "longitudinal" correlations by detecting the three atomic pseudo-spins along the Oz quantization axis. Atoms A$_1$ and A$_3$ are directly detected. For A$_2$, the $|+_2\rangle$ and $|-_2\rangle$ states (linear combinations of i and g) are mapped onto i and g, respectively, by a $\pi/2$ Ramsey pulse R_2^{gi}. Note that the complete sequence for A$_2$ amounts to a SP-QND determination of the field intensity. The three atoms should thus be detected in $\{e_1, i_2, g_3\}$ or $\{g_1, g_2, e_3\}$, with equal probabilities. Figure 14 presents the observed detection probabilities for the eight relevant channels. The longitudinal correlations (black bars) are clearly observed. The other channels (white bars) correspond to spurious effects.

These correlations, taken alone, could be explained classically as a statistical mixture

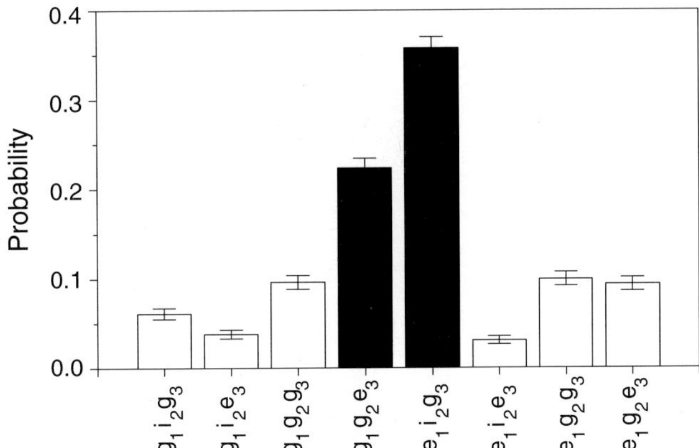

Fig. 14. – Longitudinal correlations (experiment I). Histograms of the detection probabilities for the eight relevant detection channels. The two expected channels (g_1, g_2, e_3 and e_1, i_2, g_3) (in black) clearly dominate the others (in white), populated by spurious processes. The error bars are statistical. Reprinted with permission from Rauschenbeutel *et al.*, *Science*, **288**, 2026. Copyright 2000, American Association for the Advancement of Science.

involving mainly $|e_1, i_2, g_3\rangle$ and $|g_1, g_2, e_3\rangle$. To exhibit the quantum coherence of the three-particle state superposition, we check, in experiment II, "transverse" pseudo-spin correlations. The corresponding timing is schematized in fig. 13(c). A_2 is detected directly along the pseudo-Oz direction (atom found in g or i). A_1 is detected along the Ox-axis, and A_3 along an axis in the horizontal plane making an angle ϕ with the Ox-axis. For these detections, we use the Ramsey pulses R_1^{eg} and R_3^{eg}, with a ϕ phase difference. We finally measure the Bell signal correlation $\langle \sigma_{x,1} \sigma_{\phi,3} \rangle$ conditioned by the final A_2 state. As a reference signal, we also record the A_1-A_3 Bell signal without sending the quantum phase gate atom A_2.

These signals are shown in fig. 15. The reference signal (diamonds) exhibits the EPR correlations discussed in sect. **5**. The A_1-A_3 correlations when A_2 is sent and detected in i and g are represented by circles and squares, respectively. In the first case, the phase of the A_1-A_3 EPR correlation is not changed whereas it is phase shifted by π in the second case. This shows that the phase of the EPR pair is controlled by the quantum phase gate atom (see eq. (18)).

The combined results of experiments I and II prove the entanglement of the three quantum systems. From the analysis of the data, we infer that the GHZ state is prepared with a 54% fidelity [57]. This experiment is the first one in which a controlled, tailorable entanglement has been produced between three individually addressed particles. It opens interesting perspectives towards the realization of stringent·tests of quantum non-locality.

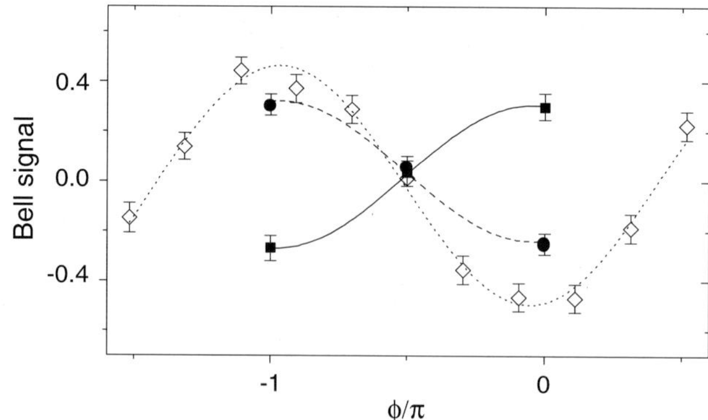

Fig. 15. – Transverse correlations (experiment II). Bell signal *versus* ϕ. Open diamonds: no A_2 atom. Solid circles: atom A_2 detected in i. Solid squares: atom A_2 detected in g. Error bars are statistical. Lines are sine fits. Reprinted with permission from Rauschenbeutel *et al.*, *Science*, **288**, 2026. Copyright 2000, American Association for the Advancement of Science.

9. – Non-resonant entanglement

9`1. *Single-atom index effect and "Schrödinger cat" states.* – Let us now consider the situation where the cavity mode is slightly detuned from the $e \to g$ transition frequency (level i does not play any role here). The atom-cavity frequency mismatch δ is greater than the vacuum Rabi frequency Ω. Since energy can no longer be conserved during elementary photon absorption and emission processes, the atom and the field cannot get entangled through resonant energy exchange. This does not mean, however, that the two subsystems do not interact. The mere presence of an atom in the cavity slightly shifts its frequency. This is a refractive index effect, produced by a single atom [48]. This effect depends upon the atom's energy state. When the cavity mode frequency is larger than the atomic transition one, the field frequency is increased or decreased, by an amount $\pm \Omega^2/4\delta$ depending upon whether the atom is in level g or e. When the atom crosses the cavity in a linear superposition of these levels, the cavity mode is then in a superposition of states corresponding to two different frequencies at once. This peculiar situation leads to the possibility of studying new kinds of entanglement between a microscopic system (the atom) and a mesoscopic field in the cavity [58]. These experiments are very briefly described here in qualitative terms.

Let us start by injecting in C, using source S, a small coherent field, with a classical amplitude α. This field decays within the cavity damping time T_r. Note that this experiment was realized with a cavity without ring. The photon storage time was 160 μs only. We then send an atom A_1 across C. The period of the field oscillations is slightly modified as the atom interacts with the field mode. This results in a kick of the field phase, after the atom has left C. Equivalently, the vector representing the field rotates in phase space by an angle $\pm \Phi = \pm \Omega^2 t_e/4\delta$, depending on the state of the atom, where t_e is the effective interaction time. If the atom is sent in the linear superposition $(|e\rangle + |g\rangle)/\sqrt{2}$, the combined atom-field system evolves into an entangled state which writes

$$(20) \qquad \frac{1}{\sqrt{2}}\left(e^{i\Phi}\left|e,\alpha e^{i\Phi}\right\rangle + \left|g,\alpha e^{-i\Phi}\right\rangle\right).$$

The situation is reminiscent of the famous "Schrödinger cat" [1] entangled with a single atom in a superposition of states corresponding to its "live" and "dead" states. We note that the vector describing the classical complex field amplitude in the Fresnel plane acts here as a kind of "meter" pointing in two different directions correlated to the atom's energy. In other words, the field plays the role of an apparatus measuring the atom's energy in a quantum non-demolition way. The situation is symmetric of the one described above (atom used to measure the field's energy non-destructively, see sect. **7**).

9`2. *A complementarity experiment*. – This leads us to an experiment demonstrating in a simple way the complementarity principle. The Ramsey fringes observed when the atom undergoes two $\pi/2$ microwave pulses resonant with the $e \Leftrightarrow g$ transition in R_1^{eg} and R_2^{eg} result from the interference of two quantum paths (see fig. 2). Interferences fringes are observed only if nothing in the apparatus allows us to distinguish these two channels. What happens then if C contains a small coherent field, non-resonant with the $e \Leftrightarrow g$ transition? Since the phase of this field is kicked by an angle depending upon the atom's state in C, this field acts as a "which path" detector [84, 85] able to reveal the atomic "path" through the Ramsey interferometer. According to the complementarity principle, the fringes then tend to vanish.

This is what we observe (see fig. 16). When the phase rotation Φ is smaller than the quantum phase fluctuations of the initial field and thus too small to permit to distinguish the two paths without ambiguity, the fringes contrast is merely reduced. The fringes disappear altogether when the phase rotation is large, removing all ambiguity on the atom's path. The field rotation Φ is adjusted by changing the detuning δ between the atomic transition and the cavity mode. A simple analysis of the experiment shows that the fringe signal is merely multiplied by the scalar product of the two final coherent states $\langle\alpha\exp[-i\Phi]|\alpha\exp[i\Phi]\rangle$. The modulus of this complex number accounts for the fringe contrast reduction, while its phase accounts for the fringe phase shifts apparent in fig. 16. This phase shift arises from the light shifts experienced by the atom inside the cavity field [59]. This phase shift provides a direct measurement of the mean photon number $\bar{n} = 9.5$ in this experiment. Note that similar complementarity experiments have been performed in quantum optics with other systems [86-90].

We have also recently performed a complementarity experiment in which the coherent field in the cavity itself replaces one of the two Ramsey pulses [91]. This field then plays the role of a "quantum beam splitter". When the coherent field contains a small photon number, it is appreciably modified by the emission of one photon by the atom. It then stores a piece of information about the atomic "path" in the Ramsey interferometer and no fringes are observed. When the photon number is large, the atom does not sensibly modify the field and fringes are visible.

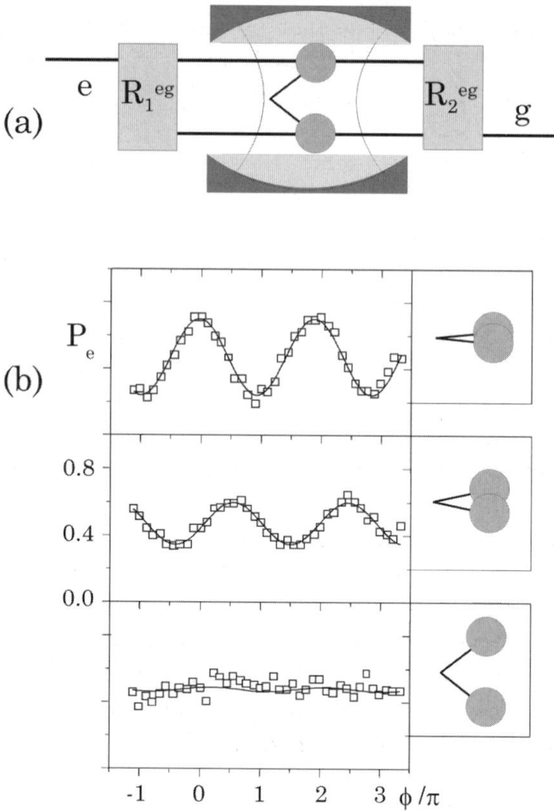

Fig. 16. – Complementarity experiment. (a) Principle of the experiment: an atom follows two interfering paths between the Ramsey zones and the phase of the field stored in the cavity provides a "which path" information. The two coherent components of the field are pictorially represented by a vector in the Fresnel plane. A small uncertainty circle on the tip of this vector depicts the field amplitude and phase quantum fluctuations. (b) Probability P_g for detecting the atom in level g is recorded, as a function of the Ramsey interferometer phase ϕ, for three different values of the field dephasing angle Φ corresponding to $\delta = 712$, 347 and 104 kHz from top to bottom. The average photon number is 9.5. The fringe contrast decreases when the separation of the field components, represented in the insets at right, is increased.

9˙3. *Decoherence caught in the act.* – Let us discuss now the quantum coherence of the Schrödinger cat state superposition described above. How long does it survive in C? To answer this question, we must analyze in more detail the nature of the field's "environment". Since the field losses are mainly due to photon scattering on mirror surface imperfections, we can describe this environment as being made of the free space around the cavity, which can be filled by the scattered photons. If the cavity contains on average $\overline{n}$ photons, a small field with about one photon escapes in the environment within a characteristic time $T_r/\overline{n}$. This microscopic field is entangled through its phase with the field left in C.

This entanglement with the environment provides a way to determine, at least in principle, the phase of the field in C. The phase of the tiny component escaping in the outside world could, in principle, be measured, reducing the field left in C to one or the other term, thereby destroying the quantum coherence[5]. This is the very essence of the decoherence phenomenon as analyzed by W. Zurek [2, 92]. Thus, after a time of the order of $T_r/\bar{n}$, the quantum coherence between the two field components in C has vanished. This explains why macroscopic fields, corresponding to huge $\bar{n}$ values, behave classically, since they experience a quasi-instantaneous decoherence process. In our experiment, however, $\bar{n}$ is of the order of 3 to 10 only. The decoherence time is then long enough to allow for the observation of transient interference signals associated to the two components of our "Schrödinger cat" state in C. Note that these conclusions do not depend on the details of the cavity relaxation. If the field losses occurred by absorption in the mirrors, instead of photon scattering, the electronic degrees of freedom in the mirrors would get entangled with the field, leading to the same dynamics for the decoherence process.

To observe decoherence in action, we send in C a delay τ after the first atom A_1 which prepares the "cat state", a second atom A_2, playing, so to speak, the role of a "quantum mouse" used to probe in C the quantum coherence of the field. This second atom kicks in turn the phase of the field components in C, recombining them partially. This recombination produces, in a correlation signal between the two atoms, an interference term sensitive to the quantum coherence between the cat-state components left by the first atom in C [58]. The A_1-A_2 correlation signal, $\eta = P(e_2/e_1) - P(e_2/g_1)$, is the difference of the conditional probabilities for finding A_2 in e provided A_1 has been detected either in e or g. The correlation η is independent of the Ramsey phase ϕ in the experimental conditions. It should be ideally 0.5 if the cavity contains a quantum superposition of coherent components and 0 when the decoherence process is completed.

The experimental correlation signal, $\eta(\tau)$, is shown in fig. 17, as a function of the time delay τ between A_1 and A_2, for two "cat" configurations pictorially depicted in the insets. As expected, the correlation decreases when τ is increased. This phenomenon occurs faster and faster when the two components of the cat are more and more separated. This provides a direct illustration of the main features of environment-induced decoherence, which acts faster and faster as the size of the systems become more and more macroscopic. Note the good agreement between the experimental points and the theoretical curves, deduced from a very simple calculation based on decoherence theory [93, 51]. This experiment, which verifies the main aspect of the decoherence theory, constitutes a step in the exploration of the quantum/classical boundary. Similar "Schrödinger cat" experiments [94] and decoherence studies [95] have been performed with an ion in a trap. In this case, however, the loss of coherence was due to the perturbing effect of a classical noise acting on the system and not to quantum entanglement with the environment.

---

[5] The fact that this measurement is unrealistic is irrelevant here. The mere fact that the information escapes in an unobserved environment is enough to destroy the quantum coherence of the "Schrödinger cat" state.

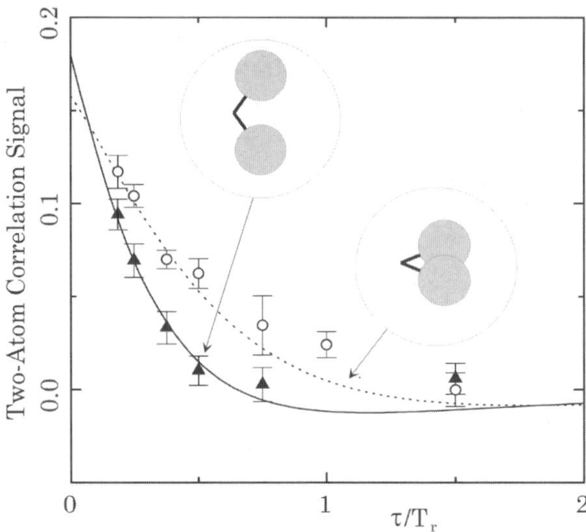

Fig. 17. – Decoherence experiment. Correlation signal η plotted *versus* the delay between the two atoms, τ. The initial coherent field contains 3.3 photons on the average. Experimental results for two different values of the field component separation are shown here (circles and triangles). The phase shifts, corresponding to $\delta/2\pi = 170$ and $70\,\mathrm{kHz}$, respectively, are depicted in the insets. The curves result from a simple analytical model.

10. – Conclusions and perspectives

We have described a very flexible and versatile set-up to generate and manipulate entanglement on a small ensemble of atoms and photons interacting either via resonant or dispersive interactions. We have prepared EPR pairs [54] of entangled atoms, demonstrated the operation of a quantum gate [55] and used it to perform the first quantum non-demolition detection of a single photon [56]. Combining up to six operations on four qubits (three atoms and one field mode), we have prepared and studied a GHZ entangled triplet [57]. In the context of quantum information processing, this experiment constitutes, to our knowledge, the most complex combination so far of successive logic operations involving individually addressable quantum systems. We have also prepared mesoscopic field state superpositions illustrating the main aspects of the "Schrödinger cat" paradox. By observing directly the evolution of these states, we have confirmed the basic features of environment-induced decoherence theories, in an experiment which provides a glimpse at the quantum/classical border [58].

The present set-up, however, suffers experimental limitations: non-ideal Ramsey and Rabi pulses limit the fidelity of complex entanglement manipulations. Cavity damping and residual thermal fields also contribute to decoherence processes. Finally, the Poisson statistics of the atomic source leads to prohibitively long data acquisition times for experiments involving more than three particle correlations. Various improvements are under way to overcome these difficulties. Better cavities without ring should allow us to manipulate atomic coherences for longer times and distances. Atomic sources based on cold atom techniques could be used to prepare single atoms on demand [96]. Finally, the detection efficiency could be boosted to nearly 100% and the thermal field background completely eliminated.

Among the possible experiments under way or under consideration, let us mention tests of quantum non-locality with massive particles (EPR, GHZ and more complex multiparticle situations [97]), quantum teleportation of atomic quantum states [98], unrestricted quantum non-demolition of photon numbers via dispersive interactions [99], measurement of the Wigner function of non-classical field states [100,101], cavity-assisted collisions between two or more Rydberg atoms, leading to other forms of qubit entanglement [102]. With deterministic single-atom sources, more complex manipulations of entanglement as well as the implementation of simple quantum algorithms or error correction codes will become possible. In these experiments the scalability implied by single-particle addressing will be an asset.

Finally, we are also considering, in the long term, experiments coupling two cavities via their interactions with a single atom. Two mesoscopic fields at macroscopic distances could thus be entangled, a situation which would marry the "strangeness" of the EPR and "Schrödinger cat" situations.

REFERENCES

[1] SCHRÖDINGER E., *Naturwiss.*, **23** (1935) 807, 823, 844.
[2] ZUREK W. H., *Phys. Today*, **44**, October issue (1991) 36.
[3] DIVINCENZO D. P., *Science*, **270** (1995) 255.
[4] EKERT A. and JOSZA R., *Rev. Mod. Phys.*, **68** (1997) 3733.
[5] EKERT A., *Phys. Rev. Lett.*, **67** (1991) 661.
[6] BENNETT C. H. *et al.*, *Phys. Rev. Lett.*, **70** (1993) 1895.
[7] SHOR P. W., in *Proceedings of the 35th Annual Symposium on the Foundations of Computer Science*, edited by S. GOLDWASSER (IEEE Computer Society) 1994.
[8] RAIMOND J. M., http://www.lkb.ens.fr/recherche/qedcav/english/englishframes.html.
[9] EINSTEIN A., PODOLSKY B. and ROSEN N., *Phys. Rev.*, **47** (1935) 777.
[10] BOHM D., *Quantum Theory* (Prentice-Hall, Englewood Cliffs, N.J.) 1951, pp. 614-623.
[11] BELL J. S., *Physics*, **1** (1964) 195.
[12] ASPECT A., DALIBARD J. and ROGER G., *Phys. Rev. Lett.*, **49** (1982) 1804.
[13] ZEILINGER A., *Rev. Mod. Phys.*, **71** (1998) S288.
[14] ASPECT A., *Nature*, **398** (1999) 189.
[15] GREENBERGER D. M., HORNE M. A. and ZEILINGER A., *Am. J. Phys.*, **58** (1990) 1131.
[16] ZUREK W. H., *Phys. Rev. D*, **24** (1981) 1516.
[17] ZUREK W. H., *Phys. Rev. D*, **26** (1982) 1862.
[18] CALDEIRA A. O. and LEGGETT A. J., *Physica A*, **121** (1983) 587.
[19] OMNÈS R., *The Interpretation of Quantum Mechanics* (Princeton University Press) 1994.
[20] VAN ENK S. J., CIRAC I. and ZOLLER P., *Phys. Rev. Lett.*, **78** (1997) 4293.
[21] STEANE A., *Phys. Rev. Lett.*, **77** (1996) 793.
[22] EKERT A. and MACCHIAVELO C., *Phys. Rev. Lett.*, **77** (1996) 2585.
[23] GROVER L. K., *Phys. Rev. Lett.*, **79** (1997) 325.
[24] LLOYD S., *Science*, **273** (1996) 1073.

[25] Haroche S. and Raimond J. M., *Phys. Today*, **49**, August issue (1996) 51.
[26] Bertoni A. *et al.*, *Phys. Rev. Lett.*, **84** (2000) 5912.
[27] Friedman J. R. *et al.*, *Nature*, **406** (2000) 43.
[28] van der Wal C. H. *et al.*, *Science*, **290** (2000) 773.
[29] Gershenfeld N. A. and Chuang I. L., *Science*, **275** (1997) 350.
[30] Jones J. A., Mosca M. and Hansen R. H., *Nature*, **393** (1998) 344.
[31] Braunstein S. L. *et al.*, *Phys. Rev. Lett.*, **83** (1999) 1054.
[32] Schack R. and Caves C. M., *Phys. Rev. A*, **60** (1999) 4354.
[33] Tittle W. *et al.*, *Phys. Rev. Lett.*, **84** (2000) 4737.
[34] Pan J.-W. *et al.*, *Nature*, **403** (2000) 515.
[35] Bouwmeester D. *et al.*, *Nature*, **390** (1998) 575.
[36] Boschi D. *et al.*, *Phys. Rev. Lett.*, **80** (1998) 1121.
[37] Furusawa A. *et al.*, *Science*, **282** (1998) 706.
[38] Rarity J. G., Owens P. C. M. and Tapster P. R., *J. Mod. Opt.*, **41** (1994) 2435.
[39] Jennewein T. *et al.*, *Phys. Rev. Lett.*, **84** (2000) 4729.
[40] Naik D. S. *et al.*, *Phys. Rev. Lett.*, **84** (2000) 4733.
[41] Monroe C. *et al.*, *Phys. Rev. Lett.*, **75** (1995) 4714.
[42] Turchette Q. A. *et al.*, *Phys. Rev. Lett.*, **81** (1998) 3631.
[43] Sackett C. A. *et al.*, *Nature*, **404** (2000) 256.
[44] Berman P., *Cavity Quantum Electrodynamics* (Academic Press) 1994.
[45] Thompson R. J., Rempe G. and Kimble H. J., *Phys. Rev. Lett.*, **68** (1992) 1132.
[46] Münstermann P. *et al.*, *Phys. Rev. Lett.*, **82** (1999) 3791.
[47] Hood C. J. *et al.*, *Science*, **287** (2000) 1447.
[48] Haroche S. and Raimond J. M., *Cavity Quantum Electrodynamics*, edited by P. Berman (Academic Press) 1994, p. 123.
[49] Raithel G. *et al.*, *Cavity Quantum Electrodynamics*, edited by P. Berman (Academic Press) 1994, p. 123.
[50] Haroche S., *Fundamental Systems in Quantum Optics, Les Houches Summer School Session LIII*, edited by J. Dalibard, J. M. Raimond and J. Zinn-Justin (North Holland, Amsterdam) 1992.
[51] Maître X. *et al.*, *J. Mod. Opt.*, **44** (1997) 2023.
[52] Raimond J. M. and Haroche S., *International Trends in Optics and Photonics, ICO IV*, edited by T. Asakura (Springer Verlag) 1999, p. 40.
[53] Brune M. *et al.*, *Phys. Rev. Lett.*, **76** (1996) 1800.
[54] Hagley E. *et al.*, *Phys. Rev. Lett.*, **79** (1997) 1.
[55] Rauschenbeutel A. *et al.*, *Phys. Rev. Lett.*, **83** (1999) 5166.
[56] Nogues G. *et al.*, *Nature*, **400** (1999) 239.
[57] Rauschenbeutel A. *et al.*, *Science*, **288** (2000) 2024.
[58] Brune M. *et al.*, *Phys. Rev. Lett.*, **77** (1996) 4887.
[59] Brune M. *et al.*, *Phys. Rev. Lett.*, **72** (1994) 3339.
[60] Maître X. *et al.*, *Phys. Rev. Lett.*, **79** (1997) 769.
[61] Hulet R. G. and Kleppner D., *Phys. Rev. Lett.*, **51** (1983) 1430.
[62] Gross M. and Liang J., *Phys. Rev. Lett.*, **57** (1986) 3160.
[63] Nussenzveig P. *et al.*, *Phys. Rev. A*, **48** (1993) 3991.
[64] Glauber R. J., *Phys. Rev.*, **131** (1963) 2766.
[65] Kim J. I. *et al.*, *Phys. Rev. Lett.*, **82** (1999) 4737.
[66] Ramsey N. F., *Molecular Beams* (Oxford University Press, New York) 1985.
[67] Rempe G., Walther H. and Klein N., *Phys. Rev. Lett.*, **58** (1987) 353.
[68] Jaynes E. T. and Cummings F. W., *Proc. IEEE*, **51** (1963) 89.
[69] Rauch H. *et al.*, *Phys. Lett. A*, **54** (1975) 425.

[70] WERNER S. A. *et al.*, *Phys. Rev. Lett.*, **35** (1975) 1053.

[71] WEIDINGER M. *et al.*, *Phys. Rev. Lett.*, **82** (1999) 3795.

[72] VARCOE B., BRATTKE S. and WALTHER H., *Nature*, **403** (2000) 743.

[73] EBERLY J. H., NAROZHNY N. B. and SANCHEZ-MONDRAGON J. J., *Phys. Rev. Lett.*, **44** (1980) 1323.

[74] ROWE M. A. *et al.*, *Nature*, **409** (2001) 791.

[75] BRAGINSKY V. B. and KHALILI F. Y., *Sov. Phys. JETP*, **46** (1977) 705.

[76] BRAGINSKY V. B. and KHALILI F. Y., *Quantum Measurement*, edited by K. S. THORNE (Cambridge University Press, Cambridge) 1992.

[77] CAVES C. M. *et al.*, *Rev. Mod. Phys.*, **52** (1980) 341.

[78] GRANGIER P., LEVENSON A. L. and POIZAT J. P., *Nature*, **396** (1998) 537.

[79] LA PORTA A., SLUSHER R. E. and YURKE B., *Phys. Rev. Lett.*, **62** (1989) 28.

[80] ROCH J. F. *et al.*, *Phys. Rev. Lett.*, **78** (1997) 634.

[81] LEVENSON M. D. *et al.*, *Phys. Rev. Lett.*, **57** (1986) 2473.

[82] BENCHEICK K., LEVENSON J. A. and LOPEZ O., *Phys. Rev. Lett.*, **75** (1995) 3422.

[83] BRUCKMEIER R., HANSEN H. and SCHILLER S., *Phys. Rev. Lett.*, **79** (1997) 1463.

[84] SCULLY M. O., ENGLERT B. E. and WALTHER H., *Nature*, **351** (1991) 111.

[85] HAROCHE S., BRUNE M. and RAIMOND J. M., *Appl. Phys.*, **54** (1992) 355.

[86] EICHMANN U. *et al.*, *Phys. Rev. Lett.*, **70** (1993) 2359.

[87] PFAU T. *et al.*, *Phys. Rev. Lett.*, **73** (1994) 1223.

[88] CHAPMAN M. S. *et al.*, *Phys. Rev. Lett.*, **75** (1995) 3783.

[89] DÜRR S., NONN T. and REMPE G., *Nature*, **395** (1998) 33.

[90] DÜRR S., NONN T. and REMPE G., *Phys. Rev. Lett.*, **81** (1998) 5705.

[91] BERTET P. *et al.*, *Nature*, **411** (2001) 166.

[92] ZUREK W. H., *Phys. World*, **10**, January issue (1997) 25.

[93] RAIMOND J. M., HAROCHE S. and BRUNE M., *Phys. Rev. Lett.*, **79** (1997) 1964.

[94] MONROE C. *et al.*, *Science*, **272** (1996) 1131.

[95] MYATT C. J. *et al.*, *Nature*, **403** (2000) 269.

[96] FRESE D. *et al.*, *Phys. Rev. Lett.*, **85** (2000) 3777.

[97] MERMIN N. D., *Phys. Rev. Lett.*, **65** (1990) 1838.

[98] DAVIDOVICH L. *et al.*, *Phys. Rev. A*, **50** (1994) R895.

[99] BRUNE M. *et al.*, *Phys. Rev. Lett.*, **65** (1990) 976.

[100] LUTTERBACH L. G. and DAVIDOVICH L., *Phys. Rev. Lett.*, **78** (1997) 2547.

[101] NOGUES G. *et al.*, *Phys. Rev. A*, **62** (2000) 054101.

[102] ZHENG S. B. and GUO G. C., *Phys. Rev. Lett.*, **85** (2000) 2392.

Optical cavity QED: Fundamentals and application as a single-photon light source

A. KUHN and G. REMPE

Max-Planck-Institut für Quantenoptik
Hans-Kopfermann-Str. 1, D-85748 Garching, Germany

1. – Introduction

Within the last few years, a major effort in the development of quantum cryptography, quantum communication and quantum information processing has been undertaken worldwide. Special attention has been paid to the realization of decoherence-free quantum systems for the storage of individual quantum bits (qubits) and to the conditional coupling between these qubits for the processing of quantum information. On the one hand, schemes coupling internal atomic states with external states of the atomic motion (*e.g.* in an ion trap) were studied, and on the other hand, the coupling of internal atomic states to quantized modes of the radiation field has been investigated. In this lecture, we concentrate on the latter. We focus our attention on an adiabatic coupling scheme between a single atom and an optical cavity, which allows one to populate either Fock states on demand, or to emit single photons into a well-defined mode of the radiation field outside the cavity. Such a deterministic single-photon source would not only enhance the security of quantum cryptography schemes [1], but would also enable optical quantum information processing with linear components [2].

The atom-cavity coupling scheme combines cavity quantum electrodynamics (CQED) with a technique known as stimulated Raman scattering by adiabatic passage (STIRAP) [3-5]. To describe the scheme, we first introduce the relevant features of CQED and the Jaynes-Cummings model [6, 7], which describes the interaction of a two-level atom with a single lossless cavity mode. Next, we discuss the transmission properties of a dissipative cavity containing an atom and the modification of the spontaneous emission properties of the atom, like the Purcell effect [8-11]. We then describe the adiabatic passage technique. It employs a three-level atom with two dipole transitions driven by two radiation fields. One of them comes from a laser, the other is that of a cavity strongly coupled to the atom. We show how to treat the atom and the two radiation fields in a dressed-state basis [12], and how the state vector of the coupled system can be changed in such a way that exactly one photon is placed into the cavity. This photon can escape through one of the mirrors, thereby forming a single-photon pulse [13]. We evaluate the adiabaticity criterion to estimate possible losses, and compare our experimental results with a numerical simulation of the process. For comparison, we also discuss a proposal [14] where the emission of a single photon from a coupled atom-cavity system is investigated in the bad-cavity regime.

2. – A short review of cavity quantum electrodynamics

In this section, we briefly review the interaction of a single two-level atom with a quantized mode of the radiation field in a cavity. First, the basic properties of optical cavities are introduced and a simple scheme to quantize the field is presented. We then introduce the interaction of a single two-level atom with a light field in free space, and extend this discussion to the full quantum-mechanical description of an atom coupled to a single mode of a cavity, leading to the Jaynes-Cummings picture. Finally, we describe the normal-mode splitting of a coupled atom-cavity system and the enhancement of the spontaneous-emission rate of an atom coupled to a cavity.

2$^{\cdot}$1. *Optical high-finesse cavities*. – We first restrict our discussion to a one-dimensional Fabry-Perot cavity with mirror separation l, mirror reflectivity, $\mathcal{R}$, close to one, mirror transmission, $\mathcal{T}$, and mirror absorption loss, $\mathcal{L}$. Light of wavelength λ and intensity I_{in} impinging on one of the cavity mirrors gives rise to a circulating intensity

$$(1) \qquad I_{\mathrm{circ}} = \frac{I_{\mathrm{in}} \mathcal{T}_{\mathrm{max}}/\mathcal{T}}{1 + 4(\mathcal{F}/\pi)^2 \sin^2(kl)},$$

where $k = 2\pi/\lambda$ is the wave number of the light, $\mathcal{T}_{\mathrm{max}} = \mathcal{T}^2/(1-\mathcal{R})^2$ the maximum transmission and

$$(2) \qquad \mathcal{F} = \frac{\pi\sqrt{\mathcal{R}}}{1-\mathcal{R}}$$

the cavity finesse. The frequency difference between two neighbouring transmission maxima is called the free spectral range, $FSR = \pi c/l$. Since we are interested in a high mirror reflectivity, the cavity transmission in the vicinity of a resonance can be described by the

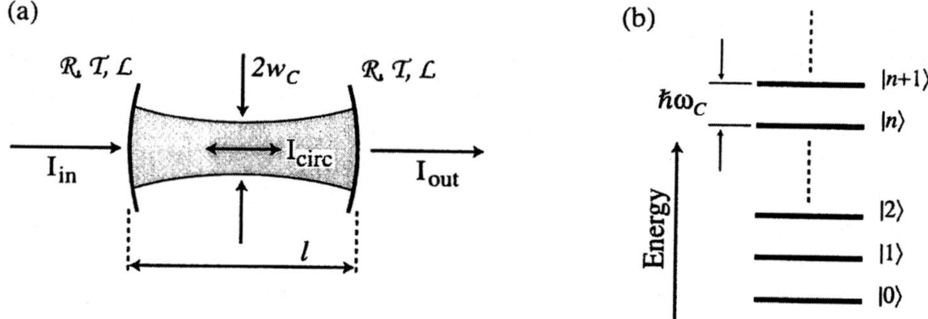

Fig. 1. – (a) One-dimensional Fabry-Perot cavity of length l with input intensity I_{in}, circulating intensity I_{circ} and transmitted intensity I_{out}. The mirrors have the reflectivity $\mathcal{R}$, transmission $\mathcal{T}$, and absorption loss $\mathcal{L}$. The waist of the $\mathrm{TEM_{00}}$ mode is w_{C}. (b) Ladder of Fock states (photon number states) showing the energy stored in a single $\mathrm{TEM_{00}}$ mode of frequency ω_{C}. As for a harmonic oscillator, the level spacing, $\hbar\omega_{\mathrm{C}}$, is equidistant, and the lowest-lying state $|0\rangle$ has a zero-point energy of $\hbar\omega_{\mathrm{C}}/2$.

Lorentzian

$$(3) \qquad \mathcal{T}_{\mathrm{cavity}} \propto \frac{\kappa^2}{\kappa^2 + \Delta_{\mathrm{L}}^2},$$

where $\Delta_{\mathrm{L}} = \omega_{\mathrm{L}} - \omega_{\mathrm{C}}$ is the difference between the laser frequency and the frequency of the cavity resonance. The linewidth (FWHM) is $2\kappa = FSR/\mathcal{F}$. Note that the decay rate of the cavity field, κ, is only determined by the mirror reflectivity and the spacing of the mirrors.

To achieve a stable single-mode operation in a real cavity, curved mirrors are used, as shown in fig. 1(a). As long as the mirror curvature, R, is large with respect to the mirror distance, all angles within the cavity remain small and the system can be treated in the paraxial approximation. In this case, Hermite-Gaussian or Laguerre-Gaussian cavity eigenmodes are obtained. We concentrate on the Gaussian $\mathrm{TEM_{00}}$ mode to simplify the further discussion. The smallest ($1/e$ field strength) radius of this mode is called its waist, w_{C}, which is given by

$$(4) \qquad w_{\mathrm{C}}^2 = \frac{\lambda}{2\pi}\sqrt{l(2R - l)}.$$

For a cavity length l shorter than the Rayleigh length, $z_{\mathrm{R}} = \pi w_{\mathrm{C}}^2/\lambda$, the variation of the mode diameter along the cavity axis can be neglected, and the spatial intensity distribution of the $\mathrm{TEM_{00}}$ mode is given by

$$(5) \qquad I_{\mathrm{sw}}(\mathbf{r}) = 4I_{\mathrm{circ}}|\psi(\mathbf{r})|^2 \qquad \text{with} \qquad \psi(\mathbf{r}) = \cos(kz)\exp\left[-\frac{x^2 + y^2}{w_{\mathrm{C}}^2}\right],$$

where the modefunction $\psi(\mathbf{r})$ reflects the standing wave in the cavity. It is normalized to one at the antinodes. The mode volume is defined by

$$(6) \qquad V \equiv \int |\psi(\mathbf{r})|^2 \, d\mathbf{r} = \frac{\pi}{4} w_{\mathrm{C}}^2 l.$$

2$\cdot$2. *Field quantization*. – In this section, we briefly discuss the quantization of the electromagnetic field in the cavity. Without loss of generality, we restrict ourselves to a single mode of the cavity with resonance frequency ω_{C}, as in the previous section. For each photon added to this mode, the energy stored in the cavity increases by $\hbar\omega_{\mathrm{C}}$, and for n photons in the mode the total energy is $\hbar\omega_{\mathrm{C}}(n + (1/2))$, where the zero-point energy of $\hbar\omega_{\mathrm{C}}/2$ has been associated with the vacuum field. The equidistant energy spacing imposes an analogue treatment of the cavity's photon-number states, the so-called Fock states, $|n\rangle$, to the energy levels of a harmonic oscillator, as shown in fig. 1(b). Consequently, creation and annihilation operators for a photon in the cavity mode, $a^\dagger$ and a, respectively, can be used to express the Hamiltonian of the cavity field,

$$(7) \qquad H_{\mathrm{C}} = \hbar\omega_{\mathrm{C}}\left(a^\dagger a + \frac{1}{2}\right).$$

Note that this Hamiltonian does not include losses. In a real cavity, all photon number states decay until the vacuum state, *i.e.* the state with no photon in the cavity, is reached.

2$\cdot$3. *Two-level atom coupled to a laser*. – Prior to the analysis of a two-level atom interacting with the quantized electromagnetic field in a cavity, we restrict our discussion to an atom being exposed to a monochromatic laser beam of frequency ω_{L}, which is treated classically. The atom is assumed to have two eigenstates, a ground state, $|g\rangle$, and an electronically excited state, $|e\rangle$, with energies $\hbar\omega_g$ and $\hbar\omega_e$, respectively. A dipole transition between these levels therefore occurs at the frequency $\omega_{eg} = \omega_e - \omega_g$. The Hamiltonian of the undisturbed atom reads

$$(8) \qquad H_{\mathrm{A}} = \hbar\omega_g|g\rangle\langle g| + \hbar\omega_e|e\rangle\langle e|.$$

The atom is exposed to an oscillating electric field $E(t) = (E_0/2)(e^{-i\omega_{\mathrm{L}} t} + \mathrm{c.c.})$, where E_0 is assumed to vary slowly in time. This is expressed by the interaction Hamiltonian

$$(9) \qquad H_{\mathrm{int}}(t) = -\frac{\hbar}{2}\left(\frac{\mu_{eg}2E(t)}{\hbar}\right)[|e\rangle\langle g| + |g\rangle\langle e|],$$

where μ_{eg} is the transition dipole moment between states $|g\rangle$ and $|e\rangle$. In a frame rotating with frequency ω_{L}, and in the rotating-wave approximation (RWA), the Hamiltonian of the coupled system, $H = H_{\mathrm{A}} + H_{\mathrm{int}}$, is given by

$$(10) \qquad H = \frac{\hbar}{2}\left(2\Delta|e\rangle\langle e| - \Omega|e\rangle\langle g| - \Omega|g\rangle\langle e|\right),$$

where $\Delta = \omega_{eg} - \omega_{\mathrm{L}}$ is the detuning of the laser from the atom's transition, and $\Omega = \mu_{eg} E_0/\hbar$ is the Rabi frequency of the driving field, assumed to be real by, e.g., an appropriate choice of the phases of the atomic wave functions. This system shows two non-degenerate eigenfrequencies,

$$(11) \qquad \omega^{\pm} = \frac{1}{2}\left(\Delta \pm \sqrt{\Omega^2 + \Delta^2}\right),$$

with the respective eigenstates

$$(12) \qquad \left|a^+\right\rangle = \cos\phi|g\rangle - \sin\phi|e\rangle \qquad \text{and} \qquad \left|a^-\right\rangle = \sin\phi|g\rangle + \cos\phi|e\rangle,$$

where the mixing angle ϕ is given by

$$(13) \qquad \tan\phi = \frac{\Omega}{\sqrt{\Omega^2 + \Delta^2} - \Delta} = \frac{\sqrt{\Omega^2 + \Delta^2} + \Delta}{\Omega}.$$

The level splitting between the two eigenstates, $\Omega_{\mathrm{eff}} = \sqrt{\Omega^2 + \Delta^2}$, is called the effective Rabi frequency. It is the rate at which the relative phase of the two eigenstates develops in time, according to $\Omega_{\mathrm{eff}} \times t$. If we now start from the assumption that both eigenstates are populated and ask for the probability to find the atom in either one of its bare states, interference terms oscillating at the frequency Ω_{eff} arise, i.e. part of the atomic population oscillates between states $|g\rangle$ and $|e\rangle$ at the effective Rabi frequency.

2`4. *Two-level atom coupled to a cavity.* – Now we assume that the two-level atom interacts with n photons in the relevant field mode of the cavity, as shown in fig. 2. In this case, the resonant Rabi frequency is

$$(14) \qquad \Omega_{\mathrm{C}} = 2g(\mathbf{r}) = 2g_0\sqrt{n}\,\psi(\mathbf{r}), \qquad \text{with} \qquad g_0 = \sqrt{\frac{\mu_{eg}^2 \omega_{\mathrm{C}}}{2\hbar\epsilon_0 V}}.$$

The description of an atom interacting with a quantized field is similar to the description of an atom interacting with a classical field of a given amplitude, with one important difference, namely conservation of energy in the absorption and emission processes. It is evident that any change of the atom's internal state must be reflected by a respective change of the cavity's photon number. It follows that the interaction Hamiltonian of the atom-cavity system now encompasses also the creation and annihilation operators, $a^\dagger$ and a, for the photons in the cavity. On the cavity axis, and in an antinode of the relevant mode, this Hamiltonian is given by

$$(15) \qquad H_{\mathrm{int}} = -\frac{\hbar}{2}(2g_0)\left[|e\rangle\langle g|a + a^\dagger|g\rangle\langle e|\right].$$

For any arbitrary excitation number, n, only the pair of product states $|g, n\rangle$ and $|e, n-1\rangle$ is coupled. Therefore, the situation corresponds to a set of independent two-level systems,

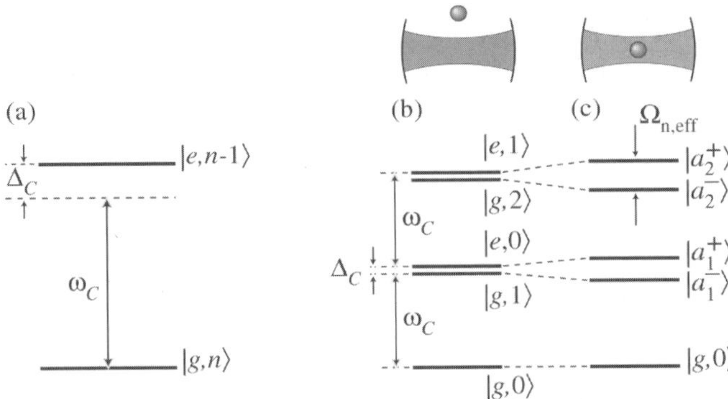

Fig. 2. – (a) A two-level atom with ground state $|g\rangle$ and excited state $|e\rangle$ coupled to a cavity containing n photons, yielding a Rabi frequency of $2g_0\sqrt{n}$. The detuning of the cavity from the atomic resonance is $\Delta_C = \omega_{eg} - \omega_C$. (b) Dressed-level scheme of the combined atom-cavity system while the atom is outside the cavity, *i.e.* for a vanishing coupling between atom and cavity. Note that the degeneracy of the state doublets is only lifted by the detuning Δ_C between atom and cavity. (c) The same as (b), but for an atom interacting with the cavity. The doublets are now split by the effective Rabi frequency, $\Omega_{n,\text{eff}} = \sqrt{\Omega^2 + \Delta_C^2} \equiv \sqrt{4ng_0^2 + \Delta_C^2}$. This scheme is named after Jaynes and Cummings, and the state doublets are the so-called Jaynes-Cummings doublets.

one for each excitation number n. Note that between the pair of states characterized by n and the neighbouring one with quantum number $n + 1$ or $n - 1$, an energy difference of $\hbar\omega_C$ results from the cavity Hamiltonian H_C.

Apart from this energy difference between neighbouring pairs of states, the analogy with single two-level systems can be used to write down the eigenfrequencies of the complete system Hamiltonian, $H = H_C + H_A + H_{\text{int}}$. In the rotating-wave approximation, they read

$$(16) \qquad \omega_n^\pm = \omega_C\left(n + \frac{1}{2}\right) + \frac{1}{2}\left(\Delta_C \pm \sqrt{4ng_0^2 + \Delta_C^2}\right),$$

where $\Delta_C = \omega_{eg} - \omega_C$ is the detuning between the atom and the cavity. The eigenstates of each coupled pair of states correspond to the ones given in (12), provided the replacements $|g\rangle \rightarrow |g,n\rangle$ and $|e\rangle \rightarrow |e, n - 1\rangle$ are made. Therefore, the atom-cavity interaction splits the photon number states into a doublet of two non-degenerate dressed states. These doublets are named after Jaynes and Cummings, who pioneered this way of describing a coupled atom-cavity system [6, 7].

It must be emphasized that there is one exception to the rule. Without excitation, both the atom and the cavity are in their ground states, $|g\rangle$ and $|0\rangle$, respectively. The only possible product state, $|g,0\rangle$, is not coupled to any other state. Therefore, its eigenfrequency, $\omega_0 = \omega_C/2$, is not altered, and no splitting occurs.

As in the case of a two-level atom coupled to a classical field, Rabi oscillations also occur in a cavity. For an atom initially in the ground state and n photons in the cavity,

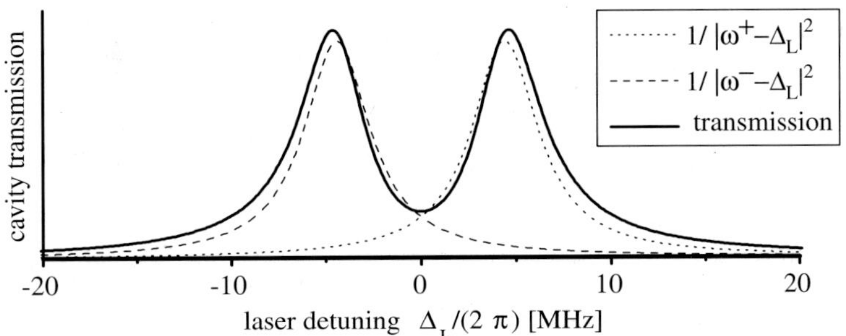

Fig. 3. – The frequency-dependent transmission of a cavity resonant with a single atom placed in its mode volume displays two peaks, indicating a normal-mode splitting. The atom-cavity parameters are $(g_0, \gamma, \kappa) = 2\pi \times (4.5, 3, 1.25)\,\mathrm{MHz}$. The cavity transmission is shown as a function of the laser-cavity detuning, $\Delta_\mathrm{L} = \omega_\mathrm{C} - \omega_\mathrm{L}$.

the probability to find the system in either one of the bare states, $|g, n\rangle$ or $|e, n - 1\rangle$, oscillates with the effective Rabi frequency $\Omega_{n,\mathrm{eff}} = \sqrt{4ng_0^2 + \Delta_\mathrm{C}^2}$. This means that the cavity field stimulates the emission of an excited atom into the cavity, thus de-exciting the atom and increasing the photon number by one. Subsequently, the atom is re-excited by absorbing a photon from the cavity field, and so forth. Since a single excited atom and a cavity containing initially only a vacuum field, *i.e.* no photon at all, are sufficient to start the oscillation between $|e, 0\rangle$ and $|g, 1\rangle$ at frequency $\sqrt{4g_0^2 + \Delta_\mathrm{C}^2}$, this oscillatory energy exchange is called vacuum-Rabi oscillation. For a resonant interaction, the oscillation frequency is $2g_0$, which is therefore called the vacuum-Rabi frequency.

2˙5. *Normal-mode splitting.* – For suitable parameters, the atom-cavity interaction leads to a splitting of the cavity transmission into two peaks, one below and the other above the unperturbed resonance frequency, ω_C, as shown in fig. 3. This effect is called normal-mode splitting. A detailed analysis goes beyond the simple Jaynes-Cummings model because the decay rates of the excited atomic level, 2γ, and the cavity field, κ, must be included, as well as the external field driving the atom-cavity system. We start our discussion by considering the relaxation rates first. These rates lead to a non-Hermitian Hamiltonian, which reads

$$(17) \qquad H' = H_\mathrm{C} + H_\mathrm{A} - \hbar g_0 \big(|e\rangle\langle g|a + a^\dagger|g\rangle\langle e|\big) - i\hbar\gamma|e\rangle\langle e| - i\hbar\kappa a^\dagger a.$$

For a cavity which is in resonance with the atom, and with n denoting the number of excitations, the eigenfrequencies of this damped system,

$$(18) \qquad \omega_n^\pm = \omega_\mathrm{C}\left(n + \frac{1}{2}\right) + \tilde{\omega}_n^\pm, \quad \text{with}$$

$$\tilde{\omega}_n^\pm = \pm\frac{1}{2}\sqrt{4ng_0^2 - (\gamma - \kappa)^2} - \frac{i}{2}\big(\gamma + \kappa(2n - 1)\big) \quad \text{for } n \geq 1,$$

are now given by complex numbers. The imaginary parts of the eigenfrequencies cause a decay of the eigenstates, $|a_n^\pm\rangle$, while the real part of $\tilde{\omega}_n^\pm$ describes the splitting. Obviously, a necessary condition for a level splitting to occur is

$$(19) \qquad 4ng_0^2 > (\gamma - \kappa)^2.$$

Otherwise, $\tilde{\omega}_n^\pm$ would be purely imaginary, describing an overdamped system. But note that condition (19) is not sufficient to resolve the level splitting. This is only possible if the real part of $\tilde{\omega}_n^\pm$ is larger than its imaginary part, which is given by the averaged damping rate $(\gamma + \kappa(2n - 1))/2$.

In case of weak excitation of the atom-cavity system, only the two states with $n = 1$ are relevant for the transmission properties of the cavity. The difference between the two corresponding eigenfrequencies amounts to

$$(20) \qquad \Omega_{\text{split}} = \sqrt{4g_0^2 - (\gamma - \kappa)^2}.$$

This splitting can be resolved for sufficiently small decay rates, $(\kappa, \gamma) \ll g_0$. This condition characterizes the strong-coupling regime of cavity QED. Note that the presence of the decay channels can lead to a significant reduction of the splitting with respect to a lossless atom-cavity system.

To calculate the cavity transmission quantitatively, we follow [15] and consider a weak laser beam impinging on one cavity mirror. The corresponding pump Hamiltonian,

$$(21) \qquad H_{\text{P}} = -i\hbar\eta\big(a - a^\dagger\big),$$

describes a coherent pumping of the cavity field with a rate η. As shown below, this leads to, on average, η^2/κ^2 photons in the resonant cavity containing no atom. With an atom present, and for sufficiently weak pumping, characterized by $\eta \ll g_0$ and $\eta < (\kappa + \gamma)/2$, the system is mainly in its ground state, $|g, 0\rangle$, and only the small fraction placed in the first excited doublet is of interest. In this case, the relevant state vector of the coupled atom-cavity system can be written as

$$(22) \qquad |\Psi(t)\rangle = c_e(t)|e, 0\rangle + c_g(t)|g, 1\rangle + c_0|g, 0\rangle.$$

With $c_0 \approx 1$, the pump Hamiltonian (21) describes a population flow into state $|g, 1\rangle$. As above, the atom and the cavity are assumed to be in resonance, whereas the laser can be detuned by an amount $\Delta_{\text{L}} = \omega_{\text{C}} - \omega_{\text{L}}$. In the rotating-wave approximation and in a frame rotating with the laser frequency, ω_{L}, the relevant terms of the Schrödinger equation now read

$$(23) \qquad i\dot{c}_e = -g_0 c_g + (\Delta_{\text{L}} - i\gamma)c_e \quad \text{and}$$

$$i\dot{c}_g = -g_0 c_e + (\Delta_{\text{L}} - i\kappa)c_g + i\eta.$$

The transmission of the cavity, $\mathcal{T}_{\text{cavity}}$, can be calculated from the average photon number, $\langle n \rangle$, which is given by the population of state $|g, 1\rangle$. Using the steady-state solution of (23) with $\dot{c}_g = \dot{c}_e = 0$, we get

$$(24) \qquad \mathcal{T}_{\text{cavity}}(\Delta_{\mathrm{L}}) \propto \langle n \rangle = c_g c_g^* = \left(\frac{\eta}{\kappa}\right)^2 \left| \frac{\kappa(\gamma - i\Delta_{\mathrm{L}})}{(\Delta_{\mathrm{L}} - \tilde{\omega}_1^+)(\Delta_{\mathrm{L}} - \tilde{\omega}_1^-)} \right|^2 .$$

The cavity transmission is shown in fig. 3. To guide the eye, two Lorentzians centred around $\Re(\tilde{\omega}_1^{\pm}) = \pm(1/2)\sqrt{4g_0^2 - (\gamma - \kappa)^2}$ with a width of $-\Im(\tilde{\omega}_1^{\pm}) = (1/2)(\gamma + \kappa)$ (HWHM) are also shown. Note that the cavity transmission cannot be decomposed into a sum of two Lorentzians.

We emphasize that the single-atom induced normal-mode spectrum as described here has not been observed yet. All experiments performed so far have either been carried out with several atoms randomly distributed over the cavity mode in such a way that their combined effect resembles that of a single atom at an antinode [16-18], or with a single atom exhibiting a complicated motional behaviour depending on the frequency of the probe laser [19].

2'6. *Enhanced spontaneous emission.* – It has been proposed by Purcell [8] and demonstrated by Heinzen *et al.* [9] and Morin *et al.* [11] that the spontaneous emission properties of an atom coupled to a cavity are significantly different from those in free space. For an analysis of the atom's behaviour, it suffices to look at the $n = 1$ doublet. As in the previous section, we include the decay rates γ and κ and make use of the Hamiltonian (17). There is no light coupled into the cavity, *i.e.* $\eta = 0$, and therefore the Schrödinger equation for the coefficients $c_{g,e}$ reads

$$(25) \qquad i\dot{c}_e = -g_0 c_g - i\gamma c_e \qquad \text{and} \qquad i\dot{c}_g = -g_0 c_e - i\kappa c_g .$$

Figure 4(a) shows the time evolution of the atom-cavity system when $\kappa \gg g_0$. The strong damping of the cavity's one-photon state inhibits any vacuum-Rabi oscillation, since the photon is emitted from the cavity before it can be reabsorbed by the atom. Therefore the transient population in state $|g, 1\rangle$ is negligible if the atom is initially (at $t = 0$) in its excited state $|e\rangle$. In this case, the adiabatic approximation $\dot{c}_g \approx 0$ can be applied, which gives

$$(26) \qquad \frac{\mathrm{d}}{\mathrm{d}t} c_e = -\gamma c_e - \frac{g_0^2}{\kappa} c_e, \qquad \textit{i.e.} \qquad c_e(t) = \exp\left[-\left[\gamma + \frac{g_0^2}{\kappa}\right] t \right].$$

It is obvious that the ratio of the emission probability into the cavity to the spontaneous emission probability into free space reads $g_0^2/(\kappa\gamma)$. This equals twice the one-atom cooperativity parameter, originally introduced in the context of optical bistability [20]. Note that the atom radiates mainly into the cavity if $g_0^2/\kappa \gg \gamma$. Together with $\kappa \gg g_0$, this condition constitutes the bad-cavity regime.

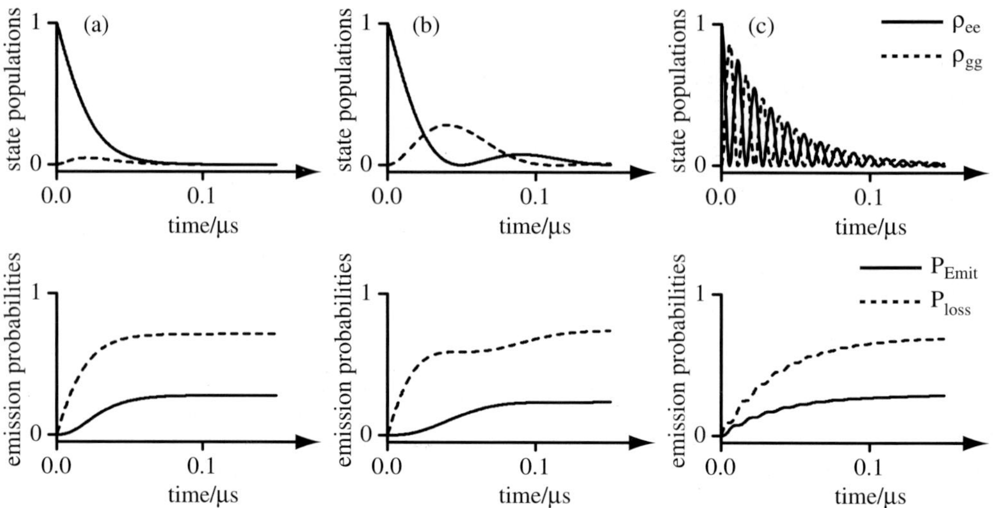

Fig. 4. – Time evolution of an excited two-level atom placed in a resonant optical cavity at $t = 0$. The initial photon number is zero. The upper row of diagrams shows the populations, ρ_{ee} and ρ_{gg}, of the product states, $|e,0\rangle$ and $|g,1\rangle$, respectively. The lower row of diagrams shows the corresponding integrated emission rate, *i.e.* the emission probabilities as a function of time, with P_{loss} denoting the probability for spontaneous emission not leading to a photon emission from the cavity, and P_{Emit} denoting the probability to emit a photon through one of the cavity mirrors. All traces in this figure result from a numerical evaluation of the master equation describing the coupled system. The three columns belong to three different coupling strengths and loss rates: (a) $(g_0, \gamma, \kappa) = 2\pi \times (4.5, 3, 12.5)\,\text{MHz}$; (b) $(g_0, \gamma, \kappa) = 2\pi \times (4.5, 3, 1.25)\,\text{MHz}$; (c) $(g_0, \gamma, \kappa) = 2\pi \times (45, 3, 1.25)\,\text{MHz}$.

In the strong-coupling regime, $g_0 \gg (\kappa, \gamma)$, however, the atom-cavity system is subject to a vacuum-Rabi oscillation between states $|e,0\rangle$ and $|g,1\rangle$, both decaying at the respective rates γ and κ. Figure 4(c) shows a situation where the atom-cavity coupling, g_0, saturates the $|e,0\rangle \leftrightarrow |g,1\rangle$ transition. On average, the probabilities to find the system in either one of these two states are equal, and the average ratio of the photon-emission probability out of the cavity, P_{Emit}, to the spontaneous photon-emission probability, P_{loss}, is given by κ/γ. The fast Rabi oscillation leads only to a small ripple on the photon emission and loss probabilities in the beginning of the interaction.

In the intermediate regime, the enhancement of the atom's spontaneous emission into the cavity cannot be written in terms of simple analytic expressions. As the bad-cavity regime and the strong-coupling regime display a dramatically different dynamical behaviour, not even an interpolation is allowed. To illustrate this statement, we assume, *e.g.*, atom-cavity parameters $(g_0, \gamma, \kappa) = 2\pi \times (4.5, 3, 1.25)\,\text{MHz}$, which lead to the time evolution shown in fig. 4(b). It is obvious that here the decay is neither exponential, as would be the case in the bad-cavity regime, nor displays ripples indicating the fast Rabi oscillation in the strong-coupling regime. Hence, if g_0 is too small to strongly saturate

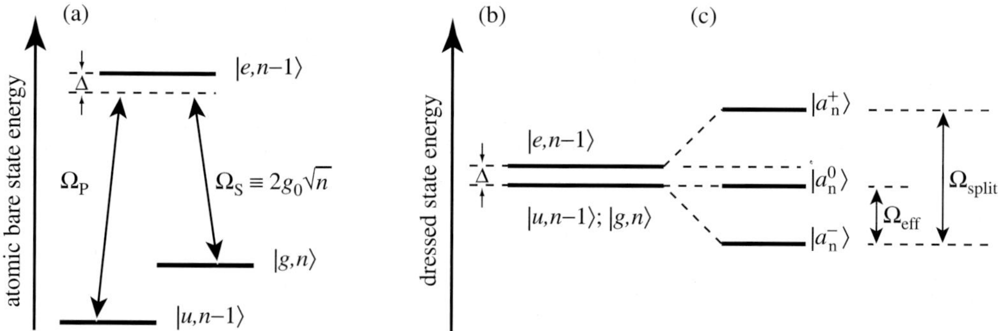

Fig. 5. – (a) A Λ-type three-level atom with two long-lived ground states $|u\rangle$ and $|g\rangle$ and excited state $|e\rangle$. The transition $|u\rangle \leftrightarrow |e\rangle$ is driven by a classical laser field with Rabi frequency Ω_{P}, and the $|g\rangle \leftrightarrow |e\rangle$ transition is either driven by a second classical laser field of Rabi frequency Ω_{S}, or coupled to a cavity containing n photons, with Rabi frequency $2g_0\sqrt{n}$. The common detuning of the cavity and the classical field(s) from their respective resonances is $\Delta = \omega_{eu} - \omega_{\mathrm{P}} = \omega_{eg} - \omega_{\mathrm{C}}$. (b) Dressed-level scheme of the combined system for vanishing coupling strengths between atom, fields and cavity. Note that the degeneracy of the state triplet is only partially lifted by the detuning Δ. (c) Same as (b), but for an atom interacting with laser field(s) and/or cavity. The state triplet is now split by $\Omega_{\mathrm{split}} = \sqrt{\Omega_{\mathrm{P}}^2 + \Omega_{\mathrm{S}}^2 + \Delta^2} \equiv \sqrt{\Omega_{\mathrm{P}}^2 + 4g_0^2 n + \Delta^2}$. Note that for a detuning Δ which is large compared to all Rabi frequencies, the Raman transition $|u\rangle \leftrightarrow |g\rangle$ is driven at the effective Rabi frequency $\Omega_{\mathrm{eff}} = (1/2)(\Omega_{\mathrm{split}} - |\Delta|) \approx (\Omega_{\mathrm{P}}^2 + \Omega_{\mathrm{S}}^2)/|4\Delta|$. This is not the usual Raman Rabi frequency, $\Omega_{\mathrm{Raman}} = \Omega_{\mathrm{P}}\Omega_{\mathrm{S}}/2\Delta$, in an "effective" two-level atom, because the differential light shift, $\Delta_{\mathrm{light}} = (\Omega_{\mathrm{P}}^2 - \Omega_{\mathrm{S}}^2)/4\Delta$, is taken into account. Our expression corresponds to $\Omega_{\mathrm{eff}} = \sqrt{\Omega_{\mathrm{Raman}}^2 + \Delta_{\mathrm{light}}^2}$. We emphasize that the Jaynes-Cummings doublets for a two-level atom-cavity system are now replaced by triplets. The dark state $|a_n^0\rangle$, which belongs to the triplet, is not shifted in energy.

the atomic transition, a simulation is mandatory to determine the system's dynamics. It reveals (see fig. 4(b)) a Rabi oscillation starting so slowly from state $|e,0\rangle$, that the initial spontaneous-emission loss from $|e,0\rangle$ dominates the photon emission probability from the cavity.

3. – Three-level atoms

Now, we consider an atom with a Λ-type three-level scheme, providing transition frequencies $\omega_{eu} = \omega_e - \omega_u$ and $\omega_{eg} = \omega_e - \omega_g$, as depicted in fig. 5. This atom is either interacting with two classical light fields of frequencies ω_{P} and ω_{S}, or, alternatively, with one classical field and a cavity mode, respectively. Assuming a Raman-resonant interaction, i.e. setting $\omega_{eu} - \omega_{\mathrm{P}} = \omega_{eg} - \omega_{\mathrm{S}} \equiv \Delta$, the behaviour of the atom is described in a rotating frame by the Hamiltonian

$$(27) \qquad H = \frac{\hbar}{2}\left[2\Delta|e\rangle\langle e| - \Omega_{\mathrm{S}}\left(|e\rangle\langle g| + |g\rangle\langle e|\right) - \Omega_{\mathrm{P}}\left(|e\rangle\langle u| + |u\rangle\langle e|\right)\right],$$

where Ω_P and Ω_S are the Rabi frequencies of the two fields, respectively. Note that we have used an extended version of the rotating-wave approximation to obtain this Hamiltonian—not only contributions rotating at twice the optical frequency were eliminated, but also all other off-resonant terms of the order of $\omega_g - \omega_u$ have been neglected, *i.e.* we assume that the pump laser (frequency ω_P) only couples to the $|u\rangle \leftrightarrow |e\rangle$ transition, and the Stokes laser (frequency ω_S) only to the $|g\rangle \leftrightarrow |e\rangle$ transition. This is justified whenever the condition $\max(\Omega_P, \Omega_S, |\Delta|) \ll |\omega_g - \omega_u|$ is met.

It is convenient to express the state vector of the atom coupled to the two radiation fields in the basis of the Hamiltonian's eigenstates, which read

$$(28) \qquad |a^0\rangle = \cos\Theta|u\rangle - \sin\Theta|g\rangle,$$

$$|a^+\rangle = \cos\Phi\sin\Theta|u\rangle - \sin\Phi|e\rangle + \cos\Phi\cos\Theta|g\rangle,$$

$$|a^-\rangle = \sin\Phi\sin\Theta|u\rangle + \cos\Phi|e\rangle + \sin\Phi\cos\Theta|g\rangle,$$

where the mixing angles Θ and Φ are given by

$$(29) \qquad \tan\Theta = \frac{\Omega_P}{\Omega_S} \quad \text{and} \quad \tan\Phi = \frac{\sqrt{\Omega_S^2 + \Omega_P^2}}{\sqrt{\Omega_S^2 + \Omega_P^2 + \Delta^2} - \Delta},$$

with Ω_P and Ω_S assumed to be real. The corresponding eigenfrequencies are

$$(30) \qquad \omega^0 = 0 \quad \text{and} \quad \omega^\pm = \frac{1}{2}\left(\Delta \pm \sqrt{\Omega_S^2 + \Omega_P^2 + \Delta^2}\right).$$

We note that the interaction with the light lifts the degeneracy of the three eigenstates as soon as the Rabi frequencies are non-zero. Furthermore, it must be emphasized that one of these states, namely $|a^0\rangle$, is neither subject to an energy shift, nor does the excited atomic state contribute to it. In the literature, $|a^0\rangle$ is therefore called a "dark state", since its population cannot be lost by spontaneous emission from the excited state $|e\rangle$, and because an atomic system prepared in $|a^0\rangle$ remains dark even when being exposed to pump and Stokes light of Rabi frequencies Ω_P and Ω_S, respectively. The lack of spontaneous emission from atoms in $|a^0\rangle$ manifests itself as a "dark resonance" in Raman spectra, when the two-photon Raman-resonance condition, $\Delta_P = \Delta_S$, is met.

The previously discussed three-level atom is now placed in the mode volume of a cavity, with the Stokes transition driven by the atom-cavity interaction in place of an externally applied field. Analogous to the case of a two-level atom discussed above, the Rabi frequency and the detuning of the Stokes transition need to be replaced by

$$(31) \qquad \Omega_S \longrightarrow 2g\sqrt{n},$$

$$(32) \qquad \Delta_S \longrightarrow \Delta_C.$$

In addition, the photon creation and annihilation operators, $a^\dagger$ and a, respectively, must be introduced in the interaction Hamiltonian. In a rotating frame, the interaction Hamiltonian now reads

$$(33) \quad H_{\text{int}} = \hbar\left[\Delta_{\text{P}}|u\rangle\langle u| + \Delta_{\text{C}}|g\rangle\langle g| - g\big(|e\rangle\langle g|a + a^\dagger|g\rangle\langle e|\big) - \frac{1}{2}\Omega_{\text{P}}\big(|e\rangle\langle u| + |u\rangle\langle e|\big)\right].$$

Note that the cavity-induced Stokes Rabi frequency is proportional to $\sqrt{n}$, since $\langle n|a^\dagger|n-1\rangle = \sqrt{n} = \langle n-1|a|n\rangle$. Given an arbitrary excitation number n, this Hamiltonian couples only the three states $|u, n-1\rangle$, $|e, n-1\rangle$, $|g, n\rangle$. For this state triplet and a Raman-resonant interaction with $\Delta_{\text{P}} = \Delta_{\text{C}} \equiv \Delta$, the eigenfrequencies of the coupled system read

$$(34) \qquad\qquad \omega_n^0 = \omega_{\text{C}}\left(n + \frac{1}{2}\right) \quad \text{and}$$

$$\omega_n^\pm = \omega_{\text{C}}\left(n + \frac{1}{2}\right) + \frac{1}{2}\left(\Delta \pm \sqrt{4ng^2 + \Omega_{\text{P}}^2 + \Delta^2}\right),$$

with eigenvectors according to (28). Of course, the replacement $|u\rangle \to |u, n-1\rangle$, $|e\rangle \to |e, n-1\rangle$ and $|g\rangle \to |g, n\rangle$ has to be made. As a consequence, the former Jaynes-Cummings doublet is now replaced by the state triplet $\{|a_n^0\rangle, |a_n^+\rangle, |a_n^-\rangle\}$ with the above-mentioned eigenfrequencies. In the limit of vanishing Ω_{P}, the states $|a_n^\pm\rangle$ correspond exactly to the Jaynes-Cummings doublet which has been discussed in the previous section, and the third eigenstate, $|a_n^0\rangle$, coincides with $|u, n-1\rangle$. Note that ω_n^0 is not affected by Ω_{P} or g. Therefore transitions between dark states of different n, $|a_n^0\rangle$, are always in resonance with the cavity frequency ω_{C}. This holds, in particular, for the transition from $|a_1^0\rangle$ to $|g, 0\rangle$, since the $n = 0$ state does not split (the corresponding states $|u, -1\rangle$ and $|e, -1\rangle$ do not exist).

4. – Adiabatic passage

Adiabatic passage techniques have been used for coherent population transfer in atoms or molecules for many years now. In two-level systems, usually a frequency chirp across the relevant resonance line is applied to a single driving field. In three-level systems, either a variation of the amplitudes of the two driving fields is used, or a frequency chirp across the relevant two-photon (Raman) resonance drives the passage. First realizations were in the rf-regime, with the aim to flip nuclear spins in NMR [21]. In the optical domain, frequency-chirped interactions have been realized for atomic or molecular samples, using either a Stark shift or the dynamical Stark effect to chirp the resonance across the fixed laser frequency [22-24]. This kind of frequency chirp has been extended to three-level systems. For example, it has been shown that population transfer can be achieved by a Stark-chirped rapid adiabatic passage (SCRAP) across the Raman resonance [25]. Moreover, driving the Raman transition with frequency-chirped laser pulses allowed a velocity-selective excitation and Raman cooling of atoms [26, 27]. Without frequency chirp, but with the Raman transition driven by two distinct pulses of variable amplitudes, effects like electromagnetically induced transparency (EIT) [28,29], slow light [30-33], and stimulated Raman scattering by adiabatic passage (STIRAP) [3-5] can be observed. We emphasize that these effects have all been demonstrated with classical light fields.

The common feature of these techniques is that the time evolution of the system's state vector,

$$(35) \qquad |\Psi(t)\rangle = \sum_i \alpha_i \exp\left[-i \int_0^t \omega_i(t')\,\mathrm{d}t'\right] |a_i(t)\rangle,$$

is only determined by the behaviour of the eigenstates, $|a_i(t)\rangle$, of the interaction Hamiltonian, and their respective eigenfrequencies, $\omega_i(t)$. The eigenstates as well as the eigenfrequencies may change in time, whereas the linear coefficients, α_i, are assumed to be non-varying in time throughout the whole interaction. In this context, the non-decaying dark state, $|a^0\rangle$, introduced above, is of enormous significance to the adiabatic passage phenomenon.

In principle, any three-level quantum system initially prepared in $|a^0\rangle$ stays in this state forever, thus allowing to control the relative population of the contributing atomic states, $|u\rangle$ and $|g\rangle$, by adjusting the values of the pump and Stokes Rabi frequencies, Ω_P and Ω_S, respectively. To show this, let us start with a system initially prepared in the state $|u\rangle$. As can be easily seen from (28), this state coincides with the dark state $|a^0\rangle$ if the condition $\Omega_\mathrm{S} \gg \Omega_\mathrm{P}$ is met in the beginning of the interaction:

$$(36) \qquad |\langle u \mid a^0\rangle|^2 = \frac{\Omega_\mathrm{S}^2}{\Omega_\mathrm{P}^2 + \Omega_\mathrm{S}^2} \xrightarrow{\Omega_\mathrm{S} \gg \Omega_\mathrm{P}} 1.$$

Similarly, for an atom interacting with a cavity in place of the stimulating Stokes field, the initial condition for the dark-state preparation reads $2\sqrt{n}g \gg \Omega_\mathrm{P}$. Once the system has been successfully prepared in the dark state, and if the state vector $|\Psi\rangle$ follows $|a^0\rangle$ during the interaction, the ratio between the populations of the contributing states always reads

$$(37) \qquad \frac{|\langle u \mid \Psi\rangle|^2}{|\langle g \mid \Psi\rangle|^2} = \frac{\Omega_\mathrm{S}^2}{\Omega_\mathrm{P}^2} \equiv \frac{4ng^2}{\Omega_\mathrm{P}^2} = \frac{|\langle u, n-1 \mid \Psi\rangle|^2}{|\langle g, n \mid \Psi\rangle|^2}.$$

Therefore, the final state of the system is determined by the relative amplitudes of the two fields when the interaction ends. In principle, this allows to establish arbitrary state superpositions of $|u\rangle$ and $|g\rangle$, or $|u, n-1\rangle$ and $|g, n\rangle$ if the atom is coupled to a cavity. In the latter case, the photon number in the cavity is affected in the same manner as the atomic states.

Up to this point, we have argued that the state vector of the system, $|\Psi\rangle$ is not only expressed in the basis of the eigenstates, but also that it *adiabatically follows* any change in the eigenstates with respect to the atom's and the cavity's bare state basis. Such a behaviour is not evident, and it holds only when the eigenstates are changing slowly. In order to analyse the criteria that must be met for "adiabatic following" in detail, we follow Messiah [34], and consider the state vector expressed by a superposition of the triplet of eigenstates,

$$(38) \qquad |\Psi\rangle = \alpha^0 |\tilde{a}^0\rangle + \alpha^+ |\tilde{a}^+\rangle + \alpha^- |\tilde{a}^-\rangle,$$

with the phase evolution included in the base vectors, $|\tilde{a}^0\rangle \equiv \exp[-i \int_{t_0}^{t} \omega^0(t')\,\mathrm{d}t']|a^0(t)\rangle$ and $|\tilde{a}^{\pm}\rangle \equiv \exp[-i \int_{t_0}^{t} \omega^{\pm}(t')\,\mathrm{d}t']|a^{\pm}(t)\rangle$. For a system prepared in the dark state $|\tilde{a}^0\rangle$ at time t_0, the probability to find it in either one of the other two eigenstates at a later time t_1 is

$$(39) \qquad P_{\pm}(t_1) = \left|\langle \tilde{a}^{\pm}(t_1) \mid \Psi(t_1)\rangle\right|^2 = \left|\alpha^{\pm}(t_1)\right|^2.$$

An adiabatic evolution of the interaction requires $P_{\pm}$ to be close to zero at all times. Using the Schrödinger equation, $\langle \tilde{a}^{\pm}|(\mathrm{d}/\mathrm{d}t)|\Psi\rangle = \langle \tilde{a}^{\pm}| - (i/\hbar)H|\Psi\rangle$, we get

$$(40) \qquad \langle \tilde{a}^{\pm}| \sum_{j=0,\pm} \left[(\dot{\alpha}^j - i\omega^j \alpha^j)|\tilde{a}^j\rangle + \alpha^j \exp\left[-i\int_{t_0}^{t} \omega^j\,\mathrm{d}t'\right]\frac{\mathrm{d}}{\mathrm{d}t}|a^j\rangle\right] = -i\omega^{\pm}\alpha^{\pm}.$$

As long as the non-adiabatic losses cause only a small perturbation, the approximations $\alpha^0 \approx 1$ and $\alpha^{\mp}\langle a^{\pm}|(\mathrm{d}/\mathrm{d}t)|a^{\mp}\rangle \approx 0$ are justified, and the transition probability to the non-dark eigenstates reads

$$(41) \qquad P_{\pm}(t_1) = \left| \int_{t_0}^{t_1} \left[\exp\left[-i\int_{t_0}^{t}(\omega^{\pm} - \omega^0)\mathrm{d}t'\right]\langle a^{\pm}|\frac{\mathrm{d}}{\mathrm{d}t}|a^0\rangle\right]\mathrm{d}t\right|^2.$$

The integrand on the right-hand side is the product of $\langle a^{\pm}|(\mathrm{d}/\mathrm{d}t)|a^0\rangle$ and an exponential oscillating at frequency $\omega^{\pm}-\omega^0$. In a sufficiently short time interval $[t_0 \dots t_1]$, we consider these two expressions to be time-independent. Hence (41) is easily integrated to give

$$(42) \qquad P_{\pm}(t_1) = \left|\frac{\langle a^{\pm}|(\mathrm{d}/\mathrm{d}t)|a^0\rangle}{\omega^{\pm} - \omega^0}\right|^2 2\left(1 - \cos\left[(\omega^{\pm} - \omega^0)(t_1 - t_0)\right]\right).$$

Now we allow $\langle a^{\pm}|(\mathrm{d}/\mathrm{d}t)|a^0\rangle$ and the difference frequency $\omega^{\pm} - \omega^0$ to vary smoothly in time. Therefore the probability, $P_{\pm}$, to loose population from the dark state will be limited by the maximum value attained by (42),

$$(43) \qquad P_{\pm} \leq \max \left|\frac{\langle a^{\pm}|(\mathrm{d}/\mathrm{d}t)|a^0\rangle}{\omega^{\pm} - \omega^0}\right|^2 \times 4.$$

From (43) we see that the condition for adiabatic following, $P_{\pm} \ll 1$, is satisfied if

$$(44) \qquad |\omega^{\pm} - \omega^0| \gg \left|\langle a^{\pm}|\frac{\mathrm{d}}{\mathrm{d}t}|a^0\rangle\right|$$

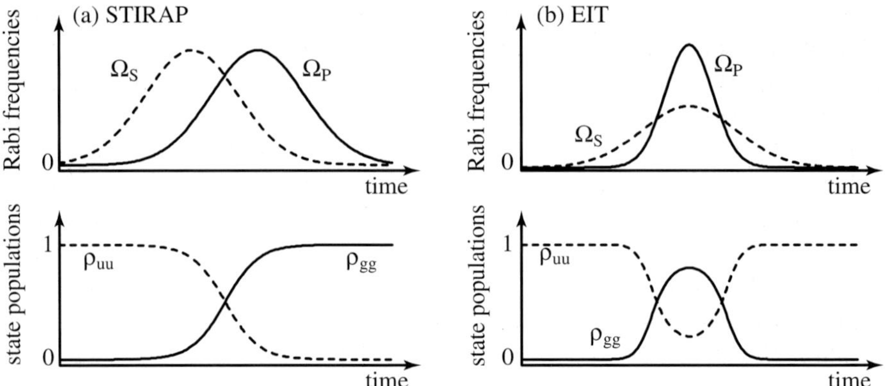

Fig. 6. – (a) Counter-intuitive interaction sequence for STIRAP realized by either two delayed Gaussian laser pulses, or by an atom travelling across two shifted Gaussian beam waists. If adiabatic following is assured, the $|u\rangle$-state population, ρ_{uu}, and the $|g\rangle$-state population, ρ_{gg}, behave as shown in the lower half of the diagram. The population is inverted, and both state populations are equal when Ω_P equals Ω_S. (b) The pump laser pulse is embedded in the stimulating Stokes laser pulse, which corresponds to the EIT (electromagnetically induced transparency) configuration. In case of adiabatic following, no net absorption and/or emission takes place, since the initial condition is re-established in the end. However, the evolution of the state populations, shown in the lower half, reveals that a transient population inversion occurs whenever $\Omega_P > \Omega_S$.

is met throughout the interaction. We emphasize that it is usually sufficient to fulfil this condition for the "most critical" moment in time, *i.e.* when the maximum of (43) is reached.

4`1. *Population transfer and Fock-state preparation*. – In Λ-type three-level systems, adiabatic passage has been successfully used for coherent population transfer between two long-lived ground states, $|u\rangle$ and $|g\rangle$, of atoms and molecules. The excitation is performed by means of a stimulated Raman transition and is known as STIRAP, first pioneered by Bergmann *et al.* [3,4]. Figure 6(a) illustrates this scheme: The atom first experiences a strong-field coupling states $|g\rangle$ and $|e\rangle$, *i.e.* $\Omega_S \gg \Omega_P$. Therefore, the initially populated state $|u\rangle$ coincides with the dark state $|a^0\rangle$. Provided adiabatic following is assured, any final condition leading to $\Omega_P \gg \Omega_S$ causes a complete population transfer to state $|g\rangle$, since $|a^0\rangle$ coincides with $|g\rangle$ in the end. This means that the interaction with the stimulating field has to start first, whereas the interaction with the pump field must be delayed. Such a sequence is often called to be counterintuitive. It can be realized by two Gaussian laser pulses of duration τ which are delayed by δ_t with respect to each other, with $\Omega_{P,S}(t) = \Omega_{P,S}^0 \exp[-(t \pm \delta_t/2)^2/\tau^2]$. Adiabaticity is best fulfilled when they are delayed by a time interval $\delta_t = -\tau$. If the two peak Rabi frequencies, Ω_P^0 and Ω_S^0, are nearly equal, the condition for adiabatic following reads

$$(45) \qquad \Omega_P^0 \approx \Omega_S^0 \gg \frac{1}{\tau}.$$

As proposed in [12], we now assume that the stimulating field, Ω_S, is replaced by the interaction of the atom with a cavity mode, populated with $n-1$ photons. If the atom is travelling through the cavity at velocity v, it experiences a Gaussian "stimulating pulse" on the $|g\rangle \leftrightarrow |e\rangle$ transition of duration $\tau = w_C/v$, where, again, w_C is the waist of the cavity mode. When the pump laser beam, coupling the $|u\rangle \leftrightarrow |e\rangle$ transition, is shifted slightly downstream, the atom is exposed to a counterintuitive interaction sequence and is therefore subject to a STIRAP-like population transfer. However, we have to take the change of the cavity's photon number into account: The coupled system evolves from $|u, n-1\rangle$ to $|g, n\rangle$, and therefore the photon number in the field mode increases by one for each atom travelling across the cavity. Once the photon is added, the atom leaves the cavity while it is still exposed to a strong pump field, with $\Omega_P \gg 2g\sqrt{n}$, and therefore the atom cannot reabsorb the photon again. If we furthermore neglect a possible decay of the photons from the cavity, *i.e.* for $\kappa = 0$, this method would allow one to prepare arbitrary Fock states [35]. Starting from an empty cavity, $|0\rangle$, and repeating the sequence with N atoms moving through the cavity one after the other, the Fock state $|N\rangle$ can be prepared.

It should be mentioned that adiabatic population transfer in a three-level atom is also possible via $|a^+\rangle$ or $|a^-\rangle$, provided the common detuning $\Delta = (\Delta_P + \Delta_S)/2$ is either positive or negative, respectively. In such a case, the contribution of $|e\rangle$ to $|a^+\rangle$ (or $|a^-\rangle$) becomes negligible, and the corresponding eigenstate is a pseudo-dark state. The adiabatic passage can be driven by a chirp of the Raman detuning, $\Delta_P - \Delta_S$, across zero. This method has been used for velocity-selective excitation of atoms and sideband cooling of neutral atoms [26, 27].

4́2. *Electromagnetically induced transparency.* – Closely related to the technique of adiabatic population transfer is the effect of electromagnetically induced transparency (EIT) [28, 29], which is illustrated in fig. 2(b). In this case, the pump laser pulse is shorter than the stimulating Stokes laser pulse, and it is "embedded" in the Stokes pulse in such a way that both the initial and the final conditions read $\Omega_S \gg \Omega_P$. Therefore, the population is transferred from state $|u\rangle$ to the dark state $|a^0\rangle$ with the rising edge of the Stokes laser pulse, and back to the initial state $|u\rangle$ with the trailing edge of the pulse. Despite a possible transient energy exchange between atom and fields during the interaction, no net energy is exchanged in the end, *i.e.* the photon numbers in the two laser pulses remain unchanged, and the system is subject to a coherent population return.

In this configuration, the stimulating Stokes laser pulse could also be replaced by the interaction of the atom with a field mode of a cavity. Provided the same timing is realized, *i.e.* the atom experiences a strong coupling to the cavity mode before and after it interacts with the pump beam, no change in the cavity's photon number takes place. It must be emphasized that a transient photon is placed in the cavity mode while the atom sees a strong pump laser pulse. Only if this photon is not emitted through one of the cavity mirrors, it is reabsorbed by the atom when the atom leaves the pump beam.

4·3. *Emission of photons*. – We now consider the special case where the cavity is empty in the beginning of the interaction with the atom, *i.e.* the initial state of the relevant field mode is $|0\rangle$. As we have seen, both the STIRAP and the EIT techniques are able to increase the photon number by one. It follows that a single photon is placed into the cavity, where it is either stored in case of STIRAP, or from where it is reabsorbed by the atom in the EIT configuration. However, a real optical cavity is always subject to photon losses because the cavity mirrors have a non-negligible transmission. Therefore a single photon will be emitted through one of the mirrors if the cavity-field decay time, κ^{-1}, is much shorter than the so-called "critical time", t_c, which is either the time needed to place a second photon into the cavity (for STIRAP), or the time to loose the photon by another mechanism, *e.g.* by photon reabsorption in the EIT configuration or by photon absorption in the mirror coatings.

5. – Single-photon sources

In subsect. 4·3, we have already pointed out that a coupled atom-cavity system could be employed as a single-photon source. The availability of such a source, which emits single photons into a well-defined mode of the radiation field, is the basis for many attempts to realize elementary quantum-logic gates, as well as feasible schemes for quantum cryptography and quantum teleportation. Moreover, all-optical quantum information processing based only on linear optics seems possible, provided highly efficient single-photon sources and single-photon detectors are available [2].

5·1. *State-of-the-art*. – So far, most schemes used for photon generation rely on spontaneous emission or parametric down-conversion and produce photons at more or less random times. Only during the last few years, different photon generation schemes have been demonstrated, like a single-photon turnstile device based on the Coulomb blockade mechanism in a quantum dot [36], the fluorescence of a single molecule [37,38], or a single colour centre (Nitrogen vacancy) in diamond [39,40], or the photon emission of a single quantum dot [41]. All these new schemes emit photons upon an external trigger event. However, the photons are spontaneously emitted into many modes of the radiation field and usually show a broad energy distribution. Only recently, cavity-enhanced spontaneous emission techniques for single-photon generation from a quantum dot have been proposed [42] and demonstrated [43,44].

5·2. *Cavity-enhanced spontaneous emission*. – Before describing a single-photon source based on the adiabatic passage scheme introduced in sect. **4**, we discuss a situation where a three-level atom is excited by a pump pulse, thereby emitting a photon into a cavity by enhanced spontaneous emission. This situation has been analysed in detail by Law *et al.* [45,14]. They consider the bad-cavity regime,

$$(46) \qquad \kappa \gg \frac{g^2}{\kappa} \gg \gamma,$$

where the loss of excitation into unwanted modes of the radiation field is small. In this regime, the cavity-field decay rate κ sets the fastest time scale, while the spontaneous-emission rate into the cavity, g^2/κ, dominates the usual incoherent decay rate, 2γ, of the population from the excited atomic level. It is assumed that any decay leads to a loss from the three-level system. Therefore the evolution of the wave vector is governed by the non-Hermitian Hamiltonian

$$(47) \qquad H' = H - i\hbar\kappa a^\dagger a - i\hbar\gamma|e\rangle\langle e|,$$

with H given in (33). To simplify the analysis, we consider only the vacuum state, $|0\rangle$, and the one-photon state, $|1\rangle$, of the cavity. Hence the state vector can be written as

$$(48) \qquad |\Psi(t)\rangle = c_u(t)|u,0\rangle + c_e(t)|e,0\rangle + c_g(t)|g,1\rangle,$$

where c_u, c_e and c_g are complex amplitudes. The time evolution of the amplitudes is given by the Schrödinger equation, $i\hbar(\mathrm{d}/\mathrm{d}t)|\Psi\rangle = H'|\Psi\rangle$, which yields

$$(49) \qquad i\dot{c}_u = \frac{\Omega_\mathrm{P}(t)}{2}c_e,$$

$$i\dot{c}_e = \frac{\Omega_\mathrm{P}(t)}{2}c_u + gc_g - i\gamma c_e,$$

$$i\dot{c}_g = gc_e - i\kappa c_g,$$

with the initial condition $c_u(0) = 1$, $c_e(0) = c_g(0) = 0$ and $\Omega_\mathrm{P}(0) = 0$. A so-called "adiabatic solution" of (49) can be derived if the exciting pump pulse satisfies $\Omega_\mathrm{P}(t) \ll g^2/\kappa$. Under these conditions, the decay is so fast that c_e and c_g are nearly time independent. This allows one to make the approximations $\dot{c}_e = 0$ and $\dot{c}_g = 0$, with the result

$$(50) \qquad c_u(t) \approx \exp\left[-\frac{\alpha}{4}\int_0^t \Omega_\mathrm{P}^2(t')\,\mathrm{d}t'\right],$$

$$c_e(t) \approx -i\frac{\alpha}{2}\Omega_\mathrm{P}(t)c_u(t),$$

$$c_g(t) \approx -i\frac{g}{\kappa}c_e(t),$$

where $\alpha = 2/(2\gamma + 2g^2/\kappa)$. Since photons are emitted out of the cavity from state $|g,1\rangle$ only, the photon emission rate is $R_\mathrm{E}(t) = 2\kappa|c_g(t)|^2$, and the photon emission probability reads

$$(51) \qquad P_\mathrm{E}(\tau) = 2\kappa\int_0^\tau |c_g(t)|^2\,\mathrm{d}t = \frac{g^2\alpha}{\kappa}\left[1 - \exp\left[-\frac{\alpha}{2}\int_0^\tau \Omega_\mathrm{P}^2(t)\mathrm{d}t\right]\right] \xrightarrow{\Omega_\mathrm{P}\tau\to\infty} \frac{g^2\alpha}{\kappa}.$$

Note that the exponential in (51) vanishes in case of a sufficiently large area, $\int_0^\tau \Omega_\mathrm{P}(t)\,\mathrm{d}t$, of the exciting pump pulse. In this limit, the photon emission probability does not depend

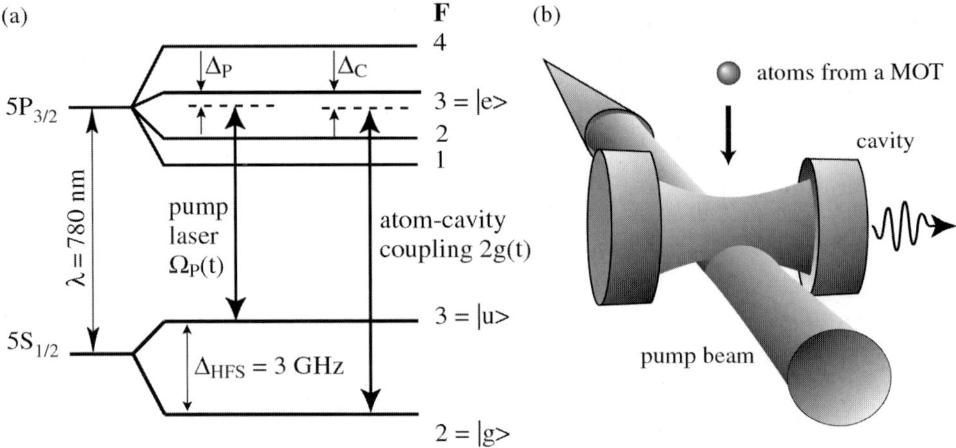

Fig. 7. – (a) Scheme of the relevant energy levels, transitions, and detunings of a ^{85}Rb atom coupled to a pump laser and a cavity. Δ_P and Δ_C denote the pump and cavity detunings from the respective atomic transitions. (b) Sketch of the experimental setup. The pump beam is displaced with respect to the cavity mode.

on the shape and amplitude of the pump pulse. As the other parameters, g, α, and κ, are in principle under the control of the experimentalist, arbitrary high photon emission probabilities could be reached. In [14], a realistic scheme using a Cs atom was proposed with $(g, \kappa, \gamma)/2\pi = (45, 45, 2.25)\,\mathrm{MHz}$. For an excitation of this system with a $3\gamma^{-1}$ long sine-squared pump pulse of peak Rabi frequency $\Omega_\mathrm{P}^0 = 2\pi \times 45\,\mathrm{MHz}$, a photon emission probability of 96.7% is found numerically.

$5^.3$. *Vacuum-stimulated Raman scattering.* – We now report on an experimental scheme [12,13] which employs a mixture of the STIRAP and EIT techniques discussed in sect. 4, and where the strong coupling of a single atom to a single cavity mode induces the Stokes transition of the Raman process.

Figure 7(a) depicts the excitation scheme for the ^{85}Rb atoms used in the experiment. A Λ-type three-level scheme is realized by the two $5S_{1/2}$ hyperfine ground states $F = 3$ and $F = 2$, which we label $|u\rangle$ and $|g\rangle$, respectively. The $F = 3$ hyperfine level of the electronically excited state $5P_{3/2}$ forms the intermediate state, $|e\rangle$. The atom interacts with a single mode of an optical cavity, whose frequency ω_C is close to the atomic transition frequency between states $|e\rangle$ and $|g\rangle$, but far off-resonance from the $|e\rangle$ to $|u\rangle$ transition. Hence, only the product states $|e, 0\rangle$ and $|g, 1\rangle$ are coupled by the cavity. For this transition, the vacuum Rabi frequency,

$$
(52) \qquad\qquad 2g(t) = 2g_0 \exp\left[-\left(\frac{t\,v}{w_\mathrm{C}}\right)^2\right],
$$

is time dependent since the atom moves with velocity v across the waist, w_C, of the

Gaussian cavity mode, as indicated in fig. 7(b).

The pump laser beam crosses the cavity axis at right angle. This beam is placed slightly downstream in the path of the atoms (by an amount $-\delta_X$ with respect to the cavity axis) and has a waist w_P, therefore causing a time-dependent Rabi frequency

$$(53) \qquad \Omega_P(t) = \Omega_P^0 \exp\left[-\left(\frac{t\,v + \delta_X}{w_P} \right)^2 \right].$$

The pump frequency is near resonant with the transition between $|u, 0\rangle$ and $|e, 0\rangle$, thereby coupling these states.

Due to the pump-beam displacement, the atom experiences the desired counterintuitive interaction sequence while travelling through the cavity. Since the cavity photon number is initially zero, and the atom is initially in state $|u\rangle$, the system's state vector $|\Psi\rangle$ should follow the dark state

$$(54) \qquad \left|a^0(t)\right\rangle = \frac{2g(t)|u, 0\rangle - \Omega_P(t)|g, 1\rangle}{\sqrt{4g^2(t) + \Omega_P^2(t)}},$$

of the $n = 1$ triplet, provided the condition for adiabatic following,

$$(55) \qquad \min\left(\frac{2g_0 w_C}{v}, \ \frac{\Omega_P^0 w_P}{v} \right) \gg 1,$$

is met. Furthermore, the detunings of the cavity, Δ_C, and of the pump pulse, Δ_P, from the corresponding atomic transition frequencies must be equal or, more precisely, the condition

$$(56) \qquad |\Delta_C - \Delta_P| < 2\kappa$$

must be fulfilled. Note that the cavity decay rate determines the linewidth of the Raman transition.

As stated in sect. **4**, adiabaticity does not assure photon emission. Hence, either the interaction time must be significantly longer than $(2\kappa)^{-1}$ to allow the emission, or the interaction with the pump beam must be strong when the atom leaves the cavity to avoid photon reabsorption. Once the photon is emitted, the empty cavity state, $|g, 0\rangle$, is reached and no further excitation is possible.

5`4. *Loss mechanisms*. – For each atom falling through the cavity, there are four possible scenarios, which either lead to no excitation at all, a loss of the excitation by spontaneous emission, or the desired emission of the photon from the cavity:

1) The atom is not excited and leaves the cavity in its initial state. Therefore, no photon is emitted from the cavity. In such a case, the system either stays in state $|u, 0\rangle$, or this state is repopulated by spontaneous emission from $|e, 0\rangle$, or is reestablished by coherent population return in the EIT configuration.

2) The atom is excited to state $|e, 0\rangle$ and emits a photon into free space outside the cavity. In this case the excitation is lost. If the atom reaches $|g, 0\rangle$, it decouples from any further interaction and leaves the cavity. Note that the atom can be "recycled" if it decays back to state $|u, 0\rangle$.

3) The atom is excited to $|e, 0\rangle$, and cavity-enhanced spontaneous decay leads to the emission of a photon into the cavity. Subsequently, the desired photon emission from the cavity takes place. In this way, the system reaches $|g, 0\rangle$ and decouples from any further interaction.

4) The population of state $|e, 0\rangle$ is always zero, and $|g, 1\rangle$ is reached by adiabatic passage via the dark state. The photon is emitted from the cavity as desired, and the system reaches $|g, 0\rangle$, thereby decoupling from any further interaction.

In the following subsection, some possible loss mechanisms are discussed in detail.

$5{\cdot}4.1.$ Non-adiabatic evolution. A partially non-adiabatic evolution of the system's state vector, $|\Psi\rangle$, to the non-dark eigenstates $|a^{\pm}\rangle$ is a possible loss channel for the photon generation process. An upper bound for these losses is given in (43), which can be evaluated further if some basic assumptions are made: First, the timing can be adjusted in such a way that the photon is most probably generated at $t = 0$, $i.e.$ when the atom is on the axis of the cavity. Hence, we can assume $\dot{g}(0) = 0$ and $g(0) = g_0$. Second, when $\Delta \gg \max(g_0, \Omega_{\rm P}^0)$, we have

$$(57) \qquad |a^+\rangle \approx -|e, 0\rangle \qquad \text{and} \qquad |a^-\rangle \approx \sin\Theta|u, 0\rangle + \cos\Theta|g, 1\rangle.$$

It can be seen from (28), (29) and (34) that the denominator of (43) is given by

$$(58) \qquad \left|\langle a^-|\frac{\mathrm{d}}{\mathrm{d}t}|a^0\rangle\right|^2 = |\dot{\Theta}|^2 \approx \left|\frac{2g_0\dot{\Omega}_{\rm P}(0)}{\Omega_{\rm P}^2(0) + 4g_0^2}\right|^2,$$

where $\Omega_{\rm P}(0)$ is the pump laser Rabi frequency on the cavity axis. From (53), we obtain for a nearly optimal situation with $2g_0 = \Omega_{\rm P}^0$ and $\delta_X = -w_{\rm P}$:

$$(59) \qquad \Omega_{\rm P}(0) = \frac{2g_0}{e} \qquad \text{and} \qquad \dot{\Omega}_{\rm P}(0) = -\frac{4v}{e w_{\rm P}}g_0,$$

where, again, $w_{\rm P}$ is the waist of the pump beam. Under these conditions, the denominator of (43) is

$$(60) \qquad \left|\langle a^-|\frac{\mathrm{d}}{\mathrm{d}t}|a^0\rangle\right|^2 = \left|\frac{2v}{w_{\rm P}e}\right|^2\left(1 + \frac{1}{e^2}\right)^{-2},$$

whereas the nominator is

$$(61) \qquad |\omega^{\pm} - \omega^0|^2 \approx \left|\frac{g_0^2}{\Delta}\right|^2\left(1 + \frac{1}{e^2}\right)^2.$$

Therefore, in case of a detuning Δ much larger than g_0, the non-adiabatic population losses to the non-dark eigenstates are limited by

$$(62) \qquad P_\pm \leq 4 \times \left| \frac{2v\Delta}{ew_\mathrm{P} g_0^2} \right|^2 \left(1 + \frac{1}{e^2} \right)^{-4} \approx 1.2 \times \left(\frac{v}{w_\mathrm{P}} \right)^2 \left(\frac{\Delta}{g_0^2} \right)^2.$$

If the detuning Δ is zero, an analogous estimation yields

$$(63) \qquad P_\pm \leq 4 \times \left| \frac{2v}{ew_\mathrm{P} g_0} \right|^2 \left(1 + \frac{1}{e^2} \right)^{-3} \approx 1.2 \times \left(\frac{v}{w_\mathrm{P}} \right)^2 \left(\frac{1}{g_0^2} \right).$$

For our experimental parameters, $v = 2\,\mathrm{m/s}$, $w_\mathrm{P} \approx w_\mathrm{C} = 35\,\mu\mathrm{m}$, and $g_0 = 2\pi \times 4.5\,\mathrm{MHz}$, non-adiabatic losses according to (63) are below 2×10^{-5} and, hence, negligible. A similarly small result holds for a detuning of $\Delta \approx 2\pi \times 10\,\mathrm{MHz}$. It follows that the Raman transition could be driven much faster than assumed here. For example, the atom could be excited by a pump pulse which matches the photon decay rate of the cavity, 2κ. In this case, one has to substitute v/w_P by 2κ in (63). For $2\kappa = 2\pi \times 2.5\,\mathrm{MHz}$, one finds that the resulting losses, $P_\pm \approx 1.2 \times (2\kappa/g_0)^2$, may reach 50% in this case.

5`4.2. Spontaneous emission. Even when the state vector of the system, $|\Psi\rangle$, follows $|a^0\rangle$ adiabatically, the excitation can possibly be lost by spontaneous emission if the excited state, $|e, 0\rangle$, makes a small contribution to state $|a^0\rangle$. In such a case, the otherwise dark state turns grey, and the time-dependent contributions of states $|e, 0\rangle$ and $|g, 1\rangle$ to state $|a^0\rangle$ determine the lost fraction. The ratio of the loss rate from the excited state to the total emission rate of the system gives the maximum spontaneous-emission loss,

$$(64) \qquad P_\mathrm{loss} \leq \max \left(\frac{2\gamma \rho_{ee}}{2\gamma \rho_{ee} + 2\kappa \rho_{gg}} \right),$$

where the populations of $|e, 0\rangle$ and $|g, 1\rangle$ are denoted by ρ_{ee} and ρ_{gg}, respectively. They are determined exclusively by the decomposition of $|a^0\rangle$ in case of an adiabatic evolution, since the other eigenstates do not contribute.

It is evident that $|a^0\rangle$ turns grey if the Raman-resonance condition, $\Delta_\mathrm{P} = \Delta_\mathrm{C}$, is not fulfilled. Moreover, it can bee seen from (47) that the cavity-field decay rate, κ, is equivalent to an imaginary "detuning" of the cavity from resonance, $\Delta_\mathrm{C} \equiv i\kappa$. This complex pseudo-detuning cannot be compensated by an appropriate complex pump detuning, since the latter should not have an imaginary part to avoid decay of the atom from the initial state $|u, 0\rangle$. A consequence of the imaginary detuning is that the excited state, $|e, 0\rangle$, contributes to state $|a^0\rangle$, even in case of a Raman-resonant excitation. It follows that part of the excitation is lost by spontaneous emission.

An analytical treatment of this effect is far beyond the scope of this lecture. However, the size of the effect can be estimated from a numerical simulation like the one shown in fig. 8(a) and (b). In this case, we estimate that the photon emission from the cavity

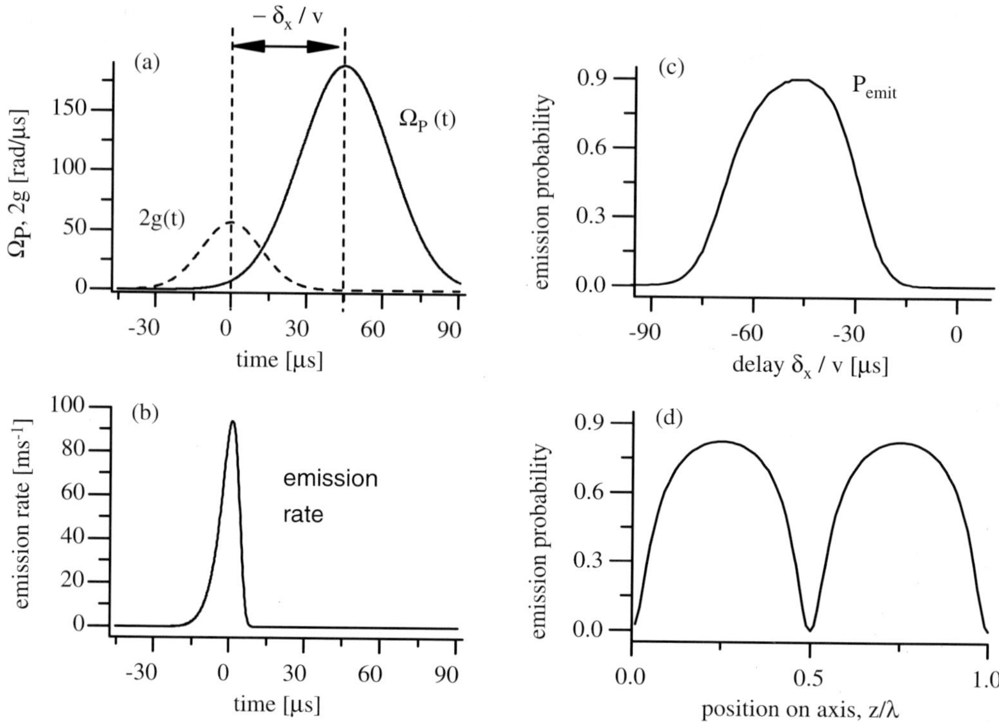

Fig. 8. – Simulation of a resonant atom-cavity interaction sequence for a cavity decay constant, $2\kappa = 2\pi \times 2.5\,\text{MHz}$, an atomic population decay rate of $2\gamma = 2\pi \times 6\,\text{MHz}$, and atoms travelling at $v = 2\,\text{m/s}$. (a) $\Omega_{\mathrm{P}}(t)$ and $2g(t)$ shown for experimental amplitudes and waists, $w_{\mathrm{C}} = 35\,\mu\text{m}$ and $w_{\mathrm{P}} = 50\,\mu\text{m}$. (b) Photon emission rate for a delay of $\delta_X/v = -45\,\mu\text{s}$. The integral of the rate yields a total photon emission probability, P_{Emit}, of 90%. (c) P_{Emit} as a function of the delay, δ_X/v, between cavity and pump interaction. (d) P_{Emit} as a function of the atomic position on the cavity axis for a delay of $\delta_X/v = -35\,\mu\text{s}$.

occurs around $t = 0$. At this moment, the atom is exposed to the Rabi frequencies $2g \approx 2g_0$ from the cavity, and $\Omega_{\mathrm{P}} \approx g_0/2$ from the pump laser. In case of a resonant excitation with $\Delta_{\mathrm{P}} = \Delta_{\mathrm{C}} = 0$, and for the relevant set of parameters at the moment of the photon emission, $(\Omega_{\mathrm{P}}, g, \gamma, \kappa) = 2\pi \times (2.25, 4.5, 3, 1.25)\,\text{MHz}$, the eigenstates of the non-Hermitian Hamiltonian (47) can be calculated numerically. The eigenstate $|a^0\rangle$, which is a dark state in the absence of dissipation, now reads

$$(65) \qquad |a^0\rangle = 0.976|u,0\rangle + i\,0.055|e,0\rangle - 0.208|g,1\rangle,$$

and the relevant populations are $\rho_{ee} = |c_e|^2 = 0.003$ and $\rho_{gg} = |c_g|^2 = 0.043$. According to (64), the spontaneous-emission losses are $P_{\mathrm{loss}} \leq 14\%$. Since we have neglected that the atom may also decay back to its initial state from where it can be recycled, the actual losses do never reach such a high value. This is evident from the numerical simulation which is discussed in the next section. It shows that only 10% of the excitation is really lost by spontaneous emission from state $|e, 0\rangle$ to state $|g, 0\rangle$, from where recycling is not possible.

5'5. *Numerical simulation.* – A numerical simulation for a single atom crossing the cavity is shown in fig. 8. To include the cavity-field decay rate, κ, and the spontaneous-emission rate of the atom, 2γ, we have employed the density-matrix formalism described in [12]. For the resonant situation, $\Delta_P = \Delta_C = 0$ shown here, the total emission probability, P_{Emit}, is expected to reach 90%. For the considered waists and amplitudes, fig. 8(c) shows that P_{Emit} reaches its maximum for $\delta_X/v = -45\,\mu$s. Note also that P_{Emit} is vanishingly small if the interaction with the pump beam coincides or precedes the interaction with the cavity mode. Figure 8(d) shows P_{Emit} as a function of the atom's position on the cavity axis for the delay realized in the experiment. Due to the standing-wave mode structure, the emission probability is zero at the nodes, and shows maxima at the antinodes. Since the dependence of P_{Emit} on the position-dependent coupling constant, g, is highly nonlinear and saturates for large g, the gaps around the nodes are much narrower than the plateaus surrounding the antinodes.

5'6. *Experimental realization.* – To realize the proposed scheme, we have chosen the setup sketched in fig. 7(b). A cloud of ^{85}Rb atoms is prepared in the $5S_{1/2}$, $F = 3$ state and released from a magneto-optical trap (MOT) at a temperature of $\approx 10\,\mu$K. A small fraction (up to 100 atoms) falls through a stack of apertures and enters the mode volume of an optical cavity at a speed of $2\,$m/s. The cavity is composed of two mirrors with a radius of curvature of $50\,$mm and a distance of $1\,$mm. The waist of the TEM$_{00}$ mode is $w_C = 35\,\mu$m, and in the antinodes the coupling coefficient is $g_0 = 2\pi \times 4.5\,$MHz. The finesse of 61000 corresponds to a linewidth $2\kappa = 2\pi \times 2.5\,$MHz (FWHM), which is significantly smaller than the natural linewidth of the ^{85}Rb atoms, $2\gamma = 2\pi \times 6\,$MHz. While one cavity mirror is highly reflective $(1 - R = 4 \times 10^{-6})$, the transmission of the other is 25 times higher, so that photons are emitted preferably in one direction. An avalanche photodiode (APD) with a quantum efficiency of 50% is used to detect them.

A reference laser is used to stabilize the cavity close to resonance with the $5S_{1/2}, F = 2 \longleftrightarrow 5P_{3/2}, F = 3$ transition with a lock-in technique. However, since an empty cavity is needed for the experiment, this laser is blocked $3.7\,$ms before the atoms enter the cavity. The pump beam is close to resonance with the $5S_{1/2}, F = 3 \longleftrightarrow 5P_{3/2}, F = 3$ transition and crosses the cavity transverse to its axis. This laser is focussed to a waist of $50\,\mu$m and has a power of $5.5\,\mu$W, which corresponds to a peak Rabi frequency $\Omega_P^0 = 2\pi \times 30\,$MHz.

To adjust the delay between the atom-cavity and the atom-pump coupling, we displace the pump beam along the vertical direction. Figure 9 shows the number of photons emitted from the cavity for a single cloud of Rb atoms released from the MOT as a function of the pump-beam displacement, δ_X, for three different values of the pump amplitude. It is obvious that a negative δ_X, *i.e.* a counterintuitive pulse sequence, is needed to reach a maximum photon emission probability. Moreover, the position of the

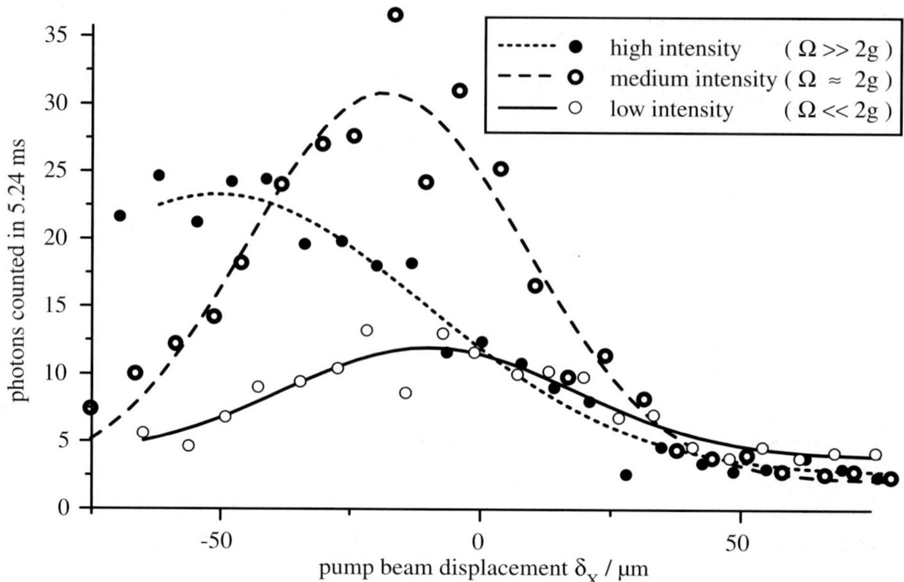

Fig. 9. – Number of photons emitted from the cavity in 5.24 ms as a function of the displacement δ_X of the pump beam (shift along the atom's trajectories) for three different intensities. A counterintuitive interaction sequence for STIRAP is realized for negative δ_X.

peak of the emission shifts to more negative values when the peak Rabi frequency, Ω_P^0, is increased. This is because the pulse amplitude needed to generate a photon is reached earlier. The highest efficiency is obtained for $\Omega_P^0 \approx 2g_0$, which is also the best condition for adiabatic following.

The left-hand side of fig. 10 depicts the case for arbitrary Δ_P and Δ_C. From this numerical simulation, it is evident that P_{Emit} is largest if the excitation is Raman resonant ($\Delta_P = \Delta_C$). However, for the delay $\delta_X/v = -35\,\mu\mathrm{s}$ chosen in the experiment, a slight reduction is expected for $\Delta_P = \Delta_C = 0$, since the waist of the pump, w_P, is larger than the waist of the cavity, w_C, and resonant excitation of the atom prior to the interaction with the cavity mode cannot be neglected. The right-hand side of fig. 10 shows the corresponding experimental results. The detunings of the cavity and the pump laser are both adjusted by means of acousto-optic modulators. To register the data, the MOT has been loaded and dropped across the cavity 50 times. The atom cloud needs 6.5 ms (FWHM) to cross the cavity mode, and within this interval, the photons emerging from the cavity are measured by the APD and recorded by a transient digitizer during 2.6 ms with a time resolution of 25 MHz. Therefore, the signal is observed for a total time of 130 ms. Due to the dark-count rate of 390 Hz of the APD, the total number of dark counts in the interval is limited to 51 ± 7.

Figure 10(a) shows the number of counted photons emerging from the cavity resonant with the atomic transition, $\Delta_C = 0$, as a function of the pump pulse detuning, Δ_P.

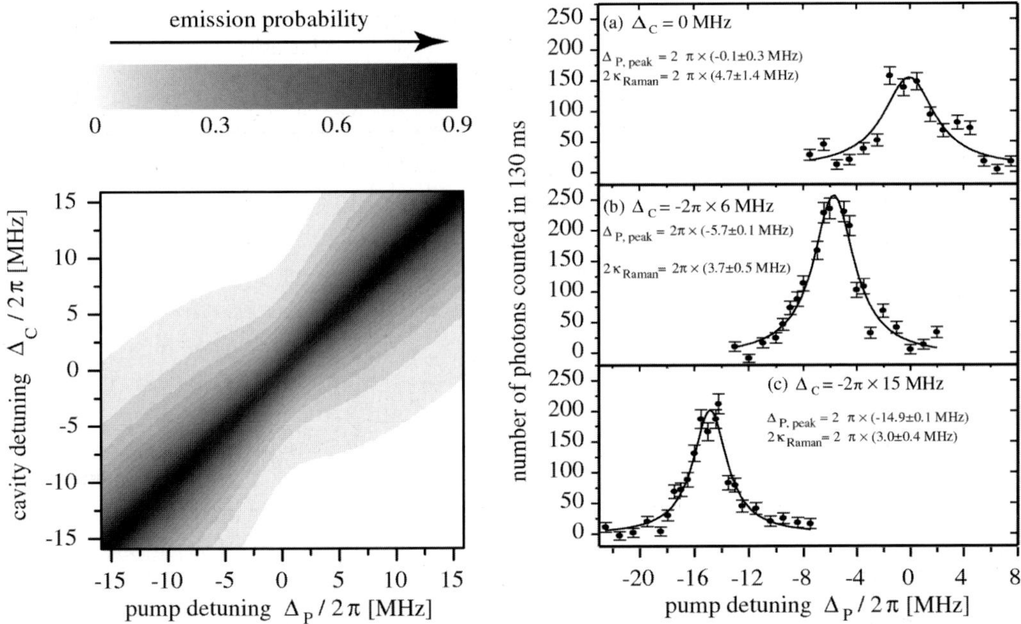

Fig. 10. – Left: Photon emission probability as a function of cavity and pump detuning, calculated for a pulse delay of $\delta_x/v = -35\,\mu$s and the parameters of fig. 8. The chosen delay gives the best fit with the experimental data shown on the right: Observed number of photons emitted from the cavity as a function of the pump laser detuning, Δ_P, for three different cavity detunings. The solid lines are Lorentzian fits to the data. The magneto-optical trap was loaded and dropped through the cavity 50 times to record the data shown here. A small background level was subtracted from the data displayed in the figure.

In this case, a small probability for atomic excitation is expected, which could lead to a small but cavity-enhanced spontaneous emission into the cavity mode, as has been shown previously [9-11]. The numerical simulation shows that an excited atom, suddenly placed at an antinode, would emit into the resonant cavity with a probability of 26% (see sect. **2˙6**), which indicates that most of the spontaneously emitted photons are lost in a random direction. This loss explains the smaller peak emission rate with respect to the off-resonant cases discussed below. Note that the cavity mode covers only a small solid angle of $\approx 4\pi \times 2.6 \times 10^{-5}$ srad, which means that the calculated spontaneous-emission rate into the cavity is enhanced by a factor 10^4. However, as can be seen in fig. 10(a), the measured linewidth is subnatural, and therefore the observed signal cannot be attributed to an excitation by the pump beam followed by enhanced spontaneous emission.

This is even more evident if the cavity is detuned (fig. 10(b, c)). The emission peak is pulled away from the atomic resonance, following the Raman resonance condition, $\Delta_P = \Delta_C$. Such a displacement proves that the light emission is not the result of a pump transition followed by enhanced spontaneous emission into the cavity. Moreover, Δ_P is too large for an electronic excitation of the atoms. Therefore, the exponentially decreasing wing of the pump beam cannot excite the atoms prior to their interaction with the cavity mode as in fig. 10(a). Hence, the losses vanish, and the peak photon emission probability is higher than for the resonant case. Note also that the observed linewidth is much smaller than the natural linewidth, $2\gamma = 2\pi \times 6\,\mathrm{MHz}$, of the atom. For $\Delta_C = -2\pi \times 15\,\mathrm{MHz}$, the line is only $3\,\mathrm{MHz}$ wide and its width approaches the cavity linewidth $2\kappa = 2\pi \times 2.5\,\mathrm{MHz}$, which limits the width of the Raman transition, since 2κ is the decay rate of the final state, $|g,1\rangle$.

In our discussion, we have assumed that the atoms interact with the cavity one-by-one. This is justified according to the following estimation: A mechanical slit restricts the atom's maximum distance from the cavity axis to $\pm 50\,\mu\mathrm{m}$. The spatial variation of g along (fig. 8(d)) and perpendicular to the cavity axis reduces the average emission probability to 37% per atom crossing the slit and the pump beam. Due to the low quantum efficiency of the APD and unavoidable cavity losses, only about 40% of the generated photons are detected. Therefore the maximum measured rate of 230 events/130 ms corresponds to a generation rate of 4.4 photons/ms, and 12 atoms/ms are needed to explain this signal. Since the photon generation takes $12\,\mu\mathrm{s}$ (FWHM, fig. 8(b)), the probability that two atoms interact with the cavity is at most 14%. This is small and, hence, negligible.

All observed features are in excellent agreement with our simulation, and we therefore conclude that the photon emission is caused by a vacuum-stimulated Raman transition, i.e. the coupling to the cavity, $g(t)$, and the Rabi frequency of the pump laser, $\Omega_P(t)$, are both high enough to assure an adiabatic evolution of the system, thus forcing the state vector $|\Psi\rangle$ to follow the dark state $|a^0\rangle$ throughout the interaction. Loss due to spontaneous emission is suppressed, and the photons are emitted into a single mode of the radiation field with well-determined frequency and direction.

6. – Summary and outlook

Vacuum-stimulated Raman scattering driven by an adiabatic passage can be used to generate single, well-characterized photons on demand, provided the Raman excitation is performed in a controlled, triggered way. In contrast to most other single-photon sources [36-41], these photons will have a narrow bandwidth and a directed emission. With a single atom trapped inside the cavity, a suitable repumping mechanism should allow to re-establish the initial condition after each photon emission. Repeated adiabatic passage and repumping cycles should therefore lead to a bit-stream of single photons. Finally, we state that the photon generation process depends on the initial state of the atom interacting with the cavity. If the atom is prepared in a superposition of states $|g,0\rangle$ and $|u,0\rangle$ prior to the interaction, this state will be mapped onto the emitted photon. A second atom placed in another cavity could act as a receiver, and with the suitable pump pulse sequence applied to the emitting and the receiving atom, a quantum teleportation of the atom's internal state could be realized [46].

* * *

This work was partially supported by the focused research program "Quantum Information Processing" of the Deutsche Forschungsgemeinschaft, and by the European Union through the IST(QUBITS) and IHP(QUEST) programs. The authors gratefully acknowledge the contributions of M. HENNRICH and T. LEGERO, who have done most of the laboratory work presented here.

REFERENCES

[1] TITTEL W., RIBORDY G. and GISIN N., *Phys. World*, March issue (1998) 41.

[2] KNILL E., LAFLAMME R. and MILBURN G. J., *Nature*, **409** (2001) 46.

[3] BERGMANN K. and SHORE B. W., *Coherent population transfer*, in *Molecular Dynamics and Stimulated Emission Pumping*, edited by H. L. DAI and R. W. FIELD (World Scientific, Singapore) 1995, pp. 315-373.

[4] BERGMANN K., THEUER H. and SHORE B. W., *Rev. Mod. Phys.*, **70** (1998) 1003.

[5] KUHN A., STEUERWALD S. and BERGMANN K., *Eur. Phys. J. D*, **1** (1998) 57.

[6] JAYNES E. T. and CUMMINGS F. W., *Proc. IEEE*, **51** (1963) 89.

[7] SHORE B. W. and KNIGHT P. L., *J. Mod. Opt.*, **40** (1993) 1195.

[8] PURCELL E. M., *Phys. Rev.*, **69** (1946) 681.

[9] HEINZEN D. J., CHILDS J. J., THOMAS J. E. and FELD M. S., *Phys. Rev. Lett.*, **58** (1987) 1320.

[10] HEINZEN D. J. and FELD M. S., *Phys. Rev. Lett.*, **59** (1987) 2623.

[11] MORIN S. E., YU C. C. and MOSSBERG T. W., *Phys. Rev. Lett.*, **73** (1994) 1489.

[12] KUHN A., HENNRICH M., BONDO T. and REMPE G., *Appl. Phys. B*, **69** (1999) 373.

[13] HENNRICH M., LEGERO T., KUHN A. and REMPE G., *Phys. Rev. Lett.*, **85** (2000) 4872.

[14] LAW C. K. and KIMBLE H. J., *J. Mod. Opt.*, **44** (1997) 2067.

[15] HECHENBLAIKNER G., GANGL M., HORAK P. and RITSCH H., *Phys. Rev. A*, **58** (1998) 3030.

[16] THOMPSON R. J., REMPE G. and KIMBLE H. J., *Phys. Rev. Lett.*, **68** (1992) 1132.

[17] CHILDS J. J., AN K., OTTESON M. S., DASARI R. R. and FELD M. S., *Phys. Rev. Lett.*, **77** (1996) 2901.

[18] MÜNSTERMANN P., FISCHER T., MAUNZ P., PINKSE P. W. H. and REMPE G., *Phys. Rev. Lett.*, **84** (2000) 4068.

[19] HOOD C. J., CHAPMAN M. S., LYNN T. W. and KIMBLE H. J., *Phys. Rev. Lett.*, **80** (1998) 4157.

[20] LUGIATO L. A., *Theory of Optical Bistability*, Vol. **XXI** (Elsevier Science Publishers) 1984, pp. 71-216.

[21] ABRAGAM A., *The Principles of Nulcear Magnetism* (Oxford University Press, Oxford) 1961.

[22] LOY M. M. T., *Phys. Rev. Lett.*, **32** (1974) 814.

[23] LOY M. M. T., *Phys. Rev. Lett.*, **36** (1976) 1454.

[24] LOY M. M. T., *Phys. Rev. Lett.*, **41** (1978) 473.

[25] YATSENKO L. P., SHORE B. W., HALFMANN T. and BERGMANN K., *Phys. Rev. A*, **60** (1999) R4237.

[26] KUHN A., PERRIN H., HÄNSEL W. and SALOMON C., *Three dimensional Raman cooling using velocity selective rapid adiabatic passage*, in *OSA TOPS on Ultracold Atoms and BEC*, edited by K. BURNETT, Vol. **7** (Optical Society of America) 1996, pp. 58-65.

[27] PERRIN H., KUHN A., BOUCHOULE I. and SALOMON C., *Europhys. Lett.*, **42** (1998) 395.

[28] HARRIS S. E., *Phys. Rev. Lett.*, **70** (1993) 552.

[29] HARRIS S. E., *Phys. Today*, **50** (1997) 36.

[30] HAU L. V., HARRIS S. E., DUTTON Z. and BEHROOZI C. H., *Nature*, **397** (1999) 594.

[31] LUKIN M. D., YELIN S. F. and FLEISCHHAUER M., *Phys. Rev. Lett.*, **84** (2000) 4235.

[32] FLEISCHHAUER M. and LUKIN M. D., *Phys. Rev. Lett.*, **84** (2000) 5094.

[33] PHILLIPS D. F., FLEISCHHAUER A., MAIR A., WALSWORTH R. L. and LUKIN M. D., *Phys. Rev. Lett.*, **86** (2001) 783.

[34] MESSIAH A., *Quantum Mechanics*, Vol. **2** (J. Wiley & Sons, New York) 1958, Chapt. 17.

[35] PARKINS A. S., MARTE P., ZOLLER P. and KIMBLE H. J., *Phys. Rev. Lett.*, **71** (1993) 3095.

[36] KIM J., BENSON O., KAN H. and YAMAMOTO Y., *Nature*, **397** (1999) 500.

[37] BRUNEL C., LOUNIS B., TAMARAT P. and ORRIT M., *Phys. Rev. Lett.*, **83** (1999) 2722.

[38] LOUNIS B. and MOERNER W. E., *Nature*, **407** (2000) 491.

[39] KURTSIEFER C., MAYER S., ZARDA P. and WEINFURTER H., *Phys. Rev. Lett.*, **85** (2000) 290.

[40] BROURI R., BEVERATOS A., POIZAT J.-P. and GRANGIER P., *Opt. Lett.*, **25** (2000) 1294.

[41] MICHLER P., IMAMOGLU A., MASON M. D., CARSON P. J., STROUSE G. F. and BURATTO S. K., *Nature*, **406** (2000) 968.

[42] BENSON O., SANTORI C., PELTON M. and YAMAMOTO Y., *Phys. Rev. Lett.*, **84** (2000) 2513.

[43] SANTORI C., PELTON M., SOLOMON G., DALE Y. and YAMAMOTO Y., *Phys. Rev. Lett.*, **86** (2001) 1502.

[44] MICHLER P., KIRAZ A., BECHER C., SCHOENFELD W. V., PETROFF P. M., ZHANG L., HU E. and IMAMOGLU A., *Science*, **290** (2000) 2282.

[45] LAW C. K. and EBERLY J. H., *Phys. Rev. Lett.*, **76** (1996) 1055.

[46] CIRAC J. I., ZOLLER P., KIMBLE H. J. and MABUCHI H., *Phys. Rev. Lett.*, **78** (1997) 3221.

Rydberg wave packets in quantum information science

Ph. H. Bucksbaum

FOCUS Center, Physics Department, University of Michigan
Ann Arbor, MI 48109-1120 USA

1. – Introduction

This series of lectures introduces the basic properties of Rydberg atoms that are relevant to quantum information science. First, we will review the definition of Rydberg atoms, and some of the basic elements of Rydberg structure. We will show how Rydberg atoms possess all the basic properties required for construction of working quantum gates. One significant problem with current quantum information science orthodoxy is that the Rydbergs have a state space that is vastly larger than the canonical 2-level "qubit". This makes it possible to store an arbitrarily large data string in a *single* Rydberg atom. Along with this apparent advantage, however, is a serious problem: the system is not easily scalable, since it is not a simple building block that can be repeated many times and coupled together. Nonetheless, entanglement exists in Rydberg states, and this will be discussed.

Rydberg state engineering is very advanced compared to most other quantum information science technologies, and this will be described in the bulk first half of these lecture notes. The construction, or "sculpting", of Rydberg states uses ultrafast laser technology, so this will be described as well.

The second half of the lecture notes describes current advances and future prospects in Rydberg Quantum Information Science. The greatest progress is in data storage, data hiding, search, and retrieval. Two methods will be discussed. The first method is quantum tomography using shaped optical pulses. The second method uses the radiation to coherently redistribute population within the Rydberg manifold. This is much more powerful, but it is also quite challenging technically. We conclude with a brief discussion of the paths to practical utilization of Rydberg sates in Quantum Information Science.

The basic properties of Rydberg atoms are discussed in a monograph by Gallagher [1]. A more complete discussion of Rydberg wave packet dynamics is in a book of essays on this subject, edited by Yaezell [2]. Finally, no introduction to this subject should fail to reference the "bible" of the quantum theory of single-electron atoms, by Bethe and Salpeter [3].

2. – Basic properties of Rydberg atoms

We begin by writing the atomic Hamiltonian in a form that may or may not be entirely familiar:

$$(1) \qquad H = \frac{p^2}{2} - \frac{1}{r} + V(\mathbf{r}, \mathbf{s})|_{r<r_0}.$$

This form uses a number of important approximations and conventions. First, the units are conventional atomic units, where $\hbar = m = |e| = 1$, and where the vacuum speed of light c is equal to $1/\alpha \approx 137$. This is not only a notational convenience; it also helps to emphasize the strong connection between the Rydberg scaling laws and the corresponding laws for Newtonian classical dynamics, as we shall illustrate. Second, since the kinetic energy term in H assumes the particle has a mass of one free electron, the dynamics is that of a single electron moving in the vicinity of an infinitely massive central potential. The potential energy has two parts: a Coulomb $1/r$ potential; and a short-range central potential V. The idea is that there is a single active electron in the atom, which spends most of its time outside the radius where the remaining electrons form a positively charged core.

This potential implies that the problem to solve is a simple one: far simpler than a *real* multi-electron atom, and nearly (but not quite!) as simple as a hydrogen atom. The eigenvalues and eigenfunctions reveal the underlying dynamics, and they vary only slightly from the solutions for hydrogen.

Quantum numbers: The single active electron moves in a centrally symmetric potential, so the "good" quantum numbers are the principal hydrogenic quantum number n, and the quantum numbers describing the angular momentum of the system. In general this could be quite complicated. The core might contain a number of "unpaired" electrons outside of its closed shells, and each of these could have spins and orbital angular momentum. The nucleus could also have a magnetic moment, which couples to the electrons via the hyperfine interaction. The general methods for sorting out the various electrostatic, spin-orbit, and spin-spin interactions are described in textbooks such as the one by Sobelman [4]. In practice, however, we will make two additional simplifying assumptions that will eliminate all of this complication with the core:

1) *The nuclear spin and the hyperfine interaction are ignored!* There is a good reason for this. The hyperfine coupling leads to splittings δE_{HFS} on the order of MHz to hundreds of MHz. This means that the *natural dynamical time scale* for the

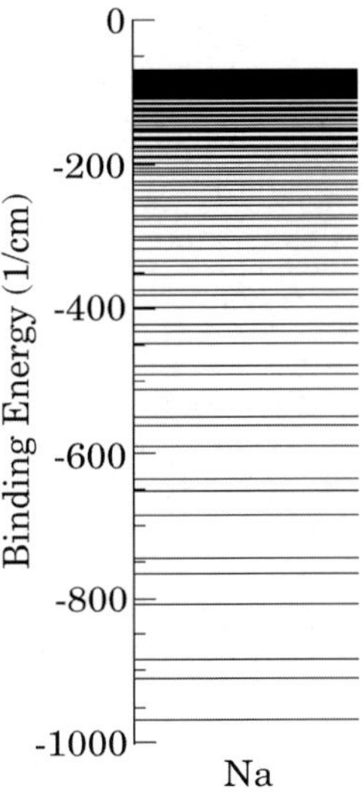

Fig. 1. – The Rydberg series for atomic sodium shows a converging pattern of sets of three eigenvalues, corresponding to the energies of states with $l = 0$, $l = 1$, and $l > 1$.

hyperfine interaction is $h/\delta E_{\mathrm{HFS}}$, which is on the order of tens of nanoseconds or longer. This is a thousand times longer than any coherent interaction that will concern us here, so we may neglect hyperfine couplings, and ignore the spin of the nucleus. The core angular momentum is then only due to the motion and spin of the core electrons, and this brings us to the second simplifying assumption:

2) *The core is an angular-momentum singlet.* This is true for the alkali atoms: lithium, sodium, rubidium, and cesium, once nuclear spin is ignored. We confine our attention to these atoms. In the standard notation of L-S (Russel-Saunders) coupling, the core is $^1S_0$. So in addition to spherical symmetry and infinite mass, the core has no angular momentum.

With these assumptions, the good quantum numbers are only those of the single electron itself. It has orbital and spin angular momentum, and these can couple as they do in hydrogen. Most of the examples in these sections will use p-states in atomic cesium, where there are two fine-structure-split series, $n^2P_{1/2}$ and $n^2P_{3/2}$. These quantum numbers are labeled $n, l = 1$, $j = 1/2$ or $3/2$, and $m_j = |j|, |j-1|, \ldots, -|j|$. One further simplification occurs if we can neglect the spin-orbit splitting. In that case, there are only three important quantum numbers: n, l, and m_l.

Eigenvalues. Rydberg never saw a Rydberg atom. The highly excited states of a single electron derive the name Rydberg state from their spectrum, which follows nearly, but not quite Rydberg's energy formula for hydrogen (fig. 1). To a good approximation

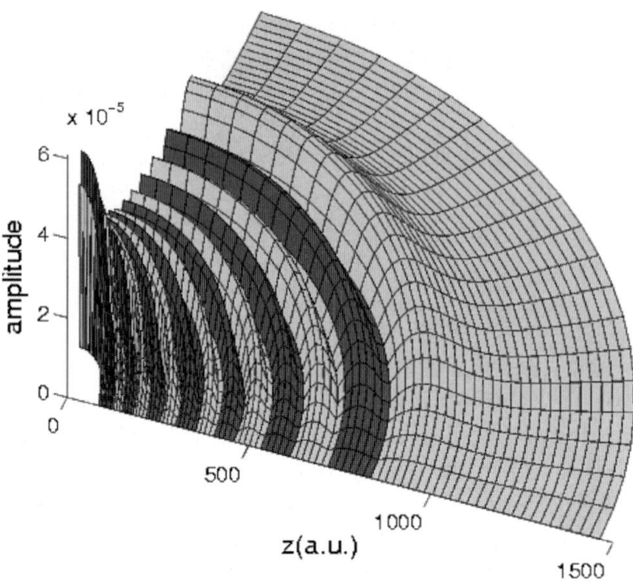

Fig. 2. – A cross-section in the (x, z)-plane of a p-state $m_l = 1$ Rydberg eigenfunction in atomic cesium. The different shadings show when the function is relatively positive real and negative real.

the state energies are as follows:

$$(2) \qquad E = -\frac{1}{2(n - \delta_l)^2} = -\frac{1}{2n^{*2}}.$$

The quantity δ_l is called the quantum defect, and it depends on the orbital-angular-momentum quantum number l. This reflects the fact that the electron sees the departure of the core potential from a simple hydrogen potential only when it is inside the core radius r_0. The fraction of time spent close to the core depends on the centrifugal barrier, which depends on the total angular momentum. The quantum defect is nearly independent of n for $n > 10$ or so. This is because high-lying Rydberg states are so weakly bound, that their momentum inside the core is essentially always that of an electron with zero energy at $r = \infty$. The quantity n^*, which incorporates the principal quantum number and the quantum defect, is called the "effective principal quantum number".

Eigenfunctions. The value of the quantum defect is related to the scattering phase shift experienced by a zero-energy electron with angular momentum l as it scatters from the core. A classical picture can help us to visualize this. As an electron drops inside the outer layers of core electrons screening the nucleus, the Coulomb force increases. The potential well is deeper and the potential gradient steeper than it would be for hydrogen. Therefore the electron speeds through this region, spends less time in the core than it would have spent in the vicinity of the +1 charge of the hydrogen nucleus, and has greater momentum while it is in the core. The electron's momentum is directly related to the wavelength of the corresponding electron eigenfunction at that place in the potential. The wave function has a shorter wavelength inside the core, and this pulls all of the nodes in the eigenfunction towards the core, and lowers the eigenvalue.

Since the potential is time-independent, the wave function is relatively real (fig. 2). The eigenstate evolves according to Schrödinger's equation $H\psi = i\dot\psi$ like a stationary state. In the configuration-space basis we write the eigenfunction

$$(3) \qquad \langle \mathbf{x} \mid \psi \rangle = \psi(\mathbf{x}, t) = \phi(\mathbf{x})e^{-i\omega t}.$$

The real part of ω is just the energy eigenvalue. Its imaginary part is the damping time constant γ. This is usually very long for Rydberg states, and a final important approximation is to neglect γ, and assume that all Rydberg eigenstates are stable.

The wave functions have a number of properties that can be understood by considering the corresponding classical system, a light particle orbiting a massively heavy central body in a $1/r$ potential. Such a system obeys Kepler's laws, and this is displayed in the scaling relations of the Rydberg states. For example, the orbital radius scales like n^{*2}. (More specifically, for hydrogen the *inverse* radius $\langle 1/r \rangle$ is equal to $1/n^2$, by the Virial Theorem.) The splitting between neighboring states must then scale as $1/n^{*3}$ by direct computation from eq. (2). Since the quantum defect is nearly constant, the time spent inside the core must also scale as $1/n^{*3}$. This then affects all properties related to the core interaction: The spin-orbit splittings, dipole matrix elements with the ground state, and photoionization cross-sections all scale like $1/n^{*3}$, because they all involve interactions between the Rydberg electron and the core.

Wave packets. Rydberg wave packets are nonstationary states that are nondegenerate coherent superpositions of eigenstates. In the notation of eq. (3), we write

$$(4) \qquad \langle \mathbf{x} \mid \psi \rangle = \psi(\mathbf{x}, t) = \sum_{n^*} a_{n^*} \phi_{n^*}(\mathbf{x}) e^{-i\omega_{n^*} t}.$$

The unit norm is maintained by the requirement that $\sum_{n^*} |a_{n^*}|^2 = 1$.

Wave packet motion is specified by the time-dependence of eq. (4), which is contained in the complex exponentials. The characteristic time scale for motion is inverse to the splitting between different occupied orbitals, *i.e.* eigenstates with nonzero a_{n^*}'s. For $\Delta n^* = 1$, this has the approximate value n^{*3}. Classically, this is the time it takes for a particle to move one radian in a circular orbit. In other words, if the states n and n' with effective quantum numbers n^* and $n^* + 1$, respectively, interfere constructively at $r = 0$ and $t = 0$, then they will interfere constructively at $r = 0$ again after one Kepler period, at

$$(5) \qquad t = \tau_{\text{Kepler}} = \frac{2\pi}{1/n^{*2} - 1/n'^{*2}} \sim 2\pi n^{*3} = 1.26 \left[\frac{n^*}{20}\right]^3 \text{ ps}.$$

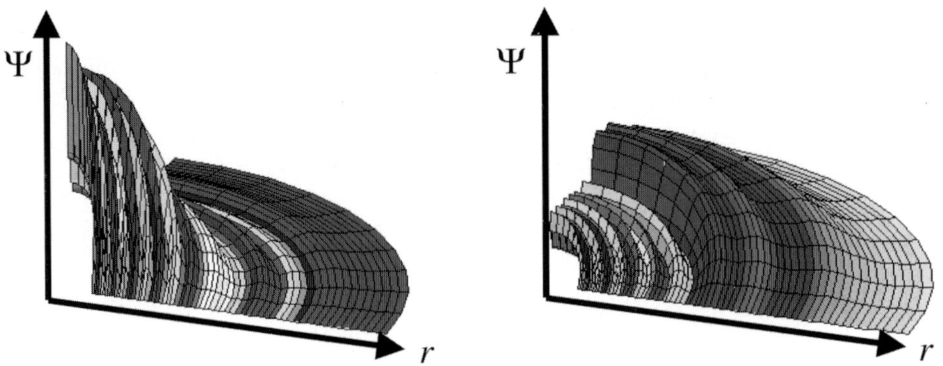

Fig. 3. – Two snapshots of a cesium Rydberg wave packet moving through one Kepler cycle. The left picture shows the wave packet at its inner turning point. The right picture shows it at the outer turning point. The gray shading represents quantum phase.

Figure 3 shows an example of the excursion of a radial wave packet through the first half of its orbit.

The instantaneous motion of a wave packet is proportional to its spatial derivative. This is just a restatement of the fact that the momentum operator is proportional to the gradient of the wave function. So in the picture above, the wave packet near its inner turning point, shown on the left, is moving much more rapidly than the wave packet at its radial outer turning point, shown on the right. The rapid change of phase with position indicates regions of high momentum.

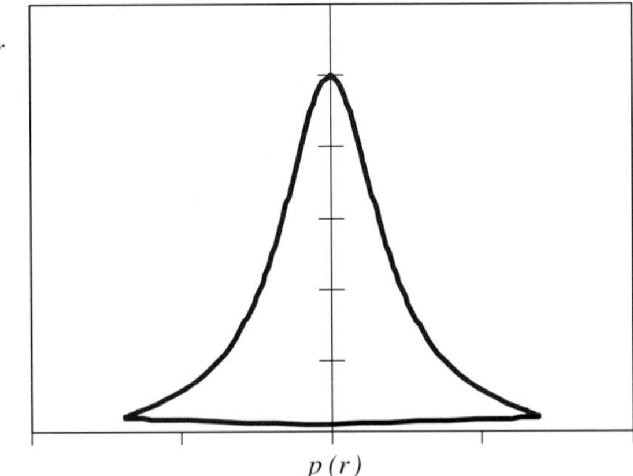

Fig. 4. – Phase space plot showing the radial position and momentum of a classical particle executing a highly elliptical Kepler orbit in a $-1/r$ potential. The radial momentum goes through zero at the outer and inner turning points. Elsewhere the trajectory has a characteristic bell shape.

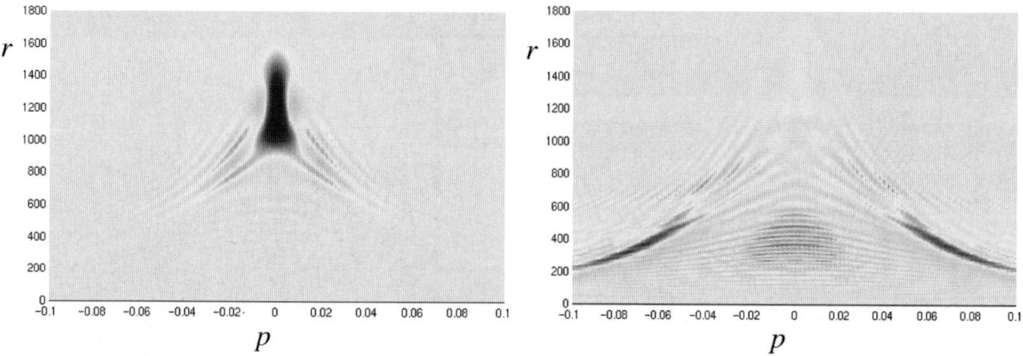

Fig. 5. – Wigner functions for a Rydberg radial wave packet, shown at the outer turning point (left) and inner turning point (right). The position and momentum scales are in atomic units. Shading shows probability density.

Classical physics can display the momentum of a particle at the same time as its position, using a phase space plot of the particle trajectory (see fig. 4).

The quantum wave packet can also be displayed using phase space, by performing a Wigner-Weyl transformation on the radial Schrödinger wave function $\Psi(r)$ to transform it to the Wigner function $\rho(r, p)$:

$$(6) \qquad \rho(r, p) = \int_{-\infty}^{\infty} \mathrm{d}s \, \exp[2ips]\psi^*(r - s)\psi(r + s).$$

The Wigner function resembles the classical phase space trajectory (see fig. 5).

In summary, Rydberg wave packets are coherent superpositions of the stationary solutions to Schrödinger's equation for a single electron orbiting a singly charged ion under the influence of the ion's Coulomb potential. Superpositions of nondegenerate orbitals exhibit time evolution similar to the trajectory of a particle in a Kepler orbit. The Schrödinger solutions display this motion in the spatial gradient of their quantum phase. The correspondence to classical motion can be seen more directly by transforming the Schrödinger wave function according to eq. (6) to form the Wigner function.

3. – Forming superposition states

This section describes the essential features of Rydberg wave packet "sculpting". Coherent quantum wave packets are produced by exciting atoms with broadband coherent laser radiation. The technology to produce optical pulses with enough bandwidth to simultaneously excite many electronic states has developed over the past twenty years in the field known as ultrafast optical science. Broadband coherent lasers can produce subpicosecond optical pulses with many scientific and technological applications, from

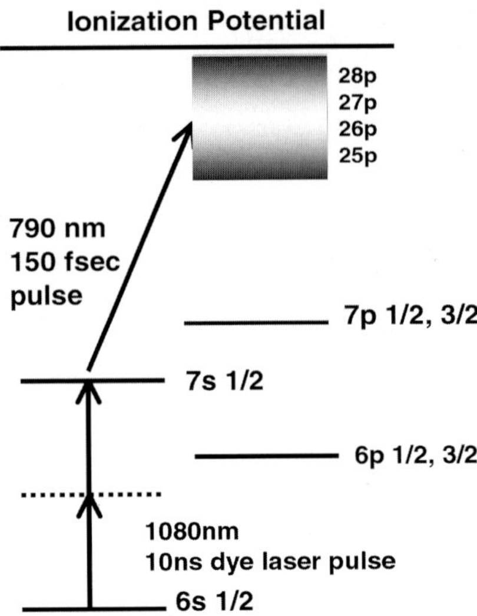

Fig. 6. – Level scheme for excitation of Rydberg wave packets in atomic Cs. The ground state is excited to the $7s$ state via a two-photon transition. $7s$ is then excited to $n = 20$–30 by $790\,\mathrm{nm}$ broadband coherent radiation.

laser fusion to optical communication. These are sufficiently broadband to excite several states for $n > 20$.

Three technological innovations have facilitated ultrafast control. The first is the Kerr-Lens Modelocked laser, developed in the early 1990s, which creates stable pulses with 10–$100\,\mathrm{nm}$ of coherent bandwidth. The mode-locking mechanism is a nonlinear focusing element in the cavity, which makes the laser optical cavity most stable if the circulating light is a single very short pulse. If the gain medium is very broadband, the pulse may have only a few optical cycles [5].

The output of a KLM laser is a continuous train of pulses about $10\,\mathrm{ns}$ apart, with about $1\,\mathrm{nJ}$ of energy per pulse. The central frequency is widely tunable, and tends to be centered around $800\,\mathrm{nm}$, or about $1.5\,\mathrm{eV}$. This photon energy is too low to excite Rydberg wave packets directly from the ground state of any atom; but there are a number of schemes where an atom can be excited to a "launch" state first, approximately $1.5\,\mathrm{eV}$ below the ionization limit for the atom. One such scheme is depicted in fig. 6. Rydberg wave packets are produced in a Cs atomic beam by first exciting ground-state atoms to the $7s$ state using a two-photon transition at $1.08\,\mu\mathrm{m}$. Ultrafast broadband laser pulses centered around $790\,\mathrm{nm}$ then excite the atoms to Rydberg states. This produces coherent superpositions of $p_{1/2}$ and $p_{3/2}$ states with principal quantum numbers between $n = 24$ and $n = 35$.

The excitation cross-section for $7s$-np transitions in atomic Cs is too small to have a high excitation probability with the pulses from the KLM oscillator, so they must be amplified. Simple multipass laser amplifiers cannot be used for this because the high

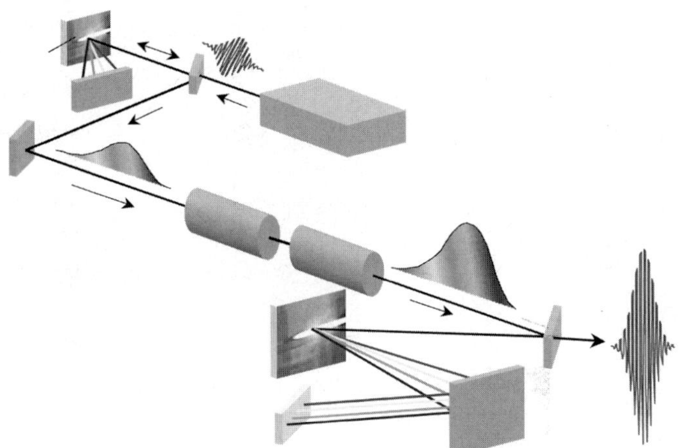

Fig. 7. – This block diagram shows the essential steps of chirped-pulse amplification (CPA). An ultrashort laser pulse is first dispersed using diffraction gratings, then amplified, and finally recompressed. CPA is essential for lasers with peak power exceeding tens of gigawatts.

peak power of an amplified femtosecond pulse is enough to cause nonlinear propagation effects that would destroy the laser mode, or the gain medium, or both; therefore, all amplified ultrafast laser systems employ a technique invented by Strickland and Mourou called chirped pulse amplification (CPA) [6]. The basic idea, which was borrowed from microwave technology, is to temporally disperse the pulse, to increase its length without destroying any of the coherent bandwidth. The longer pulse has lower peak power, and can be amplified without danger of destroying itself or the optics. The final step in CPA is recompression using a pair of diffraction gratings (see fig. 7).

The ultrashort pulses provide broadband coherent radiation, which is the raw material for pulse shaping. The final ingredient is the device that molds the different colors into the final pulse shape. This is accomplished with a Fourier filter, or "pulse shaper". A pulse shaper is essentially two spectrometers placed back-to-back, with a programmable filter in between, as shown in fig. 8 [7].

The heart of our pulse shaper is a programmable acousto-optic modulator (AOM) made of tellurium dioxide (TeO_2), which forms a Fourier filter [8]. An acoustic wave traveling through the crystal creates a transient transmission grating which diffracts the optical wave at the Bragg angle. Since the acoustic wave moves at the sound speed v_s, it hardly changes while the optical pulse travels through the crystal. Therefore, the complex amplitude of the traveling acoustic wave $A(t)\cos\omega_s t = A(y/v_s)\cos\omega_s t$ is mapped onto the optical field $E(\omega)$ as it passes through the AOM. The shaped beam then has the form

$$(7) \qquad E_{\text{shaped}}(\omega) = E_{\text{input}}(\omega) \times a(\omega) \times e^{i\varphi(\omega)t},$$

where $a(\omega)e^{i\phi(\omega)} = A(y(\omega)/v_s)$.

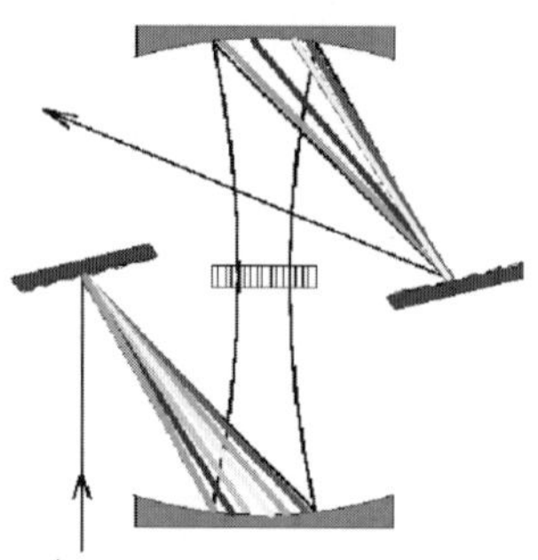 

Fig. 8. – Left: Schematic diagram of a Fourier filter. Light from an ultrafast pulse is dispersed, and then passes through a programmable amplitude and phase filter, then recombined to form any pulse shape consistent with the initial bandwidth. Right: Photograph of the Neos acousto-optic modulator.

The shaped pulses can be measured using spectral interferometry. In this technique, the pulse to be measured is combined with an unshaped reference pulse, and then analyzed in a spectrometer. The spectrally resolved interference between signal and reference is a direct measure of the spectral phase function.

4. – Measuring sculpted Rydberg wave packets

Quantum interference techniques can be used to view Rydberg wave packet sculptures [9]. The cesium Rydberg wave packet $\Psi(\mathbf{x}, t)$ is determined by the amplitude and phase of the constituent eigenstates:

$$(8) \qquad \Psi(\mathbf{x}, t) = \sum_{np} a_n e^{i\Phi_n} \phi_{npm}(\mathbf{x}) e^{-i\omega_n t} + a_0 \phi_s e^{-i\omega_s t},$$

where ϕ_s is the launch state. For weak excitation, we can use Fermi's Golden rule to calculate the coefficients:

$$(9) \qquad a_n e^{i\Phi_n} \propto E(\omega_{np} - \omega_{7s})\langle npm|z|7s\rangle \propto \frac{E(\omega_{np} - \omega_{7s})}{n^3}.$$

This expression relates the quantum state amplitude and phase to the amplitude and phase of the sculpted optical field.

Decoherence and dephasing due to Doppler effect and collisions can change the state coefficients, and so we minimize these effects by using an atomic beam. Residual dephasing mechanisms are stray electric fields, and some residual first-order Doppler shifts if the beam is not perpendicular to the laser. These contribute to coherent dephasing times on the order of 10^{-7} s. Decay of the state amplitudes can occur due to excitation by blackbody radiation, or fluorescence decay, with associated T_1 times on the order of 10^{-5} s. These times set the limits for producing and measuring wave packets.

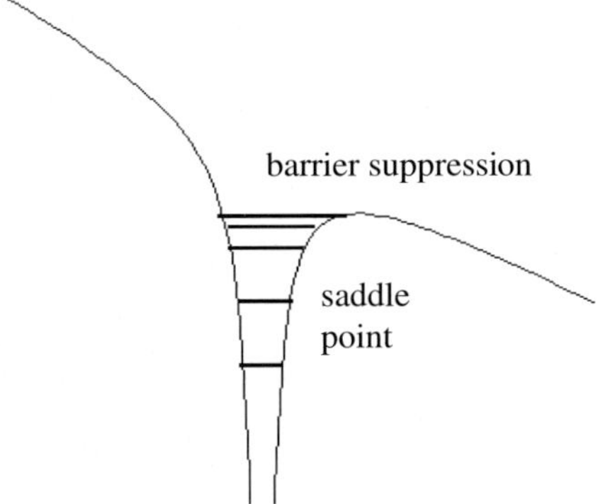

Fig. 9. – Coulomb potential in the presence of a static uniform electric field, shown in cross-section. The critical field for ionization of a state n is approximately given by $F = 1/(16n^4)$.

4'1. Amplitude measurement. – Quantum amplitudes of Rydberg wave packets can be measured through state-selective detection. The simplest state-selective method for atomic beams is the technique of ramped field ionization (fig. 9) [1]. If a uniform electric field is applied to the Rydberg wave packets, field ionization becomes possible. However, the field ionization probability is not a monotonically increasing function of field, because the energy structure of the wave packet in the field is discretized into constituent Stark eigenstates. Each energy eigenstate with energy E has its own characteristic critical ionization field F. For alkali atoms, this is roughly given by $|F| = (1/4)E^{-2}$ in atomic units.

When a wave packet experiences the critical field for one of its constituent states, the ionization occurs with probability proportional to the square of the amplitude for that state. Thus, the atomic beam ionization distribution maps the squared state amplitudes.

In some situations, state-selective field ionization is inconvenient or impossible. For example, wave packets in a dense gas or a liquid may relax so rapidly that the ramped field technique takes too long, or the ionized electrons cannot be extracted. In that case, it is often possible to use quantum wave packet interferometry to analyze the wave packet amplitudes. Quantum interferometry uses a Michelson interferometer to split the optical pulse that excites the wave packet into two parts with a variable time delay τ. Each pulse in the pair excites an identical wave packet in the atom, and the two wave packets coherently interfere, depending on the time delay:

$$(10) \qquad \Psi = (\mathbf{x}, t, \tau) = e^{i\omega_{gs}\tau} \sum_n a_n u_n(\mathbf{x}) e^{-i\omega_n t}(1 + e^{-i\omega_n \tau}).$$

The total wave function depends only on the squared amplitudes and the time delay:

$$(11) \qquad \langle \Psi \mid \Psi \rangle \propto \sum_n |a_n|^2 \cos \omega_n \tau.$$

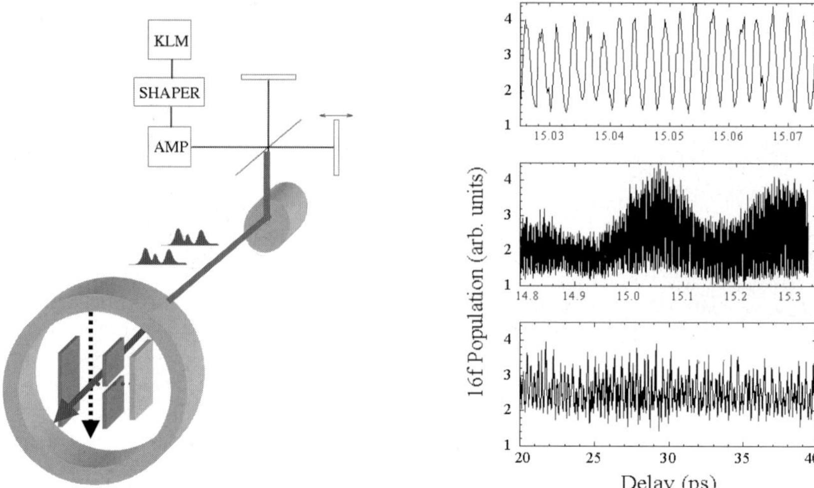

Fig. 10. – Left: Block diagram of a Michelson interferometer that creates two identical wave packets in the same atoms, with a variable time delay. Right: The wave packet excitation as a function of time delay shows structure on several time scales. The Fourier transform of this structure reveals the energies and squared amplitudes of the orbitals occupied by the wave packet.

A Fourier analysis of the ionization current $|\langle \Psi \mid \Psi \rangle|^2$ will yield the amplitudes (fig. 10).

In wave packet interferometry, the atomic wave packet is interfering with its twin, produced at a different time. Since each atomic orbital is resonant with a different frequency component of the field, the relative phases between different colors is unimportant. The phase is not measured in the interferometric process, so even temporally incoherent light could be used to make the measurement [10].

4'2. *Phase measurement*. – The phase of the state coefficients a_i can be determined using a holographic technique similar to spectral interferometry. In this measurement, the atoms are excited to a second, *reference* wave packet superposition with a time delay τ with respect to the sculpted packet. The known reference pulse gets around the problem discussed in the last paragraph of the previous section. The reference excitation has real amplitudes, produced by a transform limited optical field.

$$(12) \qquad \Psi_{\text{ref}}(\mathbf{x}, t, \tau) = e^{i\omega_{gs}\tau} \sum_n b_n u_n(\mathbf{x}) e^{-i\omega_n t}.$$

The probability for state i depends on the relative phase ϕ_t between signal and reference:

$$(13) \qquad P_i = |a_i|^2 + |b_i|^2 + 2|a_i||b_i| \cos[(\omega_i - \omega_{gs})\tau - \phi_i].$$

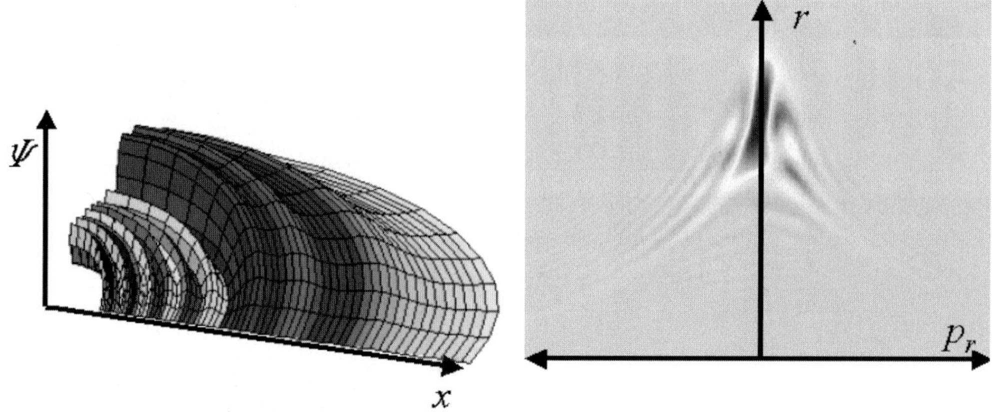

Fig. 11. – Two views of a sculpted wave packet. Left: Schrödinger wave function. The amplitude is shown as a function of x and z for a p-state wave packet oriented along x. The phase is represented by the shading of the wave packet. Right: Wigner representation of the same wave packet in the (r, p_r) phase space plane.

Since this delay time τ is not stable with respect to the optical frequency $\omega_i - \omega_{gs}$, the phases are extracted by averaging several laser shots and constructing the correlation function

$$(14) \qquad r_{ij} = \frac{\langle P_i P_j \rangle - \langle P_i \rangle \langle P_j \rangle}{(\Delta P_i)(\Delta P_j)} = \cos[(\omega_i - \omega_j) - (\phi_i - \phi_j)].$$

A reconstructed wave packet is shown in fig. 11.

5. – Quantum information applications

5˙1. *Data storage and retrieval.* – Quantum control techniques can be used to load, hide, retrieve, and manipulate information in wave packets. Consider, for example, the storage of a simple binary number, such as 000001000. A wave packet could encode this number in various ways. For example, one could load an n-state quantum wave packet with an n-bit number according to the prescription that at a specified time t, the phase is real and positive for binary 0, but real and negative for binary 1. Since every state in the wave packet has equal amplitude, ordinary spectroscopic techniques such as state-selective field ionization cannot reveal which state stores the binary 1. This bit is hidden from view. If the same data were stored in a classical binary register with n locations, one would have to search each location to find the marked bit. The search would take, on average, $n/2$ steps; however, the rules of quantum state manipulation provide some simple methods for revealing the marked bit (fig. 12).

Grover has considered efficient searches of quantum data registers [11]. He pointed out that when information is stored as phase, the superposition principle of quantum

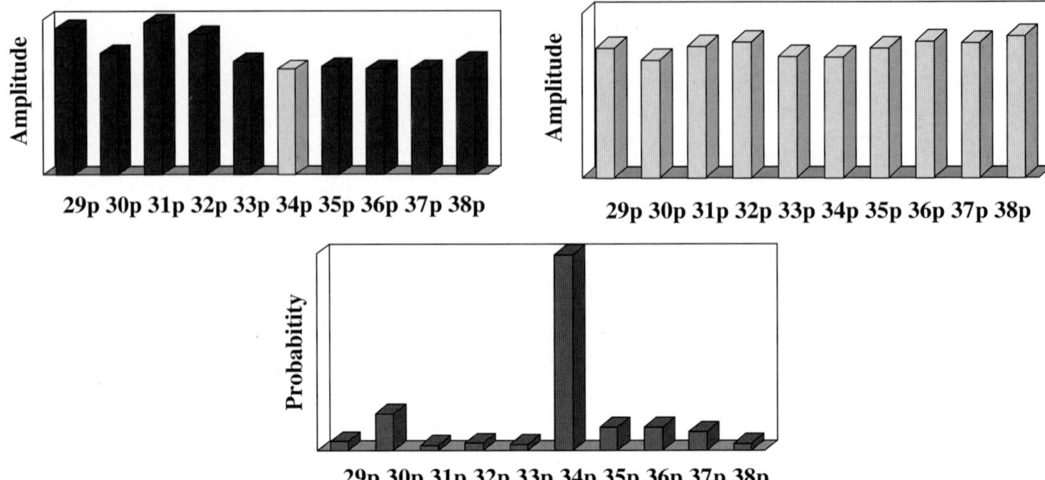

Fig. 12. – Top left: Bar graph representing a Rydberg data register for the binary number 0000010000. The binary bit is encoded into the phase of the n-state, so that binary 0 states are real and positive (dark phase), while binary 1 states are real and negative (light phase). Top right: The decoding wave packet has equal amplitudes, and all negative real phases. Bottom: The superposition of both wave packets amplifies the binary 1 bits, while destructive interference destroys the binary 0 bits.

mechanics provides an efficient method to search the database. Such a search uses wave-mechanical superposition. Wave interference is also present in some classical analog data storage media, such as optical holograms, which use the classical phases of interfering electromagnetic waves. The quantum version has an added nonclassical feature of wave function collapse after a measurement. Grover's method does not make use of nonlocal entanglement, which is another nonclassical feature of quantum systems [12].

Quantum holographic techniques provide a simple means to find the marked bit [13]. Consider a Rydberg wave packet where each occupied state has the same amplitude, but binary 1's and 0's are denoted by phase. Such a state is excited from the launch state in the atom by a sculpted laser field, which performs a unitary transformation A:

$$(15) \qquad A|\Psi\rangle = \begin{pmatrix} 1 & -\varepsilon & \varepsilon & \cdot & \varepsilon \\ \varepsilon & 1 & 0 & \cdot & 0 \\ -\varepsilon & 0 & 1 & \cdot & 0 \\ \cdot & \cdot & \cdot & \cdot & \cdot \\ -\varepsilon & 0 & 0 & \cdot & 1 \end{pmatrix} \begin{pmatrix} 1 \\ 0 \\ 0 \\ \cdot \\ 0 \end{pmatrix} = \begin{pmatrix} 1 \\ \varepsilon \\ -\varepsilon \\ \cdot \\ -\varepsilon \end{pmatrix} \begin{matrix} 7s \\ 27p \\ 28p \\ \cdot \\ 38p \end{matrix}.$$

To find the hidden information, the atom is excited by a search pulse which produces a second wave packet with all quantum phases relatively real. Such a pulse is universal: it will retrieve the information, no matter which states stored 1's and 0's. This is represented by a unitary tranformation B:

$$(16) \qquad BA|\Psi\rangle = \begin{pmatrix} 1 & -\varepsilon & -\varepsilon & \cdot & -\varepsilon \\ \varepsilon & 1 & 0 & \cdot & 0 \\ \varepsilon & 0 & 1 & \cdot & 0 \\ \cdot & \cdot & \cdot & \cdot & \cdot \\ \varepsilon & 0 & 0 & \cdot & 1 \end{pmatrix} \begin{pmatrix} 1 \\ \varepsilon \\ -\varepsilon \\ \cdot \\ -\varepsilon \end{pmatrix} = \begin{pmatrix} 1 \\ 2\varepsilon \\ 0 \\ \cdot \\ 0 \end{pmatrix}.$$

The unitary operator B is, in fact, the very same "reference" pulse that was used to measure the phases in the sculpted wave in the previous section. The superposition leads to destructive interference of any states where a binary 0 was stored, but constructive addition to any states with a binary 1. The combined wave packet then has the same information as before, but now it is encoded into quantum amplitudes rather than phases. These can then be read out using state-selective field ionization.

This is a simple example of a general class of search algorithms that make use of the properties of superposition and quantum interference, which were introduced by Grover [8]. Our particular implementation of a Grover-style search algorithm has some unresolved difficulties if the register is too large. The simple form of the decoding pulse only works in the perturbation theory limit, where almost all of the probability amplitude in the wave packet resides in the "launch" state ($7s$ for our work in Cs). If there are too many states in the Rydberg wave packet, the launch state will become depleted. The data retrieval is still possible, but now the unitary transformation that amplifies the "1" bits and suppresses the "0" bits must depend on the total pulse energy. In other words, we move into *the strong-field regime*.

5ʹ2. *Data base searches with strong THz pulses*. – The limitations on unitary transformations in the Rydberg state space that are imposed by perturbation theory can be overcome by employing shaped broadband coherent far-infrared pulses. These terahertz pulses can be produced and shaped using some of the same ultrafast technology that was developed for optical pulse shaping.

The terahertz version of a subpicosecond optical pulse is a simple half-cycle of broadband coherent radiation. In other words, the transverse electric field simply turns on and then turns off in less than one picosecond. These "half-cycle pulses" (HCPs) are not quite light and not quite radio waves. Generation mechanisms rely on ultrafast optical pulses. When these impinge on a biased semiconductor surface they can create an electron-hole plasma, which radiates a half-cycle pulse (fig. 13) [2].

HCPs can transfer significant momentum to a Rydberg system, leading to coherent state redistribution, or even field ionization. The ionization mechanism is essentially that of an impulse:

$$(17) \qquad Q = -\int_{\mathrm{HCP}} F(t)\,\mathrm{d}t.$$

HCP ionization of Rydberg atoms have been studied extensively [14], as have redis-

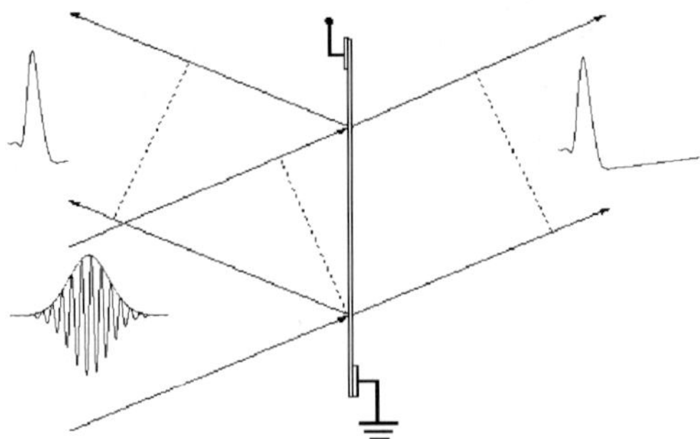

Fig. 13. – A large-aperture photoconducting switch is biased with an electric field in the surface plane. When sub-picosecond pulses illuminate the switch, far-infrared radiation is generated along the transmitted and reflected laser beam direction.

tribution of eigenstates [15], and interactions with Rydberg wave packets [16]. In the present context of data base searches, research has shown that an HCP can efficiently discover the position of a single phase-marked state in a Rydberg wave packet, and amplify that state while suppressing the rest [17].

The HCP locates the phase-flipped state because its propensity to change the properties of a wave packet depends on the location of the probability density. The momentum transfer is just constant, given by eq. (17); but the energy change implied by this momentum change is largest when the wave packet has a lot of momentum of its own. This occurs at the ion core, where the potential energy is large and negative. A wave packet localized at the origin corresponds to a coherent sum of orbitals with equal phase. A Rydberg wave packet data register with one flipped bit is described by a superposition of such a localized wave packet, plus a phase-reversed orbital:

$$(18) \qquad |\Psi_j\rangle = \sum_j |n_j lm\rangle - 2|n_0 lm\rangle.$$

The HCP thus redistributes the localized component over a large number of states, essentially depleting the probability in any one state. At the same time, the spatially delocalized eigenstate representing the single marked orbital is amplified by a factor of approximately $2^2 = 4$. This is an example of the quantum diffusion and quantum amplification of information, discussed by Grover [11]. This simple idea can also be shown in more rigorous calculations, in which the finite bandwidth and duration of the HCP is taken into account explicitly [15].

A test of the HCP quantum search was performed in a beam of atomic Cs. The Rydberg data register was loaded using wave packet sculpting techniques that were described above. Briefly, an optical pulse was sculpted to excite a Rydberg data register with one marked bit in Cs np_z states with $n = 24$ to 29. State-selective field ionization of the initial wave packet reveals only the amplitudes of the states, as shown in fig. 14a. An HCP was generated inside the atomic beam apparatus by an ultrafast optical pulse

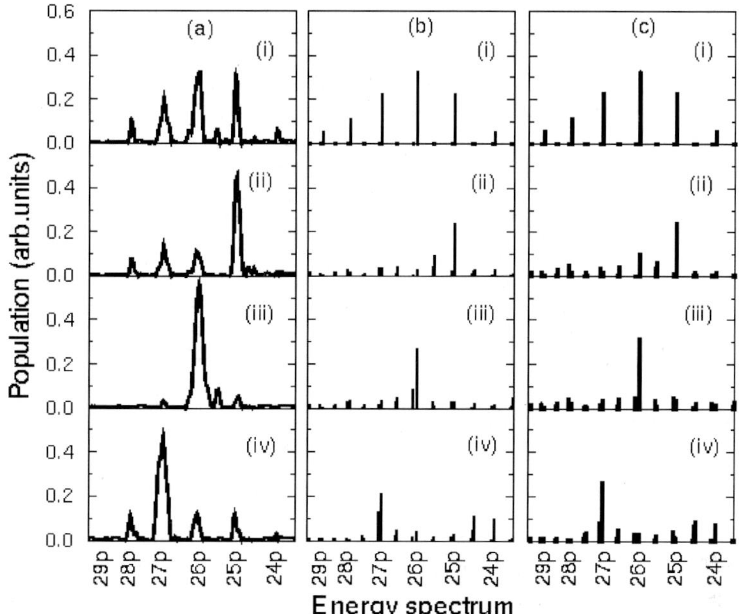

Fig. 14. – Data (a) and calculations (b, c) of the state-selective field ionization spectra of the initial wave packet (i), and the final wave packets at three different times, when state $25p$ (ii), $26p$ (iii), and $27p$ (iv) are phase-reversed from the other states in the wave packet. (b) Impulse model calculation, using the approximation of eq. (17). (c) Model calculation including the direct integration of the time-dependent Schrödinger equation.

illuminating an electrically biased GaAs wafer. The far-infrared radiation was focused on the atomic beam by an off-axis parabolic mirror. Its peak field strength was calibrated by observing the threshold for ionization of Rydberg states [14]. The HCP delivered an impulse of approximately $Q_0 = 0.0043$ atomic units, which was enough to significantly mix s, p, and d states [15]. It was polarized parallel to the Rydberg state. The HCP interrogated the system at time t, converting the wave packet to a simpler structure where for the most part, only the marked state remains. Figure 14 shows ramped field ionization data for three different data register configurations, in which a different state in the register was phase reversed.

5˙3. *Entanglement in Rydberg atoms.* – Grover's algorithm does not require more than one degree of freedom—radial motion, for example. Nonetheless, single-electron systems such as these Rydberg atoms contain several degrees of freedom, and these can be useful in performing other kinds of quantum computation. In this final section of the paper we will describe some of the degrees of freedom that could be manipulated to produce "entanglement", and that might be suitable for quantum algorithms such as Shor's algorithm, which require entanglement to achieve efficiency.

Even if we neglect all of the degrees of freedom associated with the electrons in the

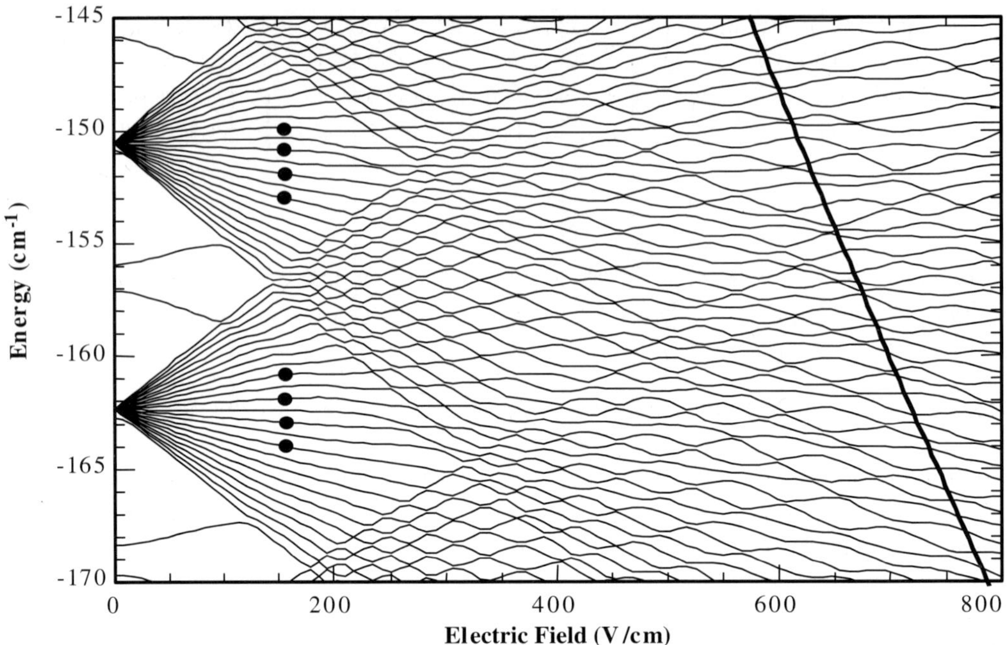

Fig. 15. – Stark manifolds for $n = 26$ and $n = 27$ in the Rydberg series for $m = 1$ states of atomic cesium. The slanted line at the right follows the critical ionization field $1/(16n^{*4})$.

ion core of a Rydberg atom, there are still three spatial degrees of freedom and two spin degree of freedom associated with the electron. Each has a quantum number describing it, as discussed in the first section of these lectures. For example, we could choose n, l, m_l, s and m_s to describe the radial, polar, azimuthal, spin, and spin projection along z for the Rydberg electron's wave function. (Alternatively, we could use the eigenstates of the spin-orbit Hamiltonian: n, l, s, j, and $m_j = m_l + m_s$.) Other examples exist for other additions to the Hamiltonian. For example, a uniform electric field produces Stark eigenstates with quantum numbers n, n_1, n_2, and m [3].

Entanglement is the formation of nonfactorizable superposition states of these different degrees of freedom. In the previous section, we have been concerned with wave packets where only the n quantum number varies. Now, to bring in another degree of freedom, let us suppose we split the degeneracy of the n-manifold by applying a uniform electric field to the atom. The n-states spit into manifolds of states, as shown in fig. 15.

The pattern of the eigenvalues reveals two different kinds of motion, associated with two degrees of freedom. The repetition of the manifolds reveals the radial motion with the Kepler period:

$$(19) \qquad \Delta E_n = \frac{2\pi}{\tau_{\text{Kepler}}} = \frac{1}{n^3}.$$

The splitting between different states in one manifold increases linearly with the electric field, although this linear Stark effect is broken up by the interaction between the manifolds once they cross. The linear Stark effect reveals the precession of angular momentum in the wave packet due to the torque exerted by the electric field:

$$(20) \qquad \Delta E_k = \frac{2\pi}{\tau_{\text{Stark}}} = 3\,\text{nF}.$$

An entangled superposition of states in the Stark manifold exhibits both types of motion. For example, consider a wave packet made of the states that are marked in fig. 15:

$$(21) \qquad |n, k\rangle = |26, -8\rangle, |27, -5\rangle, |26, -4\rangle, |27, -1\rangle, |26, 0\rangle, |27, 3\rangle, |26, 4\rangle, |27, 7\rangle.$$

Here, k is equal to the difference between the two parabolic quantum numbers n_1 and n_2. This state oscillates radially as it precesses from nearly linear motion along the radius, to nearly circular motion, and back again. (Actually, the words "linear" and "circular" do not do justice to the beauty of the motion.)

Entanglements can be useful in quantum information science when it is possible to measure one of the two entangled degrees of freedom without disturbing the amplitudes and phases associated with the other degree of freedom. Thus, in the example above, suppose it is possible to measure n without disturbing the "parabolic" quantum number k. This might be accomplished by field ionization with a field of 700 v/cm. The Rydberg wave packet subjected to this field will "collapse". This means that, according to the rules of quantum mechanics, it will field ionize with a probability of 50%, since half of the probability resides in states above the energy to ionize in the field. There is a 50% probability, however, that the atom will survive. In this case, it must be in the simpler wave packet state described by the substates that could not ionize in this field. Effectively, then, this "partial measurement" has yielded a measurement of n, without disturbing k. Experiments to demonstrate this effect are underway. Similar principles underlie the proposals for quantum measurements of entangled states in Josephson junction systems, electrons on the surface of liquid He, and electrons in quantum dots.

6. – Conclusion

In these lectures we have described some of the science and technology associated with Rydberg atoms for quantum information applications. We should end with some acknowledgements, and a disclaimer!

The ideas and experiments described in these lectures are largely the work of dedicated students and post-docs in my group. I particularly acknowledge the work of Jae Ahn, Chandra Rahman, Tom Weinacht, and Robert Jones. The funding for all of these experiments has come from the National Science Foundation.

Can we conclude that Rydberg atoms provide the best path to realizing practical quantum computers? Certainly not! Although Rydberg atoms are simple and easily manipulated, they lack a fundamental characteristic that will be needed if quantum computers are to become useful. Rydberg atom state spaces are not easy to scale to very large numbers of degrees of freedom. Therefore the state space cannot grow exponentially with the hardware resources, as is possible with other systems such as ions in traps. This problem is not fundamental, however. Entanglement of Rydberg states distributed as "Schrödinger's cats" across several atoms is absolutely essential, and has been recently demonstrated by Haroche, described elsewhere in this volume in the context of cavity QED. Similar entanglement may be possible without a cavity, as described recently by Polzik and Zoller.

REFERENCES

[1] Gallagher T., *Rydberg Atoms* (Cambridge Press, Cambridge, Mass.) 1995.

[2] Yaezell J. and Uzer T. (Editors), *The Physics and Chemistry of Wavepackets* (John Wiley, New York) 1999.

[3] Bethe H. A. and Salpeter E. E., *Quantum Mechanics of One and Two Electron Atoms* (Academic Press, New York) 1957.

[4] Sobelman I. I., *Atomic Spectra and Radiative Transitions*, second edition (Springer-Verlag, Berlin) 1991.

[5] Spielman C., Curley P. F., Brabec T. and Krausz F., *IEEE J. Quantum Electron.*, **QE-30** (1994) 1100, and references therein.

[6] Strickland D. and Mourou G., *Opt. Commun.*, **55** (1985) 447.

[7] Tull J. X., Dugan M. A. and Warren W. S., *Adv. Opt. Mag. Res.*, **20** (1990) 1; Weiner A. M., Leird D. E., Patel J. S. and Wullert J. R., *J. Quantum Electron.*, **28** (1992) 1.

[8] Tull J. X., op. cit. [7].

[9] Weinacht T. C., Ahn J. and Bucksbaum P. H., *Phys. Rev. Lett.*, **80** (1998) 5508.

[10] Jones R. R., Schumacher D. W., Gallagher T. F. and Bucksbaum P. H., *J. Phys. B*, **28** (1995) L405.

[11] Grover L. K., *Phys. Rev. Lett.*, **79** (1997) 325; 4709.

[12] Lloyd S., *Phys. Rev. A*, **60** (2000) 301.

[13] Ahn J., Weinacht T. C. and Bucksbaum P. H., *Science*, **287** (2000) 463.

[14] Jones R. R., You D. and Bucksbaum P. H., *Phys. Rev. Lett.*, **70** (1993) 1236; Reinhold C. O. *et al.*, *J. Phys. B*, **26** (1993) L659.

[15] Tielking N. E. and Jones R. R., *Phys. Rev. A*, **52** (1995) 1371.

[16] Raman C. *et al.*, *Phys. Rev. Lett.*, **76** (1996) 2436.

[17] Ahn J., Hutchinson D. N., Rangan C. and Bucksbaum P. H., *Phys. Rev. Lett.*, **86** (2001) 1179.

Cavity QED—experiments with atoms and ions(*)

H. Walther

*Sektion Physik der Universität München and Max-Planck-Institut für Quantenoptik
85748 Garching, Germany*

1. – Introduction

The many applications discussed in quantum communication and quantum cryptography require sources able to produce a preset number of photons. Single photons are, for example, a necessary requirement for secure quantum communication [1-3], for quantum cryptography [4] and in special cases also for quantum computing [5]. However, photon fields with fixed photon numbers are also interesting from the point of view of fundamental physics since they represent the ultimate non-classical limit of radiation. When the photon number state is generated by strong coupling of excited-state atoms, a corresponding number of ground-state atoms is simultaneously populated. Such a system therefore produces a fixed number of atoms in the lower state as well. This type of atom source is a long-sought-after *gedanken* device [6]. Single photons have been generated by several processes such as single-atom fluorescence [7], single-molecule fluorescence [8], two-photon down-conversion [9], Coulomb blockade of electrons [10], and one- and two-photon Fock states have been created in the micromaser [11] (see also [12]). As these sources do not produce the photons on demand, they are better described as "heralded" photon sources, because they are stochastic either in the emission direction or in the time of creation. A source of single photons or even more generally Fock states created on demand has not yet been demonstrated. Cavity quantum electrodynamics (QED) provides

(*) This work was performed in collaboration with S. Brattke, G. R. Guthöhrlein, M. Keller, W. Lange and B. Varcoe.

us with both the possibility of generating a photon at a particular time and localising its emission direction. To this end, there have been several proposals making use of high-Q cavities that are basically capable of serving as sources of single photons [3, 13-15]. The current paper reviews the work on a microwave source able to produce a preset number of photons and lower-state atoms. The principle of the source and the first experimental demonstration will be described. It is based on the One-Atom Maser and allows the generation of a specified photon Fock state ($n \geq 1$) *on demand*, without the need for conditional measurements, being therefore independent of detector efficiencies.

In the second part of the paper the work towards a new single-photon source in the visible spectral range is described. This source uses a single trapped ion which is placed into a cavity. The first progress towards the realization of this source will be reported.

Steady-state operation of the One-Atom Maser or micromaser has been studied extensively both theoretically [16] and experimentally, and has already been used to demonstrate many quantum phenomena of the radiation field such as sub-Poissonian statistics [17], the collapse and revival of Rabi oscillations [18], and entanglement between the atoms and cavity field [19]. More recently two experiments have demonstrated that Fock states (*i.e.* states with a fixed photon number) can be readily created in the normal operation of the maser, by means of either state reduction [11] or steady-state operation of the micromaser in a trapping state [12]. State reduction is possible owing to the entanglement between the state of the outgoing atoms and the cavity field; detection of a lower-state atom means that a field originally in an n photon Fock state is projected onto the $n + 1$ state [20]. As a source of single photons such a source can be compared to two-photon down-conversion, in which an idler beam is used to herald the creation of a photon in the signal beam. Both are subject to the same limitation in that the creation of the Fock state is unpredictable, and imperfect detectors further reduce the probability that a state, once created, is also detected. In contrast it is shown here that the micromaser can be used to prepare Fock states with small photon numbers in the cavity *on demand* and independent of detection efficiencies. Simultaneously the same number of ground-state atoms are produced with an efficiency of up to 98%.

Trapping states are a feature of the low-temperature operation of the micromaser, for which the steady-state photon distribution closely approximates a Fock state under certain conditions. They are typical of strongly coupled systems. They occur when atoms perform an integer number, k, of Rabi cycles under the influence of a fixed photon number n:

$$(1) \qquad \sqrt{n + 1}\, g t_{\text{int}} = k\pi,$$

where g is the effective atom-field coupling constant and t_{int} is the interaction time. Trapping states are characterised by the number of photons n and the number of Rabi cycles k. The trapping state $(n, k) = (1, 1)$ therefore refers to the one-photon, one-Rabi-oscillation trapped field state. In other words, trapping states occur when the interaction time is chosen such that the emission probability becomes zero for certain operating parameters of the maser. Therefore at some time during the steady-state operation of

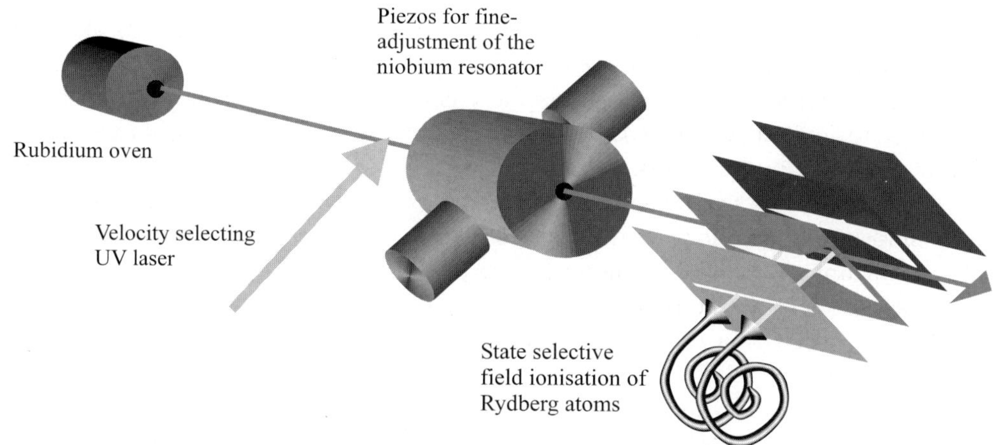

Fig. 1. – The atoms leaving the rubidium oven are excited into the $63P_{3/2}$ Rydberg state by means of a UV laser at an angle of 11°. After the cavity the atoms are detected by state-selective field ionisation. Tuning of the cavity is performed with two piezo translators. An auxiliary atomic beam (not shown) is used to stabilise the laser frequency. The laser is locked to a Stark-shifted atomic resonance of the auxiliary beam, thus allowing the velocity subgroup selected by excitation to be continuously changed within the range of the velocity distribution of the atoms.

the micromaser in a trapping state the field will enter a Fock state and become stabilised. The particular Fock state is known and is determined by the interaction time between atom and cavity as given by the trapping-state formula (eq. (1)). The Fock state, once prepared, is preserved owing to the trapping condition with a minimum probability of photon emission. Following the state preparation, the beam of pump atoms can be turned off and the Fock state remains in the cavity for the duration of the cavity decay time. For simplicity we will concentrate in the following on the preparation of a one-photon Fock state, however, the method can also be generalised to the generation of fields of higher photon numbers.

The micromaser setup used for the experiments is shown in fig. 1 and is operated in the same way as described in [12]. Briefly, a ^{3}He-^{4}He dilution refrigerator houses the closed superconducting microwave cavity. A rubidium oven provides two collimated atomic beams: the main beam passing directly into the cryostat and a second used to stabilise the laser frequency [12]. A frequency-doubled dye laser ($\lambda = 297\,$nm) was used to excite rubidium (^{85}Rb) atoms to the Rydberg $63P_{3/2}$ state from the $5S_{1/2}(F = 3)$ state. The cavity is tuned to the 21.456 GHz transition from the $63P_{3/2}$ state to the $61D_{5/2}$ state, which is the lower or ground state of the maser transition. For this experiment a cavity with a Q-value of 4×10^{10} was used; this corresponds to a field decay time of 0.6 s or a photon lifetime of 0.3 s. This Q-value is the largest ever achieved in this type of experiment and the photon lifetime is more than two orders of magnitude higher than that achieved in related experiments [21]. This cavity is used to study micromaser operation in great detail. To realise the Fock state it is necessary to switch the excitation of the Rydberg atoms on and off in a predefined pulse sequence; this was achieved by means of an intensity modulating electro-optic modulator triggered by control software. The pulse duration and pulse separation can both be tailored to the conditions required for the particular experiment.

To demonstrate the principle of this source, fig. 2 shows a sequence of twenty subsequent pulses obtained via a Monte Carlo simulation [22] of the micromaser operating in the $(1,1)$ trapping state. In each pulse there is a single emission event, producing a single lower-state atom and leaving a single photon in the cavity. In the case of the loss of a photon by dissipation, one of the next incoming excited-state atoms will restore the single-photon Fock state. This condition has been observed in [12] when sub-Poissonian atom statistics were measured with the maser operating in a trapping state. The in-

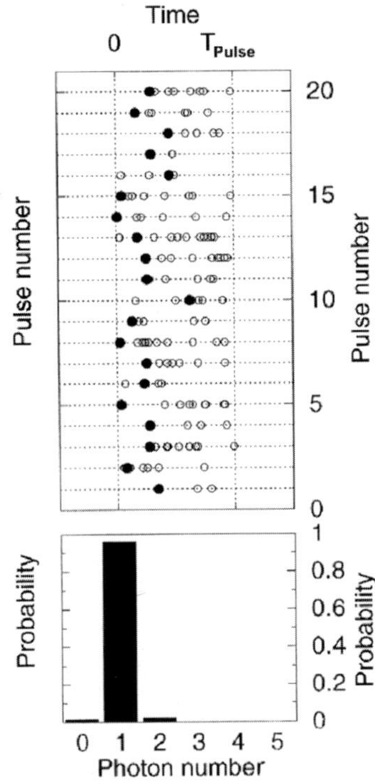

Fig. 2. – A simulation of a subset of twenty subsequent atom bunches after the cavity and the associated probability distribution for photons or lower-state atom production (filled circles represent lower-state atoms and open circles represent excited-state atoms). The start and finish of each pulse is indicated by the vertical dotted lines marked 0 and τ_{pulse}, respectively. The operating conditions are the $(1, 1)$ trapping state $(gt_{\text{int}} = 2.2)$ conditions. The size of the atoms in this figure is exaggerated for clarity. With the real atomic separation there are 0.06 atoms in the cavity on average (*i.e.*, the system operates in the one-atom regime). The other parameters are $\tau_{\text{pulse}} = 9{,}92\ \tau_{\text{cav}}$, $n_{\text{th}} = 10^{-4}$ and $N_a = 7$ (see also ref. [23] and [24]).

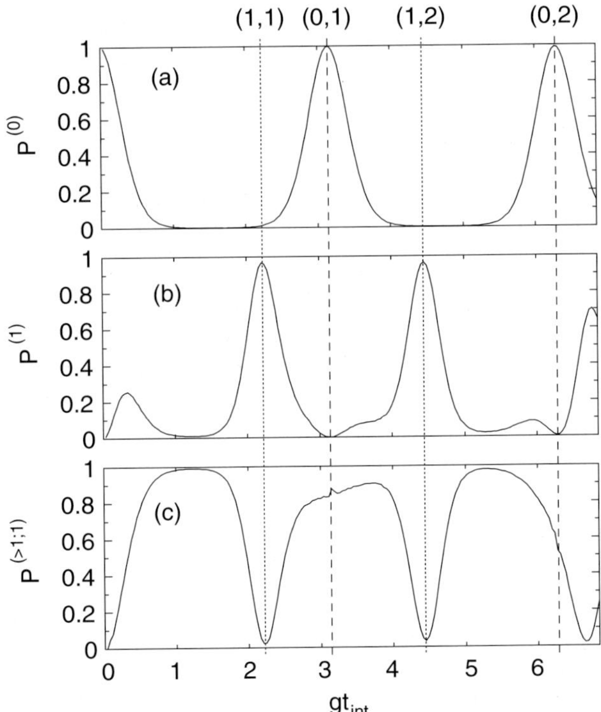

Fig. 3. – The probability of finding (a) no lower-state atoms per pulse $P^{(0)}$, (b) exactly one lower-state atom per pulse $P^{(1)}$, and (c) a second lower-state atom, if one has already been detected $P^{(>1;1)}$. The parameters are $\tau_{\text{pulse}} = 0.02\tau_{\text{cav}}$, $N_a = 7$ atoms and $n_{\text{th}} = 10^{-4}$. The maximum value of $P^{(1)}$ is 98% for the $(1,1)$ trapping state.

fluence of thermal photons and variations in interaction time or cavity tuning further complicates this picture, resulting in reduced visibility of steady-state Fock states [12]. Pulsed excitation as discussed here, however, reduces these perturbations and the Fock state maintains a high visibility [24].

Figure 3 shows three curves obtained again from a computer simulation, that illustrate the behaviour of the maser under pulsed excitation as a function of interaction time for more ideal (but achievable) experimental parameters. The simulations show the probability of finding no ground-state atom per pulse ($P^{(0)}$); finding exactly one ground-state atom per pulse ($P^{(1)}$); and the conditional probability of finding a second ground-state atom in a pulse already containing one ($P^{(>1;1)}$). It is shown below that the latter plot of the conditional probability, $P^{(>1;1)}$, has the advantage of being especially suitable for the comparison between theory and experiment as it is relatively insensitive to the detection probability for atoms in the upper and lower maser levels.

From the simulations it follows that with an interaction time corresponding to the $(1,1)$ trapping state, both one photon in the cavity and a single atom in the lower state are produced with a 98% probability. In order to maintain an experimentally verifiable quantity, most of the simulations presented relate to the production of lower-state atoms rather than to the Fock state left in the cavity. Pulse lengths on the other hand are rather short ($0.01\tau_{\mathrm{cav}} \leq \tau_{\mathrm{pulse}} \leq 0.1\tau_{\mathrm{cav}}$) so there is little dissipation and the one-photon state in the cavity following the pulse is very close to the probability of finding an atom in the lower state. Note that at no time in this process is a detector event required to project the field; the field evolves to the trapping state as a function of time, when the suitable interaction time has been chosen.

The variation of the time when an emission event occurs during an atom pulse in fig. 2 is due to the variable time spacing between the atoms as a consequence of Poissonian statistics and the stochasticity of the quantum process. The atomic rate therefore has to be high enough so that there will be a sufficient number of excited atoms per laser pulse, in order to maintain the 98% probability of an atom emitting.

To guarantee single-atom single-photon operation, the duration of the preparation pulses must be short in relation to the cavity decay time. For practical purposes, the pulse duration should be smaller than $0.1\tau_{\mathrm{cav}}$ for dissipative losses to be less than 10%. Apart from reducing the fidelity of the Fock state produced, losses increase the likelihood of a second emission event leading to larger number of lower-state atoms than photons in the field; whereby the 1:1 correspondence between both would be lost. Shorter atom pulses reduce the dissipative loss, however the number of atoms per cavity decay time (usually labelled N_{ex}) must be larger than ten times the threshold value of the atomic flux to realise the Fock source with a significant fidelity. Since a minimum atom number is required to produce the desired state, care must also be taken to avoid atom beam densities violating the one-atom-at-a-time condition.

For a large range of operating conditions, the production of Fock states of the field and single lower-state atoms is remarkably robust against the influence of thermal photons, variations of the velocity of atoms and other influences such as mechanical vibrations of the cavity. Much more so than the steady state trapping states, for which highly stable conditions with low thermal photon numbers are required [12, 23].

An obvious side effect of the production of a single photon in the mode is, as mentioned already, that a single atom in the lower state is produced. This atom is in a different state when it leaves the cavity and is therefore distinguishable from the pump atoms, hence under this operation, the micromaser also serves as a source of single atoms in a particular state, a requirement for many proposed experiments [6, 25].

Although the distribution of lower-state atoms leaving the cavity will be maximally sub-Poissonian, the arrival time of an atom within the pumping pulse still shows a small uncertainty, the upper limit of which is determined by the pump pulse duration in the range of 0.01–$0.1\tau_{\mathrm{cav}}$ for the parameters used in this paper. The separation of the pulses is $\geq 3\tau_{\mathrm{cav}}$ leading to a small relative variation in the arrival times. If one would increase the pump rate still further, the pulse lengths could be further reduced and the arrival of an atom becomes even more predictable.

The present setup of the micromaser was specifically designed for steady-state operation and is therefore not ideal for the parameter range presented here. However, the current setup does permit a comparison between theory and experiment in a relatively small parameter range. The experimental test relies on the measurement of an absolute number of atoms and although the operation of the Fock source is independent of detector efficiencies, the experimental test is blurred by the fact that the state-selective field ionisation detectors for the Rydberg atoms do not reach an efficiency of 100%. Atoms in a particular state will therefore be missed leading to wrong or misleading results. In order to circumvent this disadvantage, it is useful to measure population correlations between subsequent atoms instead. Owing to the strong coupling between the atoms and cavity, the cavity field and the state of the pumping atom are entangled following the interaction. A subsequent pumping atom will thereby also become entangled with any previous one, thus the population correlations between subsequent atoms are determined by the particular dynamics of the atom-cavity interaction. The connection between population correlations and the micromaser dynamics, has been studied in detail in previous papers [19, 26]. It is important to note that even in the presence of lost counts, the correlations between subsequently detected atoms are maintained. Conditioning the experimentally measured parameter on the detection of atom pairs that contain at least one lower-state atom, provides both a value appropriate to the existent correlation and —at the same time— directly related to the total probability of finding one atom per pulse.

By means of an extremely high cavity Q factor (4×10^{10}) an N_{ex} of approximately 60 was accessible for a short range of interaction times around the maximum in the Maxwell-Boltzmann velocity distribution, which happens to correspond to the interaction time for the $(1, 1)$ trapping state. A pulse length of $\tau_{\mathrm{pulse}} = 0.066\tau_{\mathrm{cav}}$ leading to an average of 4 atoms per pulse was chosen as a compromise between the effects of dissipation and external influences, while still providing a pump rate above the threshold for single-photon Fock-state production. Figure 4 shows the results of the comparison of theory and experiment for a scan over the $(1, 1)$ trapping state. Plotted here is the ratio of two-atom events to the total number of two-atom events detected that contain at least one lower state. That is,

$$(2) \qquad P^{(>1;1)} = \frac{N_{\mathrm{gg}}}{N_{\mathrm{gg}} + N_{\mathrm{eg}} + N_{\mathrm{ge}}},$$

where for example N_{eg} is the probability of detecting a two-atom event containing first an upper-state atom (e) and then a lower-state atom (g) in any given pulse. The number of three-atom events detected is negligible and can be ignored as a contributing factor. Constructing the parameter in this way ensures that there is a one-to-one correspondence between the correlation parameter and the maximum probability of finding exactly one atom per pulse. That is, to a good approximation for pump rates and pulse durations employed in this experiment, the difference between the upper bound P^{max} and the measured correlation $P^{(>1;1)}$ gives the probability of finding exactly one atom per pulse ($P^{(1)}$). $P^{(>1;1)}$ in eq. (2) is independent of the absolute detector efficiency and depends only on the relative detector efficiencies and the miscount probability (the probability that a given atomic level is detected in the wrong detector), each of which has been

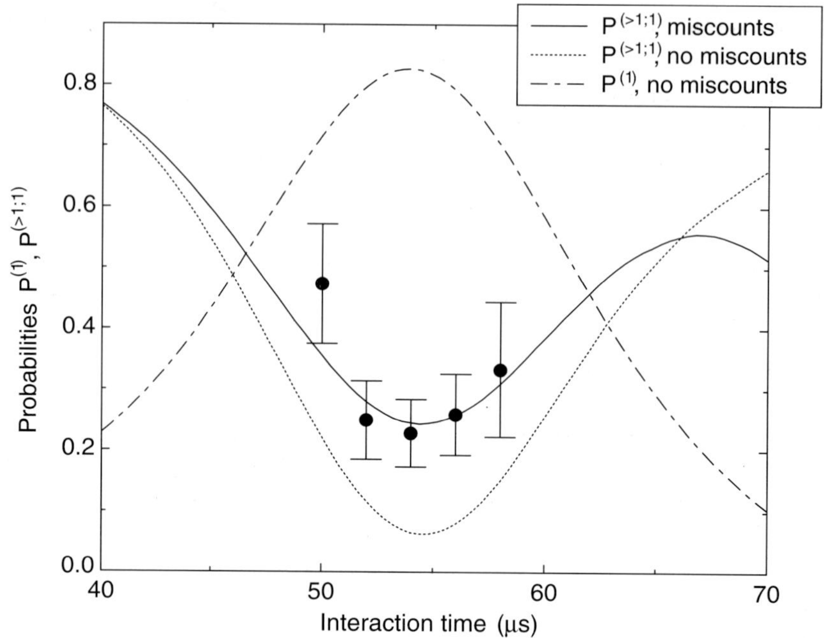

Fig. 4. – Comparison between theory and experiment. Investigated here is the behaviour shown in fig. 3. The experimental data is evaluated according to eq. (2). This ratio is to the absolute detector efficiency and is dependent only on the relative detector efficiencies and the miscount probability which can be measured experimentally. The relative detector efficiencies can be considered approximately equal. On the other hand a calculation of the probability of finding one lower-state atom per pulse is highly dependent on the absolute detector efficiency. The two theoretical curves presented are a theoretical prediction of Fock-state creation and a theoretical result which takes into account the experimentally measured miscounts of 7% in the lower-state detector and 2% in the excited-state detector. Given the extreme conditions for the operation of the apparatus, there is an excellent agreement between theory and experiment. The single-atom probability is evaluated to be 83.2%. The parameters for the experiment were $\tau_{\mathrm{cav}} = 300\,\mathrm{ms}$, $\tau_{\mathrm{pulse}} = 0.066\tau_{\mathrm{cav}}$, pulse spacing of $1\,\mathrm{s}$, $n_{\mathrm{th}} = 0.03$, $N_{\mathrm{a}} = 4$.

measured experimentally. The theoretical curves of fig. 4 represent an evaluation of the probability of finding exactly one atom per pulse and the conditional probability introduced in fig. 3. The curves are evaluated both for the ideal situation of no detector miscounts and for the measured detector miscounts of 7% in the lower-state detector and 2% in the excited-state detector. When the miscounts are incorporated into the data there is an excellent match between the experimental points and the theoretical curve.

As the present apparatus was designed for operating the micromaser in steady-state conditions, the current results were obtained under non-ideal operating parameters of the apparatus. In order to overcome this disadvantage future developments will incorporate two improvements, to increase the atomic flux and to introduce a second pulse of atoms with variable velocity to act as a field probe. With these changes it will be easier to arrive at the optimum conditions for the Fock source and, in addition, will allow direct measurements of the cavity photon number by means of Rabi oscillations of the probe atom [11] and further studies of quantised field effects.

The source presented here has the significant advantage over our previous method of Fock-state creation [11] of being unconditional and therefore significantly faster in preparing a target quantum state. Previously the state was prepared by the dynamics of the interaction of excited-state atoms with the cavity field. State reduction occurred on detection of a lower-state atom indicating the preparation of the desired cavity field. In the current experiment, however, the cavity field is correctly prepared in 83.2% of the pulses, independently of state reduction and hence atomic detection efficiency. Simply detecting the lower-state atom as it emerges from the cavity increases the fidelity of the one-photon Fock state to $\geq 95\%$ at the moment of detection (incorporating both dissipative losses and detector miscounts). Assuming 40% detector efficiency, detection of a lower-state atom within any given preparation pulse will occur with a 36.8% probability. This should be compared to the 95% fidelity of the measured Fock state in [11], in which there was a $\leq 1\%$ probability of detecting the correctly prepared state. Thus along with the first observation of the operation of a single-photon Fock-state source, we also have an order-of-magnitude improvement of Fock-state creation over our previous experiment.

In the second part of this paper, we would like to report on the progress of our work on ions in optical cavities. The interaction of a single atom with a single field mode of a high-finesse cavity has been the subject of a number of experiments in the field of cavity QED [15]. However, most of these investigations suffer from a lack of control over the position of the atom, which results in non-deterministic fluctuations of the coupling between atoms and field. In this context, the strong localisation and position control available when an ion trap is combined with an optical cavity would be a big step forward and would become a key technology for future progress in cavity QED in the optical range. We are presently implementing two experiments exploiting the localisation of an ion in a cavity. By pulsed excitation of a maximally coupled ion, single-photon wave packets may be emitted from the cavity on demand [14,15,27] (single-photon gun). Under conditions of strong coupling, a single calcium ion in the cavity provides sufficient gain to build up a laser field [13]. Analogous to a single ion in free space, which previously was shown to be an excellent source of antibunched light [28,29], radiation from a single-ion laser has non-classical photon statistics and correlations.

An equally attractive goal in the area of cavity QED is the simultaneous interaction of two or more ions with a single-cavity mode. Due to the linear geometry of our trap several ions can be stored within the mode volume. As a first test, we have placed an array of two ions in the cavity field and observed the total fluorescence. We succeeded in matching the ion crystal to the two maxima of the TEM_{01} mode of the cavity. In such a configuration the cavity field may be used to entangle the two ions [30,31]. This is a promising alternative to schemes involving the ions' motional degrees of freedom, since there is no need for cooling the vibrational modes of the string below the Doppler temperature. Using a cavity to perform quantum operations on adjacent pairs of ions in a long string is a viable route to a scalable quantum computer.

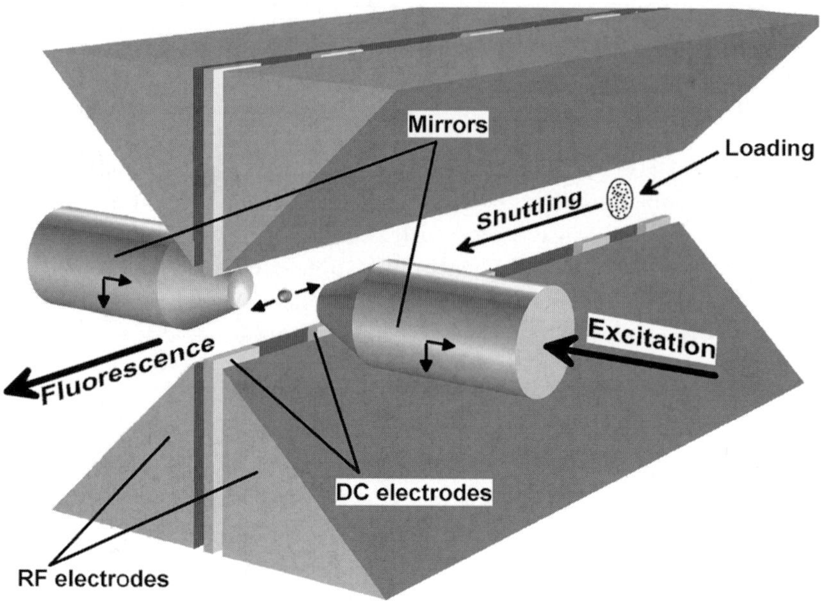

Fig. 5. – Experimental arrangement of trap electrodes and cavity mirrors. The ion is loaded at the rear end of the trap and shuttled to the mirror region. Fluorescence is observed from the side of the cavity. For scans in the direction of the trap axis, the ion is moved with DC electrodes. In all other directions, the cavity is translated relative to the ion's position, as indicated by arrows.

In the following we give a progress report of our experiment. (For details see also ref. [32].) We are using a linear trap with Ca ions (see fig. 5). As an initial test of the setup for the above-mentioned cavity QED experiments, we have used the trapped Ca ions to probe the optical field in the cavity. The Ca ion is sensitive to radiation close to the resonance line $4^2S_{1/2}$–$4^2P_{1/2}$ at a wavelength of $\lambda = 397$ nm. The fluorescent light emitted by the ion is collected with a lens (numerical aperture $= 0.17$) and detected with a photomultiplier tube (overall detection efficiency $\eta \approx 10^{-4}$). The observed fluorescence rate R is proportional to the local intensity of the optical field at the position $\vec{r}$ of the ion, $i.e.$ $R \propto I(\vec{r})$, provided there is no saturation of the atomic transition. By scanning the position of the ion in the field and detecting the fluorescence rate at each point, a high-resolution map of the optical intensity distribution is obtained. It should be noted that a single ion can also probe the amplitude distribution $\vec{E}(\vec{r})$ of the light field and hence measure its phase. To this end, heterodyne detection of the fluorescent light must be used, with the exciting laser as a local oscillator [7, 29].

With the single ion as a probe, we have investigated the eigenmodes of a Fabry-Perot resonator formed by two mirrors (radius of curvature $= 10$ mm) at a distance of $L = 6$ mm (fig. 5). The transverse-mode pattern is described by Hermite-Gauss functions with a beam waist $w_0 \approx 24\,\mu$m, while in the direction of the cavity axis a standing wave builds up. In the experiment, a particular cavity mode is excited by a laser beam with a power

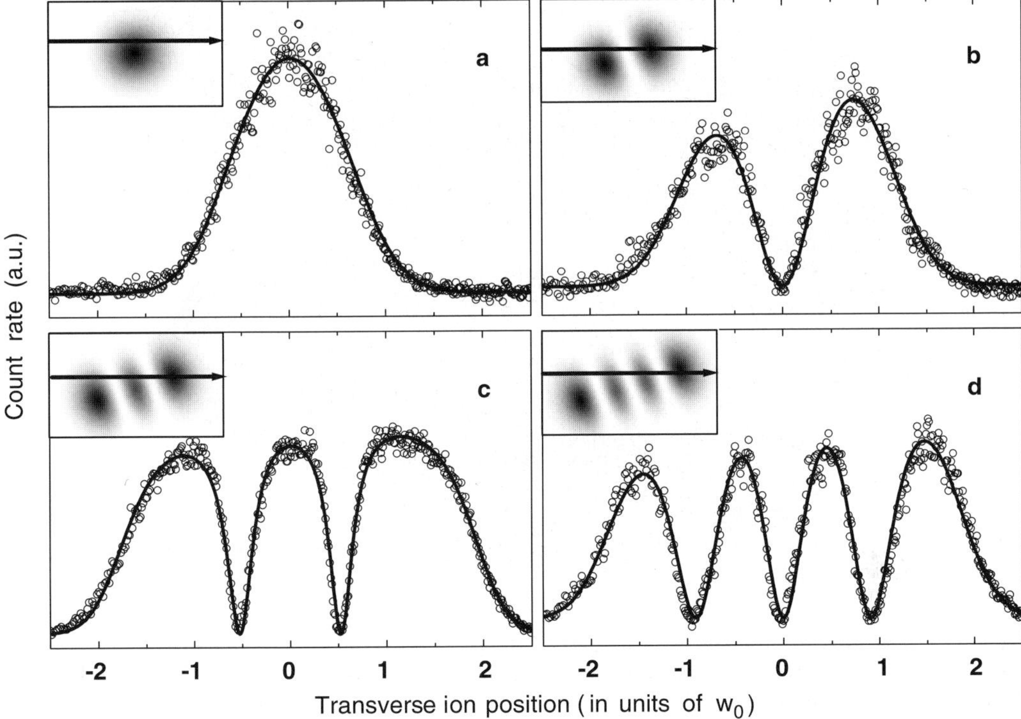

Fig. 6. – Transverse profiles of the Hermite-Gauss modes of the cavity, obtained by monitoring the ion's fluorescence while scanning over a range of $120\,\mu$m. The solid line is a fit including saturation of the transition. The inset shows the calculated intensity distribution of the mode and indicates the scan path. The modes are a) TEM_{00}, b) TEM_{01}, c) TEM_{02}, d) TEM_{03}.

of a few hundred nanowatts at 397 nm. The length of the cavity is actively stabilised to this mode.

An ion is loaded in the trap after electron-impact ionisation of calcium atoms. Since the electron beam and the calcium beam would degrade the optical mirrors and make stable trapping difficult, we use a linear trap and load it in a region spatially separated from the observation zone, as shown in fig. 5. Subsequently, DC electrodes along the axis are employed to shuttle the ion over a distance of 25 mm to the uncontaminated end of the trap, where the cavity is located, oriented at right angles to the trap axis. Residual DC fields in the radial direction must be carefully compensated with correctional DC voltages to place the ion precisely on the nodal line of the RF field (coinciding with the trap axis), to avoid the trapping field exciting the micromotion of the ion.

In the direction of the trap axis, the ion is confined in a DC potential well, which is approximately harmonic with an oscillation frequency of $\omega_z \approx 300\,\text{kHz}$. By applying asymmetric voltages, the minimum of the potential well and thus the equilibrium position of the ion is moved along the trap axis. By simultaneously monitoring the fluorescence, we

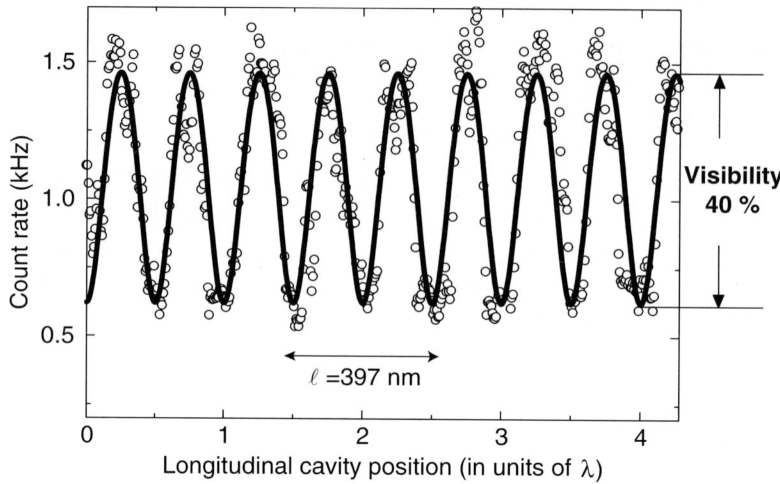

Fig. 7. – Single-ion mapping of the longitudinal structure of the cavity field. The visibility is determined by the residual thermal motion of the Doppler-cooled ion. It corresponds to a resolution of 60 nm. The localisation of the ion's wave packet in this measurement is 16 nm.

have sampled one-dimensional cross-sections of the cavity mode. The width of the ion's wave function in the axial potential well is a few hundred nanometres, which provides sufficient resolution to map the transverse-mode pattern with an intensity distribution varying on a scale given by the cavity waist w_0.

Figure 6 shows scans of the first four TEM_{0n} modes of the cavity obtained in this way. The fluorescence data is not entirely symmetric because of a small displacement and rotation of the cavity eigenmodes with respect to the trap axis. In each plot, an inset indicates the path along which the ion is scanned. The solid curves in fig. 6 are obtained from a fit using Hermite-Gauss functions and take into account saturation of the ion's transition. The influence of saturation is apparent in fig. 6c, where a slightly higher intensity was injected into the cavity. In all cases, the correspondence with the measured fluorescence is excellent.

The ion's motion must be restrained to the trap axis, since off the axis the radio-frequency field of the trap would lead to micromotion. To scan other dimensions of the field, the sample must be moved. In our experiment this is done by piezoelectrically translating the entire cavity assembly perpendicular to the trap axis.

RF confinement of the ion perpendicular to the trap axis is also harmonic, but the corresponding oscillation frequency $\omega_r \approx 1.1\,\text{MHz}$ is larger than the axial frequency so that field structures in the radial direction of the trap are better resolved. The resolution achieved with our method may be determined most accurately by probing the standing-wave field created between the cavity mirrors, which varies on a scale of $\lambda/2$. To this end, the cavity was moved parallel to its axis while keeping the ion stationary and monitoring its fluorescence. Figure 7 shows the mapping of the cavity field obtained in this way. A pronounced standing-wave pattern is observed with a visibility of 40%. For details see also ref. [32].

Taking advantage of the excellent localisation in ion traps, we have performed the most precise measurement to date of a three-dimensional spatial structure of an optical field over a range of up to $100\,\mu$m. As a demonstration we scanned modes of a low-loss optical cavity. The precise positioning we achieve implies a deterministic control of the coupling between ion and field. At the same time, the field and the internal states of the ion are not affected by the trapping potential. What we have realised, therefore, is an ideal system for cavity QED with a single particle.

2. – Conclusion

In this paper we reviewed our work on the generation of photon number states on demand using the micromaser. In addition, we describe first experiments using a single trapped ion in connection with an optical cavity. The system described in the second part of the paper shows a lot of promise and will open a series of new interesting applications.

REFERENCES

[1] Zbinden H., Gisin N., Huttner B. and Tittel W., *J. Cryptol.*, **13** (2000) 207; Lo H.-K. and Chau H. F., *Science*, **283** (1999) 2050.

[2] Gheri K. M., Saavedra C., Törmä P., Cirac J. I. and Zoller P., *Phys. Rev. A*, **58** (1998) R2627; van Enk S. J., Cirac J. I. and Zoller P., *Phys. Rev. Lett.*, **78** (1997) 4293; van Enk S. J., Cirac J. I. and Zoller P., *Science*, **279** (1998) 205.

[3] Cirac J. I., Zoller P., Kimble H. J. and Mabuchi H., *Phys. Rev. Lett.*, **78** (1997) 3221; Parkins A. S., Marte P., Zoller P. and Kimble H. J., *Phys. Rev. Lett.*, **71** (1993) 3095; Parkins S. and Kimble H. J., *J. Opt. B*, **1** (1999) 496.

[4] Jennewein T., Simon C., Weihs G., Weinfurter H. and Zeilinger A., *Phys. Rev. Lett.*, **84** (2000) 4729; Naik D. S., Peterson C. G., White A. G., Berglund A. J. and Kwiat P. G., *Phys. Rev. Lett.*, **84** (2000) 4733; Tittel W., Brendel J., Zbinden H. and Gisin N., *Phys. Rev. Lett.*, **84** (2000) 4737.

[5] The error tolerant quantum computing proposal by Gottesman D. and Chuang I. L., *Nature*, **402** (1999) 390 requires that a "quantum resource" be supplied *on demand* to facilitate computation. Such a source can be provided by the apparatus considered here. See also Preskill J., *Nature*, **402** (1999) 357; Johnathon D. and Plenio M. B., *Phys. Rev. Lett.*, **83** (1999) 3566.

[6] Sources of single atoms, such as the one described in this paper, are routinely employed for hypothetical tasks such as the creation of an atomic beam with arbitrary timing sequence or for the stabilisation of cavity states. For an example see Vitali D., Tombesi P. and Milburn G., *Phys. Rev. A*, **57** (1998) 4930.

[7] Höffges J. T., Baldauf H. W., Lange W. and Walther H., *J. Mod. Opt.*, **44** (1997) 1999.

[8] Brunel C., Lounis B., Tamarat P. and Orrit M., *Phys. Rev. Lett.*, **83** (1999) 2722.

[9] Hong C. K. and Mandel L., *Phys. Rev. Lett.*, **56** (1986) 58.

[10] Kim J., Benson O., Kan H. and Yamamoto Y., *Nature*, **397** (1999) 500.

[11] Varcoe B. T. H., Brattke S., Weidinger M. and Walther H., *Nature*, **403** (2000) 743.

[12] WEIDINGER M., VARCOE B. T. H., HEERLEIN R. and WALTHER H., *Phys. Rev. Lett.*, **82** (1999) 3795.

[13] MEYER G. M., BRIEGEL H.-J. and WALTHER H., *Europhys. Lett.*, **37** (1997) 317.

[14] LAW C. K. and EBERLY J. H., *Phys. Rev. Lett.*, **76** (1996) 1055; LAW C. K. and KIMBLE H. J., *J. Mod. Opt.*, **44** (1997) 2067; DOMOKOS P., BRUNE M., RAIMOND J. M. and HAROCHE S., *Eur. Phys. J. D*, **1** (1998) 1.

[15] KUHN A., HENNRICH M., BONDO T. and REMPE G., *Appl. Phys. B*, **69** (1999) 373; PINKSE P. W. H., FISCHER T., MAUNZ P. and REMPE G., *Nature*, **404** (2000) 365; YE J., VERNOOY D. W. and KIMBLE H. J., *Phys. Rev. Lett.*, **83** (2000) 4987; HOOD C. J., LYNN T. W., DOHERTY A. C., PARKINS A. S. and KIMBLE H. J., *Science*, **287** (2000) 1447.

[16] See, for example, SCULLY M. O. and ZUBAIRY M. S., *Quantum Optics* (Cambridge University Press) 1997.

[17] REMPE G., SCHMIDT-KALER F. and WALTHER H., *Phys. Rev. Lett.*, **64** (1990) 2783.

[18] REMPE G., WALTHER H. and KLEIN N., *Phys. Rev. Lett.*, **58** (1987) 353.

[19] ENGLERT B., LÖFFLER M., BENSON O., WEIDINGER M., VARCOE B. and WALTHER H., *Fortschr. Phys.*, **46** (1998) 897.

[20] KRAUSE J., SCULLY M. O. and WALTHER H., *Phys. Rev. A*, **36** (1987) 4547.

[21] NOGUES G., RAUSCHENBEUTEL A., OSNAGHI S., BRUNE M., RAIMOND J. M. and HAROCHE S., *Nature*, **400** (1999) 239.

[22] A detailed account of the simulations used in this paper and a comparison with ideal micromaser theory can be found in BRATTKE S., ENGLERT B.-G., VARCOE B. T. H. and WALTHER H., *J. Mod. Opt.*, **47** (2000) 2857.

[23] BRATTKE S. *et al.*, *Optics Express*, **8** (2001) 131.

[24] BRATTKE S., VARCOE B. T. H. and WALTHER H., *Phys. Rev. Lett.*, **86** (2001) 3534.

[25] Proposals such as the teleportation of an atomic state using multiple atomic beams would be substantially enhanced when atoms arrive on demand rather than by chance. See for example: DAVIDOVICH L., ZAGURY N., BRUNE M., RAIMOND J. M. and HAROCHE S., *Phys. Rev. A*, **50** (1994) R895; CIRAC J. I. and PARKINS A. S., *Phys. Rev. A*, **50** (1994) R4441; MOUSSA M. H. Y., *Phys. Rev. A*, **55** (1997) R3287.

[26] BRIEGEL H.-J., ENGLERT B.-G, STERPI N. and WALTHER H., *Phys. Rev. A*, **49** (1994) 2962.

[27] HENNRICH M., LEGERO T., KUHN A. and REMPE G., *Phys. Rev. Lett.*, **85** (2000) 4872.

[28] DIEDRICH F. and WALTHER H., *Phys. Rev. Lett.*, **58** (1987) 203.

[29] HÖFFGES J. T., BALDAUF H. W., EICHLER T., HELMFRID S. R. and WALTHER H., *Opt. Commun.*, **133** (1997) 170.

[30] PELLIZZARI T., GARDINER S. A., CIRAC J. I. and ZOLLER P., *Phys. Rev. Lett.*, **75** (1995) 3788.

[31] ZHENG S. B. and GUO G. C., *Phys. Rev. Lett.*, **85** (2000) 2392.

[32] GUTHÖHRLEIN G. R., KELLER M., HAYASAKA K., LANGE W. and WALTHER H., *Nature*, **414** (2001) 49.

Quantum information processing with neutral atoms:
Experimental challenges

S. L. ROLSTON

Atomic Physics Division, Physics Laboratory, National Institute of Standards and Technology
Gaithersburg, MD 20899, USA

1. – Introduction

Atomic physics offers an attractive approach to overcome the challenges presented by quantum computation: many-particle quantum systems that are at the same time well isolated from the environment, and yet exquisitely controllable. Two-level systems, known as qubits in the quantum computation world, have long been the playground of atomic physics and quantum optics. The knowledge that has been accumulated has led to numerous tools for the manipulation of such systems, including radio frequency and optical techniques, techniques based on adiabatic passage, spin echoes, and quantum state tomography, to name just a few. The cesium atomic clock, now operating at an accuracy of better than 2×10^{-15}, is an example of the control of a two-level system that is possible within atomic physics [1]. Many of the tools of atomic physics can be applied to neutral atoms as well as trapped ions. Neutral atoms have an advantage in that they are immune to decoherence-inducing electric fields, a current source of difficulty in the much more developed ion-trap implementations of quantum information processing. But neutral atoms also have a significant disadvantage in that the interactions necessary for entangling operations are much weaker than the Coulomb interactions available to ions. It seems likely that both strategies can make significant contributions to the field of quantum information, and both are being actively pursued.

In these notes I will deal exclusively with the challenges and opportunities presented by considering neutral atoms as the basis of quantum information processing. These notes are based on two lectures given at the Enrico Fermi Summer School on Experimental Quantum Computation and Information in July 2001. At the writing of these notes, there is little in the way of published experimental results directly concerning quantum information processing with neutral atoms. These notes therefore will concentrate on the various proposed systems, and in particular attempt to identify the difficult experimental problems that must be confronted and overcome.

A recent paper by David DiVincenzo [2] on the physical implementation of quantum computing offers an ideal framework in which to evaluate the prospects for quantum computation. He presents five requirements that are necessary for the construction of a useful quantum computer: 1) scalable qubits, 2) universal gates, 3) initialization, 4) high fidelity and small decoherence, and 5) measurement. For each of these requirements, I will summarize the relevant issues and technologies available for neutral atoms.

The topics to be covered include choices of qubits; methods of neutral atom confinement; optical lattices; light scattering as it relates to confinement and decoherence; methods for loading including laser cooling, sideband cooling, and Bose-Einstein condensates; atom-atom interactions and possible quantum logic gates; sources of decoherence and technical and fundamental noise; and addressing of the qubits.

2. – Neutral atom qubits

Any quantum information processor must be composed of individual two-level quantum systems (multi-level systems are also possible, but will not be considered here) called qubits. To be useful, such qubits must have long coherence times (*i.e.*, a coherent superposition of the two states of the qubit must remain coherent for a sufficiently long time to allow information processing to occur). They must be manipulable. For example, it should be possible to perform "single-bit" rotations—ideally an arbitrary rotation along any axis (although this is an overly restrictive requirement). They must also have some controllable interactions with other qubits in the system. This will allow the construction of quantum logic gates, and the ability to generate states with entanglement between qubits. An arbitrary quantum algorithm can be implemented with one- and two-qubit gates, which is all that is needed for universal quantum computation.

For neutral atoms, there are a number of choices for qubits. In general the ground-state structure of atoms is composed of hyperfine levels, labeled by quantum numbers F, m_F, where F is the total angular momentum (coupling the nuclear spin to the electronic angular momentum) and m_F is the projection of the angular momentum along the z-axis. All of these states have exceedingly long lifetimes, because either they are true ground states, or their decay is through magnetic-dipole emission of microwave photons, a strongly suppressed process. In the absence of a magnetic (or electric) field the m_F levels associated with a particular F value are degenerate, so one might imagine ignoring the m_F quantum number and just considering a qubit between F and F'. Because it is operationally very difficult to fully eliminate stray fields, a more useful choice is between a particular pair of Zeeman sublevels, F, m_F and F', m'_F, where $\Delta m_F = 0, 1$, or 2 (because to perform single-bit rotations we will consider at most a two-photon (Raman) coupling). If $\Delta m_F = 0$, then magnetic and electric fields will not cause any relative phase evolution between the two states of the qubit (to first order), eliminating at least one possible source of decoherence. More complex encodings are possible, and will most likely be inevitable as error correction coding and decoherence free subspace encoding [3] is implemented.

Many atoms have metastable states, so it is possible to envisage creating a qubit between two metastable states, such as the $^3P_2$ and $^3P_0$ metastable states that exist in the noble gases and akali earths. In this case the energy difference between the two states of the qubit is now an optical energy, which brings up an often overlooked challenge for quantum computation: synchronization. Throughout a quantum computation, there will be many superpositions of individual qubits, $|0\rangle + e^{i\phi}|1\rangle$, where $\phi = \phi_0 + \omega_{01}t$. The value of this phase is an integral part of the calculation. Any later operations must be synchronized with this time-evolving phase. This will be quite a challenge, in that our quantum computer will have to have atomic clock-like levels of timing precision. In the end this may favor qubits with smaller energy differences, whose phases evolve at a slower rate, or encoding schemes where the two states of the effective qubit are degenerate.

A completely different approach to the qubit is to use the external degree of freedom of atoms trapped in small wells. One can imagine a periodic optical potential (optical lattice) where the ground and first excited vibrational states of the potential wells form the qubit. This has the advantage in that we could use an atom with a $^1S_0$ ground state, which would be about as insensitive to the environment as possible. One drawback to such a scheme is that in the course of logic operations, one has to be sure to not connect to the higher-lying vibrational states, which would be outside the computational basis. If the wells were harmonic this would be hard to achieve, since the energy between 0 and 1 would be the same as between 1 and 2. Another complication is that high quantum efficiency readout would be challenging, especially in an atom without hyperfine structure. (Readout on atomic systems relies on scattering many photons from one level, while not coupling to the other. This discrimination is accomplished through polarization and/or energy discrimination, normally in the GHz range for typical hyperfine couplings. Such discrimination would not be possible if the levels are separated by less than the natural linewidth of the cycling transition, almost certainly the case if the external degrees of freedom are used for the qubits.)

There have been numerous traps demonstrated that can confine neutral atoms, but for quantum computation there will be a premium on traps with strong confinement. Many of the proposed two-qubit gates have their speed proportional to the atomic density in a well. The Lamb-Dicke effect, proportional to the extent of the ground-state wave function, suppresses inelastic scattering events. Strong confinement also means high oscillation frequencies, which aids in avoiding the inevitable acoustic and $1/f$ sources of noise and decoherence.

There are two (viable) methods to confine neutral atoms: optical and magnetic confinement (there have been other traps demonstrated, but they are unlikely to have any impact on quantum computation). Magnetic traps are quite ubiquitous, as they form the basis of almost all of the multitude of Bose-Einstein condensation experiments [4]. The vast majority of these traps are too large (*i.e.* too weak of a spring constant) to be of any use for quantum computation. Recent advances in microtraps, however, have created the possibility of producing sufficiently tightly confining magnetic traps. Two such traps produced with lithographic techniques have recently been used to create "BECs on a chip" [5,6].

Optical lattices provide a way to create periodic potentials in one, two, and three dimensions, with characteristic lattice spacings of order of the optical wavelength. A tightly focussed laser can be used to form a Far-Off-Resonance Trap (FORT) which can also optically trap atoms. Arrays of such dipole traps can be constructed and manipulated [7]. Since most of the proposed schemes for implementing quantum logic operations on neutral atoms have been based on optical confinement, and optical lattices in particular, the remainder of this section will treat lattices in some detail.

The confinement of neutral atoms for quantum information purposes must rely solely on the conservative part of the atom-light interaction, described by the light shift [8]. To assure that the conservative part (which scales as $1/\delta$, where $\delta = \omega - \omega_0$ is the detuning of the laser frequency from the natural resonant frequency of the atom) is much larger than the dissipative (spontaneous emission) component (which scales as $1/\delta^2$), the detunings of the laser light will always have to be very large compared to the linewidth of the atomic transition. Depending on the exact situation, the detuning may be large compared to the excited-state hyperfine structure, the excited-state fine structure, or even the transition itself (*e.g.* a $10.6\,\mu$m CO_2 laser trap). The light shift (in the large detuning limit) can be written as

$$(1) \qquad\qquad\qquad U = \frac{\hbar\Omega^2}{4\delta},$$

where Ω is the on-resonance Rabi frequency for a laser field (the precession frequency of the Bloch vector representing the 2-level atom when $\delta = 0$). When the detuning is large compared to the excited-state structure, it is no longer strictly a two-level system, but it can still be treated as such by introducing an effective Rabi frequency, calculated by summing over all of the dipole matrix elements between the relevant states. For example, if the detuning is large compared to the excited-state hyperfine structure, but small compared to the fine structure, one should sum over all excited states in the fine-structure manifold the laser is tuned near. When one takes into account polarization, some simple rules are apparent. If the detuning is large compared to the excited-state hyperfine structure, but less than the fine-structure splitting, *i.e.* $\epsilon_{\rm fs} \gg \delta \gg \epsilon_{\rm hfs}$, there is "no time" for the atom to feel the nuclear-electronic coupling of the hyperfine interaction, and the atom will behave as if the nuclear spin were irrelevant. For example, in the alkalis, the levels would respond as if the ground state of the atom were a $J = 1/2$ state, with only two levels. If we look at the Clebsch-Gordan coefficients that couple $J = 1/2 \rightarrow 3/2$ (the D2 transition), they are equal for π light, and different for σ light. The light shift for all hyperfine levels will be the same for π-polarized light, and different for σ light. This can be very useful, because it means that the light shift is independent of the m_F quantum number in the ground state if π-polarized light is used. These detuning-dependent results can of course be obtained quantitatively by summing over all $m_F \to m'_{F'}$ transitions, with appropriate $3j$ and $6j$ symbols necessary to describe the transition amplitudes. Using similar arguments, if one is detuned far compared to the fine structure, *i.e.* $\delta \gg \epsilon_{\text{fs}}$, the atom will act like an $L \to L'$ transition. For the alkali atoms, this would be a $0 \to 1$ transition, where all three Clebsch-Gordan coefficients are equal. This implies that the light shift is then independent of both the laser polarization and the m_F value. For such large detunings, the atom acts like a scalar particle.

A useful general expression [9] of the light shift is in terms of the atomic polarizability tensor. For large detunings ($\delta \gg \gamma$) the light shift operator can be written as

$$\hat{U}(z) = -\mathbf{E}_{\text{L}}^{*}(\mathbf{z}) \cdot \hat{\alpha} \cdot \mathbf{E}_{\text{L}}(\mathbf{z}), \tag{2}$$

where

$$\hat{\alpha} = \frac{\sum_{e_i} \hat{d}_{ge_i} \hat{d}_{e_i g}}{\hbar \delta_{ge_i}}, \tag{3}$$

and $\hat{d}_{ge_i}$ is the electric-dipole operator (which will include all Clebsch-Gordan coefficients and $6j$ symbols), δ_{ge_i} is the detuning between the ground state and each excited state e_i, and $\mathbf{E}_{\text{L}}$ is the electric field of the applied laser beams. This formulation makes it simple to calculate the m_F-dependent and polarization-dependent light shifts. In particular, for alkali atoms with $\epsilon_{\text{fs}} \gg \delta \gg \epsilon_{\text{hfs}}$ the light shift operator can be broken into a scalar component and a piece that looks like a magnetic field:

$$\hat{U}_F = U_0 \left(\frac{2}{3} |\vec{\epsilon}(z)|^2 \hat{\mathbf{I}} - \frac{i}{3} [\vec{\epsilon}^{\,*}(z) \times \vec{\epsilon}(z)] \cdot \frac{\hat{\mathbf{F}}}{F} \right), \tag{4}$$

where $\hat{\mathbf{I}}$ is the identity, U_0 is the light shift for a unit C-G coefficient, $\hat{F}$ is the angular momentum operator, and $\vec{\epsilon}$ is the polarization vector of the light. For certain implementations of quantum logic gates (discussed below) this will be important, as they will rely on being able to manipulate different spin states differently using the differential light shifts.

If δ starts to become significant with respect to ω_0, the rotating wave approximation begins to fail. The counter-rotating term that is normally ignored in calculating Ω should be included, as well as matrix elements to other electronic states. In the limit where the detuning is large, and the frequency small compared to all electronic transitions (as would be the case for a CO_2 laser) then the light shift should be calculated using the DC polarizibility of the atom, $U = -\alpha_{\text{dc}} E^2$. For the alkali atoms, where the D lines carry almost all of the oscillator strength, an approximate expression including the counter-rotating term is

$$U_0 = \frac{-\hbar \Omega^2}{4} \left(\frac{1}{\omega_0 - \omega} + \frac{1}{\omega_0 + \omega} \right), \tag{5}$$

where ω_0 is the frequency of the D transition. From this we can see that the counter-rotating term increases the light shift by a factor of two as $\omega \to 0$. This effect can begin to become significant, even when the laser is not too far away: for Rb (D2 transition at 780 nm) trapped by a Nd:YAG laser at 1064 nm, the counter-rotating term increases the light shift by $\approx 15\%$.

For all implementations of quantum computing, the confining potential should be as conservative as possible. We require the spontaneous-emission rate to be many times ($\geq 10^4$) the rate of operations in the computation. The spontaneous-emission rate for a two-level system (for $\delta \gg \gamma$) can be written as

$$(6) \qquad \gamma_s = \frac{\hbar \Omega^2}{4\delta^2} \gamma.$$

A more useful expression is obtained by comparing this with the expression for the light shift potential. The spontaneous-emission rate is simply the trap depth (in frequency units) divided by the detuning in natural linewidth units: $\gamma_s = (U_0/\hbar)(\gamma/\delta)$. In general, one is often first confronted with designing a particular trap depth, so this expression allows a quick calculation of the spontaneous-emission rates. When the detuning is large so that many excited states are contributing, this expression for the spontaneous-emission rate is still approximately valid. The matrix elements that contribute to the light shift are the same as those that contribute to the scattering rate, so Clebsch-Gordan sums will be the same. Because of the different detuning dependence of the spontaneous-emission rate compared to the light shift, this expression is only valid in the large detuning limit, where the detuning to each of the contributing states is approximately the same (this will tend to be normal operating parameters if we want a very conservative potential). If we were tuned close to one fine-structure level, for instance, then the spontaneous-emission rate must be summed for each transition. If the detuning is getting to be comparable to the optical frequency, then the counter-rotating term cannot be ignored. It adds to the light shift, but subtracts for the spontaneous-emission rate (which assures there is no spontaneous emission for a DC electric field).

There are a multitude of possible configurations of geometry and polarizations possible to create optical lattices [10-12]. It would be impossible to cover them all in these notes, but there is an extensive literature available. One word of caution, particularly with respect to polarization-gradient lattices. It is important when calculating the m_F-dependent potentials to take into account the light shift from all relevant levels, as discussed above. Much of the work that has been done with optical lattices has been with small detunings (less than typical excited-state hyperfine splittings) which are not suitable for quantum computation purposes, because of the relatively large spontaneous-emission rate. The large detunings that will be needed both restrict the types of lattices possible in terms of polarization gradients, and expand the geometries, because radiation pressure can be ignored (because it is due to spontaneous emission). An example of this is a lattice where all the beams are incident from one side [13].

The simple one-dimensional lattice for a two-level atom contains much of the relevant physics that will be necessary to use optical lattices for quantum computation purposes. We can write the optical potential for a 1D standing wave as

$$(7) \qquad U(z) = U_0 \sin^2(kz) \;=\; \frac{U_0}{2}(1 - \cos(2kz)).$$

Note that in solid-state terms, this describes a lattice with a single reciprocal lattice vector, $K = 2k$. If we expand the potential around $z = 0$ we can find the harmonic approximation to the frequency:

$$(8) \qquad \omega_{\mathrm{h}} = \sqrt{\frac{U_0 \epsilon_{\mathrm{R}}}{\hbar^2}},$$

where $\epsilon_{\mathrm{R}} = \hbar^2 K^2/(2m) = 4\hbar^2 k^2/(2m) = 4E_{\mathrm{R}}$ is the characteristic energy associated with the reciprocal lattice vector, and $k = 2\pi/\lambda$, where λ is the laser wavelength. Note that this is four times the usual recoil energy (E_{R}). Since the depth of the potential is U_0, the number of bound states can be estimated as

$$(9) \qquad N_{\mathrm{b}} \approx \frac{U_0}{\hbar\omega_{\mathrm{h}}} = \sqrt{\frac{U_0}{\epsilon_{\mathrm{R}}}}.$$

If we change the periodicity of the lattice by using non-counterpropagating beams, the expression for the frequency remains unchanged—just the value for ϵ_{R} is altered ($K = 2k/\sin(\theta/2)$). For quantum computation we will usually only have to consider the ground state, unless the qubit will be stored in the vibrational degree of freedom. Because the potential is periodic, and the barriers between wells are finite, there is the possibility of tunneling between wells. This would be bad, as our qubits could become scrambled. The amount of tunneling can be calculated by diagonalizing the Hamiltonian in a Bloch state picture. For the simple 1D lattice, the Hamiltonian can be written as

$$(10) \qquad H_{ij} = \delta_{ii}(i + q)^2 + \delta_{i,i\pm1}\frac{U_0}{\epsilon_{\mathrm{R}}},$$

where q is the quasimomentum. This describes an infinite-dimensional, tri-diagonal Hamiltonian. Its dimension can be restricted, with a sufficient number of states included to get the appropriate accuracy. Typically 10–15 states are quite sufficient to get good accuracy. The tunneling of the ground band is found by evaluating the width of the band, by calculating $\epsilon(q = 1) - \epsilon(q = 0)$. This is plotted in fig. 1 in terms of E_{R}. For our quantum computation needs, the tunneling time must be appropriately long ($\geq 10^5$ gate operations).

Another critical parameter is the extent of the ground-state wave function, z_0, since many of the gate schemes will be proportional to the atomic density in a well. If we make this dimensionless, then it is the Lamb-Dicke parameter, kz_0. This could be found by looking at the eigenfunctions from the band structure calculation, but for wells deep

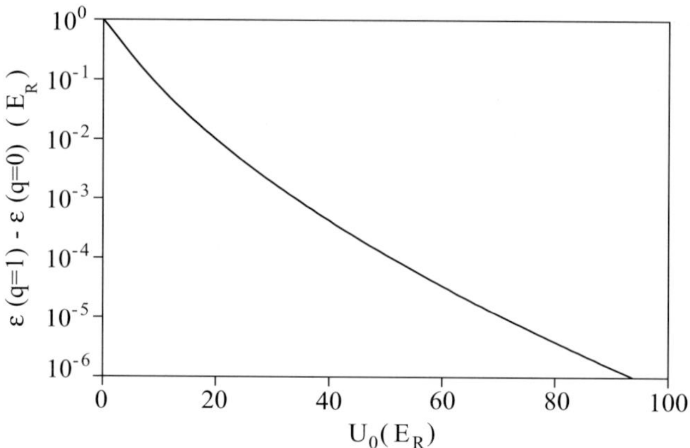

Fig. 1. – The tunneling rate in an optical lattice must be sufficiently slow so that there is no possibility of qubits tunneling from one site to another during the computation. This graph plots the width of the ground-state band (proportional to the tunneling rate) as a function of the lattice depth.

enough to make tunneling appropriately negligible, the harmonic approximation is quite accurate:

$$(11) \qquad kz_0 = \sqrt{\frac{\hbar k^2}{2m\omega_0}} = \frac{1}{2}\left(\frac{\epsilon_R}{U_0}\right)^{1/4}.$$

The canonical polarization-sensitive optical lattice is the so-called lin-angle-lin lattice, formed by counterpropagating linearly polarized beams, with an angle θ between the polarization vectors. We can decompose each of the linearly polarized beams into a superposition of σ^+ and σ^- beams. The lattice can then be seen as a superposition of two circularly polarized standing waves displaced from each other by $\Delta z = (\lambda/4)\sin\theta$. If $\theta = 0$ the two standing waves have no relative displacement, and the lattice is a pure intensity gradient, with linear polarization everywhere. If $\theta = 90°$ the two lattices are exactly out of phase; there is no intensity variation, and it is a pure polarization-gradient lattice. If we now imagine an atom with a $1/2 \rightarrow 3/2$ transition, cold atoms with $m_F = 1/2$ will be trapped at the antinodes of the σ^+ lattice (assuming a red-detuned lattice), while the atoms in $m_F = -1/2$ will be trapped in the σ^- antinodes. If θ is varied the atoms can be controllably overlapped with one another, which will be the underlying basis for some of the conditional two-qubit gates. For an atom with a more complicated structure (such as the alkalis) it is not so simple—one has to calculate the polarization-dependent light shift for each m_F state. As discussed above, depending on the detuning relative to the excited-state hyperfine and fine structure, the different m_F states may respond differently. If $\delta \gg \epsilon_{fs}$, there will be no polarization dependence, and no polarization-gradient lattices are possible.

3. – Initialization

For most of the proposed implementations of quantum logic gates in optical lattices, the atoms need to be in the ground state. Even if the qubit is encoded in hyperfine states, there may be a coupling to the external degree of freedom. For instance, if the atom-atom interaction is proportional to the atomic density, it would be significantly altered if the atom was in the first excited state of the lattice, rather than the ground state. For this reason, cooling is important. Simple laser cooling, including Sisyphus cooling, is not sufficient. The typical temperature for Sisyphus cooling [14] is $k_{\mathrm{B}}T \approx 0.2U_0$ with a lower limit at $k_{\mathrm{B}}T \approx 10E_{\mathrm{R}}$ which corresponds to about 25% population in the ground-state band. There are a number of approaches that can be taken. We can do a filtration process, discarding any atoms in excited bands. This can be done by adiabatically lowering the potential so that there is only one bound state in a well, and allowing the untrapped atoms to be removed by gravity. A somewhat quicker approach is to accelerate the lattice, and only trapped atoms will move with the lattice, allowing a spatial separation between the trapped and untrapped atoms [15].

Another approach is sideband cooling, a technique first demonstrated by the Jessen group [16] for neutral atoms in 2D, and extended to 3D in recent work by others [17,18]. In this approach one makes a stimulated Raman coupling between n, m_F and $n-1, m_F-1$, where n is the vibrational quantum number (using a magnetic field and some π-polarized light). The dissipation necessary for any cooling scheme is provided by a repumper that predominantly drives $\Delta n = 0$ transitions (because of the Lamb-Dicke effect) between $n - 1, m_F - 1$ and $n - 1, m_F$. The net effect is a reduction of the vibrational quantum number, *i.e.*, cooling. It works quite well: in 2D ref. [16] found a ground-state population $n_0 \geq 0.95$, while in 3D refs. [17,18] found $n_0 = 0.37$ and 0.80, respectively.

A Bose-Einstein condensate offers an attractive way to load the ground state. Because the condensate consists of many atoms in the ground state of the confining potential (magnetic or optical trap plus mean-field potential) it is possible to connect this ground state to the ground state of an optical lattice through an adiabatic transition. At NIST we have shown that we can load more than 99.5% of the atoms into the ground-state band of a 1D optical lattice [19]. This was done with what may seem like a surprisingly short linear ramp of the lattice potential in $20\,\mu$s. That the ramp can be this fast is best understood by invoking band structure. Because the condensate has a uniform phase profile, it will load $q = 0$ in the ground-state band (the quasimomentum q describes the phase difference between neighboring wells in a periodic potential). If we consider the free-particle dispersion relation ($\epsilon = q^2/2m$) viewed in the single Brillioun zone picture, we see that for $q = 0$ states, there is a gap of $4E_{\mathrm{R}}$, even when there is no optical potential present. This arises from the simple fact that stimulated processes in an optical lattice must change the momentum by a reciprocal lattice vector $(2k)$ which has a large energy cost associated with it. The rate at which the system parameter can be changed is related to the difference in eigenvalues of the system available to the state. Since it is large to start, we can go fast. Because of symmetry arguments, the quasimomentum cannot change, and $\delta n = 0, 2, 4, \ldots$, so that the first unwanted state populated is the $n = 2$,

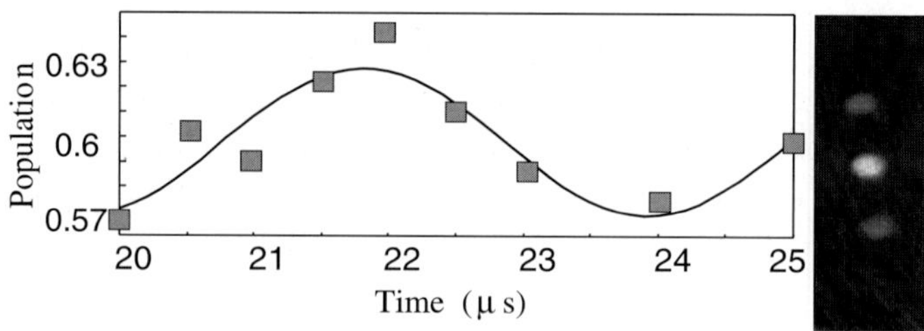

Fig. 2. – Residual oscillations in the $0\,\hbar k$ component of the plane-wave decomposition of a sodium BEC lattice state which was loaded by ramping up a 1D optical lattice ($\approx 13E_{\mathrm{R}}$) in $20\,\mu$s. The beating is a measure of the population ($\approx 0.5\%$) of the $n = 2$ band due to the non-adiabaticity of the process. The time-of-flight image of the released condensate on the right-hand side reveals the plane-wave decomposition of the lattice state after slow loading. It agrees well with the calculated plane-wave decomposition for the ground state.

$q = 0$ (which is degenerate with $n = 1$, $q = 0$ for $U_0 = 0$). Figure 2 shows a momentum state analysis of the adiabatic loading process. The experiment varies the amount of time the atoms sit in the lattice. If there is some residual amplitude in the $n = 2$ band, it will interfere with the $n = 0$ state in the projection onto plane-wave states that is made upon a sudden release from the lattice. This "heterodyne" effect gives a very sensitive measure of residual population in the excited states. Such adiabatic loading techniques should work equally well in 1, 2 or 3 dimensions, and are clearly an ideal way to load optical lattice, of course at the price of requiring a BEC!

For all of the methods discussed so far, there is nothing to assure exactly one atom in each lattice site, which in the end will of course be necessary for quantum computation (some schemes may even require exactly one atom in every other site—even more difficult). In a paper by Jaksch *et al.* [20] it was suggested that a Mott-insulator transition would be a possible solution. If we consider a lattice shallow enough so that there is some tunneling between wells (we will only consider the ground band), there is the possibility of two atoms in one well. The repulsive atom-atom interactions will provide an energy cost for this situation. It is clear that when the atom-atom interaction is included, the lowest-energy state of the system is to have exactly one atom in each well (excess atoms would still be free to tunnel). One state has now separated from the ground-state band, with an energy gap proportional to the energy cost to move one atom into an already occupied site. The strategy would then be to adiabatically load the lattice from a BEC on a slow enough timescale compared to this energy. The final state would be a product state of N single-atom Fock states, with no transport allowed—an insulator state. At this point it would be possible to continue to increase the well depth to freeze out any tunneling. The many-atom analog of this phenomenon has recently been demonstrated with a 1D optical lattice [21]. In this case they observed a significant (factor of ≈ 10–15) reduction in the number fluctuations in the ≈ 12 wells populated from a Rb BEC, when they adiabatically ramped up an optical lattice over 200 ms. In this experiment, the signature of reduced number fluctuations is the loss of contrast of the diffraction pattern that is normally observed when atoms with coherence over many lattice sites are released. The phase of each of the one-atom Fock states is undefined, so the diffraction pattern would be that of a multiple slit grating with a random phase at each slit. Using this detection scheme to demonstrate very high fidelity population of the insulator state may be problematic, as loss of contrast is not that sensitive a measure.

4. – Gates

One of the requirements for quantum computation will be to perform operations on single qubits. Since any unitary operation on a single qubit can be decomposed as a series of rotations on the Bloch sphere, a useful goal is to be able to perform an arbitrary rotation on the Bloch sphere, which will ensure the ability to construct an arbitrary single qubit gate. (This is a more stringent goal than may be necessary as certain algorithms may only require a restricted set of unitary operations, but at this stage universality seems like a laudable goal.) Single-bit operations should be rather straightforward and essentially the same as has been demonstrated in ion trap systems. We have before us the full panoply of tools that have been developed in atomic physics: microwave transitions; Raman transitions; pulse shaping techniques; polarization and magnetic field control; coherent control techniques; rapid adiabatic passage; and most of the pulse sequencing techniques from NMR. Because the details will depend entirely on the qubits used and the actual system implemented, I will not spend any more time on single qubit operations— these are among the requirements that seem most tractable.

Perhaps the most important realization about the structure of a quantum computer was that it could be constructed with gates that did not involve more than two qubits at a time. Imagine the complexity if a 100-qubit computer also required 100-qubit logic gates! There have been a number of proposals for two-qubit gates for atomic systems, too many to fully describe here [22-26]. I will summarize some of the schemes, and refer the reader to the literature to find more details.

The two-qubit gates necessary to fully implement quantum logic operations are conditional gates: the state of one qubit is conditioned on the state of the second qubit. There are a number of two-qubit gates that are universal, *i.e.* capable of being used (with one-qubit operations) to construct an arbitrary quantum algorithm. The canonical gates, which are the types being proposed for the atomic physics systems, are controlled-not or controlled-phase gates. In a controlled-not gate, the state of qubit 2 if flipped if and only if qubit 1 is in the logical state 1. A controlled-phase gate would write a phase (typically π) on qubit 2 if and only if qubit 1 were in state 1. To physically implement such conditionality requires some sort of interaction between the two atoms comprising the qubits. For neutral atoms, this can be a virtual (or real) exchange of photons, producing a dipole-dipole coupling; it can be a Van der Waals interaction between two atoms; it can be a magnetic dipole interaction. This list essentially exhausts all the possible interactions between two neutral atoms.

The first suggested gate that used the dipole-dipole interaction was proposed by Brennen *et al.* [22]. It relied on trapping atoms in different Zeeman sublevels in an

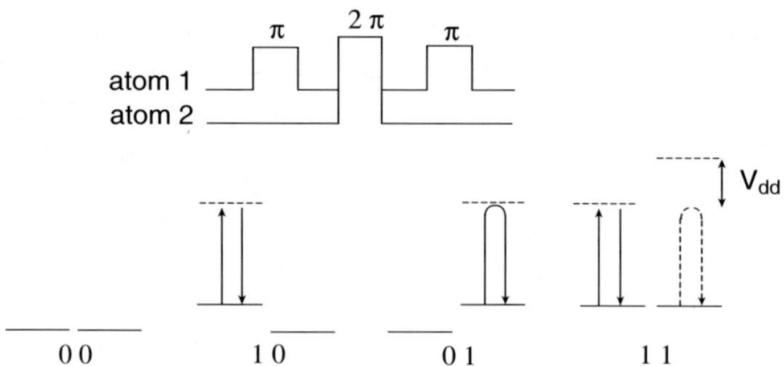

Fig. 3. – Schematic of the Rydberg gate. Atom 1 undergoes two π pulses and atom 2 undergoes a single 2π pulse as shown. Because of the large dipole-dipole interaction between two excited Rydberg atoms, if both atoms are in the $|1\rangle$ state, atom 2 does not get excited by the 2π pulse because of the large induced detuning. Except for the $|00\rangle$ state, the states accumulate a π phase shift, which yields a conditional phase gate.

optical lattice where these sublevels could be trapped independently. The gate was accomplished by shifting the interpenetrating lattices so that the atoms were close to one another, and applying a pulse of near resonant light, which would create a small excited-state population and a dipole-dipole coupling. By appropriate tuning, they arranged that only if both atoms were in the logical state 1 would there be a dipole-dipole coupling, and by leaving the light on long enough, a π phase shift would be developed, creating a controlled-phase gate. What makes this proposal at all feasible is their recognition that although both the dipole-dipole coupling and the excited-state population (and the associated decoherence associated with spontaneous emission) were proportional to the square of the dipole matrix element, the ratio scaled as kR^{-3}, where R is the separation. So there is a regime where the coherent dipole-dipole coupling used to create the gate is much larger than the decoherence-producing spontaneous-emission rates. Realistic estimates put the fidelity of such gates in the 0.99 range.

The speed of such gates based on the dipole-dipole interaction would be limited by the size of the interaction which is at best $\hbar\gamma$, limited by the excited state width, and in fact much smaller to keep coherence winning out over decoherence. A scheme that allowed for much faster gates was proposed [23] that relied on the extremely large dipole-dipole couplings available with Rydberg atoms. Because the coupling scales as n^4, where n is the principal quantum number, couplings in the GHz range are quite realistic. Then one can in principle imagine gate speeds on the ns scale. In fact, the Rydberg gates have much more than just speed to offer. In investigating such gates, the authors realized that a gate could be constructed in a way that made it much less sensitive to the motional states of the atoms, which may greatly aid in the simplicity of a quantum computer if the coupling to other degrees of freedom is weak. The proposed Rydberg gate used a blockade phenomenon and is displayed schematically in fig. 3. The qubits were assumed to be in the hyperfine states of the atoms. If we imagine individual addressing of the atoms, a gate can be constructed by applying a π-pulse to atom 1, followed by a 2π-pulse on atom 2, followed by a π-pulse on atom 1 to complete the gate. We will further assume that if the atom is in logical state $|0\rangle$, it does not get excited, and if it is in 1 it gets excited to a Rydberg level. We can examine the effect of the pulse sequence: if the two atoms are in state $|00\rangle$, nothing happens and they end up in $|00\rangle$. If they start in $|01\rangle$ or $|10\rangle$, one atom is a spectator, and the other atom sees a 2π-pulse (either a single 2π or a pair of π) and acquires a π phase shift (the property of a two-level system). If the atoms start in the $|11\rangle$ state, the first π-pulse excites atom one, but when the pulse is applied to atom 2, the dipole-dipole shift is so large that the two-atom excited state is detuned from resonance, and excitation is suppressed (the blockade effect) so nothing happens to atom 2. Then the 3rd pulse returns atom 1 to the ground state, and the $|11\rangle$ state has picked up a π phase shift just like the $|01\rangle$ and $|10\rangle$ states. Since the $|00\rangle$ state is unaffected there is a conditional phase and the appropriate gate has been achieved. The proposal goes on to evaluate more realistic schemes that do not involve addressing individual atoms, and use adiabatic passage to minimize excited-state populations. They estimate fidelities for the example in the proposal of order 0.99 (this can most likely be improved with a more extensive exploration of parameter space). The Rydberg gate offers a number of advantages: it can be fast, because the dipole-dipole coupling can be large; the spacing between atoms can be large ($\geq 1\,\mu$m) which will make individual addressing much simpler; and perhaps most importantly, coupling to motional states is suppressed—the blockade effect works if the atoms are close enough to make a shift large compared to the detuning. If the atoms positions vary slightly (or if they are in the first excited state of a lattice well instead of the ground state) this will only have a second order effect on the fidelity. The biggest challenge for the Rydberg gates will most likely be laser technology—it will require high laser power (because the matrix elements for excitation of a Rydberg level from the ground state are small). This may necessitate pulsed laser technology, an area where 10^{-4} intensity control to achieve high-fidelity gates will be demanding, to say the least.

Another atom-atom interaction that has been proposed to create conditional gates is the Van der Waals interaction. Although it may seem counterintuitive, a collision between ground-state atoms can be fully coherent if the atoms are cold (s-wave collisions). Just like the dipole-dipole interaction, a phase shift can be accumulated to create a controlled-phase gate. The advantage is that because the atoms are in the ground states, there is no source of decoherence from spontaneous emission. The disadvantage is that it is a much weaker interaction, which will require the atoms to be closer together, and will produce slower logic operations. The first proposal [24] to use ground-state interactions came from the Zoller group. They envisioned a spin-sensitive optical lattice trapping two different Zeeman sublevels. Depending on which level the atoms were in, they would be shifted in space, only overlapping (and interacting with) the second atom of the gate if it were in the other unshifted state. One complication of this particular scheme is that the lattice must have every other well occupied. The Zoller group has also considered gates developed with atoms that are set oscillating in a potential, undergoing coherent collisions with another atom, again developing a phase [25]. In all of these schemes, the conditionality appears in that there are different forces applied to the atoms depending on their state, and the atom-atom interaction allows for a gate to be constructed.

A two-qubit gate that relies on Van der Waals interactions, but uses vibrational states as the qubits has been proposed by the NIST group [27]. In this scheme the two atoms are trapped in the two wells of a double-well potential (created in an optical lattice by superimposing a lattice at frequency ω with a lattice at 2ω). The two states of the qubit are the ground and first excited states of a well. The gate is operated by lowering the barrier between the wells so that there is some appreciable tunneling between wells. If both atoms are in the excited state (the $|11\rangle$ state) they will have a much larger Van der Waals interaction than if one or both of the atoms are in the ground state. From this one can construct a conditional phase gate. The disadvantage of this gate is that it is relatively slow, because not only is the atom-atom interaction small, but one must operate the gate adiabatically so that one never ends up with two atoms in one well. This condition is set by the energy scale of the atom-atom interaction, and leads to gate times of order $1\,\mathrm{ms}$. Because this gate relies on tunneling, it is exponentially sensitive to the barrier height (and therefore laser intensity), and as described would require extraordinary laser stabilization. This should be a generic cautionary point—gates that rely on tunneling (with the exponential sensitivity that comes along with this) may be problematic because they will place undue burdens on our technical abilities. For this particular gate, there is a modification that can correct for this problem. If instead of relying on barrier tunneling, we introduce a coupling laser that connects the $|1\rangle$ state to a higher-lying (above barrier) state that has amplitude in both wells, this would introduce a state-selective coupling between the wells that is much less sensitive to the height of the barrier.

5. – Coherence

Perhaps the most significant challenge for any quantum computation implementation will be to reduce the levels of decoherence to a place where error correction techniques can be useful. Current theoretical estimates put the threshold for fault tolerant quantum computation at an error probability of 10^{-4} or so [28], but we can hope and expect that this threshold will be relaxed as more progress is made in the field. There are a number of sources of decoherence and dissipation that are possible for atoms trapped in optical lattices, but they are quite well understood. Among the possible sources are inelastic collisions, time-varying magnetic fields, laser amplitude noise, laser position noise, and spontaneous emission. Spontaneous emission is a completely calculable decoherence source that will most likely be manageable. As discussed above, the depth of the optical potential that forms the lattice is proportional to I/δ (where I is the laser intensity, and δ is the laser detuning), while the spontaneous rate scales as I/δ^2. Clearly the optical potential can be held constant while reducing the spontaneous-emission rate by simultaneously increasing the intensity and detuning. Optical lattices and single-focus dipole traps have been made with scattering rates less than one photon per second. Since gate operation speeds will inevitably be limited by oscillation frequencies within the lattice (typically 10–1000 kHz), decoherence due to spontaneous emission may already be below the fault tolerance threshold. There is a caveat, in that some of the proposed gates require specific laser detunings to create the appropriate polarization-dependent lattices, or involve some spontaneous emission during the two-qubit gate operation.

The most likely source of decoherence and dissipation will be laser noise. Heating rates were calculated [29] in an analysis of both laser amplitude noise and beam pointing noise in a single-focus dipole trap (laser frequency noise will be negligible). This analysis is applicable to optical lattices, where the case will be more favorable because of the higher oscillation frequencies in a lattice over that of a dipole trap. If we assume we want less than one heating event per 10^5 oscillation frequencies, and a 100 kHz oscillation frequency, this analysis yields fractional laser amplitude noise requirements of $S_a(200\,\mathrm{kHz})$ $\leq 10^{-13}\,\mathrm{Hz}^{-1}$ and laser beam pointing noise of $S_x(100\,\mathrm{kHz}) \leq 10^{-17}\,\mu\mathrm{m}^2/\mathrm{Hz}$. Recent work [30] on a CO_2 laser dipole trap reported an amplitude noise well below this requirement, and trap lifetimes of 300 s. The pointing noise should also be achievable, especially since the optical lattice frequency is well above acoustic frequencies, the primary source of beam pointing instabilities. We can in addition bring to bear the formidable technologies of optical interferometry to minimize these problems. If the qubit is encoded in hyperfine levels, heating can degrade gate fidelity due to coupling between external and internal degrees of freedom during the gate operation. It is probably best to treat a transition to a vibrationally excited level as the loss of the qubit from the computational basis. Cooling of this qubit could return the qubit to the basis, where error correction routines could then correct for what would now look like a bit flip (although the cooling must be done without adversely affecting neighbors—a clear challenge). If the qubit is encoded in the vibrational states, amplitude noise will produce a direct decoherence of the qubit (the noise on the lattice will cause the phase of the upper state to evolve with a random component), but preliminary estimates suggest this will be a tractable problem. It may also have a significant effect on the fidelity of two-qubit gates that are based on tunneling. One significant possibility for "dissipation" is that an atom is lost entirely from the lattice due to an inelastic collision. If we can develop a way of reloading this site, then once again this just looks like a bit flip error that can be corrected with error correction coding.

A simple calculation can reinforce the challenge of a 10^{-4} error rate target. Uniform laser intensity profiles are notoriously difficult to achieve. First, there is the problem of a Gaussian intensity profile—if we want a lattice uniform to 10^{-4} over $100\,\mu\mathrm{m}$, it would require a waist of 1.4 cm, and it is unlikely we can find sufficient laser intensity to achieve a reasonable well depth for such large beams. Of course the semiconductor industry in the course of developing optical lithography has developed very sophisticated field flatteners, so perhaps we can take advantage of their innovations. The other challenge in intensity uniformity is scattered light. Because of the effects of interference, a 100 mW beam can interfere with a 0.5 nW beam and produce 10^{-4} intensity variations. These are not insurmountable problems, but we will have an easier task if the fault-tolerant fidelity threshold is relaxed by new breakthroughs in quantum error correction methods.

6. – Measurement

One of the potential strengths of optical lattices is their inherent parallelism. Perhaps new algorithms will be developed that can exploit this. If not and individual addressing is required, this will be a major challenge. The inherent difficulty is that the spacing in a typical optical lattice is sub-wavelength, and optical resolution is limited to about a wavelength. One obvious solution is to use a long-wavelength optical lattice. This could be created with a long-wavelength laser (*e.g.* 10.6 μm CO_2 laser) [31], or a multiple beam lattice with all the angles between beams being small (possible if the detuning is large and radiation pressure can be neglected) [13]. The readout can then be done on a length scale smaller than a lattice constant. The downside to this is that most schemes for logic gates require tight confinement for good gate speed (except perhaps the Rydberg gates), which will be compromised with a long-wavelength lattice. Another possibility is to use multiple wavelengths to create "superlattices" that have both a fine period (for tight confinement) and a coarse period for large separation and individual addressing. We may also borrow techniques from MRI and apply a gradient that brings only one site into (Raman) resonance. Although this would require a very large magnetic-field gradient to achieve wavelength resolution, using the AC Stark shift of a focused laser should produce the required spatial resolution. In the end, the measurement issues will be intimately connected with the qubits, the gates used, and error coding techniques that might use blocks of qubits.

7. – Summary

It is instructive to return to DiVincenzo's rules [2] to see where neutral atom quantum computing stands. Scalable qubits are clearly available, given the large number of atoms routinely loaded in optical lattices. The primary choice of qubit is between internal and external degrees of freedom. Initialization with sideband cooling or adiabatic loading from a BEC seems quite possible, with the challenge to load exactly one atom in each desired lattice site. Decoherence seems like a tractable problem; the issue of the fidelity of gate operations and the requirements on the laser systems is most challenging. There are numerous proposals for two-qubit gates, but none have been demonstrated yet. The gates based on Rydberg atoms have much to offer, but will also be difficult in terms of laser requirements. The final rule is the ability to measure qubits, which in this case really means the ability to individually address them. There are some possible solutions in terms of lattice geometries, but their effects on the gates will have to be carefully evaluated.

This discussion has been based on our current ideas about quantum computation and the associated algorithms and error correction techniques. It is not at all clear that they are the best tools to use with optical lattices. Perhaps the strongest point of an optical lattice is that it can trap many (millions) of atoms, and quite easily get them to act in parallel. It may be that algorithms written to exploit this geometry will be both powerful, and also actually lessen the requirements (for individual addressing, for example). If we survey all the proposed schemes for quantum computers, they all have some promise with many roadblocks and uncertainties, and neutral atom computing is no different. But it is perhaps the system (along with ion traps) where the underlying physics is best understood, and for this reason may in the end prove useful, either as a quantum computing system, or a testing ground for much of what will eventually make up a quantum computer.

REFERENCES

[1] Meekhof D. M., Jefferts S. R., Stepanovic M. and Parker T. E., *IEEE Trans. Instrum. Meas.*, **50** (2001) 507.

[2] DiVincenzo D., *Fortschr. Phys.*, **48** (2000) 771.

[3] Nielsen M. A. and Chuang I. L., in *Quantum Computation and Quantum Information* (Cambridge University Press) 2000.

[4] http://amo.phy.gasou.edu/bec.html/.

[5] Hansel W. *et al.*, *Nature*, **413** (2001) 498.

[6] Ott H. *et al.*, *Phys. Rev. Lett.*, **87** (2001) 230401.

[7] Dumke R. E. *et al.*, quant-ph/0110140.

[8] Cohen-Tannoudji C., Dupont-Roc J. and Grynberg G., in *Atom-Photon Interactions* (Wiley-Interscience) 1992.

[9] Deutsch I. H. and Jessen P. S., *Phys. Rev. A*, **57** (1998) 1972.

[10] Jessen P. S. and Deutsch I. H., *Adv. Atom. Mol. Opt. Phys.*, **37** (1996) 95.

[11] Petsas K. I., Coates A. B. and Grynberg G., *Phys. Rev. A*, **50** (1994) 5173.

[12] Rolston S. L., *Phys. World*, **11**, No. 10 (1998) 27.

[13] Anderson B. P., Gustavson T. L. and Kasevich M. A., *Phys. Rev. A*, **53** (1996) R3727.

[14] Gatzke M. *et al.*, *Phys. Rev. A*, **55** (1997) R3987.

[15] Madison K. W., Fischer M. C. and Raizen M. G., *Phys. Rev. A*, **60** (1999) R1767.

[16] Hamann S. E. *et al.*, *Phys. Rev. Lett.*, **80** (1998) 4149.

[17] Han Dian-Juan *et al.*, *Phys. Rev. Lett.*, **85** (2000) 724.

[18] Kerman A. J. *et al.*, *Phys. Rev. Lett.*, **84** (2000) 439.

[19] Denschlag J., Simsarian J. E., Haeffner H., McKenize C., Browaeys A., Cho D., Helmerson K., Rolston S. L. and Phillips W. D., *J. Phys. B*, **35** (2002) 3095.

[20] Jaksch D. *et al.*, *Phys. Rev. Lett.*, **81** (1998) 3108.

[21] Orzel C. *et al.*, *Science*, **291** (2001) 2386.

[22] Brennen G. K. *et al.*, *Phys. Rev. Lett.*, **82** (1999) 1060.

[23] Jaksch D. *et al.*, *Phys. Rev. Lett.*, **85** (2000) 2208.

[24] Jaksch D. *et al.*, *Phys. Rev. Lett.*, **82** (1999) 1975.

[25] Calarco T. *et al.*, *Phys. Rev. A*, **6102** (2000) 2304.

[26] You L. and Chapman M. S., *Phys. Rev. A*, **6205** (2000) 2302.

[27] Charron E., Tiesinga E., Mies F. and Williams C., to be published.

[28] Preskill J., *Phys. Today*, **52**, No. 6 (1999) 24.

[29] Savard T. A., *Phys. Rev. A*, **56** (1997) R1095.

[30] O'Hara K. M. *et al.*, *Phys. Rev. Lett.*, **82** (1999) 4204.

[31] Scheunemann R. *et al.*, *Phys. Rev. A*, **6205** (2000) 1801.

BEC

Bose-Einstein condensation: The cure for decoherence arising from inhomogeneity

E. A. CORNELL

JILA, National Institute of Science and Technology and Department of Physics
Boulder, CO 80309-0440, USA

In a resonance experiment performed on a sample of two-level atoms, there are really two kinds of decoherence to consider. The first, call it "atom-by-atom" decoherence, results from the internal phases of the various atoms in a given sample evolving at different rates, thus washing out the transverse polarization of the sample. This process arises for instance from spatial inhomogeneity, or from inter-atom collisions (fig. 1a). The second kind, call it "shot-to-shot" decoherence, arises when the atoms of a given sample remain coherent with each other, but the overall phase becomes unpredictable from one realization of the experiment to another (fig. 1b). This could be a consequence, for instance, of a drift in the mean ambient magnetic field, or even in the frequency of the oscillator with which the experimenter defines local time.

It is a peculiar feature of Bose-Einstein condensates [1] that they are essentially immune to an entire class of decoherence processes, the "atom-by-atom" type. In response to an imposed inhomogenous potential, the condensate adjusts its internal density such that the mean-field self-repulsive energy of the condensate exactly cancels the inhomogeneity. The energy released by this relaxation is converted into a relative small number of high-energy excitations, atoms which no longer participate in the spectroscopic measurement. Thus the bulk of the condensate appears to relax in an irreversible matter in such as way as to cancel the spatial inhomogeneity. Somewhat counterintuitively, this relaxation is deterministic from the point of view of the quantum phase—the memory of the quantum phase of the bulk of the condensate is not erased in the process [2]. These notes, a summary of a lecture given at the Varenna school, describe a measurement of the relative internal phase of two components of a Bose-Einstein condensate; indeed we

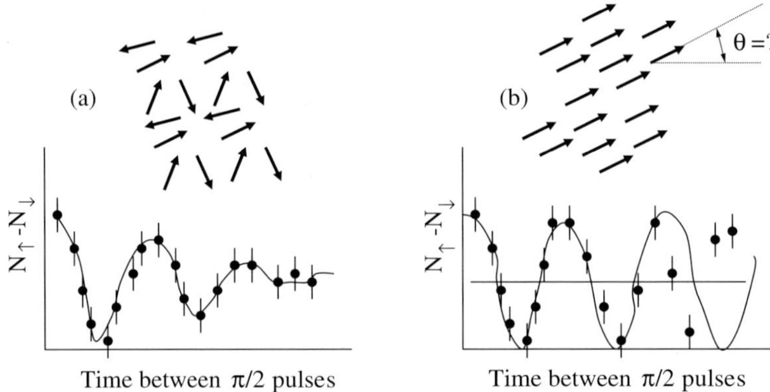

Fig. 1. – Schematic showing two varieties of decoherence. (a) In "atom-by-atom" decoherence, the internal phase of each individual atom in a collection randomizes with respect to the other atoms. In this illustration, the various atoms in a sample are represented by the arrows, and the phase of each atom is portrayed as the plotted angle of its arrow. In a hypothetical Ramsey-type experiment, the final net spin of the sample (defined as the number of atoms spin-up minus the number of spin-down atoms) gradually decays to zero. (b) In "shot-to-shot" decoherence, the sample remains internally coherent, in the sense that all the atoms in a given shot have the same phase. But for longer and longer evolution times, the value of the phase can no longer be accurately predicted. In the Ramsey experiment, for longer times the points will average to zero, but for any given shot will show the full variance from all spin up to all spin down.

see that the phase survives the relaxation process, such that we can measure a definite value for the phase well after one might expect that applied inhomogeneity might have washed it out.

The paradoxical aspect of the Rb-87 mixed-condensate system is that it combines two traits that are usually thought of as mutually exclusive: i) the $|1\rangle$ and $|2\rangle$ components of the condensate are distinguishable, in the usual senses of the term—the components do not at experimental energy scales spontaneously interconvert, and the components can be selectively detected and imaged; and ii) the components possess a measurable relative quantum phase, which evolves at a rate proportional to the difference in chemical potential between the two fluids, and which can, under the right circumstances, be read out, much as the current across a Josephson junction reads out the relative quantum phase across it. Interconversion (and phase read-out) *can* be driven via a stimulated process and therefore can be turned on and off. The system thus bears considerable resemblance to the idealized pair of condensates linked by a removable Josephson junction that Leggett [3, 4] and others have envisioned in various thought experiments.

The experiments described here all begin with a sample of approximately 5×10^5 spin-aligned, Bose-condensed Rb-87 atoms at a temperature of less than 50 nK, confined in a 3d, harmonic magnetic potential at density $1 \times 10^{14}/\text{cm}^3$. The number density of noncondensed atoms is a factor of 30 or more smaller. In this paper we will skip the details of refrigeration—the technique is a hybrid method that combines many of

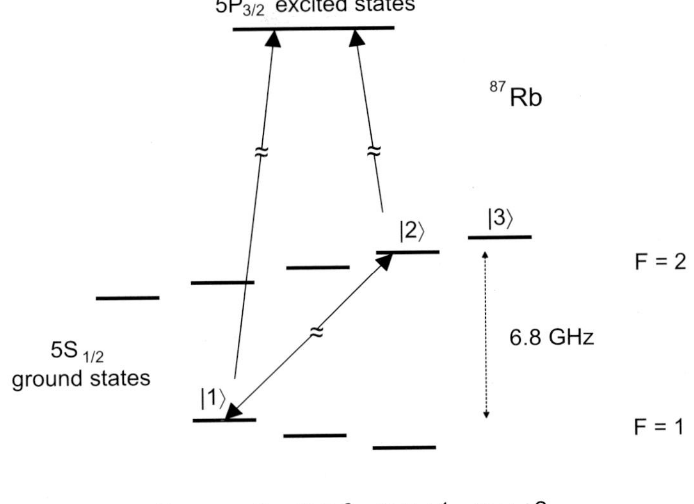

Fig. 2. – The relevant hyperfine and Zeeman structure of Rb-87 in a magnetic field. The atoms are moved coherently from state $|1\rangle$ to state $|2\rangle$ with a microwave pulse [5]. Imaging is accomplished via absorption in optical transitions to the excited states.

the notable advances [6-8] in atom cooling from the 1980s: laser cooling and trapping, magnetic trapping, and evaporative cooling. The synthesis of these various approaches was begun at JILA from 1989–1992 [9, 10], but it was only after considerable additional development [11-14] that, in 1995, the parts worked well enough together to permit the first observation of BEC in dilute atomic gases [1, 15, 16]. In the current version of our apparatus, the cooling process is automated and reliable, much like a (well-behaved) dilution refrigerator. A freshly prepared condensate is available for study about once per minute. The isolation of the sample from the 300 K environment (less than 1 cm away) is excellent—thermal equilibration times with the environment are on the order of centuries—but residual heating and the collisional decay of the condensate limit the duration of an experiment on any given condensate to a few seconds.

The ground state of an Rb-87 atom is split in a magnetic field into eight levels by hyperfine and Zeeman interactions (fig. 2). The states labelled $|1\rangle$, $|2\rangle$, and $|3\rangle$ can be trapped at a local minimum in the magnetic field; it is with states $|1\rangle$ and $|2\rangle$ ($|F = 1, m = -1\rangle$, and $|F = 2, m = 1\rangle$, respectively) that we have performed most of our mixed-fluids experiments [17, 18]. The other relevant internal structure of the Rb-87 atom is a set of excited-state levels 1.6 eV above the ground state (fig. 2). We probe the spatial distribution of the atoms by imaging the absorption on an optical transition to these levels. The natural 6 MHz linewidth of the optical transition is much less than the energy splitting between, for instance, states $|1\rangle$ and $|2\rangle$. With the correct choice of laser frequency, we can image the spatial distribution of either state $|1\rangle$ or of state $|2\rangle$, or, if we prefer, of the combined density of the two states [19].

Species $|1\rangle$ and $|2\rangle$ meet both usual criteria for calling fluids "distinguishable". One can in the literal sense distinguish one density distribution from the other, simply by adjusting the frequency of the probe laser so that one species or the other casts its shadow onto our imaging CCD array (fig. 2). This measurement, as performed in refs. [19-21], is destructive, but as has been shown [22, 23] it need not be so. In any case the fact that we can prepare condensates with a high repetition rate and a large degree of reproducibility means that the destructive aspect of the imaging is not very important to us: we observe the time evolution of the clouds simply by repeating the experiment many times with increasing dwell times. The second sense of "distinguishable" is that particles do not spontaneously interconvert. This requirement is enforced in our system by the enormous difference in internal energies between the two states. In the absence of applied electromagnetic fields, the 6.8 GHz hyperfine energy (fig. 2) is a million times larger than any other energy scale of the condensate system. The energy released by a *single* atom converting from state $|2\rangle$ to state $|1\rangle$, if that energy were to be distributed thermally through our sample, is sufficient to drive the *entire sample* out of the Bose-condensed state. The fact that we do not observe our condensates melting during our measurements is a guarantee that spontaneous interconversion is not a factor in the system.

The positions of the two fluids are determined by their respective confining magnetic potentials. The magnetic moments of the two states are nominally the same, but due to various small effects—gravity, the nuclear magnetic moment, the onset of nonlinearity in the Zeeman shifts, and subtle dynamical effects [14, 24] of our magnetic trap—the location of the two minima of the confining potentials $V_1(r)$ and $V_2(r)$ can be adjusted to be either exactly coincident or slightly offset. For the work described here, the confining potentials are axially symmetric and harmonic, with single-particle oscillation frequencies of $\omega_z/2\pi = 60\,\mathrm{Hz}$, and $\omega_r/2\pi = 21\,\mathrm{Hz}$. With the spatial offset between the two potentials set to zero, and in the absence of interspecies interactions, the two fluids would be completely interpenetrating. In fact, inter- (and intra-) species interactions are a dominant factor in determining the density distributions. The dynamics are well described in the mean-field language of coupled Gross-Pitaevskii equations for the order parameters [25-27]:

$$\text{(1)} \qquad i\hbar\dot{\Phi}_1 = \left(-\frac{\hbar^2}{2m}\nabla^2 + V_1 + u_1|\Phi_1|^2 + u_{12}|\Phi_2|^2 \right)\Phi_1,$$

and

$$\text{(2)} \qquad i\hbar\dot{\Phi}_2 = \left(-\frac{\hbar^2}{2m}\nabla^2 + V_2 + V_{\mathrm{hf}} + u_2|\Phi_2|^2 + u_{21}|\Phi_1|^2 \right)\Phi_2,$$

where m is the mass of the Rb atom, V_{hf} is the magnetic-field–dependent hyperfine splitting between the two states in the absence of interactions, $u_i = 4\pi\hbar^2 a_i/m$ and $u_{ij} = 4\pi\hbar^2 a_{ij}/m$, $|\Phi_i|^2$ is the condensate density, and the intraspecies and interspecies scattering lengths are a_i and $a_{ij} = a_{ji}$. Note that these equations do not allow for species interconversion; they conserve independently the number of atoms in each species.

Considerable theoretical work has already gone into studying ground-state solutions [25-27] and small-amplitude excitations [28-31] of the order parameters. We review here the results most relevant to the current experiments. As the total number of atoms trapped increases, the quantum kinetic energy term $\nabla^2\Phi$ becomes less significant, until it is almost negligible for the numbers used in double-condensate experiments. Most theoretical treatments neglect this term, thus making the so-called Thomas-Fermi approximation. In this case, the steady-state, single-component condensate density profile in a parabolic potential will have an inverted parabaloid density profile, with density tapering smoothly from the peak at cloud center to zero at its edge. Due to an accidental degeneracy [32, 33] in the Rb-87 scattering system, $a_1 \approx a_{12} \approx a_2$. For this special case, one can see by inspection of eqs. (1) and (2) that the steady-state total density of a two-component BEC will also have an inverted parabola form, even though the relative densities of the components may have intricate structure. It turns out that there is a critical value [25-27] for the interaction term $a_{12}^c = \sqrt{a_1 a_2}$. For $a_{12} > a_{12}^c$, two components should have little spatial overlap—they should spontaneously separate in the trap, while for $a_{12} < a_{12}^c$, there should be a relatively large region of interpenetration. In Rb-87, the scattering lengths are known at the 1% level to be in the proportion $a_1 : a_{12} : a_2 : 1.03 : 1 : 0.97$, with the average of the three being 55(3) Å [19, 33].

Since the mutual interaction term is within 1% of its critical value, anything that breaks the symmetry between the two species becomes important. For instance, the fact that $a_1 > a_2$ means it is energetically favorable for the $|1\rangle$ atoms to preferentially move towards the lower-density region at the periphery of the cloud, forming a spherical shell around $|2\rangle$ [25-27], but since the difference between a_1 and a_2 is small, even a rather minor vertical offset in the spatial centers of V_1 and V_2 will result in the components' separating up-and-down, rather than radially in-and-out. In steady state, then, one expects the clouds to be largely, although not completely [34] spatially separated. Further, because the steady-state energetics only modestly favor component separation, one would expect that in the ensuing dynamical behavior, small oscillations about the steady-state configuration [28-31] will feature a number of "soft" modes, with frequencies low compared to the excitations of the total density.

All of these qualitative theoretical expectations have been borne out in preliminary experiments [20]. Figure 3 shows the density profile [35] of two condensates after they have come to steady state. In this measurement, the potentials have been offset a distance equal to only 3% of the overall sample size, but the resulting component separation is much larger. Note however that there does remain a region of overlap, which is useful for work described below. The $|2\rangle$ component is originally created (as described below) such that it completely overlaps the $|1\rangle$ component, and we have observed [20] the time evolution of the component separation as it relaxes to the steady-state, separated condition. During this time, the overall density profile is essentially unperturbed. The period of the resulting damped oscillations of the relative positions of the two components is about 30 ms. In comparison, the period of the lowest-order nontrivial excitation of the total density profile is about 25 ms.

If we do not deliberately break symmetry of the confining potentials in the experiment

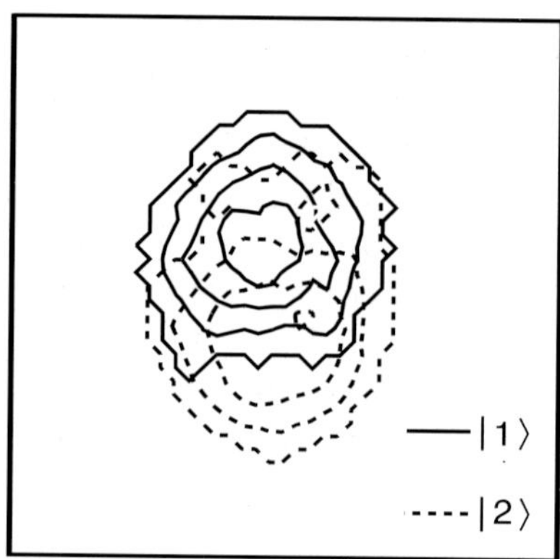

Fig. 3. – Contour plot of the expanded [35] density distribution after the two components have undergone separation and their relative motions have damped. The image is $182\,\mu\text{m}^2$. Before expansion the size of the total distribution was 40 by $14\,\mu\text{m}$. Note that a considerable region of interpenetration remains [36].

but rather keep them as perfectly overlapped as possible, we see, at early times, the onset of the expected inward-outward component separation.

In the absence of any applied coupling the 6.8 GHz energy separation between the two states is enormous, but the presence of an oscillating magnetic field tuned close to the hyperfine splitting changes the situation dramatically. An atom can readily be transferred from one state to the other—the difference in energy is absorbed from (or stimulated into) the coherent field of 6.8 GHz photons [5]. That the population transfer resulting from the application of a coupling field is sensitive to the relative phase of the two states is well known in microwave spectroscopy [37] and NMR and is easy to understand in the case of a single, two-level atom.

The Schrödinger equation [38] for the internal state of a two-level atom (or of any two-level quantum system) in the presence of a coupling drive at frequency ω_{rf} is

$$（3）\qquad i\dot{A}_1 = \omega_1 A_1 + \frac{\Omega}{2}e^{i\omega_{\text{rf}}t}A_2,$$

$$（4）\qquad i\dot{A}_2 = \omega_2 A_2 + \frac{\Omega^*}{2}e^{-i\omega_{\text{rf}}t}A_1,$$

where A_i is the probability amplitude to be in state $|i\rangle$, $\hbar\omega_i$ is the internal energy of state $|i\rangle$, and Ω is the amplitude of the coupling. We assume the detuning $\delta \equiv \omega_{\text{rf}} - \omega_2 + \omega_1$ is small, such that $|\delta| \ll \omega_{\text{rf}}$, and concentrate on the case that the coupling drive is off ($\Omega = 0$) most of the time and on only for brief pulses of duration τ, such that $1/\tau \gg |\delta|$. When we discuss "phase" in this context, we are talking about the real quantities α_1, α_2, or α_{rf} defined in the relations $A_1 = |A_1|e^{i\alpha_1}$, $A_2 = |A_2|e^{i\alpha_2}$, and $\Omega e^{-\omega_{\text{rf}}t} = |\Omega|e^{i\alpha_{\text{rf}}}$.

There is one particularly useful special-case solution of eqs. (3) and (4). A pulse with duration and amplitude such that $|\Omega|\tau = \pi/2$ is known as a "$(\pi/2)$-pulse". If the atom is initially in state $|1\rangle$ with unit probability (*i.e.*, $|A_1| = 1$ and $|A_2| = 0$), then after the application of the pulse the atom will have equal probability of being in either state (*i.e.*, $|A_1| = 1/\sqrt{2}$ and $|A_2| = 1/\sqrt{2}$). The $(\pi/2)$-pulse thus creates a 50-50 coherent superposition of the two states, also "writing" a particular initial relative phase. After the pulse is applied, the relative phase evolves at the rate $\omega_2 - \omega_1$. The accumulated phase can be measured by a second $(\pi/2)$-pulse [37]. When a $(\pi/2)$-pulse is applied to a two-level system in 50-50 coherent superposition, the population transferred by the pulse from state $|1\rangle$ to state $|2\rangle$ is proportional to $\cos(\alpha_1 - \alpha_2 - \alpha_{\mathrm{rf}})$, where the phases α refer to the total phase accumulated during the dwell time T between the two pulses. One measures the transition probability from state 1 to state 2 by repeating the experiment for different T; the result is "Ramsey fringes", a cosine at the detuning frequency δ. This technique is known as the method of separated oscillatory fields [37] or "twin-pulse" spectroscopy.

Now it becomes clear how we modify eqs. (1) and (2) to include a coupling field applied to a Bose condensate:

$$
(5) \qquad i\hbar\dot{\Phi}_1 = \left(-\frac{\hbar^2}{2m}\nabla^2 + V_1 + u_1|\Phi_1|^2 + u_{12}|\Phi_2|^2 \right)\Phi_1 + \frac{\hbar\Omega(t)}{2}e^{i\omega_{\mathrm{rf}}t}\Phi_2,
$$

and

$$
(6) \qquad i\hbar\dot{\Phi}_2 = \left(-\frac{\hbar^2}{2m}\nabla^2 + V_2 + V_{\mathrm{hf}} + u_2|\Phi_2|^2 + u_{21}|\Phi_1|^2 \right)\Phi_2 + \frac{\hbar\Omega(t)}{2}e^{-i\omega_{\mathrm{rf}}t}\Phi_1.
$$

The probability amplitudes A_i of eqs. (3) and (4) have now become field amplitudes Φ_i, and the physics has become correspondingly richer. One can still perform experiments, however, that explore simple limits.

In one such experiment, we begin with atoms in the pure $|1\rangle$, relaxed to a steady-state, near-pure condensate with spatial distribution described by Φ_0. Next we apply a $(\pi/2)$-pulse, transferring half the population to state $|2\rangle$. Immediately after the $(\pi/2)$-pulse, the two-state system can be described by $\Phi_1 = \Phi_0 e^{i\mu_1 t}/\sqrt{2}$, and $\Phi_2 = \Phi_0 e^{i\mu_2 t}/\sqrt{2}$, where μ_1 and μ_2, which differ by about the hyperfine frequency, are the chemical potentials of the two components. We know however that this perfectly overlapping state is not the steady state of the mixed BEC system; this is in fact the point of departure for the relaxation-to-equilibrium experiments described previously. Within a matter of a few tens of milliseconds the components will separate, but on shorter time scales, the condensates do not have time to realize that they are a pair of mutually repulsive quantum fluids. One could equally well describe the system at short times as a single condensate containing about a million atoms, each in an internal, coherent superposition governed by eqs. (3) and (4). Figure 4 shows the result of applying a second $(\pi/2)$-pulse immediately, and then measuring the final number of atoms in the $|2\rangle$ state. The high-contrast Ramsey fringes that result are much as one would expect from a similar experiment performed on, for instance, an atomic beam [37]. This "conventional" response of condensates at short times to coupling drives has also been seen in condensate Rabi oscillations [19, 39].

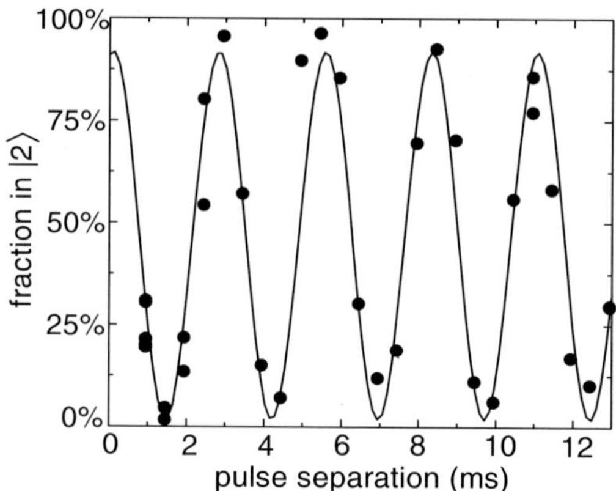

Fig. 4. – Population transfer from $|1\rangle$ to $|2\rangle$ resulting from twin $(\pi/2)$-pulse coupling pulses, as a function of delay between the two pulses. The first pulse prepares the condensate in an equal superposition of the two states, the second pulse induces further population transfer sensitive to the relative phase that has evolved during the pulse separation time. The coupling drive is detuned from the energy difference between the two condensates by about 360 Hz.

If the system is allowed to evolve longer after the first $(\pi/2)$-pulse, so that the components begin to separate, it is no longer very useful to describe the system as a single condensate of atoms in a superposition state. One should think of them instead as distinct fluids evolving their separate ways, albeit (as we shall see) with a well-defined relative phase. The effects of component separation show up clearly in twin-pulse spectroscopy: fig. 5 shows some condensate Ramsey fringes collected over longer periods. The loss of

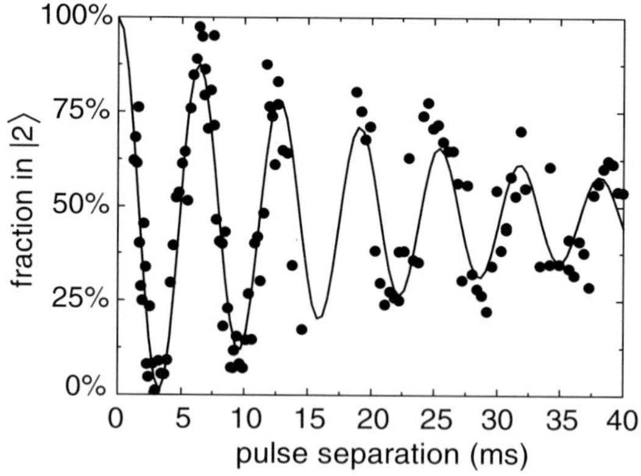

Fig. 5. – Population transfer from $|1\rangle$ to $2\rangle$ resulting from twin $(\pi/2)$-pulse coupling pulses, as a function of delay between the two pulses. Results are similar to those shown in fig. 4, except that over the longer duration the components physically separate—the loss of spatial overlap reduces the contrast ratio of the Ramsey fringes.

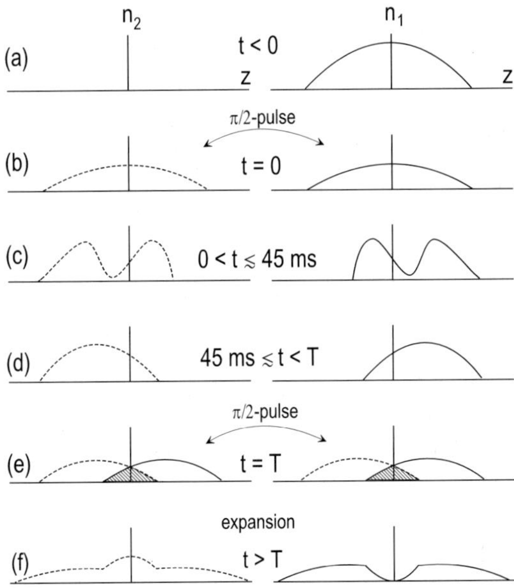

Fig. 6. – A schematic [21] of the condensate interferometer. (a) The experiment begins with all of the atoms in condensate $|1\rangle$ at steady state. (b) After the first $(\pi/2)$-pulse, the condensate has been split into two components with a well-defined initial relative phase. (c) The components begin to separate in a complicated fashion due to mutual repulsion as well as a $0.4\,\mu$m vertical offset in the confining potentials (see also fig. 3 of ref. [20]). (d) The relative motion between the components eventually damps with the clouds mutually offset but with some residual overlap. Relative phase continues to accumulate between the condensates until (e) at time T a second $(\pi/2)$-pulse remixes the components; the two possible paths by which the condensate can arrive in one of the two states in the hatched regions interfere. (f) The cloud is released immediately after the second pulse and allowed to expand for imaging. In the case shown, the relative phase between the two states at the time of the second pulse was such as to lead to destructive interference in the $|1\rangle$ state and a corresponding constructive interference in the $|2\rangle$ state.

contrast in the fringes is not due to inhomogenous broadening of the atomic transition, at least not in the conventional sense. Rather, it arises from the dwindling spatial overlap between the two components, which in turn arises from the mutual repulsion of the components, as we saw earlier. During component separation the components are of course moving with respect to each other, and therefore there is a gradient in the relative phase across the sample; it is no longer so simple to define a single quantity as the phase difference between the two clouds.

At still longer inter-pulse times, the results of the twin-pulse experiments become once again explainable in terms of a single relative phase. After the component-separation process has gone to completion, the relative motion of the condensates damps, and presumably the gradients of the relative phase vanish as well. As shown in fig. 3 above, the post-separation clouds preserve a reasonable amount of spatial overlap. As shown schematically in fig. 6, the application the second $(\pi/2)$-pulse at this time acts as a sort

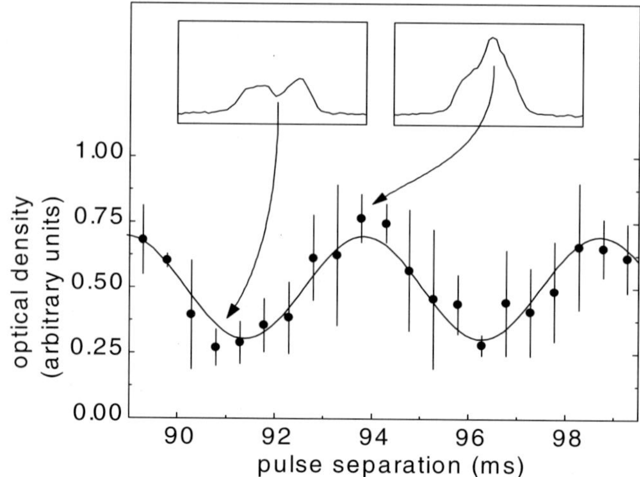

Fig. 7. – The value of the condensate density in the $|2\rangle$ state is extracted at the center of the overlap region (inset) and plotted as a function of T. Each point represents the average of 4 separate realizations and the thin bars denote the rms scatter in the measured interference for an individual realization. The thick line is a sinusoidal fit to the data, from which we extract the angular frequency μ_2-μ_1-$\omega_{\rm rf}$.

of recombiner on a separated-arm atom-interferometer [40]. Depending on the relative phase between the two components, we can see (fig. 7) destructive or constructive interference in the region where the components overlap. Thus, the double-pulse experiment with a long time between the pulses acts as a condensate interferometer that allows us to determine the relative phase between two largely separated condensate components.

We find the contrast ratio manifest in fig. 7 quite remarkable. It is evidence that the two components have a well-preserved memory of their relative phase *even after* there has been damping of the external degrees of freedom, with which the internal states are entangled. The phase of the condensate evidently has a robustness one would not expect in a single-particle experiment. This is clearly an area worthy of extensive study—how robust is the relative phase, as a function of temperature, of relative equilibrium separation, of time between pulses? How well can the relative phase survive a "nondestructive" observation [22, 23], which can take many sequential images of the same condensate? In single-atom interferometry, the connection between the availability of "which-path" information and the loss of coherence has been thoroughly examined [41]. How do these results generalize to condensates?

$$* \ * \ *$$

This research is supported by the NSF, the ONR, and NIST.

REFERENCES

[1] Anderson M. H., Ensher J. R., Matthews M. R., Wieman C. E. and Cornell E. A., *Science*, **269** (1995) 198.

[2] Sinatra A. and Castin Y., *Eur. Phys. J. D*, **8** (2000) 319.

[3] Leggett A. J., *J. Low Temp. Phys.*, **110** (1998) 719.

[4] Sols F., *Physica B*, **194** (1994) 1389.

[5] A transition from the $|1\rangle$ state ($F = 1$, $m = -1$) to the $|2\rangle$ state ($F = 2$, $m = 1$) requires two units of angular momentum. The microwave coupling drive must therefore be composed of two photons. As described in ref. [19], we use a two-photon drive well detuned from the intermediate $F = 1$, $m = 0$ state. For all practical purposes, one may model the coupling drive as a single-photon transition with a particular detuning and effective strength.

[6] Raab E. L., Prentiss M., Cable A., Chu S. and Pritchard D. E., *Phys. Rev. Lett.*, **59** (1987) 2631.

[7] Hess H. F., Kochanski G. P., Doyle J. M., Masuhara N., Kleppner D. and Greytak Th. J., *Phys. Rev. Lett.*, **59** (1987) 672.

[8] Migdall A. L., Prodan J. V., Phillips W. D., Bergeman Th. H. and Metcalf H. J., *Phys. Rev. Lett.*, **54** (1985) 2596.

[9] Monroe C., Swann W., Robinson H. and Wieman C., *Phys. Rev. Lett.*, **65** (1990) 1571.

[10] Monroe Ch., Cornell E. and Wieman C., in *Proceedings of the International School of Physics "Enrico Fermi", Course CXVIII*, edited by E. Arimondo, W. D. Phillips and F. Strumia (North-Holland) 1992, p. 361.

[11] Ketterle W., Davis K. B., Joffe M. A., Martin A. and Pritchard D. E., *Phys. Rev. Lett.*, **70** (1993) 2253.

[12] Petrich W., Anderson M. H., Ensher J. R. and Cornell E. A., in *Fourteenth International Conference on Atomic Physics (ICAP XIV)*, 1994, p. 7.

[13] Davis K. B., Mewes M.-O., Joffe M. A., Andrews M. R. and Ketterle W., *Phys. Rev. Lett.*, **74** (1995) 5202.

[14] Petrich W., Anderson M. H., Ensher J. R. and Cornell E. A., *Phys. Rev. Lett.*, **74** (1995) 3352.

[15] Davis K. B., Mewes M.-O., Andrews M. R., van Druten N. J., Durfee D. S., Kurn D. M. and Ketterle W., *Phys. Rev. Lett.*, **75** (1995) 3969.

[16] Bradley C. C., Sackett C. A., Tollett J. J. and Hulet R. G., *Phys. Rev. Lett.*, **75** (1995) 1687; **79** (1170) 1997.

[17] The exception is ref. [32], for which states $|1\rangle$ and $|3\rangle$ were used. In ref. [32] it was discovered that, due to an accidental degeneracy in scattering lengths, spin-exchange processes are suppressed in Rb-87 (see ref. [42]). In the absence of this degeneracy, one would expect condensate mixtures of different hypefine levels to have a very short lifetime.

[18] The MIT BEC group has confined condensates in optical potentials (ref. [43]) and shown that experiments can be performed on mixtures on the $m = 1$, $m = 0$ and $m = -1$ sublevels of the $F = 1$ hyperfine state of sodium.

[19] Matthews M. R., Hall D. S., Jin D. S., Ensher J. R., Wieman C. E., Cornell E. A., Dalfovo F., Minniti C. and Stringari S., *Phys. Rev. Lett.*, **81** (1998) 243.

[20] Hall D. S., Matthews M. R., Ensher J. R., Wieman C. E. and Cornell E. A., *Phys. Rev. Lett.*, **81** (1998) 1539.

[21] Hall D. S., Matthews M. R., Wieman C. E. and Cornell E. A., *Phys. Rev. Lett.*, **81** (1998) 1543.

[22] Andrews M. R., Mewes M.-O., van Druten N. J., Durfee D. S., Kurn D. M. and Ketterle W., *Science*, **81** (1996) 273.

[23] BRADLEY C. C., SACKETT C. A. and HULET R. G., *Phys. Rev. Lett.*, **78** (1997) 985.

[24] HALL D. S., ENSHER J. R., JIN D. S., MATTHEWS M. R., WIEMAN C. E. and CORNELL E. A., *Proc. SPIE*, **3270** (1998) 98.

[25] HO T.-L. and SHENOY V. B., *Phys. Rev. Lett.*, **77** (1996) 3276.

[26] ESRY B. D., GREENE C. H., BURKE J. P. jr. and BOHN J. L., *Phys. Rev. Lett.*, **78** (1997) 3594.

[27] PU H. and BIGELOW N. P., *Phys. Rev. Lett.*, **80** (1998) 1130.

[28] BUSCH TH., CIRAC J. I., PÉREZ-GARCÍA V. M. and ZOLLER P., *Phys. Rev. A*, **56** (1997) 2978.

[29] GRAHAM R. and WALLS D., *Phys. Rev. A*, **57** (1998) 484.

[30] PU H. and BIGELOW N. P., *Phys. Rev. Lett.*, **80** (1998) 1134.

[31] ESRY B. D. and GREENE C. H., *Phys. Rev. A*, **57** (1998) 1265.

[32] MYATT C. J., BURT E. A., GHRIST R. W., CORNELL E. A. and WIEMAN C. E., *Phys. Rev. Lett.*, **78** (1997) 586.

[33] BURKE J. P. jr., BOHN J. L., ESRY B. D. and GREENE C. H., *Phys. Rev. A*, **55** (1997) R2511; KOKKELMANS S. J. J. M. F., BOESTEN H. M. J. M. and VERHAAR B. J., *Phys. Rev. A*, **55** (1997) R1589; JULIENNE P. S., MIES F. H., TIESINGA E. and WILLIAMS C. J., *Phys. Rev. Lett.*, **78** (1997) 1880.

[34] Sharp inter-component boundaries carry a kinetic energy cost. In the limit of $a_1 \approx a_{12} \approx a_2$, interspecies boundaries are expected to have a finite width large compared to the condensate healing length. See TIMMERMANS E., e-print cond-mat/9709301.

[35] In order to get better spatial resolution of the clouds, we utilize a pre-imaging expansion technique. The confining potentials are discontinuously turned off, which allows the self-repulsion of the cloud to expand rapidly the density distribution. We know from both numerical simulation and laboratory experience that this process does little other than linearly expand the space the cloud occupies. In particular, the relative positions of the two components in the cloud are preserved.

[36] The figures (and some of the language) in this manuscript have previously appeared in our publications [19-21, 44].

[37] RAMSEY N. F., *Molecular Beams* (Clarendon Press) 1956.

[38] See, for instance, ref. [37], or COHEN-TANNOUDJI C., DUPONT-ROC J. and GRYNBERG G., *Atom-Photon Interactions* (John Wiley and Sons, New York) 1992.

[39] MEWES M.-O., ANDREWS M. R., KURN D. M., DURFEE D. S., TOWNSEND C. G. and KETTERLE W., *Phys. Rev. Lett.*, **78** (1997) 582.

[40] KEITH D. W., EKSTROM C. R., TURCHETTE Q. A. and PRITCHARD D. E., *Phys. Rev. Lett.*, **66** (1991) 2693.

[41] CHAPMAN M. S., HAMMOND T. D., LENEF A., SCHMIEDMAYER J., RUBENSTEIN R. A., SMITH E. and PRITCHARD D. E., *Phys. Rev. Lett.*, **75** (1995) 3783.

[42] BURKE J. P. jr., BOHN J. L., ESRY B. D. and GREENE C. H., *Phys. Rev. A*, **55** (1997) R2511.

[43] STAMPER-KURN D. M., ANDREWS M. R., CHIKKATUR A. P., INOUYE S., MIESNER H.-J., STENGER J. and KETTERLE W., *Phys. Rev. Lett.*, **80** (1998) 2027.

[44] CORNELL E. A., HALL D. S., MATTHEWS M. R. and WIEMAN C. E., *J. Low Temp. Phys.*, **113** (1998) 151.

Laser ionization of a Rb Bose-Einstein condensate towards quantum computation

E. ARIMONDO

Istituto Nazionale per la Fisica della Materia and Dipartimento di Fisica
Università di Pisa - Via F. Buonarroti 2, I-56126 Pisa, Italy

1. – Introduction

The recent development of laser cooling and trapping techniques has made possible the controlled realization of dense and cold atomic samples, thus opening the way for spectroscopic investigations in the low and ultra-low temperature regimes not accessible with conventional techniques. Magneto-optical traps (MOT), and magnetic traps (MT) where atoms are laser cooled and confined in space through the applications of counterpropagating laser beams and/or magnetic fields, produce cold atomic samples with sufficient atoms and atomic density high enough to perform accurate measurements of spectroscopy and collisional physics. Since a few years much attention has been concentrated on the photoionization of cold alkali atoms, as rubidium [1,2], cesium [3], lithium and sodium [4], magnesium [5]. More recently ultra-cold plasma has been created making use of laser cold atomic samples and laser excitation to high Rydberg states [6].

The realization of Bose-Einstein condensates (BEC) of alkali atom vapors has attracted much interest to new aspects of photon-matter interaction arising from the coherent nature of the atomic ensemble. Thus attention has been paid recently to a theoretical analysis of photoionization of a BEC by monochromatic laser light [7]. The products of the photoionization process (electrons and ions) obey Fermi-Dirac statistics. Due to the coherent nature of the initial atomic ensemble and the narrow spectral width of the laser ionization source, the occupation number of the electron/ion final states can become close to unity, especially for excitation close to threshold. Under those conditions the condensate ionization rate should be slowed down by a Pauli blockade process and finally determined by a balance between the laser light acting on the atoms and the rate of escape of the charged products from the system.

Stimulated by these theoretical predictions an experiment investigating the photoionization process within a rubidium condensate has been started. Several experimental observations have been performed, initially based on pulse laser sources and later on cw lasers tuned to the rubidium atomic resonances. The progress realized up to now with the photoionization of cold rubidium atoms and of the rubidium condensate will be reviewed. The experiment is not yet at the stage of controlling the predictions on the Pauli blockade, but it has certainly produced some interesting results. The major result is the observation that the ionization process does not destroy the condensate itself and could be used for a fine control of the production of very cold positive and negative charges, without modifying the atoms remaining in the condensate.

The motion of charged particles within a gaseous condensate should be compared to the motion of positive and negative charges within the superfluid helium, a research area quite important for the determination of the properties of superfluid liquid helium [8,9], and more recently for the spectroscopy of the impurity injected into the superfluid [10]. The study of the motion of the positive and negative particles within a gaseous condensate could probe the condensate excitations.

The interaction between negative or positive charges was never considered, to the best of our knowledge, in the framework of the quantum computing applications. Whence it is not possible to claim, *sic et simpliciter*, that the present investigation of condensate photoionization could be applied to quantum computing. However the excitation of a gaseous condensate to the ionization continuum level, in analogy to the excitation of Rydberg states in a cold gas [11-13], could find applications in the realization of two-qubit gates.

Section **2** of the present work analyses of the ion evolution and production inside a gaseous Bose-Einstein condensate. The following sections discuss the experimental apparatus and the experimental results obtained so far. Section **5** concludes presenting schemes based on the condensate photoionization for the applications to quantum gates in cold gases.

2. – Ions within a BEC

2˙**1.** *Ion motion*. – Suppose that one introduces into the condensate a charged particle of atomic size, as an ion or an electron, that can readily be manipulated by externally applied forces. Such particle could then serve as a microscopic probe of the medium in which it is immersed, for instance a superfluid liquid or a condensate. In fact this approach was used in a large series of experimental investigations with ^{4}He superfluid liquid, with the measurements of low and high electric field mobilities in order to probe the superfluid excitations [8-10]. The interaction between the charged impurities and the vortex lines and rings has produced important information on these excitations [14]. As a manifestation of superfluidity, objects travelling below a critical velocity through the superfluid should propagate without dissipation, as tested in liquid ^{4}He study dragging negative

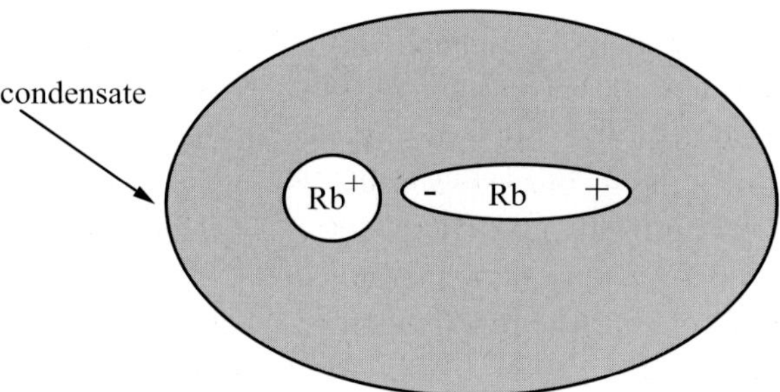

Fig. 1. – Schematic diagram indicating a Rb atom of the condensate polarized by the electric field created by a Rb$^+$ ion created within the condensate after the electron has being extracted.

ions through an applied external electric field. Similarly, in gaseous Bose-Einstein condensates, the motion of microscopic atomic impurities has recently allowed to measure the superfluidity critical velocity [15]. An additional evidence for the superfluidity in gaseous Bose-Einstein condensates could be obtained by measuring the motion of the ions produced by laser ionization in the presence of an external electric field. However charge hopping processes, with electrons jumping from a neighboring atom onto the positive ion, may also contribute to the conductivity in an ultracold gas [16].

The superfluid liquid helium experiments demonstrated that the positive and negative ions used as probes in those experiments were not the bare ions and electrons [8, 10]. Suppose that an ion is introduced into a dielectric medium. The strong electric field due to the ion charge polarizes the surrounding medium; *i.e.*, each atom of the surrounding medium close to the ion should acquire an induced dipole moment as indicated schematically in fig. 1. The electric field due to the ion exerts on the polarized atom of the surrounding medium a net force directed toward the ion, so that the surrounding medium experiences a compression with the polarizable particles attracted towards regions of higher electric field. For the case of liquid helium, the high compression leads to a helium solidification within a small sphere surrounding the ion. Therefore, a positive charge within the superfluid is really a complex consisting of a small positively charged sphere. Similar behavior, with the production of a bubble around an electron, applies to the negative charges. The structure of the charged defects produced in liquid helium was confirmed by the recent spectroscopy of the alkali and alkali-earth ions implanted by laser sputtering into superfluid helium [10].

A similar local deformation should take place also for a gaseous Bose-Einstein condensate. For a single charged particle immersed into a condensate of cold rubidium atoms the interaction potential U is given by

$$(1) \qquad U = -\frac{1}{2}\frac{1}{(4\pi\epsilon_0)^2}\frac{e^2}{r^4}\alpha,$$

with the rubidium static polarizability $\alpha_{\mathrm{Rb}} = 4\pi\epsilon_0 \times 48 \times 10^{-30}\,\mathrm{m}^3$, with the potential being repulsive for short distances due to exchange interaction of the electrons. In order to estimate the importance of the ion-neutral interaction, for a density of $10^{20}\,\mathrm{m}^{-3}$, we calculate an average interatomic distance $r = n^{-1/3}$, leading to $U/h = 3.9\,\mathrm{kHz}$, which is comparable to or larger than the typical chemical potential. This means that locally the polarization effect is important for the condensate dynamics and may even produce a destruction of the condensate phase. Estimating the global condensate importance of the ion charge is less straightforward since the cloud-averaged interaction energy is highly sensitive to the cut-off at small distances due to the singular behavior of the r^{-4} potential. Apart from the exchange interaction at small distances, this average should be also sensitive to details of the 2-body correlation function for small distances, which before the ionization is governed by the van der Waals interaction between ground-state atoms. This analysis evidences that in the experiments dealing with ion-condensate interactions within dissipative traps, the energy transfer from the charged particles to the neutral atoms of the condensate could be investigated.

$2^{.}2.$ *Pauli blockade.* – Due to the coherent nature of the initial atomic ensemble and narrow spectral width of the light source producing the photoionization process, the occupation numbers of the final states n_e for electrons and n_i for positive ions, with $n_e = n_i$, becomes close to unity. Because the ionization products (electrons and ions) obey Fermi-Dirac statistics, the rate of the laser-driven photoionization depends on the final-state occupation through $1 - n_e$. Thus, in the presence of n_e values close to unity, the ionization process is blocked by the previous electron production. The final-state occupation fractions n_e and n_i are determined by the balance between the photoionization process and the rate of escape of the electrons and ions from the photoionization final states. A crucial role in that balance is played by the interactions between the charged and neutral particles of the condensate and between the charged particles themselves. However, just after the photoionization the fermions are very far from thermodynamical equilibrium, hence the standard formulae for transport processes valid for a plasma deviating slightly from the equilibrium state cannot be applied. Moreover, owing to the wave packet spread the wave function of a particular fermion may overlap significantly with wave packets of many other fermions, beyond the standard binary-collision approach. Finally, through the polarization of the surrounding condensate atoms, a single charged particle will influence the density distribution of the cloud and thus the cloud-averaged interaction energy. The work by Mazets [7] has briefly touched all these contributions, but a correct determination of the Pauli blockade role requires a careful evaluation of the contributions.

3. – Experimental setup

In the Pisa laboratory, an apparatus for laser cooling and Bose-Einstein condensation of rubidium atoms is in operation [17]. It is based on a double MOT system, where around $5 \times 10^7\ ^{87}\mathrm{Rb}$ atoms are loaded into the lower MOT, transferred into a magnetic trap, of the triaxial Time Orbiting Potential type [18]. After evaporation with both circle-of-death and radiofrequency, we arrive at the condensation temperature with typically 4×10^4 atoms; further radiofrequency evaporation yields pure condensates with 2×10^4 atoms. The final frequency ν of the radiofrequency cut, between 8 MHz and 3.5 MHz, determines the final density of the cold cloud, whence the formation of a dense thermal cold cloud, or the creation of the rubidium condensate. Depending on the chosen sequence of cooling and trapping steps, neutral samples with temperatures ranging from a few hundred K down to tens of nK can be obtained. Typical densities (10^{10}–10^{13} atoms/cm^3) are large enough to perform reliable investigations by absorption or emission techniques. Changing the trap parameters, we can study photoionization of atomic samples with different phase space densities, reaching the condensed phase with a number of atoms that allows a dynamical analysis of the cold atom sample (temperature, density distribution). The present experimental setup does not include a charge detector inside the vacuum cell to monitor the production of positive and negative charges. Thus the action of the photoionizing laser on the cold cloud is monitored through the decrease of the atoms remaining in the trap, applying a shadow imaging detection using a near-resonant probe beam. The absorptive shadow cast by the cold atoms is imaged onto a CCD camera. Measurements on both condensates and very cold thermal clouds were performed after a few milliseconds of free fall of the released atomic clouds, when their typical dimensions were of the order of 1030 μm.

In a first stage, the photoionization of the cold atomic cloud was produced by pulsed lasers. The 4.177 eV energy required for the rubidium ionization from the ground state was provided either by the one-photon absorption at the 296 nm wavelength, or by a two-photon non-resonant transition at the 593 nm wavelength. The absorption of one 296 nm photon or two 593 nm photons by the ground-state rubidium atom produced an electron with kinetic energy around 10 meV.

The two-photon 594 nm radiation was produced by an excimer-pumped dye laser, with intensities in the tens of MW/cm^2 range and a pulse duration of 16 ns. Laser pulse repetition rates up to 12 Hz were used. The 297 nm radiation was generated from the 594 nm radiation through a frequency-doubling process within a BBO crystal, with intensity in the 1 MW/cm^2 range. The photoionizing laser was focused on the cloud of rubidium atoms, whose dimensions were in the 5 μm range for the condensate cloud and in the 100–200 μm range for the MOT cloud. The dye laser was focused down to 50 μm, while for the pulsed ultraviolet radiation, owing to the beam astigmatism, beam waists of 200 μm and 400 μm in the two directions were achieved. For a photoionization experiment with the transverse section of the photoionizing laser smaller than the cold-cloud transverse area, the ionization efficiency is reduced by the ratio between the laser transverse area and the cloud area.

In a second stage of the experiment, radiation at 420 nm wavelength, generated by frequency-doubling a diode laser at 840 nm, was focused onto the cold cloud or the atomic condensate. The 20 mW continuous source allowed us to apply in the focused area intensities up to a few kW/cm^2. The 420 nm wavelength was chosen in order to excite the second resonance line from the ground state to the $6P_{1/2}$ or $6P_{3/2}$ states, so that, in combination with a a continuous laser operating at 1010 nm, a two-photon ionization process from the ground state is produced. This two-photon ionization process has not been explored yet in the experiment. Instead Rb ionization from the excited $5P_{3/2}$ state and from the ground $5S_{1/2}$ state has been examined.

It should be pointed out that in all the experiment the motion of the charged particles within the condensate will depend critically on the presence of stray electric fields, and the presence of the magnetic-trap high currents close to the condensate could make very difficult the control the stray electric fields.

4. – Experimental results

We first irradiated the cold atoms of the Magneto Optical Trap (MOT) with the 534 nm or 296 nm ionizing lasers, studying the induced additional losses. Within the MOT the cold atoms are lost from the trap because of collisional loss mechanisms. The application of the ionizing laser induces additional losses for the atoms in the trap. These additional losses modify the exponential decay for the atoms remaining into the trap. Switching off the MOT loading mechanism, the decrease in the number N of atoms remaining in the trap is [3]

$$(2) \qquad\qquad N = N_0 e^{-(\gamma + \gamma_{\mathrm{P}})t},$$

with N_0 the initial number of cold atoms, γ the trap loss rate for collisions with the MOT background vapor and γ_{P} the photoionization additional loss produced. If p_{e} denotes the $5P_{3/2}$ excited-state occupation fraction produced by the cooling process, the one-photon photoionization loss rate can be written

$$(3) \qquad\qquad \gamma_{\mathrm{P}} = [\sigma_{\mathrm{P}}^{g}(1 - p_{\mathrm{e}}) + \sigma_{\mathrm{P}}^{e} p_{\mathrm{e}}] \frac{I_{\mathrm{P}} \lambda_{\mathrm{P}}}{\hbar c},$$

for photoionization cross-sections σ_{P}^{g} and σ_{P}^{g} from ground and excited states, respectively, in the presence of a laser with intensity I_{P} and wavelength λ_{P}.

In order to determine the γ_{P}^{g} photoionization rate as losses for atoms confined from a magnetic trap, we monitored the atoms remaining in the magnetic trap by measuring the fluorescence light emitted after excitation by a 1 μs laser pulse of 780 nm resonance light. For the pulsed laser photoionization, the laser was applied periodically to the cold atoms during the time required to empty the magnetic trap, while the time decay of the fluorescence was recorded, as shown in fig. 2. If the photoionization laser was applied synchronously with the 780 nm resonant laser, the photoionization from the first excited state represented an additional loss from the magnetic trap. For the data analysis the p_{e} excited-state fraction of eq. (3) accounted for the temporal superposition between the 780 nm laser and the photoionization laser.

The solid line in fig. 2, representing the exponential fit to the atom decay in the absence of uv photoionization light, corresponds to an exponential decay time $\gamma^{-1} = 60$ s, produced by collisions of the atoms in the magnetic trap with the background gas. In the presence of the 296 nm photoionization laser, γ_{P} ionization rates up to one-fifth of

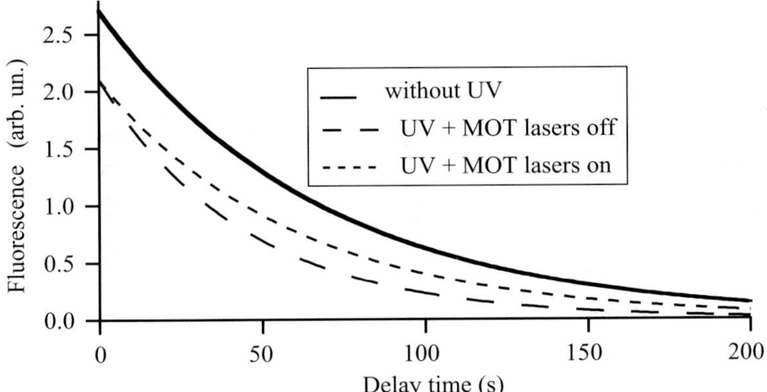

Fig. 2. – Time decay for the fluorescence from the rubidium atoms contained in the MOT in the absence (continuous line) and in the presence of a 296 nm photoionization laser light (dashed and dotted lines). For the dotted line the MOT lasers were switched on simultaneously with the uv light in order to populate the $5P_{3/2}$ state. A 12 Hz pulsed laser repetition rate was used, with 130 μJ pulse energy and $I_P = 0.5\,\mathrm{MWcm}^{-2}$ peak intensity.

the collisional γ rate were measured. The dotted and dashed lines correspond to the application of the pulsed uv light synchronously or not with the atomic excitation laser. In fig. 2 the fluorescence emission at the initial time is decreased by the presence of an additional loss mechanism in the steady rate operation, as analyzed in refs. [2, 3] for the MOT operation. Because the 2 mm MOT diameter was larger than the laser beam transverse dimension, the photoionization rate of eq. (3) was reduced by the ratio between the transverse dimension of the laser and that of the cold cloud. Analysing the fluorescence decays at different laser intensities we derived the photoionization cross-sections $\sigma^{\mathrm{g}}_{296} = 7.6 \times 10^{-20}\,\mathrm{cm}^2$ for the ground state and $\sigma^{\mathrm{e}}_{296} = 5.4 \times 10^{-18}\,\mathrm{cm}^2$ for the excited first P state. The ground-state ionization cross-section value is in agreement with the ground-state value reported in the literature, $\sigma^{\mathrm{g}}_{296} = 1 \times 10^{-19}\,\mathrm{cm}^2$ [19].

Similar ionization analyses were applied for cold atoms or condensates confined by the magnetic trap, except that each shadow imaging measurement of the atoms contained within the magnetic trap was destructive for the sample. For the 594 nm two-photon ionization of the condensate we measured losses much larger than those predicted on the basis of the ground-state two-photon cross-section theoretical value $\sigma^{2\mathrm{g}}_{594} = 5 \times 10^{-49}\,\mathrm{cm}^4\mathrm{s}$ [20]. For instance at a peak intensity $I_P = 500\,\mathrm{MW\,cm}^{-2}$ we measured a 3×10^{-2} single-pulse photoionization loss against an estimated 1.5×10^{-3} loss on the basis of the theoretical cross-section. Moreover we observed no clear evidence of a photoionization threshold behavior when the laser wavelength was scanned around the value the 593.6 nm threshold value, as shown by the data of fig. 3. There the ionization fraction f is defined as

$$(4) \qquad f = \frac{N - N_0}{N_0},$$

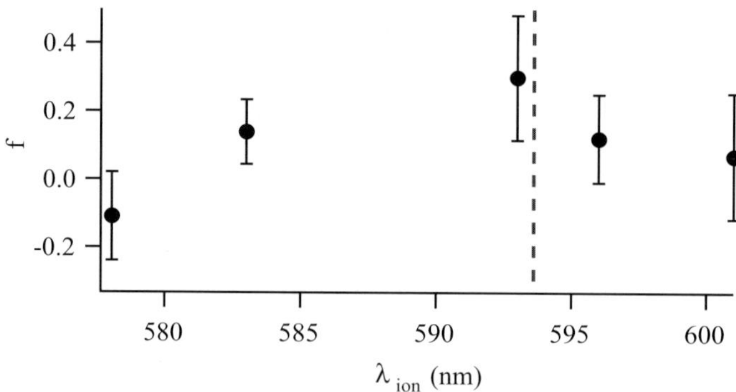

Fig. 3. – Two-photon ionization fraction f for a single laser pulse applied to the condensate *versus* the laser wavelength λ_P at the intensity $I_P = 10\,\text{MW cm}^{-2}$. The vertical line determines the threshold wavelength for the two-photon ionization from the ground state. The negative value on the first point arises from the fluctuations in the condensate number over the measurement time.

with N_0 the initial number of atoms and N the number of atoms remaining in the trap after a single photoionization laser pulse.

In the experiment with a single ionizing pulse on the BEC sample, we measured also the density profile modifications of the condensed cloud produced by the photoionization laser, with results as those in fig. 4. We noticed that the photoionization pulse ejected a secondary condensate from the original one. After its creation, the secondary condensate separated in space from the original one. The shadow image of fig. 4, obtained after 11 ms evolution of the condensates within the magnetic trap and a 18 ms free fall time of flight, presents two condensate clouds, the upper condensate at the place of the original one and the ejected secondary condensate separated in space, at a lower vertical position. By measuring the position of the secondary condensate as a function of the evolution time τ_{mag} within the magnetic trap, as shown in fig. 5, we realized that the secondary cloud is constituted by atoms in the $F = 2, m = 1$ Zeeman state. In fact the position difference $z_{m=1} - z_{m=2}$ between the secondary condensate and the initial one oscillates with τ_{mag} because the equilibrium position of the $F = 2, m = 1$ condensate within the magnetic trap is different from that of the $F = 2, m = 2$ condensate, and the $F = 2, m = 1$ condensate is initially prepared in the magnetic-trap equilibrium position of the $F = 2, m = 2$ condensate. We also discovered that a fit of the oscillating condensate position could be obtained only by supposing that the secondary condensate was created with an initial velocity $v_z(0) = -7\,\text{cm/s}$, *i.e.*, a velocity pointing downwards the vertical axis.

The process taking place within the condensate could be described as an $F = 2, m = 1$ atom laser emission from the original condensate following the excitation by the photoionizing laser. However we do not know which process has precisely produced the coherent transfer of condensate atoms from the initial Zeeman level to the final one. In effect, because of the $\Delta m = 0$ selection rule for the excitation with a linearly polarized laser field, a Raman process associated to the absorption-emission cycle with the photoionizing

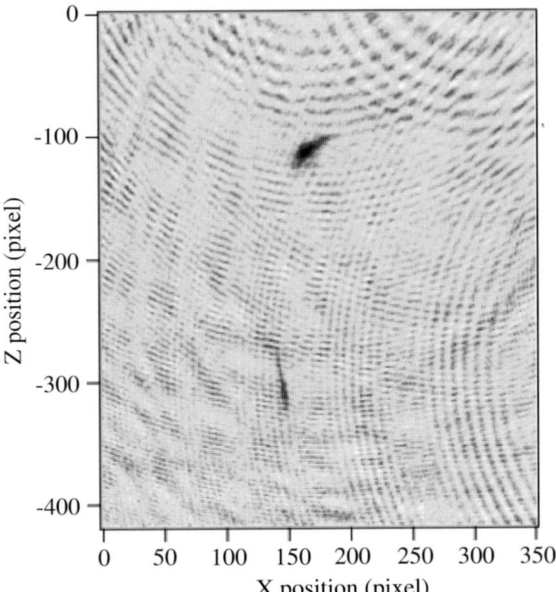

Fig. 4. – Shadow image of a condensate after a single 594 nm laser ionization pulse. The original condensate cloud, in the upper part, appears deformed with respect to the round shape observed in the absence of the ionizing laser. Moreover a secondary condensate cloud is present in the image. The periodic circular fringe patterns are an artifact of the imaging technique. At the initial time 2×10^4 rubidium atoms were present in the condensate at a 50 nK temperature, with a spatial width of 15 μm produced by the magnetic trap with frequency 23 Hz along the vertical axis. The condensates are released from the trap 11 ms after the photoionization pulse, and a 18 ms free fall takes place before shadow imaging.

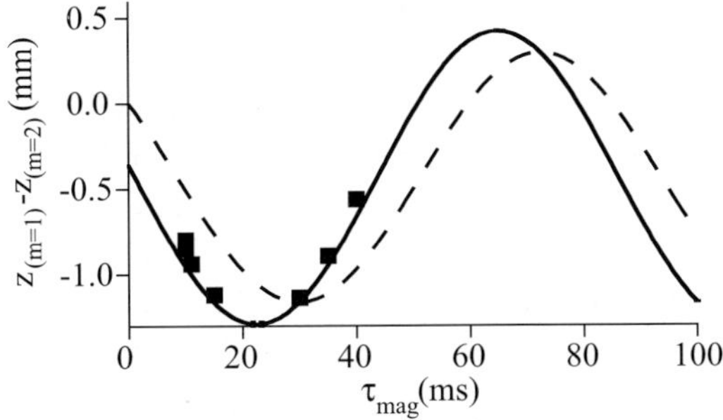

Fig. 5. – Data for the relative displacement $z_{m=1} - z_{m=2}$ of the $(F = 2, m = 1)$ secondary condensate, ejected from the $(F = 2, m = 2)$ primary one *versus* the evolution time τ_{mag} within the magnetic trap, following the ionization by a single 594 nm pulse. The oscillating fit functions correspond to different initial velocities $v_z(0)$ of the secondary cloud, $v_z(0) = 0$ for the dashed line and $v_z(0) = -1.6$ cm/s for the continuous line. The condensate positions were measured after a 18 ms free fall time, producing the vertical shift of the continuous line at $\tau_{mag} = 0$.

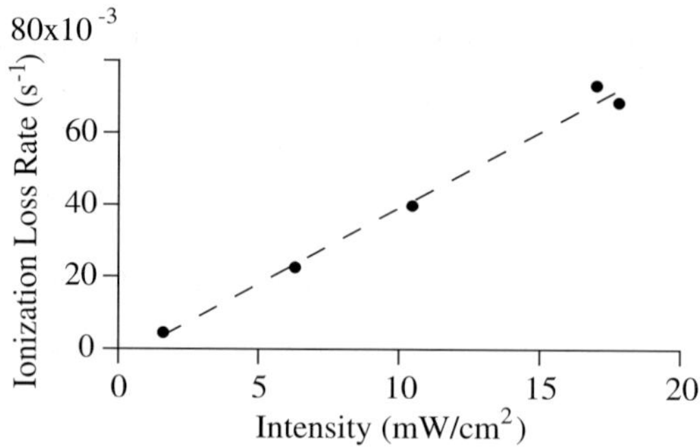

Fig. 6. – MOT ionization loss rate *versus* the intensity of a cw laser at 420 nm, producing photoionization from the $5^2P_{3/2}$ state. A cross-section value of $\sigma^e_{420} = 5 \times 10^{-49}$ cm^{4}s was derived on the basis of the excited-state occupation $p_e = 0.14$.

photons cannot transfer atoms from the initial $m = 2$ state to the $m = 1$ final one. Weak violation of the selection rule may arise either because the trap magnetic field produces a different quantization axis or because the polarization of the photoionization pulsed laser is not constant during the laser pulse. The interpretation of these results, as well the threshold absence in the data of fig. 3 and the large photoionization loss, suggests the occurrence of different phenomena. However it should be noted that the losses from the magnetic trap are not necessarily associated to the ionization of the cold atoms. In fact an escape from the magnetic trap is also produced by the processes where the atomic internal state is not modified, but the atom acquires a kinetic energy larger than the magnetic trap depth. Thus collisions between cold electrons produced by the photoionization process and the cold atoms could produce large losses from the condensate either by ionizing the condensate atoms or by imparting to them a kinetic energy large enough to escape from the magnetic trap. Kinetic energy could be gained by the condensate atoms also through the action of the dipole forces associated to the pulsed laser. Focusing and deflection of atoms by pulsed lasers has demonstrated that the atomic momentum can be modified by several units of the photon momentum [21]. We estimated that the focused pulsed laser geometry used in the photoionization experiment could provide the atoms with a momentum kick large enough to escape from the trap and to generate the oscillating motion of the secondary condensate cloud in fig. 4. An additional feature of the condensate images in fig. 4 is the strongly elongated shape presented by both the original condensate and the secondary one, while before the ionization process the condensate shape was nearly spherical [17]. Also this condensate deformation cannot be explained at the present level of our understanding of the photoionization process. It should be only noted that the electron emission by the photoionization process is asymmetric in space and the condensate deformations could be linked to the electron production.

Photoionization experiments based on the continuous 420 nm laser source performed within the MOT have allowed to measure the photoionization cross-section from the excited $^2P_{3/2}$ state, as shown by the data in fig. 6. The measured values is in very good agreement with that measured by Dinneen *et al.* [1] and estimated by Aymer *et al.* [22]. With the same source we performed within the magnetic trap, with very cold atoms, a two-photon non-resonant ionization from the ground state, whose cross-section can be estimated to be $\sigma_{420}^{2g} = 4 \times 10^{-49}$ cm^{4}s from the analysis in [20]. Anomalously large fractional losses were observed, an anomaly similar to that presented by the data of fig. 3. The comprehension of all these anomalies will require additional investigations.

5. – Toward quantum gates

The pulsed and continuous laser sources used to photoionize the condensate can be also tuned below the ionization threshold in order to excite the condensate atoms to high Rydberg states. Thus the apparatus described in the present work could be applied also to realize two-qubits quantum gates. Furthermore we have recently loaded the rubidium condensate into a one-dimensional optical lattice formed by a standing-wave laser nearly resonant with the rubidium first transition line [23]. The Rydberg excitation within the optical lattice could be used to produce a regular periodic structure of atoms excited into states with strong dipole-dipole interactions. At an $R = 390$ nm distance of a 780 nm laser optical lattice, the dipole-dipole interaction of atoms excited to Rydberg states provides the large interaction energy required to perform fast gate operations. Because our experimental results for the excitation below the ionization threshold have evidenced out an anomalously large loss of atoms, the quantum gate application of atomic high Rydberg states requires further careful studies of the process efficiency.

For quantum gate applications, an additional element provided by the present experimental investigation is the dipole blockade based on the excitation to continuum states above the atomic ionization limit. In an ensemble of cold atoms excited into Rydberg states, the level shifts associated to the atom dipole-dipole interactions can be used to block the transitions into states with more than a single atomic excitation [12]. Such a dipole blockade should take place also for atoms excited into continuum states. It should be pointed out that schemes have been proposed where the final state, the continuum one in the present case, is never populated [11]. Whence an atomic photoionization, with the creation of ions inside the condensate and their motion under the action of external electric fields, should not occur. The dipole blockade based on continuum states may gain from the strong interaction between the charged species. On the other hand the presence of a continuum of excited states could modify the excitation conditions required for the dipole blockade realization.

In conclusions we have presented experimental results for the photoionization of a condensate. The study of the ionic motion within the condensate will provide interesting tests on the condensate local deformation and the superfluid properties. Additional investigation is required on order to test the large losses presented by the condensate following the photoionization and the high Rydberg state excitation. The application of the present scheme to quantum gates requires a preliminary theoretical study.

* * *

The contributions of D. CIAMPINI, F. FUSO, J. H. MÜLLER, M. ANDERLINI, O. MORSCH and J. THOMSEN to the experimental results is acknowledged. The author is grateful to G. ALBER, W. M. FAIRBANK jr and P. ZOLLER for useful discussions. This research was supported by the Sezione A of the INFM-Italy through a PAIS Project, by the MURST-Italy through a Progetto di Ricerca Integrato, and by the EU through the Cold Quantum-Gases Network, contract HPRN-CT-2000-00125.

REFERENCES

[1] DINNEEN T. P., WALLACE C. D., TAN K. N. and GOULD P. L., *Opt. Lett.*, **17** (1992) 1706; DUNCAN B. C., SANCHEZ-VILLICAN V., GOULD P. L. and SADEGHPOUR H. R., *Phys. Rev. A*, **63** (2001) 043411.

[2] GABBANINI C., GOZZINI S. and LUCCHESINI A., *Opt. Commun.*, **141** (1997) 25; GABBANINI C., GOZZINI S. and LUCCHESINI A., in *Laser Spectroscopy*, edited by Z. J. WANG *et al.* (World Scientific, Singapore) 1998, p. 200; GABBANINI C., CECCHERINI F., GOZZINI S. and LUCCHESINI A., *J. Phys. B*, **31** (1998) 4143.

[3] MARAGÓ O., CIAMPINI D., FUSO F., ARIMONDO E., GABBANINI C. and MANSON S. T., *Phys. Rev. A*, **57** (1998) R4110; PATTERSON B. M., TAKEKOSHI T. and KNIZE R. J., *Phys. Rev. A*, **59** (1999) 2508; FUSO F., CIAMPINI D., ARIMONDO E. and GABBANINI C., *Opt. Commun.*, **173** (2000) 223; ARIMONDO E., CIAMPINI D., FUSO F. and GABBANINI C., *Appl. Surf. Sci.*, **154-155** (2000) 527.

[4] WIPPEL V., BINDER C., HUBER W., WINDHOLZ L., ALLEGRINI M., FUSO F. and ARIMONDO E., *Eur. Phys. J. D*, **7** (2001) 285.

[5] RUSCHEWITZ F., PENG J. L., DEGNER R., HINDERTHUER H., SCHELLER D., BETTERMANN D. and ERTMER W., in *Proceedings of the 1996 European Quantum Electronics Conference* (Hamburg University) 1996, p. 112; MASDEN D. N. and THOMSEN J., *J. Phys. B*, **35** (2002) 2173.

[6] KILLIAN T. C., KULIN S., BERGESON S. D., OROZCO L. A., ORZEL C. and ROLSTON S. L., *Phys. Rev. Lett.*, **83** (1999) 4776; KULIN S., KILLIAN T. C., BERGESON S. D. and ROLSTON S. L., *Phys. Rev. Lett.*, **85** (2000) 318; ROBINSON M. P., LABURTHE TOLRA B., NOEL M. W., GALLAGHER T. F. and PILLET P., *Phys. Rev. Lett.*, **85** (2000) 4466; KILLIAN T. C., LIN M. J., KULIN S., DUMKE R., BERGESON S. D. and ROLSTON S. L., *Phys. Rev. Lett.*, **86** (2001) 3759; DUTTA S. K., FELDBAUM D., WALZ-FLANNIGAN A., GUEST J. R. and RAITHEL G., *Phys. Rev. Lett.*, **86** (2001) 3993.

[7] MAZETS I. E., *Quantum Semiclass. Opt.*, **10** (1998) 675.

[8] For reviews of early experiments see CARERI G., *Progress in Low Temperature Physics*, edited by C. J. GORTER, Vol. III (North-Holland, Amsterdam) 1961, p. 58; REIF F., *Quantum Fluids*, edited by N. WISER and D. J. AMIT (Gordon and Breach, London) 1970, p. 165.

[9] For comprehensive theoretical reviews see FETTER A. L., *The Physics of Liquid and Solid Helium*, part I, edited by K. H. BENNEMANN and J. B. KETTERSON (Wiley, New York) 1974, Chapt. 3, p. 207; SCHWARS K. W., *Advances in Chemical Physics XXXIII*, edited by L. PRIGOGINE and S. A. RICE (Wiley, New York) 1975, p. 1.

[10] For a review of more recent work see KANORSKY S. I. and WEIS A., *Advances in Atomic, Molecular and Optical Physics*, edited by B. BEDERSON and H. WALTHER, Vol. **38** (Academic Press, Boston) 1997, p. 87.

[11] Jaksch D., Cirac J. I., Zoller P., Rolston S. L., Côté R. and Lukin M. D., *Phys. Rev. Lett.*, **85** (2000) 2208.

[12] Lukin M. D., Fleischhauer M., Côté R., Duan L. M., Jaksch D., Cirac J. I. and Zoller P., *Phys. Rev. Lett.*, **87** (2001) 037901.

[13] Rolston S. L., this volume, p. 101.

[14] Donnelly R. J., *Quantized Vortices in HeII* (Cambridge University, New York) 1991.

[15] Chikkatur A. P., Görlitz A., Stamper-Kurn D. M., Inouye S., Gupta S. and Ketterle W., *Phys. Rev. Lett.*, **85** (2000) 483; Ketterle W. and Inouye S., *C. R. Acad. Sci. (Paris)*, **2** (2001) 339.

[16] Côté R., *Phys. Rev. Lett.*, **85** (2000) 5316.

[17] Müller J. H., Ciampini D., Morsch O., Smirne G., Fazzi F., Verkerk P., Fuso F. and Arimondo E., *J. Phys. B*, **33** (2000) 4095.

[18] Hagley E. W., Deng L., Kozuma M., Wen J., Helmerson K., Rolston S. L. and Phillips W. D., *Science*, **283** (1999) 1706.

[19] Marr G. V. and Creek D. M., *Proc. R. Soc. London, Ser. A*, **304** (1968) 233.

[20] Bebb H. B., *Phys. Rev.*, **149** (1966) 25.

[21] Mützel M., Haubrich D. and Meschede D., *Appl. Phys. B*, **70** (2000) 689; Goepfert A., Bloch I., Haubrich D., Lison F., Schütze R., Wynands R. and Meschede D., *Phys. Rev. A*, **86** (1997) R3354.

[22] Aymar M., Robaux O. and Wane S., *J. Phys. B*, **17** (1984) 993.

[23] Morsch O., Müller J. H., Cristiani M., Ciampini D. and Arimondo E., *Phys. Rev. Lett.*, **87** (2001) 14040201, and cond-mat/0108457.

Bose-Einstein condensates in a 1D optical lattice

S. Burger, F. S. Cataliotti, F. Ferlaino, C. Fort, P. Maddaloni,
F. Minardi and M. Inguscio

Laboratorio Europeo di Spettroscopia Nonlineare (LENS)
Istituto Nazionale per la Fisica della Materia (INFM)
Dipartimento di Fisica dell' Università di Firenze
via Nello Carrara 1, I-50019 Sesto Florentino (Firenze), Italy

1. – Introduction

Bose-Einstein condensates (BECs) are quantum systems which can be easily manipulated and characterized due to their macroscopic nature [1]. The employment of BECs in atomic physics has stimulated a wealth of new experiments which is often compared to the rapid development in optics and spectroscopy after the invention of the laser [2].

Atomic BECs confined to optical lattices have been proposed for the realization of quantum-computing schemes [3, 4]. These approaches are of special interest because of the precise manipulation tools available in atomic physics.

Atoms confined in a periodic potential exhibit quantum effects known from solid-state physics, like Bloch oscillations and Wannier-Stark ladders, which have been observed by exposing cold atoms to the dipole potential of far-detuned optical lattices [5, 6]. The achievement of BEC has given the possibility to explore also macroscopic quantum effects in this context.

In a first experiment with BECs loaded into a 1D optical lattice, quantum interference could be observed, leading to the formation of the first "mode-locked" atom laser [7]. The coherent nature of a BEC governs its dynamics in optical lattices. From the well-defined phase of the macroscopically occupied wave function describing the BEC it follows that at low fluid velocities the BEC is performing a superfluid motion in the lattice [8]. In a regime of larger potential height of the optical lattice, the system is also ideally suited to study the Josephson effect: At a potential depth of the lattice sites exceeding the thermal energy BECs collectively tunnel from one site to the next, at a rate which depends on the difference in phase between the sites. Under the same conditions, thermal clouds of atoms are fixed to the wells of the optical lattice [9].

Moreover, atoms trapped in tightly confining potential wells offer a controllable way to investigate ground-state properties of the system [10] and lower-dimensional physics, *e.g.*, the condensation process in lower dimensions [11].

Recently, also the squeezing of matter waves [12] and the decoherence of BECs in 2D optical lattices [13] have been investigated. Experiments in which optical lattices are applied to the BEC on much shorter time scales have investigated Bragg diffraction as a tool for interferometry and spectroscopy [14,15], Bloch oscillations [16], and dynamical tunnelling [17].

Here we discuss recent experiments on macroscopic quantum effects of BEC dynamics in optical lattices. The lecture is organized as follows: In sect. **2** we briefly introduce the setup for the implementation of an optical lattice into a BEC experiment. Section **3** concentrates on the Bose-Einstein phase transition in the periodically modulated trap and discusses effects of two-dimensional (2D) physics. Ground-state properties of the coherent array of BECs in the optical lattice are reviewed in sect. **4**. In sect. **5** we discuss on the superfluid motion and the density-dependent breakdown of superfluidity of a BEC in an optical lattice. Section **6** concentrates on the direct observation of a coherent atomic current in an array of Josephson junctions. Section **7** concludes the paper with an outlook on future directions.

2. – Experimental setup

Techniques for the achievement of BEC in dilute atomic gases have been described in detail in ref. [1]. In the experiments discussed in the following sections we create BECs of ^{87}Rb by the combination of laser cooling in a double magneto-optical trap system and evaporative cooling in a static magnetic trap of the Ioffe type [18]. The BECs are produced in the $(F = 1, m_F = -1)$ state, with atom numbers of the order of $N \sim 10^6$. Due to the anisotropic magnetic-trapping potential, the condensates are cigar-shaped with the long axis oriented horizontally; the typical dimensions (Thomas-Fermi radii) are $R_x \sim 60\,\mu$m and $R_\perp \sim 6\,\mu$m.

We create a 1D optical lattice by superimposing to the long axis of the magnetic trap a far detuned, retroreflected laser beam with wavelength λ (see fig. 1). The resulting potential is given by the sum of the magnetic (V_B) and the optical potential (V_{opt}):

$$(1) \qquad V = V_B + V_{\mathrm{opt}} = \frac{1}{2}m\left(\omega_x^2 x^2 + \omega_\perp^2\left(y^2 + z^2\right)\right) + V_0 \cos^2 kx,$$

where m is the atomic mass, $\omega_x = 2\pi \times 9\,$Hz and $\omega_\perp = 2\pi \times 90\,$Hz are the axial and radial frequencies of the magnetic harmonic potential, and $k = 2\pi/\lambda$ is the modulus of the wave vector of the optical lattice. By varying the intensity of the laser beam (detuned typically $\Delta = 150\,$GHz to the blue of the D_1 transition at $\lambda = 795\,$nm) up to $14\,$mW/mm^2 we can

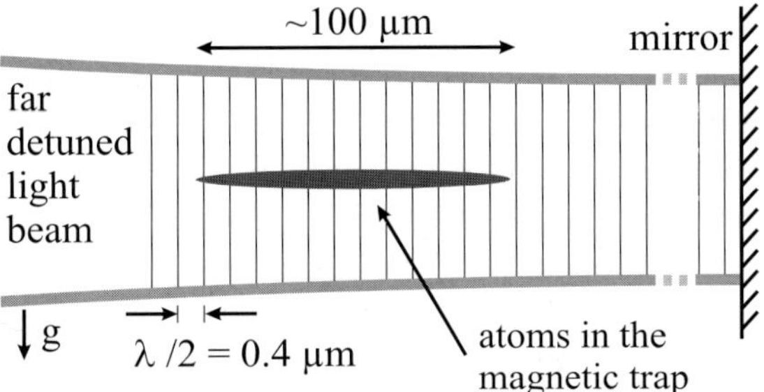

Fig. 1. – Schematic set up of the experiment.

vary the optical-lattice potential height V_0 from 0 to $V_0 \sim 5E_{\mathrm{R}}$, where E_{R} is the recoil energy corresponding to the emission or absorption of one lattice photon, $E_{\mathrm{R}} = \hbar^2 k^2 / 2m$. To calibrate the optical potential, we measure the Rabi frequency of the Bragg transition between the momentum states $-\hbar k$ and $+\hbar k$ induced by the standing wave [19]. Due to the large detuning of the optical lattice, spontaneous scattering can be neglected for the experiments on BEC dynamics which are performed typically on a timescale of $\tau \sim 2\pi/\omega_x$; nevertheless, spontaneous scattering leads to a reduction of the total atom number during the preparation of the BEC.

Bose-Einstein condensates in the combined magnetic trap and optical lattice are prepared by superimposing the optical lattice to the trapping potential already during the last hundreds of ms of the RF-evaporation ramp. Figure 2 shows the expected linear density distribution of the ground state in a weakly binding lattice ($V_0 = 1.5\,E_{\mathrm{R}}$), as obtained by numerical propagation of the Gross-Pitaevskii equation in imaginary time. In the experiment, the density modulation on the length scale of $\lambda/2$ cannot be directly resolved, due to the limited resolution of the imaging system.

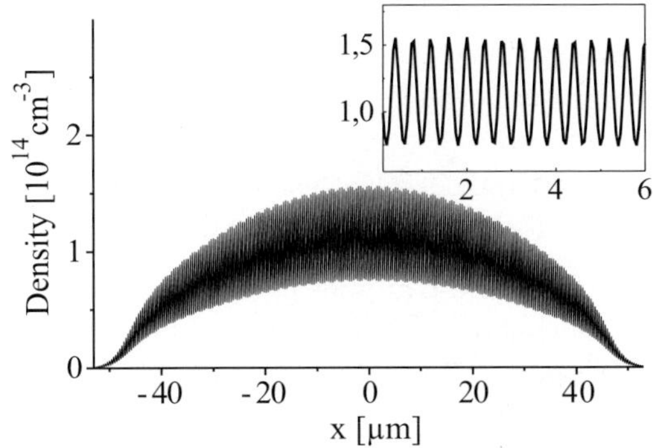

Fig. 2. – Linear density distribution of a BEC in the combined potential of the magnetic trap and optical-lattice potential, obtained from a numerical simulation with the parameters $N = 3 \times 10^5$ and $V_0 = 1.5\,E_{\mathrm{R}}$.

3. – Bose-Einstein phase transition in lower dimensions

In this section we concentrate on the Bose-Einstein phase transition of the atomic gas in the combined potential of the magnetic trap and the optical lattice which allows to identify consequences of the reduced dimensionality of the system [11].

In atomic physics, important steps towards the realization of pure 2D systems of neutral atoms have been made in different systems: Significant fractions of atomic systems could be prepared in the 2D potentials of optical lattices [20, 21] and of an evanescent wave over a glass prism [22], quasi-condensates could be observed in 2D atomic hydrogen trapped on a surface covered with liquid ^{4}He [23], and 3D condensates of ^{23}Na with low atom numbers could be transferred to the 2D regime by an adiabatic deformation of the trapping potential [24].

By using Bose-Einstein condensates trapped in optical lattices it has become possible to overcome major limitations of previous experiments: First, an optical lattice can confine a large array of 2D systems, which allows measurements with a much higher number of involved atoms with respect to a single confining potential. Second, the macroscopic population of a single quantum state (BEC) naturally transfers the whole system to a pure occupation of the 2D systems, which could so far not be realized with thermal atomic clouds.

The dimensionality of a gas of weakly interacting bosons has important consequences for its thermodynamical properties. While in 3D the gas undergoes the phase transition to BEC even in free space, in 2D systems BEC at finite temperatures can only exist in a confining potential [25]. For the ideal gas in a 2D harmonic trap with the fundamental frequency ω the analytical solutions for the condensation temperature, T_c, and the dependence of the condensate fraction N_0/N (number of particles in the ground state, N_0, and total particle number, N), are given by [26]

$$(2) \qquad k_B T_c = \hbar\omega \left(\frac{N}{\zeta(2)} \right)^{1/2}, \qquad \frac{N_0}{N} = 1 - \left(\frac{T}{T_c} \right)^2,$$

where $\zeta(s)$ is the *zeta*-function, defined as $\zeta(s) = \sum_{n=1}^{\infty} 1/n^s$. In the 3D case these dependencies are

$$(3) \qquad k_B T_c = \hbar\omega \left(\frac{N}{\zeta(3)} \right)^{1/3}, \qquad \frac{N_0}{N} = 1 - \left(\frac{T}{T_c} \right)^3.$$

In our experiment, the magnetic-trapping potential is a 3D potential which confines the atoms to an overall cigar-shaped distribution, while in the 1D optical lattice the atoms are confined to 2D planes. Therefore, by increasing the strength of the optical lattice superposed to a 3D potential it is possible to follow the transition from a 3D BEC to an array of 2D degenerate atomic clouds confined radially by the magnetic potential and assorted in the axial direction like disks in a shelf.

For reaching the quasi-2D regime, the motion of the particles has to be effectively "frozen" in the direction of the optical lattice beam [25], *i.e.*, the fundamental frequency in

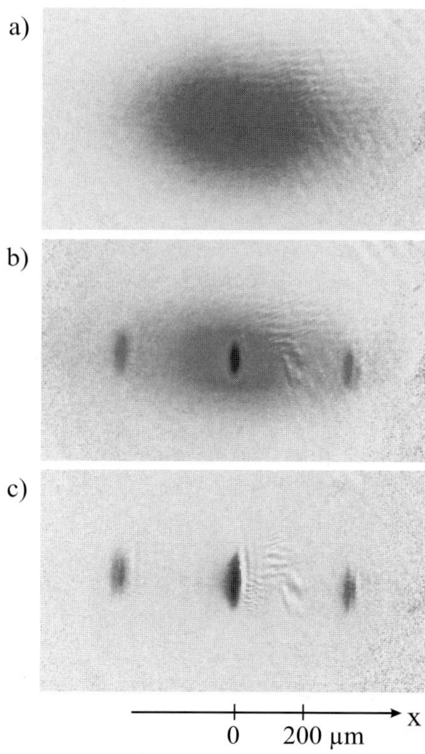

Fig. 3. – Absorption images of a thermal cloud (a), a mixed cloud (b), and a pure BEC (c), expanded for 26.5 ms from the combined magnetic trap and optical lattice ($V_0 \sim 4E_\mathrm{R}$). The corresponding temperatures and atom numbers are $T \approx 210\,\mathrm{nK}$ and $N \approx 4 \times 10^5$ (a), $T \approx 110\,\mathrm{nK}$ and $N \approx 1.5 \times 10^5$ (b), $T < 50\,\mathrm{nK}$ and $N \approx 1.5 \times 10^4$ (c).

a single lattice site, ω_l, has to fulfill $\hbar\omega_l \gg k_\mathrm{B}T$. For our experimental parameters of $T < 200\,\mathrm{nK}$ and $\omega_l \approx 2\pi 14\,\mathrm{kHz}$ (for $V_0 \approx 4\,E_\mathrm{rec}$) this relation is well satisfied. Nevertheless, due to the small width of the barriers, atoms can tunnel between the lattice sites. The low energy of thermal atoms allows them to tunnel only over a few sites during the duration of the experiment. Therefore, we expect only minor changes of the thermodynamic properties due to such processes. In contrast, tunnelling of ground-state atoms is greatly enhanced because the ground state is macroscopically occupied (see sect. **6**). As a result, the BECs at the optical lattice sites form a phase-coherent ensemble giving rise to the interference pattern in the expansion (see sect. **4**).

In order to measure the effects of dimensionality on T_c and on $N_0/N(T)$ we have adjusted the final temperature of the atomic clouds by evaporative cooling and recorded the atomic density distributions at different temperatures. Figure 3 shows absorption images of such ensembles, expanded from a combined trap with a lattice-potential height of $V_0 \approx 4\,E_\mathrm{rec}$. The interference pattern of the expanding array of BECs appears in three spatially separated peaks (in fig. 3b, c) [10].

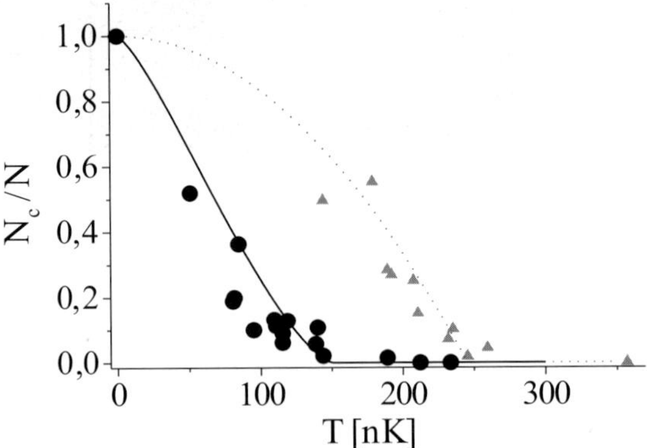

Fig. 4. – Ground-state occupation *vs.* temperature in the combined trap with $V_0 \approx 4\,E_{\text{rec}}$ (circles), and in the 3D purely magnetic trap (triangles). The solid (respectively, dotted) line gives the expected dependence for the combined (respectively, purely magnetic) trap, according to eq. (2) (respectively, eq. (3)).

By integrating over the atomic density distribution, we obtain the number of atoms in the ground state, N_0, and in the thermal cloud, $N_{\text{th}} = N - N_0$. Figure 4 shows the ground-state fraction N_0/N, as a function of the temperature of the ensemble. In the case of the 3D potential of the pure magnetic trap (triangles in fig. 4) this ratio reproduces the shape expected from eq. (3) (dotted line, using a linear fit to the measured function $N(T)$). The shape of the curve for ensembles produced in the combined trap is much smoother around the (lower) transition temperature, and mixed clouds with a relatively small condensate fraction exist in a broad temperature range well below T_c.

This behaviour can be qualitatively understood with a simplifying model [11] assuming the subsequent formation of 2D BECs at the different lattice sites: Due to the magnetic-trapping potential the central lattice wells are populated with a higher number of atoms, which—according to eq. (2)—leads to a higher critical temperature for the central clouds than for the clouds in the wings of the overall density distribution. BECs form first in the central 2D disks, lowering the temperature leads to BEC formation at more and more lattice sites. The solid line in fig. 4 shows the ground-state occupation according to this model; the curve agrees well with the experimental data points. A more sophisticated theoretical modelling of the problem, including interactions and the effect of tunnelling on the thermodynamic properties, remains to be developed.

4. – Expansion of a BEC from the combined trap

In this section we introduce ground-state properties of the fully coherent array of condensates in the optical lattice. To this aim, we explore the interference pattern in the expanded cloud, reflecting the initial geometry of the sample. The time evolution of the interference peaks, their relative population as well as the radial size of the expanding cloud are compared to theoretical results developed in [10].

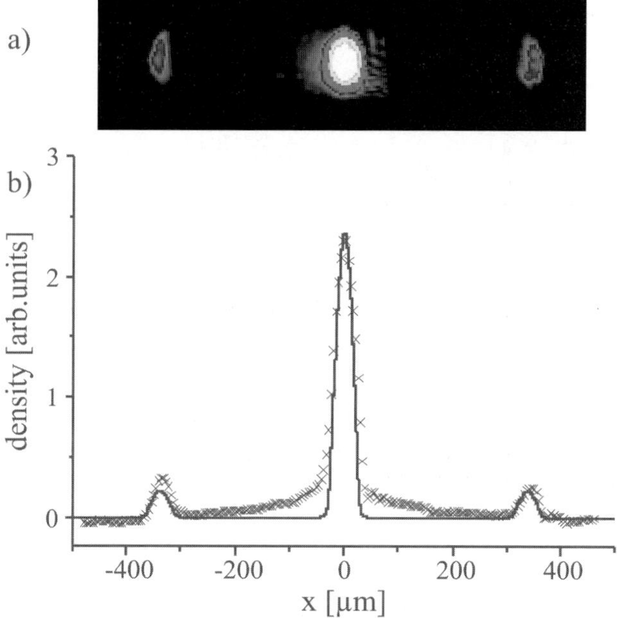

Fig. 5. – a) Absorption image of the expanded array of condensates, showing three interference peaks, $n = -1, 0, 1$. b) Density profile obtained from the absorption image (crosses) and calculated density profile for the experimental parameters $V_0 = 5E_{\mathrm{R}}$ and $t_{\mathrm{exp}} = 29.5\,\mathrm{ms}$ [10]. The wings of the central peak in the experimental density profile result from a small thermal component.

In analogy of multiple order interference fringes in light diffraction from a grating, the expansion of an array of coherent BECs leads for long expansion times (*i.e.*, in the far field) to distinct peaks in position space, reflecting the momentum distribution. The analogy is best understood considering a periodic and coherent array of condensates aligned along the x-axis. The momentum distribution of the whole system is affected in a profound way by the lattice structure and exhibits distinctive interference phenomena (see [10]). The momentum distribution is characterized by sharp peaks at the values $p_x = 2n\hbar k$ with n integer (positive or negative) whose weight is modulated by a function $n_0(p_x)$.

In fig. 5a we show a typical image of the cloud taken at $t_{\mathrm{exp}} = 29.5\,\mathrm{ms}$ with a total number of atoms $N = 20000$ and a potential height of $V_0 = 5E_{\mathrm{R}}$. The structure of the observed density profiles is well reproduced by the free expansion of the ideal gas assuming a periodic ground-state wave function Ψ describing the array of BECs [10]. A predicted result for the density distribution $n(x) = |\Psi(x)|^2$ evaluated for the experimental parameters is shown in fig. 5b (continuous line).

From the experimental images we determine the relative population of the lateral peak with respect to the central one. The results for the relative population of the first

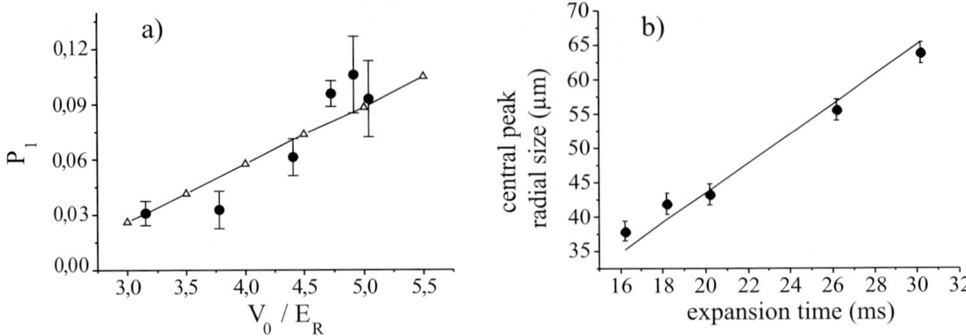

Fig. 6. – a) Experimental (circles) and theoretical values (triangles) of the relative population of the $n = 1$ peak with respect to the central one ($n = 0$) as a function of the lattice potential V_0 in units of the recoil energy E_R. b) Radial size of the central peak as a function of the expansion time. Experimental data points are compared to the expected asymptotic law $R_\perp(t) = R_\perp \omega_\perp t$.

lateral peak as a function of the optical-lattice potential V_0 are shown in fig. 6a. This figure also shows the corresponding results of the 1D theoretical model developed in [10].

More generally, it can be shown that the relative population of the $n \neq 0$ peaks with respect to the central one ($n = 0$) obeys the simple law

$$(4) \qquad P_n = \exp\left[\frac{-4\pi^2 n^2 \sigma^2}{d^2}\right],$$

where σ characterizes the width of a single condensate in a lattice site and d is the periodicity of the lattice, $d = \lambda/2$. Equation (4) holds also in the presence of a smooth modulation of the atomic occupation number N_k in each well. Result (4) shows that, if σ is much smaller than d, the intensity of the lateral peaks will be high, with a consequent important layered structure in the density distribution of the expanding cloud. The value of σ is determined, in first approximation, by the optical confinement. It can be obtained numerically giving $\sigma/d = 0.30$, 0.27, and 0.25 for $V_0 = 3$, 4 and $5\,E_R$, respectively [10].

A 3D model permits also to explain the behaviour of the radial expansion of the gas. In the presence of the density oscillations produced by the optical lattice the problem is not trivial and should be solved numerically by integrating the GP equation. However, after the lateral peaks are formed, the density of the central peak expands smoothly according to the asymptotic law $R_\perp(t) = R_\perp(0)\omega_\perp t_{\text{exp}}$, holding for a cigar configuration in the absence of the optical lattice [27]. As can be seen from fig. 6b, the experimental values of the radial size of the peak after varied expansion time compare well to this law [10].

Concluding, we remark that the well-understood behaviour of the expanding cloud can be used as a tool to investigate the coherence properties of the BEC. Deviations from the shape of the interference pattern and from the population of the interference peaks can be interpreted to result, *e.g.*, from decoherence effects. In recent experiments, the interference peaks have been used to prove the well-defined phase relation in an array of Josephson junctions [9]. The disappearance of the interference peaks has been used to study squeezing of matter waves [12]. Further studies include the possible effects of thermal decoherence in the presence of tighter optical traps [28, 29].

5. – Superfluid dynamics

Superfluidity of BECs is a direct consequence of their coherent nature [30]. It is manifested in the appearance of vortices [31, 32] and scissors modes [33] as well as in a critical velocity for the onset of dissipative processes [34]. A far-detuned optical lattice at a low potential height, $V_0 < 2\,E_R$, is well suited to study in detail the critical velocity because it acts like a medium with a microscopic roughness on the BEC moving through it, being velocity-dependent compressed and decompressed as it propagates.

In order to investigate the dynamics in the combined trap we translate the magnetic-trapping potential in the x-direction by a variable distance Δx ranging up to $300\,\mu$m in a time $t \ll 2\pi/\omega_x$. Therefore, the BEC finds itself out of equilibrium and is forced into motion by a potential gradient. After an evolution time $t_{\rm ev}$ in the displaced trap, both the magnetic trapping and the optical lattice are switched off and after a free expansion of $26.5\,$ms the cloud is imaged, giving information on the momentum distribution of the system.

In the magnetic trap the center-of-mass motion of the BEC in the displaced trap is an undamped oscillation with frequency $\omega_x = 2\pi \times 8.7\,$Hz and amplitude Δx. In the combined trap formed by the magnetic and the (weakly confining, $V_0 \sim 1.5\,E_R$) optical-lattice potentials we observe dynamics in different regimes: For small displacements, $\Delta x < 50\,\mu$m, the dynamics of the BEC resembles the "free oscillation" at the same amplitude but with a significant shift in frequency which can be explained in terms of an effective atomic mass [8]. By varying the potential height V_0 we are able to tune this effective mass. The undamped dynamics without dissipative processes in the small-amplitude regime is a manifestation of superfluid behavior of the BEC. When we further increase the initial displacement Δx and hence the velocity of the BEC, it enters a regime of dissipative dynamics. We observe a damped oscillation in the trap and dissipative processes heating the cloud.

The critical velocity in a superfluid is proportional to the local speed of sound, $c_{\rm s}$, which depends on the density n, $c_{\rm s}(r) = \sqrt{n(r)/m\,(\delta\mu/\delta n)}$, with the chemical potential μ. Therefore superfluidity breaks down first in the wings of the BEC where the density is lowest.

In order to measure the velocity- and density-dependent onset of dissipation and thereby the spectrum of critical velocities in the BEC, we have varied the displacement Δx and recorded atomic distributions after a fixed evolution time $t_{\rm ev} = 40\,$ms. For low velocities, $v < 2\,$mm/s, the sample follows the position of a freely moving BEC ("lattice off" in fig. 7); no thermal component appears.

Upon increasing the velocity of the BEC, we observe a retardation of a part of the cloud, leading to a well-detectable separation from the superfluid component after free

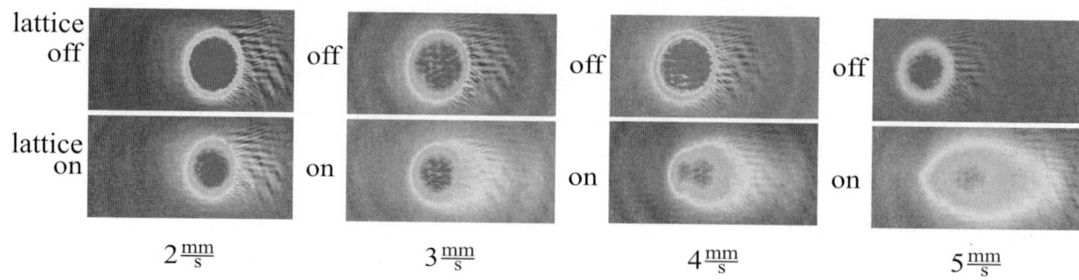

Fig. 7. – Grey scale representation of absorption images of atomic clouds after evolution with different maximum velocities (indicated) in the magnetic trap ("lattice off") and in the combined trap ("lattice on").

evolution (see fig. 7). For velocities $v \sim 4\,\mathrm{mm/s}$ we observe that only the central part of the fluid is moving without retardation. For velocities $v \sim 4\,\mathrm{mm/s}$ we observe that only a part of the fluid is moving without retardation; the central position of this part is the same as the central position of the "freely oscillating" BEC. For even higher velocities all of the atoms are retarded and form a heated cloud with a Gaussian density distribution. The spatial separation from the thermal component allows a clear demonstration of the superfluid properties of inhomogeneous Bose-Einstein condensates.

Figure 8 shows the ratio of atom number in the non-retarded component (parabolic density profile, "superfluid component"), N_s, and the total atom number, N, as a function of the maximum velocity attained during the evolution in the optical lattice [35]. The envelope function of the density distribution of the BEC is an inverted parabola in 3D (see fig. 2) and hence, by integration over the high-density region, we get an equation for the relative number of atoms in the superfluid part of the BEC for a given velocity v, $N_\mathrm{s}(v)/N = [5/2 \times (1 - v^2/v_{\mathrm{max}}^2)^{3/2} - 3/2 \times (1 - v^2/v_{\mathrm{max}}^2)^{5/2}]$, where v_{max} is the critical velocity at maximum density. This expression implies that about 90% of the atomic

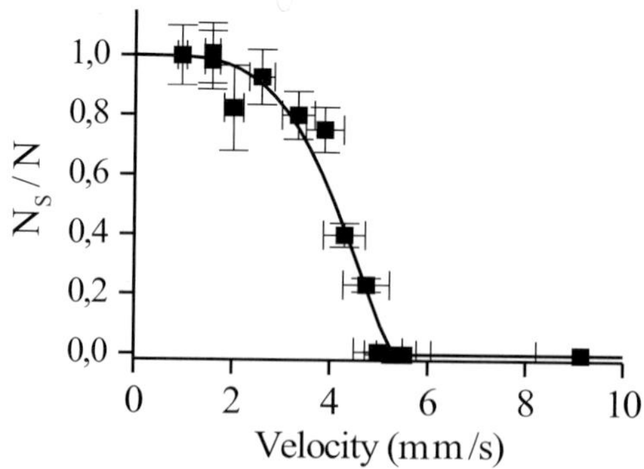

Fig. 8. – Fraction of atoms in the superfluid component *vs.* maximum propagation velocity.

probability density is localized in a region which remains superfluid up to velocities $v \simeq v_{\max}/2$. The line in fig. 8 shows that the above expression for $N_{\rm s}(v)/N$ gives a very good account of the data, the fitted value of the maximum velocity being $v_{\max} = (5.3 \pm 0.5)\,{\rm mm/s}$.

In the regime of a weak optical-lattice potential, the lattice acts with a rather small perturbation on the BEC, and the atoms are completely delocalized within the macroscopic BEC wave function. However, by increasing the potential we reach a regime where the atoms are confined to single lattice sites. Here, the ensemble is better described in terms of an array of discrete BECs, connected to each other by tunnelling. This regime is addressed in the following section.

6. – Observation of a Josephson current in an array of coupled BECs

Two macroscopic quantum systems which are coupled by a weak link produce the flow of a supercurrent I between them, driven by their relative phase $\Delta\phi$,

$$(5) \qquad\qquad I = I_{\rm c} \sin \Delta\phi\,,$$

where $I_{\rm c}$ is the critical Josephson current [36, 37]. The relative phase evolves in time proportionally to the difference in chemical potential between the two quantum fluids.

The first experimental evidence of a current-phase relation was already obtained in superconducting systems soon after Josephson's proposal [38]. Also, phase-coherent tunnelling of BEC atoms from an optical lattice to the continuum, driven by gravity, has been observed [7]. We realize a one-dimensional array of bosonic Josephson junctions (JJs) by preparing an array of BECs in the sites of the optical lattice with an interwell barrier energy V_0 which is high compared to the chemical potential of the BECs [9]. Every two condensates in neighbouring wells overlap slightly with each other due to a finite tunnelling probability, and therefore constitute a JJ, with the possibility to adjust the critical current $I_{\rm c}$ by tuning the laser intensity. By driving the system with the external harmonic potential, we investigate the current-phase dynamics and measure the critical Josephson current as a function of the interwell potential V_0.

In its ground state the system consists of spatially separated condensates; tunneling between adjacent wells leads to a constant phase over the whole array. Therefore, the condensates show an interference pattern after an expansion from the combined trap (see fig. 5).

To observe a Josephson current in the array we non-adiabatically displace the magnetic trap along the lattice axis by a distance of $\sim 30\,\mu{\rm m}$. The potential energy the atoms gain in this process is much smaller than the interwell potential barrier, but the relative phases of the BECs in the different wells are driven by this process. According to eq. (5) we expect a Josephson current. A collective motion can be established only with a well-defined phase relation between the condensates. This locking of the relative phases shows up in the expanded cloud interferogram. The expected Josephson current is observed as the collective oscillation of the atomic ensemble. In fig. 9a the positions

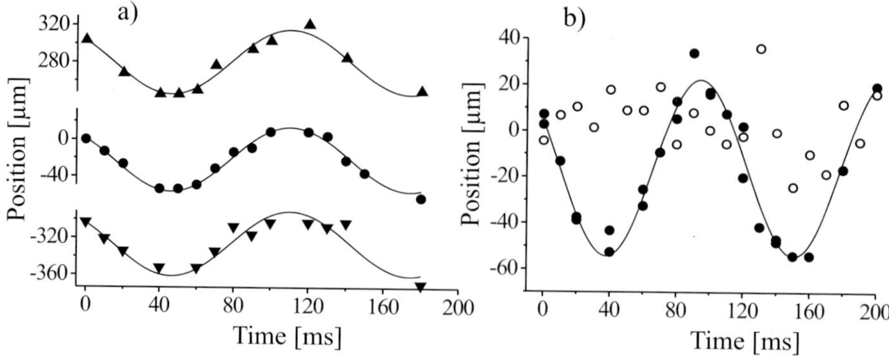

Fig. 9. – a) The three peaks of the interference pattern of an array BECs, expanded for 28 ms after the propagation in the optical lattice. b) Center-of-mass positions of thermal clouds after expansion from the magnetic trap (filled circles) and from the combined magnetic trap and optical lattice (open circles) as a function of evolution time in the displaced respective trap.

of the three peaks in the interferogram are plotted as a function of time spent in the combined trap after the displacement of the magnetic trap. The motion performed by the center of mass of the ensemble is an undamped oscillation at a frequency $\omega < \omega_x$.

The coherent nature of the oscillation is also proven by repeating the same experiment with a thermal cloud. In this case—although atoms can individually tunnel through the barriers—no macroscopic phase is present in the cloud and no motion of the center of mass is observed. The center-of-mass positions of thermal clouds in the optical lattice are shown in fig. 9b, together with the oscillation of thermal clouds in the absence of the optical potential. As can clearly be seen, the movement of thermal clouds is strongly suppressed in the presence of the optical lattice. We have also subjected mixed clouds to the displaced potential, where only the condensate fraction starts to oscillate while the thermal component remains static; the interaction of the two eventually leads to

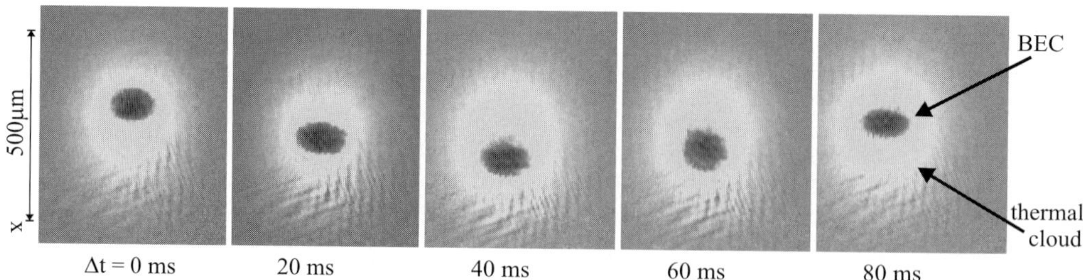

Fig. 10. – Grey scale representation of absorption images of mixed atomic clouds in the displaced magnetic trap with superimposed optical lattice. In the time evolution Δt the ground state is moving relative to the thermal cloud.

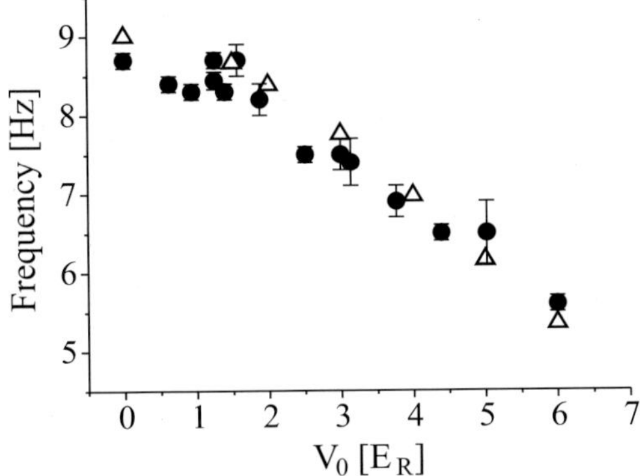

Fig. 11. – Frequency of the oscillation of a coherent ensemble in the JJ array as a function of the interwell potential height. Filled circles: experimental data, open triangles: results from a numerical simulation of the Gross-Pitaevskii equation [39].

a damping of the condensate motion and a heating of the system. Figure 10 shows absorption images of mixed clouds, where a BEC moves relative to the thermal cloud.

As can be derived from the phase-current relation of the JJ array [9], the critical Josephson current is related to the small amplitude oscillation frequency ω of the JJ array by the simple relation $I_c = (4\hbar/m\lambda^2)(\omega/\omega_x)^2$. Figure 11 shows experimental values of the oscillation frequency ω together with results from numerical solutions of the Gross-Pitaevskii equation. The possibility to precisely adjust the critical Josephson current presents a major advantage of Josephson junctions in Bose-Einstein condensates, where due to the elaborate manipulation tools of atomic physics a variety of parameters can be tuned, compared to systems realized in solid-state physics.

7. – Conclusions

In this lecture we have discussed thermodynamical and ground-state properties of a dilute gas in the potential of an optical lattice. We have also investigated macroscopic quantum effects in the dynamics of the system, such as superfluid motion, a density-dependent critical velocity for the onset of dissipation, and—in a regime of higher lattice potentials—an oscillating Josephson current.

Future directions of this work include the study of BECs in 2D and 3D optical lattices and in superlattices, which offers the possibility of a better control of the atom number per lattice site and of the manipulation of single lattice sites, and possible applications of quantum-computing schemes. We also plan to further study BEC properties in these systems, such as the dynamical behaviour of BECs in lower dimensions or the formation of bright atomic solitons in optical lattices [40].

* * *

Our understanding of the various experiments performed at LENS benefitted very much from the theoretical support in collaborations with M. L. CHIOFALO, P. PEDRI, L. PITAEVSKII, A. SMERZI, S. STRINGARI, M. TOSI, and A. TROMBETTONI. We acknowledge stimulating discussions with M. ARTONI, G. FERRARI, and G. V. SHLYAPNIKOV. We also acknowledge support by the EU under contracts HPRI-CT 1999-00111 & HPRN-CT-2000-00125, by MURST through the PRIN 1999 & PRIN 2000 Initiatives, and by INFM under the contract *"Photon Matter"*.

REFERENCES

[1] INGUSCIO M. *et al.* (Editors), *Bose-Einstein Condensation in Atomic Gases* (IOS Press, Amsterdam) 1999.

[2] MARTELLUCCI S. *et al.* (Editors), *Bose-Einstein Condensates and Atom Lasers* (Kluwer, New York) 2000.

[3] JAKSCH D., BRUDER C., CIRAC J. I., GARDINER C. W. and ZOLLER P., *Phys. Rev. Lett.*, **81** (1998) 3108

[4] BRENNEN G. K., CAVES C. M., JESSEN P. S. and DEUTSCH I. H., *Phys. Rev. Lett.*, **82** (1999) 1060.

[5] DAHAN M. B., PEIK E., REICHEL J., CASTIN Y. and SALOMON C., *Phys. Rev. Lett.*, **76** (1996) 4508.

[6] WILKINSON S. R., BHARUCHA C. F., MADISON K. W., NIU Q. and RAIZEN M. G., *Phys. Rev. Lett.*, **76** (1996) 4512.

[7] ANDERSON B. P. and KASEVICH M. A,. *Science*, **282** (1998) 1686.

[8] BURGER S., CATALIOTTI F. S., FORT C., MINARDI F., INGUSCIO M., CHIOFALO M. L. and TOSI M., *Phys. Rev. Lett.*, **86** (2001) 4447.

[9] CATALIOTTI F. S., BURGER S., FORT C., MADDALONI P., MINARDI F., INGUSCIO M., TROMBETTONI A. and SMERZI A., *Science*, **293** (2001) 843.

[10] PEDRI P., PITAEVSKII L., STRINGARI S., FORT C., BURGER S., CATALIOTTI F. S., MADDALONI P., MINARDI F. and INGUSCIO M., *Phys. Rev. Lett.*, **87** (2001) 220401.

[11] BURGER S., CATALIOTTI F. S., FORT C., MADDALONI P., MINARDI F. and INGUSCIO M., *Europhys. Lett.*, **57** (2002) 1.

[12] ORZEL C., TUCHMAN A. K., FENSELAU M. L., YASUDA M. and KASEVICH M. A., *Science*, **291** (2001) 2386.

[13] GREINER M., BLOCH I., MANDEL O., HÄNSCH T. W. and ESSLINGER T., *Phys. Rev. Lett.*, **87** (2001) 160405.

[14] KOZUMA M., DENG L., HAGLEY E. W., WEN J., LUTWAK R., HELMERSON K., ROLSTON S. L. and PHILLIPS W. D., *Phys. Rev. Lett.*, **82** (1999) 871.

[15] STENGER J., INOUYE S., CHIKKATUR A. P., STAMPER-KURN D. M., PRITCHARD D. E. and KETTERLE W., *Phys. Rev. Lett.*, **82** (1999) 4569.

[16] MORSCH O., MÜLLER J. H., CRISTIANI M., CIAMPINI D. and ARIMONDO E., *Phys. Rev. Lett.*, **87** (2001) 140402.

[17] HENSINGER W. K., HÄFFNER H., BROWAEYS A., HECKENBERG N. R., HELMERSON K., MCKENZIE C., MILBURN G. J., PHILLIPS W. D., ROLSTON S. L., RUBINSZTEIN-DUNLOP H. and UPCROFT B., *Nature*, **412** (2001) 52.

[18] FORT C., PREVEDELLI M., MINARDI F., CATALIOTTI F. S., RICCI L., TINO G. M. and INGUSCIO M., *Europhys. Lett.*, **49** (2000) 8.

     161

[19] PEIK E., DAHAN M. B., BOUCHOULE I., CASTIN Y. and SALOMON C., *Phys. Rev. A*, **55** (1997) 2989.

[20] VULETIĆ V., CHIN C., KERMAN A. J. and CHU S., *Phys. Rev. Lett.*, **81** (1998) 5768.

[21] BOUCHOULE I., MORINAGA M., PETROV D. S. and SALOMON C., e-print (2001) quant-ph/0106032.

[22] GAUCK H., HARTL M., SCHNEBLE D., SCHNITZLER H., PFAU T. and MLYNEK J., *Phys. Rev. Lett.*, **81** (1998) 5298.

[23] SAFONOV A. I., VASILYEV S. A., YASNIKOV I. S., LUKASHEVICH I. I. and JAAKKOLA S., *Phys. Rev. Lett.*, **81** (1998) 4545.

[24] GÖRLITZ A., VOGELS J. M., LEANHARDT A. E., RAMAN C., GUSTAVSON T. L., ABO-SHAEER J. R., CHIKKATUR A. P., GUPTA S., INOUYE S., ROSENBAND T. and KETTERLE W., *Phys. Rev. Lett.*, **87** (2001) 130402.

[25] PETROV D. S., HOLZMANN M. and SHLYAPNIKOV G. V., *Phys. Rev. Lett.*, **84** (2000) 2551.

[26] BAGNATO V. and KLEPPNER D., *Phys. Rev. A*, **44** (1991) 7439.

[27] CASTIN Y. and DUM R., *Phys. Rev. Lett.*, **77** (1996) 5315.

[28] PITAEVSKII L. and STRINGARI S., *Phys. Rev. Lett.*, **87** (2001) 180402.

[29] CUCCOLI A., FUBINI A., TOGNETTI V. and VAIA R., e-print (2001) cond-mat/0107387.

[30] DALFOVO F., GIORGINI S., PITAEVSKII L. P. and STRINGARI S., *Rev. Mod. Phys.*, **71** (1999) 463.

[31] MATTHEWS M. R., ANDERSON B. P., HALJAN P. C., HALL D. S., WIEMAN C. E. and CORNELL E. A., *Phys. Rev. Lett.*, **83** (1999) 2498.

[32] MADISON K. W., CHEVY F., WOHLLEBEN W. and DALIBARD J., *Phys. Rev. Lett.*, **84** (2000) 806.

[33] MARAGÓ O. M., HOPKINS S. A., ARLT J., HODBY E., HECKENBLAIKNER G. and FOOT C. J., *Phys. Rev. Lett.*, **84** (2000) 2056.

[34] RAMAN C., KÖHL M., ONOFRIO R., DURFEE D. S., KUKLEWICZ C. E., HADZIBABIC Z. and KETTERLE W., *Phys. Rev. Lett.*, **83** (1999) 2502.

[35] Due to spontaneous scattering during the preparation of the BEC in the combined trap the total atom number is reduced by a factor ~ 0.7.

[36] BARONE A. and PATERNO G., *Physics and Applications of the Josephson Effect* (Wiley, New York) 1982.

[37] SMERZI A., FANTONI S., GIOVANNAZZI S. and SHENOY S. R., *Phys. Rev. Lett.*, **79** (1997) 4950.

[38] JOSEPHSON B. D., *Phys. Lett.*, **1** (1962) 251.

[39] TROMBETTONI A., PhD-thesis, SISSA, Trieste.

[40] ZOBAY O., PÖTTING S., MEYSTRE P. and WRIGHT E. M., *Phys. Rev. A*, **59** (1999) 643.

IONS

Trapped ions and quantum information processing

D. J. Wineland

National Institute of Standards and Technology - Boulder, CO 80305-3328, USA

1. – Introduction

Efforts to realize experimentally the elements of quantum computation (QC) using trapped atomic ions have been stimulated in large part by a 1995 paper by Cirac and Zoller [1]. In this scheme, ions confined in a linear RF (Paul) trap are cooled to a point where they are strongly coupled through their Coulomb interaction and form a stable spatial array. In this condition, the ions' motion is best described by normal modes. Two stable or metastable internal levels in each ion form a qubit. For typical experimental conditions, ion spacings are large enough ($> 1\,\mu$m) that the direct coupling of internal states of separate ions is negligible, thereby precluding logic gates based on internal-state interactions. Cirac and Zoller [1] suggested cooling the ions to their motional ground state and then using the ground and first excited state of a particular motional mode as a qubit (hereafter called the motion qubit). For most motional modes, every ion has nonzero amplitude; therefore, the mode can act as a data bus to transfer information between ions. For example, to make a universal logic gate between qubits formed from the internal states of two ions (hereafter referred to as ion qubits), the superposition state of a particular ion qubit is first mapped onto the selected motion qubit by using a laser beam focused onto that ion. After this step, a gate operation is performed between the motion qubit and a second ion qubit, also selected by laser beam focusing. Finally, the motion qubit state is mapped back onto the first ion qubit resulting in a gate between the first and second ion qubits. Being able to perform (universal) gates between any two selected ion qubits coupled with the ability to perform single-ion qubit rotations provides the basis for general quantum computation [2]. The ion trap scheme satisfies the main requirements for a quantum computer as outlined by DiVincenzo [3]: 1) a scalable system

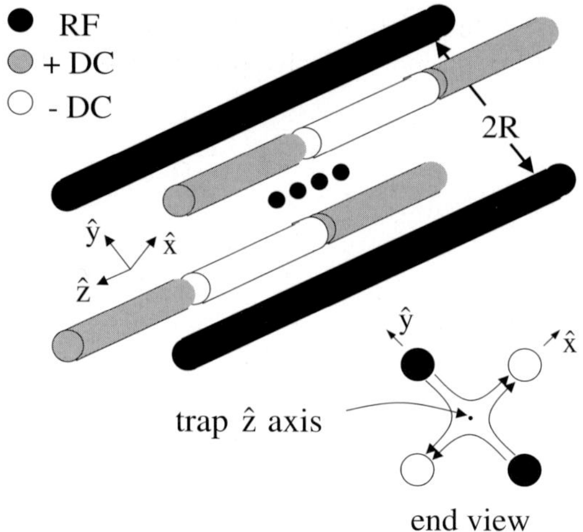

Fig. 1. – Electrode configuration for a linear RF (Paul) trap. A common RF potential $V_0 \cos(\Omega_T t)$ is applied to the dark electrodes; the other electrodes are held at RF ground through capacitors (not shown) connected to ground. The lower-right portion of the figure shows the x, y electric fields from the applied RF potential at an instant when the RF potential is positive relative to ground. A static electric potential well is created (for positive ions) along the z-axis by applying a positive potential to the outer segments (grey) relative to the center segments (white).

of well-defined qubits, 2) a method to reliably initialize the quantum system, 3) long coherence times, 4) existence of universal gates, and 5) an efficient measurement scheme. Because of this, it is studied in several laboratories. Below, we briefly describe single-ion qubit rotations and gate operations realized in ion traps. We focus on experiments carried out at NIST but note that similar work is being carried out at several other laboratories including Aarhus, Almaden (IBM), Hamburg, Innsbruck, Los Alamos (LANL), Michigan, Munich (MPI-Garching), and Oxford.

Although a few operations can now be strung together, these experiments and those at other laboratories are currently limited by various sources of decoherence. Decoherence can result from rather mundane causes, such as fluctuating magnetic fields, or from sources more difficult to control such as spontaneous emission. Studies of decoherence are also of enduring fundamental interest because of the contradictions between what quantum mechanics predicts and what we observe in our classical world [4, 5]. Even if we avert our eyes from these fundamental issues, decoherence is "the" important problem in quantum information processing. Although it appears that there is no fundamental reason why decoherence cannot be sufficiently reduced in order to make a useful quantum computer, it will clearly occupy most of our attention as the field progresses.

2. – Trapology

Ion traps come in various forms [6]; for brevity, we restrict our discussion to the linear RF (Paul) trap shown schematically in fig. 1. This trap is essentially a quadrupole mass filter plugged on the axis by superimposing a static potential well. In the (x, y)-plane of the figure, ions are bound by a ponderomotive pseudopotential

$$(1) \qquad U_{x,y}(\mathbf{r}) = \frac{q^2}{2m\Omega_T^2} \langle E^2(\mathbf{r}) \rangle \simeq \frac{q^2 V_0^2}{4m\Omega_T^2 R^4}(x^2 + y^2),$$

where q is the ion's charge, m its mass, E is the RF electric field (resulting from a potential $V_0 \cos(\Omega_T t)$ applied to the dark electrodes of fig. 1), $\mathbf{r}$ is the radial distance from the trap axis, and R is equal to the distance between the trap axis and the nearest electrode surface. The oscillation frequency of an ion in this pseudopotential is given by

$$(2) \qquad \omega_{x,y} \simeq \frac{qV_0}{\sqrt{2}\Omega_T m R^2},$$

where we assume the pseudopotential approximation ($\omega_{x,y} \ll \Omega_T$) and assume that $\omega_z \ll \omega_{x,y}$ so that the static radial forces are much smaller than the pseudopotential forces. From the form of eq. (1), we see that the ions seek the region of minimum $|E(\mathbf{r})|$. Typical conditions in the NIST ^{9}Be$^+$ experiments are: $q = 1e$, $V_0 = 500\,\text{V}$, $\Omega_T/2\pi \simeq 100\,\text{MHz}$, $R = 200\,\mu\text{m}$, so that $\omega_{x,y}/2\pi \simeq 24\,\text{MHz}$. Note that optical dipole traps work similarly. In this case, Ω_T is the optical frequency and the (spatially inhomogeneous) laser electric field traps the electron to which the atomic core is attached. However, since the electron response is dispersive about its resonance frequency, atoms in optical traps seek minimum $|E(\mathbf{r})|$ for laser tuning blue of the electron's resonance and maximum $|E(\mathbf{r})|$ for red detuning.

For the purposes of quantum computing, to a good approximation, we can view the linear trap as providing a 3D harmonic well where the strength of the well in two directions (x and y in fig. 1) is much stronger than in the third direction (z). When a small number of ions is trapped and cooled, each ion seeks the bottom of the trap well, but the mutual Coulomb repulsion between ions results in an equilibrium configuration in form of a linear array, like beads on a string. To give an idea of array size, two ions in such a trap are spaced by $2^{1/3}s$; three ions are spaced by $(5/4)^{1/3}s$, where $s \equiv q^2/(4\pi\epsilon_0 m\omega_z^2)^{1/3}$. Equivalently, for singly charged ions, the spacing parameter in μm is $s\,(\mu\text{m}) = 15.2(M(u)\nu_z^2(\text{MHz}))^{-1/3}$, where the ion mass is expressed in a.m.u. and the axial z-frequency in MHz. For $\nu_z = 5\,\text{MHz}$, two ^{9}Be$^+$ ions are separated by $3.15\,\mu\text{m}$. The basic operation of linear RF traps is described in ref. [6]; these traps with particular application to QC are also discussed in refs. [7-10].

2$^{\cdot}$1. *RF micromotion*. – Although the linear trap can be viewed as providing a 3D harmonic well, in practice, we must also pay attention to the ions' "RF-micromotion". Classically, the micromotion is the forced oscillatory motion at the trap drive frequency Ω_T when the ion is displaced from the trap axis. When the confinement in the (x, y)-plane is dominated by ponderomotive forces, the micromotion amplitude is small and approximately equal to $\sqrt{2}\omega_{x,y}/\Omega_T$ times the displacement of the ion from the trap axis. In a quantum-mechanical treatment of the motion, the wave packet breathes at frequency

Ω_T with an amplitude related to its displacement from the axis by approximately the same ratio (see, for example, refs. [11, 12]). Since typically, $\omega_{x,y} \ll \Omega_T$, we can usually neglect these effects and treat the "secular" motion (the motion in the ponderomotive well) as a quantized oscillator in a static harmonic well of frequency $\omega_{x,y}$.

One effect of the RF micromotion is to smear out the ion's wave function which, in turn, reduces the interaction with optical radiation. This latter effect can be compensated by increasing the radiation intensity. A more serious issue concerning micromotion is the effect of static electric fields which displace the mean position of an ion from the trap axis. These static fields might arise from the Coulomb repulsion between ions if, for example, the ions are forced into a zig-zag pattern around the z-axis by an axial potential that is too strong compared to the x, y potential [13]. However, we usually operate with an x, y potential strong enough to confine ions on the axis; in fact, the primary reason for choosing the linear trap for QC is to minimize the effects of micromotion. A more insidious source of micromotion is displacement from the trap axis caused by stray static electric fields. These fields can shift the x, y trap potential minimum to a location away from the z-axis where the force from the stray static field must be compensated by the ponderomotive force. In this case, a steady RF micromotion ensues even when the ions' secular motion is cold. If the micromotion amplitude is too large, it can significantly alter an ion's absorption spectrum with important consequences for laser cooling [14, 15]. In principle, the extra micromotion should not cause heating if the stray fields are uniform over the entire ion sample since the common mode RF micromotion does not resonantly couple to the secular motion. However, in practice, RF heating is a problem with multiple trapped ions so it is important to compensate these fields. This is probably related to the RF heating mediated by the nonlinear Coulomb interaction which has been extensively studied at Garching (MPI) and Almaden (IBM) for two ions in a spherical Paul trap; for a summary, see ref. [16].

Often, the origin of these stray fields is not known in specific experiments. They might be caused by potentials on surfaces outside of the immediate trap area. They might also originate from contact potential (work function) variations on trap electrode surfaces caused by different crystal planes, extraneous material adsorbed on the electrode surfaces, or perhaps by electrons which reside on insulating layers attached to the electrode surfaces. The adverse effects of adsorbed layers and electrons can be mitigated by the use of heaters in the electrodes [17, 18]; however, it is usually necessary to compensate for these stray fields by applying static correction potentials to the trap electrodes or auxiliary electrodes outside of the trap structure.

Different methods can be used to sense micromotion; for a summary see, for example, ref. [19]. A common method is to directly sense the micromotion through its effect on (near-resonant) laser light scattering on an allowed transition, usually taken to be the Doppler-cooling transition. Two regimes are important to distinguish, depending on the ratio γ/Ω_T, where γ is the radiative linewidth of the allowed transition. When $\gamma/\Omega_T \gg 1$, the micromotion adiabatically frequency-modulates the ions' absorption profile through the micromotion-induced first-order Doppler shift. By tuning a laser beam to the side of the absorption profile, the fluorescence is modulated at Ω_T. By minimizing this modulation with the additional correction potentials, the component of micromotion along the direction of the laser beam is minimized. When $\gamma/\Omega_T \ll 1$, the absorption spectrum develops discrete RF sidebands on both sides of the carrier; by minimizing these, the component of micromotion along the direction of the laser beam is minimized. Alternatively, for $\gamma/\Omega_T \ll 1$, the time-averaged strength of the carrier fluorescence is a good indicator of the micromotion. For laser intensities below saturation, this fluorescence is proportional to $J_0^2(k_i x_{\mathrm{RF}})$, where J_0 is the zeroth-order Bessel function, k_i is the component of the laser beam k-vector along the direction of motion, and x_{RF} is the amplitude of RF micromotion. The condition for RF micromotion nulling is readily identified since the maximum value of $J_0^2 = 1$ can be distinguished from the first auxiliary maximum where $J_0^2 \simeq 0.16$ for $k_i x_{\mathrm{RF}} \simeq 3.83$. For any value of γ/Ω_T, the motion must be sensed with three non-coplanar laser beam directions to insure the absence of micromotion in all directions.

Not all consequences of RF micromotion are bad. The $J_0^2(k x_{\mathrm{RF}})$ suppression noted above can actually be used as a form of differential selection of two-ion qubits when the ions are so close together that ion selection by laser-beam focusing is precluded. This selection method has been used to entangle two-ion qubits [20] and might be used in more general schemes of quantum computation [21].

3. – Ion entanglement

The key entangling operation in the Cirac/Zoller scheme for QC is one that couples an ion qubit with the motion qubit. For simplicity, assume that the ion qubit has a ground state (labeled $|\downarrow\rangle$) and a higher metastable state (labeled $|\uparrow\rangle$) which are separated in energy by $\hbar\omega_0$. We assume transitions between these levels can be excited with a laser beam. The interaction between an ion and the electric field of the laser beam can be written as

$$(3) \qquad H_{\mathrm{I}}(t) = -\widetilde{d}E = -\widetilde{d}E_0 \cos(k\widetilde{z} - \omega_{\mathrm{L}}t + \phi),$$

where $\widetilde{d}$ is the electric dipole operator, $\widetilde{z}$ is the ion position operator, k is the laser beam k-vector (taken to be parallel to $\hat{z}$), ω_{L} is the laser frequency, and ϕ is the laser field phase at the position of the ion. E is the laser electric field which is assumed to be classical. The dipole operator $\widetilde{d}$ is proportional to $\sigma^+ + \sigma^-$, where $\sigma^+ \equiv |\uparrow\rangle\langle\downarrow|, \sigma^- \equiv |\downarrow\rangle\langle\uparrow|$, and $\widetilde{z} \propto a + a^\dagger$, where a and $a^\dagger$ are the lowering and raising operators for the harmonic oscillator of the selected motional mode which has frequency ω_z. (Here, we assume all other z-modes are cooled to and remain in their ground states and for simplicity have neglected them in $\widetilde{z}$.) In the Lamb-Dicke limit, where the extent of the ion's motion is much less than $\lambda/2\pi = 1/k$, in the interaction frame, and making the rotating wave approximation [10], we can write eq. (3) as

$$(4) \qquad H_{\mathrm{I}} \simeq \hbar\Omega\sigma^+ e^{-i(\omega_{\mathrm{L}}-\omega_0)t+\phi}[1 + i\eta(ae^{-i\omega_z t} + a^\dagger e^{i\omega_z t})] + \mathrm{h.c.}$$

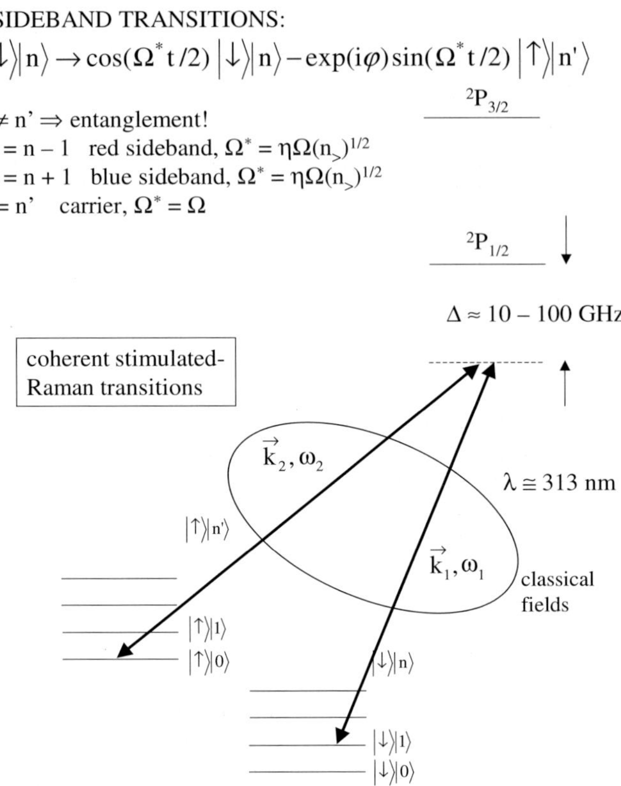

Fig. 2. – Entanglement of internal and motional quantum states using coherent stimulated Raman transitions. The figure indicates the situation for $^9\mathrm{Be}^+$ ions where $|\downarrow\rangle = |F = 2, m_F = \pm 2\rangle$ and $|\uparrow\rangle = |F = 1, m_F = \pm 1\rangle$. The Lamb-Dicke limit is assumed. The laser beams are applied for time t. When driving sideband transitions, $\Omega^* = \eta\Omega\sqrt{n_>}$, where $n_>$ is the larger of n or n'. When driving carrier transitions ($n = n'$), $\Omega^* = \Omega$.

Here, $\Omega \equiv Ed_{\downarrow\uparrow}/(2\hbar)$, where $d_{\downarrow\uparrow}$ is the electric dipole matrix element between $|\downarrow\rangle$ and $|\uparrow\rangle$ and $\eta = kz_0$ is the Lamb-Dicke parameter ($\ll 1$ in the Lamb-Dicke limit) where $z_0 \equiv \sqrt{\hbar/2m\omega_z}$ is the extension of the ground-state wave function.

For certain choices of ω_{L}, H_{I} is resonant and the spin and motion can be entangled efficiently. For example, when $\omega_{\mathrm{L}} = \omega_0 - \omega_z$, $H_{\mathrm{I}} \simeq i\eta\hbar\Omega\sigma^+ a e^{i\phi} + \mathrm{h.c.}$ This is usually called the "red-sideband" coupling and is formally equivalent to the Jaynes-Cummings Hamiltonian [22] from quantum optics. Here, $|\downarrow\rangle \to |\uparrow\rangle$ transitions are accompanied by $n \to n-1$ motional mode transitions. When $\omega_{\mathrm{L}} = \omega_0 + \omega_z$, $H_{\mathrm{I}} \simeq i\eta\hbar\Omega\sigma^+ a^\dagger e^{i\phi} + \mathrm{h.c.}$, the "blue sideband" where $|\downarrow\rangle \to |\uparrow\rangle$ transitions are accompanied by $n \to n + 1$ transitions. When $\omega_{\mathrm{L}} = \omega_0$, $H_{\mathrm{I}} \simeq \hbar\Omega\sigma^+ e^{i\phi} + \mathrm{h.c.}$ and $|\downarrow\rangle \to |\uparrow\rangle$ transitions are driven that do not change n ("carrier" transitions).

In the $^9\mathrm{Be}^+$ experiments at NIST, two laser beams are used to drive stimulated Raman transitions between two ground-state hyperfine levels which serve as ion qubit levels. Here, k must be replaced by the difference between k-vectors for the two beams, and $\Omega \propto E_1 E_2/\Delta$, where $E_{1,2}$ are the electric fields of the two beams and Δ is the nominal detuning of the beams from an allowed transition [10], as indicated in fig. 2. Since η must be kept large enough so that the time to carry out an entangling operation is not too long, we must add corrections (suppression factors) to the expressions for the Rabi frequencies Ω^*, but the basic picture remains the same. These corrections, which can be called Debye-Waller factors [10], have been observed in ref. [23]. They are rather direct evidence of the wave packet nature of the ions because to obtain the correct interaction with the laser, we must average the laser field over the extent of the ion wave packet rather than assuming they are point particles.

As a function of the time that H_{I} is applied, the system sinusoidally oscillates, or "Rabi-flops" between states $|\!\downarrow\rangle|n\rangle$ and $|\!\uparrow\rangle|n'\rangle$, as indicated in fig. 2 and as observed in experiments [23, 24]. If the ion qubit starts in the state $|\!\downarrow\rangle|n\rangle$ or $|\!\uparrow\rangle|n\rangle$, then for values of t where $\eta\Omega\sqrt{n_>}t \neq m\pi/2$ (m an integer and $n_>$ the greater of n or n'), we create entanglement between the ion and motion qubits. This is the basic source of entanglement in ion experiments from which universal logic gates have been constructed. For example, a controlled-not and π-phase gate between the motion and ion qubit for a single ion has been realized at NIST [10, 25]. As described by Cirac and Zoller [1], to use this controlled-not gate to realize a controlled-not between two ion qubits (1 and 2), we would first start the motion in the ground state. We then map the state of ion qubit number 1 onto the motion qubit with the red sideband so that we carry out the operation $(\alpha|\!\downarrow\rangle + \beta|\!\uparrow\rangle) \otimes |n = 0\rangle \rightarrow |\!\downarrow\rangle \otimes (\alpha|0\rangle + \beta|1\rangle)$. If we now carry out the controlled-not between the motion qubit and ion qubit 2, followed by mapping the motion qubit back onto ion qubit 1, we realize a controlled-not between ion qubits 1 and 2. As discussed in ref. [1], this requires the ability to focus laser beams onto ions 1 and 2 separately. This reveals conflicting requirements because for high operation speed, we need high ω_z. (Basically the ion must move in order to drive sideband transitions; therefore, the limit on sideband transition times is on the order of $2\pi/\omega_z$.) On the other hand, high values of ω_z imply close spacing and therefore strong focusing in order to achieve individual ion addressing.

More recently, using the scheme suggested by Sørensen and Mølmer [26,27] and Solano et $al.$ [28], the NIST group has realized a universal gate between two ion qubits [29,30] which takes the form

$$(5) \qquad |\!\downarrow\rangle|\!\downarrow\rangle \longrightarrow \frac{1}{\sqrt{2}}(|\!\downarrow\rangle|\!\downarrow\rangle + i|\!\uparrow\rangle|\!\uparrow\rangle),$$

$$|\!\uparrow\rangle|\!\uparrow\rangle \longrightarrow \frac{1}{\sqrt{2}}(|\!\uparrow\rangle|\!\uparrow\rangle + i|\!\downarrow\rangle|\!\downarrow\rangle),$$

$$|\!\uparrow\rangle|\!\downarrow\rangle \longrightarrow \frac{1}{\sqrt{2}}(|\!\uparrow\rangle|\!\downarrow\rangle + i|\!\downarrow\rangle|\!\uparrow\rangle),$$

$$|\!\downarrow\rangle|\!\uparrow\rangle \longrightarrow \frac{1}{\sqrt{2}}(|\!\downarrow\rangle|\!\uparrow\rangle + i|\!\uparrow\rangle|\!\downarrow\rangle).$$

Here, laser beams are tuned to resonantly excite the transitions $|\downarrow\rangle|\downarrow\rangle \rightarrow |\uparrow\rangle|\uparrow\rangle$ and $|\downarrow\rangle|\uparrow\rangle \rightarrow |\uparrow\rangle|\downarrow\rangle$ through states excited non-resonantly, or virtually, with simultaneously applied red and blue sidebands [26,27]. This gate has the advantages that 1) laser beam focusing (for individual ion addressing) is not required, 2) it can be carried out in one step in a time approximately equal to the mapping operation above, 3) it does not require use of an additional internal state as in the scheme of Cirac and Zoller and 4) it does not require precise control of the motional state (as long as the Lamb-Dicke limit is satisfied). From this gate, a controlled-not gate can be constructed [26]. To do this, addressing is required on each ion but only on internal-state (carrier) rotations. These single-bit operations can be carried out with the ions separated where individual addressing is easier. Alternatively, even when the ions are equally illuminated, Ωt rotations (fig. 2) can be implemented on either ion selectively by dividing the carrier rotations into two parts [30]. In the first, a $\Omega t/2$ pulse is driven on both ions. In the second part, the ions are first shifted and separated so that the position of the first ion is unchanged but the second ion is shifted by a distance $\lambda/2$. Therefore, after the second $\Omega t/2$ pulse is applied, the first ion has rotated by Ωt but the second ion has rotated back to its original state. Differential ion addressing has also been achieved by inducing differential micromotion as noted in sect. **2**·1 [20].

In general, to initialize a group of ion qubits for each experiment, we use a combination of internal-state optical pumping (to pump to the $|\downarrow\rangle$ state of fig. 2) and laser cooling to optically pump the motional modes to their ground states [24, 31-33]. As in many atomic physics experiments, the observable in the ion trap experiments is the ion qubit state, or spin. For example, we can efficiently detect the $|\downarrow\rangle$ internal state of the ion by applying circularly polarized laser light resonant with the transition $|\downarrow\rangle \leftrightarrow |e\rangle$, where $|e\rangle$ is a short-lived excited electronic state that, by selection rules, can only decay back to $|\downarrow\rangle$ upon emitting a photon. In contrast, the transition $|\uparrow\rangle \leftrightarrow |e\rangle$ is out of resonance, and an ion in the state $|\uparrow\rangle$ scatters negligible light. For a summary of this "quantum jump" detection technique, see ref. [34].

Although the basic elements of quantum computation using ions have been demonstrated, the fidelity and number of operations have been limited by the effects of decoherence. In the following, we discuss how some fundamental aspects of decoherence can be demonstrated and also discuss some of the sources of decoherence which, in practice, limit the fidelity and coherence of the trapped-ion system.

4. – Motion decoherence

For brevity, we will assume that we are interested in characterizing the decoherence of a superposition state $\alpha|\psi_1\rangle + \beta|\psi_2\rangle$, where $|\psi_1\rangle$ and $|\psi_2\rangle$ are states of a single mode of motion for trapped ions. This mode of motion we call the quantum "system" under investigation. For simplicity, we take the mode to be the z-mode of a single trapped ion. In a basic model of decoherence [4,5], it is assumed that this state couples to the outside world or "environment" whose initial state is $|\phi_e\rangle$. Therefore, the initial state of the motion and environment can be written $\psi_0 = (\alpha|\psi_1\rangle + \beta|\psi_2\rangle) \otimes |\phi_e\rangle$. After the motional state couples to the environment, the system and environment evolve to

$$(6) \qquad \psi_0 = (\alpha|\psi_1\rangle + \beta|\psi_2\rangle) \otimes |\phi_e\rangle \longrightarrow \psi_{\text{final}} = \alpha|\psi_1'\rangle|\phi_{e1}\rangle + \beta|\psi_2'\rangle|\phi_{e2}\rangle.$$

For a specific type of system/environment coupling, we can assume $|\psi_1'\rangle = e^{i\varsigma_1}|\psi_1\rangle$ and $|\psi_2'\rangle = e^{i\varsigma_2}|\psi_2\rangle$; therefore, after the interaction, $\psi_{\text{final}} = \alpha|\psi_1\rangle|\phi_{e1}\rangle + e^{i(\varsigma_2-\varsigma_1)}\beta|\psi_2\rangle|\phi_{e2}\rangle$ and we see how the initial states become correlated with the environment. Now, if $\langle\phi_{e2}|\phi_{e1}\rangle \simeq 0$, and if the final environment states are uncontrolled and unmeasured (or unmeasurable), we must ignore these degrees of freedom (mathematically trace over them) such that the pure state ψ_{final} becomes a statistical mixture expressed by the density matrix $\rho_{\text{final}} = |\alpha|^2|\psi_1\rangle\langle\psi_1| + |\beta|^2|\psi_2\rangle\langle\psi_2|$. Hence, the off-diagonal terms of the density matrix (or "coherences") from the pure state $\alpha|\psi_1\rangle + \beta|\psi_2\rangle$ are lost to the environment.

It is sometimes useful to incorporate a quantum measuring device or quantum "meter" into the scheme above using a "von Neumann chain" [4]. Here we assume that the quantum system, initially given by $\alpha|\psi_1\rangle + \beta|\psi_2\rangle$, is first coupled to a quantum meter and, in general, this combination is then coupled to the environment. In the first stage of coupling, we have

$$(7) \qquad (\alpha|\psi_1\rangle + \beta|\psi_2\rangle) \otimes |\psi_M\rangle \otimes |\phi_e\rangle \longrightarrow (\alpha|\psi_1'\rangle|\psi_{M1}\rangle + \beta|\psi_2'\rangle|\psi_{M2}\rangle) \otimes |\phi_e\rangle.$$

Upon coupling to the environment, the evolution is

$$(8) \qquad (\alpha|\psi_1'\rangle|\psi_{M1}\rangle + \beta|\psi_2'\rangle|\psi_{M2}\rangle) \otimes |\phi_e\rangle \longrightarrow \alpha|\psi_1''\rangle|\psi_{M1}'\rangle|\phi_{e1}\rangle + \beta|\psi_2''\rangle|\psi_{M2}'\rangle|\phi_{e2}\rangle,$$

and, as before, if we assume $\langle\phi_{e2}|\phi_{e1}\rangle \simeq 0$ and that the final environment states are uncontrolled and unmeasured (or unmeasurable), then the final state of the system and meter is expressed by the density matrix

$$(9) \qquad |\alpha|^2|\psi_1''\rangle\langle\psi_1''||\psi_{M1}'\rangle\langle\psi_{M1}'| + |\beta|^2|\psi_2''\rangle\langle\psi_2''||\psi_{M2}'\rangle\langle\psi_{M2}'|.$$

Therefore the correlation between the system and meter states is established yielding the expected classical result and the coherences are lost to the environment. In the first experiment described below we will consider a case that is described by expression (8) but we have a situation where $|\phi_{e1}\rangle \simeq |\phi_{e2}\rangle \simeq |\phi_e\rangle$. However since, in practice, we can not measure $|\phi_e\rangle$, and because $|\phi_e\rangle$ changes from experiment to experiment, we must average over the distribution of $|\phi_e\rangle$ states which leads to decoherence very similar to the case where $\langle\phi_{e2}|\phi_{e1}\rangle \simeq 0$.

Including the quantum meter more closely describes the ion experiments discussed below because the "system" is the ion motion which is not directly measured, but can be coupled to the ion spin which is the meter that is measured (by laser scattering.) However, by introducing the meter, we also begin to see some of the problems that emerge in trying to describe decoherence at a fundamental level. For example, if the couplings of the system and meter to the environment are very weak or can be turned off, then it seems more natural to include the meter within the original quantum system. In this case, we see from expression (7) that the system plus meter is described by the entangled state $\alpha|\psi_1'\rangle|\psi_{M1}\rangle + \beta|\psi_2'\rangle|\psi_{M2}\rangle$ whose off-diagonal density matrix elements do not decohere. Here, the natural dividing line separating the quantum and classical world is between the quantum system plus meter and the environment. On the other hand, if the system is not coupled to the environment and the meter is weakly coupled to the system but strongly coupled to the environment, it is more natural to include the meter with the environment and make the quantum/classical separation between the system and meter. It appears that this choice of dividing line cannot be made on fundamental grounds, but is rather made for convenience, depending on experimental circumstances. Nevertheless, if we avert our eyes from this issue, some experiments on ions have been able to exhibit the features of conventional descriptions of decoherence. We consider two examples of this below.

4`1. *High-temperature amplitude reservoir.* – Decoherence of harmonic oscillator superposition states, with coupling to a variety of reservoir environments, has been investigated extensively in theory; see for example refs. [4, 35-42]. The model in these studies is a system harmonic oscillator coupled to a bath of environment quantum oscillators. The interaction Hamiltonian for this situation can be modelled as

$$(10) \qquad H_{\mathrm{I}} = \hbar \sum_{i=0}^{\infty} \Gamma_i ab_i^{\dagger} + \mathrm{h.c.},$$

where a is the lowering operator for the system oscillator, $b_i^{\dagger}$ is the raising operator for the i-th environment oscillator, and Γ_i gives the strength of coupling between the system oscillator and the i-th environment oscillator. The case where the system oscillator is a single mode of the electromagnetic field has been investigated extensively in the experiments of Haroche *et al.* [43]. This and other models of decoherence in the specific context of trapped-ion experiments are discussed theoretically in refs. [10, 44-47]. One model assumes that the motion of a trapped ion couples to the (noisy) uniform electric field $\mathbf{E}$ caused by the environment oscillators through the potential $U = -q\mathbf{x} \cdot \mathbf{E}$, where $\mathbf{x}$ is the displacement of the ion from its equilibrium position (proportional to the amplitude of motion). Such a model corresponds to the case of a noisy electric field due to a resistor coupled between the trap electrodes [10, 48] and is described by the Hamiltonian in eq. (10).

To speed up decoherence for experimental convenience, we can simulate a hot resistor coupled between trap electrodes by applying random uniform electric fields along the axis of the trap, that have a spectrum which overlaps the ion's axial-motion frequency ω_z [49, 50]. We generate axial fields in the trap by applying voltages to one of the trap electrodes. A commercial function generator produces pseudo-random voltages which are applied through a band-pass filter centered near ω_z, defining the frequency spectrum of the reservoir. In the experiments, we measured the initial coherence and subsequent

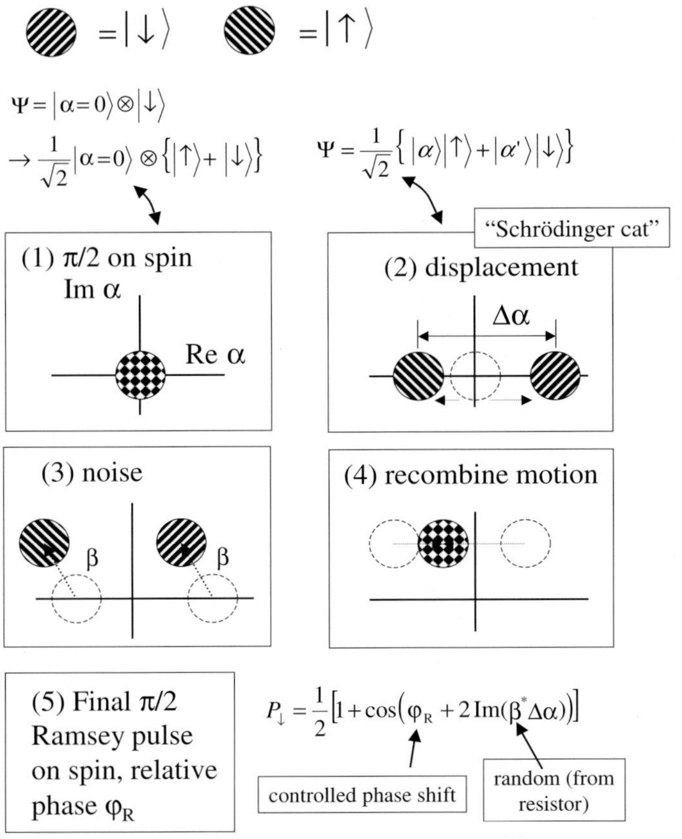

Fig. 3. – Pictorial representation in phase space of a Ramsey interferometer [51] experiment designed to observe decoherence of coherent state superpositions caused by coupling to an amplitude reservoir (from refs. [49, 50]). Starting from the initial system/meter state $|\alpha = 0\rangle|\downarrow\rangle$, we create the Schrödinger-cat state shown in panel (2) by applying a $\pi/2$ pulse on the spin followed by a spin-dependent optical-dipole-force displacement. Noise is then applied to the motion simulating coupling to a hot resistor. A (random) displacement β in phase space results as shown in panel (3). Subsequently we reverse the steps used to create the cat state but with a phase shift ϕ_R on the final Ramsey $\pi/2$ pulse. Finally, we measure the probability $P_\downarrow$ of finding the ion in the $|\downarrow\rangle$ internal state. In the absence of noise ($\beta = 0$), we obtain the characteristic sinusoidal oscillations as a function of ϕ of Ramsey interferometry; with noise, the Ramsey fringe contrast is reduced because we must average β over a distribution of values characteristic of the applied amplitude noise.

decoherence of the motion state superpositions with single-atom interferometry, as described in detail in refs. [49, 50]. Briefly, we first create the entangled system/meter state shown in the left-hand side of expression (8) with $|\psi_1'\rangle = |\alpha\rangle, |\psi_{M1}\rangle = |\uparrow\rangle, |\psi_2'\rangle = |\alpha'\rangle$, $|\psi_{M2}\rangle) = |\downarrow\rangle$ ($\alpha = \beta = 1/\sqrt{2}$ in expression (8) as indicated in part (2) of fig. 3). Here, $|\alpha\rangle$ and $|\alpha'\rangle$ are coherent states of amplitude α and α'. This state is usually called a

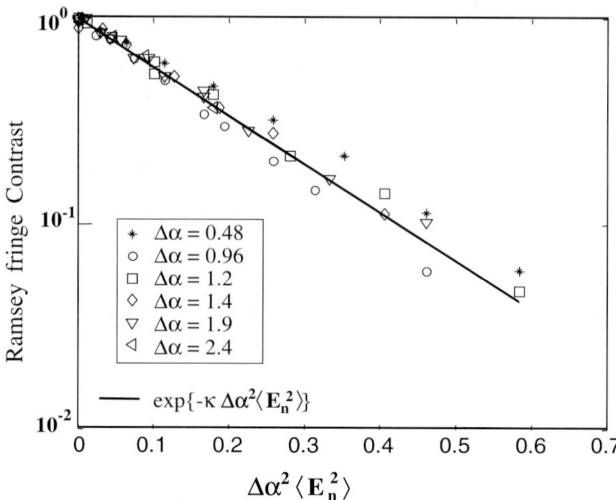

Fig. 4. – Loss of coherence of a Schrödinger-cat state caused by coupling to an amplitude reservoir (from ref. [49]). The amplitude noise was applied for a fixed time ($3\,\mu$s) but with varying amplitude. The contrast for all experiments is normalized to the contrast observed in the absence of applied noise. The expected scaling was observed showing that the decoherence rate is proportional to the square of the phase-space separation $\Delta\alpha$ of the cat-state components.

"Schrödinger-cat" state. The superposition is then coupled to the reservoir (the applied fluctuating fields) for a fixed time. Finally, the components of the motional-state superposition are overlapped by reversing the steps which initially created it and the internal state is measured. Repeating the experiment many times, the internal state of the ion is measured as a function of the relative phase of the creation and reversal steps, and the contrast of the resulting interference fringes characterizes the amount of coherence remaining after coupling to the reservoir. The results are displayed in fig. 4. We observe that the contrast of the interferometer fringes decays as $\exp\left[-|\alpha-\alpha'|^2\langle E_n^2\rangle\right]$, where $\sqrt{\langle E_n^2\rangle}$ is the r.m.s. value of the applied fluctuating fields. This scaling was also observed for the ambient fluctuating fields in the experiment [49,50]. This exponential dependence of the decoherence on the "size" of the cat-state $\Delta\alpha \equiv |\alpha-\alpha'|$ agrees with theoretical predictions and indicates why it is so difficult to preserve superpositions except on a microscopic scale.

The decoherence observed here fits into the general scheme outlined above with some notable differences. After interaction with our reservoir (right-hand side of expression (8)), the environment has acted on the system (motion state) and not the meter (spin) so that $|\psi'_{M1}\rangle = |\psi_{M1}\rangle$ and $|\psi'_{M2}\rangle = |\psi_{M2}\rangle$. (Of course, when we finally make a measurement of the spin, we couple to the environment via laser scattering, but this is not the environment coupling we study here.) Also, even though we must have $|\phi_{e1}\rangle \neq |\phi_{e2}\rangle$ because of the difference in back-action from the states $|\alpha\rangle$ and $|\alpha'\rangle$ to the environment, to a very good approximation the back-action of the system on the environment is negligible so that, for all practical purposes, $\langle\phi_{e2}|\phi_{e1}\rangle = 1$ and $|\phi_{e1}\rangle = |\phi_{e2}\rangle = |\phi_e\rangle$. Here we have the situation that we could, in principle, measure $|\phi_e\rangle$ without disturbing it appreciably and apply an operation which undoes its effect. In the case of purposely applied noise, we could actually do this, but for the case of the ambient noise, this is, in practice, impossible and we must average over the ensemble of environment states $\{|\phi_e\rangle\}$ leading to the expected decoherence exhibited in fig. 4.

In related experiments discussed elsewhere [49, 50], we have characterized the decoherence when we apply the amplitude reservoir to superpositions of two Fock states and also when we apply a phase reservoir, characterized by the interaction Hamiltonian

$$(11) \qquad H_{\mathrm{I}} = \hbar a a^\dagger \sum_{i=0}^{\infty} \Gamma_i b_i^\dagger + \mathrm{h.c.}$$

to the Schrödinger-cat state and to superpositions of two Fock states.

$4^{\cdot}2.$ *"Engineered" zero-temperature amplitude reservoir.* – In the cavity-QED decoherence experiments of ref. [43], where the system oscillator frequency is around $50\,\mathrm{GHz}$ and the ambient temperature is $0.6\,\mathrm{K}$, the equilibrium oscillator quantum number is around 0.05 and the amplitude reservoir temperature can be assumed to be $0\,\mathrm{K}$ to a good approximation. In the ion experiments, the oscillator frequency is around $5\,\mathrm{MHz}$ and the ambient temperature is around room temperature implying that the equilibrium oscillator number is very large. (Actually, the ambient fluctuating fields behave as if the reservoir were much hotter as described below.) Nevertheless, we can simulate a $T = 0$ reservoir as suggested by Poyatos *et al.* [41]. The basic process is laser cooling as indicated in fig. 5. Coherent Raman beams couple the states $|\downarrow\rangle|n\rangle \leftrightarrow |\uparrow\rangle|n-1\rangle$ with Rabi rate Ω^* (fig. 2). At the same time, an optical pumping beam causes spontaneous Raman transitions from $|\uparrow\rangle|n\rangle$ to $|\downarrow\rangle|n\rangle$ through the excited P state at rate γ. From the diagram in fig. 5, we see that all populations are driven towards the state $|\downarrow\rangle|0\rangle$ thereby simulating a $T = 0$ reservoir. By varying the strength of the Raman and optical pumping couplings and time, we have control over the reservoir parameters.

In a NIST experiment [49, 50], we examined the time evolution of the coherence of the Fock state superposition $\psi = (1/\sqrt{2})(|0\rangle + |2\rangle)|\downarrow\rangle$ for varying lengths of reservoir interaction time. To observe this coherence, we make an interferometer similar to one described in the previous section. (Note that in this experiment, the spin serves two functions: it allows us to construct the interferometer, make measurements and, in the middle of any particular interferometer experiment, it also acts as part of the environment.) The resulting data are shown in fig. 6. Each data point is the contrast of the interference fringes after the system interacts with the reservoir for a time τ. We show two cases, $\gamma > \Omega^*$ and $\gamma < \Omega^*$. In the first case, we observed the decay of the coherence due to coupling to the simulated $T = 0$ reservoir. In this case, it is natural to assume that the reservoir is the spin plus the rest of the environment which includes the spontaneously scattered photons. (Since we do not rigorously satisfy $\gamma \gg \Omega^*$, we also see some initial non-exponential decay evident of the Zeno effect [49, 50].) In the second

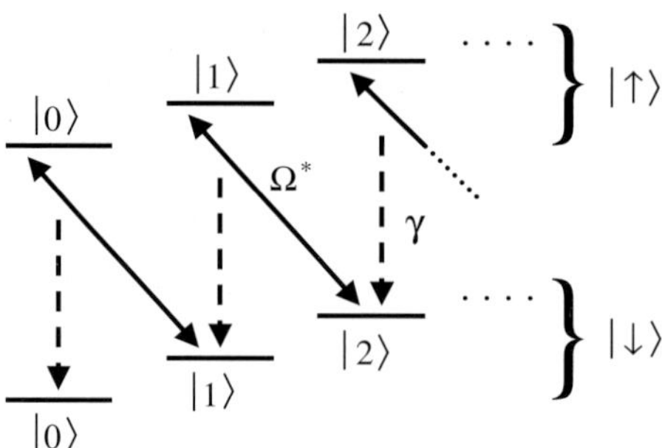

Fig. 5. – Schematic diagram depicting an "engineered" $T = 0$ reservoir. A red sideband (re: fig. 2) drives transitions $|\downarrow\rangle|n\rangle \leftrightarrow |\uparrow\rangle|n-1\rangle$, while spontaneous Raman scattering is used to make transitions $|\uparrow\rangle|n\rangle$ to $|\downarrow\rangle|n\rangle$ at rate γ. When $\gamma \gg \Omega^*$, the system acts as a $T = 0$ reservoir for motional states.

case, the coherence between the $|0\rangle$ and $|2\rangle$ state disappears and reappears over time, with an overall decay of the fringe contrast. The underlying effect is population transfer back and forth (Rabi flopping) between the states $|\downarrow\rangle|2\rangle$ and $|\uparrow\rangle|1\rangle$ with a coupling of the $|\uparrow\rangle|1\rangle$ state to the outside environment through spontaneous Raman scattering. If we make $\gamma \to 0$, then we see near perfect revival of the fringe contrast as expected [50]. In this last instance, we have restricted the size of the environment to the internal spin states so that we can reverse the effects of decoherence (of the $(1/\sqrt{2})(|0\rangle + |2\rangle)$ state). (A scheme for observing a similar reversal in the context of cavity-QED is discussed in ref. [52].) In this experiment, when the atom scatters a photon through spontaneous Raman scattering, no measurement we do will allow us to restore the initial superposition. That is, the emission of a spontaneous photon projects the atom into a definite state ($|\downarrow\rangle|0\rangle$ or $|\downarrow\rangle|1\rangle$ in the context of fig. 5) and phase information is lost. (This is the same situation as in the experiments of Brune $et\ al.$ [43].) The decoherence in this experiment is to be contrasted with the Schrödinger-cat experiment where we could, in principle, detect the environment and reverse its effects. Note that although the data for $\gamma < \Omega^*$ illustrate how coherence transferred to the environment can be recovered, an alternative explanation would say that by transferring the $|\downarrow\rangle|2\rangle$ component of the superposition to the $|\uparrow\rangle|1\rangle$ state, we gain "which-path" information in our interferometer—the paths being the $|\downarrow\rangle|0\rangle$ and $|\downarrow\rangle|2\rangle$ parts of the superposition. The oscillation in which-path information is analogous to that illustrated by the experiments of Chapman $et\ al.$ [53], Dürr $et\ al.$ [54], and Bertet $et\ al.$ [55].

This second experiment also illustrates a fundamental problem in explaining decoherence. If we restrict the size of the environment to be the spin ($\gamma \to 0$), then the information is not lost and can be recovered, as indicated by the revival of the contrast. Even in the case of photon emission, the information need not be lost if the photon is emitted into a lossless cavity where it can be recovered, as illustrated in the cavity-QED experiments [56, 57]. Therefore it seems that decoherence is useful to describe situations

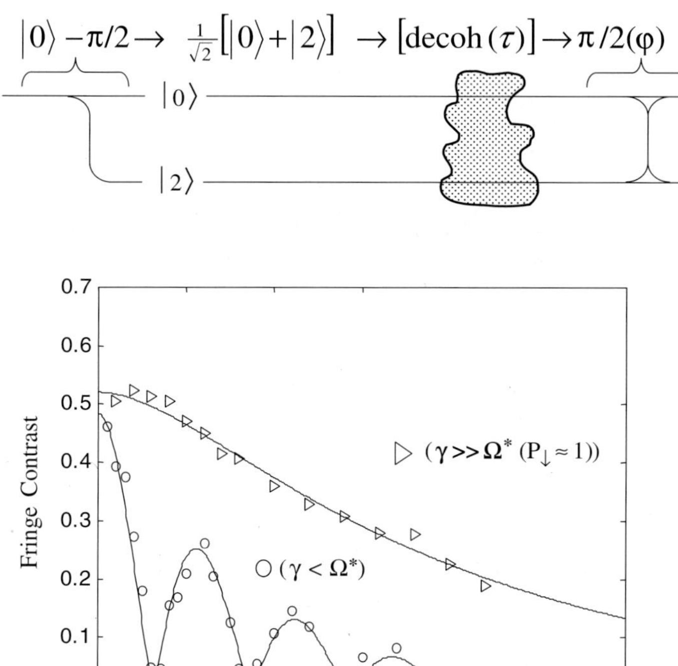

Fig. 6. – Data for decoherence of motional superpositions $(1/\sqrt{2})(|0\rangle + |2\rangle)$ coupled to an engineered $T = 0$ reservoir (from refs. [49,50]). In the upper part of the figure we show schematically how the motional-state Ramsey interferometer works, where the reservoir is applied during the time indicated by the shaded area. In the lower part of the figure, we show the interferometer contrast as a function of time for two cases of the relative values of Ω^{*} and γ. The initial contrast (without decoherence) is not unity due to imperfections in the motional-state Ramsey interferometer.

where, for practical and technical reasons, we cannot retain the overall system information. However, these limitations seem only to be practical and not fundamental *unless* something is missing in quantum mechanics. For a summary of alternatives, see, for example, ref. [58].

4'3. *Motion decoherence in practice.* – The ambient motion decoherence observed in the NIST experiments can be characterized by the model described in the first case above; its characteristics are the same as that caused by thermal electronic noise in the resistance of the electrodes or resistance coupled to the electrodes. At the relatively low ion oscillation frequencies used, where the characteristic wavelength is much larger than the electrode spacing, this could be described by Johnson noise associated with lumped circuit elements attached to the electrodes. Typical heating rates (expressed as quanta per second from the motional ground state) are observed to be around $\langle dn/dt \rangle \simeq 10^3$–$10^4\,\mathrm{s}^{-1}$ for $\nu_z \simeq 10\,\mathrm{MHz}$ and $R \simeq 150\,\mu\mathrm{m}$ [59]. The problem is that given our best estimates of the electrode resistance [10] and attached circuit elements, the electrodes would have to be at a temperature of $10^6\,\mathrm{K}$ or higher to explain most of the heating results [59]. The principle cause of the anomalously large fluctuating fields and resultant heating is still not understood at this time, but some of its characteristics are known. It appears to be rather broad band (no sharp resonances observed from 2 to 20 MHz) and is likely caused by electric-field noise emanating from the electrodes. For example, heating from collisions due to background gas or an electron field emission source would tend to heat the internal modes of two ions at about the same rate as the center-of-mass (COM) modes. However, we observe that the internal mode heating is negligible compared to the heating of the COM modes [60] indicating that the fields at the site of the ions are approximately uniform spatially. (We note that the experiments of Rohde *et al.* [61] disagree with this finding.) A general trend is that the heating increases as $1/R^\kappa$ where $\kappa > 2$. For thermal electronic noise from circuit resistance, $\kappa = 2$ [10]. If the noise were due to fluctuating patch fields, $\kappa = 4$ [59]. The scatter of the data using $^9\mathrm{Be}^+$ ions is rather large but is consistent with $\kappa = 4$. Data on other traps [31,61] appear to also be consistent with κ equal to 4 or more.

In the experiments of ref. [59], there was some indication that beryllium deposition onto the electrodes caused a higher heating rate. (Beryllium ions are created in the trap by ionizing neutral beryllium atoms that pass near the center of the trap after being emitted from a wide angle source. Some of the beryllium atoms from the source are deposited on electrodes.) By masking the electrodes from direct beryllium deposition, preliminary evidence indicates a significant drop in heating rate (M. Rowe *et al.*, unpublished). In any case, we conjecture that for clean metal electrode surfaces, free from oxides or adsorbed gases (which could support potentially mobile electrons), the heating should approach that predicted by thermal electronic noise $\langle dn/dt \rangle \simeq 1\,\mathrm{s}^{-1}$.

5. – Entanglement-enhanced quantum measurement: "spin squeezing"

In addition to the potential application of ion traps to quantum computation, the techniques developed in the pursuit of this goal might be used to increase the precision of quantum-limited measurements. In this context, the subject of atomic-state squeezing or "spin-squeezing" has received increased attention in the last couple of years with experimental observations of spin-squeezing being reported in refs. [62-64]. In related work, atomic number squeezing has been reported in a Bose-Einstein-condensate/optical-lattice experiment [65]. As in the case of photon squeezing, squeezing of atomic ensembles should find application in interferometry [66-69] such as in Mach-Zehnder interferometers [70,71] and Ramsey spectroscopy [72-74]. Before describing an example of spin squeezing and how it can lead to enhanced measurement capabilities, we first recall some elements of quantum-limited measurement uncertainty in a more general context.

One is perhaps most familiar with measurement uncertainty relations for non-commuting observables. For example, let $\tilde{O}_1$ and $\tilde{O}_2$ be non-commuting operators for the quantum system under consideration. If we make measurements of these operators on identically prepared systems, the Schwartz inequality [75] implies that

$$\Delta \tilde{O}_1 \Delta \tilde{O}_2 \geq \frac{|\langle [\tilde{O}_1, \tilde{O}_2] \rangle|}{2}, \tag{12}$$

where brackets $\langle \; \rangle$ denote an ensemble average and $(\Delta \tilde{O}_i)^2 \equiv \langle \tilde{O}_i^2 \rangle - \langle \tilde{O}_i \rangle^2$, $i \in \{1,2\}$ are the variances of measurements of $\tilde{O}_1$ and $\tilde{O}_2$ around their mean values. As a simple example, when $\tilde{O}_1 = \tilde{x}$, the position operator, and $\tilde{O}_2 = \tilde{p}$, the momentum operator, the Schwartz inequality leads to the Heisenberg uncertainty relation for measurements of the position and momentum of a particle.

We can also speak about uncertainty relations involving operators and system parameters in the following way. Suppose we have a system operator $\tilde{O}(\zeta)$ which depends on a classical system parameter ζ. Now if ζ fluctuates from measurement-to-measurement by an amount characterized by r.m.s. fluctuations $\delta\zeta$, this implies fluctuations in $\tilde{O}$ given by $\Delta\tilde{O} = |d\langle\tilde{O}\rangle/d\zeta|\delta\zeta$. Conversely, we can write this expression as

$$\delta\zeta = \frac{\Delta\tilde{O}}{|d\langle\tilde{O}\rangle/d\zeta|}. \tag{13}$$

Therefore, if we use measurements of $\tilde{O}(\zeta)$ to determine ζ, our precision is limited by fluctuations in $\tilde{O}$ through eq. (13). As an example, suppose in eq. (12) that $\tilde{O}_1 = \tilde{O}$ (a particular, but unspecified system operator) and $\tilde{O}_2 = \mathbf{H}$, the system Hamiltonian. $[\tilde{O}, \mathbf{H}]$ is related to the time derivative of $\tilde{O}$ through Heisenberg's equation $\hbar d\tilde{O}/dt = -i[\tilde{O}, \mathbf{H}] + \hbar\partial\tilde{O}/\partial t$. If $\tilde{O}$ does not explicitly depend on time, then eq. (12) implies $\Delta\tilde{O}\Delta\mathbf{H} \geq \hbar|d\langle\tilde{O}\rangle/dt|/2$. From eq. (13), we find $\Delta\mathbf{H}\delta t \geq \hbar/2$, the time-energy uncertainty relation.

As a concrete example of the use of spin squeezing, we examine the case of eq. (13) when ζ is the angle of rotation of a system of spins. Measurement of spin rotation angle is essentially the problem of Ramsey spectroscopy (see, for example, refs. [51, 72-74, 76]) or particle interferometry [66, 67]. A recent experiment is described in ref. [64]; for brevity, we examine only one case treated there. We consider a system of N spin-$(1/2)$ particles with total angular momentum $\mathbf{J} = \sum_{i=1}^{N} \mathbf{S}_i$, where $\mathbf{S}_i$ is the spin of the i-th particle. For each spin, $|\downarrow\rangle_i$ and $|\uparrow\rangle_i$ are spin eigenstates with respect to a chosen axis. Rotations of the entire system are characterized by the operator $R = \exp[-i\zeta\mathbf{J} \cdot \hat{\mathbf{u}}]$ for ζ an angle and $\hat{\mathbf{u}}$ the axis of rotation (throughout this section, angular momentum is expressed in units of $\hbar$). To make the best estimate of ζ, we want to prepare an input state and choose an observable $\tilde{O}$ that will minimize $\delta\zeta$ in eq. (13).

For non-entangled spin-$(1/2)$ particles, the states which minimize $\delta\zeta$ are angular-momentum coherent states [77]. Such coherent states can be obtained from the state $|\Psi_0\rangle = |\downarrow\rangle_1|\downarrow\rangle_2 \cdots |\downarrow\rangle_N = |J = N/2, m_J = -N/2\rangle$ by an overall rotation. In this case $\tilde{O} = \tilde{J}_\perp$ minimizes $\delta\zeta$, where $\tilde{J}_\perp$ is the angular-momentum operator perpendicular to $\langle\mathbf{J}\rangle$ in the plane of rotation and we find $\delta\zeta = \Delta J_\perp/|\langle\mathbf{J}\rangle|$. Assuming no additional sources

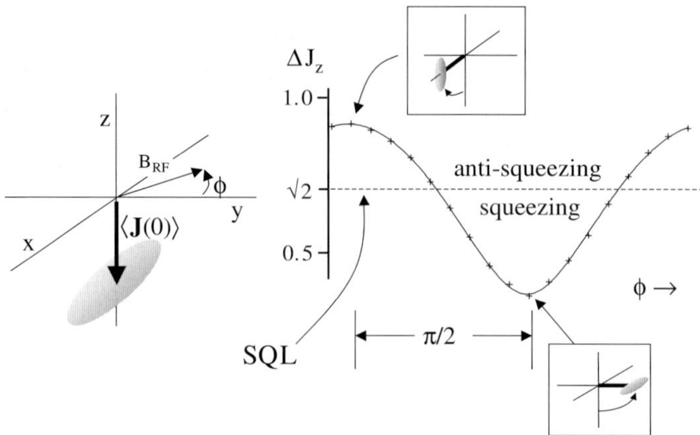

Fig. 7. – Spin squeezing. In the left part of the figure we depict the net angular-momentum vector $\langle \mathbf{J} \rangle$ of a system of N spin-(1/2) particles which at time $t = 0$ points along the negative z-axis in a frame which rotates at the spin precession frequency. The extent of the spheroid at the end of the vector is meant to indicate the square roots of the variances of the angular momenta measured along different axes. For a coherent spin state ($\psi(0) = |\downarrow\rangle_1 |\downarrow\rangle_2 \cdots |\downarrow\rangle_N = |J = N/2, m_J = -N/2\rangle$), we have $\Delta J_z(0) = 0$ and $\Delta J_x(0) = \Delta J_y(0) = \sqrt{J/2} = \sqrt{N}/2$. These uncertainties would be indicated by a thin disk of radius $\sqrt{J/2}$ in the (x, y)-plane at the end of the angular-momentum vector. If the spins are entangled or squeezed, then in general $\Delta J_z(0) \neq \Delta J_x(0) \neq \Delta J_y(0) \neq 0$. When the spin is squeezed, the uncertainty in one direction is reduced below the value for a coherent state. In the figure, the spin is squeezed in the y-direction. To measure the x, y squeezing of the initial state, we rotate the spin about an equivalent RF field (B_{RF}) by an angle $= \pi/2$ and measure ΔJ_z as indicated in the right part of the figure. Data is shown for the squeezed spin state of two $^9\mathrm{Be}^+$ ions. Error bars for the measurements are smaller than the size of the data-point marker. (From ref. [64].)

of error, for coherent states we obtain $\delta\zeta = \delta\zeta_c = 1/\sqrt{N}$, the standard quantum limit (SQL) [67-69, 72, 73, 76] (see fig. 7). This limit has been observed in experiments [76, 78]. To reduce $\delta\zeta$ below the SQL we can use a method which employs states that are particularly well described by "spin squeezing" [67, 68, 73]. Such states are depicted in fig. 7. We quantify the improvement from squeezing with the parameter $\xi_R = \delta\zeta/\delta\zeta_c$ [73]. It can be shown that the minimum possible value of ξ_R is $1/\sqrt{N}$ ($\delta\zeta = 1/N$), the Heisenberg limit.

The squeezing experiment was carried out on two $^9\mathrm{Be}^+$ ions [64]. Squeezing was created by the scheme of Sørensen and Mølmer [27] where we realized a generalization of eq. (5) which (ideally) carries out the transformation

$$(14) \qquad |\downarrow\downarrow\rangle \longrightarrow |\Psi\rangle \equiv \cos\alpha|\downarrow\downarrow\rangle + i\sin\alpha|\uparrow\uparrow\rangle.$$

For the data of fig. 7, $\alpha = \pi/6$. In general, $\Delta J_\perp$ of the state $|\Psi\rangle$ depends on ϕ, as indicated in fig. 7. To determine $\Delta J_\perp(\phi)$, we rotate $|\Psi\rangle$ into the (x, y)-plane by driving a $\pi/2$-pulse on both ions ($\theta = \pi/2$) about the vector $B_{\mathrm{RF}(\phi)}$ and measure ΔJ_z for different values of ϕ. This operation preserves the expectation values of $\langle \mathbf{J} \rangle$ and $\Delta J_\perp$ in the rotated frame. Therefore, we can regard the experiment as measuring the precision of a $\zeta = \theta = \pi/2$ rotation of the spins for the initial state $|\Psi\rangle$; this precision will be optimized for certain values of ϕ. From the figure, the squeezing (and anti-squeezing) of $\Delta \tilde{J}_\perp$ is clearly evident. However, from eq. (14), $|\langle \mathbf{J} \rangle| \equiv |\langle J_z(0) \rangle| = |\cos(2\alpha)|$; that is, the length of the net angular momentum shrinks as we induce the squeezing which, in turn, reduces the improvement in angular precision. In the experiment, we extracted $|\langle \mathbf{J} \rangle|$ by applying the carrier rotation of fig. 2 for various values of $\theta = \Omega^* t$ and recording $\langle J_z \rangle$ ("Rabi flopping"). For the ideal case, we have

$$(15) \qquad \delta\theta(\phi) = \frac{\Delta J_\perp(\phi)}{|\langle \mathbf{J} \rangle|} = \frac{\sqrt{(1/2)(1 - \sin(2\alpha)\sin(2\phi))}}{|\cos(2\alpha)|}.$$

For $\alpha = \pi/10$, we measured $|\langle \mathbf{J} \rangle|$ to be 0.768(2) (ideally, for $\alpha = \pi/10$, we expect 0.809). The highest measured sensitivities achieved were $\delta\theta_{\min} = 0.65(1)$, or $\xi_R = 0.92(1)$. (Ideally, for $\alpha = \pi/10$, $\delta\theta_{\min} = 0.561$ ($\xi_R = 0.794$) [64].) We therefore saw an increase in sensitivity over the SQL for two spins ($\delta\theta_c = 1/\sqrt{2}$). The discrepancy with the theoretically expected minimum values was caused primarily by imperfect entangled-state preparation. Note that ideally, from eq. (15), for two spins, $\delta\theta_{\min} \to 1/2$ (the Heisenberg limit) as $\alpha \to \pi/4$. However, then $\langle \mathbf{J} \rangle \to 0$ so that any added noise prevents us from achieving this limit. In ref. [64], we also performed a measurement of spin angle where we used input states of the form $|\Psi\rangle = (1/\sqrt{2})[|J, -J\rangle + e^{i\beta}|J, +J\rangle] = (1/\sqrt{2})[|{\downarrow}\rangle_1|{\downarrow}\rangle_2 \cdots |{\downarrow}\rangle_N + e^{i\beta}|{\uparrow}\rangle_1|{\uparrow}\rangle_2 \cdots |{\uparrow}\rangle_N]$ and measured the parity operator $\tilde{O} = \tilde{P} = \prod_{i=1}^{N} \sigma_{zi}$, where $\sigma_{zi} = 2S_{zi}/\hbar$ is the Pauli spin operator in the z-direction for the i-th particle. We used this strategy to then perform a Ramsey spectroscopy experiment along the lines suggested in ref. [74], achieving an improvement over the case where the spins are not entangled.

Will spin squeezing really be useful? For example, one might argue that we can always improve atomic clock stability by increasing the number of atoms and do without squeezing. However, in atomic clock experiments on ions, we usually want to keep the number of ions fairly small in order to obtain high accuracy. This is so, for example, because as the number of ions increases, the sample spreads out due to the Coulomb interaction and therefore the ions can sample different regions of magnetic field giving rise to systematic clock frequency shifts which are hard to determine. Therefore, for reasons like this, in an atomic clock based on the 40.5 GHz hyperfine transition in ^{199}Hg$^+$ ions [79], only seven ions were used. If ideal squeezing is obtained on N atoms in an atomic clock, the time to reach a given measurement precision would be reduced by a factor of N compared to the case of unentangled atoms. This would be important even for relatively small N because in experiments on atomic clocks, the output frequency is typically averaged for many weeks or months in order to obtain the highest precision.

6. – Towards a quantum computer

To go beyond a demonstration of the basic elements of quantum computing using trapped ions, we require a significant increase in operation fidelity and a way to scale the system to an arbitrarily large size. To improve operation fidelity, we must significantly reduce all sources of decoherence. Below we discuss some of the most important sources of decoherence and possible cures. As a possible way of scaling up, we discuss one possibility in which arrays of interconnected traps are used.

6′1. *Improving motional coherence.* – Since motion decoherence (dominated by anomalous heating so far [59]) has usually been the primary limitation on logic operation fidelity, we hope that we can find its physical cause and eliminate or significantly reduce it as described briefly above. At first, one might think that there is an easy way out of this problem since the heating observed so far (sect. 4′3) is dominated by uniform (noisy) fields which, to a high degree, heat only the COM motion. Therefore, in an array of ions, we could use one of the internal modes, and be free of the effects of COM heating. This argument is true so long as the COM motion stays in the Lamb-Dicke limit (extent of the motion wave packet $\ll \lambda/2\pi$). Unfortunately, we are never strictly in the Lamb-Dicke limit and second-order effects ("Debye-Waller factors" [10,60]) due to COM motion quickly reduce overall operation fidelity. One way to overcome the effects of heating is sympathetic laser cooling [80]. In the context of QC, one or more ions in the array of ions is laser cooled and through the Coulomb interaction the other ions are cooled (see, for example, refs. [16,61,81,82]). Of course, laser cooling will cause decoherence of the mode that is cooled. Therefore, we cool modes that couple to uniform fields and, for logic, use modes where this coupling is absent. The basic idea for this scheme is described in ref. [83] where the center ion in a string of ions (N odd) is cooled, thereby sympathetically cooling the other ions. (The mode structure for linear arrays of ions for $N \leq 10$ with equal mass and charge has been calculated by James [84].)

6′2. *Phase coherence.* – Phase decoherence can occur because a desired superposition state acquires an unknown phase shift ϕ; for internal states, $\alpha|\downarrow\rangle + \beta|\uparrow\rangle \rightarrow \alpha|\downarrow\rangle + \beta e^{i\phi}|\uparrow\rangle$. Such fluctuations can originate from magnetic-field fluctuations or Stark-shift fluctuations. Also, an unknown shift can occur in the laser phase (or the phase difference between lasers in the stimulated-Raman case). This might be intrinsic to the laser, but can also occur from beam path fluctuations such as those caused by unstable optics mounts or the trap vibrating relative to the optical beam paths. Laser beam path fluctuations can be minimized by rigid mounting but, ultimately, may have to be eliminated with servos [85-87]. Intrinsic laser phase fluctuations can be very important when ion qubit levels are separated by optical energies, requiring careful attention to laser frequency stability. However, the state of the art is improving so that visible lasers with linewidths around 0.1 Hz for 20 s averaging times have been reported [88]; this gives phase coherence times of around 10 s.

A common cause of internal-state-superposition phase fluctuations in trapped-ion experiments are fluctuations of the magnetic field **B**. The energy between ion qubit states can depend on the magnetic field; for example, the transition frequency between $|\downarrow\rangle$ and $|\uparrow\rangle$ states in the $^9\text{Be}^+$ example of fig. 2 has a magnetic-field dependence of $2.1\,\text{MHz G}^{-1}$. Therefore if the field fluctuates, the ion qubit superposition acquires a corresponding phase shift. In experiments where ion qubit rotations are induced by stimulated-Raman transitions, phase fluctuations can also be caused by fluctuations in laser-induced Stark shifts (proportional to laser beam intensities). However, the loss of fidelity from fluctuating Stark shifts is on the order of the loss of fidelity due to the rotation angle errors caused by laser intensity fluctuations so that in either the single-photon or stimulated-Raman case, the laser intensity must be stabilized to about the same degree [10]. (In certain favorable cases, due to cancellation effects, Stark shifts in the Raman case are reduced further by the ratio of the hyperfine frequency to the Raman laser detuning Δ.)

Phase fluctuations arising from fluctuating magnetic fields can be made very small by using ion qubit levels whose dependence on field vanishes to first order. This condition can be satisfied for ion qubit transitions $|F, m_F = 0\rangle \rightarrow |F', m_F = 0\rangle$ for $\mathbf{B} \rightarrow 0$. Unfortunately, a finite magnetic field is typically needed to define polarization directions and break the degeneracy of Zeeman sublevels; however, depending on the circumstances, this field can be small enough that the linear dependence on $|\mathbf{B}|$ is small. This strategy is often used for atomic clocks based on hyperfine transitions (see, for example, ref. [89]). Finally, in certain cases, transition frequency extrema are obtained at finite field. Such cases can provide very high immunity from field fluctuations. For atoms with relatively small hyperfine frequencies, certain hyperfine transitions will become field-independent at modest field strengths. For example, $^9\text{Be}^+$ and $^{25}\text{Mg}^+$ both have hyperfine transitions that are field-independent at magnetic fields around $100\,\text{G}$. As a specific example, the $|F = 2, m_F = 0\rangle \rightarrow |F = 3, m_F = 1\rangle$ transition in $^{25}\text{Mg}^+$ is field-independent for $|\mathbf{B}| \equiv B_0 = 109.6\,\text{G}$ and has a frequency dependence of $26[(B - B_0)/B_0]^2\,\text{MHz}$ for deviations from this field. Therefore a change of magnetic field of $0.01\,\text{G}$ from B_0 will cause a frequency shift of about $0.2\,\text{Hz}$ leading to a phase shift of 1 radian in about $0.8\,\text{s}$. This is to be contrasted with the case of the $|F = 2, m_F = \pm 2\rangle \rightarrow |F = 1, m_F = \pm 1\rangle$ transitions ($2.1\,\text{MHz G}^{-1}$) that have a shift of $21\,\text{kHz}$ for a $0.01\,\text{G}$ change leading to a 1 radian shift in $7.6\,\mu\text{s}$! Field-independent transitions can also be found at much higher fields ($\gg 100\,\text{G}$) [10, 90]; but the required field homogeneity may be hard to maintain over the entire ensemble of ions.

Phase fluctuations can also occur in motional state superpositions through a coupling of the form given in eq. (11). Suppression of these effects requires careful attention to the stability of trap potentials (and suppression of the fluctuating fields that cause heating). However, we gain some relative immunity from these kinds of shifts because they are important only during gate operations.

$6{}^{\cdot}3$. *Internal-state amplitude coherence*. – By amplitude decoherence we mean a random process of the form $\cos(\alpha)|\downarrow\rangle + e^{i\phi}\sin(\alpha)|\uparrow\rangle \rightarrow \cos(\alpha')|\downarrow\rangle + e^{i\phi}\sin(\alpha')|\uparrow\rangle$. A straightforward source of such errors is given by intensity fluctuations on laser beams which can, for example, result in the transformation $|\downarrow\rangle \rightarrow \alpha'|\downarrow\rangle + \beta'|\uparrow\rangle$ when, in fact, we desire the transformation $|\!\downarrow\rangle \to \alpha|\!\downarrow\rangle + \beta|\!\uparrow\rangle$ [10]. This is a very important practical problem but in many laser systems, intensity noise can be suppressed to near the shot-noise limit which would be sufficient in the context of QC. Because intensity noise may be much harder to suppress in some laser systems than others, this may eventually lead to a choice of one ion species over another.

Spontaneous emission [91,92,10] is a more troublesome, fundamental source of internal-state decoherence. For optical transitions, the radiative decay from the excited state limits the time before decoherence occurs, but for very narrow, weakly allowed transitions, the intensity of the laser must be so high to maintain a certain Rabi rate that spontaneous emission can occur from non-resonant, but much more strongly allowed transitions [92]. Spontaneous emission is also a concern when using stimulated-Raman transitions to drive qubit transitions based on ground-state hyperfine levels (assumed to be separated in energy by $\hbar\omega_0$). The basic problem arises from two conflicting requirements. When the transitions are driven by stimulated-Raman processes through a fine-structure level (the usual case) the overall Rabi rate for the process (see fig. 2) is approximately equal to $\Omega^* \simeq g_1 g_2/\Delta$, where g_1 and g_2 correspond to the resonant Rabi rates to the fine-structure level (from each qubit level) and Δ is the detuning from this level. This approximation is valid as long as $\omega_0 \ll \Delta \ll \omega_{\mathrm{FS}}$, where ω_{FS} is the fine-structure splitting (the splitting between the $^2P_{3/2}$ and $^2P_{1/2}$ levels of the first excited electronic levels in ions having a single unpaired outer electron). At the same time, the rate of spontaneous emission from coupling to this level is approximately equal to $R_s \simeq \gamma(g_1^2 + g_2^2)/\Delta^2$, where γ is the rate of spontaneous decay from this level. From these expressions, we see that we want to increase Δ to maximize the ratio of Rabi rate to spontaneous emission rate Ω^*/R_s. Unfortunately there is a limit to this approach because, as Δ becomes larger than ω_{FS}, Ω^* approaches zero. The basic effect is that the Rabi rate amplitude from coupling to the other fine-structure level cancels that due to coupling to the first level. This results in a net Rabi rate given approximately by $\Omega^* \simeq (g_1 g_2/\Delta)(\omega_{\mathrm{FS}}/(\Delta + \omega_{\mathrm{FS}})$. At the same time, the rate of spontaneous emission from both fine-structure levels add together so that for $\Delta \gg \omega_{\mathrm{FS}}$, $\Omega^*/R_s \simeq (g_1 g_2/(g_1^2 + g_2^2))(\omega_{\mathrm{FS}}/2\gamma)$ which is equal to $Q_s \equiv \omega_{\mathrm{FS}}/4\gamma$ for $g_1 = g_2$. Therefore, to maximize the number of logic operations before spontaneous emission decoherence, we want to find an ion qubit where Q_s is as large as possible. For several ions of interest for quantum computing we find $Q_s \simeq 2.6 \times 10^3$ ($^9\mathrm{Be}^+$), 1.7×10^4 ($^{25}\mathrm{Mg}^+$), 7.5×10^4 ($^{43}\mathrm{Ca}^+$), 2.8×10^5 ($^{88}\mathrm{Sr}^+$), 9.1×10^4 ($^{67}\mathrm{Zn}^+$), 4.2×10^5 ($^{113}\mathrm{Cd}^+$), 9.8×10^5 ($^{199}\mathrm{Hg}^+$). The precise values of Ω^* and Q_s must take into account the polarizations used for the Raman beams. This could lower Q_s by a factor of two or three depending on the qubit hyperfine transition and laser polarization used; however, if specific laser polarizations are used, this reduction can be avoided. Also, the stated values of Q_s are for carrier transitions. If travelling-wave beams are used for the Raman transitions, then Q_s must be further reduced for sideband transitions by approximately the Lamb-Dicke parameter. To satisfy the conditions for fault tolerance, we require $Q_s > 10^4$–10^5 [9]. Based on these considerations it appears that $^9\mathrm{Be}^+$ must be ruled out as the ultimate ion qubit; however, it still serves as a useful test-bed for many studies (and could eventually be used as the cooling ion for sympathetic cooling).

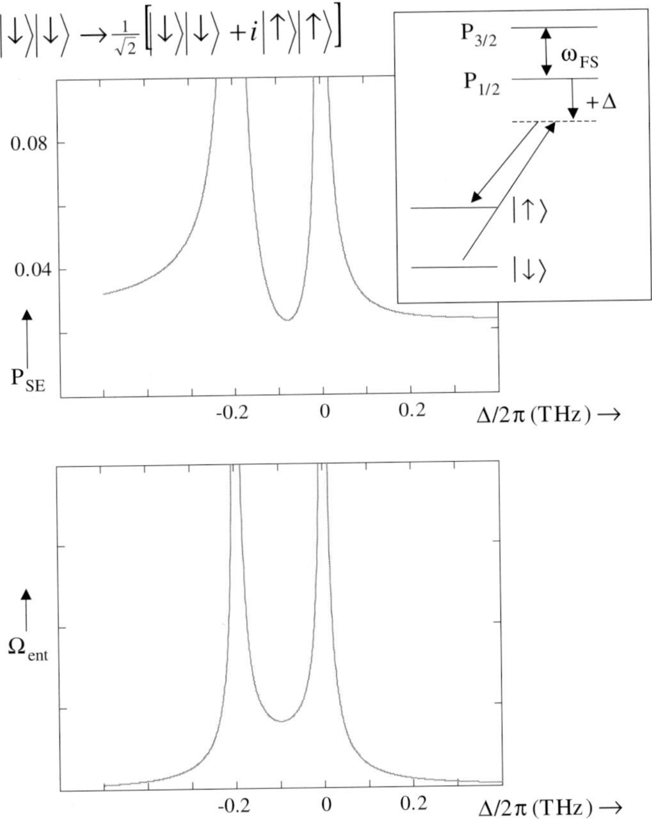

Fig. 8. – In the upper part of the figure we plot the theoretical probability of spontaneous emission P_{SE} during the gate operation $|\downarrow\rangle|\downarrow\rangle \rightarrow (1/\sqrt{2})[|\downarrow\rangle|\downarrow\rangle + i|\uparrow\rangle|\uparrow\rangle]$ of Sørensen and Mølmer [27] for $^9\mathrm{Be}^+$ ions as a function of detuning Δ. For $^9\mathrm{Be}^+$, $\omega_{\mathrm{FS}}/2\pi = 0.198\,\mathrm{THz}$ and conditions are taken to be those of the experiments of ref. [29]. In the lower part of the figure, the overall Rabi rate Ω_{ent} for this process is plotted for the same values of Δ.

To further illustrate the problem of spontaneous emission, in fig. 8, we plot the probability of spontaneous emission P_{SE} during the gate operation described by the first line of eq. (5) (realized in the experiment of ref. [29]). From this figure, we see that Q_s is maximized by either tuning the Raman laser frequencies between the fine-structure levels ($\Delta/2\pi \simeq -0.080\,\mathrm{THz}$) or by making $|\Delta| \gg \omega_{\mathrm{FS}}$. However for the latter condition Ω^* becomes very small (or the required laser intensity becomes very large), indicating that we want to tune between the fine-structure levels instead (as was done in the experiment). For this experiment, fig. 8 indicates that the probability of spontaneous emission was quite high ($\simeq 2\%$); however, we did not use optimum polarization, had extraneous laser lines from unused electro-optic-modulator sidebands, so that in a properly designed system, P_{SE} could be reduced by about a factor of five. Nevertheless, this case illustrates why $^9\mathrm{Be}^+$ will not be a good choice for an ion qubit in the long term.

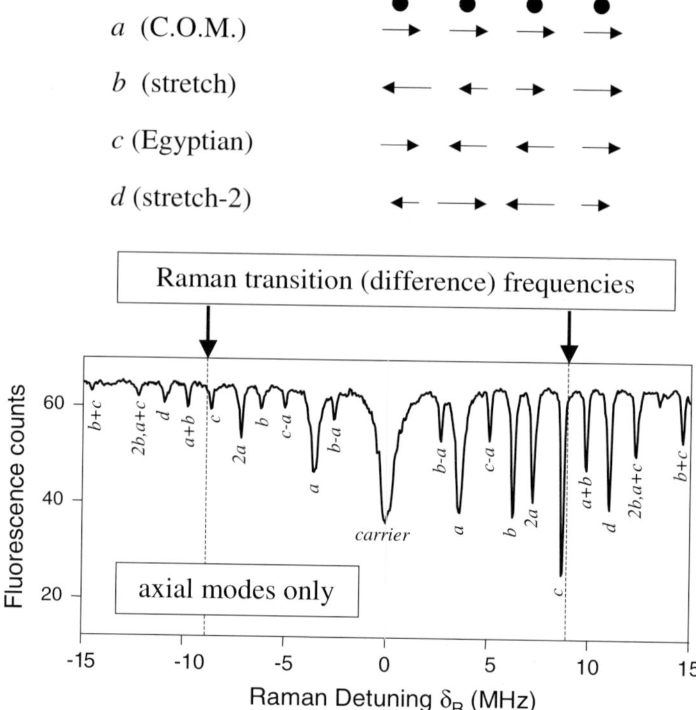

Fig. 9. – Absorption spectrum observed on four $^9\text{Be}^+$ ions. The effective k-vector is parallel to the trap axis so that the laser beams couple only to axial modes. According to the scheme of ref. [27], in order to realize the entangling operation $|\downarrow\downarrow\downarrow\downarrow\rangle \rightarrow (1/\sqrt{2})[|\downarrow\downarrow\downarrow\downarrow\rangle + |\uparrow\uparrow\uparrow\uparrow\rangle]$, we want to simultaneously, but non-resonantly, excite the ions near symmetric red and blue sidebands which, in the experiment, were taken to be at the frequencies indicated by the vertical arrows and dotted lines in the bottom part of the figure. We choose mode "c" because for this mode, the mode amplitudes are equal for all ions as required if the ions are equally illuminated. The horizontal scale is the detuning δ_R of the Raman (difference) frequency from the carrier transition. The vertical scale is the average fluorescence recorded after each application of the Raman laser beams (proportional to the number of ions in the $|\downarrow\rangle$ state. The mode amplitudes (see ref. [84]) for each of the ions for all z-axis modes are represented pictorially in the upper part of the figure. The absorption spectrum has many features including the desired ones; for example, the feature labeled b-a indicates $|\downarrow\rangle \rightarrow |\uparrow\rangle$ transitions accompanied by a single quantum excitation of the "b" mode with a simultaneous de-excitation of the "a" mode. From ref. [99] (See also refs. [61,82,100].)

$6^{.}4$. *Multiplexing.* – In principle, a quantum computer could be constructed by storing a sufficiently large number of ion qubits in a single linear trap as in the original Cirac/Zoller proposal [1]. However, as the number of ions increases, it becomes very difficult to spectrally isolate the logic motional mode from extraneous "spectator" modes [10]. As an illustration of this, in fig. 9 we show a Raman-transition spectrum for four trapped ions that was observed in conjunction with the experiments of Sackett *et al.* [29], where

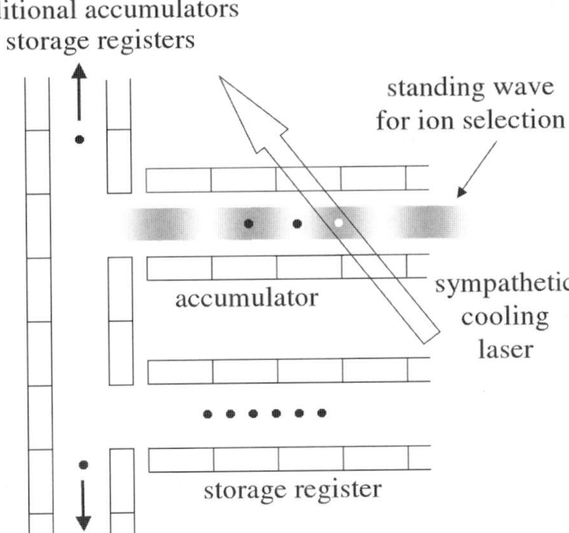

Fig. 10. – Basic elements of a multiplexed trap array. Logic ions (indicated by black dots) are held in accumulators or storage registers. Selected logic ions are moved into accumulators by applying time-sequenced potentials to segmented electrodes that are shown schematically in the figure. Since the ions will likely heat during this process, they will be re-cooled through sympathetic cooling with a "cooling ion" (shown as a white dot in the accumulator). Logic ions could be individually addressed with focused laser beams [1, 101], with the techniques used in refs. [30, 102] or, perhaps with standing waves as indicated in the figure.

the operation $|\downarrow\downarrow\downarrow\downarrow\rangle \rightarrow (1/\sqrt{2})[|\downarrow\downarrow\downarrow\downarrow\rangle + |\uparrow\uparrow\uparrow\uparrow\rangle]$ was implemented. The vertical arrows and dotted lines indicate the (two) Raman-beam difference frequencies to be applied according to the scheme of ref. [27]. The trap well depth and Zeeman shifts were carefully chosen to avoid overlapping spectral lines, but the nature of the problem is evident from the figure; as the number of ions and motional modes increases, it will become very difficult to avoid overlaps with unwanted spectral lines or, alternatively, the operations must become prohibitively slow to achieve the desired spectral resolution. This can be offset by making the trap oscillation frequencies higher, but from eq. (1), to do this we must increase V_0 (which currently leads to arcing across the electrodes) or reduce R, which apparently causes severe heating [59].

These problems could be avoided if, instead of using the motion as a data bus, photon cavity modes are used as proposed in ref. [93]. This scheme is also attractive because the gate speed would not be limited to the ion motion frequencies. However, it appears that operation fidelity would currently be limited by cavity losses. Another possibility is to multiplex the system by putting ions in arrays of distinct traps. Ions in different traps could be coupled optically [94-96] or by means of an ion held in a special trap that is moved from trap to trap to spread entanglement [97]. In an approach being pursued at NIST [10, 98], we will investigate the transfer of ion qubits between different interconnected traps in an array. The basic elements are illustrated in fig. 10. Ions are held either in memory locations or in accumulators that are shielded from one another. In the accumulators, two or more logic qubits are held along with at least one "cooling ion". Logic ions are moved between accumulators and memory locations. Transfer of ions could be accomplished with interconnected linear traps with segmented electrodes that are indicated schematically in the figure. An individual ion or groups of ions could be transferred by generating a local (moving) potential well that is configured with appropriate time-dependent potentials on the segmented electrodes. Construction of these traps could be along the lines of the lithographic traps described in ref. [59]. However, the trap geometry shown there is not compatible with "T" or "X" junctions, so that for these functions a geometry like that of fig. 1 could be used where, for example, the "DC" electrodes are in the plane of the T or X, and the (common) RF electrodes lie outside of this plane above and below it. Alternatively, a three-layer electrode geometry similar to that described in ref. [103] could be employed.

Moving an ion or groups of ions between two locations can, in principle, be done arbitrarily fast, without any heating. To see this, consider a 1D geometry where we want to move an ion from the initial position $z = 0$ to final position z_f. We assume the ion is first trapped in a harmonic well located at $z = 0$ (that is, the well is centered at position $z = 0$) and cooled to its ground state. We suddenly shift the harmonic well to a position z_1 where, ideally, $z_1 \gg z_f$. We now let the ion fall down the potential until it reaches the position $z_f/2$ at which point we shift the well to a position $z_f - z_1$. The ion now climbs the potential hill and stops at position z_f at which point we suddenly switch the well to position z_f. As long as the strength of the harmonic well is preserved during this process, the ion's wave packet does not change from that corresponding to the zero-point state. For example, after the first shift of the well, as the ion falls down the potential hill, it maintains its shape since it is a coherent state. This process can, in principle, be made arbitrarily fast by making z_1/z_f arbitrarily large. Of course, in practice, speeds will be limited by how fast the potentials can be switched. Here, a trade-off is encountered because we desire to filter the electrodes from uncontrolled external fluctuating voltages, which limits the speed of switching. The best compromise will be determined by experiment, where we must include the ambient external noise.

More difficult is the separation of a group of ions located in one trap followed by transfer of the subgroups to separate traps to combine them with other ions. As a model (1D) problem, consider two ions in a trap which are initially cooled to their ground states (more precisely, their center-of-mass and stretch modes are cooled to the ground state). We want to separate them and place them in traps at different locations without appreciable heating. The problem is very similar to that of moving an ion from one location to the other with some important practical differences. Basically, we want to drive a potential hill or "wedge" between the ions, and keep the local well depth for an individual ion strong. However, for electrode sizes currently used, we cannot make this wedge narrow enough to maintain this condition. In practice, the width of the wedge will be approximately equal to the distance from the ions to the electrodes (100–200 μm in current traps) whereas the distance between two $^9\mathrm{Be}^+$ ions from the Coulomb repulsion in a 5 MHz trap is only about $3\,\mu$m. Two approaches might be considered to overcome this difficulty.

The most straightforward solution is to use very small traps for the separation so that a narrow wedge can separate the ions. At the present time, this seems like an ominous prospect since the heating that has been observed appears to increase strongly as the trap dimensions are reduced [59]. However, we are optimistic that the source of anomalous heating will eventually be eliminated so we keep this option open. A second approach is to optimize the separation speed *vs.* heating for the current trap sizes. Classically, this is a straightforward problem. After cooling the ions, we relax the well to let the Coulomb repulsion push the ions apart. After a sufficient separation has taken place we drive the wedge between the ions and, by appropriate timing of external potentials (as in the single-ion transfer problem), we then catch the ions in separate wells at rest. However, in the quantum case, the wave packet will spread during this process so that even though the center of the wave packet will be at rest after this process, it will not correspond to the zero-point state of the final well. If the separation of the ions is done adiabatically, the ions can be caught in motional states near their ground states, but this may be too slow. Optimization in this case will require a combination of quantum modelling coupled with experiment. Since we do not anticipate catching the ions in their respective ground states after this process, we will rely on sympathetic cooling to provide a final cooling.

In addition to the heating incurred in ion transfer, a number of sources of internal-state decoherence might be incurred. If the magnetic field varies in time (for example, from currents generated during switching of "static" potentials) and/or over the trap array, phase decoherence will likely occur. This can be mitigated by the use of field-independent transitions as described in subsect. **6**·**2**, or the use of logical ion qubits encoded into a (phase) decoherence-free subspace (DFS) using pairs of physical qubits [30] (with or without the use of field-independent transitions). The latter method will suppress decoherence due to magnetic-field spatial variations if the length scale of spatial variations is long compared to the spacing between ion pairs in the DFS. The use of such encoding also nearly eliminates the sensitivity of a quantum computer to phase variations over the duration of the computation and between different regions of the trap array (D. Kielpinski *et al.*, unpublished).

7. – Summary

It is probably fair to say that building a quantum computer will be extremely difficult in any currently proposed implementation. We have briefly discussed some of the features and main problems in trying to construct a quantum computer based on atomic ion qubits that are confined in a trap. In spite of the many problems encountered with this scheme, it in principle satisfies all of the required criteria [3] and the most pressing problems (motional heating, laser intensity fluctuations, spontaneous emission, ion transfer) have straightforward solutions. Certainly the quest to try to overcome these problems will be interesting and challenging.

* * *

The NIST work is supported by the U.S. National Security Agency (NSA) and the Advanced Research and Development Activity (ARDA) under contract MOD-7171.00, the U.S. Office of Naval Research (ONR), the U.S. National Reconnaissance Organization (NRO) and NIST. This paper is a contribution of the National Institute of Standards and Technology and is not subject to U.S. copyright. The author is indebted to many collaborators at NIST whose work provide the basis for these lecture notes. These people include A. BEN-KISH, J. BEALL, J. BERGQUIST, J. BOLLINGER, J. BRITTON, B. DEMARCO, W. ITANO, D. KIELPINSKI, B. KING, C. LANGER, D. LEIBFRIED, D. MEEKHOF, V. MEYER, C. MYATT, T. ROSENBAND, M. ROWE, C. SACKETT, Q. TURCHETTE, C. WOOD, and particularly C. MONROE who has now established a separate ion trap quantum computation project at the University of Michigan. The author also thanks J. BRITTON, C. LANGER, D. LEIBFRIED, and V. MEYER for helpful comments and suggestions on the manuscript.

REFERENCES

[1] CIRAC J. I. and ZOLLER P., *Phys. Rev. Lett.*, **74** (1995) 4091.

[2] DIVINCENZO D. P., *Phys. Rev. A*, **51** (1995) 1015.

[3] DIVINCENZO D. P., *The physical implementation of quantum computation*, in *Scalable Quantum Computers*, edited by S. L. BRAUNSTEIN and H. K. LO (Wiley-VCH, Berlin) 2001, pp. 1-13.

[4] ZUREK W. H., *Phys. Today*, **44**, December issue (1991) 36.

[5] ZUREK W. H., *Decoherence, einselection, and the quantum origins of the classical*, quant-ph/0105127, 2001.

[6] GHOSH P. K., *Ion Traps* (Clarendon Press, Oxford) 1995.

[7] Special focus issue: *Quantum computing/quantum cryptography - message coding with the help of quantum mechanics, Fortsch. Phys.*, **46**, No. 4-5 (1998) 325-592.

[8] Special focus issue: *Fundamental problems in quantum theory, Fortsch. Phys.*, **46**, No. 6-8 (1998) 593-928.

[9] BRAUNSTEIN S. L. and LO H. K., *Scalable Quantum Computers*, 1st edition (Wiley-VCH, Berlin) 2001.

[10] WINELAND D. J., MONROE C., ITANO W. M., LEIBFRIED D., KING B. E. and MEEKHOF D. M., *J. Res. Natl. Inst. Stand. Tech.*, **103** (1998) 259.

[11] BARDROFF P. J., LEICHTLE C., SCHRADE G. and SCHLEICH W. P., *Phys. Rev. Lett.*, **77** (1996) 2198.

[12] BARDROFF P. J., LEICHTLE C., SCHRADE G. and SCHLEICH W. P., *Act. Phys. Slov.*, **46** (1996) 231.

[13] ENZER D. G., SCHAUER M. M., GOMEZ J. J., GULLEY M. S., HOLZSCHEITER M. H., KWIAT P. G., LAMOREAUX S. K., PETERSON C. G., SANDBERG V. D., TUPA D., WHITE A. G., HUGHES R. J. and JAMES D. F. V., *Phys. Rev. Lett.*, **85** (2000) 2466.

[14] DEVOE R. G., HOFFNAGLE J. and BREWER R. G., *Phys. Rev. A*, **39** (1989) 4362.

[15] BLÜMEL R., KAPPLER C., QUINT W. and WALTHER H., *Phys. Rev. A*, **40** (1989) 808.

[16] WALTHER H., *Adv. At. Mol. Opt. Phys.*, **31** (1993) 137.

[17] YU N., NAGOURNEY W. and DEHMELT H., *Am. J. Phys.*, **69** (1991) 3779.

[18] MILLER J. D., BERKELAND D. J., CRUZ F. C., BERGQUIST J. C., ITANO W. M. and WINELAND D. J., *A cryogenic linear ion trap for* ^{199}Hg$^+$ *frequency standards, IEEE International Frequency Control Symposium* (1996), pp. 1086-1088.

[19] BERKELAND D. J., MILLER J. D., BERGQUIST J. C., ITANO W. M. and WINELAND D. J., *J. Appl. Phys.*, **83** (1998) 5025.

[20] TURCHETTE Q. A., WOOD C. S., KING B. E., MYATT C. J., LEIBFRIED D., ITANO W. M., MONROE C. and WINELAND D. J., *Phys. Rev. Lett.*, **81** (1998) 1525.

[21] LEIBFRIED D., *Phys. Rev. A*, **60** (1999) R3335.

[22] JAYNES E. T. and CUMMINGS F. W., *Proc. IEEE*, **51** (1963) 89.

[23] MEEKHOF D. M., MONROE C., ITANO W. M., KING B. E. and WINELAND D. J., *Phys. Rev. Lett.*, **76** (1996) 1796.

[24] ROOS CH., ZEIGER TH., ROHDE H., NÄGERL H. C., ESCHNER J., LEIBFRIED D., SCHMIDT-KALER F. and BLATT R., *Phys. Rev. Lett.*, **83** (1999) 4713.

[25] MONROE C., MEEKHOF D. M., KING B. E., ITANO W. M. and WINELAND D. J., *Phys. Rev. Lett.*, **75** (1995) 4714.

[26] SØRENSEN A. and MØLMER K., *Phys. Rev. Lett.*, **82** (1999) 1971.

[27] SØRENSEN A. and MØLMER K., *Phys. Rev. A*, **62** (2000) 02231-1.

[28] SOLANO E., DE MATOS FILHO R. L. and ZAGURY N., *Phys. Rev. A*, **59** (1999) 2539.

[29] SACKETT C. A., KIELPINSKI D., KING B. E., LANGER C., MEYER V., MYATT C. J., ROWE M., TURCHETTE Q. A., ITANO W. M., WINELAND D. J. and MONROE C., *Nature*, **404** (2000) 256.

[30] KIELPINSKI D., MEYER V., ROWE M. A., SACKETT C. A., ITANO W. M., MONROE C. and WINELAND D. J., *Science*, **291** (2001) 1013.

[31] DIEDRICH F., BERGQUIST J. C., ITANO W. M. and WINELAND D. J., *Phys. Rev. Lett.*, **62** (1989) 403.

[32] MONROE C., MEEKHOF D. M., KING B. E., JEFFERTS S. R., ITANO W. M., WINELAND D. J. and GOULD P., *Phys. Rev. Lett.*, **75** (1995) 4011.

[33] ROOS C. F., LEIBFRIED D., MUNDT A., SCHMIDT-KALER F., ESCHNER J. and BLATT R., *Phys. Rev. Lett.*, **85** (2000) 5547.

[34] BLATT R. and ZOLLER P., *Eur. Phys. J.*, **9** (1988) 250.

[35] CALDEIRA A. O. and LEGGETT A. J., *Phys. Rev. A*, **31** (1985) 1059.

[36] WALLS D. F. and MILBURN G. J., *Phys. Rev. A*, **31** (1985) 2403.

[37] COLLETT M. J., *Phys. Rev. A*, **38** (1988) 2233.

[38] WALLS D. F. and MILBURN G. J., *Quantum Optics*, 1st edition (Springer, Berlin) 1994.

[39] VOGEL W. and WELSCH D. G., *Quantum Optics*, 1st edition (Akademie Verlag, Berlin) 1994.

[40] BUŽEK V. and KNIGHT P. L., *Prog. Opt.*, **34** (1995) 1.

[41] POYATOS J. F., CIRAC J. I. and ZOLLER P., *Phys. Rev. Lett.*, **77** (1996) 4728.

[42] SCHLEICH W. P., *Quantum Optics in Phase Space*, 1st edition (Wiley-VCH, Berlin) 2001.

[43] BRUNE M., HAGLEY E., DREYER J., MAÎTRE X., MAALI A., WUNDERLICH C., RAIMOND J. M. and HAROCHE S., *Phys. Rev. Lett.*, **77** (1996) 4887.

[44] SCHNEIDER S. and MILBURN G. J., *Phys. Rev. A*, **57** (1998) 3748.

[45] MURAO M. and KNIGHT P. L., *Phys. Rev. A*, **58** (1998) 663.

[46] SCHNEIDER S. and MILBURN G. J., *Phys. Rev. A*, **59** (1999) 3766.

[47] BONIFACIO R., OLIVARES S., TOMBESI P. and VITALI D., *Phys. Rev. A*, **61** (2000) 053802-1.

[48] WINELAND D. J. and DEHMELT H. G., *J. Appl. Phys.*, **46** (1975) 919.

[49] MYATT C. J., KING B. E., TURCHETTE Q. A., SACKETT C. A., KIELPINSKI D., ITANO W. M., MONROE C. and WINELAND D. J., *Nature*, **403** (2000) 269.

[50] TURCHETTE Q. A., MYATT C. J., KING B. E., SACKETT C. A., KIELPINSKI D., ITANO W. M., MONROE C. and WINELAND D. J., *Decoherence and decay of motional quantum states of a trapped atom coupled to engineered reservoirs*, Phys. Rev. A, **62** (2000).

[51] RAMSEY N. F., *Molecular Beams* (Oxford University Press, London) 1963.

[52] RAIMOND J. M., BRUNE M. and HAROCHE S., Phys. Rev. Lett., **79** (1997) 1964.

[53] CHAPMAN M. S., HAMMOND T. D., LENEF A., SCHMIEDMAYER J., RUBENSTEIN R. A., SMITH E. and PRITCHARD D. E., Phys. Rev. Lett., **75** (1995) 3783.

[54] DÜRR S., NONN T. and REMPE G., Phys. Rev. Lett., **81** (1998) 5705.

[55] BERTET P., OSNAGHI S., RAUSCHENBEUTEL A., NOGUES G., AUFFEVES A., BRUNE M., RAIMOND J. and HAROCHE S., Science, **411** (2001) 166.

[56] MAÎTRE X., HAGLEY E., NOGUES G., WUNDERLICH C., GOY P., BRUNE M., RAIMOND J. M. and HAROCHE S., Phys. Rev. Lett., **79** (1997) 769.

[57] VARCOE B. T. H., BRATTKE S., WEIDINGER M. and WALTHER H., Nature, **403** (2000) 743.

[58] LEGGETT A. J., *Phys. World*, December issue (1999) 73.

[59] TURCHETTE Q. A., KIELPINSKI D., KING B. E., LEIBFRIED D., MEEKHOF D. M., MYATT C. J., ROWE M. A., SACKETT C. A., WOOD C. S., ITANO W. M., MONROE C. and WINELAND D. J., Phys. Rev. A, **61** (2000) 063418-1.

[60] KING B. E., WOOD C. S., MYATT C. J., TURCHETTE Q. A., LEIBFRIED D., ITANO W. M., MONROE C. and WINELAND D. J., Phys. Rev. Lett., **81** (1998) 1525.

[61] ROHDE H., GULDE S. T., ROOS C. F., BARTON P. A., LEIBFRIED D., ESCHNER J., SCHMIDT-KALER F. and BLATT R., J. Opt. B, **3** (2001) S34.

[62] HALD J., SØRENSEN J. L., SCHORI C. and POLZIK E. S., Phys. Rev. Lett., **83** (1999) 1319.

[63] KUZMICH A., MANDEL L. and BIGELOW N. P., Phys. Rev. Lett., **85** (2000) 1594.

[64] MEYER V., ROWE M. A., KIELPINSKI D., SACKETT C. A., ITANO W. M., MONROE C. and WINELAND D. J., Phys. Rev. Lett., **86** (2001) 5870.

[65] ORZEL C., TUCHMAN A. K., FENSELAU M. L., YASUDA M. and KASEVICH M. A., Science, **291** (2001) 2386.

[66] YURKE B., Phys. Rev. Lett., **56** (1986) 1515.

[67] YURKE B., McCALL S. L. and KLAUDER J. R., Phys. Rev. A, **33** (1986) 4033.

[68] KITAGAWA M. and UEDA M., Phys. Rev. A, **47** (1993) 5138.

[69] SANDERS B. C. and MILBURN G. J., Phys. Rev. Lett., **75** (1995) 2944.

[70] JACOBSON J., BJÖRK G., CHUANG I. and YAMAMOTO Y., Phys. Rev. Lett., **74** (1995) 4835.

[71] BOUYER P. and KASEVICH M. A., Phys. Rev. A, **86** (1997) R1083.

[72] WINELAND D. J., BOLLINGER J. J., ITANO W. M., MOORE F. L. and HEINZEN D. J., Phys. Rev. A, **46** (1992) R6797.

[73] WINELAND D. J., BOLLINGER J. J., ITANO W. M. and HEINZEN D. J., Phys. Rev. A, **50** (1994) 67.

[74] BOLLINGER J. J., ITANO W. M., WINELAND D. J. and HEINZEN D. J., Phys. Rev. A, **54** (1996) R4649.

[75] MESSIAH A., *Quantum Mechanics* (John Wiley and Sons, Inc.) 1961.

[76] ITANO W. M., BERGQUIST J. C., BOLLINGER J. J., GILLIGAN J. M., HEINZEN D. J., MOORE F. L., RAIZEN M. G. and WINELAND D. J., Phys. Rev. A, **47** (1993) 3554.

[77] ARECCHI F. T., COURTENS E., GILMORE R. and THOMAS H., Phys. Rev. A, **6** (1972) 2211.

[78] SANTARELLI G., LAURENT PH., LEMONDE P., CLAIRON A., MANN A. G., CHANG S., LUITEN A. N. and SALOMON C., Phys. Rev. Lett., **82** (1999) 4619.

[79] BERKELAND D. J., MILLER J. D., BERGQUIST J. C., ITANO W. M. and WINELAND D. J., *Phys. Rev. Lett.*, **80** (1998) 2089.

[80] LARSON D. J., BERGQUIST J. C., BOLLINGER J. J., ITANO W. M. and WINELAND D. J., *Phys. Rev. Lett.*, **57** (1986) 70.

[81] WINELAND D. J., BERGQUIST J. C., BERKELAND D. J., BOLLINGER J. J., CRUZ F. C., ITANO W. M., JELENKOVIĆ B. M., KING B. E., MEEKHOF D. M., MILLER J. D., MONROE C. and TAN J. N., *Application of laser-cooled ions to frequency standards and metrology*, in *Proceedings of the 5th Symposium on Frequency Standards and Metrology*, edited by J. C. BERGQUIST (World Scientific, Singapore) 1996, pp. 11-19.

[82] WINELAND D. J., BERGQUIST J. C., ITANO W. M., BOLLINGER J. J. and MANNEY C. H., *Phys. Rev. Lett.*, **59** (1987) 2935.

[83] KIELPINSKI D., KING B. E., MYATT C. J., SACKETT C. A., TURCHETTE Q. A., ITANO W. M., MONROE C. and WINELAND D. J., *Phys. Rev. A*, **61** (2000) 032310-1.

[84] JAMES D. F. V., *Appl. Phys. B*, **66** (1998) 181.

[85] BERGQUIST J. C., ITANO W. M. and WINELAND D. J., *Laser stabilization to a single ion*, in *Frontiers in Laser Spectroscopy, Proceedings of the International School of Physics "Enrico Fermi", Course CXX*, edited by T. W. HÄNSCH and M. INGUSCIO (North Holland, Amsterdam) 1994, pp. 359-376.

[86] MA L. S., JUNGNER P., YE J. and HALL J. L., *Opt. Lett.*, **19** (1994) 1777.

[87] YOUNG B. C., RAFAC R. J., BEALL J. A., CRUZ F. C., ITANO W. M., WINELAND D. J. and BERGQUIST J. C., Hg$^+$ *optical frequency standard: recent progress*, in *Laser Spectroscopy*, edited by R. BLATT, J. ESCHNER, D. LEIBFRIED and F. SCHMIDT-KALER, Vol. XIV (World Scientific, Singapore) 1999, pp. 61-70.

[88] YOUNG B. C., CRUZ F. C., ITANO W. M. and BERGQUIST J. C., *Phys. Rev. Lett.*, **82** (1999) 3799.

[89] FISK P. T. H., *Rep. Prog. Phys.*, **60** (1997) 761.

[90] BOLLINGER J. J., HEINZEN D. J., ITANO W. M., GILBERT S. L. and WINELAND D. J., *IEEE Trans. Instr. Meas.*, **40** (1991) 126.

[91] PLENIO M. B. and KNIGHT P. L., *Phys. Rev. A*, **53** (1996) 2986.

[92] PLENIO M. B. and KNIGHT P. L., *Proc. R. Soc. London, Ser. A*, **453** (1997) 2017.

[93] PELLIZZARI T., GARDINER S. A., CIRAC J. I. and ZOLLER P., *Phys. Rev. Lett.*, **75** (1995) 3788.

[94] CIRAC J. I., ZOLLER P., KIMBLE H. J. and MABUCHI H., *Phys. Rev. Lett.*, **78** (1997) 3221.

[95] VAN ENK S. J., CIRAC J. I. and ZOLLER P., *Phys. Rev. Lett.*, **78** (1997) 4293.

[96] DEVOE R. G., *Phys. Rev. A*, **58** (1998) 910.

[97] CIRAC J. I. and ZOLLER P., *Nature*, **404** (2000) 579.

[98] WINELAND D. J., MONROE C., MEEKHOF D. M., KING B. E., LEIBFRIED D., ITANO W. M., BERGQUIST J. C., BERKELAND D., BOLLINGER J. J. and MILLER J., *Entangled states of atomic ions for quantum metrology and computation*, in *Atomic Physics 15*, edited by H. B. VAN LINDEN VAN DEN HEUVELL, J. T. M. WALRAVEN and M. W. REYNOLDS, Vol. **15** (World Scientific, Singapore) 1997, pp. 31-46.

[99] MONROE C., SACKETT C. A., KIELPINSKI D., KING B. E., LANGER C., MEYER V., MYATT C., ROWE M., TURCHETTE Q., ITANO W. M. and WINELAND D. J., *Scalable entanglement of trapped ions*, in *Atomic Physics 17*, edited by E. ARIMONDO, P. DE NATALE and M. INGUSCIO, *AIP Conf. Proc.*, Vol. **551** (Melville, New York) 2001, pp. 173-186.

[100] SCHMIDT-KALER F., ROOS C. H., NÄGERL H. C., ROHDE H., GULDE S., MUNDT A., LEDERBAUER M., THALHAMMER G., ZEIGER T. H., BARTON P., HORNEKAER L., REYMOND G., LEIBFRIED D., ESCHNER J. and BLATT R., *J. Mod. Opt.*, **47** (2000) 2573.

[101] NÄGERL H. C., LEIBFRIED D., ROHDE H., THALHAMMER G., ESCHNER J., SCHMIDT-KALER F. and BLATT R., *Phys. Rev. A*, **60** (1999) 145.

[102] ROWE M. A., KIELPINSKI D., MEYER V., SACKETT C. A., ITANO W. M., MONROE C. and WINELAND D. J., *Nature*, **409** (2001) 791.

[103] SCHRAMA C. A., PEIK E., SMITH W. W. and WALTHER H., *Opt. Commun.*, **101** (1993) 32.

Towards quantum computation with trapped Ca^+ ions

F. Schmidt-Kaler, D. Leibfried(*), J. Eschner and R. Blatt

Institut für Experimentalphysik, Universität Innsbruck
Technikerstraße 25, A-6020 Innsbruck, Austria

1. – Introduction

Computing and processing of information using the laws of quantum mechanics was proposed already in the eighties by Feynman and Deutsch [1,2]. They thought about the storage of information in states of a single two-level quantum system. Thus, information cannot only be represented by a bit in a state 0 or 1, but must be accounted for in superposition states $|\psi\rangle$

$$(1) \qquad\qquad |\psi\rangle = \alpha|0\rangle + \beta|1\rangle, \qquad \text{with} \qquad |\alpha|^2 + |\beta|^2 = 1.$$

Such quantum states are to be considered as being simultaneously in the logical states $|0\rangle$ or $|1\rangle$ and only a (quantum) measurement will find either one of the states $|0\rangle$ or $|1\rangle$ with the probabilities $|\alpha|^2$ and $|\beta|^2$. This elementary storage of quantum information was coined a quantum bit or qubit as a generalization of the classical bit. It seems an odd proposition at first to consider information as possibly being in a superposition, since it appears impossible to predictably handle and process such an elusive quantum state. Therefore, and since quantum algorithms did not seem to be advantageous for computational purposes, until a few years ago quantum computing has been widely regarded as a curiosity by the information sciences.

(*) Present address: NIST, 325 Broadway, Boulder, CO 80305, USA.

This situation changed dramatically in 1994 when Shor [3] came up with a quantum algorithm which allows one to factorize large numbers much faster than with any classical computation. With this at hand, a quantum computer technology would endanger all of today's powerful cryptosystems since they are all based on the fact that factorization of large numbers is a difficult computational problem. Therefore, the discovery of a powerful quantum algorithm for fast factoring and the subsequently found fast algorithm for a data basis search by Grover [4] spawned a worldwide search for systems to implement and build a quantum computer.

The storage and processing of quantum information requires systems with long-lived quantum states in order to keep the information in the quantum memory during the computational process. The necessity to manipulate the thus stored quantum information needs single quantum systems well isolated from interacting with the environment and which can be initialized, measured and individually handled. Logical operations between qubits can be implemented with a controllable interaction of the single quantum systems [5]. During the last years, therefore, a large variety of physical systems have been proposed and investigated for their use in quantum information processing. NMR-based experiments [6], ion trap realizations [7], quantum dot systems [8] and Josephson junctions [9] are among some of the most promising techniques currently being investigated.

In the following, quantum computation is discussed using trapped ions with particular emphasis on a realization using trapped Ca^+ ions. This paper is organized as follows: In sect. **2** the general scheme of an ion trap quantum computer will be presented. Linear ion traps are described in sect. **3** and the implementation of a quantum computer with trapped Ca^+ ions is described in sect. **4**. Initialization of the quantum register by laser cooling of single ions and ion strings is summarized in sect. **5**. Information processing requires addressing of individual ions which is described in sect. **6** and the manipulation of the stored quantum information is shown in sect. **7**. We conclude by summarizing the current state of the experiments on quantum computation with trapped Ca^+ ions.

2. – Scheme of an ion trap quantum computer

Strings of trapped ions have been proposed for quantum computation by Cirac and Zoller [7]. They showed that all requirements [5] of a quantum system can be met and they described how a universal quantum gate can be implemented. Ions in Paul traps have been used for many years now in precision measurements, especially for time and frequency standards applications [10]. They are therefore ideal elements to carry quantum information, and the qubits can be realized with either narrow optical transitions, long-lived hyperfine states or their Zeeman sub-states. In linear Paul traps strings of ions can be stored [11, 12] and can thus serve as the equivalent to a classical register, *i.e.* as a quantum register. For the initialization of the qubits the internal states of the ions must be preset in well-defined states which can be achieved by individually addressing ions in a string with an interacting laser beam.

Figure 1 shows schematically how quantum gate operations between the ions can be achieved. The corresponding procedure is described in detail in ref. [7]: In a first step the

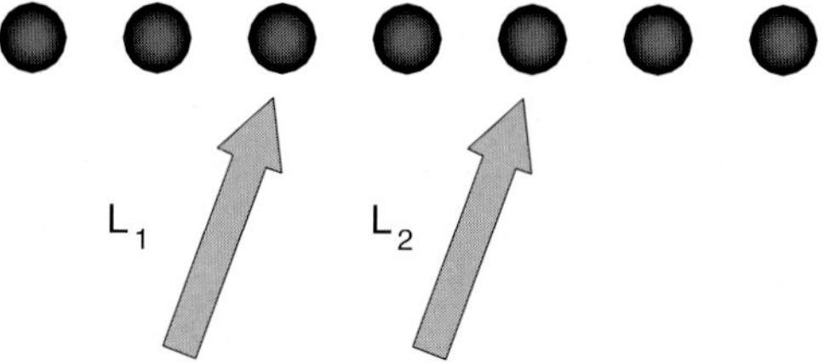

Fig. 1. – Cirac-Zoller scheme of a quantum computer: A string of trapped ions is individually addressed and the universal gate operations are realized by mapping the control qubit with laser L$_1$ to the motion of the string and then introducing a conditional sign change on the wave function of the target qubit with laser L$_2$.

entire ion string (*i.e.* the quantum register) is cooled to the ground state of its motion in the ion trap. Then, the initial quantum information is stored in the internal states of the different ions (*i.e.* the individual qubits) which can be considered as independent. However, since the mutual Coulomb repulsion spatially separates the ions, any induced motion will eventually couple the ions. Therefore a universal quantum gate can be achieved by addressing an ion (which the algorithm identifies as the controlling qubit), in the string and exciting it with the interacting laser beam in such a way that its internal excited-state amplitude is mapped to a (single-phonon) quantum motion of that ion. This phonon, however, is now carried by the entire string, and an operation on one of the other ions (chosen according to the underlying algortithm as the target qubit), which depends on whether or not there is motion in the string, allows one to realize the conditional gate operation required.

After the initialization stage, the addressing laser will act on a selected ion out of the string (given by the chosen quantum algorithm) and excite the motion according to its internal excited-state amplitude. Then, a second pulse on the target qubit can introduce a 2π rotation (on another internal auxiliary transition) and will therefore introduce a minus sign in the wave function if and only if the ion string was put in motion by the first pulse. Eventually, a final laser pulse on the first ion maps the motion back to its internal state which then concludes the conditional operation. For a full C-NOT gate operation (*i.e.* the quantum equivalent to the classical XOR gate operation) also arbitrary individual qubit rotations are required. Thus, the general scheme of an ion trap quantum computation can be visualized as indicated in fig. 2.

Any algorithm can be broken down in a series of such one- and two-qubit operations and therefore this set of instructions constitutes a universal quantum gate [13]. Thus, the realization of these quantum gates allows one to build and operate a quantum computer.

During the last years several other techniques have been proposed to implement gate operations with trapped ions. Sørensen and Mølmer [14,15], and with a different formulation Milburn [16], proposed a scheme for "hot" quantum gates, *i.e.* their procedures for gate operations do not require ground-state cooling of an ion string. Although successfully applied to trapped Be$^+$ ions [17], with the trapping parameters currently available,

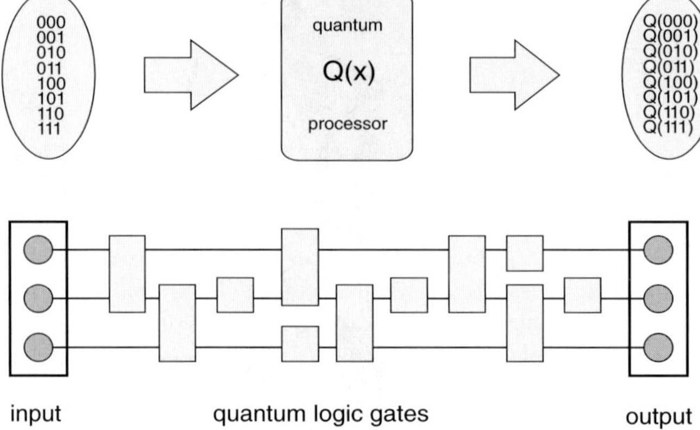

Fig. 2. – General scheme of a quantum computer: Quantum information processing takes place from left to right. Initially stored information is manipulated with single bit rotations (square shapes on the lines representing individual quantum bits) and with two-qubit gates which are indicated by the rectangular shapes connecting two qubits. During the entire process, superpositions are maintained and processed until a final measurement reads the outcome of the computation.

these gate procedures seem not applicable to Ca^+ ions. Other gates based on ac Stark shifts have been suggested by Jonathan *et al.* [18] and holonomic quantum gates (using geometric phases) have been proposed by Luming *et al.* [19]. For the latter proposals so far no experiments are currently available. In the following we will concentrate on experiments to realize the Cirac-Zoller proposal [7] with trapped Ca^+ ions.

3. – Linear ion traps

For a quantum computation it is imperative to preserve coherence during the logic operations, hence techniques with long coherence times are preferentially considered. These are in particular available in systems which have so far been used in precision experiments such as for time and frequency standards and metrology. Especially, ions trapped in a ultrahigh vacuum have been shown to provide extremely long coherence times. Similarly, nuclear spins of atoms and molecules are only weakly interacting with the surrounding environment and are hence suitable candidates for quantum information processing.

The necessity to implement quantum registers with trapped ions naturally leads to the use of linear traps for storing ion strings [20]. The linear variant of the Paul trap is based on the quadrupole mass-filter potential [21]. This potential provides confining forces in the two directions perpendicular to the z-axis, but the motion along the z-axis is not affected. For axial confinement, additional electrodes must be attached. Radial confinement of the ions is created by a dc-voltage U_{dc} and an ac-voltage $V_{ac}\cos(\Omega t)$

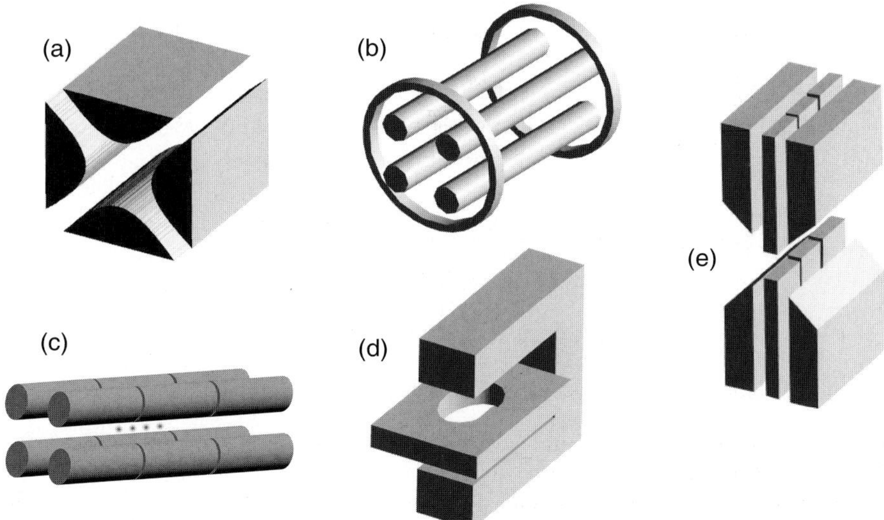

Fig. 3. – Electrode configurations used in linear ion traps. a) Paul mass filter, b) linear ion trap with rods and rings, c) segmented ion trap, d) elongated Paul trap, e) linear endcap trap [22].

applied to the electrodes. Near the trap axis this creates a potential of the form

$$(2) \qquad \Phi = \frac{U_{dc} + V_{ac}\cos(\Omega t)}{2r_0^2}\left(x^2 - y^2\right),$$

where r_0 denotes the distance from the trap axis to the surface of one of the electrodes. As is inferred from eq. (2), the potential is ideally created using hyberbolically shaped electrodes (see fig. 3a). For simplicity they are usually approximated by cylindrical rods as in fig. 3b or more elaborate shapes (fig. 3c-e), depending on the requirements for laser access and diagnostics.

Axial confinement is provided by an additional static potential U_{cap} applied along the z-axis using additional ring electrodes (cf. fig. 3b) or segmented parts of the rod electrodes (cf. fig. 3c). This creates a static harmonic well in the z-direction which is characterized by the longitudinal trap frequency

$$(3) \qquad \omega_z = \sqrt{\frac{2\kappa q U_{cap}}{m z_0^2}}.$$

Here, m and q denote the ion mass and charge, z_0 is half the length between the axially confining electrodes and κ is an empirically determined geometry factor of order unity which accounts for the particular electrode configuration. In principle, exact values of κ can be obtained either numerically or, in some cases, analytically. From a practical point of view, however, using a measured value of κ suffices to describe the experimental data.

The resulting ion motion in the (x, y)-plane consists of a harmonic secular motion (macromotion) at frequencies $\omega_{x,y}$. Superimposed is always the so-called micromotion at the trap's drive frequency Ω. For small rf-amplitudes V_{ac} and for $U_{dc} = 0$ (which is usually chosen), this micromotion is negligible and a confined ion oscillates as if trapped in a harmonic pseudopotential Ψ in the radial direction, given by

$$(4) \qquad q\Psi = q\frac{|\nabla\Phi|^2}{4m\Omega^2} = \frac{1}{2}m\omega_{\mathrm{r}}^2\left(x^2 + y^2\right)$$

with the radial secular frequency $\omega_{\mathrm{r}} \approx qV_{ac}/(\sqrt{2}m\Omega r_0^2)$.

A major advantage of a linear Paul trap (compared to a three-dimensional Paul trap used for the storage of single ions) is that the micromotion completely vanishes for a string of ions confined to the z-axis. The motion is then a pure harmonic oscillation in the static potential providing axial confinement. Furthermore, radial and axial trap frequencies can be chosen independently in a wide range, as is seen from eqs. (2), (3).

Although the use of linear traps for quantum registers with ions seems favorable, an elongated version of the three-dimensional Paul trap can nevertheless be used as well to provide small strings of two and three ions [23]. Such a device consists of an elliptically shaped ring electrode and two endcap electrodes (see fig. 3d) and the ion string is oriented along the long axis of the ring electrode. With this geometry, much higher trap frequencies are possible than with a linear trap, which is of advantage for optical cooling. On the other hand, in this trap there is always residual micromotion present which could eventually lead to a loss in the fidelity of the coherent manipulations.

The motion of an ion string can be described by common harmonic-oscillator modes with frequencies which can be readily calculated [24,25]. The lowest frequency ω_z, corresponding to center-of-mass oscillation of the string, is just the single ion secular frequency and the higher frequencies are approximately scaling with $\sqrt{2N-1}\,\omega_z$, where N denotes the number of ions in the string. Cirac and Zoller considered using the lowest-order mode frequency, however, higher-order modes could be used as well. The higher-order modes are harder to excite but this could even be favorable since any spurious heating of these modes is then suppressed as well.

4. – Ca$^+$ for quantum computation

The relevant levels and transitions of the ^{40}Ca$^+$ ion are shown in fig. 4. Excitation and laser cooling (see sect. **5** below) is achieved using the allowed dipole transitions at 397 nm and 866 nm. A well-defined quantization axis is produced with a magnetic field of about 4 Gauss which split the states into their Zeeman sublevels. With an additional σ^+ polarized beam at 397 nm the pure electronic state $S_{1/2}$, $m = 1/2$ of the ion(s) can be prepared.

The two qubit levels are realized with the $S_{1/2}$ ground state and the metastable $D_{5/2}$ state which has a natural lifetime of about 1 s (fig. 4). Between these levels quadrupole transitions are driven with a Ti:Sapphire laser at 729 nm, stabilized to a high-finesse

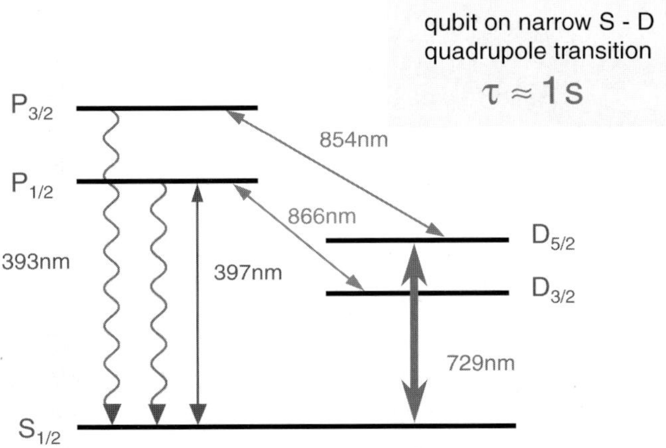

Fig. 4. – Level scheme for trapped Ca$^+$ ions.

cavity with the Pound-Drever-Hall technique. To maintain the coherence necessary for qubit manipulations this laser has to be highly stable. Doppler cooling and highly efficient state detection using the quantum jump technique [26] is achieved with the laser beams at 397 nm and 866 nm. When the ion is in the ground state many photons are scattered and observed at 397 nm. If the ion is in $D_{5/2}$, the electron is decoupled from the excitation and no fluorescence photons can be emitted. Although only a small fraction (less than 10^{-2}) of these fluorescence photons is collected by a lens and imaged onto a photomultiplier with a quantum efficiency of about 15%, one can distinguish the two qubit states within 2 ms of detection time. This allows us to measure the state of our qubit with nearly 100% efficiency.

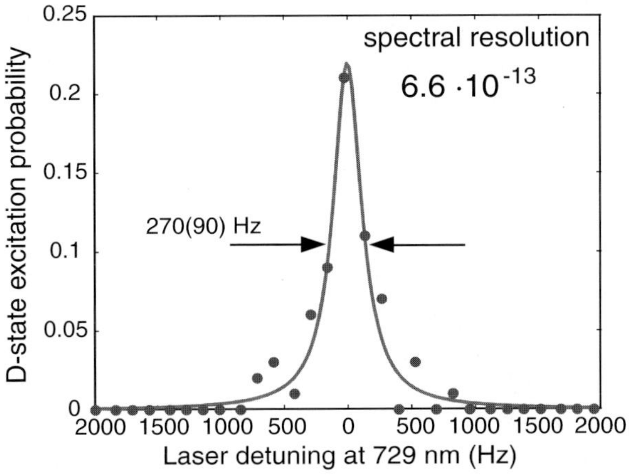

Fig. 5. – High-resolution spectroscopy on the $S_{1/2} \leftrightarrow D_{5/2}$ quadrupole transition.

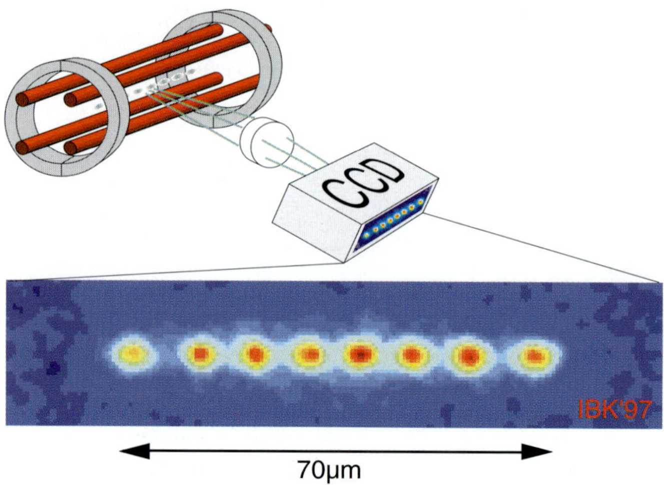

Fig. 6. – String of 8 Ca$^+$ ions in a linear ion trap.

Using this quantum jump (or electron shelving) technique, high-resolution spectroscopy is performed on the $S_{1/2} \leftrightarrow D_{5/2}$ quadrupole transition. As can be seen from fig. 5, the spectral resolution obtained on the $S_{1/2} \leftrightarrow D_{5/2}$ transition is about $7 \cdot 10^{-13}$ which already gives an upper bound for the obtainable coherence time for the qubit manipulations. The observed linewidth of 270(90) Hz is due to the residual laser linewidth and to residual magnetic fields fluctuating at the 50 Hz line frequency. The latter influence can be reduced by further magnetic shielding. We have determined an upper bound of 76(5) Hz (FWHM) for the effective linewidth of our laser system, by observing the fringe contrast in high-resolution Ramsey spectroscopy on the $S_{1/2}$-$D_{5/2}$ transition as a function of the time delay between the two excitation pulses [27].

5. – Laser cooling of trapped ions

For an implementation of the Cirac-Zoller proposal, ground-state cooling of the ions is indispensable. For this, we use a sequence of cooling steps, starting with Doppler cooling and then followed by an appropriate ground-state cooling procedure described below.

5$^{.}$1. *Doppler cooling.* – After the ions are loaded to the trap, resonance fluorescence is continuously generated by shining the two laser wavelengths at 397 nm and 866 nm on the ions. Doppler cooling [28] is achieved by tuning the 397 nm laser below the resonance (while keeping the 866 nm laser on resonance) and this leads to crystallization of the ions and the formation of an ion string. An example of an 8-ion string is shown in fig. 6.

Doppler cooling eventually leaves the ions in a common harmonic motion with a residual vibrational excitation number of about $n_z \simeq 20$–40 in the axial direction and $n_{x,y} \simeq 7$–15 in the radial direction for trap frequencies of a few MHz.

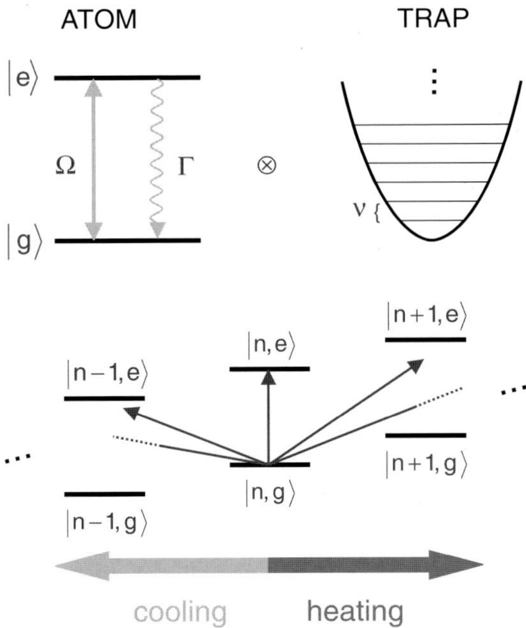

Fig. 7. – Two-level atom in a harmonic trap.

5‘2. *Sideband cooling*. – The major requirement for the realization of the Cirac-Zoller proposal is, however, that a string of ions can be initialized in the ground state of its common motion inside a trap [7]. Therefore, we first consider the quantum structure of the moving trapped ion. For a two-level atom confined in a harmonic trap, fig. 7 shows the appropriate level scheme and the corresponding transitions.

Due to the harmonic motion, each electronic level splits up into a manifold of levels between which transitions can be excited by the laser light. With an upper-state decay constant Γ smaller than the trap frequency ν, an exciting laser could be selectively tuned to any of these transitions, *e.g.* $|g, n\rangle$ to $|e, n-1\rangle$. The subsequent decay is dominant on transitions which do not involve a change in $|n\rangle$ and, therefore, for a laser detuning to the lower sideband (*i.e.* for $|g, n\rangle \leftrightarrow |e, n-1\rangle$ transitions), optical pumping of the vibrational states eventually results in the population of the lowest state (*i.e.* the ground state of the harmonic motion in the trap) only. This sideband cooling technique [29, 30] requires that the exciting laser can be tuned selectively to the different vibrational levels, *i.e.* $\Gamma < \nu$ is necessary condition. This cannot be achieved on the $S_{1/2} \leftrightarrow P_{1/2}$ transition in Ca$^+$ since $\Gamma(P_{1/2}) \simeq 20\,\mathrm{MHz}$ and the trap frequencies are several MHz at best if individual optical addressing of the ions is still to be assured (see below). Instead, sideband cooling in Ca$^+$ is achieved by strongly saturating the $S_{1/2} \leftrightarrow D_{5/2}$ transition on the lower sideband (*i.e.* for a detuning of the 729 nm laser of $\Delta = -\nu$) and simultaneously shining in laser light at the 854 nm transition which quickly quenches the $D_{5/2}$-state population via the $P_{3/2}$-state. The intensity of this quenching laser at 854 nm is adjusted for optimum cooling

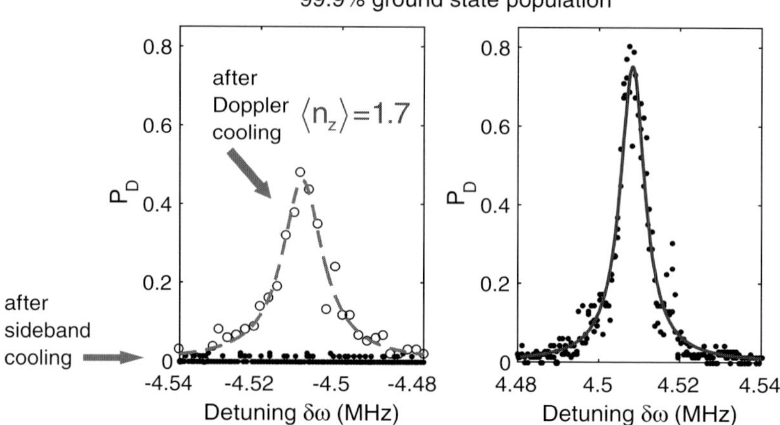

Fig. 8. – Red and blue sidebands of the axial mode (at 4.51 MHz) of one ion after Doppler cooling (dashed) and after resolved sideband cooling (solid). From the ratio of sideband strengths we determine 99.9% ground-state occupation.

during the experiment. Optical pumping to the $S_{1/2}(m = -1/2)$ level is prevented by occasional short laser pulses of σ^+-polarized light at 397 nm. The duration of those pulses is kept at a minimum to prevent unwanted heating. The ground-state occupation is found by comparison of the on-resonance excitation probability for red and blue sideband transitions [30]. For a single ion in a spherical Paul trap we obtained up to 99.9% of motional ground-state occupation within 6 ms (see fig. 8).

Application of the sideband cooling radiation to the corresponding mode frequencies of an ion string in a similar way results in ground-state cooling of two and more ions in a similar way [27]. The resulting cooling times required to reach the ground state are on the order of 1 ms. It has been observed, though, that without the acting cooling lasers, the ion motion heats up in time due to residual noise fields [31]. The source of this heating is not entirely understood, although there is some evidence that this heating scales proportional to $\sim d^{-4}$, where d denotes the average distance between the ions and the nearest electrode. In our case we observe a heating time of 190 ms/phonon for the axial motion and 70 ms/phonon for the radial motion [30], where one phonon denotes the addition of one vibrational quantum. Similar results are obtained for two ions in a linear trap [27].

5˙3. *EIT cooling*. – Very low temperatures are achieved with resolved sideband cooling only if the red sideband is excited with a narrow excitation bandwidth. Otherwise nearby nonresonant transitions (*e.g.*, carrier transitions) will lead to unwanted excess heating and severely increase the final temperature of the ion(s). Unless sideband frequencies are degenerate, this limits resolved sideband cooling to one motional sideband at a time, resulting in involved cooling procedures for ion strings. Moreover the phonons scattered in the process of cooling one motional mode will reheat the other modes. Although for

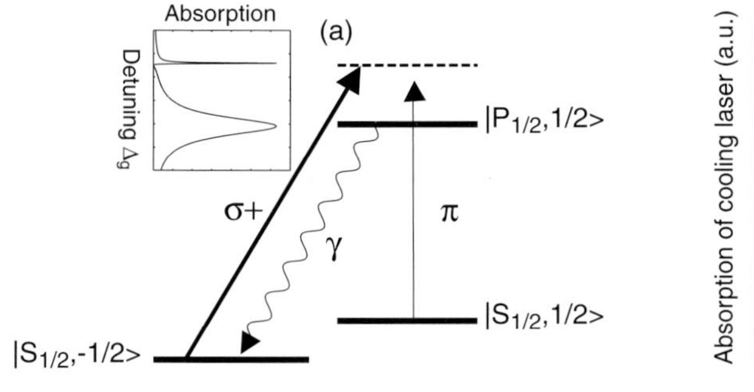
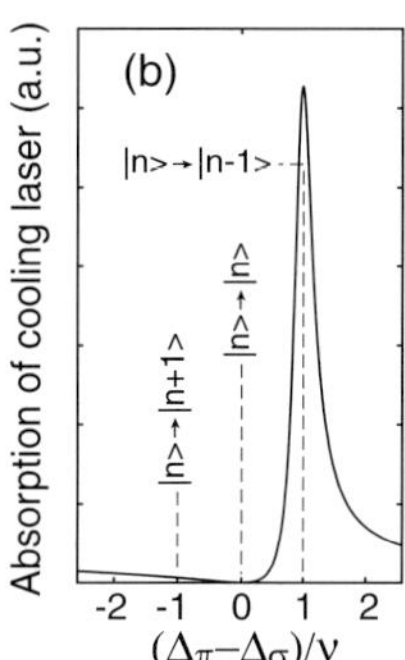

Fig. 9. – (a) Level scheme of the $S_{1/2}$-$P_{1/2}$ manifold and lasers used for EIT cooling. A strong σ^+-polarized beam creates a Fano-type absorption profile for a weak π-polarized beam. (b) By tuning the π-polarized beam to $\Delta_\pi = \Delta_\sigma$ and choosing the ac Stark shift equal to the trap frequency, cooling red sideband transitions are much more probable than heating blue sideband transitions, and the carrier is completely suppressed.

a Cirac-Zoller gate only the motional mode that is used as the "quantum-bus" has to be cooled to a very high degree, the other motional modes must be cooled into the Lamb-Dicke regime [32]. To reach this regime by Doppler cooling requires trap frequencies of 10 MHz or higher and results in an ion spacing that is hard to optically resolve. This leads to a trade-off between addressing as well as detecting individual ions and sufficiently cooling all vibrational modes of a string. In our experiments with two or more ions in the linear trap we decided to maintain good conditions for individual addressing and limited our axial center-of-mass (COM) frequency to 700 kHz. Under these conditions it was desirable to find a cooling technique that is not as narrow-band as resolved sideband cooling but has a lower cooling limit than Doppler cooling. A recent proposal to use electromagnetically induced transparency (EIT) for the cooling of trapped particles [33] promised to cool the ion deeply into the Lamb-Dicke regime for all motional degrees of freedom simultaneously. We adapted this cooling scheme for the case of the $[S_{1/2}, P_{1/2}]$ four-level system in Ca$^+$ that we also use for Doppler cooling [34]. The manifold is dressed with a σ^+-polarized beam at 397 nm, blue detuned by $\Delta_\sigma \simeq 70$ MHz (3.5 linewidths of the S-P transition), that connects the $S_{1/2}$, $m = -1/2$ with the $P_{1/2}$, $m = 1/2$ level. Under these circumstances a second low-intensity π-polarized beam experiences an absorption (Fano-) profile as depicted in fig. 9(a).

In addition to the usual line profile around $\Delta_\pi = 0$, a dark resonance (EIT) is created at $\Delta_\pi = \Delta_\sigma$, and a bright resonance appears at $\Delta_\pi = \Delta_\sigma + \delta$, where δ is the ac Stark shift due to the σ^+-polarized beam. For cooling the π-polarized beam is tuned to $\Delta_\pi = \Delta_\sigma$ and δ is adjusted to match the trap frequency. This creates an asymmetry in absorption for carrier and sidebands: The carrier is almost completely suppressed due to the dark resonance, the blue sideband is in the shallow wing of the profile, but the red sideband is greatly enhanced by the bright resonance, see fig. 9(b). When we tuned the Stark-shift δ to be equal to one of the motional modes at 3.34 MHz we were able to cool this mode

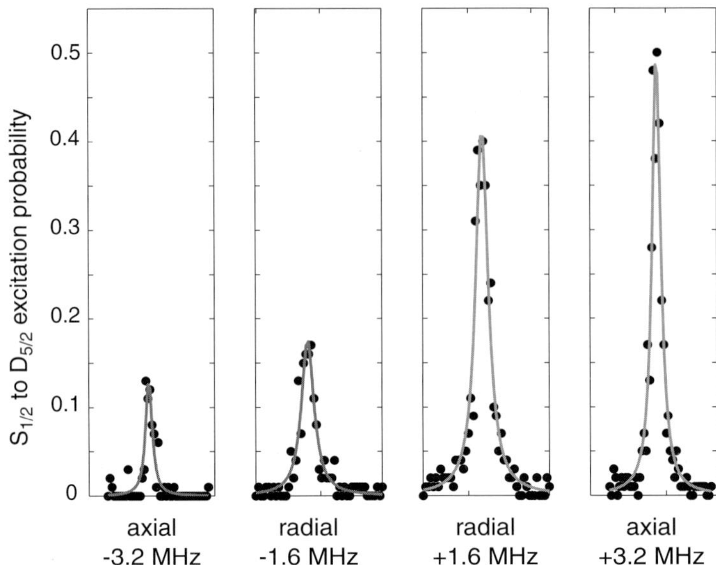

Fig. 10. – Simultaneous cooling of two vibrational modes with an oscillation frequency difference of 1.73 MHz. The sideband asymmetry corresponds to 58% ground-state occupation ($\bar{n} = 0.85$) in the mode at 1.61 MHz and 74% ($\bar{n} = 0.35$) at 3.34 MHz.

to 90% ground-state occupation or $\bar{n} = 0.1$ [34]. Moreover, as sketched in fig. 9(b) the bright resonance can have a substantial width and the red sideband need not necessarily coincide exactly with the maximum of the bright resonance to get a cooling effect. This opens the possibility to cool several motional modes *simultaneously*, as long as they are not too far apart in frequency.

For a demonstration of simultaneous cooling of two vibrational modes with this method we chose the two modes at 1.61 MHz and 3.34 MHz. The ac Stark shift δ of the σ^+-beam was adjusted halfway between the two mode frequencies. With this settings we achieved 58% ground-state occupation ($\bar{n} = 0.85$) in the mode at 1.61 MHz and 74% ($\bar{n} = 0.35$) at 3.34 MHz (see fig. 10).

5′4. *EIT-cooling of linear ion strings*. – Since EIT-cooling allows simultaneous cooling of several modes at different frequencies, it seems to be particularly suited for ion strings in linear traps. The frequencies of the axial vibrational eigenmodes of a linear string have been calculated [24, 25] and measured [35]. For a 10-ion string trapped in a linear trap with a center-of-mass axial frequency of 0.7 MHz, the closest inter-ion spacing is found to be $3.0\,\mu$m ($4.5\,\mu$m for $N = 5$). The axial vibration frequencies are 0.7 MHz, 1.22 MHz, 4.6 MHz.

We now estimate the performance of EIT-cooling for a 10-ion string. The result is displayed in fig. 11: For a 10-ion string indeed *all* vibrational modes are cooled to a mean phonon number $\bar{n} < 1$. This is promising for the application of cold ion strings for quantum information processing [7, 14, 15, 20]. For high-fidelity gate operations it

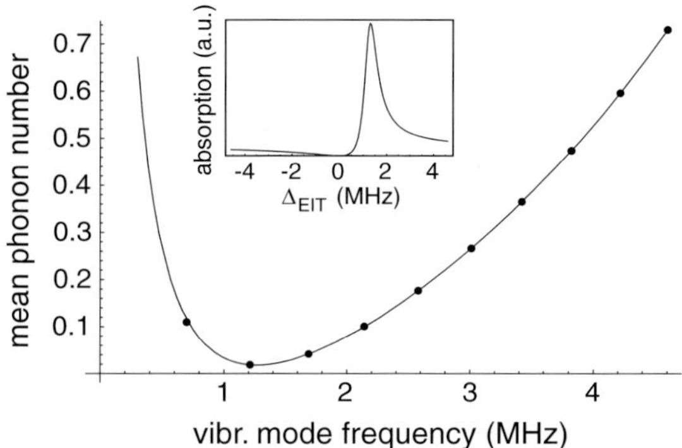

Fig. 11. – EIT-cooling of the axial modes of a linear string based on the $S_{1/2}$-$P_{1/2}$ transition with $\Omega_\sigma = 30\,\mathrm{MHz}$, $\Omega_\pi = 0.5\,\mathrm{MHz}$, $\Gamma_P = 20\,\mathrm{MHz}$, and a detuning of $\Delta_\sigma = \Delta_\pi = 75\,\mathrm{MHz}$. The axial trap frequency is 0.7 MHz. For the calculation we have chosen the light intensity such that the bright state is ac Stark shifted by $\sim 3\,\mathrm{MHz}$ (see inset). The bright resonance with a width of $\sim 0.5\,\mathrm{MHz}$ leads to cooling for all axial modes. The mean phonon numbers (black dots) of all axial modes are plotted *versus* the mode frequencies.

is required that all modes which can couple to the laser light (the so-called *spectator modes*) must be cooled well into the Lamb-Dicke regime $\eta^2 \equiv k^2 \langle x^2 \rangle \ll 1$, where the spatial extension of the motional wave packet $\sqrt{\langle x^2 \rangle}$ is smaller than the laser wavelength $\lambda = 2\pi/k$. This is due to the fact that thermally excited spectator modes cause a blurring of the Rabi frequency Ω which disturbs the precision of quantum operations. For the case of $3N-1$ spectator modes i with $\bar{n}^i$ and Lamb-Dicke factors η_i, the initial state is a mixed state. Each time the experiment will happen with slightly different initial conditions and the Rabi frequency for carrier or sideband transitions will differ. The relative blurring reads like $\Delta\Omega/\Omega = \sqrt{\sum \eta_i^4 \bar{n}^i (\bar{n}^i + 1)/(3N-1)}$ (see eq. (126) in ref. [36]) which we find as small as $3 \cdot 10^{-4}$ for our specific example.

6. – Addressing of individual ions

Implementing universal gate operations implies that one is also able to do individual qubit operations, *i.e.* it is necessary to individually access single ions out of the string and manipulate their quantum information. Therefore, addressing of individual ions with a laser is required. Alternative schemes to address individual ion-qubits have been proposed and used for a 2-ion string [37], however, the full implementation of the Cirac-Zoller gates makes individual addressing indispensable.

Since we had already shown that the imaging system has a sufficient resolution (see fig. 6 above) to resolve the individual ion in a string, we used the same lens to address the ions [38]. As is indicated in fig. 12, we employ an electro-optic deflector to laterally steer the exciting laser beam at 729 nm.

This beam traces the observation channel back to the ions using a dichroic mirror and allows us to spatially select any ion out of the string. The $3.7(3)\,\mu\mathrm{m}$ spatial resolution

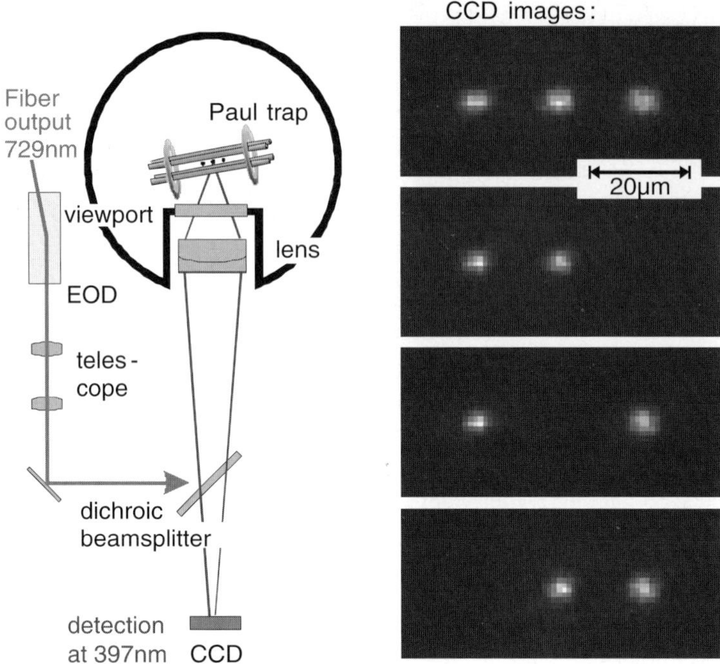

Fig. 12. – Scheme of the experimental setup to address single ions in the string. Laser light at 729 nm from an optical fiber is super-imposed with the fluorescence at 397 nm on a dichroic beam splitter. Thus, the laser is focused by the objective lens onto the ion string. A two-lens telescope transforms the fiber output for optimum focusing. The electro-optic deflector (EOD) can be used to shift the direction of the laser beam to address different ions in the string.

of the addressing beam was directly observed on the ions by measuring the light shift caused by the addressing beam. With this technique, single ions can be individually manipulated as is shown by the images in fig. 12, where an addressing laser pulse was used to excite the $D_{5/2}$-state. The resulting electron-shelving is readily observed by the missing fluorescence. If we apply a pulse to the ion addressed, the probability of exciting a neighboring ion in the ground state and 5 μm away would be about 1%.

7. – Manipulation of the quantum information

For quantum information processing, it is important to know for how long coherent interaction with the ion(s) is possible. For this we cooled one ion to the ground state and then irradiated the ion with light at the blue sideband frequency (this interaction is used in a quantum gate to transfer the internal state of a qubit into the motion) [30]. We then monitored the occupation probability of the D-state *versus* the pulse length on the blue sideband. The same interaction was also used after preparing the ion in the $n = 1$ motional Fock state by a π-pulse on the blue sideband followed by a repumping pulse on the $D_{5/2}$-$P_{3/2}$ transition.

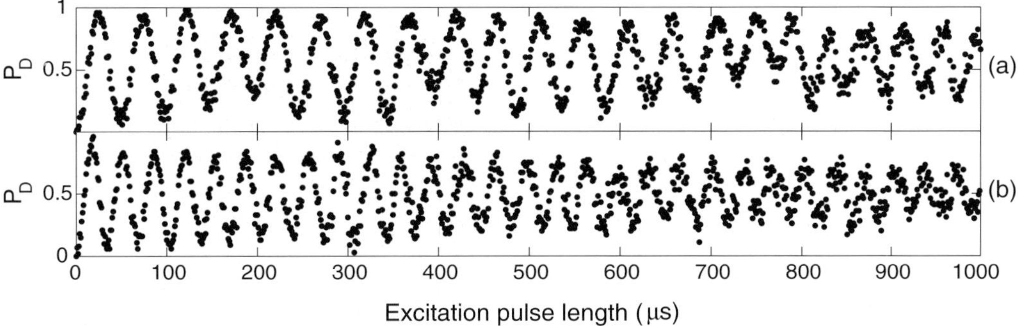

Fig. 13. – (a) Rabi oscillations on the blue sideband for the initial motional state $|n = 0\rangle$. (b) Rabi oscillations as in (a), but for $|n = 1\rangle$.

As fig. 13 shows, we were able to observe Rabi flops for both initial motional states with a contrast of better than 50% for 1 ms. The ratio of Rabi frequencies is $\sqrt{2}$ as expected for this kind of interaction in the Lamb-Dicke regime. These results make us confident that we should be able to apply multiple gate pulses equivalent to at least 40 π-pulses before the fidelity of the total operation drops below 0.5. In our system motional heating is too slow to be the prime source of the observed decoherence, we rather attribute it to time-dependent magnetic-field fluctuations that shift the levels, to slow vibrations in our setup that introduce a fluctuating Doppler shift of the 729 nm beam, and to laser intensity fluctuations.

8. – Conclusions and outlook

With the results obtained so far all necessary ingredients to perform a two-bit quantum logic gate are available with trapped ions. The time scales in our system are well separated (cf. fig. 14). The fastest characteristic time of about 1 µs is given by the harmonic motion of the ion(s). A π-pulse on a motional sideband takes about 20 µs [39] and, as stated above, the fidelity of coherent manipulations remains above 0.5 for times smaller than 1 ms. Our laser system would allow for coherent manipulations for at least 15 ms. Motional heating starts to play a role for times around 100 ms. The ultimate source of decoherence in our experiment is the 1 s lifetime of the $D_{5/2}$-state. In the near

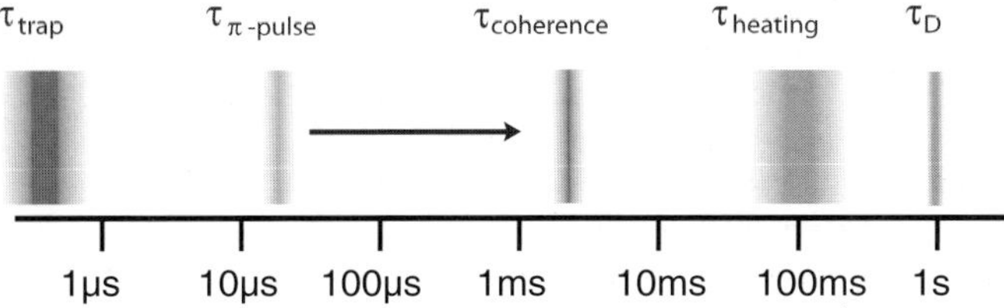

Fig. 14. – Timescales for the current experiments with trapped Ca$^+$ ions.

future we plan to use our ability to individually address ions to demonstrate a C-NOT quantum logic gate with two ions and to create maximally entangled states with 2 and more ions. Apart from the spontaneous decay, all limitations discussed above are of a technical nature. On the long run, it will be decisive whether or not logical gates can be operated with a fidelity of $\sim 10^{-5}$. At this level of precision error correction codes promise stable large-scale quantum computation [40, 41]. In the near future, we envision to implement small quantum algorithms and to perform first experiments on error corrections with up to 5 ions.

* * *

This work was supported by the Fonds zur Förderung wissenschaftlicher Forschung (FWF) within the special research grant SFB 15, by the European Commission within the TMR networks "Quantum Information" (ERB-FMRX-CT96-0087) and "Quantum Structures" (ERB-FMRX-CT96-0077), the IHP network "QUEST" (HPRN-CT-2000-00121), the IST network "QUBITS" (IST-1999-13021) and by the Institut für Quanteninformation GmbH.

REFERENCES

[1] Feynman R. P., *Int. J. Theor. Phys.*, **21** (1982) 467.

[2] Deutsch D., *Proc. R. Soc. London, Ser. A*, **400** (1985) 97.

[3] Shor P., in *Proceedings, 35th Annual Symposium on Foundations of Computer Science* (IEEE Press, Los Alamitos, CA) 1994; Shor P., *SIAM J. Comp.*, **26** (1997) 1484; see also Ekert A. and Josza R., *Rev. Mod. Phys.*, **68** (1996) 733.

[4] Grover L. K., *Phys. Rev. Lett.*, **79** (1997) 325.

[5] DiVincenzo D. P. and Loss D., *Superlatt. Microstruct.*, **23** (1998) 419; Bennett Ch. H. and DiVincenzo D. P., *Nature*, **404** (2000) 247.

[6] Chuang I. L., Gershenfeld N., Kubinec M. G. and Leung D. W., *Proc. R. Soc. London, Ser. A*, **454** (1998) 447.

[7] Cirac J. I. and Zoller P., *Phys. Rev. Lett.*, **74** (1995) 4091.

[8] Loss D. and DiVincenzo D. P., *Phys. Rev. A*, **57** (1998) 120.

[9] Mooij J. E., Orlando T. P., Levitov L., Tian L., van der Waal C. H. and Lloyd S., *Science*, **285** (1999) 1036.

[10] Blatt R., in *Atomic Physics 14, Proceedings of the 14th International Conference on Atomic Physics*, edited by D. J. Wineland, C. E. Wieman and S. J. Smith (AIP Press, New York) 1995, p. 219.

[11] Waki I., Kassner S., Birkl G. and Walther H., *Phys. Rev. Lett.*, **68** (1992) 2007; Birkl G., Kassner S. and Walther H., *Nature*, **357** (1992) 310.

[12] Raizen M. G. *et al.*, *Phys. Rev. A*, **45** (1992) 6493.

[13] DiVincenzo D. P., *Phys. Rev. A*, **51** (1995) 1015.

[14] Mølmer K. and Sørensen A., *Phys. Rev. Lett.*, **82** (1999) 1971; Sørensen A. and Mølmer K., *Phys. Rev. Lett.*, **82** (1999) 1835.

[15] Mølmer K. and Sørensen A., *Phys. Rev. A*, **62** (2000) 022311.

[16] Milburn J., quant-ph/9908037.

[17] Sackett C. A., Kielpinski D., King B. E., Langer C., Meyer V., Myatt C. J., Rowe M., Turchette Q. A., Itano W. M., Wineland D. J. and Monroe C., *Nature*, **404** (2000) 256.

[18] JONATHAN D., PLENIO M. B. and KNIGHT P., *Phys. Rev. A*, **62** (2000) 042307; JONATHAN D. and PLENIO M. B., *Phys. Rev. Lett.*, **87** (2001) 127901.

[19] DUAN L. M., CIRAC I. and ZOLLER P., *Science*, **292** (2001) 1695.

[20] See articles in: BOUWMEESTER D., EKERT A. and ZEILINGER A. (Editors), *The Physics of Quantum Information* (Springer-Verlag, Berlin) 2000.

[21] PRESTAGE J., DICK G. J. and MALEKI L., *J. Appl. Phys.*, **66** (1989) 1013; JANIK G. R. and MALEKI L., *J. Appl. Phys.*, **67** (1990) 6050; RAIZEN M. G., GILLIGAN J. M., BERGQUIST J. C., ITANO W. M. and WINELAND D. J., *J. Mod. Opt.*, **39** (1992) 233.

[22] NÄGERL H. C., SCHMIDT-KALER F., ESCHNER J., BLATT R., LANGE W., BALDAUF H. and WALTHER H., in *The Physics of Quantum Information*, edited by D. BOUWMEESTER, A. EKERT and A. ZEILINGER (Springer-Verlag, Berlin) 2000, p. 163.

[23] JEFFERTS S. R., MONROE C., BELL E. W. and WINELAND D. J., *Phys. Rev. A*, **51** (1995) 3112.

[24] STEANE A., *Appl. Phys. B*, **64** (1997) 623.

[25] JAMES D. F. V., *Appl. Phys. B*, **66** (1998) 181.

[26] NAGOURNEY W., SANDBERG J. and DEHMELT H., *Phys. Rev. Lett.*, **56** (1986) 2797; SAUTER TH., NEUHAUSER W., BLATT R. and TOSCHEK P. E., *Phys. Rev. Lett.*, **57** (1986) 1696; BERGQUIST J. C., HULET R., ITANO W. M. and WINELAND D. J., *Phys. Rev. Lett.*, **57** (1986) 1699.

[27] ROHDE H., GULDE S. T., ROOS C. F., BARTON P., LEIBFRIED D., ESCHNER J., SCHMIDT-KALER F. and BLATT R., *J. Opt. B: Quant. Semiclass. Opt.*, **3** (2001) 34.

[28] STENHOLM S., *Rev. Mod. Phys.*, **58** (1986) 699.

[29] DIEDRICH F., BERGQUIST J. C., ITANO W. M. and WINELAND D. J., *Phys. Rev. Lett.*, **62** (1989) 403.

[30] ROOS C., ZEIGER T., ROHDE H., NÄGERL H. C., ESCHNER J., LEIBFRIED D., SCHMIDT-KALER F. and BLATT R., *Phys. Rev. Lett.*, **83** (1999) 4713.

[31] TURCHETTE Q. A., KIELPINSKI D., KING B. E., LEIBFRIED D., MEEKHOF D. M., MYATT C. J., ROWE M. A., SACKETT C. A., WOOD C. S., ITANO W. M., MONROE C. and WINELAND D. J., *Phys. Rev. A*, **61** (2000) 063418-1.

[32] WINELAND D. J., MONROE C., ITANO W. M., LEIBFRIED D., KING B. and MEEKHOF D. M., *J. Res. Nat. Inst. Stand. Tech.*, **103** (1998) 259.

[33] MORIGI G., ESCHNER J. and KEITEL C., *Phys. Rev. Lett.*, **85** (2000) 4458.

[34] ROOS C. F., LEIBFRIED D., MUNDT A., SCHMIDT-KALER F., ESCHNER J. and BLATT R., *Phys. Rev. Lett.*, **85** (2000) 5547.

[35] NÄGERL H. C., BECHTER W., ESCHNER J., SCHMIDT-KALER F. and BLATT R., *Opt. Exp.*, **3** (1998) 89.

[36] WINELAND D. J., MONROE C., ITANO W. M., LEIBFRIED D., KING B. E. and MEEKHOF D. M., *J. Res. Natl. Inst. Stand. Technol.*, **103** (1998) 259.

[37] TURCHETTE Q. A., WOOD C. S., KING B. E., MYATT C. J., LEIBFRIED D., ITANO W. M., MONROE C. and WINELAND D. J., *Phys. Rev. Lett.*, **81** (1998) 3631; LEIBFRIED D., *Phys. Rev. A*, **60** (1999) 3335; MINTERT F. and WUNDERLICH C., *Phys. Rev. Lett.*, **87** (2001) 257904.

[38] NÄGERL H. C., LEIBFRIED D., ROHDE H., THALHAMMER G., ESCHNER J., SCHMIDT-KALER F. and BLATT R., *Phys. Rev. A*, **60** (1999) 145.

[39] STEANE A., ROOS C. F., STEVENS D., MUNDT A., LEIBFRIED D., SCHMIDT-KALER F. and BLATT R., *Phys. Rev. A*, **62** (2000) 042305.

[40] KNILL E., LAFLAMME R. and ZUREK W. H., *Proc. R. Soc. London, Ser. A*, **454** (1998) 365.

[41] GOTTESMAN D., *J. Mod. Opt.*, **47** (2000) 333, and references therein.

PHOTONS

Experimental quantum communication

H. ZBINDEN, N. GISIN, G. RIBORDY, D. STUCKI and W. TITTEL

Gap-Optique, Université de Genève - 20 rue de l'Ecole-de-Médecine
1211 Genève 4, Switzerland

1. – Introduction

Loosely speaking, quantum communication is about sending (or teleporting) quantum objects, preferably single photons and entangled photons, from Alice to Bob. Possible applications of quantum communication include quantum key distribution (QKD) and also connecting quantum computers in distributed networks. Even if quantum computers are a long-term goal, it remains interesting to play around with entangled photons and to determine how far entanglement can be transported in optical fibers. With current technology however, it has been shown that QKD is feasible and the first prototypes are constructed.

Firstly though, we start with the essential tools of quantum optical experiments, photon counters and entangled photon pairs sources (sect. **2**). We then show that energy-time entanglement is very robust and that the Bell inequality can be violated over more than 10 km (sect. **3**). In sect. **4**, finally, we give a short overview over Cryptography and QKD with faint laser pulses as well as entangled photons. This paper gives only a rough overview of the subject. For more detailed information refer to the cited papers or to one of the recent books and review articles [1-5].

2. – Tools

2˙1. *Photon counters*. – Optical fibers provide the most practical means of distributing photons for quantum communication. They feature extremely low loss in the so-called telecom windows at 1300 nm ($\leq 0.35\,\mathrm{dB/km}$) and 1550 nm ($\leq 0.25\,\mathrm{dB/km}$). Unfortunately, commercial photon counters based on silicon avalanche photo diodes (Si-APD) are only efficient at wavelengths below 1000 nm. For longer wavelengths materials with smaller band-gaps like Ge (up to 1400 nm) and InGaAs (up to 1700 nm) must be used.

The available Ge and InGaAs APDs are designed for telecom applications and not optimized for photon counting. Nevertheless, by selecting the appropriate models, photon counting can be achieved in the so-called Geiger mode. By applying a negative bias voltage above the breakdown, a single photon triggers an avalanche of electron-hole pairs, a macroscopic current pulse that can be detected. To avoid destruction of the APDs, the avalanche must be stopped using a quenching circuitry. The simplest way is to introduce a large resistor (in the order of $100\,\mathrm{k\Omega}$) in series, so that the voltage over the diode drops below threshold as soon as a current is passing (passive quenching). Other possibilities are to detect the output pulse and to actively lower the voltage (active quenching) or to bias the diode only for a short time, when the arrival of a photon is expected (gated mode).

The key features of photon counting APDs, the quantum efficiency, timing jitter, dark count rate and afterpulse probability, are dependent on the bias voltage and temperature. Afterpulses are due to released electron-hole pairs that were trapped during the preceding avalanche. Higher voltage leads to higher efficiency, lower jitter, higher dark counts rates and afterpulse probability. Lower temperature leads to lower dark count rates but more afterpulses. Therefore, for each application the appropriate parameters have to be chosen. Note that the rather high dark count probabilities of today's photon counters at the telecom wavelengths are the limiting factor of the reach of fiber-based QKD.

The photon counting properties of many APDs have been described before [6-9](^{1}). Figure 1 illustrates the performance of different InGaAs APDs in a temperature range that can be obtained by thermoelectric cooling. Since InGaAs APDs do not work properly with passive quenching, the gated mode is used.

$2\!\cdot\!2$. *Entangled photon pair sources.* – In his seminal test of Bell inequalities in the early '80s, Aspect [11] used a complicated two-photon source based on radiative cascade of calcium. Fortunately, current photon sources are much simpler. They are based on a process called spontaneous parametric downconversion (SPDC) [12], the spontaneous splitting of a pump photon into two daughter photons (by convention called signal and idler) within a non-linear crystal. As with frequency doubling (which is the inverse process, where two photons generate one photon) or optical parametric oscillation (OPO, which is the same process, but stimulated), we need a high second-order ($\chi^{(2)}$) non-linear coefficient. Moreover, energy and momentum must be conserved:

$$(1) \qquad \hbar\omega_p = \hbar\omega_s + \hbar\omega_i,$$

$$(2) \qquad \vec{k}_p = \vec{k}_s + \vec{k}_i = \frac{n(\omega_s)\omega_s}{c} + \frac{n(\omega_i)\omega_i}{c}.$$

(^{1}) Now also commercially available [10].

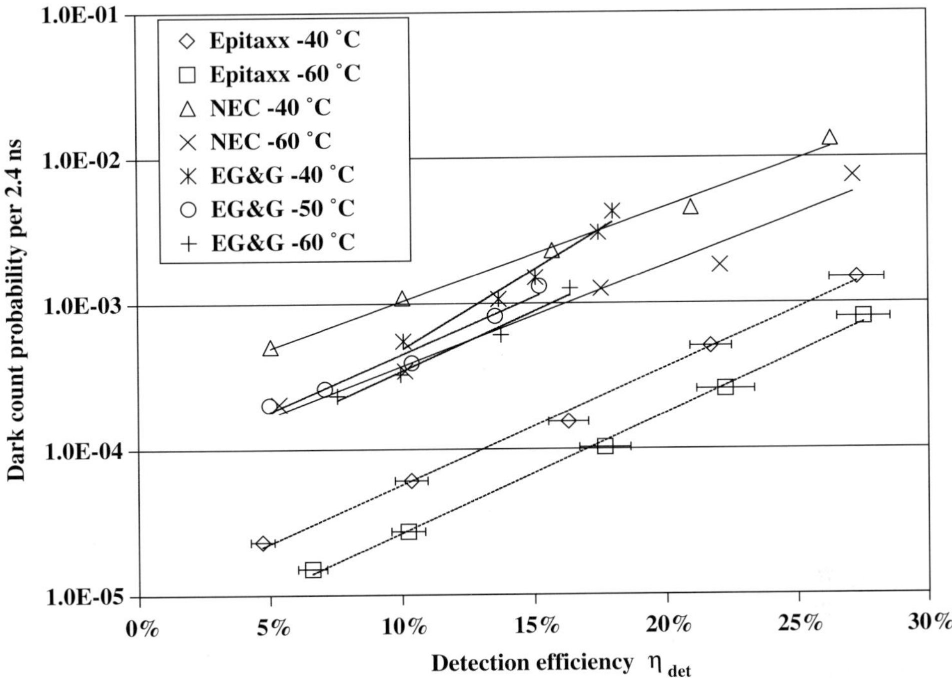

Fig. 1. – Dark count probability per gate *vs.* detection efficiency for different types of InGaAs at different temperatures (from [9]).

Taking profit of birefringence, eqs. (1) and (2) can be satisfied at the same time in spite of chromatic dispersion (phase-matching). We distinguish two types of phase-matching: in type-I phase-matching, the downconverted photons have the same polarization, which is orthogonal to the one of the pump; in type-II phase-matching signal and idler have orthogonal polarizations. Due to the phase-matching the downconverted photons are emitted on cones, each cone corresponding to a different wavelength. Collinear emission happens exactly for one combination of signal and idler wavelength, depending on the angle of incidence of the pump beam. The spectrum of the downconverted photons in a single spatial mode is very broadband (depending on the phase-matching) and varies between 1–100 nm, this means that coherence times are very short ($< 1\,\mathrm{ps}$) and one can say that the photons are emitted "simultaneously". The conversion efficiencies are very small, in the order of 10^{-10} in a single spatial mode. This efficiency can be increased considerably by using a waveguide in periodically poled $LiNbO_3$ (PPLN) [13]. Table I gives an overview of different photon pair source realised by different groups.

3. – Bell experiments

Quantum theory predicts correlation between outcomes of distant measurements on entangled states. This was pointed out by Einstein, Podolski, and Rosen already in

Table I. – *Overview of different photon pairs sources (from [13]). R_{single} and R_{coinc} are the single and coincidence count rates, respectively. p_{pdc} is the probability that a pump photon splits into two daughter photons that are emitted in one spatial mode (except*). Note that the bandwidth of the downconverted photons is not the same in all setups.*

	PPLN waveguide [13]	KNbO$_3$ [14]	BBO cascade type I [15]	BBO type II [16]	QPM-PPSF [17]
P_p (mW)	0.001	10	150	150	300
$\lambda_p/\lambda_{s,i}$	657/1314	655/1310	351/702	351/702	766/1332
APD/η_{det}	Ge/10%	Ge/10%	Si/65%	Si/35%	InGaAs/10%
R_{single} (kHz)	150	250	435	100	36
R_{coinc} (kHz)	1.6	5	21	20	0.5
p_{PDC}	$2.2 \cdot 10^{-6}$	$1.9 \cdot 10^{-10}$	$3.4 \cdot 10^{-11}$ *	$8.1 \cdot 10^{-13}$	$1.1 \cdot 10^{-12}$

1935 [18]. John Bell showed with his inequality that this feature is incompatible with any theory involving local variables [19]. In this section, we present a long-distance Bell experiment using energy-time entangled photons.

3'1. *Energy-time entanglement.* – The experiment of Aspect [11] as well as more recent experiments of Weihs [20] *et al.* used polarization entangled photons to demonstrate violation of the Bell inequalities. Polarization entanglement is conceptually easy to understand and the analyzers, polarizers or polarizing beamsplitters, are particularly simple. However, polarization entanglement is severely restricted in optical fibers. Optical fibers are slightly birefringent, there is the so-called polarization mode dispersion (PMD) of the order of 0.1–0.5 ps/$\sqrt{\text{km}}$. This means that the two components of an arbitrary polarized light pulse will separate in time, and if this separation becomes larger than the coherence time, the pulse gets depolarized. For downconverted photons with bandwidths of the order of 50 nm, depolarization occurs already after several hundred meters. Therefore, polarization entangled photons are not suited for transmission through long-optical fibers.

The photon pairs created by SPDC are entangled in energy and time: on the one hand their individual energy is not sharply defined (large bandwidth) in contrast to the sum of the energies (narrow bandwidth of the pump laser); on the other hand, although their emission time is unknown (large coherence time of the pump photon), they are emitted simultaneously (short coherence time). The analysis of energy-time entangled photons is done with so-called Franson interferometers [21]. The pairs of downconverted photons are split, and one photon is sent to each party through quantum channels. Both Alice and Bob possess identically unbalanced Mach-Zehnder interferometers, with photon counting detectors connected to the outputs (see fig. 2). Locally, if Alice or Bob change the phase of their interferometer, no effect on the count rates is observed (no

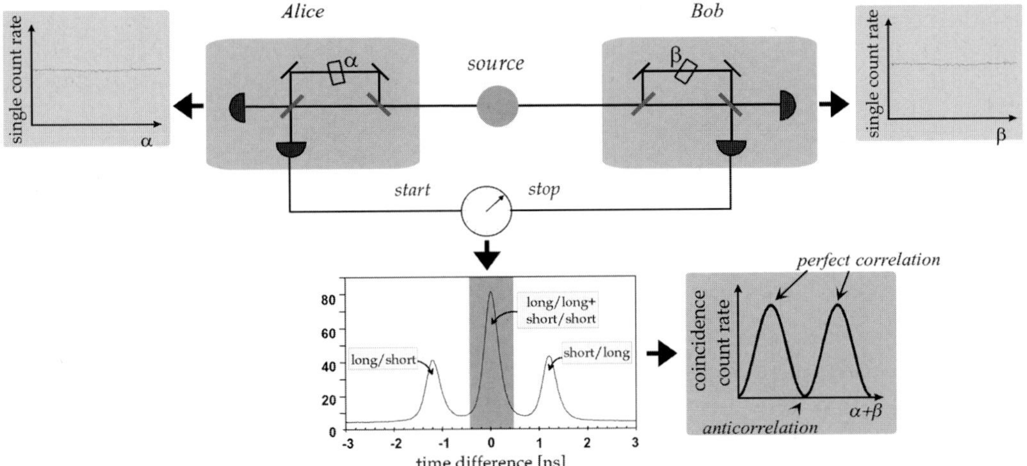

Fig. 2. – Schemeatic of non-local correlations of energy-time entangled photons using Franson-interferometers.

single-photon interference), since the arm length difference is larger than the coherence length of the photons. Looking at Bob's detection-time with respect to the arrival time at Alice's detector, three different values are possible for each combination of detectors. The different possibilities in a time spectrum are shown in fig. 2. One photon can take the long arm at Alice's, while the other one takes the short one at Bob's; this leads to the left peak. The opposite is also possible and leads to the right peak. Finally, the photons can propagate both through the long arms or both through the short arms, leading to the central peak. When the path differences of the interferometers are matched within a fraction of the coherence length of the downconverted photons, the short-short and the long-long processes are indistinguishable, provided that the coherence length of the pump photon is larger than the path-length difference. We can write the corresponding quantum state in the following way:

$$(3) \qquad |\psi\rangle = \frac{1}{\sqrt{2}} \left[|s\rangle_1 |s\rangle_2 + e^{i(\alpha+\beta)} |l\rangle_1 |l\rangle_2 \right].$$

Conditioning detection only on the central time peak, one observes two-photon interferences which depend on the sum of the relative phases in Alice's and Bob's interferometers (α, β)-non-local quantum correlation (see fig. 2). The phase in the interferometers at Alice's and Bob's can, for example, be adjusted so that both photons always emerge from the same output port. Hence, though the outcome on each side is completely random, they can be perfectly correlated.

We can assign the output ports with $+$ and $-$ such that the R_{+-} denotes the coincidence count rate between the $+$ labeled detector at apparatus 1 and the $-$ labeled one at apparatus 2 and similarly for the other permutations. This enables to calculate the so-called correlation coefficient

$$(4) \qquad E(\alpha, \beta) := \frac{R_{+,+}(\alpha, \beta) - R_{+,-}(\alpha, \beta) - R_{-,+}(\alpha, \beta) + R_{-,-}(\alpha, \beta)}{R_{+,+}(\alpha, \beta) + R_{+,-}(\alpha, \beta) + R_{-,+}(\alpha, \beta) + R_{-,-}(\alpha, \beta)}.$$

The Bell inequalities give an upper limit for a combination of four such correlation coefficients with different analyzer settings α, β under an assumption of local hidden variable theories (LHVT). One of the most often used forms, known as the Clauser-Horne-Shimony-Holt (CHSH) Bell inequality, is

$$(5) \qquad S = \left| E(\alpha, \beta) + E(\alpha, \beta') + E(\alpha', \beta) - E(\alpha', \beta') \right| \leq 2.$$

Quantum mechanics predicts the correlation function (V denotes the fringe visibility, that is in practice lower than the theoretical value of 1):

$$(6) \qquad E^{\mathrm{QM}}(\alpha, \beta) = V \cos(\alpha + \beta).$$

Using the settings

$$(7) \qquad \alpha = \frac{-\pi}{4}, \qquad \alpha' = \frac{\pi}{4}, \qquad \beta = 0, \qquad \beta' = \frac{\pi}{2},$$

and assuming $V = 1$, one gets

$$(8) \qquad S^{\mathrm{QM}} = 2\sqrt{2}.$$

This value predicted by quantum mechanics is higher than the limit for LHVT. The violation thus shows that the description of nature as provided by quantum mechanics is unreconcilable with the assumptions leading to Bell inequalities. If it is found that the correlation coefficient E is described by a sinusoidal function of the form of eq. (6), then the CHSH inequality can be reduced to

$$(9) \qquad S = \frac{4}{\sqrt{2}} V \leq 2.$$

Hence, observing a visibility V greater than

$$(10) \qquad V \geq \frac{1}{\sqrt{2}} \approx 0.71$$

will directly show that the correlations under test cannot be explained by LHVT.

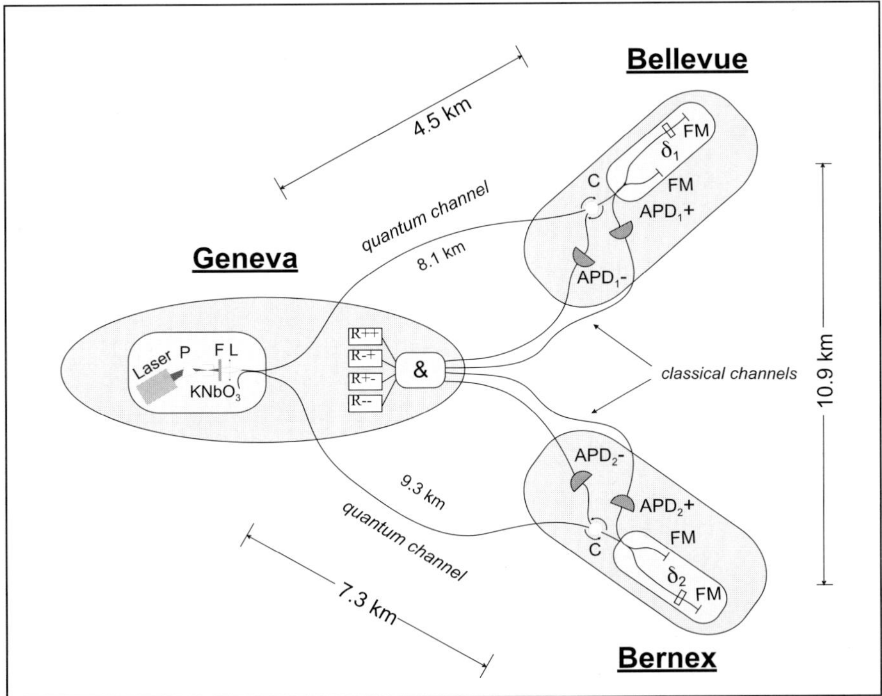

Fig. 3. – Experimental setup for a long-distance Bell experiment.

3'2. *A long-distance Bell experiment.* – The schematic setup of the experiment is given in fig. 3, for more details see [14,22]. The source of pairs of energy-time entangled photons is placed at a telecommunication station in the center of Geneva. The photons of 1310 nm wavelength are created by type-I, degenerate, collinear SPDC in a KNbO3 crystal, using the output of 655 nm diode laser with an external cavity. The photons travel through installed standard telecom fiber to two little villages Bellevue and Bernex, separated by 10.9 km. The two analyzers consist of all-fiber optical Michelson interferometers made of standard 3 dB fiber couplers and so-called Faraday mirrors (FM). The FMs make sure that the polarization transformation is always identical in both arms, in fact, the output state is always orthogonal to the input state. The optical path-length differences of about 24 cm are equal within 10 μm in all interferometers. To change the phase the temperature of the devices can be varied. In order to have access to the second output port of the analyzer, coinciding with the input arm of the Michelson interferometer, we used 3-port optical circulators C. The detectors are liquid-nitrogen–cooled, passively quenched Ge APDs. The classical information about detection time and detector number is transmitted by laser pulses via supplementary telecommunication fibers back to the source.

Figure 4 shows the four different coincidence count rates $R_{+,+}$, $R_{+,-}$, $R_{-,+}$ and $R_{-,-}$ as a function of the phase. Figure 5 gives the correlation coefficient E as a function of phase. This time, the scan range is larger and we can observe a reduction in the

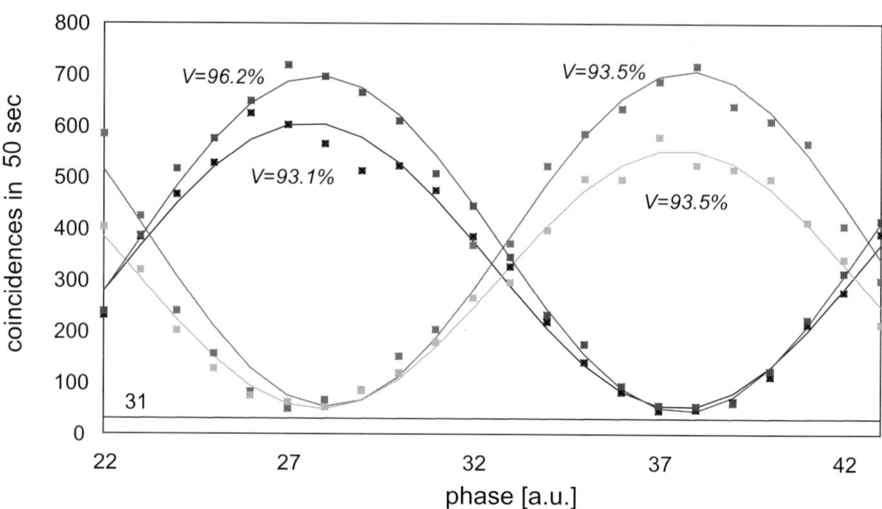

Fig. 4. – Coincidence count rates *vs.* phase $\alpha + \beta$ (arbitrary units).

visibility if the difference in the imbalance becomes significant. The width of the envelope is roughly 10λ and corresponds to the coherence length of the downconverted photons. Fitting a cosine function to the central part of the curve, we can calculate a visibility of $85.3 \pm 0.9\%$, *i.e.* a violation of Bell inequalities of 16σ. If the noise floor is subtracted the visibility, V_{net} is even higher: 95.5%. This is the same kind of visibility we obtain in an

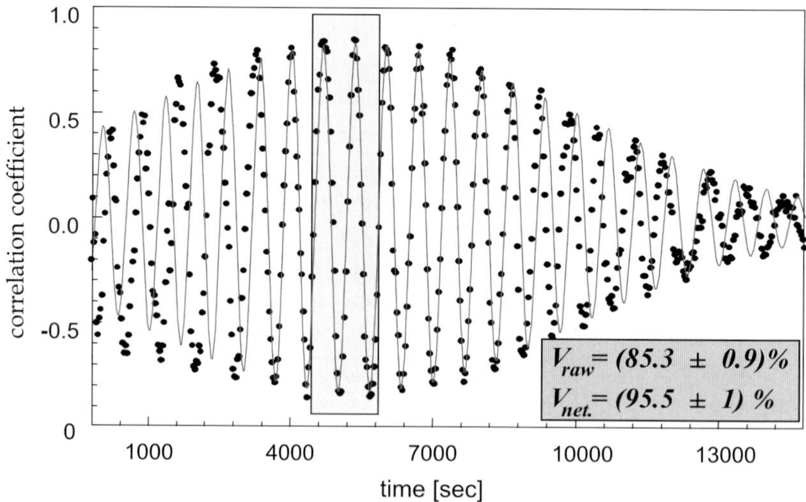

Fig. 5. – The correlation coefficient E as a function of phase $\alpha + \beta$. The phase is varied by changing the temperature of one interferometer as a function of time, therefore the scale of the x-axis is in seconds.

experiment in the laboratory over a few meters, hence no decoherence can be measured over 10 km. The photons may be absorbed, certainly, but if they arrive at the other end, they remain maximally entangled.

4. – QKD

4'1. *Introduction.* – Cryptography is the art of hiding information in a string of bits which is meaningless to any unauthorized party. To achieve this goal, one uses encryption: a message is combined according to an algorithm with some additional secret information—the key—to produce a cryptogram. In the traditional terminology, Alice is the party encrypting and transmitting the message, Bob is the one receiving it, and Eve is the malevolent eavesdropper. For a crypto-system to be considered secure, it should be impossible to unlock the cryptogram without Bob's key. In practice, this demand is often softened, and one requires only that the system is sufficiently difficult to unlock. The idea is that the message should remain protected as long as the information it contains is valuable.

There are two main classes of crypto-systems, the public-key and the secret-key crypto-systems.

Public key systems are based on so-called one-way functions: given a certain x, it is easy to compute $f(x)$, but difficult to do the inverse, *i.e.* compute x from $f(x)$. "Difficult" means that the task shall take a time that grows exponentially with the number of bits of the input. The RSA (Rivest, Shamir, Adleman) crypto-system, *e.g.*, is based on the factorizing of large integers. Anyone can compute 137×53 in a few seconds, but it may take a while to find the prime factors of 28907. To receive a message Bob chooses a private key (based on two large prime numbers) and computes from it a public key (based on the product of these numbers) which he discloses publicly. Now Alice can encrypt her message using this public key and transmit it to Bob, who decrypts it with the private key. Public key systems are very convenient and became very popular over the last 20 years, however, they suffer from two potential major flaws. To date, nobody knows for sure whether or not factorizing is indeed difficult. For known algorithms, the time for calculation increases exponentially with the number of input bits, and one can easily improve the safety of RSA by choosing a longer key. However, a fast algorithm for factorization would immediately annihilate the security of the RSA system. Although it has not been published yet, there is no guarantee that such an algorithm does not exist. Second, problems that are difficult for a classical computer could become easy for a quantum computer. If one of these two possibilities came true, RSA would become obsolete. One would then have no choice, but to turn to secret-key crypto-systems.

Crypto-systems based on a public algorithm and a relatively short secret key are very convenient and widely used. The AES (Advanced Encryption Standard, 2000), *e.g.*, uses a 128-bit key and the same algorithm for coding and decoding. The secrecy of the cryptogram, however, depends again on the eavesdropper's calculating power and time. The only crypto-system providing proven, perfect secrecy is the "one-time pad" proposed by Vernam in 1935. With this scheme, a message is encrypted using a random key of equal length, by simply "adding" each bit of the message to the corresponding bit of the key. The scrambled text can then be sent to Bob, who decrypts the message by "subtracting" the same key. The bits of the ciphertext are as random as those of the key and consequently do not contain any information. Although perfectly secure, the problem with this system is that it is essential for Alice and Bob to share a common secret key, at least as long as the message they want to exchange, and use it only for a single encryption. This key must be transmitted by some trusted means or personal meeting, which turns out to be complex and expensive.

At this point quantum cryptography or better quantum key distribution (QKD), developed by Bennett and Brassard in 1984 [23], enters the scene (see [4] for a review). It allows two physically separated parties to create a random secret key without resorting to the services of a courier, and to verify that the key has not been intercepted. This is due to the fact that any measurement of incompatible quantities on a quantum system will inevitably modify the state of this system. This means that an eavesdropper, Eve, might get information out of a quantum channel by performing measurements, but the legitimate users will detect her and hence not use the key. For convenience the quantum system is in practice a single photon propagating through an optical fiber, and the key can be encoded either by its polarization or by its phase. Several groups presented realizations over distances of a few tens of km (see [24-27] for recent experiments). QKD can also be performed with entangled photon pairs [28]; first experimental results have been published recently [29, 30, 16].

In the following subsections, we would like to explain QKD and the BB84 protocol [23] by the example of a standard polarization-based setup. We then present a scheme, utilizing entanglement and based on the BB84 protocol, that is optimized for long-distance transmission in optical fibers.

4'2. *How QKD works*. – To understand how quantum cryptography works, consider the BB84 communication protocol [23] by Bennett and Brassard (fig. 6). Alice and Bob must be connected by a quantum channel and a classical public channel. Normally single photons are used to carry the information and the quantum channel is usually an optical fiber. The public channel can be any communication link, such as a phone line or an internet connection (which is often an optical fiber, too, but this time with macroscopic pulses). The procedure consists of 4 steps.

First, Alice sends single photons in one of the four following polarization states: vertical, horizontal, $+45°$ and $-45°$. She chooses randomly one of the polarization states for each bit and records her choice. Bob has two analyzers available and selects one randomly before recording each photon. One of the analyzers allows him to distinguish between horizontally and vertically polarized photons, whereas the second one distinguishes between $+45°$ and $-45°$ photons. Bob records the analyzers used and the outcomes.

Second, after exchanging a sufficient number of photons, Bob announces on the public channel the sequence of analyzers he used, but not the results he obtained. Alice compares this sequence with the list of bits she sent, and tells Bob for which photons compatible polarization state and analyzers were used, but not which polarization states

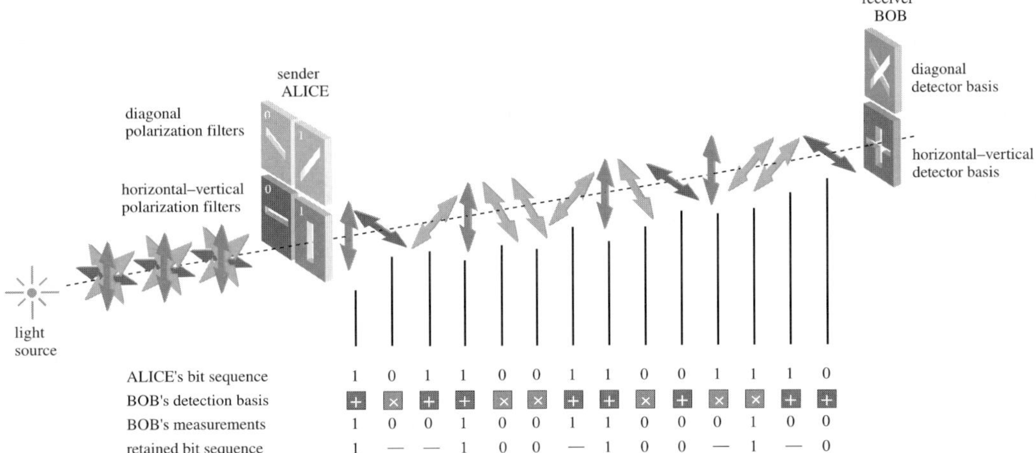

ALICE's bit sequence	1	0	1	1	0	0	1	1	0	0	1	1	1	0
BOB's detection basis	+	×	+	+	×	×	+	+	×	+	×	×	+	+
BOB's measurements	1	0	0	1	0	0	1	1	0	0	0	1	0	0
retained bit sequence	1	—	—	1	0	0	—	1	0	0	—	1	—	0

Fig. 6. – The principle of QKD according the BB84 protocol. Alice sends down an optical fiber photons polarized randomly either horizontally, vertically, at $+45°$ or at $-45°$ (row 1), Bob randomly chooses one of his analyzer basis (row 2) and records his result (row 3). Then they compare the used basis and retain all results with compatible basis (row 4).

she sent. In the cases where incompatible states and analyzers were used, the bits are simply discarded. For the remaining bits, Alice and Bob know that they have the same values, at least if nothing—no eavesdropper—perturbed the transmission. If, for example, Alice send a vertically polarized photon and Bob measures it with the vertical-horizontal analyzer, he should always get a "vertical" result.

Third, Alice and Bob select a random subset of their key and compare it over the public channel to assess the secrecy of their communication. The consequence of the interception of their key by an eavesdropper would be a reduction of the correlation between the values of their bits. Let us suppose that Eve cuts the fiber, performs a measurement on each photon with an equipment similar to Bob's, and sends a photon to Bob, prepared according to her result (intercept-resend strategy). In 50% of cases she will choose the wrong analyzer and send a photon polarized in the wrong basis that will result with 50% chance in a wrong count at Bob's. Hence 25% of Alice's and Bob's bit values will disagree, so that the eavesdropper can easily be detected.

Fourth, also when no eavesdropper is disturbing the bit exchange, there will be in practice some errors in the transmission and Alice and Bob's string will not coincide perfectly. The remaining errors are removed by standard error correction methods, which in turn reduces the length of the key. The determined quantum bit error rate (QBER, normally in the order of a few percent) is attributed to Eve. The corresponding information that Eve might gain (using the optimum strategy and technology) can be reduced to an arbitrarily low value, by a procedure called privacy amplification, again at the expense of the length of the usable key. Therefore it is very important to keep the experimental error rate as low as possible. The remaining key can now be used with total confidence to encrypt a message.

Table II. – *Comparison of QKD experiments with faint laser pulses. Note that R_{raw} (the raw bit rate, before sifting, error correction and privacy amplification) and QBER are proportional to μ.*

Authors	λ (nm)	Setup	Fiber length (km)	μ	R_{raw} (Hz)	QBER (%)
Ribordy *et al.* [27]	1300	plug&play	23	0.1	486	5.4
D. Bethune *et al.* [31]	1550	plug&play	10	0.3	3000	1.5
Marand *et al.* [32]	1300	phase	30	0.2	260	4
R. Hughes *et al.* [25]	1550	phase	48	0.63	10	9.3

4˙3. *QKD with faint laser pulses.* – Let us consider how, in practice, a polarization coding scheme could be implemented. In most experimental setups faint laser pulses are used instead of real single-photon states. The number of photons in a laser pulse are Poisson-distributed, thus the average number of photon per pulse μ must be well below 1 ($\mu = 0.1$, say), to limit the probability of obtaining more than 1 photon per pulse. Alice's light source is hence a pulsed and strongly attenuated semiconductor laser. Four lasers with polarizers oriented at $0°$, $90°$, $45°$ and $135°$ are used, followed by three passive couplers. The lasers fire at random at a given pulse rate. Bob randomly selects the polarization analyzer basis. This is most easily done by introducing a passive optical coupler that directs the photon to one of the two polarizing beam splitters (PBS) oriented at $45°$. The photons are then detected with a photon counter and acquisition electronics collect the data.

The axis of the polarizers at Alice and Bob must be aligned and kept aligned. This is the main difficulty of a fiber optic implementation of the polarization scheme. For this reason a polarization controller with an automated feedback must be introduced, to compensate polarization fluctuations due to mechanical or thermal changes in the fiber link. Realignment may be necessary every few seconds or after an hour depending on the stability of the link. There are schemes using two interferometers and phase-modulators to code the bits [32, 25]. However, conventional phase coding setups also need polarization control and moreover an automated balancing of the interferometers. There is one phase coding setup, that prevents any alignments and is for this reason called plug&play setup. The trick is to send the pulses forth and back, using only one (auto-stabilized) interferometer and a Faraday mirror that compensates all birefringence in the fiber [27].

Table II gives a non-exhaustive review over QKD experiments realized by different groups.

4˙4. *QKD with entangled photon pairs.* – QKD systems relying on photon pairs offer two advantages over those using faint laser pulses (typically 0.1 photon per pulse). With the latter systems, most pulses are empty. Since Bob does not know which ones contain a photon, he must always "open" his detectors, increasing thus the overall error probability. Thus, for a given attenuation, photon pairs systems feature lower error rates, since Bob

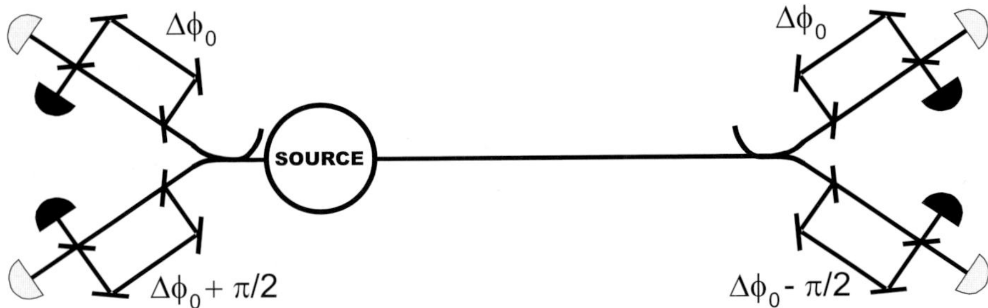

Fig. 7. – Passive choice of the basis: the photon chooses at Alice's and Bob's couplers to go to the one or the other interferometer.

only opens his detector, when Alice detected a photon. Also, photon pair systems allow passive state preparation by Alice. In this case, even when two (or more) photon pairs are created within a gate time, one photon of a pair is not at all correlated with a photon of another pair. The so-called photon number splitting attack, which has an important impact on the security of faint pulses systems [33], is thus not possible with systems using entangled photons and passive state preparation.

It is then possible to exchange bits by associating values to the two ports. This is, however, not sufficient and a second basis must be implemented (analogous to BB84), to ensure security against eavesdropping attempts. This can, for example, be done by adding a second interferometer to the analyzer (see fig. 7). In this case, a photon chooses randomly to go to one or the other interferometer. The second set of interferometers can be adjusted to also yield perfect correlations between output ports. The relative phase between their arms should, however, be chosen so that when the photons go to interferometers not associated, the outcomes are completely uncorrelated. Such a system features passive state preparation by Alice and also a passive basis choice by Bob. This constitutes an elegant solution: neither a random number generator, nor an active modulator are necessary.

This scheme has been realized by our group at Geneva University [34]. It constitutes the first experiment in which the space-like separation between Alice and the source, needed for Bell tests, was replaced by an asymmetric setup, optimized for QKD (see fig. 8). The two-photon source (a $KNbO_3$ crystal pumped by a doubled Nd-YAG laser) provides energy-time entangled photons at non-degenerate wavelengths—one around 810 nm, the other one centered at 1550 nm. This choice allows the use of high-efficiency silicon-based single-photon counters featuring low noise to detect the photons of the lower wavelength. To avoid the high transmission losses at this wavelength in optical fibers, the distance between the source and the corresponding analyzer is very short, of the order of a few meters. The other photon, at the wavelength where fiber losses are minimal, is sent via an optical fiber to Bob's interferometer and is then detected by InGaAs APDs. Implementing the BB84 protocol in the way discussed above, with a total of four interferometers, is difficult. They must be aligned and their relative phase kept

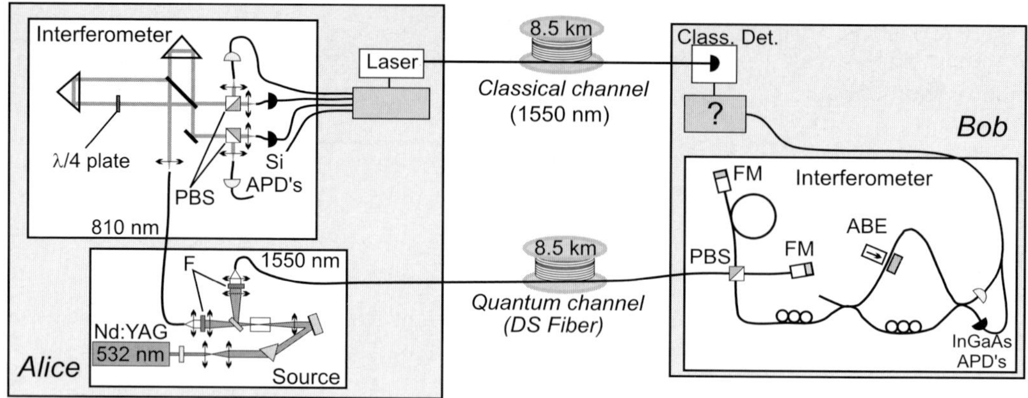

Fig. 8. – Asymmetric system for QKD utilizing photon pairs. (PBS: polarizing beamsplitter, FM: Faraday mirror, ABE: adjustable birefringent element, F: spectral filters.)

accurately stable during the whole key distribution session. To simplify this problem, we use birefringent interferometers and polarization multiplexing. The photons impinge on a polarizing beamsplitter and choose randomly between vertical or horizontal polarization. Due to the birefringent element in the interferometer they experience a different phase shift, corresponding to a different basis. We demonstrated QC over a transmission distance of 8.5 km in a laboratory setting using a fiber on a spool and generated several Mbits of key in hour-long sessions [34]. This is the largest span realized to date for QC with photon pairs. Some recent QKD experiments with entangled photon pairs are reviewed in table III.

4'5. *Faint laser pulses vs. entangled photons*. – The presented QKD system exploiting photon pairs is, contrary to faint laser pulses systems, immune to the photon number splitting attack thanks to passive state preparation. However, its operation is unquestionably more complex than that of the self-aligned system using faint pulses. The main difficulty is the need to stabilize both interferometers. In practice, an increase in the security of a system is always accompanied by an increase in cost and complexity. One

TABLE III. – *Comparison of QKD_ experiments with entangled photon pairs.*

Authors	λ (nm)	Coding	Fiber length	R_{raw} (Hz)	QBER (%)
Ribordy *et al.* [34]	1550	phase	8450 m	133	8.6
Jennewein *et al.* [16]	702	polarization	500 m	850	2.5
Naik *et al.* [30]	702	polarization	table-top	10	3.0
Tittel *et al.* [29]	1310	phase	table-top	33	4

can distinguish three levels of security. First, an entanglement-based system could be immune to all attacks performed with existing, as well as future technology, including photon number splitting. Its cost and complexity may however be too high for real applications. Second, one can consider systems based on faint pulses. They are immune to existing technology [33], but not in all cases to future technological developments. On the other hand, they are easy to operate. Finally, one can use public key cryptography, which is considered to offer sufficient security, when implemented with suitable key lengths. It is also convenient to apply, as it does not require a dedicated channel, and has been in use for many years. Nevertheless, it is vulnerable to future developments in computer power, while all QKD systems can only be attacked with technology existing at the time of the key exchange. In addition, the security of public key cryptography can be jeopardized overnight by theoretical advances. In this event, QKD with faint pulses would constitute the only realistic replacement technology.

5. – Conclusions

We gave a short overview on some aspects of quantum communication. QKD is the technology that is by far the most advanced in the field of quantum information. Prototypes of QKD-apparatuses are already available [10], whereas quantum computers will not be available for many years, if at all.

$$* \; * \; *$$

HZ would like to thank for the invitation of the Italian Physical Society. The presented work was financially supported by the European projects EQCSPOT, QUCOMM and the Swiss OFES.

REFERENCES

[1] NIELSEN M. A. and CHANG I. L., *Quantum Computation and Quantum Information* (Cambridge University Press) 2000.
[2] LO H. K., POPESCU S. and SPILLER T. (Editors), *Introduction to Quantum Computation and Information* (World Scientific) 1998.
[3] BOUWMEESTER D., EKERT A. and ZEILINGER A. (Editors), *The Physics of Quantum Information* (Springer) 2000.
[4] GISIN N., RIBORDY G., TITTEL W. and ZBINDEN H., *Quantum Cryptography*, quant-ph/0101098, *Rev. Mod. Phys.*, **74** (2002) 145.
[5] TITTEL W. and WEIHS G., *Quantum Information and Computation*, **1** (2001) 3.
[6] LACAITA A., FRANCESE P., ZAPPA F. and COVA S., *Appl. Opt.*, **30** (1994) 6902.
[7] RIBORDY G., GAUTIER J. D., ZBINDEN H. and GISIN N., *Appl. Opt.*, **37** (1998) 2272.
[8] RARITY J., WALL T., RIDLEY K., OWENS P. and TAPSTER P., *Appl. Opt.*, **39** (2000) 6746.
[9] STUCKI D., RIBORDY G., STÉFANOV A., ZBINDEN H., RARITY J. and WALL T., *J. Mod. Opt.*, **48** (2001) 1967.
[10] http://www.idquantique.com
[11] ASPECT A., GRANGIER P. and ROGER G., *Phys. Rev. Lett.*, **47** (1981) 460.

[12] BURNHAM D. C. and WEINBERG D. L., *Phys. Rev. Lett.*, **25** (1970) 84.

[13] TANZILLI S., DE RIEDMATTEN H., TITTEL W., ZBINDEN H., BALDI P., DE MICHELI M., OSTROWSKI D. B. and GISIN N., *Electron. Lett.*, **37** (2001) 26.

[14] TITTEL W., BRENDEL J., ZBINDEN H. and GISIN N., *Phys. Rev. Lett.*, **81** (1998) 3563.

[15] KWIAT P., WAKS E., WHITE A., APPELBAUM I. and EBERHARD P., *Phys. Rev. A*, **60** (1999) 773.

[16] JENNEWEIN T., SIMON C., WEIHS G., WEINFURTER H. and ZEILINGER A., *Phys. Rev. Lett.*, **84** (2000) 4729.

[17] BONFRATE G., PRUNERI V., KAZANSKI P. G., TAPSTER P. and RARITY J., *Appl. Phys. Lett.*, **75** (1999) 2356.

[18] EINSTEIN A., PODOLSKI B. and ROSEN N., *Phys. Rev.*, **47** (1935) 777.

[19] BELL J. S., *Physics*, **1** (1964) 195.

[20] WEIHS G., JENNEWEIN T., SIMON C., WEINFURTER H. and ZEILINGER A., *Phys. Rev. Lett.*, **81** (1998) 5039.

[21] FRANSON J. D., *Phys. Rev. Lett.*, **62** (1989) 2205.

[22] TITTEL W., BRENDEL J., ZBINDEN H. and GISIN N., *Phys. Rev. A*, **59** (1999) 4150.

[23] BENNETT C. and BRASSARD G., in *Proceedings of IEEE International Conference on Computers, Systems and Signal Processing, Bangalore* (IEEE, New York) 1984, p. 175.

[24] TOWNSEND P., *Opt. Fiber Tech.*, **4** (1998) 345.

[25] HUGHES R., MORGAN G. and PETERSON C., *J. Mod. Opt.*, **47** (2000) 533.

[26] MÉROLLA J.-M., MAZURENKO Y., GOEDGEBUER J.-P. and RHODES W., *Phys. Rev. Lett.*, **82** (1999) 1656.

[27] RIBORDY G., GAUTIER J.-D., GISIN N., GUINNARD O. and ZBINDEN H., *J. Mod. Opt.*, **47** (2000) 517.

[28] EKERT A., RARITY J., TAPSTER P. and PALMA M., *Phys. Rev. Lett.*, **69** (1992) 1293.

[29] TITTEL W., BRENDEL J., ZBINDEN H. and GISIN N., *Phys. Rev. Lett.*, **84** (2000) 4737.

[30] NAIK D., PETERSON C., WHITE A., BERGLUND A. and KWIAT P., *Phys. Rev. Lett.*, **84** (2000) 4733.

[31] BETHUNE D. and RISK W., *Appl. Opt.*, LP**41** (2002) 1690, quant-ph/0104089.

[32] MARAND C. and TOWNSEND P., *Opt. Lett.*, **10** (1995) 1695.

[33] FÉLIX S., GISIN N., STEFANOV A. and ZBINDEN H., *J. Mod. Opt.*, **48** (2001) 2009, quant-ph/0102062.

[34] RIBORDY G., BRENDEL J., GAUTIER J. D., GISIN N. and ZBINDEN H., *Phys. Rev. A*, **63** (2001) 012309.

Delayed-choice entanglement— Swapping with vacuum one-photon quantum states

F. Sciarrino, E. Lombardi and F. De Martini

Istituto Nazionale di Fisica della Materia, Dipartimento di Fisica
Università "La Sapienza" - Roma, 00185 Italy

State entanglement, the most distinctive, fundamental feature of modern physics, is at the heart of the essential non-locality of the quantum world, *i.e.* of the irremovable property of nature first discovered in 1935 by Einstein-Podolsky-Rosen (EPR) and later formally analyzed by J. S. Bell [1] and recently by L. Hardy [2]. In the context of the modern fields of quantum information and computation, entanglement lies at the core of several important protocols and methods as, for instance, the quantum state teleportation (QST), a fundamental process that has been implemented by different experimental approaches [3, 4]. Very recently quantum teleportation with an unprecedented large "fidelity" has been experimentally demonstrated by adoption of the new concept of "entanglement of one photon with the vacuum" by which each quantum superposition state, *i.e.* "qubit", is physically implemented by a two-dimensional subspace of Fock states of a mode of the electromagnetic field, specifically the state spanned by the QED "vacuum" and the 1-photon state [5]. This method requires, as we shall see, an entirely new re-formulation of the Hilbert space framework supporting the evolution of quantum information, and then the conception of new devices and methods to implement the transformation algebra of states and operators. In view of a further clarification of the new method in the perspective of future more complex applications we investigate in the present letter the procedure called "entanglement swapping" in which the teleported state itself is entangled, *i.e.* where the teleported system does not even enjoy its own state [6].

Let us first outline the swapping process in the new perspective. It is well known that the establishment of entanglement between two (or more) distant "quantum systems" does not necessarily require, as generally believed after the original EPR approach, a

direct original interaction between these ones, but can be realized by merely projecting by an appropriate joint measurement the independent entangled states pertaining to the separated systems, even in the absence of any previous mutual interaction. According to this scenario two separate observers, Alice (A) and Bob (B), independently prepare two sets of entangled "singlets". They perform on one "component" of each singlet an appropriate test of EPR non-locality, *e.g.* a standard Bell-inequality test [1], a Hardy's "no-inequality ladder" test [2,3] or a "continuous variables" homodyne detection test [7]. The other two components of the singlets are sent to a third party, Eve (E), who performs a joint test at its choice on the components he received, one from A and one from B. By doing that Eve projects (*i.e.* "swaps") the state of the two originally non-entangled distant components in the hands of A and B onto an entangled state. Recently it has been argued by Asher Peres that the swapping process could be completed by Eve not necessarily at the time at which the two distant systems were tested by A and B but at any retarded "delayed choice" time [8]. Indeed, according to Eve's choice at a later time a fourth "verification party", Victor (V), can sort the samples already tested by A and B into subsets and can verify that each subset behaves as if it consisted of entangled pairs of distant systems that have never communicated in the past even indirectly via other systems. This may appear a paradoxical result, as we shall see.

In the present work we report the experimental demonstration of Peres's "delayed choice" process by applying the new concept of "entanglement of one photon with the vacuum" [5,9]. Because of its novelty let us outline here the rationale of this approach. The concept of "non-locality of a single photon", first introduced by Albert Einstein in 1927 [10], has been thoroughly analyzed in the last decade by S. M. Tan *et al.* [11], by L. Hardy [12] and others in connection with the superposition state emerging from a beam splitter (BS) excited by a single photon at one of its input ports. In our view this state should indeed be interpreted as an *entangled state* by considering that in the domain of optics the *modes* of the electromagnetic field (e.m.) rather that the photons must be taken as the "systems", or "components" to be entangled. This is consistent with the content of two recent comprehensive theoretical works by E. Knill *et al.* [13] and by M. Duan *et al.* [14]. Thus, any single-particle superposition state expressed in the form $\Sigma_A = (2)^{-1/2}(|1\rangle_A|0\rangle_{A'} - |0\rangle_A|1\rangle_{A'})$ must be interpreted as a "singlet" entangling the mode pair $(k_A, k_{A'})$ which is excited by the Fock states $|1\rangle$ and $|0\rangle$, this last one expressing the QED vacuum state. If the same states $|0\rangle$, $|1\rangle$ are interpreted, respectively, as the logic "zero", "one" information states, the singlet Σ_A is viewed as an *E-bit*, *i.e.* an entangled bit of quantum information [15]. Of course in order to make use of the entanglement present in this picture we need to use the second quantization procedure of creation and annihilation of particles and/or use states which are superpositions of states with different numbers of particles. Another puzzling aspect of this second quantized picture is the need to define and measure the relative phase between states with different number of photons, such as the relative phase between the vacuum and one-photon states appearing in eq. (2), below. That we can associate a relative phase between the *vacuum* and anything else seems most surprising, but it is less so if we recall the more familiar case of a coherent state, where the relative phase between the different photon number states in the superposition is

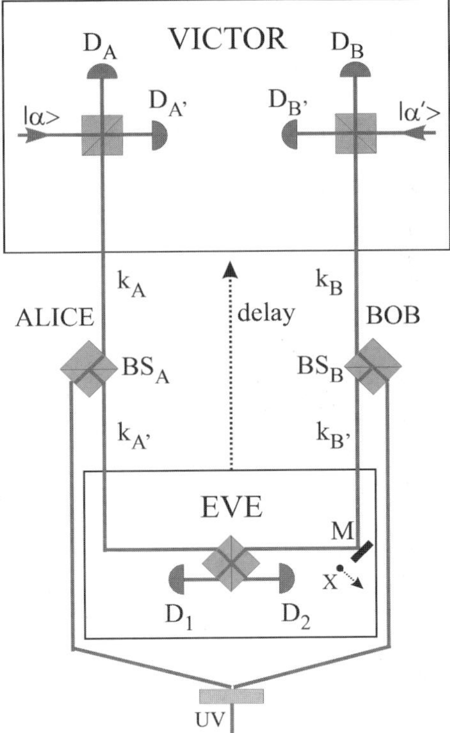

Fig. 1. – Layout of an experimental demonstration of the delayed-choice entanglement swapping process. In the actual experiment the 2-homodyne apparatus was replaced by the 2-detector set shown in fig. 2, inset.

reflected physically in the phase of the classical electric field. To be able to control these relative phases we need in general, and in analogy with classical computers, to supply all gates and all sender/receiving stations of any quantum information network with a common synchronizing *clock* signal, *e.g.* provided by an ancillary photon or by an ancillary multi-photon, Fourier-transformed coherent state. Optionally in simple cases, as in the present work, an *ad hoc* clock generator is not needed as the mutual phase information can be retrieved by a linear two-mode superposition in a beam splitter (BS).

An example concerning the present experiment is illustrated by fig. 1 which shows how the non-locality implied by the quantum state of the overall system could be tested by the distant parties A and B via two coherent states $|\alpha\rangle \equiv ||\alpha|\exp[i\theta]\rangle$ and $|\alpha'\rangle \equiv ||\alpha|\exp[i\theta']\rangle$ that can operate at the same time as *clock states* and as *local oscillators* (LO) of the corresponding homodyne detectors performing the same test. The feasibility of a similar single-photon homodyne technique has been demonstrated recently [16].

Figure 1 shows the basic layout of the delayed-choice entanglement swapping experiment. Pair of photons were generated by Spontaneous Parametric Down-Conversion

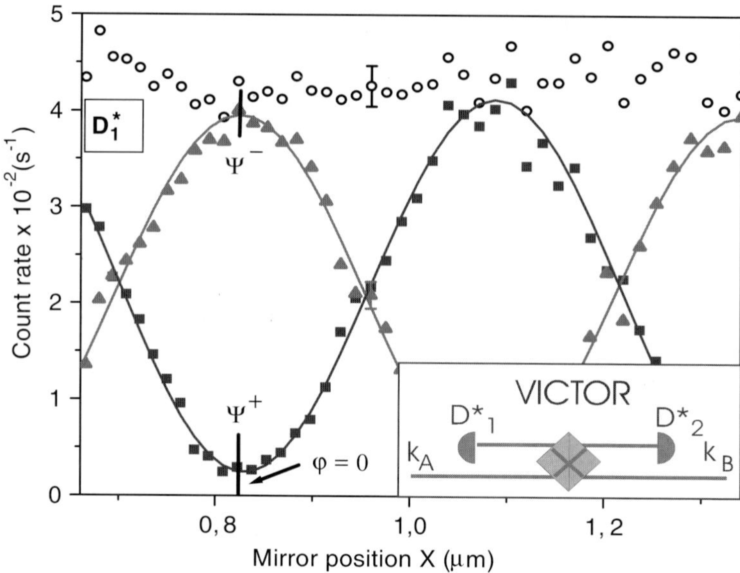

Fig. 2. – Experimental results of the measurement of the count rate by detector D_1^* as a function of delayed settings of the phase φ determined by micrometric displacements X of the mirror M (open circles). A verification party, Victor, can sort at a later time the recorded pattern in two subsets showing two sinusoidal fringe patterns with opposite phases corresponding to the Bell states $\Psi^{\pm}$, for $\varphi = 0$. The visibility of the fringe patterns is $V = 91 \pm 2\%$. The inset shows the 2-detector apparatus that has been adopted to perform experimentally the EPR non-locality test and that for that purpose replaces, in a fully equivalent fashion, the double homodyne apparatus shown in fig. 1.

(SPDC) excited by a single mode UV CW argon laser in a Type-I $LiIO_3$ crystal with the same wavelengths (wl) $l = 727.6\,\mathrm{nm}$ and with the same linear polarizations (π). Each pair of photons, each of which associated with an ultrashort optical pulse characterized by a *coherence time* $\tau_c = 0.1\,\mathrm{ps}$, was injected into 2 equal 50 : 50 beam splitters, BS_A and BS_B, characterized by equal *real* transmittivity and reflectivity parameters, $t = r = 2^{-1/2}$. Precisely, each BS consisted of a 45° π-rotator followed by a calcite crystal. As is well known [1, 17], the *product state* character of each pair, $|\Phi\rangle = |1\rangle_A \otimes |1\rangle_B$, did not imply any inter-particle EPR correlation, in agreement with the data reported in fig. 2 (open circles). In other words, as far as the dynamics of the overall system is concerned, each photon pair could have been supplied equally well by any pair of distant sources. The state $|\Phi\rangle$ was transformed by the BSs into the product of two singlets defined over the pairs of output modes $(k_A, k_{A'})$ and $(k_B, k_{B'})$:
$|\Phi\rangle = \Sigma_A \otimes \Sigma_B = (1/2)(|1\rangle_A|0\rangle_{A'} - |0\rangle_A|1\rangle_{A'}) \otimes (|1\rangle_B|0\rangle_{B'} - |0\rangle_B|1\rangle_{B'})$. The pure state $|\Phi\rangle$ may be expressed as a sum of products of Bell states defined in the two 2-dimensional Hilbert subspaces spanned by the state eigenvectors to be measured, respectively, by the

couple (Alice, Bob) and by Eve:

$$(1) \qquad |\Phi\rangle = \Sigma_A \otimes \Sigma_B = \frac{1}{2}\left[\Phi^+ \otimes \Phi^+_E - \Phi^- \otimes \Phi^-_E - \Psi^+ \otimes \Psi^+_E + \Psi^- \otimes \Psi^-_E\right],$$

and the Bell states defined in the corresponding 2-d Hilbert sub-spaces are [5]

$$(2) \qquad \Phi^\pm = \frac{1}{\sqrt{2}}(|0\rangle_A|0\rangle_B \pm |1\rangle_A|1\rangle_B), \qquad \Psi^\pm = \frac{1}{\sqrt{2}}(|0\rangle_A|1\rangle_B \pm |1\rangle_A|0\rangle_B),$$

$$\Phi^\pm_E = \frac{1}{\sqrt{2}}(|0\rangle_{A'}|0\rangle_{B'} \pm |1\rangle_{A'}|1\rangle_{B'}), \qquad \Psi^\pm_E = \frac{1}{\sqrt{2}}(|0\rangle_{A'}|1\rangle_{B'} \pm |1\rangle_{A'}|0\rangle_{B'}).$$

Equation (1) shows how the original entanglement condition existing within the two separated systems $(k_A, k_{A'})$ and $(k_B, k_{B'})$ can be swapped to the "extreme" modes k_A and k_B by any joint Bell-type measurement made by Eve on the "intermediate" modes $(k_{A'}, k_{B'})$. In the absence of such a measurement the overall state $|\Phi\rangle$ is a superposition while the one reaching the (A + B) sector is a mixed state.

Suppose that one of the two detectors D_j of the Eve sector "clicks", *i.e.* measures the state Ψ^-_E, say. A sudden state reduction occurs that projects the overall system onto the corresponding entangled Bell state: $|\Phi\rangle \Rightarrow \Psi^-$. Eve's apparatus, consisting of a $50:50$ beam splitter and of a φ-phase shifter, is apt to perform this task with a 50% efficiency. Indeed it can be easily found by applying the standard BS theory [5] that the realization of the 1-photon Bell state Ψ^-_E (or Ψ^+_E) over the input modes $k_{A'}$, $k_{B'}$ determines a click by D_1 (or D_2). It is also well known that the states $\Phi^\pm_E$ corresponding to a 2-photon excitation of Eve's sector cannot be discriminated by any linear device [18]. Note however that the present experiment is *noise free* since a 2-photon excitation of Eve's sector implies no detections by the (A + B) sector, an event easily discarded by the electronic apparatus. An additional degree of freedom under Eve's control, indeed an optional "delayed choice", was provided by the micrometric displacement ΔX of the mirror M, activated by a piezoelectric transducer. This one induced a corresponding phase shift $\Delta\varphi = (2)^{3/2}\pi\lambda^{-1}\Delta X$ between the modes $(k_{A'}, k_{B'})$. Optionally, the same task can be accomplished by fast Electro-Optic (EO) phase modulator, as we shall see. The 4 detectors adopted in the experiment were equal Si-avalanche EG&G-SPCM200 modules with quantum efficiencies $QE \approx 0.45$.

Suppose that a complete EPR non-locality test is performed by Alice and Bob by means of the two optical homodyne devices shown in fig. 1, according to the scheme by Tan *et al.* [11]. Assume that the eigenvalues of the clock-LO coherent states are $\alpha = |\alpha|\exp[\theta]$, $\alpha' = |\alpha|\exp[\theta']$. By a simple extension of a previous analysis [11] it can be shown that if Eve's detector D_1 clicks, *i.e.* Ψ^- is realized, the probability of a coincidence involving the detectors D'_A and D_B is

$$(3) \qquad \langle\Psi^-|I_{A'}I_B|\Psi^-\rangle = \frac{1}{4}|\alpha|^2\left\{|\alpha|^2 + [1 + \cos(\theta' - \theta + \varphi)]\right\}.$$

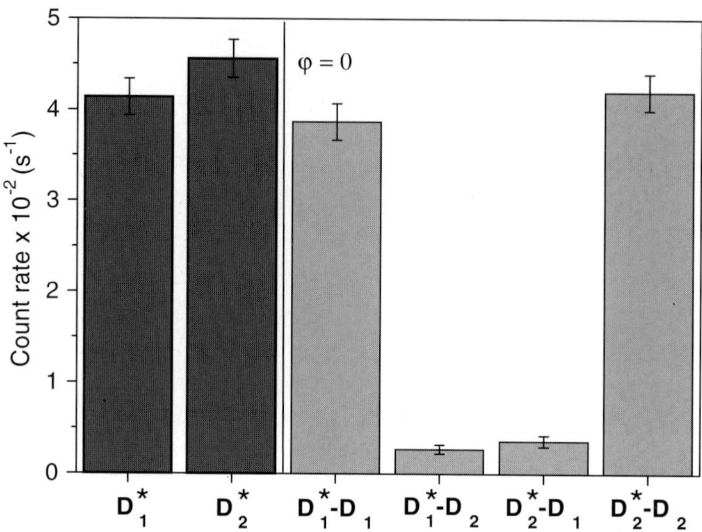

Fig. 3. – Histograms showing the measured detection count rates by D_1^* and D_2^* and the accuracy affecting the experimental determination of the Bell states $\Psi^\pm$ obtained by the delayed coincidence rates involving all detector pairs: D_i^*-D_j $(i, j = 1, 2)$, for $\varphi = 0$.

Rather than performing the difficult double homodyne experiment, in our case Alice and Bob carried out an equally significant EPR non-locality test by mixing the modes (k_A, k_B) by a 50 : 50 BS coupled to the detector pair D_1^*, D_2^*: fig. 2, inset. In analogy with Eve's apparatus, this device may be thought to perform a test on the Bell states $\Psi^\pm$ spanning the Hilbert subspace pertaining to the 2-d manifold (k_A, k_B). At the same time it also provides the necessary synchronizing *clock* effect, as said. Consider for instance the photo-detection by D_1^*. Note first that the coincidence probability of simultaneous clicks by D_1^* and D_1 is found by the standard theory to be expressed by [5]

$$(4) \qquad \langle \Psi^- | I_D I_{D^*} | \Psi^- \rangle = \frac{1}{2}[1 + \cos \varphi],$$

proportional to the expression (3) obtained for the homodyne devices by setting $\theta = \theta'$ and $|\alpha|^2 \ll 1$. Similar results are found for the other three coincidence combinations involving D_i^* and D_j $(i, j = 1, 2)$. Let Alice and Bob carry an experiment aimed at the measurement of the rate of detection by D_1^*: fig. 2. Since the two systems to be tested (k_A, k_B) lack any original non-local character, it is natural to expect a total insensitivity to any change of local parameters acting on remote parts of the apparatus, as for instance the phase shift $\Delta\varphi$. This is indeed shown by the experimental data (open circles) given in fig. 2. However, had the "verification party", Victor, kept the record of the individual outcomes of both pairs (D_1, D_2) and (D_1^*, D_2^*), at a later time he could sort into two subsets the already tested samples detected by Alice and Bob. Figures 2 and 3 show that indeed each subset behaves as if it consisted of entangled pairs of distant systems. Note that these ones have never communicated in the past even indirectly via other systems. Furthermore, as pointed out by Peres, after Alice and Bob have recorded the results of all their measurement, Eve has still the freedom of deciding which experiment she will perform [8]. This one may consist of a standard Bell measurement, or a joint measurement with a $\Delta\varphi$ shift, or a POVM measurement [18] or one of the exotic, interesting single-photon non-locality tests suggested by L. Hardy [12]. Indeed in the present experiment, owing to a spatial displacement of the corresponding detector sets, Eve's action could take place with a time delay $\Delta\tau \approx 3\,\mathrm{ns} \gg \tau_\mathrm{c}$ with respect to the time of the state reduction event determined by the test performed by Alice and Bob. In other words, since in our case $\Delta\tau$ was about 3×10^3 larger than τ_c, the photon *"coherence time"*, the "swapping" process was completed by Eve's apparatus long after the complete annihilation of the particle measured by (A + B). In order to offer an even more convincing demonstration, a sophisticated $\Delta\tau = 20\,\mathrm{ns}$ delay apparatus has been realized allowing delayed fast $\Delta\varphi$ changes by a randomly driven EO Phase Modulator (Inrad 621-040 with $\Delta\varphi = \pi \equiv \Delta\varphi_{\lambda/2}$ driven by $400\,\mathrm{V}$ rectangular pulses) triggered by the (A + B) detection apparatus, *i.e.* long *after* the completion of the A + B test. Note in fig. 2 that shifts $\pm\Delta\varphi_{\lambda/2}$ correspond to the detection interchanges: $\Psi^- \rightleftarrows \Psi^+$. This makes the original Peres's argument, conceived for standard Bell-inequality tests of $(2 \otimes 2)$-d Hilbert photon π-states, fully consistent with the present experiment [1,8].

How could then Eve's delayed choice determine data already irrevocably recorded? According to Peres, it is meaningless to assert that two quantum systems are entangled without specifying their state, or to assert that a system is in a pure state without specifying that state or to attribute an objective meaning to the quantum state of a single system. If these prescriptions are forgotten one may encounter paradoxes as the one seen here: a past event may sometimes appear to be determined by future actions [8]. For better clarification it is perhaps worth reminding here that: "a phenomenon is not a phenomenon until is a measured phenomenon" (J. A. Wheeler) asserting the inanity of any intellectual speculation involving mental modelling of the inner evolution of a quantum superposition process. Furthermore, as pointed out by Richard Jozsa [15], any apparent retrodictive process, *e.g.* associated with the quantum evolution in the presence of the EPR non-locality in a teleportation process, cannot finally lead to paradoxes or contradictions of causality because of the inherent inaccessibility of the quantum information.

In conclusion, we have illustrated experimentally an enlightening aspect of quantum EPR non-locality. We have accomplished that by implementing a new method of photon quantum entanglement that is expected to play a significant role in the field of modern quantum information as well in future studies on EPR quantum non-locality. For instance, the application of our new methods of high-"fidelity" quantum teleportation and entanglement swapping to modern "quantum repeaters" will certainly improve in the near future the technology of quantum communication at large distances [14,19].

* * *

We are indebted with the FET European Network on Quantum Information and Communication (Contract IST-2000-29681-ATESIT) and with MURST for funding.

REFERENCES

[1] Bell J. S., *Speakable and Unspeakable in Quantum Mechanics* (Cambridge University Press, Cambridge) 1987.
[2] Hardy L., *Phys. Rev. Lett.*, **71** (1993) 1665; Boschi D., Branca S., De Martini F. and Hardy L., *Phys. Rev. Lett.*, **79** (1997) 2755.
[3] Bennett C., Brassard G., Crepeau C., Jozsa R., Peres A. and Wootters W. K., *Phys. Rev. Lett.*, **70** (1993) 1895.
[4] Bouwmeester D., Pan J., Weinfurter H. and Zeilinger A., *Nature*, **390** (1997) 575; Boschi D., Branca S., De Martini F., Hardy L. and Popescu S., *Phys. Rev. Lett.*, **80** (1998) 1121; Furusawa A. *et al.*, *Science*, **282** (1998) 706.
[5] Lombardi E., Sciarrino F., Popescu S. and De Martini F., quant-ph/0109160 and *Phys. Rev. Lett.*, **88** (2002) 70402; Sciarrino F., Lombardi E., Milani G. and De Martini F., *Phys. Rev. A*, **66** (2002) 024309.
[6] Zukowski M., Zeilinger A., Horne M. A. and Ekert A., *Phys. Rev. Lett.*, **71** (1993) 4287; Bose S., Vedral V. and Knight P. L., *Phys. Rev. A*, **57** (1998) 822; Pan J., Bouwmeester D., Weinfurter H. and Zeilinger A., *Phys. Rev. Lett.*, **80** (1998) 3891.
[7] Ou Z. Y., Pereira S., Kimble J. and Peng K., *Phys. Rev. Lett.*, **68** (1992) 3663.
[8] Peres A., *J. Mod. Opt.*, **47** (2000) 139.
[9] Jennewein T., Weihs G., Pan J. and Zeilinger A., *Phys. Rev. Lett.*, **88** (2002) 017903.
[10] Niels Bohr, in *Albert Einstein: Philosopher-Scientist*, edited by P. A. Schlipp (Evantson, Ill.) 1949.
[11] Tan S. M., Walls D. F. and Collett M. J., *Phys. Rev. Lett.*, **66** (1991) 252.
[12] Hardy L., *Phys. Rev. Lett.*, **73** (1994) 2279; Albert D., Aharonov Y. and D'Amato S., *Phys. Rev. Lett.*, **54** (1985) 5; Oliver B. and Stroud C., *Phys. Lett. A*, **135** (1989) 407; Vaidman L., *Phys. Rev. Lett.*, **75** (1995) 2063.
[13] Knill E., Laflamme R. and Milburn G., *Nature*, **409** (2001) 46.
[14] Duan M., Lukin M., Cirac J. and Zoller P., *Nature*, **414** (2001) 413.
[15] Popescu S. and Jozsa R., in *Introduction to Quantum Computation and Information*, edited by H. Lo, S. Popescu and T. Spiller (World Scientific, London) 1998.
[16] Kuzmich A., Walmsley I. A. and Mandel L., *Phys. Rev. Lett.*, **85** (2000) 1349.
[17] Peres A., *Quantum Theory: Concepts and Methods* (Kluwer, Dordrecht) 1993, Chapt. 9.
[18] Casamiglia J. and Lutkenhaus N., *Appl. Phys. B*, **72** (2001) 67.
[19] Briegel H., Dur W., Cirac J. and Zoller P., *Phys. Rev. Lett.*, **81** (1998) 5932.

Frequency hopping in quantum interferometry:
Efficient up-down conversion for qubits and ebits

P. Mataloni, G. Giorgi and F. De Martini

Dipartimento di Fisica and Istituto Nazionale per la Fisica della Materia
Università di Roma "La Sapienza" - Roma, 00185, Italy

Interferometry of quantum particles is rooted at the core of modern physics as it provides a unique tool of investigation and a direct demonstration of fundamental properties of nature as complementarity, nonlocality and quantum nonseparability [1]. In modern times all methods and protocols of quantum information and quantum computation involve interferometry through the very definition of the conceptual cornerstones of this science, *viz.* the *qubit* and the *ebit* [2]. In this framework it is well-known that the 1*st-order* interference of particles, *e.g.* optical photons in the present work, is an utterly fragile property that can be easily spoiled by de-coherence when the overall system exhibits a certain degree of complexity. More fundamentally it is a common notion that the interfering particles cannot be substantially disturbed by collisions, let alone by the *hardest* of all possible collisions, *i.e.* the ones implying the *annihilation* or the *creation* of the particles themselves.

In the present paper we demonstrate experimentally that the last seemingly obvious condition is not a necessary one. In facts, there exits a class of *information-preserving* unitary transformations of parametric type allowing a nonlinear (NL) frequency conversion of all quantum interferometric dynamical structures via a QED particle annihilation or/and creation process. Conceptually the overall process may be considered as the dynamical "reversal" of the *quantum injected* NL parametric particle amplification/squeezing process that has been realized recently (QIOPA) [3]. Furthermore the frequency conversion process may be noiseless and may be easily realized with a "quantum efficiency" close to its maximum value, $QE = 1$. These are indeed very useful properties that will be of large technological interest in the domain of quantum information and computation.

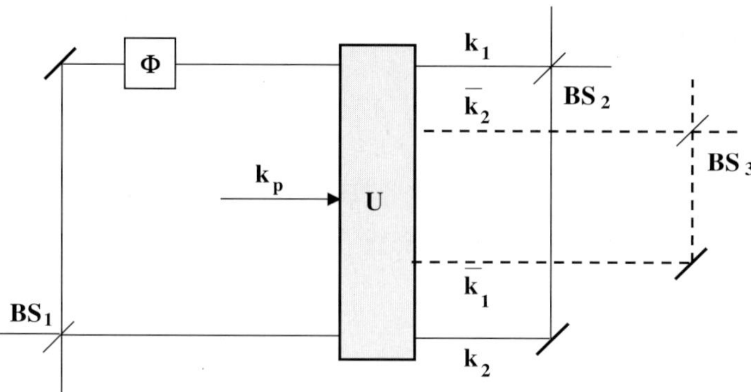

Fig. 1. – Schematic diagram of the nonlinear Mach-Zehnder interferometer (MZ-IF) terminated at two different correlated frequencies.

Figure 1 shows the schematic diagram of a new kind of *single-photon* Mach-Zehnder (MZ) 1st-order interferometer (IF). The optical structure consists of an input 50/50 beam splitter (BS) coupled to two-photon wave vector (wv) modes k_j ($j = 1, 2$) at the wavelength (wl) λ, in our case lying in the infrared (IR) spectral region. Before standard mode recombination by the output BS_2, this simple mode structure is interrupted by a device **U** providing a unitary nonlinear (NL) transformation **U** on the input single-photon superposition state, *i.e.* the *qubit* defined in a 2-dimensional Hilbert **k**-space:

$$(1) \qquad |\Phi\rangle = \alpha|1\rangle_1|0\rangle_2 + \beta|0\rangle_1|1\rangle_2,$$

where the state labels 1, 2 refer to the IF modes k_1 and k_2, respectively. Assume now for simplicity and with no lack of generality: $\beta = \alpha \exp[i\Phi]$ and $\alpha \equiv 2^{-1/2}$. Let us consider the frequency up-conversion process. The NL frequency-conversion unitary evolution operator,

$$(2) \qquad \mathbf{U} \equiv \exp\left[\widetilde{\mathbf{g}} \sum_{j=1,2} \left(\widehat{\mathbf{a}}_j \widehat{\overline{\mathbf{a}}}_j^{\dagger}\right) + \text{h.c.}\right],$$

provides the QED *annihilation* of the input qubit (1) defined by the momenta $\hbar k_j$, phase Φ and wl λ, and the simultaneous QED *creation* of a new qubit $|\Psi\rangle$ defined by the momenta $\hbar \overline{k}_j$ ($j = 1, 2$), phase Ψ and wl $\overline{\lambda} = (\lambda^{-1} + \lambda_p^{-1})^{-1}$ in our case lying in the ultraviolet (UV) spectral region. The *2nd-order* tensor parameter $\widetilde{\mathbf{g}}$ is proportional to the interaction time t, to the 2nd-order susceptibiliy $d^{(2)}$ of the NL medium and to the "pump" field $\mathbf{E}_p$, with wl λ_p, wv k_p, phase Θ. The pump field is assumed to be a single plane-wave coherent "classical field" undepleted by the interaction. The wl's and wv's $\overline{k}_j$, k_j, k_p are mutually connected by the energy conservation $\overline{\lambda}^{-1} = (\lambda^{-1} + \lambda_p^{-1})$ and by the *phase-matching condition* (PMC) for the 3-wave interaction: $k_j + \overline{k}_j +$

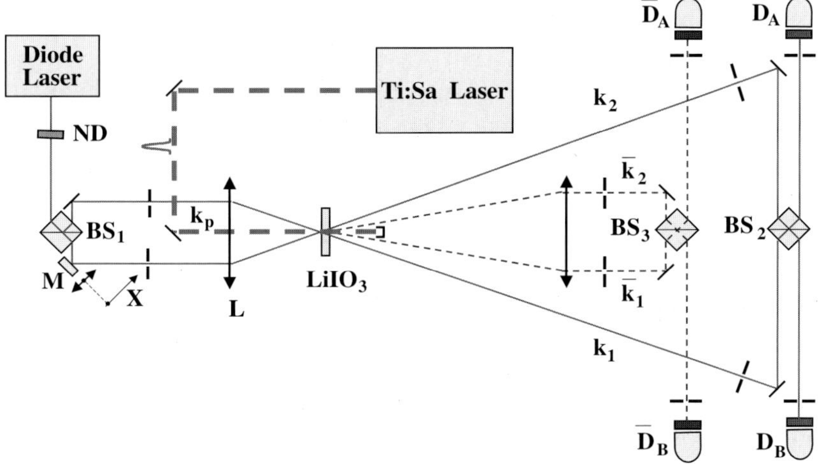

Fig. 2. – Lay-out of the single-particle MZ-IF experiment.

$\mathbf{k}_\mathrm{p} = 0$, leading in our simple plane case to an equation with two solutions: $j = 1, 2$. Similarly, the qubit phases at different wl's are also connected by $\Phi - \Psi + \Theta = 0$. Assume $\Theta = 0$ for convenience. We see that by introduction of the device $\mathbf{U}$ and of an additional output BS$_3$ the standard MZ-IF is transformed into a new kind of interferometer terminated at two *different* wavelengths λ and $\overline{\lambda}$. Correspondingly two sets of *interference fringes* can be retrieved upon changes $\Delta\Phi = 2^{3/2}\pi X/\lambda$ of the mutual phase of the modes $\mathbf{k}_j$ via displacements X of an optical mirror M activated by a piezo-transducer. Note that since in the present experiment PMC couples *deterministically* each mode $\mathbf{k}_j$ to a corresponding $\overline{\mathbf{k}}_j$, no additional quantum interference effects arise in the overall NL coupling process [4].

Let us now venture in a more detailed account of the experiment shown in fig. 2. The input field was generated by a cw linearly polarized, single transverse mode diode laser (RLT8810MG), operating at the IR wl $\lambda = 876.1$ nm with a FWHM linewidth $\Delta\lambda \approx 1$ nm. The laser beam was highly attenuated by a set of neutral density filters (ND) to the approximate *single photon* level [5]. The achievement of this important condition was tested in agreement with the following considerations. Assuming a single-mode coherent field with mean photon number m the probability of n photon excitation is given by the Poisson statistics: $P(n, m) = \frac{m^n}{n!}e^{-m}$. Accordingly, the ratio of the probability of 2 photon *vs.* 1 photon mode excitation is: $\sigma \equiv \frac{P(2,m)}{P(1,m)} = \frac{m}{2}$ and the "single photon" condition implies $\sigma \ll 1$. Because of the strong beam attenuation, in our experiment a value as small as $\sigma = 4.1 \times 10^{-7}$ at wl $\lambda = 876.1$ nm has been evaluated on the basis of the number of coincidence pulses detected simultaneously by the photodiodes D_A and D_B coupled with the two output modes of the MZ interferometer set with the IF phase $\Phi = \pi/2$. In the context of our experiment, this condition also implies that any up-converted single photon on mode $\overline{\mathbf{k}}_j$ was associated with the vacuum field on $\mathbf{k}_j$ after the NL interaction, and viceversa.

All detectors operating at the IR wl λ were equal Si avalanche single-photon SPCM-200PQ diodes with Quantum Efficiency $QE \approx 30\%$ while the two detectors operating at the UV wl $\overline{\lambda}$ were photomultipliers: Philips-56DUVP ($QE \approx 23\%$) and Hamamatsu-R943-02 ($QE \approx 21\%$). A computer interfaced Stanford Research 400 counter was adopted for counting and averaging the detected signals.

The input single-photon field with wl λ was injected into the input 50/50 beam splitter BS_1 of the double MZ-IF with output modes $\mathbf{k}_j$, $j = 1, 2$. These modes were mutually Φ-dephased by a piezoelectrically driven mirror (M) and the associated fields were brought by a $f = 3.5\,\text{cm}$ lens (L) into a common focal region, with diameter $\varphi \approx 20\,\mu\text{m}$, within a NL LiIO$_3$, $l = 1\,\text{mm}$ thick crystal slab, cut for Type-I phase matching. Here a strong NL 3-wave interaction took place between the input field associated with $\mathbf{k}_j$, the up-converted field associated with the $\overline{\mathbf{k}}_j$ and a single-mode high-intensity "pump" field associated with the ultrashort pulses emitted with wl $\lambda_\text{p} = 795\,\text{nm}$ by a mode-locked 76 MHz Coherent MIRA 900 Ti-Sa femtosecond laser. Finally, the output beams $\mathbf{k}_j$ emerging from the NL crystal were again superimposed by a 50/50 beam splitter BS_2 thus completing the usual MZ-IF scheme at the input wl λ. In a similar way the two up-converted UV output beams at the UV wl $\overline{\lambda} = 416.8\,\text{nm}$ were superimposed on an independent 50/50 BS_3, thus completing the MZ-IF scheme at the up-converted wl $\overline{\lambda}$. The difficult task of filtering the very weak beam at wl $\overline{\lambda}$ against the very strong UV beam at wl $\lambda_\text{p}/2$ was overcome by spatial discrimination after the NL crystal and by the adoption of two interference filters at 416.8 nm with bandwidth 10 nm.

Note that the up-conversion unitary transformation of the quantum superposition state (or "qubit") at the IR wl λ into the "UV qubit" with wl $\overline{\lambda}$ is a *noise-free* process since energy conservation does not allow any amplification of the input vacuum state. Of course, the inverse transformation process is also possible as an input qubit with wl $\overline{\lambda} > \lambda_\text{p}$ can be frequency "down-converted" into a corresponding one with wl $\lambda = \overline{\lambda}\lambda_\text{p}(\overline{\lambda} - \lambda_\text{p})^{-1} > \overline{\lambda}$. This "optical parametric amplifying" (OPA) process is nevertheless affected by a *squeezed vacuum* noise due to the amplification of the input vacuum state at wl λ [3].

The expression of the quantum efficiency (QE) of the up-conversion process, defined as the ratio between the average numbers of scattered and input photons on the modes $\overline{\mathbf{k}}_j$, $\mathbf{k}_j$ as a function of the peak intensity of the pump pulse I_p and for perfect collinear interaction, can be obtained by previous evaluation of the field at the output of the nonlinear interaction: $U^\dagger \hat{a} U$ and $U^\dagger \hat{\overline{a}} U$ [6]. This leads to the expression

$$(3) \qquad QE = \sin^2\left(\frac{\pi}{\varepsilon_0}\sqrt{\frac{I_\text{p}}{\lambda\lambda_\text{p}n^\circ\overline{n}^\text{e}(\vartheta_m)}}\,d^{(2)}l\right).$$

where n° is the "ordinary ray" refraction index at wl λ while $\overline{n}^\text{e}(\vartheta_m)$ is the "extraordinary ray" refraction index at wl $\overline{\lambda}$ for the angle ϑ_m determined by $\mathbf{k}_\text{p}$ and by the optic axis of the crystal (phase matching angle) [7].

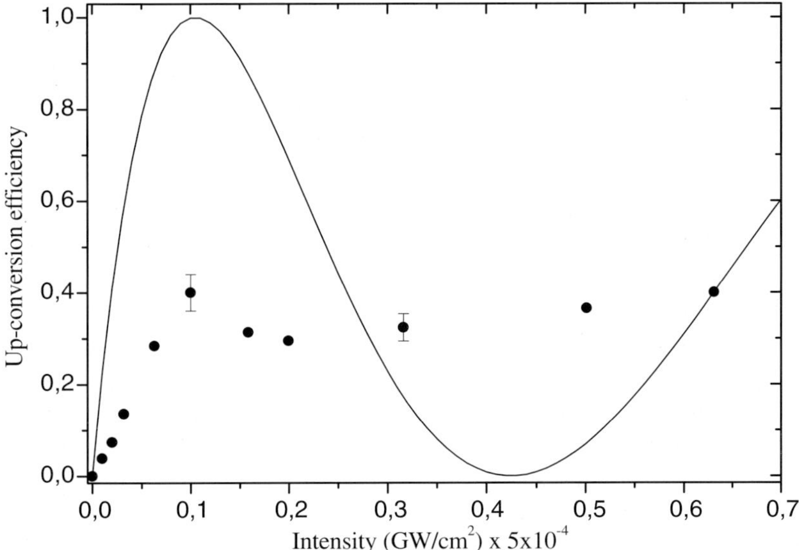

Fig. 3. – Up-conversion quantum efficiency QE as a function of the laser pump intensity I_p: theoretical (continuous line) and experimental results.

The expression of QE just given in eq. (3) refers to the case of a collinear interaction, a condition which can be approximated by improvement of the spatial superposition of the beams in the NL interaction region. We report in fig. 3 the theoretical value of QE as a function of the pump intensity I_p, calculated for the 1 mm thick LiIO$_3$ crystal adopted in the experiment. The theoretical value of I_p corresponding to the limit condition $QE = 1$ is found $I_\mathrm{p} = 200\,\mathrm{GW/cm^2}$. This figure may be compared with the experimental result $QE \approx 0.4$ we obtained in a side experiment by focusing a low repetition rate 100 fs, $I_\mathrm{p} = 200\,\mathrm{GW/cm^2}$ pulse on the same crystal: fig. 3 [6]. This discrepancy is attributed to the nonperfect realization of the plane wave collinear interaction in the active focal spot of the focusing lens. In the present MZ-IF experiment the peak intensity of each Ti-Sa laser pulse was $\approx 1\,\mathrm{GW/cm^2}$, and the corresponding measured value of the quantum efficiency was $QE \simeq 3 \cdot 10^{-3}$.

The persistence of the quantum superposition condition within the IR $\to$ UV frequency hopping process is demonstrated in fig. 4 by the two correlated interference fringe patterns showing an equal periodicity upon changes of the mutal dephasing $\Delta\Phi$ of the IR modes $\mathbf{k}_1$, $\mathbf{k}_2$. As previously emphasized in a different context [3], this is but one aspect of a very general *information preserving* transformation of all unitary NL parametric up- (or down-) conversion transformations. By these ones any input qubit at wl λ and expressed by eq. (1) is generally transformed into another at wl $\overline{\lambda} \neq \lambda$ keeping the *same* complex parameters α, β of the original one, *i.e.* fully reproducing its *quantum information* content. In addition and most importantly, the present work shows that these transformations can be *noise-free* and can be realized with a quantum efficiency close to its maximum value.

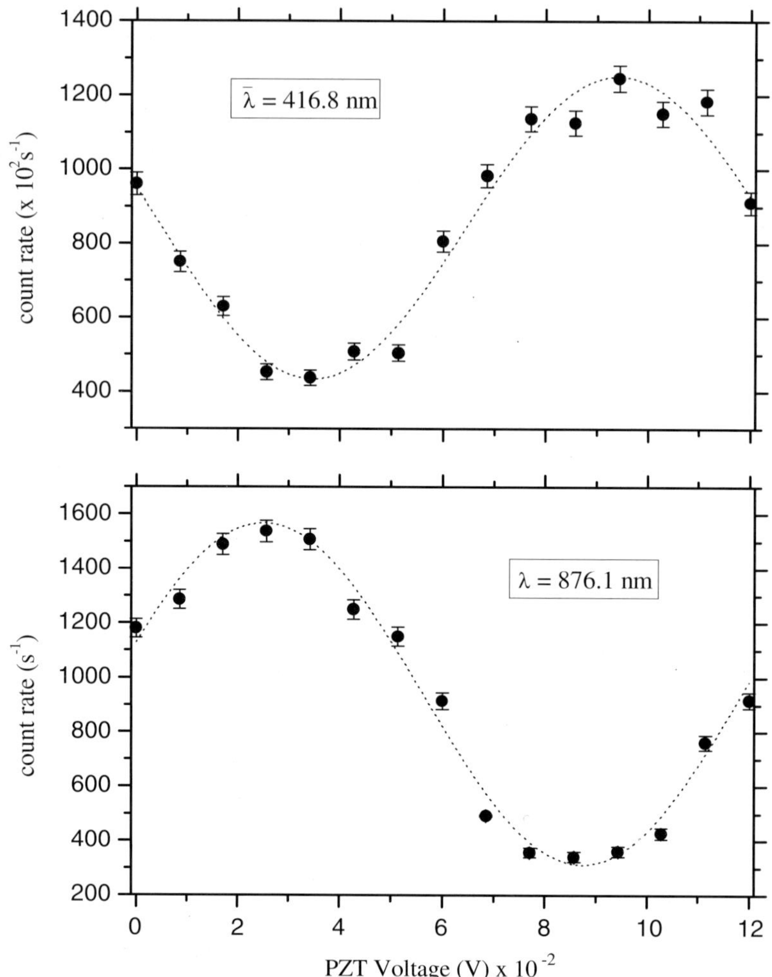

Fig. 4. – Single-photon interference fringes obtained within the same experiment at the different correlated wavelengths λ and $\overline{\lambda} = (\lambda^{-1} + \lambda_{\mathrm{p}}^{-1})^{-1}$. The phase period of the fringing patterns has been found to be in agreement with the figure of merit (0.7 nm/V) of the piezoelectrical transducer activating the mirror M.

Apart from the fundamental relevance of these results due to the peculiar paradigmatic and historical status of single-particle interferometry in the quantum-mechanical context, the present work is expected to have a large impact on modern quantum information technology. This can be illustrated by the following example. Consider a case in which quantum information is encoded on a single microwave photon with wl λ, *e.g.* within the cavity of a micromaser. If we want to transfer conveniently this information at a large distance we need to use an optical fiber exhibiting its low loss behaviour in the IR spectral region, at wl λ'. This can be done in a "information lossless" manner in a NL waveguide by the up-conversion $\lambda \to \lambda'$. If now this information is to be transferred to a set of trapped atoms in an optical cavity, we may need a further lossless up-conversion into the visible: $\lambda' \to \lambda''$, etc. This scenario may generally represent an appealing alternative to other linear methods, but in some cases it may indicate the *only available* solution to sort quantum information out of a nanostructure quantum device and, most importantly, to interconnect it efficiently within a large information network made of heterogeneous components. This may be the case of a NMR quantum gate operating at a radiofrequency wl [8] or of a superconducting quantum dot gate or a SQUID device operating at still lower electromagnetic (e.m.) frequencies [9].

So far we have been dealing only with conversion of "qubits". The extension of our NL method to a two-photon entangled state, or specifically to elementary entangled information carriers, *i.e. ebits*, can be easily realized in several ways depending on the nature of the entanglement. Consider for instance a linear-polarization (π) entangled 2-photon state emitted over two spatial modes $\mathbf{k}_1$, $\mathbf{k}_2$ by a Spontaneous Parametric Down-Conversion (SPDC) process in a NL crystal: $|\Phi\rangle = \alpha|\uparrow\rangle_1|\uparrow\rangle_2 + \beta|\downarrow\rangle_1|\downarrow\rangle_2$ [3, 10]. With reference to a technique recently adopted by P. Kwiat *et al.* [10] the modes $\mathbf{k}_j$ and the strong coherent pump beam with wv $\mathbf{k}_\mathrm{p}$ could be focused by the common lens L beam into a combination of two equal thin Type-I NL plane crystal slabs, *e.g.* LiIO$_3$, $l = 1\,\mathrm{mm}$ thick, and placed in mutual contact along their plane orthogonal to $\mathbf{k}_\mathrm{p}$. If these slabs are mutually rotated around the axis parallel to $\mathbf{k}_\mathrm{p}$ by an angle $\phi = \pi/2$ and the linear polarization $\boldsymbol{\pi}_\mathrm{p}$ of the pump beam is also rotated by $\phi = \pi/4$, both nonlocally correlated orthogonal π-state components of the injected entangled state undergo equal NL transformations given by [2], thus realizing an overall, *information-preserving* up- (or down-) frequency conversion of $|\Phi\rangle$.

By recent works [11, 12] a new conceptual and formal perspective has been introduced in quantum information according to which the optical *field's modes* rather than the photons are taken as the carriers of quantum information and entanglement. Furthermore in that picture any *qubit* is physically implemented by a two-dimensional subspace of Fock states of the e.m. field, specifically the state spanned by the vacuum state and the 1-photon state. According to this perspective the class of information-preserving NL transformations of the state given by eq. (1) investigated in the present work should be more correctly referred to as *entangled states* and may indeed provide a useful new set of unitary transformations for the Hilbert space evolution of these new information states.

$$* \; * \; *$$

This work has been supported by the FET European Network on Quantum Information and Communication (Contract IST-2000-29681: ATESIT) and by PAIS-INFM 2002 (QEUPCO).

REFERENCES

[1] Wootters W. K. and Zurek W. H., *Phys. Rev. D*, **19** (1979) 473; Bartell L. S., *Phys. Rev. D*, **21** (1980) 1698; Jammer M., *The Philosophy of Quantum Mechanics* (Wiley, New York) 1974.

[2] Popescu S., in *Introduction to Quantum Computation and Information*, edited by H. Lo, S. Popescu and T. Spiller (World Scientific, New York) 1998, p. 29.

[3] De Martini F., *Phys. Rev. Lett.*, **81** (1998) 2842; *Phys. Lett. A*, **250** (1998) 15; De Martini F., Mussi V. and Bovino F., *Opt. Commun.*, **179** (2000) 581; De Martini F., Di Giuseppe G. and Padua S., *Phys. Rev. Lett.*, **87** (2001) 150401; De Martini F., Buzek V., Sciarrino F. and Sias C., *Nature*, **419** (2002) 815.

[4] It is possible to conceive a NL 3-wave interaction in which the PMC is relaxed, *e.g.* in a very thin crystal, and the input modes are coupled *nondeterministically* to the up-converted ones. The Feynman path indistinguishability implied by the additional quantum chance condition may lead to additional interference phenomena that interplay with the dominant behaviour of the nonlinear MZ-IF.

[5] Loudon R., *The Quantum Theory of Light* (Clarendon Press, Oxford) 1983, Chapt. 5.

[6] Mataloni P., Jedrkiewicz O. and De Martini F., *Phys. Lett. A*, **243** (1998) 270.

[7] Boyd R. W., *Nonlinear Optics* (Academic Press, San Diego) 1992, Chapt. 2.

[8] Chuang I. L., in *Introduction to Quantum Computation and Information*, edited by H. Lo, S. Popescu and T. Spiller (World Scientific, New York) 1998, p. 311.

[9] Nakamura Y., Pashkin Y. and Tsai J. S., *Nature*, **398** (1999) 786; Loss D. and Di Vincenzo D., *Phys. Rev. A*, **57** (1998) 120.

[10] Kwiat P. G., Waks E., White A. G., Appelbaum I. and Eberhard P. H., *Phys. Rev. A*, **60** (1999) R773.

[11] Knill E., Laflamme R. and Milburn G., *Nature (London)*, **396** (1998) 52; Lombardi E., Sciarrino F., Popescu S. and De Martini F., quant-ph/0109160 and *Phys. Rev. Lett.*, **88** (2002) 70402.

[12] Duan M., Lukin M., Cirac J. and Zoller P., *Nature (London)*, **414** (2001) 413; Sciarrino F., Lombardi E. and De Martini F., quant-ph/0201019 v1; *Phys. Rev. A*, **66** (2002) 024309.

QUANTUM COMPUTATION – QUANTUM MEASUREMENT

Quantum logic gates with semiconductor quantum dots

D. P. DiVincenzo

IBM Research Division, Thomas J. Watson Research Center
PO Box 218, Yorktown Heights, NY 10598, USA

1. – Introduction

In this written lecture I would like to concentrate on the practical philosophy that has guided us in our thinking about quantum computing using semiconductor quantum dots. It is a strange enterprise. On the one hand, it employs tools and language that are traditional to theoretical solid state physics: we calculate Green functions, transition rates, and spectra, and we predict the results of magnetization, transport, and noise measurements. On the other hand, our motivation for the enterprise is very, very different from that which has typically driven theoretical work; this motivation is very practical, worldly, and audacious—to build a new kind of machine.

This philosophy has strengths and weaknesses. I think that it has enabled us to propose many novel and interesting things to do in the laboratory, which would never have occurred to us if we had been pursuing traditional approaches. But it also has a serious danger: some of our audacious proposals are the result of a hopeful extrapolation of existing, known physics into new regimes. We may find that our extrapolations are naive, and the resulting predictions are the worst sort of nonsense. We have not even completed that first cycle of experimental confrontation of this new theoretical approach, and I face that trial with some trepidation and humility.

But, I said I was going to be practical, so let me get down to details!

2. – Quantum dots for quantum computing

What do we need to have, or to create, a qubit? There are a set of attributes that a physical system should have if it is to embody a qubit: 1) it should be a few-level quantum system; 2) it should be possible to isolate this system from its surroundings; 3) its parameters (energy eigenstates, transition matrix elements) should be very accurately known; 4) the Hamiltonian of the system should be subject, at least to some extent, to accurate time-dependent external control; 5) the state of the qubit should, ideally, be readable by a projective quantum measurement.

It goes without saying that no physical system of any kind, solid state or not, satisfies all these requirements perfectly. But this enumeration is very useful, in that it shows why almost all quantum phenomena seen in solid state systems are *not* suitable for use as qubits. We fail on every point: 1) It has been typical in solid state physics to have quantum states forming a continuum or band. 2) The lifetime of any of these states is limited because of scattering or interaction with the environment. 3) Physical parameters such as effective masses or g-factors are not known to ultra-high precision. 4) The Hamiltonian of the system is usually a given, with no opportunity for modification. 5) Measurements in solid state physics are typically on ensembles; single quantum measurements are not feasible. So, it seems that on every point, solid state physics is incompatible with the possibility of qubits.

But it appears that each of these conditions of textbook solid state physics is subject to change, such that we can hope that qubits are possible in the solid state setting. The enumeration above indicates that we must be very selective in choosing particular, unusual solid state settings in which the conditions for qubits become satisfied.

The development of the science of solid state quantum dots [1,2] provides hope that all the difficulties can, perhaps, be overcome. This development has no well-defined beginning, as its growth is clearly related to studies of electron tunnelling, and this subject is as old as solid state physics itself. But starting about fifteen years ago, workers studying tunnel junctions realized that it was possible to construct junctions of small enough area that the discrete nature of the current passing through the junction, the fact that it consists of the passage of individual electrons, would begin to have some new consequences. In particular, it was understood that the presence of one electron in the junction region would, because of Coulomb repulsion, prevent other electrons from being in the neighborhood at the same time. Many phenomena result from this simple scenario, which are described under the heading of "Coulomb blockade" effect.

The work of the group of Kastner [1] showed the first dramatic manifestation of the Coulomb blockade effect: their work on small Si MOSFET structures showed that, if electrons in the tunneling region could be trapped temporarily (in islands in the MOSFET channel, or in grains in the tunneling barrier), then conduction of the channel could be cut off altogether by Coulomb repulsion. Small semiconductor structures were soon built in which this effect could be intentionally enhanced and controlled by gating; thus the "single electron transistor" (SET) was born.

The SET is still a very long way from a qubit, but in fact it already takes several crucial steps in this direction. It makes the creation of a discrete, controllable quantum system possible. Further developments continued to take it in this direction. In designing structures in which the SET effects were enhanced, workers soon understood that what they were doing was recreating the Millikan oil-drop experiment in the solid state: These structures consisted of a small island—the quantum dot—separated from the source and drain by tunneling barriers. The chemical potential of the island could be varied continuously by a voltage gate. At most values of the chemical potential, the dot, because of its electrical isolation, has a well-defined integer number N of electrons. In this state, the Coulomb blockade effect prevents tunneling of electrons from the source, via the dot, to the drain. The transistor is "off". But at particular values of the chemical potential, when the favored number of electrons on the dot passes from N to $N+1$, this tunneling becomes energetically possible: the transistor turns "on".

Thus did solid state physics approach an atomic-like scenario, having (at least approximately) an isolated entity with a stable, discrete number of electrons. And even though this electron number is enormously large in most cases (*e.g.*, $N = 10^{11}$, say), the *relevant* number of electons in the dot, namely the ones associated with the conduction band or "outer shell", could be quite small. After a number of generations of technical advances in the fabrication of quantum dots, any value of this effective N, from 1 to essentially infinity, are now possible [3]. There are cases where it is difficult to say exactly what the value of N is, but to know that it is close to, say, 20; in addition, there are new techniques (involving the Kondo effect [1]) which enable one to determine whether N is odd or even. This particular capability is of interest for quantum-dot qubits, as we will see shortly.

So, we have an externally tunable pseudo-atom. But where is the qubit? There remain various possible ways to proceed, many of which are hampered by the "pseudoness" of these pseudo-atoms. For one thing, quantum dots do not really have fixed physical parameters as atoms do—uncontrolled motions of charged impurities inside the semiconductor, even far from the quantum dot, cause undesired changes in the dot properties. This has long been recognized as a detrimental state of affairs, and a lot of technical work has been devoted to minimizing this effect. Less susceptible to technical improvement is the inevitably greater degree of decoherence in the solid state environment as compared with an atom in vacuum: contributors to decoherence include phonons, nuclear spins, and the excitiation of impurities, surface states, or interface states. So, many of the coherences of quantum dots are expected to be short lived; many of them have not yet been measured, but those that have [4] indicate that many quantum-dot states will be unsuitable for representing a qubit.

But things are not hopeless. I mention these difficulties to explain the initial motivation, in the work of Loss and myself [5], for using spin as the principal carrier of quantum information in the quantum dot. In atomic physics, there are instances of coherences involving only changes of the spin quantum number that have immeasurably long coherence times. So, while the spin coherence times in quantum dots will, for all the reasons mentioned above, undoubtedly be enormously shorter than in isolated atoms, we believe that these spin coherences have a good chance of being coherent enough to

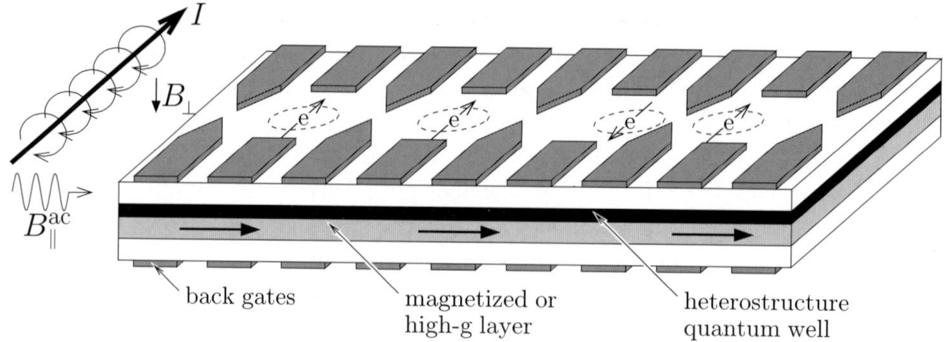

Fig. 1. – Layout for a quantum-dot quantum computer.

form a workable qubit. We are hearteneed by the existing measurement of T_2 lifetimes of spins in semiconductors; in these experiments, lifetimes approaching 1 ms have been seen in Si [6], and approaching 1 μs in GaAs [7].

But we cannot be sure that these experiments are relevant for the coherence time in the kind of quantum-dot structures that we will describe in more detail below. Coherence times are very dependent on the details; the presence of impurities, the exact state of interfaces and surfaces, the electromagnetic environment, all influence the degree of decoherence. Our work is based on the hope that the "best case" coherence times of today can become the "typical case" coherence times of the future.

Given this situation with the spin state of electrons in semiconductor structures, Loss and I developed a proposal for a quantum computing architecture as shown in fig. 1. It is informed by the existing state of experiment, but it has several elements that take us beyond the current state of the art. I will discuss several of these now: first, our spin exchange scheme, which I think of as "somewhat beyond" the state of art; and then, our ideas for single-spin operations and measurements, which I consider "far beyond".

$2\dot{}1.$ *Exchange coupling.* – In the structure shown, the coupling between neighboring dots can be controlled by varying the potential barrier between them. This is done by raising and lowering the potential on the two pointy-ended gates in the diagram between pairs of dots. In the literature these are referred to as the "point contact gates"; this nomenclature is historical, and refers to the fact that these gates have typically been used to turn on and off a conducting path for electrons between the two quantum dots. Since we will never impose an electric current flow on our quantum-dot system, this designation is not especially informative for us. But the fact that these point-contact experiments have previously been done is heartening for us, because it shows that it is possible to change the coupling between dots. This will provide the basis for our basic two-qubit gate.

What we propose is really the most basic molecular physics, once you view the quantum dot as a pseudo-atom. In effect, we propose to engineer a collision between two of these "atoms". When the ground-state orbitals of the two dots begin to overlap, then by the usual reasoning the atomic levels begin to split. In the usual, simple scenario (no relativistic effects) the "bonding" (singlet) level goes down in energy, while the "antibonding" triplet level goes up.

What does this mean as a quantum gate? We consider [8, 9] a quantum gate to be a controlled time-dependent Hamiltonian acting, in this case, in the space of two qubits. Clearly there is a time-dependent Hamiltonian in this situation, as the voltage V of the two point-contact gates has been changed as a function of time. This changing voltage changes the overlap of the two pseudo-atoms, so that the singlet-triplet splitting is changed. If we construct a model Hamiltonian that describes only this variable singlet-triplet splitting, then the Hamiltonian can only be of the form

$$(1) \qquad H = J(t)\vec{S}_1 \cdot \vec{S}_2.$$

That is, the functional form is completely fixed, with only an overall scale $J(t)$ that is variable. Here $\vec{S}$ is the vector of spin-1/2 operators (proportional to the vector of Pauli matrices). The interaction described by this functional form is, in many contexts, called the *Heisenberg exchange* interaction (or just the exchange interaction), and the scalar J is called the exchange constant. Of course here it is not constant, as it varies in time through the "collision". Actually, in the cases of interest to us, it is more appropriate to think of J as being a function of V (because we consider the changes of the Hamiltonian to occur adiabatically [10]), so to be complete we might write the prefactor of the equation as $J(V(t))$. The functional form of $J(V)$ depends on the details of the confining potential, the semiconductor band structure, and other details; but we can make several important general observations. Since J is nonzero only if there is overlap between the atoms, we expect that $J(V)$ goes very rapidly (exponentially) to zero when V is raised high enough to significantly separate the two dots.

We have also estimated [10] (using a little bit more than a pencil and paper) the order of magnitude that J may reach when the dot coupling is turned on. Our calculations indicate that the maximum value that J can conveniently attain is of the order of magnitude of 0.1 meV. This number is helpful in that it sets the scale of the gate operation time that can be striven for. This is given by $\hbar/J_{\mathrm{max}}$, and is of the order of tens of picoseconds (10^{-11} s).

It is worth emphasizing that at this point we have definitely entered the realm of plausible, theoretically sound, but experimentally completely untested schemes. No one has tried to pulse the voltage gates in this sort of structure and see the quantum state evolve under the exchange coupling. There are several reasons why this has not been done: the task is technically hard, on top of all the technically hard things that are required to have the quantum-dot structure. The added hardness, as I see it, is connected with changing the exchange coupling *fast*. No one is quite prepared to launch picosecond pulses down into their cryostats (although it is not so far off the agenda, either). There are other technical hurdles related to this—no one knows how reproducible these quantum-dot structures will be. Perhaps with picosecond pulses being launched at it (corresponding to GHz electromagnetic radiation), the structures will become very "noisy", with individual atoms in the crystal hopping around and messing things up. These phenomena are known, and some things are known about how to keep them under control, but they may get worse under the conditions of the experiment that we contemplate.

2˙2. *Quantum measurement.* – But besides these technical difficulties, there is another problem with doing the proposed experiment that is, in some sense, more fundamental: there is no way, presently, of measuring that the desired effect (the exchange coupling) has taken place. Let us examine a little further how the desired experiment is supposed to go [5]. We should first set the spins on the two dots to some "interesting" initial condition, so that we can watch the state evolve as a function of time under the exchange interaction. Both spins up (a simple initial condition to achieve by placing the whole system in a large magnetic field) will not do, as this state does not change under the exchange coupling (since it is already a pure triplet). So we need something else: we proposed [5] that a state where spin 1 is up and spin 2 is in a mixture (perhaps thermal) of up and down. I can think of three ways of achieving this initial state, each of which is rather hard. 1) Arrange that the magnetic field or g-factor on dot number 2 be zero, either using materials variations, or a big magnetic field gradient. 2) Use a resonant pulse that is selective for dot 2 to rotate its state. 3) Entangle dot 2 with a third dot by exchange coupling them. If this pair drops into their own (singlet) ground state, dot 2 is left in a completely mixed state. This last is perhaps the most elegant alternative, and most useful for future developments of the architecture—but it does require a functioning third dot.

So, the initialization is hard enough, but the measurement of the state after pulsing the dot is even harder. No magnetometry or magnetic resonance measurement is sensitive enough to sense the state of individual spins. One could imagine (as we did previously [5]) that one could assemble a large ensemble of dots and measure them as an ensemble. This is possible, but very demanding of device fabrication (to achieve a high degree of uniformity among, say, a million dot structures), and not so useful in the further development of quantum computing hardware. Much better is to develop some new scheme for sensing the state of individual dot spins. Again, there is no shortage of ideas for how to proceed, but all these ideas require new embellishments and introduce yet further difficulties for the experiment. Here are some of them.

The first approach which we introduced involves the idea of a "spin filter" [5]. This refers to an electron tunneling structure in which the tunneling barrier is sensitive to spin. The most straightforward way (conceptually) to make this happen is to choose the barrier material to be ferromagnetic. If a geometry is designed so that the electron in the quantum dot can be introduced to such a tunnel barrier by some sort of gating scheme, then the occurrence of tunneling of that single electron through the barrier can serve as a measurement of the spin of that electron. The measurement of the resulting single-electron "current" is a difficult but well-understood metrological problem—the "single-electron transistor" has the necessary sensitivity to detect the arrival of the tunnelled electron, under the right circumstances.

At least, this approach seems to be much superior to any attempt to detect the electron spin directly, that is, by sensing the magnetic field that it produces. Such single-spin magnetometry has been under invenstigation for many years in the atomic-force-microscope community, but no system of sufficient sensitivity has been developed. A clear problem with the spin-filtering scheme is that it requires a magnetic barrier material that is well matched to the semiconductors that comprise the quantum-dot device. Various magnetic barrier materials are well known, and their mating to high-quality semiconductor crystals is under investigation, but it is not a mature art.

The second approach [11] for sensing the qubit spin has the advantage that it would not require any magnetic materials. It exploits differences between the singlet and triplet states of coupled dot pairs. As a result of the Pauli exclusion principle, the binding energy of the triplet state of two electrons occupying the same dot is typically less than the binding energy of the singlet state; in extreme cases, the triplet may not be bound at all. This situation provides a means of measurement, if not of a single electron spin, at least of the relative spin of a pair of electrons. (This is almost equally useful in quantum computation.) If the two electrons in neighboring single-electron dots are in a singlet state, then an applied potential difference between the two dots can cause an electron to transfer from one dot to the other; the electrons remain in a singlet state. If the electrons are in a triplet state, this transfer is less energetically favorable (or it is impossible). So, again, we have a situation where we have to distinguish the occurrence, or not, of a "one-electron current", and the same techniques as we just discussed would be used.

The third approach, developed recently by Recher *et al.* [12], shows that the singlet-triplet splitting arising from the Pauli principle can be used in a more direct way to accomplish the desired single-spin measurement. Instead of a two-step process, in which first the spin state is converted to a charge difference, which then must be read by other means, they show how to make the spin detection doable directly as an electrical transport measurement. The idea is that in the process of two-terminal transport of electrons through the dot, electrons must be temporarily (or virtually) added or subtracted from the dot. This intermediate state is either a singlet or a triplet. If the source and drain reservoirs are spin polarized, then, depending on whether the dot spin is up or down, the intermediate state will be only a singlet or only a triplet. The differences in the intermediate state will cause the level of steady state current to be different for the two spin states.

Each of these three approaches has promise, and each has difficulties. The prospect of a single-step measurement provided by Recher *et al.* [12] is very appealing, compared with the difficult two-step method of the other two. This third scheme has its own problems: several dot parameters have to be tuned into the right range to make the scheme work. In addition, it too requires magnetic materials, in the form of a spin-polarized source and drain. This might be easier than the magnetic barrier of the first technique, since a low-density electron gas in a magnetic field can be spin polarized, so that different crystalline materials may not be required. Despite its greater material complexity, the first method might ultimately be favored on account of the potentially greater robustness of the effect: the energy scales for the magnetic tunneling phenomenon can be relatively large, so less fine tuning of parameters might be necessary than with the other two techniques.

But we have to admit that all of this is quite speculative at the moment. Quite a bit of new laboratory work will be necessary before we can see which of these approaches, or some other one, will lead to a feasible technique for spin detection. Before the single-quantum measurement capability is in place, we cannot really foresee how further progress will be made towards semiconductor quantum computing.

2˙3. *Single-spin rotations.* – I am pretty sure that the capabilities that I have just discussed are absolutely necessary for quantum-dot quantum computing to progress. Although I would never rule out the possibility of some new, ingenious scheme for doing quantum computing in a completely new way in these sorts of solid state systems, every possible approach that we can think of today will use two-spin exchange and one- or two-spin measurements. But there is one other capability that we have discussed a lot: the capability of doing single-spin unitary operations. We now understand (see below) that this capability is *not* necessary for quantum computation; but if it could be done well, it would be a desirable supplement to the capabilities we have already discussed.

Ideally, any single-spin operation should be doable, that is, any rotation of the Bloch sphere of any spin in the device should be possible. Almost as good would be any such rotation around two fixed axes, say the x-axis and the z-axis. Rotations around only one axis, on the other hand, are much less useful. The point is that compositions of rotations around two axes can be used to generate any rotation—for orthogonal axes, this is the Euler-angle construction.

I want to quickly review the schemes that have been advanced for doing these one-spin rotations. They are nearly as numerous as, and they bear some relation to, the approaches than have been developed for single-spin measurement. They all involve hard work and difficult extensions of the current state of the art. Here is the list, in no special order:

1) In our original work [5], we suggested achieving one-bit gates by the incorporation of magnetic materials. Suppose, for example, that some of the semiconducting dots were ferromagnetic. Then by transferring electrons in and out of these dots, the spins of these electrons would precess around the magnetization vector of the dot. If the in and out times are controlled, then desired one-qubit rotations can be effected. Note that this requires magnetic dots with at least two different magnetization directions (some can be one, some the other), which could be obtained by variation of the shape anisotropies. We have subsequently [13] mentioned several different variants of this idea: One involves a magnetic barrier layer underneath the dots (just as in the single-spin measurement discussion above). Pushing the electron wave function partially into this barrier by gating could also accomplish the desired precession.

2) If materials with different g-factors are available and controllable, then the same procedures can be accomplished without the need for magnetized materials. The III-V semiconductors offer a complete variety of g-factors from 0 to large positive or negative, and with various anisotropies. Suppose the layer containing the quantum dots has $g = 0$, the barrier layer above has one g tensor, and the barrier layer below another g tensor.

Then, again, by pushing the electrons in the dot up or down with top and/or bottom gates in the presence of a uniform B field, controlled single-spin rotations can be effected. At least two different g tensors (with different anisotropies) would be needed to produce all needed rotations.

3) One can conceive of doing the job "directly", that is, by applying a highly inhomogeneous magnetic field to the sample, so that the field is strong at one dot and weak at all other dots. Field gradients of the necessary strength have been demonstrated in nanofabricated structures, but the necessary shaping of the field has not been done, and I believe that there is little prospect for doing so.

4) All of the above techniques can be supplemented with the tools of pulsed magnetic resonance. For example, the third approach might be made more feasible in that a simple linear magnetic field gradient might suffice: by making the resonant frequency of the spin of each quantum dot unique, then an rf field of a chosen frequency can rotate just a particular spin. The flexibility of pulsed magnetic resonance permits any desired spin rotation. The first two schemes could also be used in the same way.

So, there is a wealth of ideas for doing one-spin rotations. All of them are hard; one could argue that they are comparable in difficulty to doing spin measurements. They are certainly not easier, and they are sufficiently different from measurement that they would require at least as much in additional effort to achieve. The first two schemes posit very difficult materials science, in the creation and marrying of poorly understood and very disparate crystalline structures. The third scheme requires a miracle of nano-electromagnetic engineering. The rf-assisted scheme has been very prominent in discussions of single-spin rotations, but it too has real problems. It requires dumping significant amounts of electromagnetic power into a system at cryogenic temperatures. In addition, the rotation rates in this scheme are generally not impressive compared with those that could be achieved directly. Rotations in a Tesla-scale field would give speeds comparable with those hopefully achievable in the two-spin exchange gate. But effective precession fields in pulsed magnetic resonance are generally in the Gauss range. (An exception is the recent all-optical spin tipping technique of Gupta *et al.* [14].)

2'4. *Exchange-only quantum computing.* – Fortunately, quantum gate theory provides various routes for the implementation of quantum algorithms. As I implied above, single-spin rotations can be entirely dispensed with, at a cost. As this result has been thoroughly discussed in a very recent paper of ours [15], I will only review it briefly here.

The cost is that it is no longer possible to think of one spin as a qubit. This is clear, since the Heisenberg coupling is incapable of flipping one spin, so it is certainly unable to do one-qubit gates in this setting. The particular constraint arising from using only the Heisenberg interaction is that it commutes with any rotation of the system; as a consequence, it conserves the total spin quantum numbers of the complete spin system. Given this constraint, the way to use the Heisenberg interaction is to utilize multispin states, all with the *same* spin quantum numbers, to represent all the states in the qubit manifold. Explicit choices guided by this general principle are not hard to find. The simplest solution is to encode the qubit state in the state of three spins. One uses the elementary fact from angular momentum theory that three spin-1/2 particles have two different (*i.e.*, orthogonal) states of total spin 1/2, and projected spin +1/2. These two states can represent the qubit.

We have shown recently [15] that the Heisenberg interaction is "efficiently" universal in this setting, in the sense that there are short sequences of exchange interactions that can implement the cNOT gate and any one bit gate. In particular, any one-bit gate is obtainable by just four exchange interactions among the three spins of the coded qubit, while the cNOT is obtained by a particular (and unique) sequence of 19 exchanges acting between nearest neighbors in a chain of six spins embodying neighboring qubits.

I might comment on why I used the phrase "efficient universality" a moment ago. We have known since 1995 that virtually any entangling quantum gate suffices, in theory, for quantum computation. However, for most of them we would have no sensible notion of what it would mean to program with them. I would assert that the approach I have just mentioned, of showing an exact logical mapping to the canonical one-bit gates and cNOT, is very desirable, although it may not be the only way to go. One could quibble about whether a 19-interaction sequence is efficient; I think that it is, in light of the potential great benefit obtained by dispensing with the physical complexity of implementing direct one-spin rotations.

2˙5. *Complications and prospects.* – What might impede our progress towards quantum-dot quantum gates? Many, many things, certainly. Excessive decoherence is always a great concern. The degree of decoherence is very dependent on details of the system—on the exact crystalline structure, including the degree of strain, the presence or state of interfaces and surfaces, on the presence and nature of impurities and defects in the materials, on the isotopic composition and the state of the nuclear spins in the system, on the temperature, and on the state of the electromagnetic environment. So, the fact that satisfactory decoherence times have been measured in samples related to the ones we envision for quantum computation does not entirely reassure me. It is hard to anticipate ahead of time whether these satisfactory decoherence times will still be present in structures of interest.

There is another, rather less well defined concern that I have, that can be put under the heading of "complexity". We are envisioning an extremely intricate, nanoscale set of devices with a large amount of control apparatus, all trying to manipulate something extremely delicate—a multispin quantum coherence. Doing a very small part of what we propose would normally be considered quite enough for a good physics experiment: either the investigation of coherence, or performing spin resonance on spins in a semiconducting layer, or looking for spin filtering effects. The assembly of all these things is daunting. The calibration problem alone, the determination of the different (unique) physical parameters of each qubit (its g factor, degree of coupling to neighboring qubits) is awesome. If this is much charge switching or $1/f$ noise that will affect these parameters, the task becomes even harder.

Theory can only hope to assist in attacking this complexity little by little. We have recently been considering one small corner of this vast parameter space, by considering some of the possible effects that might arise due to spin-orbit interaction in the quantum-dot system [16]. We have shown that in a (hopefully) general setting, spin-orbit coupling, if it is reasonably weak, should not present any problem for quantum computation. I expect that there will have to be many, many more theoretical efforts of this sort before we can rationally design a working quantum computer.

I will close at this point, leaving a vast amount to be done, perhaps by new students like yourselves. I believe that the prospects are bright. Quantum information processing is too good an idea not to be made real someday. I think that we have hardly begun to understand how significant a development this could be.

$$* \quad * \quad *$$

I am grateful for the support of the National Security Agency and the Advanced Research and Development Activity through Army Research Office contract numbers DAAG55-98-C-0041 and DAAD19-01-C-0056.

REFERENCES

[1] Kouwenhoven L. P., Schöoen G. and Sohn L. L., in *Mesoscopic Electron Transport*, edited by L. Sohn, L. Kouwenhoven and G. Schön, *NATO ASI Ser. E*, Vol. **345** (Kluwer, Dordrecht) 1997, p. 1.
[2] Jacak J., Hawrylak P. and Wojs A., *Quantum Dots* (Springer-Verlag, Berlin) 1998.
[3] Ciorga M. *et al.*, *Phys. Rev. B*, **61** (2000) R16315.
[4] Buks E. *et al.*, *Nature*, **391** (1998) 871.
[5] Loss D. and DiVincenzo D. P., *Phys. Rev. A*, **57** (1998) 120.
[6] Gordon J. P. and Bowers K. D., *Phys. Rev. Lett.*, **1** (1958) 368.
[7] Awschalom D. D. and Kikkawa J. M., *Phys. Today*, **52**, June issue (1999) 33.
[8] DiVincenzo D. P., in *Mesoscopic Electron Transport*, edited by L. Sohn, L. Kouwenhoven and G. Schön, *NATO ASI Ser. E*, Vol. **345** (Kluwer, Dordrecht) 1997, p. 657 (cond-mat/9612126).
[9] DiVincenzo D. P., *Fortsch. Phys.*, **48** (2000) 771.
[10] Burkard G., Loss D. and DiVincenzo D. P., *Phys. Rev. B*, **59** (1999) 2070.
[11] Kane B. E., *Nature*, **393** (1998) 133.
[12] Recher P., Sukhorukov E. V. and Loss D., *Phys. Rev. Lett.*, **85** (2000) 1962.
[13] DiVincenzo D. P., Burkard G., Loss D. and Sukhorukov E. V., *Quantum Mesoscopic Phenomena and Mesoscopic Devices in Microelectronics*, *NATO ASI Ser. E*, Vol. **559** (Kluwer, Dordrecht) 2000, Chapt. 27, pp. 399-428 (cond-mat/9911245).
[14] Gupta J. A., Knobel R., Samarth N. and Awschalom D. D., *Science*, **292** (2001) 2458.
[15] DiVincenzo D. P., Bacon D. A., Kempe J., Burkard G. and Whaley K. B., *Nature*, **408** (2000) 339.
[16] Bonesteel N. E., Stepanenko D. and DiVincenzo D. P., quant-ph/0106161.

Quantum optical implementation of quantum information processing

J. I. Cirac, Luming Duan and P. Zoller

Institute for Theoretical Physics, University of Innsbruck
Technikerstrasse 25-2, A-6020 Innsbruck, Austria

1. – Introduction

It is generally recognized that all the microscopic phenomena that we observed can be described and explained by the principles of Quantum Mechanics. These principles have been extensively tested, and some of them are commonly used in several technological applications. Other principles, like the ones related to the superposition principle and the measurement process, and which are in the realm of most of the paradoxes and strange phenomena related to Quantum Mechanics, have only recently become important in some applications. In particular, they form the basis of a new theory of information which may revolutionize the fields of communication and computation [1].

The basic ideas behind quantum communication and computation are very simple [1]. Quantum communication deals with sending quantum states from one place to another one in such a way that they arrive intact. The most important application so far in this field is the one in which a sender (traditionally called Alice) tries to convey a secret message to a receiver (traditionally called Bob). The message is encoded in the state $|\Psi\rangle$ of a quantum system. Due to the fact that the quantum state of a system is distorted if somebody performs a measurement, Bob will receive a wrong state if a third (malevolent) party (Eve) tries to read the message. This way of secret communication is usually called quantum cryptography, and it is the only provably secure way in which two partners can share secret messages. In the context of quantum computation, the existence of entangled states of several particles offers the possibility of performing certain computational tasks in times much shorter than the ones taken by common (classical) computers. By acting on a system entangled to other systems, one modifies the state of the whole system at the same time, which leads to an important speed-up in several computations. In particular, if one could build a quantum computer one would be able to decompose very large numbers (of $n \gg 1$ digits) into prime factors in a time that scales polynomially with n ($t \simeq an^b$, with a and b constants) [2], in contrast to the exponential dependence of classical computers [3]. A quantum computer would therefore allow to break all the classical cryptographic protocols which are based on the impossibility of factorizing large numbers in relatively short times. There are also other algorithms which make use of the superposition principle in Quantum Mechanics and they are more efficient that the classical counterparts.

The experimental situation in quantum computing and quantum communication is very different. Whereas in the second case it is already a mature field, close to reaching the commercial level, the first one is still in its infancy. For the moment, it is possible to construct very small prototypes of quantum computers which, of course, do not offer any advantage over nowadays classical computers. In fact, it seems almost impossible that a *useful* quantum computer can be created in the next twenty or thirty years. Nevertheless, pursuing research in this field does not have the only goal of being useful in the near future, but there are several other goals which can be attained in the way. In particular, creating a quantum computer means that we can manipulate the quantum state of an enormous system at will, which, apart from being capable of bringing some surprises to our present knowledge of Quantum Mechanics, paves the way for some other applications based on this theory which may be discovered in the future and do not require a large system. On the other hand, the first experiments on quantum cryptography took place between location separated by few centimeters. At present, quantum cryptography over distances of the order of 50 km is possible. Other experiments on quantum communication achieve basically the same distances. Their extension to longer distances does not seem to be straightforward though, since the systems carrying the quantum states (*i.e.* photons) are eventually absorbed and therefore the quantum states are distorted before they arrive at their destination. A way to overcome this problem is to use quantum repeaters [4], in which small quantum computers amplify in a sense the quantum states so that they arrive safely at their destination.

There are very few systems in which one can implement a small quantum computer. Many of the ideas come from the field of Quantum Optics. The reason is the spectacular experimental development of this field during the last years, which has allowed, so to say, to dominate the quantum world. In particular, the internal quantum levels of atoms and ions can be manipulated very efficiently using lasers. One can basically stop them (*i.e.* cool them) with laser cooling techniques. It is also possible to manipulate their quantum state of motion by pushing them with laser light. These methods, when combined appropriately, allow, at least in principle, to perform quantum computations and to build quantum repeaters. On the other hand, very recently it has been recognized that these goals can also be achieved using atomic ensembles, instead of single atoms. The idea is to manipulate some collective degrees of freedoms of atomic ensembles using lasers, the main advantage being that the atoms do not need to be manipulated one by one, and they can be at room temperature.

The aim of this paper is to review some of the quantum optical systems that have been proposed to perform quantum computations, both using single atoms and using atomic ensembles. In the next section we will give a brief introduction to some of the main topics in quantum information theory, with particular emphasis on the ones that are needed in the next sections. In the third section we will show how to use single atoms (ions) and photons in order to perform several tasks related to quantum information processing, whereas in the fourth section we will introduce several methods to deal with atomic ensembles.

2. – Basic concepts in quantum information theory

$2^.1$. *Introduction.* – Most of the counter-intuitive predictions of Quantum Mechanics are related to the superposition principle. For example, according to Quantum Mechanics the properties of one object are generally not defined when we do not observe it. This principle, when applied to more than one system, may lead to very intriguing phenomena related to non-locality (actions in some system may affect in a special way some other systems) which have called the attention of philosophers and physicists since the advent of Quantum Mechanics. The basic ingredient of such phenomena is *entanglement, i.e.* the possibility of having two or more systems in a state which displays (quantum) correlations. Apart from its fundamental interest, entanglement plays an important role in most of the applications in the field of quantum information [1]. In particular, entangled states are crucial for quantum communication and computation. In this section we will review the concept of entanglement, as well as some of the basic concepts in quantum communication and computation.

The characterization of this intriguing property of Quantum Mechanics, entanglement, is one of the central theoretical issues in quantum information theory. In fact, there are still many open questions regarding the entanglement properties of two or more quantum systems. Although for pure states of two systems, entanglement is well understood, for more systems we do not know yet how to quantify this property. The situation becomes much more complicated if the state of the system is mixed. In that case, we do not even know in general how to determine whether two systems are entangled or not. This problem has important consequences in current experiments in this field, since after preparing a quantum state of a system one would like to determine whether it is entangled or not (as we will explain, if it is not entangled this would mean that we could have created the same state in a much cheaper and straightforward way). In the next subsection of this section we will review this property, entanglement, both for pure and mixed states and will give some of the known criteria to determine whether a mixed state is entangled or not.

Mixed entangled states are not directly useful for quantum information purposes. However, there exist some methods, known under the name purification protocols, that allow us to make those states useful. We will review some of the basic purification protocols in the third subsection.

Quantum computation and communication are the most visible applications in the field of quantum information. In the last subsections of this section we will review these two topics paying special attention to the physical properties that a quantum system must possess in order to be useful for these applications. In particular, we will give a set of requirements to build a quantum computer which will serve us in the next chapters to show that several quantum optical systems serve for this purpose. We will also review one of the main tools in the field of quantum communication, teleportation, which combined with certain purification protocols allow to extend quantum communication over arbitrarily long distances.

2`2. *Entanglement*

2`2.1. Entanglement of pure states. The superposition principle is one of the basic concepts in Quantum Mechanics: If a system can be in two different states (associated to the vectors $|0\rangle, |1\rangle \in H$) then it can also be in the state described by a linear superposition $\alpha|0\rangle + \beta|1\rangle$. This implies the existence of states in which properties are not well defined. The situation is even more intriguing when we have a composite system. For example, let us consider two subsystems A and B whose states are associated to the elements of two Hilbert spaces, H_A and H_B, respectively. We will assume that these systems are located at different places, although for most of our treatment this condition is not required. Let us consider states of the whole system in which one subsystem is in certain state $|i\rangle_A$ and the other in $|j\rangle_B$. One denotes those states as $|i\rangle_A \otimes |j\rangle_B \in H = H_A \otimes H_B$, or simply $|i, j\rangle \in H$. According to the superposition principle, any superposition of these states must also be possible, *i.e.* the state represented by $|\Psi\rangle \equiv \alpha|0, 0\rangle + \beta|1, 1\rangle \in H$. A state of this form cannot be described as a certain state for system A and some other state for system B; that is, there exists no pair of vectors $|\phi_{1,2}\rangle_{A,B} \in H_{A,B}$ such that $|\Psi\rangle = |\phi_1, \phi_2\rangle$. States of this form are called entangled states and play a fundamental role in quantum information. Note that their existence arises from the fact that the states of the whole system must be described as elements of a Hilbert space themselves. One says that H is the tensor product of H_A and H_B, *i.e.* $H = H_A \otimes H_B$. Thus entangled states are a direct consequence of the tensor product structure of the Hilbert space describing composite systems.

In the following, we will denote by $\{|k\rangle\}_{k=1}^{d_{A,B}}$ an orthonormal basis in $H_{A,B}$, respectively. Although the definitions and results apply for general dimensions, for most of the examples we will consider qubits, *i.e.* systems where $d_A = d_B = 2$. In that case we will take as a basis $\{|0\rangle, |1\rangle\}$. We will use the Pauli operators

$$\sigma_x = |1\rangle\langle 0| + |0\rangle\langle 1|,$$

$$\sigma_y = -i(|1\rangle\langle 0| - |0\rangle\langle 1|),$$

$$\sigma_z = |1\rangle\langle 1| - |0\rangle\langle 0|.$$

For qubits, there are some entangled states which play a very important role in quantum information, the so-called Bell states. They are

$$|\Psi^{\pm}\rangle = \frac{1}{\sqrt{2}}(|0,1\rangle \pm |1,0\rangle),$$

$$|\Phi^{\pm}\rangle = \frac{1}{\sqrt{2}}(|0,0\rangle \pm |1,1\rangle).$$

The most important properties of entangled states is that they carry correlations. That is, if we measure an observable in A and another in B, the outcomes will be, in general, correlated. For example, if we have the state $|\Psi^{-}\rangle$ and measure the observable σ_z in both systems we will obtain the opposite result. Actually, if we measure any observable $\vec{\sigma} \cdot \vec{n}$ we will always obtain opposite results in A and B, the reason being that $|\Psi^{-}\rangle$ is invariant under global rotations (*i.e.* $U \otimes U|\Psi^{-}\rangle \propto |\Psi^{-}\rangle$ for all unitary operators $U \in SU(2)$). Note that for all entangled states there always exist some correlations. For product vectors, however, the outcomes in A are independent of the outcomes in B. This can be also viewed by noting that if A and B are two observables, then $\langle A \otimes B \rangle = \langle A \rangle \langle B \rangle$ for product vectors, but not (in general) for entangled states. The existence of correlations, by itself, is not a property of entangled states. For example, if somebody provides with two boxes A and B in which there are either two black or two white balls, when we open the box we will see correlations. However, the correlations carried by entangled states are, in some sense, different to those, since they occur for any pair of observables. In fact, classical correlations like the ones displayed by the balls in the boxes are restricted by Bell's inequalities [5], whereas the ones corresponding to entangled states may violate them. This is why with the correlations contained in entangled states we can perform things that are not possible using classical correlations.

In order to create entangled states out of product states we need interactions. This can be easily understood as follows. If we do not have interactions, the Hamiltonian describing the evolution of systems A and B will be written as $H = H_A \otimes 1_B + 1_B \otimes H_B$, where 1 is the identity operator. Since $H_A \otimes 1_B$ and $1_B \otimes H_B$ commute with each other, we have that the evolution operator can always be written as $U(t) = U_A(t) \otimes U_B(T)$, and the product state $|\Psi(0)\rangle = |\phi_1(0)\rangle_A \otimes |\phi_2(0)\rangle_B$ will evolve into $|\Psi(t)\rangle = |\phi_1(t)\rangle_A \otimes |\phi_2(t)\rangle_B$ which is a product state. Operators of the form $U = U_A \otimes U_B$ are called local operators. Similarly, we cannot get entangled states by measuring observables in A and B independently, since the state after the measurement will be changed by local operators. One says that entanglement cannot be created by local operations (operations meaning any action on the systems). Note, however, that product states can be obtained by local operations (in particular, by measurements).

Let us show now how we can tell whether a state is entangled or not, and how much. We consider a state of the form

$$\tag{1} |\Psi\rangle = \sum_{i=1}^{d_A} \sum_{j=1}^{d_B} c_{i,j} |i,j\rangle.$$

All the information about the state is in the coefficients $c_{i,j}$ which form a $d_A \times d_B$ matrix that we will call C. Note that we could have chosen another orthonormal bases in $H_{A,B}$ to express this state. In fact, there is a particular basis in which the matrix of the coefficients is diagonal and positive. If we choose such a basis to write the state, it will have the simple form

$$(2) \qquad |\Psi\rangle = \sum_{k=1}^{d} d_k |u_k, v_k\rangle,$$

where $d = \min(d_A, d_B)$, and $\sum_{k=1}^{d} d_k^2 = 1$, with $d_k \geq 0$. This form is called Schmidt decomposition. Its existence directly follows from the singular value decomposition of the matrix C, i.e., the existence of two unitaries U and V and a diagonal one D whose diagonal elements are d_k such that $C = UDV$. The d_k are called Schmidt coefficients and the bases $\{|u_k\rangle\} \in H_A$ and $\{|v_k\rangle\} \in H_B$ are called Schmidt bases. Once we have expressed the state in the Schmidt decomposition, it is very simple to obtain some other information. For example, if we are interested in predicting expectation values or probabilities of outcomes if we only measure system A (or B), all the information about them is in the reduced density operator $\rho_A = \text{tr}_B(|\Psi\rangle\langle\Psi|)$ (analogously for ρ_B). We obtain

$$(3) \qquad \rho_A = \sum_{k=1}^{d} d_k^2 |u_k\rangle\langle u_k|, \qquad \rho_B = \sum_{k=1}^{d} d_k^2 |v_k\rangle\langle v_k|.$$

Conversely, the Schmidt coefficients and the corresponding bases can be easily found by simply diagonalizing both reduced density operators.

For a product state $|\phi_1, \phi_2\rangle$, the reduced density operators are rank-one projectors, i.e. $\rho_{A,B} = |\phi_{1,2}\rangle\langle\phi_{1,2}|$. This means that there is only one Schmidt coefficient which is different than zero. Conversely, if we have a state with only one Schmidt coefficient then it must be a product state. Equivalently, $|\Psi\rangle$ is a product state if and only if the corresponding reduced density operators correspond to pure states. This means that if we have an entangled state, the corresponding reduced density operators must correspond to mixed states, or, equivalently, that there must be more than one non-zero Schmidt coefficients.

Thus, we see that the entanglement of a state is directly related to the mixedness of the reduced density operators. This is intuitively clear since, as we mentioned above, entangled states give rise to correlations and if we only observe one of the systems we lose information about these correlations which results in the fact that we will effectively have a mixed state. This suggests that we can measure the degree of entanglement by the degree of mixedness of the reduced density operators. There are several measures of mixedness of density operators; perhaps the most popular one is the von Neumann entropy $S(\rho) = -\text{tr}(\rho \ln(\rho))$. For a pure state this entropy is zero, whereas for a maximally mixed state (described by the identity operator, properly normalized) it gives $\log_2 d$, where d is the dimension of the Hilbert space. The entropy is convex, i.e. for $p \in [0,1]$,

$S[p\rho_1 + (1-p)\rho_2] \geq pS(\rho_1) + (1-p)S(\rho_2)$, which means that it always increases by mixing (*i.e.* by losing information). This motivates the following definition: Given a state $|\Psi\rangle$, we define the *entropy of entanglement*, $E(\Psi)$ as the von Neumann entropy of the reduced density operator [6]. Thus, we have

$$(4) \qquad E(\Psi) = S(\rho_A) = S(\rho_B) = -\sum_{k=1}^{d} d_k^2 \log_2(d_k^2).$$

The entropy of entanglement only depends on the Schmidt coefficients, but not on the corresponding basis. This means that it is invariant under local unitary operations. That is, if $|\Psi'\rangle = (U_A \otimes U_B)|\Psi\rangle$, then $E(\Psi') = E(\Psi)$. On the other hand, one can show that it cannot increase on average by local operations [7]. That is, if we perform (independent) measurements in A and B and obtain the state $|\Psi_k\rangle$ after the measurement with probability p_k, we have that

$$(5) \qquad E(\Psi) \geq \sum_k p_k E(\Psi_k).$$

Note, however, that the previous inequality does not imply that none of the $E(\Psi_k)$ can be larger than $E(\Psi)$, or even the maximum allowed $\log_2 d$. States $|\Psi\rangle \in H_{AB} = C^d \otimes C^d$ for which $E(\Psi) = \log_2(d)$ are called *maximally entangled states* in d dimensions.

Let us now consider more systems $A_1, A_2, \ldots, A_N$. Now, we can have entangled states and product states of the different systems [8]. For example, we can have a state of the form $|\Psi\rangle = |\phi_1\rangle_{A_1 A_3} \otimes |\phi_2\rangle_{A_2 A_5 A_6} \otimes |\phi_3\rangle_{A_4}$, where $|\phi_{1,2,3}\rangle$ cannot be written as product states. It is clear that in a state like that, the parties A_1 and A_3 are entangled with each other, but not to the rest; similarly, the parties A_2, A_5, and A_6 are entangled among them, and the party A_4 is completely disentangled.

In general we can consider all possible partitions of those systems in which we group some of them. For example, we can consider the partition $(A_1 A_3)$, $(A_2 A_5 A_6)$, (A_4). We can classify the entangled states according to the different partitions. That is, a state is entangled according to some partition if it can be written as a product state of the corresponding disjoint elements of the groups, but not within each of the groups. In order to determine the partition corresponding to a particular state we can calculate all possible reduced density operators and look whether they correspond to mixed states or not.

The quantification of the multipartite entanglement is a more complicated question which can be illustrated by the following example. Let us consider three parties and the states [9]

$$(6) \qquad |\text{GHZ}\rangle = \frac{1}{\sqrt{2}}(|0,0,0\rangle - |1,1,1\rangle),$$

$$(7) \qquad |\text{W}\rangle = \frac{1}{\sqrt{3}}(|0,0,1\rangle + |0,1,0\rangle + |1,0,0\rangle).$$

Those are entangled states according to the partition $(A_1 A_2 A_3)$. However it is hard to say which one is more entangled. Certainly, the first one possesses a sticking non-local behavior, in the sense that it can be used to prove Bell's theorem without using inequalities [9]. However, it is very weak in the sense that if one party does not participate in the measurement (or is lost), then all the entanglement disappears. However, the second one retains some entanglement even if one particle is lost (in fact it is the most robust against particle losses) [10].

$2{\cdot}2.2.$ **Entanglement of mixed states.** The states that we have considered in the previous subsection are idealized. In reality, all systems interact with some sort of environment. Thus, we should include the state of the environment in our description in order to be consistent. In fact, due to the interaction between system and environment they will become entangled even if initially they were in a product state: $|\Psi_S(0)\rangle_A \otimes |\Psi_E(0)\rangle_E \to |\Psi(t)\rangle_{SE}$. Since we are only interested in our system, all the information that we can acquire (without performing measurements in the environment) are contained in the reduced density operator $\rho_S(t) = \mathrm{tr}_E[|\Psi(t)\rangle_{SE}\langle\Psi(t)|]$, which will correspond to a mixed state. Actually, this process in which a pure state is converted into a mixed state via its interaction with the environment is sometimes called decoherence. The term decoherence comes from a process which makes the coherence (non-diagonal elements of the density operator in a given basis) to vanish. However, since this definition depends on the basis some authors prefer to call decoherence any process which is not describable by a unitary operator, *i.e.* which comes from the interaction of the system with some other systems.

Note that density operators can always be written in the form

$$(8) \qquad \rho = \sum_k p_k |\phi_k\rangle\langle\phi_k|,$$

where the p_k are positive and add up to one. One particular decomposition of this form is the spectral decomposition, in which, additionally, the vectors $|\phi_k\rangle$ form an orthonormal basis. In general, except for pure states (rank-one density operators) there are infinitely many decompositions. Note that a decomposition like (8) tells us one way of creating a state described by ρ. We simply have to prepare the system in state $|\phi_k\rangle$ with probability p_k. The fact that there exist infinitely many decompositions of a state means that it can be prepared in infinitely many different forms. For example, the state $\rho = I/2$ of one qubit can be prepared by choosing randomly one among the states $\{|0\rangle, |1\rangle\}$ or one among the states $\{|+\rangle, |-\rangle\}$, where $|\pm\rangle = (|0\rangle \pm |1\rangle)/\sqrt{2}$. It is also worth stressing that even though the systems are prepared in different forms, they are completely indistinguishable. The reason is that the probability of any outcome after a measurement is completely determined by the density operator, so that if two systems have the same density operator they cannot be distinguished by performing any measurement (and therefore by any means). Density operators are linear and self-adjoint ($\rho = \rho^\dagger$), have trace one ($\mathrm{tr}(\rho) = 1$)), and are positive ($\rho \geq 0$).

We thus have to define entanglement for mixed states [11]. Following what happens with pure states, it makes sense to define entangled states as those that require inter-actions between the systems in order to be prepared, and non-entangled (or separable) as the ones that can be created without interactions. More specifically, in Quantum Information a state is called entangled if it cannot be prepared by local operations (and classical communication) out of a product state. As we will see, this definition is equivalent to imposing that mixtures of product states are not entangled. Let us give some examples with two qubits: The state described by $\rho = |0,0\rangle\langle 0,0|$ is not entangled since it is already a product state. Any density operator of the form $\rho = \rho_A \otimes \rho_B$ is not entangled since the states $\rho_{A,B}$ can be prepared locally out of the state $|0\rangle$. To see this, if we write ρ as in (8) we can simply transform the state $|0\rangle$ into $|\phi_k\rangle$ with probability p_k. The state

$$\rho = \frac{1}{2}(|0,0\rangle\langle 0,0| + |1,1\rangle\langle 1,1|)$$

is not entangled since it can be locally prepared as follows. We choose randomly 0 or 1. If we have 0, we prepare both A and B in state $|0\rangle$ and otherwise in $|1\rangle$. Obviously, in this way we do not need any interaction between the systems. We just need classical communication between the location of A and B so that the corresponding preparers can agree on the state they prepare. With these examples we see that the definition of entanglement is equivalent to the following mathematical characterization: ρ is separable if and only if there exist $p_k \geq 0$ and $\{|a_k\rangle\} \in H_A$ and $\{|b_k\rangle \in H_B$ such that

$$\rho = \sum_k p_k |a_k, b_k\rangle\langle a_k, b_k|.$$

Otherwise it is entangled. It turns out that it is very hard in practice to determine whether a given state is entangled or not. There exists, however, an important suffi-cient criterion that may be useful in some occasions. It states [12] that if ρ^{T_A} has a negative eigenvalue (*i.e.* it is not positive) then ρ is entangled. Here ρ^{T_A} stands for the partial transpose of ρ with respect to the first system in the basis $\{|k\rangle\}_{k=1}^{d_A} \in H_A$, *i.e.* $_A\langle k|\rho^{T_A}|k'\rangle_A =_A \langle k'|\rho|k\rangle$. In general, the converse of this criterion is not true [13]. That is, there exist entangled states fulfilling $\rho^{T_A} \geq 0$. However, in low dimensions (if $d_A \times d_B \leq 6$) this criterion (called Peres-Horodecki criterion [12, 14]) gives a necessary and sufficient condition: ρ is separable if and only if $\rho^{T_A} \geq 0$. For other separability criteria see [15].

Given some state, sometimes we would like to know "how close" it is to some pure state, like for example a Bell state. In order to measure this quantity we define the fidelity of ρ with respect to some state $|\Phi\rangle$ as

$$F = \langle\Phi|\rho|\Phi\rangle.$$

A fidelity $F \simeq 1$ means that our state is very close to the desired one. Note that a completely random state has a fidelity $F = 1/d$, where d is the dimension of the total Hilbert space.

A particularly useful family of mixed states are the so-called Werner-like states. Let us consider two systems A and B with corresponding Hilbert spaces of dimension d both. Given a state ρ we depolarize it locally by applying the same random unitary operator to system A and system B. One can show that the state after this process has the form

$$(9) \qquad \rho_F = F\frac{P_a}{d_a} + (1 - F)\frac{P_s}{d_s},$$

where $P_s = (1 + \pi_{AB})/2$ and $P_a = (1 - \pi_{AB})/2$ are the projector onto the symmetric and antisymmetric subspaces (π_{AB} is the permutation operator), and $d_s = d(d + 1)/2$ and $d_a = d(d - 1)/2$ the corresponding dimensions. Here, $F = \mathrm{tr}(P_a\rho)$. ρ_F is called Werner-like state, since Werner [11] was the first one who introduced them for the case of qubits. One can easily show that ρ_F is entangled if and only if $\rho_F^{T_A}$ is not positive.

2˙3. *Purification*. – Most of the applications in the field of Quantum Information are based on the use of superpositions of pure states. However, in practice, the state that one has at disposal are mixed. For example, if one would like to perform quantum cryptography over long distances using entangled photons, when they arrive at the final location their state will also be entangled to the environment and therefore mixed. The longer the distance the photons have to travel, the more mixed they will become. Unfortunately, if they are significantly mixed, the security of the corresponding cryptographic protocol will no longer be ensured. This fact considerably limits the distances over which one can perform secure quantum cryptography. Fortunately, there is a method that allows to make the states purer, and even more entangled. The idea is to use several copies of a state which is not useful for the applications of Quantum Information, but that it is still entangled [16, 17]. Using local operations and classical communication it is sometimes possible to obtain fewer copies of particles in a state which is closer to a maximally entangled states, for example the state $|\Phi^+\rangle$. This process is called entanglement purification (or distillation), and will be the subject of the present subsection.

Let us consider the two-qubit Werner state

$$\rho_F = F|\Psi^-\rangle\langle\Psi^-| + \frac{(1 - F)}{3}(|\Psi^+\rangle\langle\Psi^+| + |\Phi^+\rangle\langle\Phi^+| + |\Phi^-\rangle\langle\Phi^-|),$$

where all the vectors appearing here are Bell states. In the present scenario, Alice and Bob share two pairs of qubits in that state. Let us denote by A_1 and A_2 Alice's particles and by B_1 and B_2 Bob's, so that their state is $\rho_F \otimes \rho_F$. The distillation procedure presented in ref. [16] is as follows. First, Alice applies the unitary transformation σ_y to her two qubits. This transforms in each pair $|\Psi^\pm\rangle \rightarrow |\Phi^\mp\rangle$. Thus, the resulting state will have now the maximum contribution coming from $|\Phi^+\rangle$. Then, both apply locally a controlled-NOT operation to their two particles, where A_1 and B_1 act like sources, and

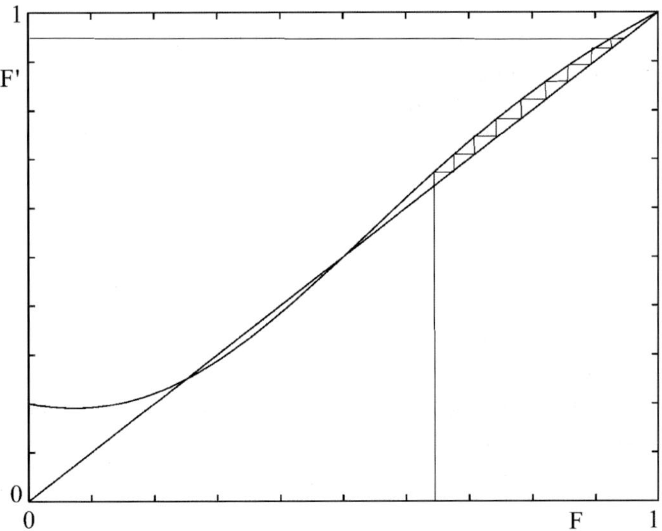

Fig. 1. – New fidelity in terms of the old fidelity for the purification protocol. Successive applications lead to a fidelity as close to one as one wishes.

the other two (A_2 and B_2) as targets. The controlled-NOT operation acts as follows:

$$(10) \qquad |0\rangle_{A_1}|0\rangle_{A_2} \longrightarrow |0\rangle_{A_1}|0\rangle_{A_2},$$

$$(11) \qquad |0\rangle_{A_1}|1\rangle_{A_2} \longrightarrow |0\rangle_{A_1}|1\rangle_{A_2},$$

$$(12) \qquad |1\rangle_{A_1}|0\rangle_{A_2} \longrightarrow |1\rangle_{A_1}|1\rangle_{A_2},$$

$$(13) \qquad |1\rangle_{A_1}|1\rangle_{A_2} \longrightarrow |1\rangle_{A_1}|0\rangle_{A_2},$$

and similarly with B. Alice and Bob then measure the state of their target particle in the basis $\{|0\rangle, |1\rangle\}$ (*i.e.*, they measure $\sigma_z^{A_2}$ and $\sigma_z^{B_2}$) and broadcast their results. If the results are the same, they keep the source particles, and otherwise they discard them. One can easily see that this is equivalent to projecting the initial states onto the subspace in which either both the sources and the targets are Φ states or both are Ψ states. The probability of having at the end of the process the state $|\Phi^+\rangle$ is

$$(14) \qquad F' = \frac{F^2 + (1-F)^2/9}{F^2 + 2F(1-F)/3 + 5(1-F)^2/9}.$$

For $1 > F > 1/2$, we have that $F' > F$. Therefore, the fidelity after this operation increases. To finish the process, the output states should be left in a Werner state, so that this process can be continued. Alice applies the operation σ_y to his source particle, which transforms $|\Phi^+\rangle \to |\Psi^-\rangle$ and then they depolarize. In summary, if the process is

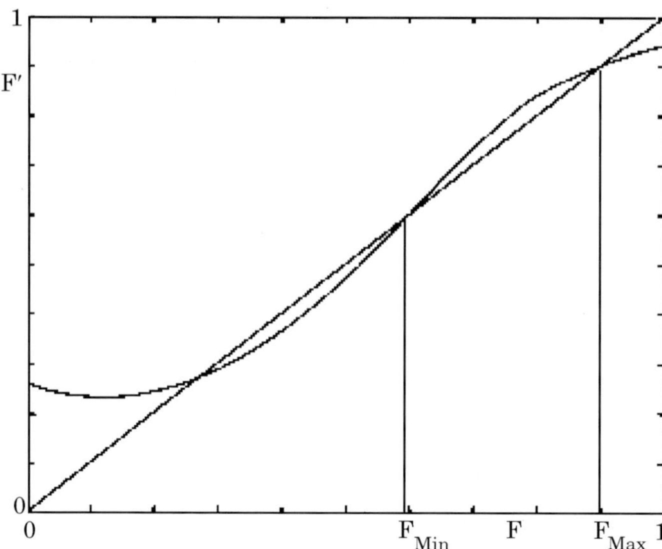

Fig. 2. – Same as in the previous figure, but with imperfections.

successful, Alice and Bob are left with a single pair in a Werner state but with fidelity F'. Then, they can take two successful pairs and repeat the same procedure to obtain a higher fidelity. By proceeding in this way they can reach a fidelity as close to one as they wish, but at the expenses of wasting many pairs. In fig. 1 we have plotted F' as a function of F and show how the fidelity increases as one repeats it with the successful pairs.

So far we have assumed that the operations that take place during the purification protocol (controlled-NOT, measurements, etc.) are perfect. In reality there will be imperfections in all these operations. One can take them into account by using some explicit models [4] or by studying the worst case scenario [18]. The result is schematized in fig. 2. Now, there is a minimum value of the original fidelity of the state F_{min} for which purification is possible. Apart from that, there is a maximum achievable fidelity F_{max} due to the imperfections.

2˙4. *Quantum computing*

2˙4.1. What is a quantum computer? A computation can be considered as a physical process that transforms an input into an output. A classical computation is that in which the physical process is based on classical laws (without coherent quantum phenomena). A quantum computation is that based on quantum laws (and in particular on the superposition principle). In quantum computation, inputs and outputs are represented by states of the system. For example, enumerating the state of a given basis as $\{|1\rangle, |2\rangle, \ldots\}$, the number N would be represented by the N-th state of this basis. A quantum computation consists of evolving the system with a designed Hamiltonian interaction, such that the states are transformed as we want. Note that the operation that transforms input into outputs has to be unitary. For example, the operation that gives 1 if a number is odd and 2 if it is even could not be implemented: $|2n+1\rangle \to |1\rangle$ and $|2n\rangle \to |2\rangle$ (where n is an integer). This operation cannot be unitary since it does not conserve the scalar product (*i.e.* $\langle 1|3\rangle = 0$ but the corresponding mapped states are not orthogonal). One can, however, use an auxiliary system so that the output is written in that system while keeping the unitarity of the operation: $|2n+1\rangle \otimes |0\rangle \to |2n+1\rangle \otimes |1\rangle$ and $|2n\rangle \otimes |0\rangle \to |2n\rangle \otimes |2\rangle$ (where n is an integer). In general, if our algorithm consists of evaluating a given function f, we can design an interaction Hamiltonian such that the evolution operator transforms the input states according to the following table:

$$|1\rangle \otimes |0\rangle \longrightarrow |1\rangle \otimes |f(1)\rangle,$$

$$|2\rangle \otimes |0\rangle \longrightarrow |2\rangle \otimes |f(2)\rangle,$$

$$\dots$$

$$|N\rangle \otimes |0\rangle \longrightarrow |N\rangle \otimes |f(N)\rangle.$$

Note that using this transformation we can, at least, do the same computations with quantum computers as with classical computers. However, with quantum computers we can do even more. We can prepare the input state that is a superposition

$$|\Psi\rangle = \frac{1}{\sqrt{N}} \sum_{k=1}^{N} |k\rangle \otimes |0\rangle \longrightarrow \frac{1}{\sqrt{N}} \sum_{k=1}^{N} |k\rangle \otimes |f(k)\rangle$$

after a single run. In principle, all the values of f are present in this superposition. Note, however, that we do not have access to this information since if we perform a measurement we will only obtain a result (with certain probability). Nevertheless, we see that with a quantum computer we can do at least the same as with a classical computer, ... and even more. This property of using quantum superpositions to run only once the computer was termed by Feynman quantum parallelism.

2`4.2. Requirements. A quantum computer consists of a quantum register (a quantum system) that can be manipulated and measured in a controlled way. In order to build a quantum computer, one needs the following elements (see also ref. [19]):

1) **A set of qubits**: These are two-level systems perfectly identified and which form the quantum register. We denote by $\{|0\rangle_k, |1\rangle_k\}$ two orthogonal states of the k-th qubit, so that the state of all the qubits (the quantum register) can be written as

$$|\Psi\rangle = \sum_{k_1,k_2,\dots k_N = 0}^{1} c_{k_1 k_2 \dots k_N} |k_1\rangle_1 \otimes |k_2\rangle_2 \otimes \dots \otimes |k_N\rangle_N.$$

In the following, we will simplify the notation and write $|k_1, k_2, \dots, k_N\rangle$ instead of the cumbersome notation that uses tensor products. Note that these qubits can be in superposition and entangled states, which gives extraordinary power to the quantum computer. Note that the state of the qubits must be kept almost pure since otherwise the power of the superpositions would not be effective. This means that the qubits must be well isolated from the environment in such a way that the process of decoherence is sufficiently slow.

2) **Universal set of quantum gates:** The controlled manipulation of the qubits means that we can perform any unitary operation U on the qubits so that $|\Psi\rangle \to U|\Psi\rangle$. In principle, if we want to perform general operations we should be able to engineer arbitrary interactions between the qubits. Fortunately, this task is enormously simplified given the fact that any U can be decomposed as a product of gates belonging to a small set, the so-called universal set of gates. This means that if we are able to perform the gates of this set we will be able to perform any computation (unitary operations on the register) by simply applying a sequence of them. There are many sets of universal gates, and of course, they are all equivalent. The most convenient set is the one that contains one two-qubit gate (an operation acting on two qubits only) plus a set of single-qubit gates. Let us recall here some of the gates of this sort ($\sigma_{x,y,z}$ are the Pauli operators acting on a qubit):

a) Single-qubit gates: act on a single qubit.

  i) Phase gate: $U_z^{(1)}(\varphi) = e^{-i\varphi\sigma_z}$,

$$|0\rangle \longrightarrow e^{i\varphi/2}|0\rangle,$$

$$|1\rangle \longrightarrow e^{-i\varphi/2}|1\rangle.$$

  ii) Excitations: $U_x^{(1)}(\theta) = e^{-i\theta\sigma_x}$,

$$|0\rangle \longrightarrow \cos\theta|0\rangle - i\sin\theta|1\rangle,$$

$$|1\rangle \longrightarrow -i\sin\theta|0\rangle + \cos\theta|1\rangle.$$

b) Two-qubit gates: act on two qubits.

  i) Controlled-not: $U_{\text{CNOT}}^{(2)} = |0\rangle\langle 0| \otimes 1 + |1\rangle\langle 1| \otimes \sigma_x$

$$|0,0\rangle \longrightarrow |0,0\rangle,$$

$$|0,1\rangle \longrightarrow |0,1\rangle,$$

$$|1,0\rangle \longrightarrow |1,1\rangle,$$

$$|1,1\rangle \longrightarrow |1,0\rangle.$$

This gate changes the state of the second qubit conditioned to the state of the first one. The first qubit is therefore called control qubit, whereas the second one is called target.

ii) Controlled-phase: $U_\pi^{(2)} = |0\rangle\langle 0| \otimes 1 - |1\rangle\langle 1| \otimes \sigma_z$

$$|0,0\rangle \longrightarrow |0,0\rangle,$$
$$|0,1\rangle \longrightarrow |0,1\rangle,$$
$$|1,0\rangle \longrightarrow |1,0\rangle,$$
$$|1,1\rangle \longrightarrow -|1,1\rangle.$$

Two different universal sets of gates are [1]

$$S_1 = \left\{ U_{\mathrm{CNOT}}^{(2)}, U_x^{(1)}\left(\frac{\pi}{4}\right), U_z^{(1)}(\varphi), \varphi \in [0, 2\pi) \right\},$$

$$S_2 = \left\{ U_\pi^{(2)}, U_z^{(1)}\left(\frac{\pi}{4}\right), U_x^{(1)}(\varphi), \varphi \in [0, 2\pi) \right\}.$$

Note that the two-qubit gates require interactions between the qubits, and therefore are the more difficult ones in practice. The fact that the operations are unitary (and therefore reversible) means that the uncontrolled interaction with any other part of the quantum computer must be avoided.

3) **Detection**: One should be able to measure σ_z on each of the qubits (or, equivalently, to detect whether they are in state $|0\rangle$ or $|1\rangle$). Note that this process requires the interaction with a measurement apparatus in an irreversible way.

4) **Erase**: We must be able to prepare the initial state of the system, for example the state $|0, 0, \ldots, 0\rangle$. Actually, this is not an extra requirement since if one is able to detect and to apply the single-qubit gate $U_x^{(1)}(\pi)$ this is enough.

5) **Scalability**: The difficulty of performing gates, measurements, etc., should not grow (exponentially) with the number of qubits. Otherwise, the gain in the quantum algorithms would be lost.

For the moment, we know very few systems which fulfill the requirements to implement a quantum computer with them. Perhaps, the most important problem is related to the necessity of finding a quantum system which is sufficiently isolated, and for which the required controlled interactions can be produced. For the moment, there exist three kind of physical systems that fulfill, at least, most of the requirements (see ref. [20]):

1) **Quantum-optical systems**: Qubits are atoms (ions), and the manipulation takes place with the help of a laser. These systems are very clean in the sense that with them it is possible to observe quantum phenomena very clearly. In fact, with them several groups have managed to prepare certain states which lead to phenomena that present certain analogies with the Schrödinger cat paradox, Zeno effect, etc.

Moreover, those systems are currently used to create atomic clocks, and with them one can perform the most precise measurements that exist nowadays. For the moment, experimentalist have been able to perform certain quantum gates, and to entangle 3 or 4 atoms. The most important difficulty with those systems is to scale up the models so that one can perform computations with many atoms.

2) **Solid-state systems**: There have been several important proposals to construct quantum computers using Cooper pairs or quantum dots as qubits. The highest difficulty in these proposals is to find the proper isolation of the system, since in a solid it seems hard to avoid interactions with other atoms, impurities, phonons, etc. For the moment, only single quantum gates have been experimentally reported. However, these systems possess the advantage that they are easily scalable.

3) **Nuclear-magnetic-resonance systems**: In this case the qubits are represented by atoms within the same molecule, and the manipulation takes place using the NMR technique. Initially, these systems seemed to be very promising for quantum computation, since it was thought that the cooling of the molecules was not required, which otherwise would make the experimental realization very difficult. However, it seems that without cooling, these systems lose all the advantages of quantum computation.

2˙4.3. Error correction. In any computation (classical and quantum) or during storing of information there will be errors. One way to fight against these errors is to improve the hardware and make it better. However, this is expensive, and not always possible. Shannon realized that instead of trying to avoid the errors it is much better to correct them. This is done by giving redundant information, and using this extra information to find out if an error occur. One can distinguish two kind of errors:

Memory errors: Those that occur to the information that is stored, regardless of whether an operation takes place or not.

Operation errors: Those that occur during an operation.

Here we will concentrate on memory errors, since the corresponding correction procedures are easier to understand. On the other hand, they play an important role not only in quantum computing, but also in quantum communication and information. Once one knows how memory errors can be corrected (with some modifications), one can understand how to correct operation errors. We will first revise the most straightforward way of correcting errors in a classical computer, and then we will show how to do it in a quantum computer.

Classical error correction

Imagine that one wants to store a single bit for a time t (we will call this bit a *logical bit*). Let us denote by P_τ the probability that one error occurs in a time interval τ; that is, the probability that the bit flips (if it was 0 then it changes to 1 and vice versa). If $P_\tau \simeq 1$ there will be problems in achieving the goal. One way to correct the errors is based on what is called *redundant coding*. This consists of using three bits to store the logical bit. That is, we encode the information such that if the logical bit is 0 the three bits are 0, and if it is 1, the three bits are 1: $0_L \equiv 000$, and $1_L \equiv 111$. These logical qubits are called *code words*. After a time τ, we will have:

- Probability of no errors: $(1 - P_\tau)^3$ (for example, if we had initially 000, after the time τ it is 000).

- Probability of error in one bit: $3P_\tau(1 - P_\tau)^2$ (for example, if we had initially 000, after the time τ it is 100, 010 or 001).

- Probability of two or more errors: $3P_\tau^2(1 - P_\tau) + P_\tau^3$.

The error correction consists of measuring if the three bits are in the same state or not. If they are in the same state, then we do nothing. If they are in a different state, we use majority vote to change the bit that is different. For example, if we have that the first and the third bit are equal and the second is different (010 or 101), we flip the second bit (000 and 111, respectively). After the correction we will have the correct state with a probability $P_\tau^c = 1 - 3P_u^2 + 2P_\tau^3$. Thus, one gains if $P_\tau^c < 1 - P_\tau$, that is, if (roughly) $P_\tau < 1/3$. If one wants to keep the state for very long times t, one has to perform many measurements. More precisely, assume that $P_\tau = 1 - e^{-\gamma\tau} \simeq \gamma\tau$ for times τ sufficiently short. Let us divide t into N intervals of duration $\tau = t/N$. For N sufficiently large, the probability of having the correct state after performing the correction after the time t will be

$$(15) \qquad P_t^c \geq \left[1 - 3\left(\frac{\gamma t}{N}\right)^2\right]^N.$$

For N large, this probability can be made as close to one as desired. One can generalize this method to the case in which one wants to store k logical bits and allow for errors in l bits. For example, encoding $0_L \equiv 00000$, $1_L \equiv 11111$, one can allow for two errors.

Quantum error correction

Imagine that one wants to store a single quantum bit in an unknown state $|\Psi\rangle = c_0|0\rangle + c_1|1\rangle$ for a time t (we will call this qubit a *logical qubit*). Let us assume that after a time τ with a probability $1 - P_\tau$ the qubit remains intact and that with a probability P_τ it changes to $|\Psi'\rangle = c_0|1\rangle + c_1|0\rangle$. This error is called spin flip, and it can be represented by the action of σ_x onto the state of the qubit. One can correct the above error by using *redundant coding* [21,22]. For example, one can encode the state of the logical qubit in 3 qubits as $|0\rangle_L \equiv |000\rangle$ and $|1\rangle_L \equiv |111\rangle$ (code words). The subspace spanned by these states is called subspace of code words. After a time τ, we will have:

- Probability of no errors: $(1 - P_\tau)^3$ (the state will be $|\Psi\rangle_L$).

- Probability of error in one bit: $3P_\tau(1 - P_\tau)^2$ (the state will be $\sigma_x^1|\Psi\rangle_L$, or $\sigma_x^2|\Psi\rangle_L$, or $\sigma_x^3|\Psi\rangle_L$).

- Probability of two or more errors: $3P_\tau^2(1 - P_\tau) + P_\tau^3$.

Note that in order to correct the errors, we cannot do the same as in the classical case, since measuring the state of the qubit will collapse it in a different state (for example $|000\rangle$), and therefore the superposition will be destroyed. What we can do is to detect whether the three bits are in the same state or not, without disturbing the state. If the qubits are in the same state, then we do nothing. If they are a different state, we use majority vote to change the bit that is different. All these measurements have to be performed without destroying the superposition. This can be done as follows: first we measure the projector $P = |000\rangle\langle000| + |111\rangle\langle111|$ (which corresponds to an incomplete measurement). If we obtain 1, then we leave the qubits as they are. If we obtain 0 then we measure the projector $P_1 = |100\rangle\langle100| + |011\rangle\langle011|$: if we obtain 1 we apply the local unitary operator σ_x^1 and if not we proceed. We measure $P_2 = |010\rangle\langle010| + |101\rangle\langle101|$; if we obtain 1 we apply the local unitary operator σ_x^2 and if not we apply the operator σ_x^3 (note that if we measure the operator P_3 we would obtain 1 with probability 1). As a result, if there was either no error or one error, it will be corrected. If there were two or more errors, they will not be corrected. Using this method, we achieve the same results as in the classical correction method, namely, by correcting very often we can keep the unknown state of a qubit as long as we want. The idea of the method for quantum error correction is based on designing the code words in such a way that every possible error (in the first, second, or third qubit) transforms the subspace of code words into another subspace which is orthogonal to it, but without modifying its internal structure. Then, by performing an incomplete measurement, we can detect in which subspace our state is, and therefore we know how to correct the error. This method can be generalized to the case in which other kinds of errors can occur. For example, imagine that with a small probability we can have errors consisting of applying the operator σ_α ($\alpha = x, y, z$) to a qubit. We want to preserve the state of k qubits against arbitrary errors in t different qubits. We will denote by E the possible operators corresponding to the errors that we want to correct. For example, $\sigma_x^1 \otimes \sigma_y^4$. We encode the k logical qubits in n qubits. The subspace of code words H_L has dimension 2^k, whereas the Hilbert space H of all the qubits has dimensions 2^n. Each of the possible error operators (consisting of up to t tensor products of Pauli operators) transform H_L into a subspace of dimension 2^k (note that the E's are unitary and therefore they conserve the dimension of the subspace on which they are applied). The subspace of code words has to be such that all these subspaces are mutually orthogonal. This condition imposes a minimum bound (the quantum Hamming bound) to the number of qubits needed, since all these orthogonal subspaces have to fit in H. One can easily show that this bound implies that

$$(16) \qquad 2^k \sum_{l=0}^{t} 3^l \binom{n}{l} < 2^n.$$

For $k = 1$ the minimum n is 5. Methods have been devised to construct codewords for each of these cases [23, 24]. On the other hand, one can take into account the errors that are produced while errors are being corrected, as well as the ones produced during operations. There is a whole theory dealing with the so-called fault-tolerant error correction [25], which basically shows that this is always possible provided the error per gate is smaller than some error threshold, which lies between 10^{-4}–10^{-6} depending on the error model. This result implies that if the error per gate is smaller than this threshold, then quantum computation is possible by using fault-tolerant error correction.

The above error correction schemes work in the presence of (undesired) coupling to the environment which leads to decoherence. In order to show that, one can expand the operator that describes the evolution of the i-th qubit with its local environment as

$$(17) \qquad U^i = \alpha^i 1^i \otimes E_0^i + \epsilon_1^i \sigma_x^i \otimes E_1^i + \epsilon_2^i \sigma_y^i \otimes E_2^i + \epsilon_3^i \sigma_z^i \otimes E_3^i,$$

where the E's are operators acting on the environment, and α^i and $\epsilon_{1,2,3}^i$ are constant numbers. Note that we can always use this expansion given the fact that the Pauli operators (plus the identity) form a basis in the space of operators acting on a qubit. We will consider that the time is sufficiently short so that all $\alpha^i \simeq 1$ and $\epsilon_{1,2,3}^i \ll 1$. The state of all the qubits after some interaction time $U|\psi\rangle|E\rangle = \otimes_{i=1}^n U^i|\psi\rangle|E\rangle$ can be expanded in terms of the epsilon keeping only the lowest orders. The error correction explained above will project the state onto only one of the terms of the resulting expression. The state of the environment will therefore factorize, and all the analysis made before remains valid.

2˙5. *Quantum communication.* – The situation one has in mind in quantum communication is the following: Alice wants to send Bob an unknown state $|\Psi\rangle$. One way of doing this is to send the particle carrying the state directly. However, the particle will very likely interact with the environment which may result in a different state, generally mixed. There are some ways of avoiding this. In the following we will describe a basic tool in quantum communication which allows to send one quantum state from one place to another provided one has a maximally entangled state shared between the two places, and is able to communicate classically without errors.

2˙5.1. Teleportation. By teleportation we define to transfer an intact quantum state from one place to another, by a sender who knows neither the state to be teleported nor the location of the intended receiver [26]. The term teleportation comes from Science Fiction meaning to make a person or object disappear while an exact replica appears somewhere else. The first teleportation experiments have recently taken place. Consider two partners, Alice and Bob, located at different places. Alice has a qubit in an unknown state $|\phi\rangle$, and she wants to teleport it to Bob, whose location is not known. Prior to the teleportation process, Alice and Bob share two qubits in a Bell state $|\Psi^-\rangle = \sqrt{1/2}(|0,1\rangle - |1,0\rangle)$. The idea is that Alice performs a joint measurement of the two-level system to be teleported and her particle. Due to the non-local correlations contained in the Bell state, the effect of the measurement is that the unknown state appears instantaneously in Bob's hands, except for a unitary operation which depends on the outcome of the measurement. If Alice communicates to Bob the result of her measurement, then Bob can perform that operation and therefore recover the unknown state (for experiments, see [27-29]).

Let us call particle 1 that which has the unknown state $|\phi\rangle_1$, particle 2 the member of the EPR that Alice possesses and particle 3 that of Bob. We write the state of particle 1 as $|\phi\rangle_1 = a|0\rangle_1 + b|1\rangle_1$ where a and b are (unknown) complex coefficients. The state of particles 2 and 3 is the Bell state $|\Psi^-\rangle$. The complete state of particles 1, 2 and 3 is therefore

$$|\Psi\rangle_{123} = \frac{a}{\sqrt{2}}(|0,0,1\rangle - |0,1,0\rangle) + \frac{b}{\sqrt{2}}(|1,0,1\rangle - |1,1,0\rangle).$$

In order to teleport the state, Alice and Bob follow this procedure:

1) *Alice measurement*: Alice makes a joint measurement of her particles (1 and 2) in the Bell basis

$$\left|\Psi^{\pm}\right\rangle = \frac{1}{\sqrt{2}}(|0,1\rangle \pm |1,0\rangle),$$

$$\left|\Phi^{\pm}\right\rangle = \frac{1}{\sqrt{2}}(|0,0\rangle \pm |1,1\rangle).$$

2) *Alice broadcasting*: Then she broadcasts (classically) the outcome of her measurement. That is, she has to send Bob two bits of classical information which indicate the outcome of the measurement.

3) *Bob restoration*: Bob then applies a unitary operation to his particle to obtain $|\phi\rangle_3$. According to the state of the particles the possible outcomes are:

- With probability 1/4, Alice finds $|\Psi^-\rangle_{12}$. The state of the third particle is automatically projected onto $a|0\rangle_3 + b|1\rangle_3$. Thus, in this case Bob does not have to perform any operation.

- With probability 1/4, Alice finds $|\Psi^+\rangle_{12}$. The state of the third particle is automatically projected onto $-a|0\rangle_3 + b|1\rangle_3$. Teleportation occurs if Bob applies σ_z to his particle.

- With probability 1/4, Alice finds $|\Phi^-\rangle_{12}$. The state of the third particle is automatically projected onto $a|1\rangle_3 + b|0\rangle_3$. Teleportation occurs if Bob applies σ_x to his particle.

- With probability 1/4, Alice finds $|\Phi^+\rangle_{12}$. The state of the third particle is automatically projected onto $a|1\rangle_3 - b|0\rangle_3$. Teleportation occurs if Bob applies σ_z to his particle.

Note that Alice ends up with no information about her original state so that no violation of the no-cloning theorem occurs. In this sense, the state of particle 1 has been transferred to particle 3. On the other hand, there is no instantaneous propagation of information. Bob has to wait until he receives the (classical) message from Alice with her outcome. Before he receives the message, his lack of knowledge prevents him from having the state. Note that no measurement can tell him whether Alice has performed her measurement or not. Since teleportation is a linear operation applied to a state, it will also work for statistical mixtures, or in the case in which particle 1 is entangled with other particles.

2˙6. *Quantum repeaters.* – We have now all necessary tools available to introduce the concept of the quantum repeater. Our goal is to create an EPR pair of high fidelity between two distant locations. Since non-local entanglement between distant particles cannot be created using only local operations, this involves the usage of a quantum channel, which is noisy in general. The bottleneck for communication over large distances is the scaling of the error probability with the length of the channel. When using, for example, optical fibers and single photons as a quantum channel, both the absorption losses and the depolarization errors scale exponentially with the length of the channel. The state of the photon or the photon itself will therefore be destroyed with almost certainty if the channel is longer than a few half-lengths of the fiber.

To overcome this problem, one can use quantum repeaters. The idea of such a repeater is to divide a long quantum channel into shorter segments, which are purified separately, before they are connected. Connecting two segments of a channel means here to build up quantum correlations across the compound channel from correlations that exist across the individual segments. This can be done by teleportation of entanglement. A quantum repeater must therefore combine the methods of entanglement purification and teleportation. Although the combination of these methods should, in principle, allow to create entanglement over arbitrary distances, it is another question how much this "costs" in terms of resources needed for purification. Resources means here the number of low-fidelity entangled pairs that have to be provided for purification of each channel segment. This quantity is related to the number of particles that have to be manipulated locally (at the connection points between the segments) in a coherent fashion. If the resources grow too fast with the length of the channel, not much will be gained by the whole procedure. A further important quantity is the error tolerance for the local operations. In every real situation, the local operations applied to one or more particles will bear some imperfections. Since such operations are the building blocks for any entanglement purification protocol, their imperfections will limit the maximum attainable fidelity for an EPR pair and the efficiency of the protocol. In the context of the quantum repeater, a maximum fidelity $F < 1$ corresponds to a residual amount of noise for each segment. When the segments are connected, this noise accumulates.

To overcome this limitation [4], we can divide the long channel into N smaller segments and create less distant entangled pairs across each segment. The number of segments N is thereby chosen in such a way that it is possible to create entangled pairs with sufficiently high initial fidelity $F > F_{\min}$ over the distance of such a segment, such that they can be purified, according to our previous discussion. In a next step, we connect these "elementary" pairs by using teleportation. For example, if we have an entangled purified pair between the nodes A_1 and A_2, and another one between A_2 and A_3, we teleport the state of the first particle in A_2 to A_3 by using this second pair. The result of this teleportation will be that the nodes A_1 and A_3 will now share an entangled state. Of course, due to imperfections during the teleportation procedure, as well as the fact that the pairs used were not perfectly pure, the new entangled state will not be pure. But as long as its fidelity is larger than $F_{\min}$, it will be possible to purify it to a value close to $F_{\max}$. Thus, the crucial point is that, on the one hand, the operations that are performed cannot be too noisy since otherwise they could decrease the fidelity below $F_{\min}$, which would make the process impossible; on the other hand, the distance between nodes has to be such that purification be possible. This limits the number of pairs one can connect before purification becomes impossible. We therefore connect a smaller number $L \ll N$ of pairs so that the resulting fidelity F_L stays above the threshold value for purification ($F_L \geq F_{\min}$) and purification is possible.

The general strategy will be to design an alternating sequence connection and (re-)purification procedures in such a way that the number of resources needed remains as small as possible, and in particular does not grow exponentially with N and thus with l. This is possible, in principle, using a nested purification protocol [4].

3. – Quantum information processing with single atoms and photons

3˙1. *Introduction.* – This section discusses various schemes of quantum information processing with single trapped atoms and photons, *i.e.*, manipulation of atoms and photons on the level of the single quantum level. Experimental realization includes laser-cooled trapped ions, either in a linear trap or in arrays of micro traps, and neutral atoms stored in far-off-resonance traps or optical lattices. Single atoms can be stored in high-Q cavities, providing an interface between atoms and photons. The models discussed below share the feature that long-lived internal atomic states, such as atomic hyperfine ground states or metastable states, serve as quantum memory to store the qubits. Furthermore, we assume that single-qubit rotations can be performed by coupling the qubit states to laser light for an appropriate time period. In general, this requires that single atoms can be addressed by laser light. The discussion during the last few years has focused on developing various schemes for two-qubit gates. The models discussed in the literature can be classified in two categories. The first version relies on the concept of a quantum data bus: in this case the qubits are coupled to a collective auxiliary quantum mode, and entanglement of qubits is achieved by swapping qubits to excitations of the collective mode. Examples for such systems are the collective phonon modes in ion traps [30], and photons in cavity QED [31,32]. Requirements often include the initialization of the quantum data bus in a pure initial state, *e.g.* laser cooling to the motional ground state in ion traps. However, recently specific protocols for "hot gates" have been developed which loosen these requirements [33,34]. The second concept for performing the two-qubit gate is controllable internal-state–dependent two-body interactions between atoms. Examples for this latter scheme are coherent cold collisions of atoms in optical traps and optical dipole-dipole interactions [35,36]. A third example is the "fast" two-qubit gate based on large permanent dipole interactions between laser-excited Rydberg atoms in static electric fields [37]. We note that the unitary operations, which can be decomposed in a series of single- and two-qubit operations on the qubits, can either be performed *dynamically,* *i.e.* based on the time evolution generated by a specific Hamiltonian, or *geometrically* as in holonomic quantum computing [38, 39]. Finally, a common feature of the quantum optical models is that the readout of the atomic qubit is performed using the method of quantum jumps.

This section is arranged as follows: we start with a detailed description of trapped ions as a physical system to implement quantum computing in subsect. **3**'2. This is followed by a subsection on cavity QED which discusses optical interconnects between atoms as quantum memory and photons for transmission of quantum information. Finally, subsect. **3**'4 discusses examples of two-qubit gates with neutral atoms based on cold coherent collisions and interaction between laser excited Rydberg atoms in electric fields.

3'2. *Trapped ions.* – In this subsection we give a theoretical description of quantum state engineering [40] and entanglement engineering [30, 41] in a system of trapped and laser-cooled ions. The development of the theory starts with the description of Hamiltonians, state preparation, laser cooling and state measurements first for single ions, which is then generalized to the case of many ions. This serves as the basis of our discussion of quantum computer models [30, 34, 33, 38].

3'3. *The model.*

3'3.1. Single trapped ion. *Motional degrees of freedom:* We consider a single ion confined in a harmonic trap and interacting with laser light [40]. We assume that the lasers are directed along one of the principal axes of the harmonic potential, which allows us to consider ion motion in only one dimension. Hence, the Hamiltonian describing the free motion of the ion in the trap is

$$(18) \qquad H_{0T} = \frac{\hat{p}^2}{2M} + \frac{1}{2}M\nu^2\hat{x}^2.$$

Here $\hat{x}$ and $\hat{p}$ are the position and momentum operators, respectively, M is the ion mass, ν is the oscillation frequency (fig. 3). We can rewrite this Hamiltonian in the familiar form $H_{0T} = \nu(a^\dagger a + 1/2)$ with raising and lowering operators a and $a^\dagger$, defined according to $\hat{x} = \sqrt{1/2M\nu}(a + a^\dagger)$ and $\hat{p} = i\sqrt{M\nu/2}(a^\dagger - a)$ (we set $\hbar = 1$).

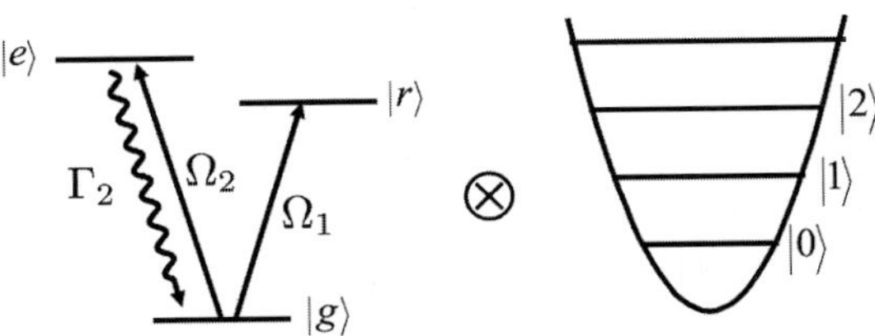

Fig. 3. – Energy levels of an ion trap. Left: internal level structure with $|g\rangle \to |r\rangle$ a metastable transition, and $|g\rangle \to |e\rangle$ a strong dissipative transition coupled by Rabi frequencies Ω_1 and Ω_2, respectively. Right: quantized energy levels in the harmonic trapping potential.

Internal degrees of freedom: We assume that the internal electronic structure of the ion is modelled by a three-level system, with levels $|g\rangle$, $|e\rangle$ and $|r\rangle$, where the first transition $|g\rangle \to |r\rangle$ is dipole-forbidden, and $|g\rangle \to |e\rangle$ is dipole-allowed (fig. 3). In our model system we will employ the transition to the metastable state $|r\rangle$ for *quantum state engineering,* while the strongly dissipative transition coupling $|e\rangle$ to the ground state will be used for *laser cooling* and *state measurement.* These transitions can be excited by laser beams of frequencies close to the corresponding resonance frequencies. Obviously, emission or absorption of laser photons will modify the atomic motion. We confine our discussion to the Lamb-Dicke limit (LDL), *i.e.,* to the limit where the ion motion is restricted to a region much smaller than the wavelength of the laser light exciting a given transition [48]. This allows us to expand the Hamiltonian describing the interaction of the ion with the laser light in terms of the Lamb-Dicke parameter $\eta_i = 2\pi a_0/\lambda_i$, where $a_0 = 1/(2M\nu)^{1/2}$ is the size of the ground state of the harmonic potential, and λ_i is the wavelength of the laser light exciting transition i. We will now write out in details the Hamiltonians describing the coupling of the ion to laser light in the LDL.

Dipole forbidden transition $|g\rangle \to |r\rangle$ *transition:* We first consider the situation in which only the laser driving the dipole-forbidden transition $|g\rangle \to |r\rangle$ is on. We assume for the moment that the interaction time with the laser beam is much shorter than the lifetime of level $|r\rangle$, so that we can neglect dissipation. The corresponding Hamiltonian is

$$(19) \qquad H_1 = H_{0T} + H_{0A_1} + H_{A_1 L},$$

where the first two terms are the bare trap and atomic Hamiltonian, and $H_{A_1 L}$ describes the interaction with the laser. In a frame rotating with the laser frequency, we have $H_{0A_1} = \delta_1 |g\rangle\langle g|$ and

$$(20) \qquad H_{A_1 L} = \frac{1}{2}\Omega_1 \sin\left[\eta_1(a + a^\dagger) + \phi_1\right](|r\rangle\langle g| + |g\rangle\langle r|),$$

$$(21) \qquad H_{A_1 L}^{\pm} = \frac{1}{2}\Omega_1\{|r\rangle\langle g| \exp[\pm i\eta_1(a + a^\dagger)] + \text{h.c.}\},$$

for standing-wave and travelling-wave configurations, respectively. Here, $\delta_1 = \omega_{L_1} - \omega_{rg}$ is the laser detuning from the internal transition, Ω_1 is the Rabi frequency, and η_1 is the Lamb-Dicke parameter for this particular transition. The index $+$ $(-)$ denotes that the laser plane wave propagates in the positive (negative) x direction, while ϕ_1 defines the position of the center of the trap in the laser standing wave.

In lowest order in the Lamb-Dicke expansion we have

$$(22) \qquad H_{A_1 L} = \frac{1}{2}\Omega_1\{|r\rangle\langle g|[\alpha_0 + \alpha_\pm(a + a^\dagger) + \mathcal{O}(\eta_1^2)] + \text{h.c.}\},$$

where $\alpha_0 = 1$, $\alpha_\pm = \pm i\eta_1$ and $\alpha_0 = \sin(\phi_1)$, $\alpha_\pm = \eta_1\cos(\phi_1)$ for a travelling- and standing-wave configuration, respectively. The Hamiltonian (22) can be further simplified if the laser field is sufficiently weak so that only pairs of bare atom $+$ trap levels are

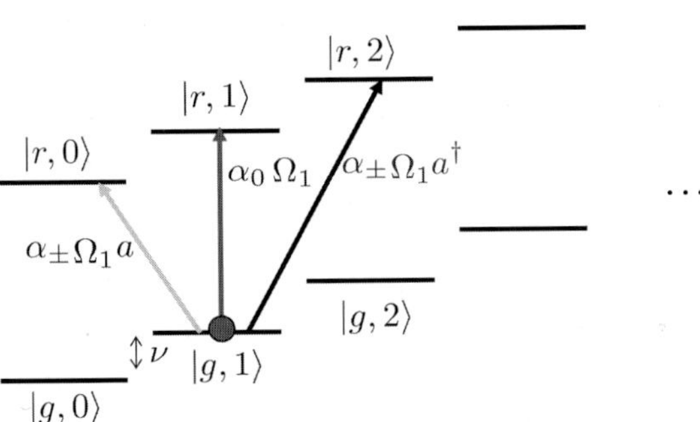

Fig. 4. – Coupling to the atom + trap levels according to the Hamiltonians (23), (24) and (25), respectively, in lowest-order Lamb-Dicke expansion.

coupled resonantly. We denote by $|n, g\rangle$ and $|n, r\rangle$ the eigenstates of the bare Hamiltonian $H_{0T} + H_{0A_1}$, where the internal two-level system is in the ground (excited) state and n is the excitation number of the harmonic oscillator. These states are degenerate for $\omega_{L_1} - \omega_{rg} = k\nu$ ($k = 0, \pm 1, \ldots$), i.e., whenever the laser is tuned to one of the "motional sidebands", corresponding to a degeneracy between $|n, g\rangle$ and $|n + k, r\rangle$. In the presence of the laser these degeneracies become avoided crossings, and for sufficiently weak laser excitation these avoided crossings will be isolated (non-overlapping). For example, for $|\omega_{L_1} - \omega_{rg}| \ll \nu$, i.e. $k = 0$, transitions changing the harmonic oscillator quantum number n are off-resonance and can be neglected. In this case the Hamiltonian (19) can be approximated by

$$(23) \qquad H_0 = \nu a^\dagger a - \frac{1}{2}\delta_1 \sigma_z + \frac{1}{2}\Omega_1(\alpha_0 \sigma_+ + \text{h.c.}),$$

where we have used the spin-(1/2) notation $\sigma_+ = (\sigma_-)^\dagger = |r\rangle\langle g|$, $\sigma_z = |r\rangle\langle r| - |g\rangle\langle g|$. For laser frequencies close to the lower motional sideband resonance $|\omega_{L_1} - (\omega_{rg} - \nu)| \ll \nu$ ($k = -1$), only transitions decreasing the quantum number n by one are important, and H_1 can be approximated by a Hamiltonian of the Jaynes-Cummings type:

$$(24) \qquad H_{\text{JC}\pm} = \nu a^\dagger a - \frac{1}{2}\delta_1 \sigma_z + \frac{1}{2}\Omega_1(\alpha_\pm \sigma_+ a + \text{h.c.}).$$

Similarly, for $|\omega_{L_1} - (\omega_{rg} + \nu)| \ll \nu$, only transitions increasing the quantum number n by one ($k = +1$) contribute, so that H_1 can be approximated by the *anti*-Jaynes-Cummings Hamiltonian

$$(25) \qquad H_{\text{AJC}\pm} = \nu a^\dagger a - \frac{1}{2}\delta_1 \sigma_z + \frac{1}{2}\Omega_1(\alpha_\pm \sigma_+ a^\dagger + \text{h.c.})$$

(see fig. 4). For the above approximations to be valid we require that the effective Rabi frequencies to the non-resonant states have to be much smaller than the trap frequency

$(\alpha_i\Omega_1/\nu)^2 \ll 1$ ($i = 0, \pm$). Note in particular that for an ion at the node of a standing light wave corrections to the JC Hamiltonian (24) are of the order $(\eta_1\Omega_1/\nu)^2 \ll 1$, *i.e.* the conditions of validity are greatly relaxed.

Eigenstates of the Hamiltonians H_0, $H_{\mathrm{JC}\pm}$ and $H_{\mathrm{AJC}\pm}$ are the dressed states familiar from cavity QED, which are obtained by diagonalizing the 2×2 matrices of nearly degenerate states. Applying a laser pulse on resonance, $\omega_{L_1} = \omega_{rg}$, will according to (23) induce Rabi flopping between the states $|n, g\rangle$ and $|n, r\rangle$, while a laser tuned for example to the lower motional sideband $\omega_{L_1} = \omega_{rg} - \nu$ will lead to Rabi oscillations coupling $|n, g\rangle$ and $|n - 1, r\rangle$. The above Hamiltonians are the basic building blocks to engineer quantum states. As an example, refs. [45,46] give a protocol to build the general motional superposition states $\sum_{n=0}^{M} c_n|g, n\rangle$ starting from the (pure) ground state $|g, 0\rangle$. The unitary operation effecting this transformation can be decomposed into unitary operations generated by the Hamiltonians (24) and (23), *i.e.* by applying a sequence of laser pulses with proper detunings and duration.

The dipole-allowed transition $|g\rangle \to |e\rangle$ *transition:* For a laser beam exciting the dipole-allowed transition $|g\rangle \to |e\rangle$ spontaneous emission is expected to play a significant role, and thus the dynamics must be described in terms of a master equation [47-49],

$$(26) \qquad \dot{\rho} = -i[H_2, \rho] + \mathcal{L}_2\rho.$$

In a rotating frame the Hamiltonian is $H_2 = \nu a^\dagger a + \delta_2|g\rangle\langle g| + H_{A_2L}$ with $\delta_2 = \omega_{L_2} - \omega_{eg}$ laser detuning from the internal transition. The laser atom coupling H_{A_2L} has a structure analogous to (19) with the replacements $|r\rangle \to |e\rangle$, a Rabi coupling Ω_2, position of the center of the trap in the laser standing wave ϕ_2, and Lamb-Dicke parameter η_2. The dissipative part of the master equation (26) can be written as [49]

$$(27) \qquad \mathcal{L}_2\rho = \Gamma_2|g\rangle\langle g|\langle e|\tilde{\rho}|e\rangle - \frac{1}{2}\Gamma_2(|e\rangle\langle e|\rho + \rho|e\rangle\langle e|),$$

where Γ_2 is the spontaneous emission rate from level $|e\rangle$, and

$$(28) \qquad \tilde{\rho} = \int_{-1}^{1} du N(u) e^{-i\eta_2 u(a+a^\dagger)} \rho e^{i\eta_2 u(a+a^\dagger)},$$

with $N(u)$ the dipole emission pattern for spontaneous emission from $|e\rangle$ to $|g\rangle$. For example, for a $\Delta m_J = \pm 1$ transition, $N(u) = 3/8(1 + u^2)$. Master equations of this type have been derived and studied in the context of laser cooling [47-49]. Physically speaking, eqs. (26), (27) describe the excitation of the atomic electron by the laser, which can either return to the ground state by either a laser-induced process or by spontaneous emission. The emission of the spontaneous photon according to the angular distribution $N(u)$ is accompanied by a momentum transfer to the atom, as described by the recycling term (28).

Spontaneous emission and laser cooling to the motional ground state: A prerequisite for many of the schemes for quantum engineering of non-classical states of motion, or entangled atomic states is that the initial motional state of the ion is prepared in a well-defined pure state, *e.g.*, the ground state $|0\rangle$ [40,30]. The standard approach to preparing such a pure state of the atomic motion is sideband cooling [42].

The theoretical description is particularly simple in the Lamb-Dicke limit, where the separation of the time scale for the internal and external dynamics allows the adiabatic elimination of the internal degree of freedom [48]. For details of the calculations we refer to [49] and references cited therein. The physical picture of laser cooling in the Lamb-Dicke limit is as follows: in the rest frame of the ion, the ion "sees" a laser field consisting of a carrier at frequency ω_{L_2} and small (motional) sidebands at frequencies $\omega_{L_2} \pm \nu$. Cooling occurs when absorption of laser photons from the upper laser sideband (at $\omega_{L_2}+\nu$) is stronger than from the lower sideband, since the former absorption reduces the external energy, whereas the latter increases the external energy. Hence, laser cooling is particularly efficient in the strong-confinement limit when one is able to tune the laser frequency in such a way that the upper laser sideband is on resonance with the two-level transition, since in this case only the photons of the upper sideband are absorbed, and consequently the ion ends up in the ground state of the harmonic trapping potential. This is the basic mechanism of *sideband cooling*. For many of the ions currently in use in Paul traps, the strong-confinement condition $\nu \gg \Gamma_2$ is not fulfilled for dipole-allowed transitions, and hence sideband cooling is not possible. However, employing auxiliary internal atomic levels and additional laser excitation allows one to effectively "design" two-level atoms for sideband cooling [42,50].

State measurement by the quantum jump technique: Implementation of quantum computing and communication protocols require measurement of the internal state of the atom [30]. In an ion trap this can be achieved with essentially 100% efficiency using the method of quantum jumps [42,47]. The theoretical understanding of quantum jumps is based on the continuous measurement theory, and we refer to [47] for a detailed mathematical description of the underlying theory. For our purpose it suffices to summarize the results as follows. Consider a *single* ion prepared initially in a superposition state on the metastable transition, $\alpha|g\rangle + \beta|r\rangle$. Switching on the laser on the strongly dissipative transition will give with probability $|\alpha|^2$ a burst of photon emissions $|e\rangle \to |g\rangle$ on the time scale $1/\Gamma_2$, or with probability $|\beta|^2$ the appearance on an emission window on the strong line. Measuring an emission window, or no window thus corresponds to a projective measurement of $|r\rangle$ or $|g\rangle$.

3˙3.2. Ions in a linear trap. The above model is readily extended to describe a string of N ions in a linear trap [42] which is the basis of the ion trap '95 quantum computer proposal [30,41] described below. A linear trap corresponds to a confinement of the motion along the x, y and z directions in an (anisotropic) harmonic potential of frequencies $\nu \equiv \nu_x \ll \nu_y, \nu_z$. The equilibrium position of the ions will be given by the confining forces of the trapping potential balancing the Coulomb repulsion between the ions. If the ions have previously been laser cooled in all three dimensions they undergo small oscillations around these equilibrium position. In this case, the motion of the ions is described in terms of normal modes.

As an example, a 1D model for two ions in a linear trap with internal levels $|g\rangle$ and $|r\rangle$ and coupled to laser light is given by the following Hamiltonian:

$$(29) \qquad H = \nu a_{\mathrm{cm}}^{\dagger} a_{\mathrm{cm}} + \sqrt{3}\nu a_{\mathrm{r}}^{\dagger} a_{\mathrm{r}} - \delta_1 |r\rangle_{11}\langle r| - \delta_2 |r\rangle_{22}\langle r| +$$

$$(30) \qquad + \frac{1}{2}\Omega_1(t)[|r\rangle_{11}\langle g| e^{-i\eta_{\mathrm{cm}}(a_{\mathrm{cm}}+a_{\mathrm{cm}}^{\dagger})} e^{-i\eta_{\mathrm{r}}(a_{\mathrm{r}}+a_{\mathrm{r}}^{\dagger})} + \mathrm{h.c.}] +$$

$$(31) \qquad + \frac{1}{2}\Omega_1(t)[|r\rangle_{22}\langle g| e^{-i\eta_{\mathrm{cm}}(a_{\mathrm{cm}}+a_{\mathrm{cm}}^{\dagger})} e^{+i\eta_{\mathrm{r}}(a_{\mathrm{r}}+a_{\mathrm{r}}^{\dagger})} + \mathrm{h.c.}].$$

Here, the first line are the Hamiltonians for the harmonic collective oscillations of center-of-mass and stretch mode with oscillation frequency ν and $\sqrt{3}\nu$, respectively, and the bare atomic Hamiltonian for the first and second ion in the rotating frame with $\delta_{1,2}$ the laser detunings. The second and third line are the laser couplings to the ions with $\Omega_{1,2}$ Rabi frequencies of the laser acting on each ion, respectively, and $\eta_{\mathrm{cm,r}}$ are the corresponding Lamb-Dicke parameters. In the general case of N ions there will be a set of N collective modes of ion motion, where again the minimum frequency of collective mode oscillations is that of the center-of-mass (CM) mode in the x-direction $\nu_x \equiv \nu$, and the next frequency is the stretch mode $\sqrt{3}\nu_x$, and all the others are larger. It is an important feature for the quantum computer proposal below that the frequency spacing of the low-lying modes is essentially *independent* of the number of ions N in the trap.

Our previous discussion of the Lamb-Dicke expansion of the Hamiltonian is readily extended to the string of N ions. In a similar way, we can model spontaneous emission of the ions by a master equation. In a typical experimental situation, the distance between the ions will be much larger than the optical wavelength, so that the spontaneous emission of the ions is independent, and the master equation will contain a sum of the independent spontaneous emission terms of the ions of the form (27).

3`3.3. Ion trap quantum computer '95. As first proposed in ref. [30], N cold ions interacting with laser light and moving in a linear trap provide a realistic physical system to implement a quantum computer. The distinctive features of this system are: i) it allows the implementation of a complete set of quantum gates between any set of (not necessarily neighboring) ions; ii) decoherence is comparatively small, and iii) the final readout can be performed with essentially unit efficiency [41, 51-58].

Figure 5 illustrates the basic setup. The qubits are represented by the long-lived internal states of the ions, with $|g\rangle_j \equiv |0\rangle_j$ representing the ground state, and $|r_0\rangle_j \equiv |1\rangle_j$ a metastable excited state ($j = 1, \ldots, N$). (In addition, we assume that there is a second metastable excited state $|r_1\rangle$ which plays the role of an auxiliary state.) In this system independent manipulation of each individual qubit is accomplished by addressing the ions with individual laser beams and inducing a Rabi rotation. The heart of the proposal is the implementation of a two-qubit gate between two (or more) arbitrary ions in the trap by exciting the collective quantized motion of the ions with lasers, *i.e.* the collective phonon mode plays the role of a quantum data bus. For this we assume that the collective phonon modes have been cooled to the ground state [52, 53].

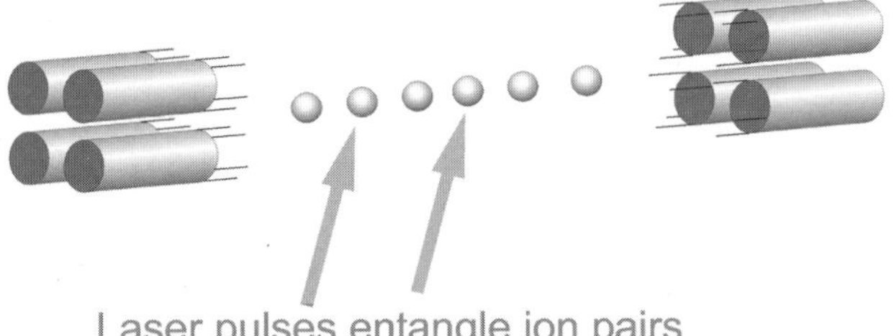

Fig. 5. – Ion trap quantum computer (schematic).

Single-qubit rotations can be performed tuning a laser on resonance with the internal transition ($\delta_j = 0$) with polarization $q = 0$. In an interaction picture the corresponding Hamiltonian is

$$(32) \qquad \hat{H}_j = \left(\frac{\Omega}{2}\right)\left[|r_0\rangle_j\langle g|e^{-i\phi} + |g\rangle_j\langle r_0|e^{i\phi}\right].$$

For an interaction time $t = k\pi/\Omega$ (*i.e.*, using a $k\pi$ pulse), this process is described by the following unitary evolution operator:

$$(33) \qquad \hat{V}_j^k(\phi) = \exp\left[-ik\frac{\pi}{2}\left(|e_0\rangle_j\langle g|e^{-i\phi} + \text{h.c.}\right)\right],$$

so that we achieve a Rabi rotation

$$|g\rangle_j \longrightarrow \cos\left(\frac{k\pi}{2}\right)|g\rangle_j - ie^{i\phi}\sin\left(\frac{k\pi}{2}\right)|r_0\rangle_j,$$

$$|r_0\rangle_j \longrightarrow \cos\left(\frac{k\pi}{2}\right)|r_0\rangle_j - ie^{-i\phi}\sin\left(\frac{k\pi}{2}\right)|g\rangle_j.$$

If the laser addressing the j-th ion is tuned to the lower motional sideband of, for example, the center-of-mass mode, we have in the interaction picture the Hamiltonian

$$(34) \qquad H_{j,q} = \frac{\eta}{\sqrt{N}}\frac{\Omega}{2}\left[|r_q\rangle_j\langle g|ae^{-i\phi} + |g\rangle_j\langle r_q|a^\dagger e^{i\phi}\right].$$

Here $a^\dagger$ and a are the creation and annihilation operator of CM phonons, respectively, Ω is the Rabi frequency, ϕ the laser phase, and η is the Lamb-Dicke parameter. The subscript $q = 0, 1$ refers to the transition excited by the laser, which depends on the laser polarization. Equation (34) follows from the Hamiltonian for the case of a linear trap, similar to the derivation of (24). The factor $\sqrt{N}$ appears since the effective mass of the CM motion is NM, and the amplitude of the mode scales like $1/\sqrt{NM}$ (Mössbauer effect).

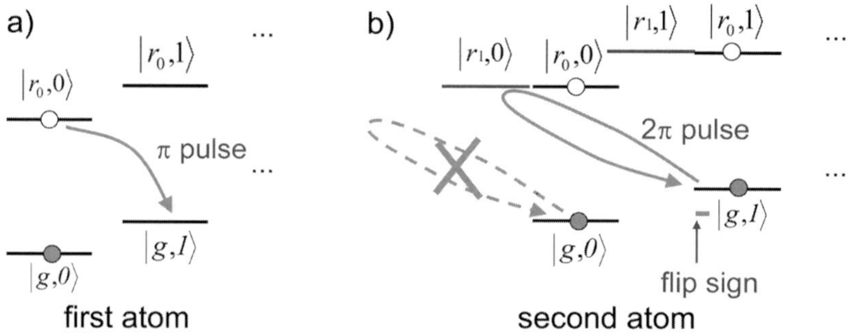

Fig. 6. – The two-qubit quantum gate. a) First step according to (36): the qubit of the first atom is swapped to the photonic data bus with a π-pulse on the lower motional sideband. b) Second step: the state $|g,1\rangle$ acquires a minus sign due to a 2π-rotation via the auxiliary atomic level $|r_1\rangle$ on the lower motional sideband.

If this laser beam is on for a certain time $t = k\pi/(\Omega\eta/\sqrt{N})$ (*i.e.*, using a $k\pi$ pulse), the evolution of the system will be described by the unitary operator

$$(35) \qquad \hat{U}_j^{k,q}(\phi) = \exp\left[-ik\frac{\pi}{2}(|r_q\rangle_j\langle g|ae^{-i\phi} + \text{h.c.})\right].$$

It is easy to prove that this transformation keeps the state $|g\rangle_j|0\rangle$ unaltered, whereas

$$|g\rangle_j|1\rangle \longrightarrow \cos\left(\frac{k\pi}{2}\right)|g\rangle_j|1\rangle - ie^{i\phi}\sin\left(\frac{k\pi}{2}\right)|r_q\rangle_j|0\rangle,$$

$$|r\rangle_j|0\rangle \longrightarrow \cos\left(\frac{k\pi}{2}\right)|r_q\rangle_j|0\rangle - ie^{-i\phi}\sin\left(\frac{k\pi}{2}\right)|g\rangle_j|1\rangle,$$

where $|0\rangle$ $(|1\rangle)$ denotes a state of the CM mode with no (one) phonon.

Let us now show how a two-bit gate can be performed using this interaction. We consider the following three-step process (see fig. 6): i) A π laser pulse with polarization $q = 0$ and $\phi = 0$ excites the m-th ion. The evolution corresponding to this step is given by $\hat{U}_m^{1,0} \equiv \hat{U}_m^{1,0}(0)$ (fig. 6a). ii) The laser directed on the n-th ion is then turned on for a time of a 2π-pulse with polarization $q = 1$ and $\phi = 0$. The corresponding evolution operator $\hat{U}_n^{2,1}$ changes the sign of the state $|g\rangle_n|1\rangle$ (without affecting the others) via a rotation through the auxiliary state $|e_1\rangle_n|0\rangle$ (fig. 6b). iii) Same as i). Thus, the unitary operation for the whole process is $\hat{U}_{m,n} \equiv \hat{U}_m^{1,0}\hat{U}_n^{2,1}\hat{U}_m^{1,0}$ which is represented diagrammatically as follows:

$$
\begin{array}{ccccccc}
 & \hat{U}_m^{1,0} & & \hat{U}_n^{2,1} & & \hat{U}_m^{1,0} & \\
 & |g\rangle_m|g\rangle_n|0\rangle & \longrightarrow & |g\rangle_m|g\rangle_n|0\rangle & \longrightarrow & |g\rangle_m|g\rangle_n|0\rangle & \longrightarrow & |g\rangle_m|g\rangle_n|0\rangle, \\
(36) & |g\rangle_m|r_0\rangle_n|0\rangle & \longrightarrow & |g\rangle_m|r_0\rangle_n|0\rangle & \longrightarrow & |g\rangle_m|r_0\rangle_n|0\rangle & \longrightarrow & |g\rangle_m|r_0\rangle_n|0\rangle, \\
 & |r_0\rangle_m|g\rangle_n|0\rangle & \longrightarrow & -i|g\rangle_m|g\rangle_n|1\rangle & \longrightarrow & i|g\rangle_m|g\rangle_n|1\rangle & \longrightarrow & |r_0\rangle_m|g\rangle_n|0\rangle, \\
 & |r_0\rangle_m|r_0\rangle_n|0\rangle & \longrightarrow & -i|g\rangle_m|r_0\rangle_n|1\rangle & \longrightarrow & -i|g\rangle_m|r_0\rangle_n|1\rangle & \longrightarrow & -|r_0\rangle_m|r_0\rangle_n|0\rangle.
\end{array}
$$

The effect of this interaction is to change the sign of the state only when both ions are initially excited. Note that the state of the CM mode is restored to the vacuum state $|0\rangle$ after the process. Equation (36) is phase gate $|\epsilon_1\rangle|\epsilon_2\rangle \to (-1)^{\epsilon_1\epsilon_2}|\epsilon_1\rangle|\epsilon_2\rangle$ $(\epsilon_{1,2} = 0, 1)$ which together with single qubit rotations becomes equivalent to a controlled-NOT.

Final readout of the quantum register (state measurement of the individual qubits) at the end of the computation can be accomplished using the quantum jumps technique with unit efficiency [42, 47].

The above proposal for the implementation of quantum computing with trapped ions has been the basis and has provided the stimulus for a significant body of experimental and theoretical work over the last few years [41, 51-58]. Highlights of the series of experimental work in particular by Chris Monroe and Dave Wineland at NIST Boulder [51, 52, 55-58], and Rainer Blatt at the University of Innsbruck [53, 54], are the implementation of a 2-qubit quantum gate with a single ion [51], ground-state cooling of a string of ions [52, 53], addressing of single ions to perform single-qubit operations [54], the generation of Bell states [58], and—as the most remarkable achievement—preparation of a maximally entangled state of four ions by the NIST Boulder group [56]. For a detailed discussion of the experimental situation we refer to the lecture notes by D. Wineland and R. Blatt in this volume. On the theory side various extensions to finite-temperature gates have been given, as well as schemes which might allow faster gate operation times [34, 33, 59]. One of the most interesting contributions is a scheme by Mølmer and Sørensen [33] which allows preparation of a maximally entangled state of N ions with a single pulse without addressing the individual ions, as employed in the four-ion NIST experiment [56]. A quantum computer model based on geometric variants of the gate [60, 39, 61] was recently proposed by Duan *et al.* [38]. Error correction of the phonon data bus was studied in [62].

3˙3.4. Ion trap quantum computer 2000. While in the ion trap '95 scheme a two-qubit gate was realized using the collective phonon mode as an auxiliary quantum degree of freedom, we describe now briefly a version on an ion trap computer where entanglement is achieved by designing an *internal-state–dependent two-body interaction* between the ions [43, 44]. This proposal has the advantage of being conceptually simpler (*e.g.*, there is no zero-temperature requirement), and obviously scalable. The model assumes that ions are stored in an *array of microtraps* (fig. 7). Similar to the ion trap '95 proposal, it is assumed that long-lived internal states of the ions serve as carriers of

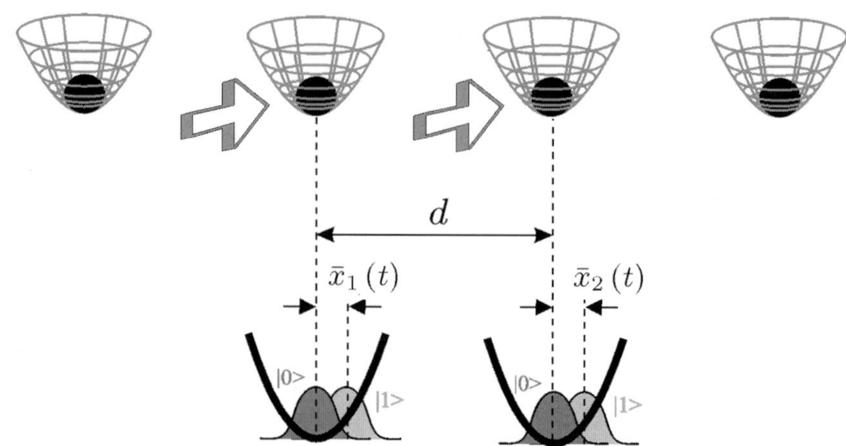

Fig. 7. – Ions stored in an array of microtraps. By addressing two adjacent ions with an external field the ion wave packet is displaced conditional to its internal state.

the qubits, and that single-qubit operations can be performed by addressing ions with a laser.

The model assumes a set of N ions confined in independent harmonic potential wells separated by some constant distance d, where d is large enough so that: i) the Coulomb repulsion is not able to excite the vibrational state of the ions; ii) the ions can be individually addressed. Two-qubit gates between two neighboring ions can be performed by slightly displacing them for a short time T if they are in a particular internal state, say $|1\rangle$. In that case, and provided the ions come back to their original motional state after being pushed, the Coulomb interaction will provide the internal wave function (quantum register) with different phases depending on the internal states of the ions. Choosing the time appropriately, the complete process will give rise to the two-qubit gate $|\epsilon_1\rangle|\epsilon_2\rangle \to e^{i\epsilon_1\epsilon_2\phi}|\epsilon_1\rangle|\epsilon_2\rangle$. In order to analyze this in a more quantitative way, we consider two ions 1 and 2 of mass m confined by two harmonic traps of frequency ω in one dimension (fig. 7). We denote by $\hat{x}_{1,2}$ the position operators of the two ions. The potential see by the ions is

$$(37) \qquad V = \sum_{i=1,2} \frac{1}{2} m\omega^2 \left(\hat{x}_i - \bar{x}_i(t)|1\rangle_i\langle 1| \right)^2 + \frac{e^2}{4\pi\epsilon_0} \frac{1}{|d + \hat{x}_2 - \hat{x}_1|},$$

where $\bar{x}_i(t)$ is the state-dependent displacement induced by the force. We are interested in the limit, where the displacements are much smaller than the distance between the traps, $|\hat{x}_{1,2}| \ll d$, and where the Coulomb energy is small compared with the trapping potentials, $\epsilon|\hat{x}_1\hat{x}_2|/a_0^2 \ll 1$. Furthermore, we assume that the motional state of the pushed ions will change adiabatically with the potential. Expansion of the Coulomb term in powers of $\hat{x}_{1,2}/d$ gives rise to a term $-m\omega^2\epsilon\hat{x}_1\hat{x}_2$ in the potential (37). It is this term which is responsible for entangling the atoms, giving rise to a conditional phase shift,

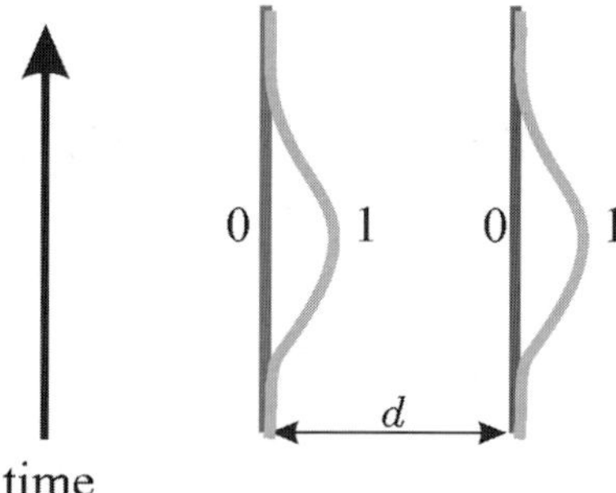

Fig. 8. – Trajectories of the qubits as a function of time. Depending on the internal state different phases are accumulated.

which can be simply interpreted as arising from the energy shifts due to the Coulomb interactions of atoms accumulated on different trajectories according to their internal states (fig. 8),

$$(38) \qquad \phi = -\frac{e^2}{4\pi\epsilon_0} \int_0^T dt \left[\frac{1}{d + \bar{x}_2 - \bar{x}_1} - \frac{1}{d + \bar{x}_2} - \frac{1}{d - \bar{x}_1} + \frac{1}{d} \right],$$

where the four terms are due to atoms in $|1\rangle_1|1\rangle_2$, $|1\rangle_1|0\rangle_2$, $|0\rangle_1|1\rangle_2$ and $|0\rangle_1|0\rangle_2$, respectively. The expression (38) depends only on mean displacement of the atomic wave packet and thus is insensitive to the temperature (the width of the wave packet) which will appear only in the problem in higher orders in $x_{1,2}/d$ of our expansion of the potential (37), or in cases of non-adiabaticity. A detailed theory of this proposal including an analysis of imperfections has been given by Calarco $et\ al.$ [44].

3˙**4.** $Cavity\ QED.$ – Cavity QED (CQED) realizes a situation where one or a few atoms interact strongly with a single quantized high-Q cavity mode, where the light field can be either in the optical or in the microwave domain (see the lecture notes by G. Rempe, S. Haroche and H. Walther). The coupling of atoms via the cavity mode can be used to engineer entanglement between the atoms [31,32,63-69]). The underlying physics is described by the Jaynes-Cummings Hamiltonian [47] which for a single two-level atom has the form

$$(39) \qquad H = \frac{1}{2}\hbar\omega_0\sigma_z + \hbar\omega_c a^\dagger a + \hbar\frac{1}{2}\left[\Omega_{eg}(\mathbf{r})a^\dagger\sigma_- + \Omega_{eg}(\mathbf{r})^*\sigma_+ a \right],$$

where the σ's are the Pauli spin operators describing the two-level atom, and $a^\dagger$ and a are the boson creation and annihilation operators of the field mode, respectively. The term proportional to $\Omega_0(\mathbf{r})$ in (39) describes the coherent oscillatory exchange of energy between the atom and the field mode. In optical cavity QED dissipation is due to spontaneous emission of the atom in the excited state and cavity decay, which can be modeled by a master equation.

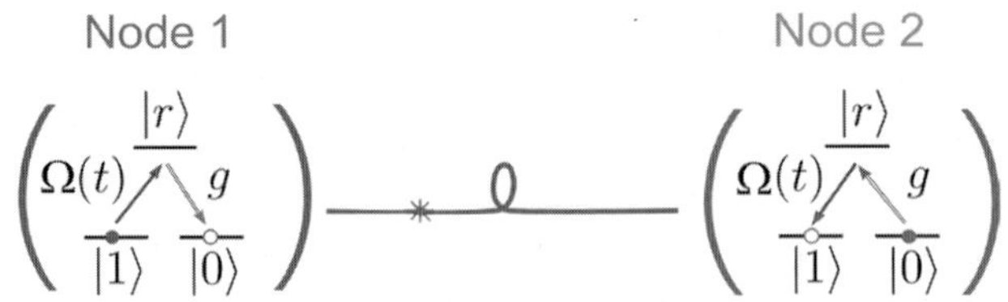

Node 1   Node 2

Fig. 9. – Transmission of a qubit from an atom at the first node to an atom at the second node according to (40) and (42).

From a formal point of view there is a close analogy between the ion trap models discussed in the previous subsection and the Jaynes-Cummings type models of CQED, where the role of the collective phonon modes of trapped ions is now taken by photons in the high-Q cavity mode. Thus, in many cases schemes for quantum state and entanglement engineering proposed in the ion traps are readily translated to their CQED counterparts. A first example of a CQED model of a quantum computer is the scheme by Pellizzari *et al.* [31]. In this scheme atoms representing the qubits are stored in a high-Q cavity, and the photon mode in the cavity plays the role of an auxiliary quantum degree of freedom. This allows entanglement of pairs of atoms, where the specific feature of ref. [31] is that photon exchange between atoms is performed using adiabatic passage along (the decoherence free subspace of) dark states, so that spontaneous emission as a decoherence channel is completely eliminated. On the other hand, CQED provides a natural interface between atoms as carriers of qubits and photons, which—once they leave the cavity—can be guided by optical fibers to transport qubits to distant locations. CQED thus serves as the paradigm for an *optical interconnect* between atoms as quantum memory and photons as carriers of qubits for quantum communication [32, 63].

3`4.1. Optical interconnects. The goal of quantum communications is to transmit an unknown quantum state from a first to a second node in a quantum network. We consider a situation where the state of a qubit is stored in the internal state of atoms. The task is to transmit the qubit according to

$$(40) \qquad (\alpha|0\rangle_1 + \beta|1\rangle_1) \otimes |0\rangle_2 \rightarrow |0\rangle_1 \otimes (\alpha|0\rangle_2 + \beta|1\rangle_2)$$

from the first to the second atom. Below we study a model of an optical interconnect based on storing atoms in high-Q optical cavities (see fig. 9). By applying laser beams, one first transfers the internal state of an atom (qubit) at the first node to the optical state of the cavity mode. The generated photons leak out of the cavity, propagate as a wave packet along the transmission line, and enter an optical cavity at the second node. Finally, the optical state of the second cavity is transferred to the internal state of an atom. Multiple-qubit transmissions can be achieved by sequentially addressing pairs of atoms (one at each node), as entanglements between arbitrarily located atoms are preserved by the state-mapping process.

The distinguishing feature of the protocol described below [32] is that by controlling the atom-cavity interaction, one can absolutely avoid the reflection of the wave packets from the second cavity, effectively switching off the dominant loss channel that would be responsible for decoherence in the communication process. For a physical picture of how this can be accomplished, let us consider that a photon leaks out of an optical cavity and propagates away as a wave packet. Imagine that we were able to "time reverse" this wave packet and send it back into the cavity; then this would restore the original (unknown) superposition state of the atom, provided we would also reverse the timing of the laser pulses. If, on the other hand, we are able to drive the atom in a transmitting cavity in such a way that the outgoing pulse were already symmetric in time, the wave packet entering a receiving cavity would "mimic" this time-reversed process, thus "restoring" the state of the first atom in the second one.

The simplest possible configuration of quantum transmission between two nodes consists of two three-level atoms 1 and 2 which are strongly coupled to their respective cavity modes (see fig. 9). The qubit is stored in a superposition of the two degenerate ground states $|g\rangle \equiv |0\rangle$ and $|e\rangle \equiv |1\rangle$. The states $|e\rangle$ and $|g\rangle$ are coupled by a Raman transition, where a laser excites the atom from $|e\rangle$ to $|r\rangle$ with to a time-dependent Rabi frequency, followed by a transition $|r\rangle \rightarrow |e\rangle$ which is accompanied by emission of a photon into the corresponding cavity mode. In order to suppress spontaneous emission from the excited state during the Raman process, we assume that the laser is strongly detuned from the atomic transition. In such a case, one can eliminate adiabatically the excited states $|r\rangle$. The Hamiltonian for the dynamics of the two ground states becomes, in a rotating frame for the cavity modes at the laser frequency, is of the Jaynes-Cummings form with time-dependent coupling $g_i(t)$

$$(41) \qquad \hat{H}_i = -\delta \hat{a}_i^\dagger \hat{a}_i - ig_i(t)\big[|e\rangle_i \, _i\langle g|a_i - \text{h.c.}\big] \quad (i = 1, 2),$$

where $\hat{a}_i$ are the destruction operators for the cavity modes $i = 1, 2$, and δ denotes the Raman detuning between the ground-state levels. For simplicity we ignore here AC-Stark shifts of the ground states due to the cavity mode and laser field, which can be easily included in a complete model [32]. The last term is a Jaynes-Cummings interaction, with an effective time-dependent coupling constant $g_i(t)$.

The goal is now to select a laser pulse shape $g_i(t)$ to accomplish *ideal quantum transmission*

$$(42) \qquad \big(c_g|g\rangle_1 + c_e|e\rangle_1\big)|g\rangle_2 \otimes |0\rangle_1|0\rangle_2|\text{vac}\rangle \longrightarrow |g\rangle_1\big(c_g|g\rangle_2 + c_e|e\rangle_2\big) \otimes |0\rangle_1|0\rangle_2|\text{vac}\rangle,$$

where $c_{g,e}$ are complex numbers. In (42), $|0\rangle_i$ and $|\text{vac}\rangle$ represent the vacuum state of the cavity modes and the free electromagnetic modes connecting the cavities. Transmission will occur by photon exchange via these modes.

It is useful to formulate this problem in the language of cascaded quantum systems. A cascaded quantum system consists of a quantum *source* driving a quantum *system* in a unidirectional coupling. In our case the source is the first node emitting a photon

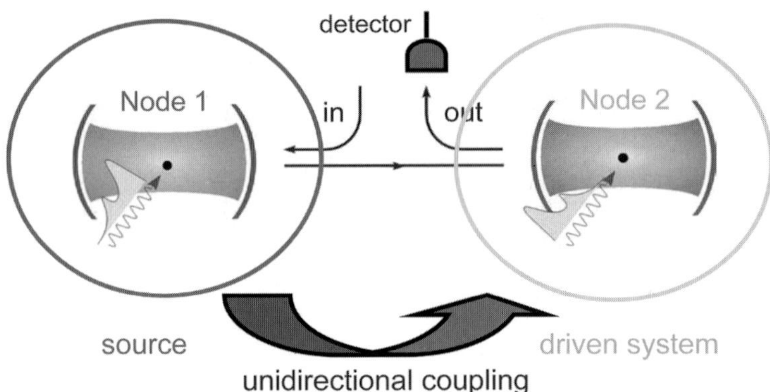

Fig. 10. – Transmission of a qubit between two atoms as a cascaded quantum system.

while the system is the second node (fig. 10). In case of perfect transmission of the qubit we require that the photon is not reflected from the second cavity, and thus there is no back reaction on the first node [70], $i.e.$ the coupling becomes unidirectional. A theory of cascaded quantum systems has been developed independently by Gardiner and Carmichael [47, 71, 72]. In the present context it is convenient to use a quantum trajectory formulation [47] of cascaded quantum systems [72]. To this end, we consider a fictitious experiment where the output field of the second cavity is continuously monitored by a photodetector (fig. 10). The evolution of the quantum system under continuous observation, conditional to observing a particular trajectory of counts, can be described by a pure state wave function $|\Psi_c(t)\rangle$ in the system Hilbert space of the two nodes (where the radiation modes outside the cavity have been eliminated). During the time intervals when no count is detected, this wave function evolves according to a Schrödinger equation with non-$Hermitian$ effective Hamiltonian

$$(43) \qquad \hat{H}_{\mathrm{eff}}(t) = \hat{H}_1(t) + \hat{H}_2(t) - i\kappa\left(\hat{a}_1^\dagger\hat{a}_1 + \hat{a}_2^\dagger\hat{a}_2 + 2\hat{a}_2^\dagger\hat{a}_1\right).$$

The detection of a count at time t_r is associated with a quantum jump according to $|\Psi_c(t_r + \mathrm{d}t)\rangle \propto \hat{c}|\Psi_c(t_r)\rangle$, where $\hat{c} = \hat{a}_1 + \hat{a}_2$, while the probability density for a jump (detector click) to occur during the time interval from t to $t+\mathrm{d}t$ is $\langle\Psi_c(t)|\hat{c}^\dagger\hat{c}|\Psi_c(t)\rangle\,\mathrm{d}t$ [47].

We wish to design the laser pulses in both cavities in such a way that ideal quantum transmission condition (42) is satisfied. A necessary condition for the time evolution is that a quantum jump (detector click, see fig. 10) never occurs, $i.e.$ $\hat{c}|\Psi_c(t)\rangle = 0 \ \forall t$, and thus the effective Hamiltonian will become a Hermitian operator. In other words, the system will remain in a $dark$ state of the cascaded quantum system. Physically, this means that the wave packet is not reflected from the second cavity. We expand the state of the system as

$$(44) \qquad |\Psi_c(t)\rangle = |c_g|gg\rangle|00\rangle_c +$$

$$+ |c_e\Big[\alpha_1(t)|eg\rangle|00\rangle_c + \alpha_2(t)|ge\rangle|00\rangle_c +$$

$$+ \beta_1(t)|gg\rangle|10\rangle_c + \beta_2(t)|gg\rangle_c|01\rangle_c\Big].$$

Ideal quantum transmission (42) will occur for $\alpha_1(-\infty) = \alpha_2(+\infty) = 1$. We can now easily derive evolution equations for the amplitudes $\alpha_i(t)$, $\beta_i(t)$ $(i = 1, 2)$. Starting from these equations Cirac *et al.* derive a class of solutions for pulses shapes in analytical form, guided by the physical expectation that the time evolution in the second cavity should reverse the time evolution in the first one, *i.e.* one looks for solutions satisfying the *symmetric pulse condition* $g_2(t) = g_1(-t)$. Numerical examples and a discussion of imperfection due to photon loss and spontaneous emission can be found in the quoted references.

Motivated by this CQED scheme various transmission protocols have been discussed based on this *photonic channel* [74, 75], including a protocol for a *quantum repeater* [4].

3˙5. *Collisional interactions for neutral atoms*. – In atomic physics with *neutral atoms* recent advances in cooling and trapping have led to an exciting new generation of experiments with Bose condensates, experiments with optical lattices, and atom optics and interferometry (see the lecture notes by E. Cornell and S. Rolston). The question therefore arises, to what extent these new experimental possibilities and the underlying physics can be adapted to the field of experimental quantum computing. In the previous subsection we have outlined possibilities of entangling neutral atoms in cavity QED schemes. Here we will discuss two examples of entangling atoms directly using controlled two-body interactions. The specific examples to be discussed are cold coherent collisions between ground-state atoms [35], and interaction of atoms via large dipole-dipole coupling of Rydberg atoms [37]. A scheme for entangling atoms with laser-induced dipole-dipole interactions was proposed by G. K. Brennen *et al.* [36, 77, 78].

3˙5.1. Entanglement via coherent ground-state collisions. Below we study coherent cold collisions [79] as the basic mechanism to entangle neutral atoms [35]. The picture of *atomic collisions* as *coherent interactions* has emerged in the field of Bose-Einstein condensation (BEC) of (ultracold) gases. In a field-theoretic language these interactions correspond to Hamiltonians which are quartic in the atomic field operators, analogous to Kerr non-linearities between photons in quantum optics. By storing ultracold atoms in arrays of microscopic potentials provided, for example, by optical lattices, these collisional interactions can be controlled via laser parameters [35]. Furthermore, these non-linear atom-atom interactions can be large [35], even for interactions between individual pairs of atoms, thus providing the necessary ingredients to implement quantum logic.

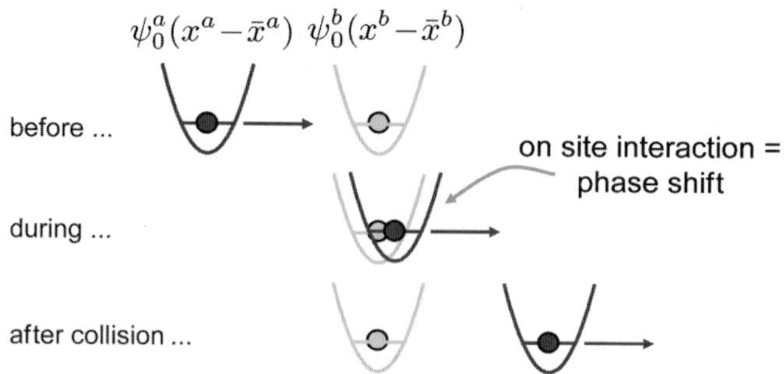

Fig. 11. – We collide a first atom in the internal state $|a\rangle$ with a second atom in state $|b\rangle$. In the collision the wave function accumulates a phase according to (46).

Consider a situation where two atoms with electrons populating the internal states $|a\rangle$ and $|b\rangle$, respectively, are trapped in the ground states $\psi_0^{a,b}$ of two potential wells $V^{a,b}$ (fig. 11). Initially, these wells are centered at positions $\bar{x}^a$ and $\bar{x}^b$, sufficiently far apart (distance $d = \bar{x}_b - \bar{x}_a$) so that the particles do not interact. The positions of the potentials are moved along trajectories $\bar{x}^a(t)$ and $\bar{x}^b(t)$ so that the wave packets of the atoms overlap for certain time, until finally they are restored to the initial position at the final time. This situation is described by the Hamiltonian

$$(45) \qquad H = \sum_{\beta=a,b} \left[\frac{(\hat{p}^\beta)^2}{2m} + V^\beta\big(\hat{x}^\beta - \bar{x}^\beta(t)\big) \right] + u^{\mathrm{ab}}(\hat{x}^a - \hat{x}^b).$$

Here, $\hat{x}^{a,b}$ and $\hat{p}^{a,b}$ are position and momentum operators, $V^{a,b}(\hat{x}^{a,b} - \bar{x}^{a,b}(t))$ describe the displaced trap potentials and u^{ab} is the atom-atom interaction term. Ideally, we would like to implement the transformation from before to after the collision,

$$(46) \qquad \psi_0^a(x^a - \bar{x}^a)\psi_0^b(x^b - \bar{x}^b) \longrightarrow e^{i\phi}\psi_0^a(x^a - \bar{x}^a)\psi_0^b(x^b - \bar{x}^b),$$

where each atom remains in the ground state of its trapping potential and preserves its internal state. The phase ϕ will contain a contribution from the interaction (collision) (fig. 12). The transformation (46) can be realized in the *adiabatic limit*, whereby we move the potentials slowly on the scale given by the trap frequency, so that the atoms

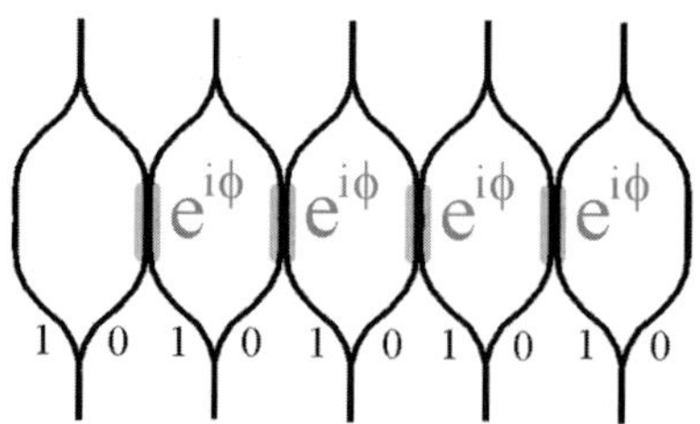

Fig. 12. – By moving an optical lattice in a state-dependent way neighboring atoms collide and acquire a phase shift.

remain in the ground state. Moving non-interacting atoms will induce single-particle kinetic phases. In the presence of interactions ($u^{\mathrm{ab}} \neq 0$), we define the time-dependent energy shift due to the interaction as

$$(47) \qquad \Delta E(t) = \frac{4\pi a_s \hbar^2}{m} \int \mathrm{d}x \left| \psi_0^a \left(x - \bar{x}^a(t)\right) \right|^2 \left| \psi_0^b \left(x - \bar{x}^b(t)\right) \right|^2,$$

where a_s is the s-wave scattering length. We assume that $|\Delta E(t)| \ll \hbar\nu$ with ν the trap frequency so that no sloshing motion is excited. In this case, (46) still holds with $\phi = \phi^a + \phi^b + \phi^{\mathrm{ab}}$, where in addition to (trivial) single-particle kinetic phases ϕ^a and ϕ^b arising from moving the potentials, we have a *collisional phase shift*

$$(48) \qquad \phi^{\mathrm{ab}} = \int_{-\infty}^{\infty} \frac{\mathrm{d}}{t} \Delta E \frac{t}{\hbar}.$$

The assumption behind the *colliding atoms by hand*, as described above, is that different internal states of the atom see a different trapping non-dissipative potential. In practice, this can be achieved by trapping atoms in a far-off-resonant trap or optical lattice, where the lasers are tuned in such a way that the atomic states $|a\rangle$ and $|b\rangle$ couple differently to the excited atomic states, so that there AC Stark shifts differ. A specific laser configuration achieving this state-dependent trapping has been analyzed in ref. [35] for alkali atoms, based on tuning the laser between the fine-structure excited states. The trapping potentials can be moved by changing the laser parameters. Such trapping potentials could also be realized with magnetic and electric microtraps [76].

So far, we have argued that one can use cold collisions as a coherent mechanism to induce phase shifts in two-atom interactions in a controlled way [80]. We can use these interactions to implement conditional dynamics. By the above arguments, the atomic wave packet in a superposition of the two internal levels can be "split" by moving the state-dependent potentials, very much like a beam splitter in atom interferometry. Thus we can move the potentials of neighboring atoms such that only the $|a\rangle$ component of the first atom "collides" with the state $|b\rangle$ of the second atom

$$(49) \qquad |a\rangle_1 |a\rangle_2 \longrightarrow e^{i2\phi^a} |a\rangle_1 |a\rangle_2,$$

$$|a\rangle_1 |b\rangle_2 \longrightarrow e^{i(\phi^a + \phi^b + \phi^{\mathrm{ab}})} |a\rangle_1 |b\rangle_2,$$

$$|b\rangle_1 |a\rangle_2 \longrightarrow e^{i(\phi^a + \phi^b)} |b\rangle_1 |a\rangle_2,$$

$$|b\rangle_1 |b\rangle_2 \longrightarrow e^{i2\phi^b} |b\rangle_1 |b\rangle_2,$$

where the motional states remain unchanged in the adiabatic limit, and ϕ^a and ϕ^b are single-particle kinetic phases. The transformation (49) corresponds to a fundamental two-qubit gate. The fidelity of this gate is limited by non-adiabatic effects, decoherence due to spontaneous emission in the optical potentials and collisional loss to other unwanted states, or collisional to unwanted states. According to ref. [35] the fidelity of this gate operation is remarkably close to one in a large parameter range. We note that loading of single atoms in laser traps has been achieved recently [81], and movable arrays of trapping potential have been created with arrays of microlenses [82]. Finally, by filling the lattice from a Bose condensate, and using the ideas related to Mott transitions in optical lattices [83] it is possible to achieve uniform lattice occupation ("optical crystals") or even specific atomic patterns, as well as the low temperatures necessary for performing the experiments proposed above.

It is interesting to generalize these ideas to the dynamics of a two-component Bose gas in an optical lattice. For example, in the adiabatic regime we denote by a_i and b_i the annihilation operators for a particle in the ground state of the potential centered at the position i, and corresponding to the internal levels $|a\rangle$ and $|b\rangle$, respectively. The effective Hamiltonian in this regime is a time-dependent Bose-Hubbard model [83],

$$(50) \qquad H = \sum_i \left[\omega^a(t) a_i^\dagger a_i + \omega^b(t) b_i^\dagger b_i + u^{\mathrm{aa}}(t) a_i^\dagger a_i^\dagger a_i a_i + \right.$$

$$\left. + u^{\mathrm{bb}}(t) b_i^\dagger b_i^\dagger b_i b_i \right] + \sum_{i,j} u_{ij}^{\mathrm{ab}}(t) a_i^\dagger a_i b_j^\dagger b_j,$$

where the ω's and u's depend on the specific way the potentials are moved. In quantum optics this Hamiltonian corresponds to a quantum non-demolition situation [47], whereby the particle number can be measured non-destructively.

3˙5.2. R y d b e r g a t o m s. Entanglement via cold coherent collisions, as described above, requires moving atoms and accumulating a phase due to (comparatively small) onsite interactions, which makes this a slow process. For neutral atoms there is a difficulty in identifying *strong and controllable long-range two-body interactions*, which is required to design a gate. Furthermore, the strength of two-body interactions does not necessarily translate into a useful fast quantum gate: large interactions are usually associated with strong mechanical forces on the trapped atoms. Thus, internal states of the trapped atoms (the qubits) may become entangled with the motional degrees of freedom during the gate, resulting effectively in an additional source of decoherence. This leads to the typical requirement that the process is *adiabatic* on the time scale of the oscillation period of the trapped atoms in order to avoid entanglement with motional states. As a result, extremely tight traps and low temperatures are required. A fast two-qubit phase gate for neutral trapped atoms, which addresses these problems, was proposed in ref. [35]. The scheme is based on i) the very large interactions of permanent dipole moments of laser-excited Rydberg states in a constant electric field to entangle atoms, and ii) allows gate operation times set by the time scale of the laser excitation or the two-particle interaction energy, which can be significantly shorter than the trap period. Among the attractive features of the gate are the insensitivity to the temperature of the atoms and to the variations in atom-atom separation.

Rydberg states [84] of a hydrogen atom within a given manifold of a fixed principal quantum number n are degenerate. This degeneracy is removed by applying a constant electric field $\mathcal{E}$ along the z-axis (linear Stark effect). For electric fields below the Ingris-Teller limit the mixing of adjacent n-manifolds can be neglected, and the energy levels are split according to $\Delta E_{nqm} = 3nqea_0\mathcal{E}/2$ with parabolic and magnetic quantum numbers $q = n - 1 - |m|, n - 3 - |m|, \ldots, -(n - 1 - |m|)$ and m, respectively, e the electron charge, and a_0 the Bohr radius. These Stark states have permanent dipole moments $\boldsymbol{\mu} \equiv \mu_z\mathbf{e}_z = 3/2\,nqea_0\mathbf{e}_z$. In alkali atoms the s- and p-states are shifted relative to the higher angular momentum states due to their quantum defects, and the Stark maps of the $m = 0$ and $m = 1$ manifolds are correspondingly modified [84].

Let us consider two atoms 1 and 2 initially prepared in Rydberg-Stark eigenstates, with a dipole moment along z and a given m, as selected by the polarization of the exciting laser. They interact and evolve according to the dipole-dipole potential

$$(51) \qquad V_{\mathrm{dip}}(\mathbf{r}) = \frac{1}{4\pi\epsilon_0}\left[\frac{\boldsymbol{\mu}_1 \cdot \boldsymbol{\mu}_2}{|\mathbf{r}|^3} - 3\frac{(\boldsymbol{\mu}_1 \cdot \mathbf{r})(\boldsymbol{\mu}_2 \cdot \mathbf{r})}{|\mathbf{r}|^5}\right],$$

with $\mathbf{r}$ the distance between the atoms. We are interested in the limit where the electric field is sufficiently large so that the energy splitting between two adjacent Stark states is much larger than the dipole-dipole interaction. For two atoms in the given initial Stark eigenstate, the diagonal terms of V_{dip} provide an energy shift whereas the non-diagonal terms couple $(m, m) \to (m \pm 1, m \mp 1)$ adjacent m manifolds with each other. We will assume that these transitions are suppressed by an appropriate choice of the initial Stark eigenstate. For a hydrogen state $|r\rangle = |n, q = n - 1, m = 0\rangle$, for example, and we find for a fixed distance $\mathbf{r} = R\mathbf{e}_z$ of the two atoms a dipole-dipole interaction $u(R) = \langle r| \otimes \langle r|V_{\mathrm{dip}}(R\mathbf{e}_z)|r\rangle \otimes |r\rangle$ with $u(R) = -9[n(n - 1)]^2(a_0/R)^3(e^2/8\pi\epsilon_0a_0) \propto n^4$. In alkali atoms we have to replace n by the effective quantum number ν [84]. We will use this large energy shift to entangle atoms.

We study a configuration where two atoms are for the moment assumed to be at fixed positions $\mathbf{x}_j$ (with $j = 1, 2$ labelling the atoms), at a distance $R = |\mathbf{x}_1 - \mathbf{x}_2|$. We store qubits in two internal atomic ground states denoted by $|g\rangle_j \equiv |0\rangle_j$ and $|e\rangle_j \equiv |1\rangle_j$. The ground states $|g\rangle_j$ are coupled by a laser to a given Stark eigenstate $|r\rangle_j$. The internal dynamics is described by a model Hamiltonian

$$H^i(t, \mathbf{x}_1, \mathbf{x}_2) = u|r\rangle_1\langle r| \otimes |r\rangle_2\langle r|+$$

$$+ \sum_{j=1,2}\left[(\delta_j(t) - i\gamma)|r\rangle_j\langle r| - \frac{1}{2}\Omega_j(t, \mathbf{x}_j)\left(|g\rangle_j\langle r| + \mathrm{h.c.}\right)\right].$$

Here $\Omega_j(t, \mathbf{x}_j)$ are Rabi frequencies, $\delta_j(t)$ are the detunings of the exciting lasers, and γ accounts for loss from the excited states $|r\rangle_j$ (fig. 13).

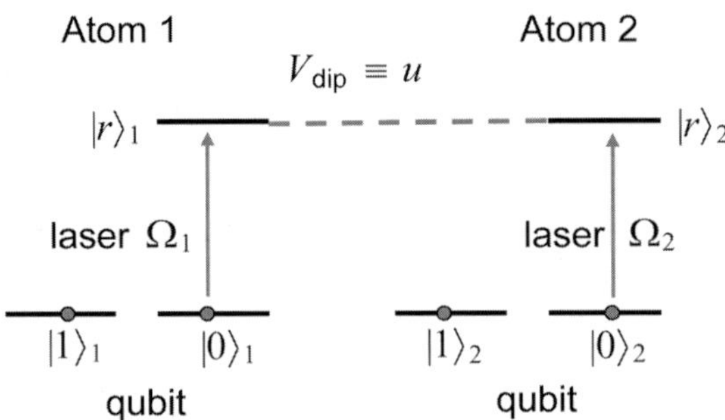

Fig. 13. – Atomic level scheme of the two-qubit gate. The laser excites the atoms in state $|1\rangle$ to Rydberg states in an electric field. The Rydberg states interact via a dipole-dipole interaction.

When we include the atomic motion, the complete Hamiltonian has the structure

$$(52) \qquad H(t, \hat{\mathbf{x}}_1, \hat{\mathbf{x}}_2) = H^T(\hat{\mathbf{x}}_1, \hat{\mathbf{x}}_2) + H^i(t, \hat{\mathbf{x}}_1, \hat{\mathbf{x}}_2),$$

$$(53) \qquad \equiv H^e(t, \hat{\mathbf{x}}_1, \hat{\mathbf{x}}_2) + H^i(t, \mathbf{x}_1, \mathbf{x}_2),$$

where H^T describes the motion of the trapped atoms, and $\hat{\mathbf{x}}_j$ are the atomic position operators, and we define $\hat{\mathbf{r}} = \hat{\mathbf{x}}_1 - \hat{\mathbf{x}}_2$. Our goal is to design a phase gate for the internal states with a gate operation time Δt with the internal Hamiltonian $H^i(t, \mathbf{x}_1, \mathbf{x}_2)$ in eq. (52), where (the c-numbers) $\mathbf{x}_j$ now denote the centers of the initial atomic wave functions as determined by the trap, while avoiding motional effects arising from $H^e(t, \hat{\mathbf{x}}_1, \hat{\mathbf{x}}_2)$. This requires that the gate operation time Δt is short compared to the typical time of evolution of the external degrees of freedom, $H^e \Delta t \ll 1$. Under this condition, the initial density operator of the two atoms evolves as $\rho_e \otimes \rho_i \to \rho_e \otimes \rho_i'$ during the gate operation. Thus the motion described by ρ_e does not become entangled with the internal degrees of freedom given by ρ_i. Typically, the Hamiltonian H^T will be the sum of the kinetic energies of the atoms and the trapping potentials for the various internal states. We will assume that the potentials are harmonic with a frequency ω for the ground states, and ω' for the excited state.

Physically, for the splitting of the Hamiltonian according to eq. (52) to be meaningful we require the initial width of the atomic wave function a, as determined by the trap, to be much smaller than the mean separation between the atoms R. We expand the dipole-dipole interaction around R, $V_{\text{dip}}(\hat{\mathbf{r}}) = u(R) - F(\hat{\mathbf{r}}_z - R) + \ldots$, with $F = 3u(R)/R$. Here the first term gives the energy shift if both atoms are excited to state $|r\rangle$, while the second term contributes to H^e and describes the mechanical force on the atoms due to V_{dip}. Other contributions to H^e arise from the photon kick in the absorption $|g\rangle \to |r\rangle$, but these terms can be suppressed in a Doppler-free two-photon absorption, for example. We obtain $H^e(\hat{\mathbf{x}}_1, \hat{\mathbf{x}}_2) = H^T - F(\hat{\mathbf{r}}_z - R)|r\rangle_1\langle r| \otimes |r\rangle_2\langle r|$.

We can study our model for a phase gate according to dynamics induced by H^i in various parameter regimes. First, in the limit $\Omega_j \gg u$, it is possible to perform a gate by exciting the Rydberg atoms with π-pulses, so that the state $|r\rangle_1|r\rangle_2$ picks up the extra

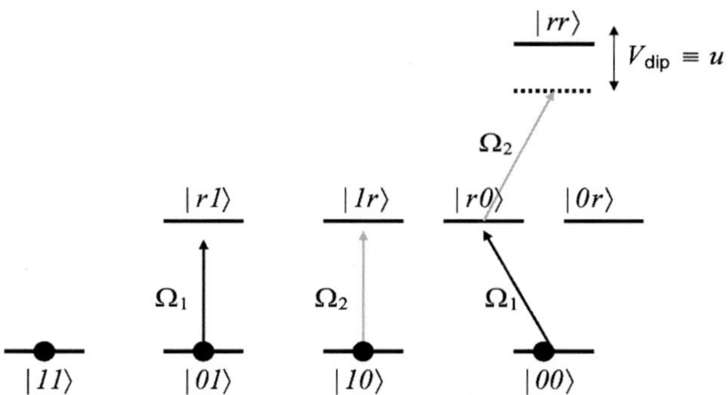

Fig. 14. – Laser excitation sequence of the dipole-dipole gate with Rydberg atoms. Qubits are stored in two internal atomic ground states denoted by $|0\rangle_j$ and $|1\rangle_j$.

phase $\varphi = u\Delta t$ from the dipole-dipole interaction. Thus, this scheme realizes a fast phase gate operating on the time scale $\Delta t \propto 1/u$. We note, however, that the accumulated phase depends on the precise value of u, *i.e.* is sensitive to the atomic distance. In addition, during the gate operation there are large mechanical effects due to the force F.

These problems can be avoided by operating the gate in the parameter regime $u \gg \Omega_j$. In the simplest case we assume that the two atoms can be addressed individually, *i.e.* $\Omega_1(t) \neq \Omega_2(t)$. We set $\delta_j = 0$ and perform the gate operation in three steps: i) We apply a π-pulse to the first atom, ii) a 2π-pulse (in terms of the unperturbed states, *i.e.* it has twice the pulse area of pulse applied in i)) to the second atom, and, finally, iii) a π-pulse to the first atom. As can be seen from fig. 14, the state $|ee\rangle$ is not affected by the laser pulses. If the system is initially in one of the states $|ge\rangle$ or $|eg\rangle$ the pulse sequence i)-iii) will cause a sign change in the wave function. If the system is initially in the state $|gg\rangle$ the first pulse will bring the system to the state $i|rg\rangle$, the second pulse will be *detuned* from the state $|rr\rangle$ by the interaction strength u, and thus accumulate a *small* phase $\tilde{\varphi} \approx \pi\Omega_2/2u \ll \pi$. The third pulse returns the system to the state $e^{i(\pi-\tilde{\varphi})}|gg\rangle$, which realizes a phase gate with $\varphi = \pi - \tilde{\varphi} \approx \pi$ (up to trivial single-qubit phases). The time needed to perform the gate operation is of the order $\Delta t \approx 2\pi/\Omega_1 + 2\pi/\Omega_2$. It is possible to formulate an *adiabatic* version of this gate with the advantage that individual addressing of the two atoms is not required, $\Omega_{1,2}(t) \equiv \Omega(t)$ and $\delta_{1,2}(t) \equiv \delta(t)$.

A remarkable feature of the second model is that, in the ideal limit, the doubly excited state $|rr\rangle$ is never populated, because the doubly excited state is shifted by the large dipole-dipole interaction, *i.e.* we have a dipole-blockade mechanism (reminiscent of the Coulomb blockade in quantum dots). Hence, the mechanical effects due to atom-atom interaction are greatly suppressed. Furthermore, this version of the gate is only weakly sensitive to the exact distance between the atoms, since the distance-dependent part of the entanglement phase $\tilde{\varphi} \ll \pi$.

We now turn to a discussion of decoherence mechanisms, which include spontaneous emission, transitions induced by black-body radiation, ionization of the Rydberg states due to the trapping or exciting laser fields, and motional excitation of the trapped atoms. While the dipole-dipole interaction increases with ν^4, the spontaneous emission and ionization of the Rydberg states by optical laser fields decreases proportional to ν^{-3}. The Rabi frequency coupling the ground to the Rydberg states scales as $n^{-3/2}$. For $\nu < 20$ the black-body radiation is negligible in comparison with spontaneous emission, and similar conclusions hold for typical ionization rates from the Rydberg states. For typical numbers one expects errors on the percent level [35].

4. – Quantum information processing with atomic ensembles

$4^{\cdot}1$. *Introduction.* – In the previous section, all the schemes for quantum information processing are based on laser manipulation of single trapped particles. In this section, we will show that many quantum information protocols can also be implemented simply by laser manipulation of atomic ensembles containing a large number of identical neutral atoms (see also the lecture notes by M. Fleischhauer). The experimental candidate systems for the atomic ensembles can be either laser-cooled atoms confined in a magnetic optical trap [85-88], or room-temperature atoms contained in a glass cell with coated walls to avoid bad collisions [89-91]. Quantum information is stored in the ground-state manifold of the atoms, such as in Zeeman sublevels with different atomic spins, or in some hyperfine atomic levels which are stable or metastable under optical transitions. Long coherence time of the relevant states has been observed in both kinds of the experimental systems mentioned above [85-91]. The motivation of using atomic ensembles instead of single particles for quantum information processing is three fold: First, laser manipulation of atomic ensembles without separate addressing of individual atoms is much easier than the laser manipulation of single particles; secondly, the quantum information encoded in atomic ensembles is robust against some practical noise (for instance, the lost of few atoms in a large atomic ensemble has negligible influence on its carried quantum information); finally and perhaps most importantly, the use of the atomic ensembles provides a novel way for enhancing the signal-to-noise ratio with some collective effects existing in this system. We know that in physical implementations of quantum information protocols, the central problem is to enhance the signal-to-noise ratio, that is, to enhance the ratio of the magnitudes between the coherent information processing and the decoherence process caused by the noisy coupling to the environment. For instance, in cavity-QED schemes, one needs to build a high-finesse cavity around the atoms to achieve strong light-atom coupling, and needs to enter the challenging strong-coupling regime for a high signal-to-noise ratio. However, we will see that due to the collective enhancement induced by the many-atom coherence, the signal-to-noise ratio in atomic ensembles can be greatly increased by encoding quantum information into some collective excitations of the ensembles. As a result of this effect, quantum information processing is made possible with much simplified experimental systems, such as atomic ensemble in weak-coupling cavities or even in free space.

This section is arranged as follows: In subsect. $4^{\cdot}2$, we describe the interaction of light with atomic ensembles with various level configurations. Our attention is focused on collective enhancement of the signal-to-noise ratio in these systems. For simple demonstration of the collective enhancement, in the first two level configurations we assume that there is a weak-coupling cavity around the atomic ensemble, following the approaches in refs. [92-95]; and in the last level configuration, we directly describe the interaction of light with a free-space atomic ensemble using a one-dimensional light propagation model, following the approaches in refs. [96-98]. Some discussions on the three-dimensional light propagation effects can be found in ref. [99]. In these two different approaches, one can find basically the same kind of collective enhancement of the signal-to-noise ratio. The systems discussed in subsect. 4˙2 provide the basic elements for physical implementation of many quantum information protocols. In subsect. 4˙3, we show how to use atomic ensembles combined with linear optical elements to realize scalable long-distance quantum communication, such as quantum key distribution, quantum teleportation, and Bell-inequality detections. Long-distance quantum communication is necessarily based on the use of photonic channels. However, due to losses and decoherence in the channel, the communication fidelity decreases exponentially with the channel length. To overcome the problem associated with the exponential fidelity decay, one needs to use quantum repeaters [4], which provide the only known way for robust long-distance quantum communication. In subsect. 4˙3, we review the recent scheme proposed in ref. [95] for physical implementation of quantum repeaters and robust quantum communication over long lossy channels. The scheme involves laser manipulation of atomic ensembles, beam splitters, and single-photon detectors with moderate efficiencies, and therefore well fits the status of the current experimental technology. The communication efficiency in this scheme scales polynomially with the channel length thereby facilitating scalability to very long distances. In subsect. 4˙4, we investigate other applications of atomic ensembles in quantum information processing. In particular, we show that atomic ensembles can be used to realize quantum light memory [100, 93, 94, 97] and single-photon sources with controllable emission time, direction, and pulse shape. Quantum light memory and a controllable single-photon source provide the basic elements for realization of a recent quantum computation scheme [101]. Another quantum computation scheme using atomic ensembles was proposed in ref. [102], which combines laser manipulation of atomic ensembles with the idea of dipole blockades from Rydberg atoms. In that scheme, one needs to exploit direct dipole-dipole interactions between atoms which are induced by exciting atoms to high Rydberg levels. This scheme will not be investigated in this review since the concept of direct atom-atom interaction is outside the scope of the current section. In subsect. 4˙5, we review the applications of atomic ensembles in continuous variable quantum information processing. Laser manipulation of atomic ensembles provides a simple way for realizing continuous variable quantum teleportation between distant atomic ensembles [98]. Probabilities of using atomic ensembles for realizing continuous variable quantum computation are also briefly remarked.

4˙2. *Interaction of light with atomic ensembles and collective enhancement of the signal-to-noise ratio.* – We consider in the following the interaction of propagating light with atomic ensembles. In particular, we demonstrate that when a sample of identical atoms with suitable level configurations are shined by a propagating optical signal, the signal will only interact with the (symmetric) collective atomic mode, whereas atomic

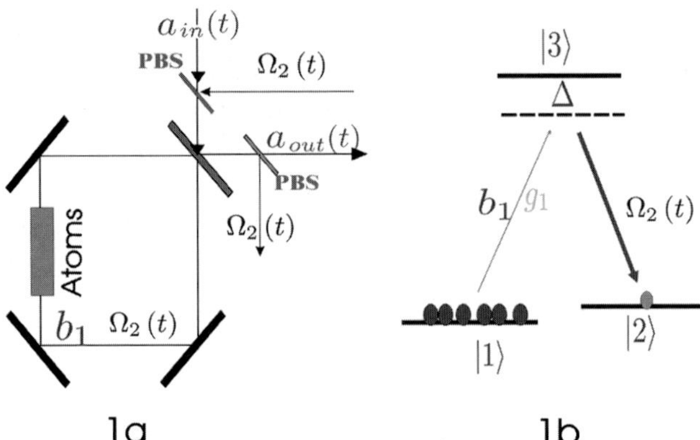

Fig. 15. – (a) An atomic ensemble in a weak coupling cavity. (b) The ΛI-level configuration.

spontaneous emissions caused by the coupling to vacuum noise fields are distributed over all the modes. As a result of this property, the signal-to-noise ratio for the collective atomic mode can be greatly enhanced. We consider three types of level configurations for the atoms: two Λ-level configurations with different kinds of coupling to the signal light and one four-level configuration. All these level schemes are useful in subsequent sections for different purposes of quantum information processing. In all these schemes, besides the propagating quantum signal, there is some driving field provided by a classical laser. We always assume that the classical driving field and the quantum signal are copropagating to assure a collective coupling condition. For a simple demonstration of the collective enhancement of the signal-to-noise ratio, first in the two Λ-level configurations we assume that there is a low-Q ring cavity around the ensemble, and a ring cavity mode, driven by the input and output quantum light signal, is then coupled to a desired atomic transition. Due to the collective enhancement, we can get a large signal-to-noise ratio even without a cavity to enhance the light-atom coupling. As an example to explicitly show this, in the four-level configuration we assume interaction of quantum light with free-space atomic ensembles under a one-dimensional light propagation model. The result confirms that we have the same kind of collective enhancement in the signal-to-noise ratio.

4˙2.1. ΛI-level configuration. In the first level configuration, we assume that the atoms have two ground states $|1\rangle$, $|2\rangle$ and one excited state $|3\rangle$ (see fig. 15). All the atoms initially occupy the ground level $|1\rangle$. The transition $|1\rangle \to |3\rangle$ is coupled with a coupling coefficient g_1 to a ring cavity mode b_1, which is then driven by the input and output quantum signals $a_{in}(t)$ and $a_{out}(t)$. The transition $|3\rangle \to |2\rangle$ is driven by a classical laser field with a Rabi frequency Ω_2 (Ω_2 can be time-dependent). The driving laser and the ring cavity mode (quantum signal) are copropagating to satisfy the collective coupling condition $(k_l - k_s)L_a \ll \pi$, where k_l and k_s are, respectively, the wave vectors of the laser and the quantum signal, and L_a is the length of the atomic ensemble. We assume off-resonant coupling with a large detuning Δ as shown in fig. 15. This level scheme has been considered in refs. [100,93,94,97] for quantum light memory (refs. [93,97] considered this level scheme with resonant coupling). Here we follow ref. [94] for a simple theoretical description. A description of the resonant coupling based on dark states can be found in refs. [93, 103], and a free-space description of this level scheme with the one-dimensional light propagation model can be found in ref. [97].

In the case of a large detuning Δ, we can adiabatically eliminate the excited level $|3\rangle$, and under the collective coupling condition, the interaction shown in fig. 15 is described by the following Hamiltonian in the rotating frame:

$$(54) \qquad H = \hbar\left(\frac{\Omega_2 g_1}{\Delta}\right) b_1^\dagger \sum_{i=1}^{N_a} \sigma_{12}^i + \text{h.c.},$$

where N_a is the total atom number, and $\sigma_{12}^i = |1\rangle_i\langle 2|$ is the atomic lowering operator for the i-th atom. We have neglected the light shift terms such as $(\Omega_2/\Delta)^2 \sum_{i=1}^{N_a} \sigma_{22}^i$ in eq. (54), which can be trivially cancelled by refining the laser frequency. We assume that the quantum signal is weak so that nearly all the atoms still remain in the level $|1\rangle$ with $\langle|1\rangle_i\langle 1|\rangle \simeq 1$. Under this weak-excitation condition, we introduce an effective bosonic operator $s = \sum_{i=1}^{N_a} \sigma_{12}^i/\sqrt{N_a}$ with $[s, s^\dagger] \simeq 1$ for the (symmetric) collective atomic mode. With the collective atomic operator, the Heisenberg-Langevin equations corresponding to the Hamiltonian (1) have the form [47]

$$(55) \qquad \dot{s} = -i\left(\frac{\sqrt{N_a}\Omega_2^* g_1^*}{\Delta}\right) b_1,$$

$$(56) \qquad \dot{b}_1 = -i\left(\frac{\sqrt{N_a}\Omega_2 g_1}{\Delta}\right) s - \left(\frac{\kappa}{2}\right) b_1 - \sqrt{\kappa} a_{\text{in}}(t),$$

where κ is the cavity decay rate, and $a_{\text{in}}(t)$ is the input quantum signal with the properties $[a_{\text{in}}(t), a_{\text{in}}^\dagger(t')] = \delta(t-t')$. The output quantum signal $a_{\text{out}}(t)$ from the cavity is connected with the input by the input-output relation $a_{\text{out}}(t) = a_{\text{in}}(t) + \sqrt{\kappa} b_1$. In the bad-cavity limit with the cavity decay rate κ much larger than the coupling rate $\sqrt{N_a}\Omega_2 g_1/\Delta$, we can adiabatically eliminate the cavity mode b_1, and obtain a direct Langevin equation for the collective atomic mode s,

$$(57) \qquad \dot{s} = -\frac{\kappa'}{2} s - \sqrt{\kappa'} a_{\text{in}}(t),$$

where the effective coupling rate $\kappa' = 4N_a|\Omega_2 g_1|^2/(\Delta^2\kappa)$, and without loss of generality we have assumed that the phase of the laser is chosen in such a way as to make $i\Omega_2 g_1 = |\Omega g_c|$. The output quantum signal, expressed by the atomic operator s, has the form $a_{\text{out}}(t) = -a_{\text{in}}(t) - \sqrt{\kappa'} s$. Equation (57) serves as the basic equation in dealing with quantum light memory in subsect. 4˙4.

To show that we have a collective enhancement of the signal-to-noise ratio, let us consider the atomic spontaneous emissions. There are two spontaneous emission processes:

first, the (about N_a) atoms in the level $|1\rangle$ can absorb photons from the quantum signal to go up to the level $|3\rangle$, and then go down through spontaneous emissions; and second, the very few atoms in the level $|2\rangle$ can absorb photons from the classical driving laser to go up and then go down through spontaneous emissions. We assume that the atomic ensemble is dilute with $k_s/\sqrt[3]{\rho_n} \geq 1$ (where ρ_n is the atomic number density) so that there are no superradiant effects for spontaneous emissions which go to all the possible directions. In this case, the total spontaneous emission rate is given by $\gamma_{t1} = N_a|g_1|^2\gamma_s/\Delta^2$ for the first process, and by $\gamma_{t2} = |\Omega_2|^2\gamma_s/\Delta^2$ for the second process, where γ_s is the resonant spontaneous emission rate (the natural bandwidth of the level $|3\rangle$). When the classical driving laser is strong with $|\Omega_2|^2 \succeq N_a|g_1|^2$, the signal-to-noise ratio R_{sn} in this configuration is estimated by $R_{sn} \sim \kappa'/\gamma_{t2} \sim 4N_a|g_1|^2/(\kappa\gamma_s)$. This result should be compared with the corresponding one in the case with single atoms trapped in a high-Q cavity, where the signal-to-noise ratio is given by $|g_1|^2/(\kappa\gamma_s)$ [104]. Therefore, for atomic ensembles with the ΛI-level configuration, the signal-to-noise ratio is greatly enhanced by the large factor of the atom number N_a due to the many-atom collective effects in coupling to the copropagating quantum signal.

4'2.2. ΛII-level configuration. In the second level configuration, each atom still has three levels $|1\rangle$, $|2\rangle$ and $|3\rangle$. The difference is now that the classical driving laser is coupled to the transition $|1\rangle \to |3\rangle$, and the quantum signal to the transition $|3\rangle \to |2\rangle$ (see fig. 16). This level configuration was considered before for a quantum description of the Raman stimulating process [96], and recently it has been shown in [95] to be useful for physical implementation of long-distance quantum communication. Here, we follow ref. [95] for a simple description and demonstration of the collective enhancement of the signal-to-noise ratio. In this level configuration, the effective Hamiltonian after adiabatic elimination of the upper level $|3\rangle$ has the following form:

$$(58) \qquad H = \hbar\left(\frac{\sqrt{N_a}\Omega_1 g_2}{\Delta}\right) s^\dagger b_2^\dagger + \text{h.c.},$$

where s is the collective atomic annihilation operator defined as before, and we have neglected the trivial light shift terms. Similar to the configuration I, we can write the

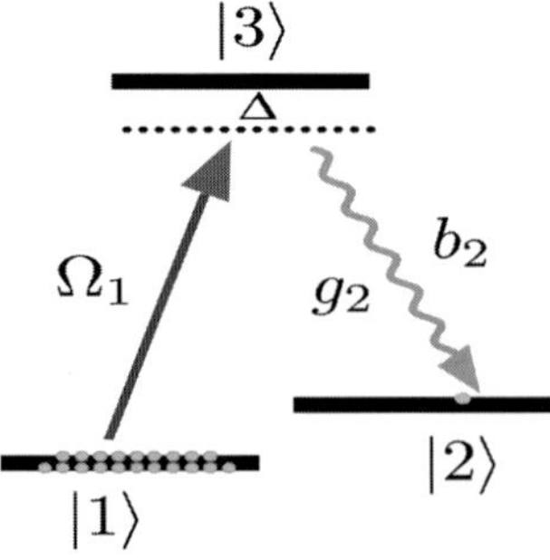

Fig. 16. – The ΛII-level configuration.

Heisenberg-Langevin equations corresponding to the Hamiltonian (58), and then adiabatically eliminate the mode b_2 in the bad-cavity limit to obtain a direct Langevin equation for the collective atomic mode s,

$$(59) \qquad \dot{s}^\dagger = \frac{\kappa'}{2} s^\dagger - \sqrt{\kappa'} a_{\mathrm{in}}(t),$$

where the effective coupling rate $\kappa' = 4N_a |\Omega_1 g_2|^2/(\Delta^2 \kappa)$, and without loss of generality we have assumed $i\Omega_1^* g_2^* = |\Omega_1 g_2|$. The output quantum signal $a_{\mathrm{out}}(t)$ is connected with the input $a_{\mathrm{in}}(t)$ by the input-output relation $a_{\mathrm{out}}(t) = -a_{\mathrm{in}}(t) + \sqrt{\kappa'} s^\dagger$.

The atomic ensembles with the ΛII-level configuration described before provide us the basic elements for implementing quantum repeaters and long-distance quantum communication which will be detailed in the next subsection. For the applications there, we explicitly solve the basic equation (59) with a vacuum input quantum signal $a_{\mathrm{in}}(t)$ which satisfies the properties $\langle a_{\mathrm{in}}^\dagger(t) a_{\mathrm{in}}(t') \rangle = 0$ and $\langle a_{\mathrm{in}}(t) a_{\mathrm{in}}^\dagger(t') \rangle = \delta(t - t')$. Equation (59) is linear and has the simple solution

$$(60) \qquad s^\dagger(t) = s^\dagger(0) e^{\kappa' t/2} - \sqrt{\kappa'} \int_0^t e^{\kappa'(t-\tau)/2} a_{\mathrm{in}}(\tau)\, \mathrm{d}\tau.$$

What we are interested in the quantum repeater scheme is some measurable quantity constructed from the output signal $a_{\mathrm{out}}(t)$. If we use photon detectors for measurement, what we detect is the integration of the output photon current during the detection time interval t_Δ, which is proportional to the intensity integration of the output field $a_{\mathrm{out}}(t)$. So in fact we measure the operator $Q_m = \int_0^{t_\Delta} a_{\mathrm{out}}^\dagger(\tau) a_{\mathrm{out}}(\tau)\, \mathrm{d}\tau$. One can find an explicit expression for Q_m by substituting the solution of $a_{\mathrm{out}}(t)$ from the input-output relation. To simplify the expression of Q_m, we define an effective single-mode bosonic operator a from the continuous field $a_{\mathrm{in}}(t)$ by the form

$$(61) \qquad a \equiv -\frac{\sqrt{\kappa'}}{\sqrt{e^{\kappa' t_\Delta} - 1}} \int_0^{t_\Delta} e^{\kappa'(t_\Delta - \tau)/2} a_{\mathrm{in}}(\tau)\, \mathrm{d}\tau.$$

With this operator, the measured quantity is expressed as

$$(62) \qquad Q_m = a_{t_\Delta}^\dagger a_{t_\Delta} + \int_0^{t_\Delta} a_{\mathrm{in}}^\dagger(\tau) a_{\mathrm{in}}(\tau)\, \mathrm{d}\tau,$$

where $a_{t_\Delta} = a \cosh r_c + s^\dagger(0) \sinh r_c$, a Bogoliubov transformation of a and $s^\dagger(0)$ with $\cosh r_c \equiv e^{\kappa' t_\Delta/2}$. The last term of eq. (62) is a trivial integration of the intensity of the vacuum field, which has no contribution to the measurement result. So what we measure is in fact the photon number in the effective single mode a_{t_Δ}. Note that eq. (60) can also be written in the Bogoliubov form $s^\dagger(t_\Delta) = s^\dagger(0) \cosh r_c + a \sinh r_c$. Both of $s^\dagger(0)$ and a are initially in vacuum states, which will be denoted by $|0_a\rangle$ and $|0_p\rangle$, respectively, in the following (the subscripts "a" for atoms, and "p" for photons). Transferring the solution to the Schrödinger picture, we conclude that after time t_Δ the collective atomic mode s and the effective single mode for the output quantum signal are in a two-mode squeezed state

$$(63) \qquad |\phi\rangle = \mathrm{sech}\, r_c \sum_n \frac{\left(S^\dagger a \tanh r_c\right)^n}{n!} |0_a\rangle |0_p\rangle.$$

This is the basic result which will be used in the next subsection.

Now let us take into account the atomic spontaneous emissions and calculate the signal-to-noise ratio for this level configuration. The current situation is quite different from the ΛI-level configuration in the sense that if one directly calculates the total spontaneous emission rate in this system, one would find that the total rate is given by $N_a(\Omega_1^2/\Delta^2)\gamma_s$ since almost all the atoms remain in the level $|1\rangle$. If one takes this rate as the noise rate, there will be no enhancement of the signal-to-noise ratio compared with the single-atom case. However, in our scheme we only concern about the collective atomic mode s since the output quantum signal is only entangled with this mode as shown by eq. (63). So instead of the calculation of the total spontaneous emission rate, we need to calculate the noise rate for the mode s. The spontaneous emissions are independent for different atoms without superradiance, and they introduce a coherence decay term to the Langevin equation of each individual atomic operator $s_i = \sigma_{12}^i$

$$(64) \qquad \dot{s^\dagger}_i = -\left(\frac{\gamma_s'}{2}\right) s_i^\dagger + \text{noise},$$

where the decay rate $\gamma_s' = (\Omega_1^2/\Delta^2)\gamma_s$. The last term in eq. (64) represents the corresponding fluctuation from the noise field which results in heating, and we have left out the coherent interaction term from the Hamiltonian (58). By taking summation of eq. (64) over all the atoms, we immediately see that there is a coherence decay term to the Langevin equation (59) of the collective atomic operator $s^\dagger$ with the decay rate still given by γ_s'. The signal-to-noise ratio R_{sn} for the collective atomic mode is thus given by $R_{sn} = \kappa'/\gamma_s' = 4N_a|g_2|^2/(\kappa\gamma_s)$. So compared with the single-atom case, the signal-to-noise ratio is still greatly enhanced by the large factor of the atom number N_a. This collective enhancement comes from the fact that the coherent interaction producing the output quantum signal involves only the collective atomic mode s, whereas the independent spontaneous emissions distribute over all the atomic modes, and thus only have small influence on the interesting mode s.

To better understand this point, it is helpful to also have a look at the master equation for the atomic density operator. The whole density operator ρ_w for the atomic states and the cavity mode obeys the following master equation [47]:

$$(65) \qquad \dot{\rho}_w = i\left[\rho_w, H\right] + \kappa \widehat{L}[b]\rho_w + \gamma_s' \sum_i \widehat{L}\left[s_i^\dagger\right]\rho_w,$$

where the Liouville superoperators $\widehat{L}[X]$ ($X = b, s_i^\dagger$) are defined as $\widehat{L}[X]\rho_w \equiv X\rho_w X^\dagger - (X^\dagger X\rho_w + \rho_w X^\dagger X)/2$. In eq. (65), the first term of the right hand side (r.h.s.) comes from the coherent Hamiltonian interaction, the second term represents the cavity output coupling, and the last term describes independent spontaneous emissions for individual atomic operators. In the bad-cavity limit, after adiabatically eliminating the cavity mode, we get from eqs. (65) and (58) the following master equation for the traced atomic density operator ρ_a:

$$(66) \qquad \dot{\rho}_a = \kappa' \widehat{L}[s^\dagger]\rho_a + \gamma'_s \sum_i \widehat{L}[s_i^\dagger]\rho_a,$$

where $\widehat{L}[s^\dagger]$ is the Liouville superoperator for the collective atomic mode. The above equation can be further simplified if we introduce the Fourier transformation to the individual atomic operators s_j ($j = 0, 1, \ldots, N_a - 1$) with the form $s_\mu \equiv \sum_j S_j e^{ij\mu/N_a}/\sqrt{N_a}$, where $s_{\mu=0}$ gives exactly the collective atomic operator s. In terms of the operators s_μ, the master equation has the form

$$(67) \qquad \dot{\rho}_a = \left(\kappa' + \gamma'_s\right) \widehat{L}[s^\dagger]\rho_a + \gamma'_s \sum_{\mu \neq 0} \widehat{L}[s_\mu^\dagger]\rho_a,$$

Under the weak-excitation condition $\langle s_j^\dagger s_j \rangle \ll 1$, the operators s_μ ($\mu = 0, 1, \ldots, N_a - 1$) commute with each other, so they represent independent atomic modes. We are only interested in the collective atomic mode s, and the populations in all the other modes s_μ with $\mu \neq 0$ have no influence on the state of the mode s. So we can trace over the modes s_μ ($\mu \neq 0$) and eliminate the last term in eq. (67). There are two contributions to the population in the collective atomic mode s: the one with a rate κ' produces a coherent output signal, and the one with a rate γ'_s emits photons to other random directions. The signal-to-noise ratio for the mode s is thus given by $R_{sn} = \kappa'/\gamma'_s \sim 4N_a|g_c|^2/(\kappa\gamma_s)$, and we get exactly the same result as before. It is interesting to note from eq. (67) that the total spontaneous emission rate of all the modes is $N_a\gamma'_s$, which could be much larger than the coherent interaction rate κ'; however, the spontaneous emission rate for the collective atomic mode is N_a times smaller than the total rate. This is why we still have collective enhancement and a large signal-to-noise ratio for this level configuration.

4'2.3. Four-level configuration. In the above two configurations of the light-atom interaction, the signal-to-noise ratio is greatly enhanced by the many-atom collective effects. Due to the collective enhancement, we by no means need a good cavity in these schemes. In fact, we can even assume to continuously decrease the cavity finesse down to 1, which corresponds to the free-space limit. In the free-space limit, the cavity decay rate κ is estimated by c/L_a, the inverse of the travelling time of the optical pulse in the ensemble. With the well-known expressions for the coupling coefficient g_1 (or g_2) and the resonant spontaneous emission rate γ_s [47], one can estimate the signal-to-noise ratio in the free-space limit by $R_{sn} \sim 4N_a|g_c|^2/(\kappa\gamma_s) \sim 3\rho_n L_a/k_s^2 \sim d_o$, where d_o denotes the on-resonance optical depth of the atomic ensemble which can be quite large

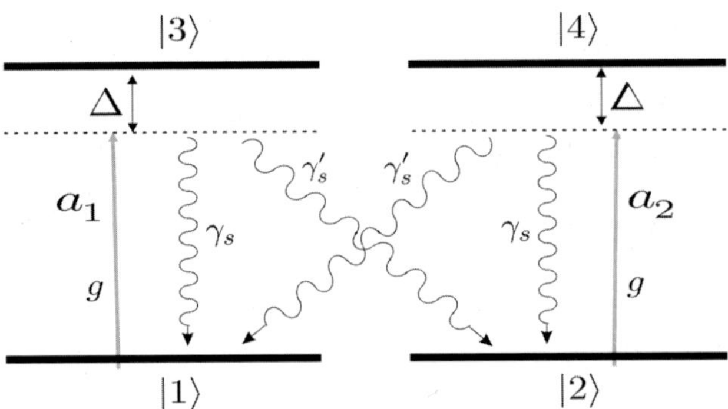

Fig. 17. – The four-level configuration.

with the current experimental technology [85-91]. So we can have a considerably large signal-to-noise ratio even without a cavity. This demonstrates that collective effects in many-atom ensembles provide us another way besides high-Q cavities to achieve strong coherent light-atom coupling. To show this more directly, we consider another light-atom interaction configuration with four levels, and in this level scheme, we solve directly the interaction of light with free-space atomic ensembles by assuming a one-dimensional light propagation model.

The relevant atomic level structure is shown by fig. 17. Each atom has two degenerate ground states and two degenerate excited states. The transitions $|1\rangle \to |3\rangle$ and $|2\rangle \to |4\rangle$ are coupled with a large detuning Δ to different circularly polarized propagating light due to the angular-momentum selection rule. This kind of interaction has been analyzed semiclassically in [105, 106], and recently shown to be applicable for quantum non-demolition measurements [92, 107-109] and for continuous variable quantum teleportation [98, 110]. Here, we follow ref. [98] for a free-space quantum description of the light-atomic-ensemble interaction. The atomic spontaneous emissions are included in the description to demonstrate the collective enhancement of the signal-to-noise ratio.

We assume a one-dimensional model for the propagating light field. As shown in ref. [99], this is justified if the atomic ensemble is of a pencil shape with Fresnel number $F = A/\lambda_0 L_a \approx 1$. Here A and L_a are the cross-section and the length of the ensemble, respectively, and λ_0 is the optical wavelength. The input laser pulse is linearly polarized and expressed as $E^{(+)}(z,t) = \sqrt{(\hbar\omega_0/4\pi\epsilon_0 A)} \sum_{i=1,2} a_i(z,t)e^{i(k_0 z - \omega_0 t)}$, where $\omega_0 = k_0 c = 2\pi c/\lambda_0$ is the carrier frequency, and i denotes two orthogonal circular polarizations, with the standard commutation relations $[a_i(z,t), a_j(z',t)] = \delta_{ij}\delta(z - z')$. The light is weakly focused with cross-section A to match the atomic ensemble. For the input of a strong coherent light with linear polarization, the initial condition is expressed as $\langle a_i(0,t)\rangle = \alpha_t$, with the total photon number over the pulse duration T satisfying $2N_p = 2c\int_0^T |\alpha_t|^2 \, dt \gg 1$. The Stokes operators are introduced for the free-space input and output light (light before entering or after leaving the atomic ensemble) by $S_x^p = (c/2)\int_0^T (a_1^\dagger a_2 + a_2^\dagger a_1)\, d\tau$, $S_y^p = (c/2i)\int_0^T (a_1^\dagger a_2 - a_2^\dagger a_1)\, d\tau$,

$S_z^p = (c/2)\int_0^T (a_1^\dagger a_1 - a_2^\dagger a_2)\, d\tau$. In free space, $a_i(z,t)$ only depends on $\tau = t - z/c$, and in this case one can check that the Stokes operators satisfy the spin commutation relations $[S_y^p, S_z^p] = iS_x^p$. For our coherent input, we have $\langle S_x^p \rangle = N_p$ and $\langle S_y^p \rangle = \langle S_z^p \rangle = 0$. With a very large N_p, the off-resonant interaction with atoms is only a small perturbation to S_x^p, and we can treat S_x^p classically by replacing it with its mean value $\langle S_x^p \rangle$. Then, we define two canonical observables for light by $X^p = S_y^p/\sqrt{\langle S_x^p \rangle}$, $P^p = S_z^p/\sqrt{\langle S_x^p \rangle}$ with a standard commutator $[X^p, P^p] = i$. These operators are the quantum variables we are interested in. Similar operators can be introduced for atoms. For an atomic ensemble with many atoms, it is convenient to define the continuous atomic operators $\sigma_{\mu\nu}(z,t) = \lim_{\delta z \to 0}(1/\rho A\delta z)\sum_i^{z \le z_i < z+\delta z} |\mu\rangle_i\langle\nu|$ ($\mu,\nu = 1,2,3,4$) with the commutation relations $[\sigma_{\mu\nu}(z,t), \sigma_{\nu'\mu'}(z',t)] = (1/\rho_n A)\delta(z-z')(\delta_{\nu\nu'}\sigma_{\mu\mu'} - \delta_{\mu\mu'}\sigma_{\nu'\nu})$. In the definition, z_i is the position of the i atom, and ρ_n is the number density of the atomic ensemble with the total atom number $2N_a = \rho AL_a \gg 1$. The collective spin operators are introduced for the ground states of the atomic ensemble by $S_x^a = (\rho A/2)\int_0^{L_a}(\sigma_{12} + \sigma_{12}^\dagger)\, dz$, $S_y^a = (\rho A/2i)\int_0^{L_a}(\sigma_{12} - \sigma_{12}^\dagger)\, dz$, $S_z^a = (\rho A/2)\int_0^{L_a}(\sigma_{11} - \sigma_{22})\, dz$. All the atoms are initially prepared in the equal superposition of the two ground states $(|1\rangle + |2\rangle)/\sqrt{2}$, which is an eigenstate of S_x^a with a very large eigenvalue N_a. As before, we treat S_x^a classically, and define the canonical operators for atoms by $X^a = S_y^a/\sqrt{\langle S_x^a \rangle}$, $P^a = S_z^a/\sqrt{\langle S_x^a \rangle}$ with $[X^a, P^a] = i$ and an initial vacuum state.

With introduction of the continuous atomic operators, the interaction between atoms and the propagating light $E^{(+)}(z,t)$ is described by the following Hamiltonian in the rotating frame:

$$(68) \qquad H = \hbar \sum_{i=1,2} \int_0^L \left[\Delta\sigma_{i+2,i+2}(z,t) + \right.$$

$$\left. + \left(ge^{ik_0 z} a_i(z,t)\sigma_{i+2,i}(z,t) + \text{h.c.}\right)\right]\rho A\, dz,$$

where the coupling constant $g = \sqrt{\omega_0/4\pi\hbar\epsilon_0 A}\, d$ and d is the dipole moment of the $|i\rangle \to |i+2\rangle$ transition. Corresponding to this Hamiltonian, the Maxwell-Bloch equations are written as [47]

$$(69) \qquad \left(\frac{\partial}{\partial t} + c\frac{\partial}{\partial z}\right)a_i(z,t) = -ig^* e^{-ik_0 z}\rho A\sigma_{i,i+2}(z,t),$$

$$\frac{\partial}{\partial t}\sigma_{\mu\nu} = -\frac{i}{\hbar}[\sigma_{\mu\nu}, H] - \frac{\gamma_{\mu\nu}}{2}\sigma_{\mu\nu} +$$

$$+ \sqrt{\gamma_{\mu\nu}}(\sigma_{\nu\nu} - \sigma_{\mu\mu})F_{\mu\nu}\ (\mu < \nu),$$

where the spontaneous emission rates (see fig. 17) are, respectively, $\gamma_{13} = \gamma_{24} \equiv \gamma_s = (\omega_0^3 |d|^2/3\pi\epsilon_0\hbar c^3)$, $\gamma_{14} = \gamma_{23} \equiv \gamma_{s'}$, and $\gamma_{12} = 0$ (the ground state has a long coherence time). The Doppler broadening caused by the atomic motion is negligible, since it is eliminated for off-resonant interactions with the collinear input and output lights.

Assuming that the spontaneous emission is independent for different atoms, the vacuum noise operators $F_{\mu\nu}$ satisfy the δ-commutation relations $[F_{\mu\nu}(z,t), F^\dagger_{\mu'\nu'}(z',t')] = (1/\rho A)\delta_{\mu\mu'}\delta_{\nu\nu'}\delta(z-z')\delta(t-t')$. To simplify eq. (69), first we change the variables by $\tau = t - z/c$, and then adiabatically eliminate the excited states $|3\rangle$ and $|4\rangle$ of atoms in the case of a large detuning, $i.e.$, $\Delta \gg g\langle a_i(z,t)\rangle \sim g\sqrt{N_p/(cT)}$. The resultant equations read

$$(70) \qquad \frac{\partial}{\partial z}a_i(z,\tau) = \frac{i|g|^2\rho A\sigma_{ii}}{\Delta c}a_i(z,\tau) - \frac{|g|^2\rho A\gamma_s\sigma_{ii}}{2\Delta^2 c}a_i(z,\tau)+$$

$$+\frac{g^* e^{-ik_0 z}\rho A\sqrt{\gamma_s}\sigma_{ii}}{\Delta c}F_{i,i+2}(z,\tau),$$

$$\frac{\partial}{\partial\tau}\sigma_{12} = \frac{i|g|^2\left(a_2^\dagger a_2 - a_1^\dagger a_1\right)}{\Delta}\sigma_{12} - \frac{|g|^2\gamma_{s'}\left(a_2^\dagger a_2 + a_1^\dagger a_1\right)}{2\Delta^2}\times$$

$$\times\,\sigma_{12} + \frac{\sqrt{\gamma_{s'}}}{\Delta}\left(g^* e^{-ik_0 z}a_2^\dagger\sigma_{11}F_{14} + ge^{ik_0 z}a_1\sigma_{22}F_{23}^\dagger\right).$$

The physical meaning of the above equation is quite clear: The first term on the right hand side is the phase shift caused by the off-resonant interaction between light and atoms, and the second and the third terms represent the damping and the corresponding vacuum noise caused by the spontaneous emission, respectively. In eq. (70), the σ_{ii} and $a_i^\dagger a_i$ are approximately constant operators, only with a small damping caused by the spontaneous emission. To consider the spontaneous emission noise to the first order, it is reasonable to assume constant σ_{ii} and $a_i^\dagger a_i$ for eq. (70). Then, this equation can be easily solved by integrating over z, τ on both sides. The result, expressed by the canonical atomic and optical operators $X^{p,a}$ and $P^{p,a}$ introduced before, has the following simple form:

$$(71) \qquad X^{p\prime} = \sqrt{1-\varepsilon_p}\left(X^p - \kappa_c P^a\right) + \sqrt{\varepsilon_p}X_s^p,$$

$$P^{p\prime} = \sqrt{1-\varepsilon_p}P^p + \sqrt{\varepsilon_p}P_s^p,$$

$$X^{a\prime} = \sqrt{1-\varepsilon_a}\left(X^a - \kappa_c P^p\right) + \sqrt{\varepsilon_a}X_s^a,$$

$$P^{a\prime} = \sqrt{1-\varepsilon_a}P^a + \sqrt{\varepsilon_a}P_s^a,$$

where the operators with (without) a prime denote the quantities after (before) the light pulse goes through the atomic ensemble, and $X_s^a, P_s^a, X_s^p, P_s^p$ represent the standard vacuum noise operators with variance $1/2$, defined from the integration of $F_{\mu\nu}(z,t)$, $X_s^p = \sqrt{c/4N_pN_a|g|^2}\int_0^T\int_0^L \rho A[ig^* e^{-ik_0 z}(a_2^\dagger\sigma_{11}F_{13} - a_1^\dagger\sigma_{22}F_{24}) + \text{h.c.}]\,dz\,dt$ for instance. The interaction and damping coefficients $\kappa_c, \varepsilon_p, \varepsilon_a$ are given, respectively, by $\kappa_c = -(2\sqrt{N_pN_a}|g|^2/\Delta c) = (3\sqrt{N_pN_a}\gamma\lambda_0^2/8\pi^2\Delta A)$, $\varepsilon_p = (N_a|g|^2\gamma_s/\Delta^2 c)$, $\varepsilon_a = (N_p|g|^2\gamma_{s'}/\Delta^2 c)$. The solution (71) is obtained under the conditions of weak excitation $\kappa_c \ll \sqrt{N_{p,a}}$ and small noise $\varepsilon_{p,a} \ll 1$. For simplicity, we assume $N_p \sim N_a$

and $\gamma_s \sim \gamma_{s'}$ so that $\varepsilon_p \sim \varepsilon_a \sim \varepsilon$. The interaction parameter κ_c can be rewritten as $\kappa_c = (3\rho_n \lambda_0^2 L_a \gamma_s)/(8\pi^2 \Delta)$ with $N_p = N_a$. For an atomic sample of density $\rho_n \sim 5 \times 10^{12}$ cm^{-3} and of length $L_a \sim 2$cm, $\kappa_c \sim 5$ is obtainable with the choice $\Delta \sim 300\gamma$, and at the same time the loss $\varepsilon_p \sim \varepsilon_a \sim \varepsilon < 1\%$. The signal-to-noise ratio for this interaction scheme is quantified by $R_{sn} = \kappa_c^2/\varepsilon \sim 3\rho_n L_a/k_0^2 \sim d_o$, which is more than 10^3 for the above example of parameters. We will see in subsect. 4˙5 that these numbers are good enough for realizing high-fidelity continuous variable quantum teleportation. By directly solving the free-space problem, we get exactly the same signal-to-noise ratio as in the two previous two-level schemes after taking the free-space limit of the cavity finesses. This clearly shows that collective enhancement of the signal-to-noise ratio is present for all these types of light-atom interactions, independent of the presence or absence of the optical cavities. The collective enhancement of the signal-to-noise ratio is an important feature of these systems, which facilitate various kinds of quantum information processing detailed below.

4˙3. *Scalable long-distance quantum communication.* – Quantum communication is an essential element required for constructing quantum networks, and it also has the application for absolutely secret transfer of classical messages by means of quantum cryptography [111]. The central problem of quantum communication is to generate nearly perfect entangled states between distant sites. Such states can be used, for example, to implement secure quantum cryptography using the Ekert protocol [111], and to faithfully transfer quantum states via quantum teleportation [26]. All the known realistic schemes for quantum communication are based on the use of the photonic channels. However, the degree of entanglement generated between two distant sites normally decreases exponentially with the length of the connecting channel due to the optical absorption and other channel noise. To regain a high degree of entanglement, purification schemes can be used [16]. However, entanglement purification does not fully solve the long-distance quantum communication problem. Due to the exponential decay of the entanglement in the channel, one needs an exponentially large number of partially entangled states to obtain one highly entangled state, which means that for a sufficiently long distance the task becomes nearly impossible.

To overcome the difficulty associated with the exponential fidelity decay, the concept of quantum repeaters can be used [4]. In principle, it allows to make the overall communication fidelity very close to unity, with the communication time growing only polynomially with the transmission distance. In analogy to a fault-tolerant quantum computing [112, 113], the quantum repeater proposal is a cascaded entanglement purification protocol for communication systems. The basic idea is to divide the transmission channel into many segments, with the length of each segment comparable to the channel attenuation length. First, one generates entanglement and purifies it for each segment; the purified entanglement is then extended to a longer length by connecting two adjacent segments through entanglement swapping [26, 114]. After entanglement swapping, the overall entanglement is decreased, and one has to purify it again. One can continue the rounds of the entanglement swapping and purification until nearly perfect entangled states are created between two distant sites.

To implement the quantum repeater protocol, one needs to generate entanglement between distant quantum bits (qubits), store them for sufficiently long time and perform local collective operations on several of these qubits. The requirement of quantum memory is essential since all purification protocols are probabilistic. When entanglement purification is performed for each segment of the channel, quantum memory can be used to keep the segment state if the purification succeeds and to repeat the purification for the segments only where the previous attempt fails. This is essentially important for polynomial scaling properties of the communication efficiency since with no available memory we have to require that the purifications for all the segments succeed at the same time; the probability of such event decreases exponentially with the channel length. The requirement of quantum memory implies that we need to store the local qubits in the atomic internal states instead of the photonic states since it is difficult to store photons for a reasonably long time. With atoms as the local information carriers it seems to be very hard to implement quantum repeaters since normally one needs to achieve the strong coupling between atoms and photons with high-finesse cavities for atomic entanglement generation, purification, and swapping [32, 75], which, in spite of the recent significant experimental advances [104, 115, 116], remains a very challenging technology.

To overcome this difficulty, ref. [95] proposes a very different scheme to realize quantum repeaters based on the use of atomic ensembles with the ΛII-level configuration. The laser manipulation of the atomic ensembles, together with some simple linear optics devices and moderate single-photon detectors, do the whole work for long-distance quantum communication. The setup is much simpler compared with the single-atom and high-Q cavity approach discussed in the previous section. To achieve this, the scheme makes significant advances in each step of entanglement generation, connection, and applications, with each step having built-in entanglement purification and resilient to the realistic noise. As a result, the scheme circumvents the realistic noise and imperfections, and at the same time keeps the overhead in the communication time increasing with the distance only polynomially. In this subsection, we will review the realization of quantum repeaters and long-distance quantum communication following the approach in ref. [95].

4'3.1. *Entanglement generation*. To realize long-distance quantum communication, first we need to entangle two atomic ensembles within the channel attenuation length. The entanglement generation scheme described here is based on single-photon interference at photodetectors, and is fault-tolerant to realistic noise. This scheme is an extension of a proposal first proposed in [117, 73] to entangle single-atoms. The extension was made in [95] to entangle atomic ensembles with significant improvements in the communication efficiency thanks to the collective enhancement of the signal-to-noise ratio for many-atom ensembles.

The system is a sample of atoms prepared in the ground state $|1\rangle$ with the ΛII-level configuration (see fig. 18). It has been shown in the previous subsection that one can define an effective single-mode bosonic annihilation operator a for the cavity output signal (it is called the forward-scattered Stokes signal in the free space case).

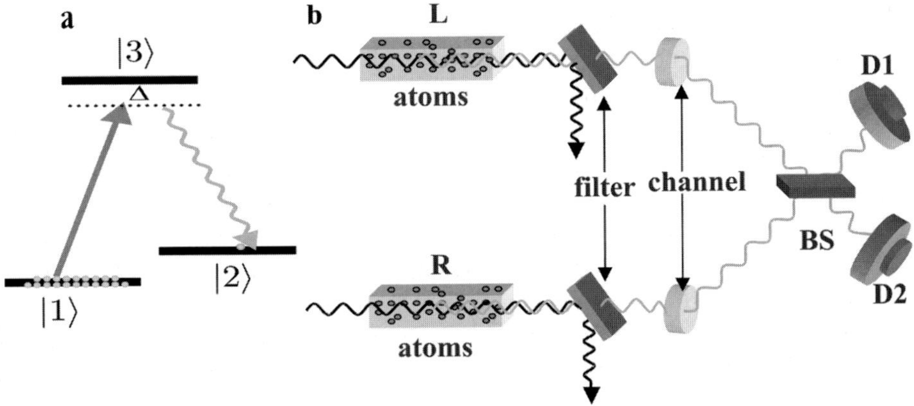

Fig. 18. – (4a) The relevant level structure of the atoms in the ensemble with $|1\rangle$, the ground state, $|2\rangle$, the metastable state for storing a qubit, and $|3\rangle$, the excited state. The transition $|1\rangle \to |3\rangle$ is coupled by the classical laser with the Rabi frequency Ω, and the forward-scattering Stokes light comes from the transition $|3\rangle \to |2\rangle$. For convenience, we assume off-resonant coupling with a large detuning Δ. (4b) Schematic setup for generating entanglement between the two atomic ensembles L and R. The two ensembles are pencil shaped and illuminated by the synchronized classical laser pulses. The forward-scattering Stokes pulses are collected after the filters (polarization and frequency selective) and interfered at a 50%–50% beam splitter BS after the transmission channels, with the outputs detected, respectively, by two single-photon detectors D1 and D2. If there is a click in D1 *or* D2, the process is finished and we successfully generate entanglement between the ensembles L and R. Otherwise, we first apply a repumping pulse to the transition $|2\rangle \to |3\rangle$ on the ensembles L and R to set the state of the ensembles back to the ground state $|0\rangle_a^L \otimes |0\rangle_a^R$, then the same classical laser pulses as the first round are applied to the transition $|1\rangle \to |3\rangle$ and we detect again the forward-scattering Stokes pulses after the beam splitter. This process is repeated until finally we have a click in the D1 *or* D2 detector.

After the light-atom interaction, the signal mode a and the collective atomic mode $s \equiv (1/\sqrt{N_a}) \sum_i |1\rangle_i \langle 2|$ are in a two-mode squeezed state with the squeezing parameter r_c proportional to the interaction time t_Δ (see eq. (63)). If the interaction time t_Δ is very small, the whole state of the collective atomic mode and the signal mode can be written in the perturbative form

$$(72) \qquad |\phi\rangle = |0_a\rangle|0_p\rangle + \sqrt{p_c}S^\dagger a^\dagger|0_a\rangle|0_p\rangle + o(p_c),$$

where $p_c = \tanh^2 r_c$ is the small excitation probability and $o(p_c)$ represents the terms with more excitations whose probabilities are equal to or smaller than p_c^2. The $|0_a\rangle$ and $|0_p\rangle$ are, respectively, the atomic and optical vacuum states with $|0_a\rangle \equiv \bigotimes_i |1\rangle_i$. There is also a fraction of light from the transition $|3\rangle \to |2\rangle$ emitted in other directions which contributes to spontaneous emissions. We have shown in the previous subsection that the contribution to the population in the collective atomic mode s from the spontaneous emissions is very small for many-atom ensembles due to the collective enhancement of the signal-to-noise ratio for this mode.

Now we show how to use this setup to generate entanglement between two distant ensembles L and R using the configuration shown in fig. 18. Here, two laser pulses excited both ensembles simultaneously, and the whole system is described by the state $|\phi\rangle_L \otimes |\phi\rangle_R$, where $|\phi\rangle_L$ and $|\phi\rangle_R$ are given by eq. (72) with all the operators and states distinguished by the subscript L or R. The forward-scattered Stokes signal from both ensembles is combined at the beam splitter and a photodetector click in either D1 *or* D2 measures the combined radiation from two samples, $a_+^\dagger a_+$ or $a_-^\dagger a_-$ with $a_\pm = (a_L \pm e^{i\varphi} a_R)/\sqrt{2}$. Here, φ denotes an unknown difference of the phase shifts in the two-side channels. We can also assume that φ has an imaginary part to account for the possible asymmetry of the setup, which will also be corrected automatically in our scheme. But the setup asymmetry can be easily made very small, and for simplicity of expressions we assume that φ is real in the following. Conditional on the detector click, we should apply a_+ or a_- to the whole state $|\phi\rangle_L \otimes |\phi\rangle_R$, and the projected state of the ensembles L and R is nearly maximally entangled with the form (neglecting the high-order terms $o(p_c)$)

$$(73) \qquad |\Psi_\varphi\rangle_{LR}^\pm = \frac{\left(S_L^\dagger \pm e^{i\varphi} S_R^\dagger\right)}{\sqrt{2}} |0_a\rangle_L |0_a\rangle_R.$$

The probability for getting a click is given by p_c for each round, so we need repeat the process about $1/p_c$ times for a successful entanglement preparation, and the average preparation time is given by $T_0 \sim t_\Delta/p_c$. The states $|\Psi_r\rangle_{LR}^+$ and $|\Psi_r\rangle_{LR}^-$ can be easily transformed into each other by a simple local phase shift. Without loss of generality, we assume in the following that we generate the entangled state $|\Psi_r\rangle_{LR}^+$.

As will be shown below, the presence of the noise modifies the projected state of the ensembles to

$$(74) \qquad \rho_{LR}(c_0,\varphi) = \frac{1}{c_0 + 1}\left(c_0 |0_a 0_a\rangle_{LR}\langle 0_a 0_a| + |\Psi_\varphi\rangle_{LR}^+ \langle \Psi_\varphi|\right),$$

where the "vacuum" coefficient c_0 is determined by the dark count rates of the photon detectors. It will be seen below that any state in the form of eq. (74) will be purified automatically to a maximally entangled state in the entanglement-based communication schemes. We therefore call this state an effective maximally entangled (EME) state with the vacuum coefficient c_0 determining the purification efficiency.

4'3.2. Entanglement connection through swapping. After the successful generation of the entanglement within the attenuation length, we want to extend the quantum communication distance. This is done through entanglement swapping with the configuration shown in fig. 19. Suppose that we start with two pairs of the entangled ensembles described by the state $\rho_{LI_1} \otimes \rho_{I_2R}$, where ρ_{LI_1} and ρ_{I_2R} are given by eq. (74). In the ideal case, the setup shown in fig. 19 measures the quantities corresponding to operators $S_\pm^\dagger S_\pm$ with $S_\pm = (S_{I_1} \pm S_{I_2})/\sqrt{2}$. If the measurement is successful (*i.e.*, one of the detectors registers one photon), we will prepare the ensembles L and R into another EME state. The new φ-parameter is given by $\varphi_1 + \varphi_2$, where φ_1 and φ_2 denote the

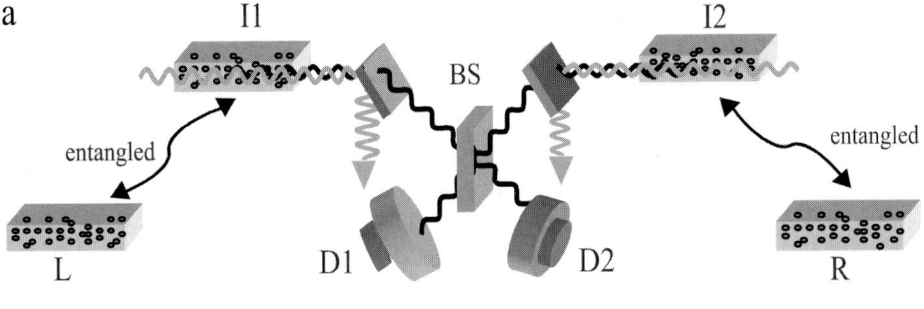

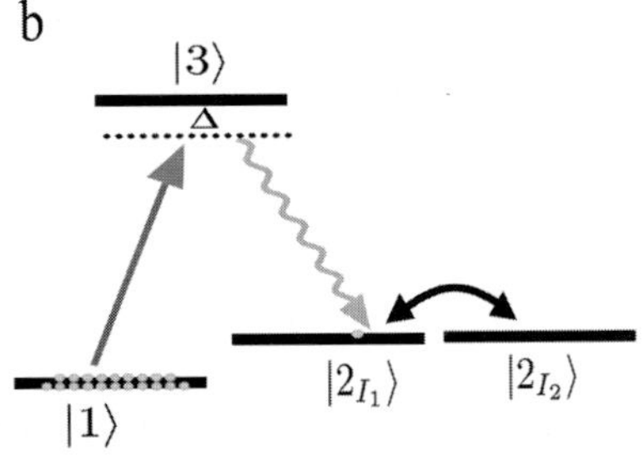

Fig. 19. – (a) Illustrative setup for the entanglement swapping. We have two pairs of ensembles L, I_1 and I_2, R distributed at three sites L, I and R. Each of the ensemble pairs L, I_1 and I_2, R is prepared in an EME state in the form of eq. (56). The excitations in the collective modes of the ensembles I_1 and I_2 are transferred simultaneously to the optical excitations by the repumping pulses applied to the atomic transition $|2\rangle \rightarrow |3\rangle$, and the stimulated optical excitations, after a 50%–50% beam splitter, are detected by the single-photon detectors D1 and D2. If either D1 *or* D2 clicks, the protocol is successful and an EME state in the form of eq. (56) is established between the ensembles L and R with a doubled communication distance. Otherwise, the process fails, and we need to repeat the previous entanglement generation and swapping until finally we have a click in D1 or D2, that is, until the protocol finally succeeds. (b) The two intermediated ensembles I_1 and I_2 can also be replaced by one ensemble but with two metastable states I_1 and I_2 to store the two different collective modes. The 50%–50% beam splitter operation can be simply realized by a $\pi/2$ pulse on the two metastable states before the collective atomic excitations are transferred to the optical excitations.

old φ-parameters for the two segment EME states. As will be seen below, even in the presence of the realistic noise and imperfections, an EME state is still created after a detector click. The noise only influences the success probability to get a click and the new vacuum coefficient in the EME state. In general we can express the success probability p_1 and the new vacuum coefficient c_1 as $p_1 = f_1(c_0)$ and $c_1 = f_2(c_0)$, where the functions f_1 and f_2 depend on the particular noise properties.

The above method for connecting entanglement can be cascaded to arbitrarily extend the communication distance. For the i-th ($i = 1, 2, \ldots, n$) entanglement connection, we first prepare in parallel two pairs of ensembles in the EME states with the same vacuum coefficient c_{i-1} and the same communication length L_{i-1}, and then perform the entanglement swapping as shown in fig. 19, which now succeeds with a probability $p_i = f_1(c_{i-1})$. After a successful detector click, the communication length is extended to $L_i = 2L_{i-1}$, and the vacuum coefficient in the connected EME state becomes $c_i = f_2(c_{i-1})$. Since the i-th entanglement connection need be repeated on average $1/p_i$ times, the total time needed to establish an EME state over the distance $L_n = 2^n L_0$ is given by $T_n = T_0 \prod_{i=1}^{n}(1/p_i)$, where L_0 denotes the distance of each segment in the entanglement generation.

4'3.3. Entanglement-based communication schemes. After an EME state has been established between two distant sites, we would like to use it in the communication protocols, such as quantum teleportation, cryptography, and Bell-inequality detection. It is not obvious that the EME state (74), which is entangled in the Fock basis, is useful for these tasks since in the Fock basis it is experimentally hard to do certain single-bit operations. In the following we will show how the EME states can be used to realize all these protocols with simple experimental configurations.

Quantum cryptography and the Bell-inequality detection are achieved with the setup shown by fig. 20a. The state of the two pairs of ensembles is expressed as $\rho_{L_1 R_1} \otimes \rho_{L_2 R_2}$, where $\rho_{L_i R_i}$ $(i = 1, 2)$ denote the same EME state with the vacuum coefficient c_n if we have done n times entanglement connection. The φ-parameters in $\rho_{L_1 R_1}$ and $\rho_{L_2 R_2}$ are the same provided that the two states are established over the same stationary channels. We register only the coincidences of the two-side detectors, so the protocol is successful only if there is a click on each side. Under this condition, the vacuum components in the EME states, together with the state components $S_{L_1}^{\dagger} S_{L_2}^{\dagger}|\text{vac}\rangle$ and $S_{R_1}^{\dagger} S_{R_2}^{\dagger}|\text{vac}\rangle$, where $|\text{vac}\rangle$ denotes the ensemble state $|0_a 0_a 0_a 0_a\rangle_{L_1 R_1 L_2 R_2}$, give no contributions to the experimental results. So, for the measurement scheme shown by fig. 17, the ensemble state $\rho_{L_1 R_1} \otimes \rho_{L_2 R_2}$ is effectively equivalent to the following "polarization" maximally entangled (PME) state (the term "polarization" comes from an analogy to the optical case):

$$(75) \qquad |\Psi\rangle_{\text{PME}} = \frac{S_{L_1}^{\dagger} S_{R_2}^{\dagger} + S_{L_2}^{\dagger} S_{R_1}^{\dagger}}{\sqrt{2}|\text{vac}\rangle}.$$

The success probability for the projection from $\rho_{L_1 R_1} \otimes \rho_{L_2 R_2}$ to $|\Psi\rangle_{\text{PME}}$ (*i.e.*, the probability to get a click on each side) is given by $p_a = 1/[2(c_n + 1)^2]$. One can also check that in fig. 20, the phase shift ψ_{Λ} $(\Lambda = L\,\text{or}\,R)$ together with the corresponding beam splitter operation are equivalent to a single-bit rotation in the basis $\{|0\rangle_{\Lambda} \equiv S_{\Lambda_1}^{\dagger}|0_a 0_a\rangle_{\Lambda_1 \Lambda_2},\ |1\rangle_{\Lambda} \equiv S_{\Lambda_2}^{\dagger}|0_a 0_a\rangle_{\Lambda_1 \Lambda_2}\}$ with the rotation angle $\theta = \psi_{\Lambda}/2$. Now, it is clear how to do quantum cryptography and Bell-inequality detection since we have the PME state and we can perform the desired single-bit rotations in the corresponding basis. For instance, to distribute a quantum key between the two remote sides, we simply choose ψ_{Λ} randomly from the set $\{0, \pi/2\}$ with an equal probability, and keep the measurement results (to be 0 if D_1^{Λ} clicks, and 1 if D_1^{Λ} clicks) on both sides as the shared secret key if the

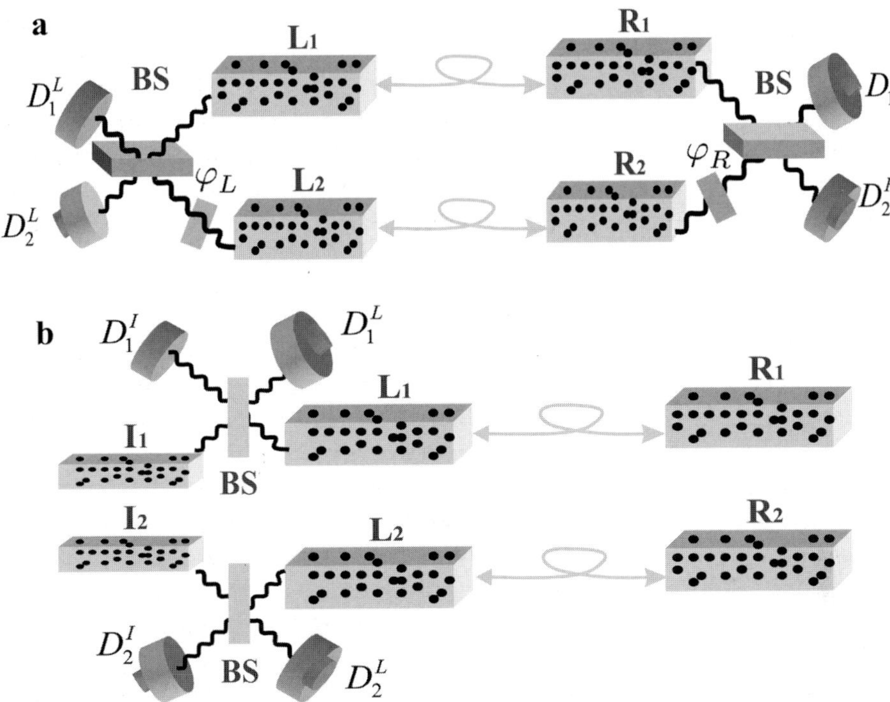

Fig. 20. – (a) Schematic setup for the realization of quantum cryptography and Bell-inequality detection. Two pairs of ensembles L_1, R_1 and L_2, R_2 (or two pairs of metastable states as shown by fig. 16b)) have been prepared in the EME states. The collective atomic excitations on each side are transferred to the optical excitations, which, respectively after a relative phase shift φ_L or φ_R and a 50%–50% beam splitter, are detected by the single-photon detectors D_1^L, D_2^L and D_1^R, D_2^R. We look at the four possible coincidences of D_1^R, D_2^R with D_1^L, D_2^L, which are functions of the phase difference $\varphi_L - \varphi_R$. Depending on the choice of φ_L and φ_R, this setup can realize both the quantum cryptography and the Bell-inequality detection. (b) Schematic setup for probabilistic quantum teleportation of the atomic "polarization" state. Similarly, two pairs of ensembles L_1, R_1 and L_2, R_2 are prepared in the EME states. We want to teleport an atomic "polarization" state $(d_0 S_{I_1}^\dagger + d_1 S_{I_2}^\dagger)|0_a 0_a\rangle_{I_1 I_2}$ with unknown coefficients d_0, d_1 from the left to the right side, where $S_{I_1}^\dagger, S_{I_2}^\dagger$ denote the collective atomic operators for the two ensembles I_1 and I_2 (or two metastable states in the same ensemble). The collective atomic excitations in the ensembles I_1, L_1 and I_2, L_2 are transferred to the optical excitations, which, after a 50%–50% beam splitter, are detected by the single-photon detectors D_1^I, D_1^L and D_2^I, D_2^L. If there are a click in D_1^I or D_1^L and a click in D_2^I or D_2^L, the protocol is successful. A π-phase rotation is then performed on the collective mode of the ensemble R_2 under the condition that the two clicks appear in the detectors D_1^I, D_2^L or D_2^I, D_1^L. The collective excitation in the ensembles R_1 and R_2, if appearing, would be found in the same "polarization" state $(d_0 S_{R_1}^\dagger + d_1 S_{R_2}^\dagger)|0_a 0_a\rangle_{R_1 R_2}$.

two sides become aware that they have chosen the same phase shift after the public declare. This is exactly the Ekert scheme [111] and its absolute security follows directly from the proofs in [118, 119]. For the Bell-inequality detection, we infer the correlations

$E(\psi_L, \psi_R) \equiv P_{D_1^L D_1^R} + P_{D_2^L D_2^R} - P_{D_1^L D_2^R} - P_{D_2^L D_1^R} = \cos(\psi_L - \psi_R)$ from the measurement of the coincidences $P_{D_1^L D_1^R}$ etc. For the setup shown in fig. 20a, we would have $|E(0, \pi/4) + E(\pi/2, \pi/4) + E(\pi/2, 3\pi/4) - E(0, 3\pi/4)| = 2\sqrt{2}$, whereas for any local hidden variable theories, the CHSH inequality [120] implies that this value should be below 2.

We can also use the established long-distance EME states for faithful transfer of unknown quantum states through quantum teleportation, with the setup shown by fig. 20b. In this setup, if two detectors click on the left side, there is a significant probability that there is no collective excitation on the right side since the product of the EME states $\rho_{L_1 R_1} \otimes \rho_{L_2 R_2}$ contains vacuum components. However, if there is a collective excitation appearing from the right side, its "polarization" state would be exactly the same as the one input from the left. So, as in the Innsbruck experiment [27], the teleportation here is probabilistic and needs posterior confirmation; but if it succeeds, the teleportation fidelity would be nearly perfect since in this case the entanglement is equivalently described by the PME state (75). The success probability for the teleportation is also given by $p_a = 1/[2(c_n + 1)^2]$, which determines the average number of repetitions for a successful teleportation.

4˙3.4. Noise and built-in entanglement purification.

We next discuss noise and imperfections in the schemes for entanglement generation, connection, and applications. In particular we show that each step contains built-in entanglement purification which makes the whole scheme resilient to the realistic noise and imperfections.

In the entanglement generation, the dominant noise is the photon loss, which includes the contributions from the channel attenuation, the spontaneous emissions in the atomic ensembles (which results in the population of the collective atomic mode s with the accompanying photon going to other directions), the coupling inefficiency of the Stokes signal into and out of the channel, and the inefficiency of the single-photon detectors. The loss probability is denoted by $1 - \eta_p$ with the overall efficiency $\eta_p = \eta_p' e^{-L_0/L_{att}}$, where we have separated the channel attenuation $e^{-L_0/L_{att}}$ (L_{att} is the channel attenuation length) from other noise contributions η_p' with η_p' independent of the communication distance L_0. The photon loss decreases the success probably for getting a detector click from p_c to $\eta_p p_c$, but it has no influence on the resulting EME state. Due to this noise, the entanglement preparation time should be replaced by $T_0 \sim t_\Delta/(\eta_p p_c)$. The second source of noise comes from the dark counts of the single-photon detectors. The dark count gives a detector click, but without population of the collective atomic mode, so it contributes to the vacuum coefficient in the EME state. If the dark count comes up with a probability p_{dc} for the time interval t_Δ, the vacuum coefficient is given by $c_0 = p_{dc}/(\eta_p p_c)$, which is typically much smaller than 1 since the Raman transition rate is much larger than the dark count rate. The final source of noise, which influences the fidelity to get the EME state, is caused by the event that more than one atom is excited to the collective mode S whereas there is only one click in D1 or D2. The conditional probability for that event is given by p_c, so we can estimate the fidelity imperfection $\Delta F_0 \equiv 1 - F_0$ for the entanglement generation by

$$\Delta F_0 \sim p_c. \tag{76}$$

Note that by decreasing the excitation probability p_c, one can make the fidelity imperfection closer and closer to zero at the price of a longer entanglement preparation time T_0. This is the basic idea of the entanglement purification. So, in this scheme, the confirmation of the click from the single-photon detector generates and purifies entanglement at the same time.

In the entanglement swapping, the dominant noise is still the losses, which include the contributions from the detector inefficiency, the inefficiency of the excitation transfer from the collective atomic mode to the optical mode [88,90], and the small decay of the atomic excitation during the storage [88,90]. Note that by introducing the detector inefficiency, we have automatically taken into account the imperfection that the detectors cannot distinguish the single and the two photons. With all these losses, the overall efficiency in the entanglement swapping is denoted by η_s. The loss in the entanglement swapping gives contributions to the vacuum coefficient in the connected EME state, since in the presence of loss a single detector click might result from two collective excitations in the ensembles I_1 and I_2, and in this case, the collective modes in the ensembles L and R have to be in a vacuum state. After taking into account the realistic noise, we can specify the success probability and the new vacuum coefficient for the i-th entanglement connection by the recursion relations $p_i \equiv f_1(c_{i-1}) = \eta_s(1 - (\eta_s/2(c_{i-1}+1)))/(c_{i-1}+1)$ and $c_i \equiv f_2(c_{i-1}) = 2c_{i-1} + 1 - \eta_s$. The coefficient c_0 for the entanglement preparation is typically much smaller than $1 - \eta_s$, then we have $c_i \approx (2^i - 1)(1 - \eta_s) = (L_i/L_0 - 1)(1 - \eta_s)$, where L_i denotes the communication distance after i times entanglement connection. With the expression for the c_i, we can easily evaluate the probability p_i and the communication time T_n for establishing a EME state over the distance $L_n = 2^n L_0$. After the entanglement connection, the fidelity of the EME state also decreases, and after n times connection, the overall fidelity imperfection $\Delta F_n \sim 2^n \Delta F_0 \sim (L_n/L_0)\Delta F_0$. We need fix ΔF_n to be small by decreasing the excitation probability p_c in eq. (76).

It is important to point out that our entanglement connection scheme also has built-in entanglement purification function. This can be understood as follows: Each time we connect entanglement, the imperfections of the setup decrease the entanglement fraction $1/(c_i + 1)$ in the EME state. However, the entanglement fraction decays only linearly with the distance (the number of segments), which is in contrast to the exponential decay of the entanglement for the connection schemes without entanglement purification. The reason for the slow decay is that at each time of the entanglement connection, we need repeat the protocol until there is a detector click, and the confirmation of a click removes part of the added vacuum noise since a larger vacuum component in the EME state results in more times of repetitions. The built-in entanglement purification in the connection scheme is essential for the polynomial scaling law of the communication efficiency.

As in the entanglement generation and connection schemes, our entanglement application schemes also have built-in entanglement purification which makes them resilient to the realistic noise. Firstly, we have seen that the vacuum components in the EME states are removed from the confirmation of the detector clicks and thus have no influence on the fidelity of all the application schemes. Secondly, if the single-photon detectors and the atom-to-light excitation transitions in the application schemes are imperfect with the overall efficiency denoted by η_a, one can easily check that these imperfections only influence the efficiency to get the detector clicks with the success probability replaced .by $p_a = \eta_a/[2(c_n + 1)^2]$, and have no effects on the communication fidelity. Finally, we have seen that the phase shifts in the stationary channels and the small asymmetry of the stationary setup are removed automatically when we project the EME state onto the PME state, and thus have no influence on the communication fidelity.

The noise not correctable by our scheme includes the detector dark count in the entanglement connection and the non-stationary channel noise and set asymmetries. The resulting fidelity imperfection from the dark count increases linearly with the number of segments L_n/L_0, and form the non-stationary channel noise and set asymmetries increases by the random-walk law $\sqrt{L_n/L_0}$. For each time of entanglement connection, the dark count probability is about 10^{-5} if we make the typical choice that the collective emission rate is about $10\,\text{MHz}$ and the dark count rate is $10^2\,\text{Hz}$. So this noise is negligible even if we have communicated over a long distance (10^3 the channel attenuation length L_{att}, for instance). The non-stationary channel noise and setup asymmetries can also be safely neglected for such a distance. For instance, it is relatively easy to control the non-stationary asymmetries in local laser operations to values below 10^{-4} with the use of accurate polarization techniques [121] for Zeeman sublevels (as in fig. 19b).

4'3.5. Scaling of the communication efficiency. We have shown that each of our entanglement generation, connection, and application schemes has built-in entanglement purification, and as a result of this property, we can fix the communication fidelity to be nearly perfect, and at the same time keep the communication time increasing only polynomially with the distance. Assume that we want to communicate over a distance $L = L_n = 2^n L_0$. By fixing the overall fidelity imperfection to be a desired small value ΔF_n, the entanglement preparation time becomes $T_0 \sim t_\Delta/(\eta_p \Delta F_0) \sim (L_n/L_0)t_\Delta/(\eta_p \Delta F_n)$. For an effective generation of the PME state (75), the total communication time $T_{\text{tot}} \sim T_n/p_a$ with $T_n \sim T_0 \prod_{i=1}^n (1/p_i)$. So the total communication time scales with the distance by the law

$$
(77) \qquad T_{\text{tot}} \sim \frac{2\left(L/L_0\right)^2}{\eta_p p_a \Delta F_T \Pi_{i=1}^n p_i},
$$

where the success probabilities p_i, p_a for the i-th entanglement connection and for the entanglement application have been specified before. The expression (77) has confirmed that the communication time T_{tot} increases with the distance L only polynomially. We show this explicitly by taking two limiting cases. In the first case, the inefficiency $1 - \eta_s$ for the entanglement swapping is assumed to be negligibly small. One can deduce from eq. (77) that in this case the communication time

$T_{\text{tot}} \sim T_{\text{con}}(L/L_0)^2 e^{L_0/L_{\text{att}}}$, with the constant $T_{\text{con}} \equiv 2t_\Delta/(\eta_p'\eta_a\Delta F_T)$ being independent of the segment and the total distances L_0 and L. The communication time T_{tot} increases with L quadratically. In the second case, we assume that the inefficiency $1 - \eta_s$ is considerably large. The communication time in this case is approximated by $T_{\text{tot}} \sim T_{\text{con}}(L/L_0)^{[\log_2(L/L_0)+1]/2+\log_2(1/\eta_s-1)+2} e^{L_0/L_{\text{att}}}$, which increases with L still polynomially (or sub-exponentially in a more accurate language, but this makes no difference in practice since the factor $\log_2(L/L_0)$ is well bounded from above for any reasonably long distance). If T_{tot} increases with L/L_0 by the m-th power law $(L/L_0)^m$, there is an optimal choice of the segment length to be $L_0 = mL_{\text{att}}$ to minimize the time T_{tot}. As a simple estimation of the improvement in the communication efficiency, we assume that the total distance L is about $100L_{\text{att}}$, for a choice of the parameter $\eta_s \approx 2/3$, the communication time $T_{\text{tot}}/T_{\text{con}} \sim 10^6$ with the optimal segment length $L_0 \sim 5.7L_{\text{att}}$. This result is a dramatic improvement compared with the direct communication case, where the communication time T_{tot} for getting a PME state increases with the distance L by the exponential law $T_{\text{tot}} \sim T_{\text{con}}e^{L/L_{\text{att}}}$. For the same distance $L \sim 100L_{\text{att}}$, one needs $T_{\text{tot}}/T_{\text{con}} \sim 10^{43}$ for direct communication, which means that for this example the present scheme is 10^{37} times more efficient.

In summary, in this subsection we explained the recent atomic ensemble scheme for implementation of quantum repeaters and long-distance quantum communication. The proposed technique allows to generate and connect the entanglement and use it in quantum teleportation, cryptography, and tests of Bell inequalities. All of the elements of the scheme are within the reach of current experimental technology, and have the important property of built-in entanglement purification which makes them resilient to the realistic noise. As a result, the overhead required to implement the scheme, such as the communication time, scales polynomially with the channel length. This is in remarkable contrast to direct communication where the exponential overhead is required. Such an efficient scaling, combined with a relative simplicity of the proposed experimental setup, opens up realistic prospectives for quantum communication over long distances.

4′4. *Other applications: Quantum light memory and single-photon source.* – In this subsection, we investigate two other significant applications of atomic ensembles in quantum information processing: laser manipulation of atomic ensembles provides a simple experimentally feasible way to realize quantum light memory and single-photon source with controllable emission time, direction, and pulse shape. The realization of quantum light memory and controllable single-photon source constitute two important steps towards implementation of a recently proposed quantum computation scheme [101]. In [101], a potentially scalable fault-tolerant quantum computation scheme is proposed based on the use of single-photon source, linear optics devices, and single-photon detectors. To realize that scheme, one is required i) to have the ability of storing qubits (that is, quantum light memory is required to store optical qubits involved in that scheme), ii) to have the desired single-photon source with controllable emission time and direction, iii) and to maintain noise and imperfections in the involved physical setups below the percent level. Except the last requirement, which is still very challenging with the current experimental technology, atomic ensembles provide an ideal system for the realization of the first two elements. Besides this potential application in quantum computation, quantum light memory and a single-photon source are also important by themselves. For instance, a single-photon source is important in the BB84 scheme for quantum key distribution to assure the absolute security and to increase the distribution efficiency [122]; and quantum light memory provides a powerful tool for eavesdropping in quantum cryptography.

4'4.1. *Quantum light memory*. It is very hard to directly store photons for a reasonably long time. However, we know that coherence of atomic internal states can be maintained for a quite long time with the current technology. The basic idea of quantum light memory is to transfer the photonic excitation to the excitation in atomic internal states so that it can be saved, and afterwards, it should be possible to restore the excitation to photons without change of its quantum state. Quantum light memory has been investigated theoretically in refs. [100, 93, 94, 97], and its experimental realization has been recently reported with either an ultracold Bose-Einstein condensate [88] or a hot atomic ensemble [90] as the storing medium. Reference [100] described a method for irreversible mapping of the light state to the atomic state, and refs. [93, 94] proposed the first quantum light memory schemes with revisable mapping between photonic and atomic states. Both schemes use the ΛI-level configuration with a weak-coupling cavity around the ensemble as described in sect. **2**. The difference is that ref. [93] is based on resonant coupling through adiabatic passages of dark states, and ref. [94] is based on off-resonant coherent Raman absorption. Reference [97] described the first scheme for storing light in a free-space atomic ensemble, which has been realized in the recent experiments [88, 90]. Here, to be consistent with other subsections in this section, we will follow the off-resonant approach in ref. [94] to review the principle for implementing quantum light memory. The readers interested in the schemes based on adiabatic passages are referred to refs. [93, 97].

We consider an atomic ensemble with the ΛI-level configuration as shown in subsect. 4'2 (see fig. 15). The input quantum optical signal is described by a continuous operator $a_{\mathrm{in}}(t)$, with $[a_{\mathrm{in}}(t), a_{\mathrm{in}}^{+}(t')] = \delta(t - t')$. We assume that the input signal has a definite pulse shape $f_{\mathrm{in}}(t)$. This is the case in most of the applications in quantum information processing. For instance, if we know that the signal comes from the output of a free-cavity mode, the signal pulse has the shape $f_{\mathrm{in}}(t) = (\sqrt{\eta}/\sqrt{1 - e^{-\kappa T}})e^{-\kappa t/2}$, $(0 \leq t \leq T)$, where κ is the cavity decay rate and T denotes the pulse duration. For the input pulse with a definite shape, we can define an effective single-mode bosonic operator $c_{\mathrm{in}} = \int_0^T f_1(t)a_{\mathrm{in}}(t)\,\mathrm{d}t$ with $[c_{\mathrm{in}}, c_{\mathrm{in}}^{\dagger}] = 1$ and a normalized shape function $f_{\mathrm{in}}(t)$. Similarly for the output optical signal $a_{\mathrm{out}}(t)$ with a definite shape $f_{\mathrm{out}}(t)$, we can also define a single-mode operator $c_{\mathrm{out}} = \int_0^T f_{\mathrm{out}}(t)a_{\mathrm{out}}(t)\,\mathrm{d}t$. The purpose of quantum light memory is to faithfully transfer the quantum state of the input optical mode c_{in} to the state of the collective atomic mode $s \equiv (1/\sqrt{N_a})\sum_{i=1}^{N_a} |1\rangle_i\langle 2|$. The state can be stored in the atomic mode s, and afterwards we need to have the ability to read out this state again to the output optical mode c_{out} with an arbitrary intentionally chosen pulse shape $f_{\mathrm{out}}(t)$.

The light-atom interaction is described by the basic Langevin equation (57), with the effective coupling rate $\kappa'(t) = 4N_a|\Omega_2(t)g_1|^2/(\Delta^2\kappa)$ adjustable by controlling the time dependence of the Rabi frequency $\Omega_2(t)$ through change of the classical laser intensity. Equation (57) is linear and has the following simple solution:

$$(78) \qquad s(T) = s(0)e^{-\int_0^T \kappa'(t)\,dt/2} -$$

$$-\int_0^T e^{-\int_t^T \kappa'(\tau)\,d\tau/2} a_{\text{in}}(t)\sqrt{\kappa'(t)}\,dt.$$

To map the state of the input optical mode c_{in} to the atomic mode s, we choose the form of $\Omega_2(t)$ so that the coupling rate $\kappa'(t)$ satisfies the differential equation

$$(79) \qquad \dot{\kappa}' = \frac{2\dot{f}_{\text{in}}}{f_{\text{in}}}\kappa' - \kappa'^2,$$

where $f_{\text{in}}(t)$ is the shape of the input pulse. This differential equation is sometimes called the impedance matching condition [93, 103]. Under this condition, it can be seen from eq. (78) that the collective atomic operator s_T at time T is given by

$$(80) \qquad s_T = s(T) = \sqrt{M}s(0t) - \sqrt{1-M}c_{\text{in}}$$

with the mapping inefficiency $M = e^{-\int_0^T \kappa'(t)\,dt}$. Since $\kappa'(t)$ is positive, the mapping inefficiency M quickly tends to zero after a sufficiently long time T. With a zero M, the photonic state is faithfully mapped to the atomic state with $s_T = -c_{\text{in}}$. For a non-zero M, there will be a small inherent loss to the state mapping caused by the vacuum noise $s(0)$.

After mapping of the photonic state to the atomic state, the classical laser is turned off and the information can be stored in the atomic internal mode s_T. Then, after some time we want to read out this state to an output optical pulse with an intentionally chosen pulse shape $f_{\text{out}}(t)$, that is, we would like to map the state of the atomic mode s_T to the output optical mode $c_{\text{out}} = \int_0^T f_{\text{out}}(t)a_{\text{out}}(t)\,dt$. To attain this goal, we turn on the classical laser to the transition $|2\rangle \to |3\rangle$ (see fig. 15), and control its intensity so that the effective coupling rate $\kappa'(t)$ satisfies the equation $\dot{\kappa}' = (2\dot{f}_{\text{out}}/f_{\text{out}})\kappa' + \kappa'^2$, which is the time reverse of the impedance matching condition (79) for write-in of the photonic state. With this condition, one can deduce from eq. (78) and the input-output relation $a_{\text{out}}(t) = -a_{\text{in}}(t) - \sqrt{\kappa'}s$ that after time T, the outgoing optical mode is given by $c_{\text{out}} = -\sqrt{M}c_T - \sqrt{1-M}s_T$, where c_T is a single-mode vacuum noise operator. The matching inefficiency M has the same form as before, and tends to zero for a sufficiently long time T. In this case, the atomic state is faithfully read out with $c_{\text{out}} = -s_T$.

The physical setup of quantum light memory discussed above could have other applications. We note that the output pulse shape $f_{\text{out}}(t)$ need not be the same as the input pulse shape $f_{\text{in}}(t)$. This means that the setup can work as a pulse shape modulator to change the shape of an optical pulse without alteration of its quantum state. The pulse

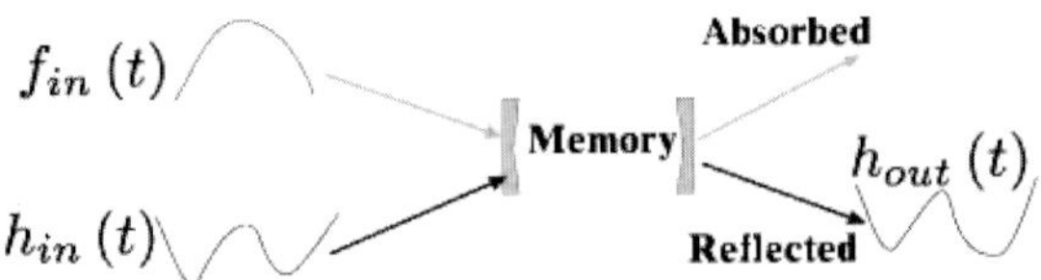

Fig. 21. – Schematic setup to illustrate the pulse shape splitter.

shape modulator could be useful in quantum communication between two cavities [32]. For instance, if one wants to input a pulse from the free decay of a cavity to another cavity with the same decay rate, the pulse will be nearly completely reflected at the mirror of the second cavity [32]. However, if between the two cavities we insert a pulse shape modulator to change the pulse shape to be its time reversal, the pulse will be nearly completely absorbed by the second cavity since the input process is exactly the time reversal of the output process. Besides the application as a pulse shape modulator, the quantum light memory setup can also be used as a pulse shape splitter illustrated by fig. 21. Consider that we have two independent pulse modes superposed in the same time window $[0, T]$, with the shapes denoted by $f_{\rm in}(t)$ and $h_{\rm in}(t)$, respectively. The pulse shape functions are orthogonal to each other with $\int_0^T f_{\rm in}^*(t) h_{\rm in}(t)\,{\rm d}t = 0$ for independent modes. Now we want to split these two modes by selectively transferring one of the optical modes $c_{\rm in} = \int_0^T f_{\rm in}(t) a_{\rm in}(t)\,{\rm d}t$ to the atomic mode s with the impedance matching condition. In this case, one can show from eq. (78) and the input-output relation that the other mode $d_{\rm in} = \int_0^T h_{\rm in}(t) a_{\rm in}(t)\,{\rm d}t$ is completely reflected. In fact, one has $d_{\rm out} \equiv \int_0^T h_{\rm out}(t) a_{\rm out}(t)\,{\rm d}t = d_{\rm in}$, where $d_{\rm out}$ is the reflected optical mode with its pulse shape changed to $h_{\rm out}(t) = h_{\rm in}(t) - (e^{R(T)} - 1)e^{-R(t)} f_{\rm in}(t) \int_0^t f_{\rm in}^*(\tau) h_{\rm in}(\tau)\,{\rm d}\tau$, where $R(t) \equiv \int_0^t \kappa'(\tau)\,{\rm d}\tau$. In this way, we obtain a pulse shape splitter to separate different shapes, just like a polarization beam splitter to separate different polarizations.

In the above we have shown how to store an effectively one-mode optical field in an atomic ensemble. It is also possible to store many optical modes in the same atomic ensemble with a step-by-step method to increase its memory capacity. For this we consider a one-dimensional atomic ensemble with length L_a. The coordinate of the j atom is denoted by x_j. We introduce a Fourier transformation to the individual atomic operator $\sigma_{12}^j = |1\rangle_j\langle 2|$ with the form $s_\mu \equiv \sum_j \sigma_{12}^j e^{i\mu x_j/L_a}/\sqrt{N_a}$ $(\mu = 0, 1, \ldots)$. Under the weak-excitation condition $\langle|2\rangle_j\langle 2|\rangle \ll 1$, the new modes s_μ are independent and satisfy bosonic commutation relations $[s_\mu, s_{\mu'}^\dagger] = \delta_{\mu\mu'}$. We can induce a transition $s_\mu \to s_{\mu+1}$ $(\mu = 0, 1, \ldots)$ between the new modes by applying an electric field with spatial gradient for a suitable time to give a coordinate-dependent phase kick $\sigma_{12}^j \to \sigma_{12}^j e^{ix_j/L_a}$. To store in one atomic ensemble many optical pulse modes which come one after another, we transfer the first optical mode to the collective atomic mode $s = s_0$ with the method described above, and then induce a transition $s_\mu \to s_{\mu+1}$ by applying a phase kick. After the phase kick, the mode s is free to be used for storing the second optical mode. This process can be continued until many optical modes are stored in the atomic modes $s, s_1, s_2, \ldots$. For releasing these atomic modes, we can simply reverse the above process.

In the real experiments, one cannot realize ideal quantum light memory with perfect state mapping between photons and atoms. As has been shown above, there is some inherent loss if the interaction time T is not much longer than the average effective coupling rate $\overline{\kappa'}$. Additional to this, atomic spontaneous emissions will also contribute to loss with a signal-to-noise ratio given by $R_{sn} \sim 4N_a|g_1|^2/(\kappa\gamma_s)$ as described in subsect. 4'2. Assume that the overall inefficiency for all the loss effects in quantum light memory is denoted by η. In this case, if one inputs an optical mode in a coherent state $|\alpha\rangle$ (which is commonly taken in experimental demonstrations), the readout state from the memory is given by $|\sqrt{1-\eta}\alpha\rangle$ with a state fidelity $F = (\langle\alpha||\sqrt{1-\eta}\alpha\rangle)^2 = e^{-x_\eta|\alpha|^2}$, where $x_\eta = (1-\sqrt{1-\eta})^2$. With this imperfection, one would like to ask how to experimentally verify that one indeed realizes a quantum light memory, which should be better than any classical memory protocol. For instance, in a classical protocol, one can measure the input optical state to get some classical information, and then according to this information prepare a similar state after some time. The problem here is very similar to that in continuous variable quantum teleportation [124, 29], and we can use the result there to provide an experimentally testable criterion for quantum light memory. This criterion is given by measuring the input-output state fidelity F. Assume that the input state to the memory is randomly chosen from the set of coherent states $\{|\alpha\rangle\}$ with a probability distribution $p(\alpha) = (\lambda/\pi)e^{-\lambda|\alpha|^2}$, where λ is a positive parameter, quantum light memory is verified with the confirmation to be better than any classical memory protocol if the measured average state fidelity $F > F_{\mathrm{cri}} = (1+\lambda)/(2+\lambda)$ [125]. With an overall loss inefficiency η for quantum light memory, the calculated average state fidelity in this case is given by $F_{\mathrm{cal}} = \lambda/(\lambda + x_\eta)$. One can see that it is always possible to make $F_{\mathrm{cal}} > F_{\mathrm{cri}}$ by choosing a large parameter λ, that is, by choosing the input coherent states close enough to the vacuum state. In a real experiment, there might be some technique noise which induces an additional fidelity decrease F_{tec} to the calculated value F_{cal}. In this case, one can verify that to meet the criterion $F = F_{\mathrm{cal}} - F_{\mathrm{tec}} > F_{\mathrm{cri}}$, one has an optimal choice of the parameter λ to be $\lambda = (1 - x_\eta)/(2F_{\mathrm{tec}}) - 1 - x_\eta/2$, and the technique fidelity decrease F_{tec} needs to be approximately below $F_{\mathrm{tec}} < (1-x_\eta)^2/4$ for a successful demonstration of quantum light memory. In the experimental demonstrations [88, 90], the above criterion has not been checked, but it seems possible to meet this criterion with the current experimental technology.

4'4.2. *Single-photon source*. As has been mentioned before, a single-photon source has many applications in quantum information processing. It is desirable to have a single-photon source with controllable emission time, direction, and pulse shape. This is required in some quantum information processing schemes since one needs to interfere two single-photon pulses, and this is generally available only when we can control the emission time, direction, and shape of the pulses. It is possible to produce single photons with the setup of optical spontaneous parametric down-conversion, which has been commonly used now in quantum communication experiments [123]. In this setup, photons are always generated in pairs due to the non-linearity in the optical crystal. If we measure one output beam with a single-photon detector, conditional on a detector click the other output beam will be projected onto a single-photon state. However, in this setup single photons will be produced at random times which are not controllable due to the randomness of spontaneous emissions. There are also proposals and experiments of using a blockade mechanism in semiconductors or other solid-state materials to produce a source of single photons [126-128], with the emission time fully controllable. To require the emitted single-photon pulses to be also directional, it seems that one needs to build a good cavity around the material. A fully controllable single-photon source is achievable if one could trap single atoms in high-Q cavities for a sufficiently long time [129]. However, as we have mentioned before, this is possible but it is an experimentally challenging work. We show here that atomic ensembles can provide another way to achieve a fully controllable single-photon source, which seems to be much easier for experimental demonstrations.

The principle of using an atomic ensemble to produce a single-photon source can be easily understood with the ideas explained in this section. The atomic ensemble working in the ΛII-level configuration generates a correlated state in the form of eq. (72) in the perturbative limit, which is an exact analogy of the optical spontaneous parametric down-conversion process. We can measure the forward-scattered signal with a single-photon detector, and conditional on a detector click, the collective atomic mode is projected to a single-excitation state. Since excitations can be stored for a reasonably long time in the ground-state manifold of the atoms, we can transfer the single-atomic excitation to the single-photonic excitation at any desired time with the method described in the previous subsection. The emission time is controllable by when the repumping pulse is applied, and the emitted single-photon pulse is directed to the forward direction. The pulse shape is controllable by changing the time-dependence of the Rabi frequency of the repumping pulse as in quantum light memory. This shows that the relatively simple system of an atomic ensemble can be used to produce single-photon pulses with fully controllable properties.

4`5. *Applications in continuous variable quantum information processing*. – In quantum information protocols, quantum information is normally carried by qubits, that is, by two-dimensional quantum systems. Similarly to the classical case, quantum information can also be carried by some observables with continuous values, the canonical observables X and P for instance. The bosonic field (such as the optical field) provides a natural physical system to carry the continuous variable quantum information. Because of this, continuous variable quantum information processing has recently aroused a lot of interest. There have been proposals for continuous variable quantum teleportation [124, 29], cryptography [131], computation [130], error correction [134], entanglement purification [132], cloning [133], etc., and continuous variable teleportation has been experimentally demonstrated by using single-mode optical fields [29]. We have seen from subsect. 4`2 that one can define a pair of canonical continuous-valued observables for atomic ensembles with suitable level schemes (the four-level scheme, for instance). This property, combined with the ability of storing (qubit or continuous variable) quantum information in the ground-state manifold of atomic ensembles and the collectively enhanced coupling of the

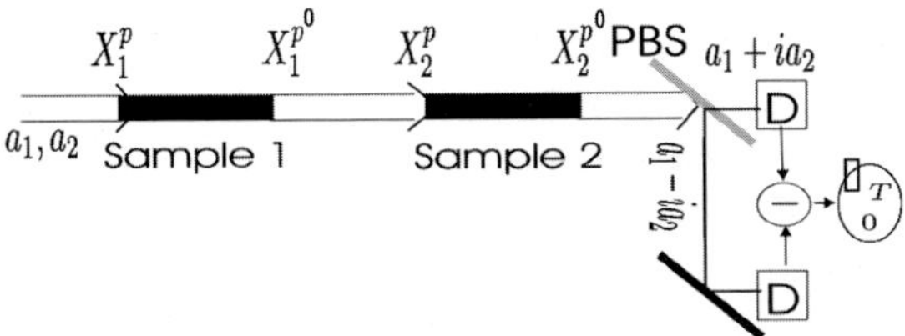

Fig. 22. – Schematic setup for Bell measurements. A linearly polarized strong laser pulse (decomposed into two circular polarization modes a_1, a_2) propagates successively through the two atomic samples. The two polarization modes $(a_1 + ia_2)/\sqrt{2}$ and $(a_1 - ia_2)/\sqrt{2}$ are then split by a polarizing beam splitter (PBS), and finally the difference of the two photon currents (integrated over the pulse duration T) is measured.

atomic mode in the ensemble to the optical mode, provides many possibilities for using atomic ensembles in continuous variable quantum information processing.

A good existing example to show these possibilities is given by the recent scheme for realizing continuous variable atomic quantum teleportation with laser manipulation of atomic ensembles [98]. Quantum teleportation of atomic states has not been realized yet due to the difficulty of achieving strong light-atom coupling. As we have seen, collective enhancement of the signal-to-noise ratio in atomic ensembles provides a possible way to go around this problem. There are two proposals to realize continuous variable teleportation with free-space atomic ensembles. Reference [110] is based on the use of an external source of entanglement (non-classical light). Reference [98] eliminates this requirement, and proposes a quantum teleportation scheme with the use of only coherent light. Based on the method in ref. [98], a recent experiment has successfully generated entanglement between two distant macroscopic atomic ensembles [91], which is an important first step towards the final realization of atomic quantum teleportation. The following of this subsection is mainly devoted to a review of the scheme proposed in ref. [98], and we will also briefly remark at the end of this subsection the possibilities of using atomic ensembles for realization of other continuous variable quantum information protocols.

The scheme in ref. [98] is based on the four-level scheme described and analyzed in details in subsect. 4'2. The transformations (71) of the canonical atomic and optical observables induced by the light-atom interaction serve as the basic equations for understanding this scheme. To teleport continuous variable states from one atomic ensemble to the other, first we need to generate entanglement between the continuous observables X_1^a, P_1^a and X_2^a, P_2^a of two distant ensembles 1 and 2. This is done through a non-local Bell measurement of the EPR operators $X_1^a - X_2^a$ and $P_1^a + P_2^a$ with the setup depicted by fig. 22. This setup measures the Stokes operator $X_2^{p'}$ of the output light. Using eq. (71) and neglecting the small loss terms, we have $X_2^{p'} = X_1^p + \kappa_c(P_1^a + P_2^a)$, so we get a collective measurement of $P_1^a + P_2^a$ with some inherent vacuum noise X_1^p. The efficiency $1 - \eta$ of this measurement is determined by the parameter κ_c with $\eta = 1/(1 + 2\kappa_c^2)$. After this round of measurements, we rotate the collective atomic spins around the x axis to get the transformations $X_1^a \to -P_1^a$, $P_1^a \to X_1^a$ and $X_2^a \to P_2^a$, $P_2^a \to -X_2^a$.

The rotation of the atomic spin can be easily obtained with negligible noise by applying classical laser pulses with detuning $\Delta \gg \gamma$. After the rotation, the measured observable of the first round of measurement is changed to $X_1^a - X_2^a$ in the new variables. We then make another round of collective measurement of the new variable $P_1^a + P_2^a$. In this way, both the EPR operators $X_1^a - X_2^a$ and $P_1^a + P_2^a$ are measured, and the final state of the two atomic ensembles is collapsed into a two-mode squeezed state with variance $\delta(X_1^a - X_2^a)^2 = \delta(P_1^a + P_2^a)^2 = e^{-2r}$, where the squeezing parameter r is given by

$$(81) \qquad\qquad r = \frac{1}{2}\ln\left(1 + 2\kappa_c^2\right).$$

Thus, using only coherent light, we generate continuous variable entanglement [135] between two non-local atomic ensembles. With the interaction parameter $\kappa_c \approx 5$, a high squeezing (and thus a large entanglement) $r \approx 2.0$ is obtainable. Note that entanglement generation is the key step for many quantum information protocols, and as an example we show in the following how to use it to achieve quantum teleportation.

To achieve quantum teleportation, first the ensembles 1 and 2 are prepared in a continuous variable entangled state using the non-local Bell measurement described above. Then, a Bell measurement with the same setup as shown by fig. 22 on the two local ensembles 1 and 3, together with a straightforward displacement of X_3^a, P_3^a on the sample 3, will teleport an unknown collective spin state from the atomic ensemble 3 to 2. The teleported state on the ensemble 2 has the same form as that in the original proposal of continuous variable teleportation using squeezing light [124], with the squeezing parameter r replaced by eq. (81) and with an inherent Bell detection inefficiency $\eta = 1/(1+2\kappa_c^2)$. The teleportation quality is best described by the fidelity, which, for a pure input state, is defined as the overlap of the teleported state and the input state. For any coherent input state of the sample 3, the teleportation fidelity is given by

$$(82) \qquad\qquad F = 1 \left/ \left(1 + \frac{1}{1 + 2\kappa_c^2} + \frac{1}{2\kappa_c^2}\right)\right. .$$

Equation (82) shows that, if there is no extra noise, a high fidelity $F \approx 96\%$ would be possible for the teleportation of the collective atomic spin state with the interaction parameter $\kappa_c \approx 5$.

Finally, we need to incorporate several sources of noise in this scheme and analyze their influence on the teleportation fidelity. The noise includes the spontaneous emission noise described by eq. (71), the detector inefficiency, and the transmission loss of the light from the first ensemble to the second one. The spontaneous emission noise can be included partly in the transmission loss and partly in the detector efficiency, so we do not analyze it separately. The effect of the detector inefficiency η_d is to replace κ_c^2 in eqs. (81) and (82) with $\kappa_c^2(1 - \eta_d)$, and the teleportation fidelity is decreased by a term η_d/κ_c^2, which is very small and can be safely ignored. The most important noise comes from the transmission loss. The transmission loss is described by $X_2^p = \sqrt{1 - \eta_t}X_1^{p'} + \sqrt{\eta_t}X_s^t$ (see fig. 22), where η_t is the loss rate and X_s^t is the standard vacuum noise. The transmission loss changes the measured observables to be $\sqrt{1-\eta_t}X_1^a - X_2^a$ and $\sqrt{1-\eta_t}P_1^a + P_2^a$. These two observables do not commute, and the two rounds of measurements influence each other. To minimize the influence on the teleportation fidelity, we choose the following configuration (for simplicity, we assume we have the same loss rate η_t from the sample 1 to 2 and from 1 to 3): In the non-local Bell measurements on the samples 1 and 2 (the entanglement generation process), we choose a suitable interaction coefficient κ_{c2} (where its optimal value will be determined below) for the second round measurement, whereas κ_{c1} for the first round of measurement is large with $\kappa_{c1}^2 \gg \kappa_{c2}^2$ (the interaction coefficient can be easily adjusted, for instance, by changing the detuning). In the local Bell measurement, we choose the same κ_{c2} for the first round of measurement and the large κ_{c1} for the second round of measurement. For a coherent input state of the ensemble 3, the teleported state on the ensemble 2 is still a Gaussian state, and the teleportation fidelity F is found to be

$$(83) \qquad F \approx 2 / \left(2 + \frac{1}{\kappa_{c2}^2} + \kappa_{c2}^2 \eta_t \right) \leq \frac{1}{1 + \sqrt{\eta_t}},$$

which is independent of the coherent input state with suitable gain for the displacements [124]. The optimal value for κ_{c2} is thus given by $\kappa_{c2} = 1/\sqrt[4]{\eta_t}$. Even with a significant transmission loss rate $\eta_t \sim 0.2$, quantum teleportation with a remarkable high fidelity $F \sim 0.7$ is still achievable, which well exceeds the fidelity criterion $1/2$ for continuous variable quantum teleportation with arbitrary coherent input states [125].

We have described the scheme in ref. [98] for achieving continuous variable quantum teleportation of atomic spin state by laser manipulation of several atomic ensembles. The proposed scheme is within the reach of the current experimental technology, and has been partially demonstrated by the first-step experiment [91]. In general, one concerns about the possibilities of using laser manipulation of atomic ensembles to realize other continuous variable quantum information schemes. The four-level configuration used here and the ΛII-level configuration used in subsect. 4`3 have the ability to produce any squeezing operation. The Λ1-level configuration can be used to realize any beam-splitter–like operation. One can also use the phase-kick technique discussed in quantum light memory subsection to manipulate many bosonic modes in one atomic ensemble, and the number of controllable modes can be further extended and is well scalable by connecting many atomic ensembles through optical pulses making use of the collective enhancement of the desired light-atom coupling in the ensembles. With these abilities, in principle one can use laser manipulation of atomic ensembles to realize any scheme which is based on the following physical requirements, that is, a series of well-controllable bosonic modes, and the ability of performing any desired squeezing or beam-splitter–like operation on these modes. The schemes belong to this class include some continuous variable quantum cryptography scheme [131], the continuous variable quantum error correction scheme in ref. [134], and the scheme in ref. [133] for quantum cloning of Gaussian continuous variable states. To realize the continuous variable quantum computation scheme in ref. [130] and the continuous variable entanglement purification scheme in ref. [132], one needs another kind of operation, *i.e.*, the Kerr operation, which is described by the single-mode Hamiltonian $H_k = \chi_k (a^\dagger]a)^2$. There have been proposals of some level configurations to realize the Kerr operation in atomic ensembles using the phenomenon of electromagnetic-field–induced transparency (EIT) [136, 137]. Unfortunately, unlike the three configurations discussed in subsect. 4·2, the signal-to-noise ratio for the Kerr operation is not collectively enhanced [138], and as a result one basically still needs to build a high-Q cavity around the ensemble and has to enter the strong-coupling regime for getting a good signal-to-noise ratio, which is experimentally very challenging. Therefore, it seems to be hard to realize continuous variable quantum computation and entanglement purification solely based on laser manipulation of atomic ensembles, but studies are going on in this direction, and possibly one can realizes it with a combination of some other techniques, using direct interactions between atoms as in refs. [102], for instance.

5. – Summary

This lecture has reviewed the recent advances of using atomic ensembles for quantum information processing. We put emphasis on the collective enhancement of the signal-to-noise ratio for the coupling between light and atomic ensembles with suitable level configurations. Due to the collectively enhanced coupling, we can do various kinds of interesting quantum information processing simply by laser manipulation of atomic ensembles in weak-coupling cavities or even in free space, which greatly simplifies their experimental demonstration. All the theoretical schemes for quantum information processing reviewed in this lecture are within the scope of the near-future experiments. We hope that these example schemes have shown the promising future of using atomic ensembles for quantum information processing, with a combination of the advantages of a long coherence time and a collectively enhanced coupling to light. The progress in this area is fast, and the important open questions include a full understanding of the interaction of light with hot and cold atomic ensembles in three-dimensional free space with various level configurations, and applications of these interaction configurations in physical realization of more quantum information protocols.

REFERENCES

[1] A very good introduction to Quantum Information and its applications can be found in: Peres A., *Quantum Theory: Concepts and Methods* (Kluwer Academic, Dordrecht) 1993; Nielsen M. and Chuang I., *Quantum Computation and Quantum Information* (Cambridge University Press) 2000; *The Physics of Quantum Information: Quantum Cryptography, Quantum Teleportation, Quantum Computation*, edited by Bouwmeester D., Ekert A. and Zeilinger A. (Springer–Verlag) 2000. On the other hand, most of the articles on the field can be found in `http://xxx.lanl.gov/archive/quant-ph`.

[2] Shor P. W., *Proceedings of the 3rd Annual Symposium on the Foundations of Computer Science* (IEEE Computer Society Press, Los Alamitos, Ca) 1994, p. 124.

[3] Ekert A. and Josza R., *Rev. Mod. Phys.*, **68** (1995) 733.

[4] BRIEGEL H. J., DUR W., CIRAC J. I. and ZOLLER P., *Phys. Rev. Lett.*, **81** (1998) 5932; DUR W., BRIEGEL H. J., CIRAC J. I. and ZOLLER P., *Phys. Rev. A*, **59** (1999) 169.

[5] BELL J. S., *Physics*, **1** (1964) 195.

[6] BENNETT C. H., BERNSTEIN H. J., POPESCU S. and SCHUMACHER B., *Phys. Rev. A*, **53** (1996) 2046.

[7] POPESCU S. and ROHRLICH D., *Phys. Rev. A*, **56** (1997) 3319.

[8] LINDEN N. and POPESCU S., *Fortsch. Phys.*, **46** (1998) 567.

[9] GREENBERGER D. M., HORNE M. and ZEILINGER A., in *Bell's Theorem, Quantum Theory and Conceptions of the Universe*, edited by M. KAFATOS (Kluwer Academic, Dordrecht) 1989, p. 69.

[10] DUR W., VIDAL G. and CIRAC J. I., *Phys. Rev. A*, **62** (2000) 62314.

[11] WERNER R. F., *Phys. Rev. A*, **40** (1989) 4277.

[12] PERES A., *Phys. Rev. Lett.*, **77** (1996) 1413.

[13] HORODECKI P., *Phys. Lett. A*, **232** (1997) 333.

[14] HORODECKI R., HORODECKI P. and HORODECKI M., *Phys. Lett. A*, **210** (1996) 377.

[15] LEWENSTEIN W. *et al.*, *J. Mod. Opt.*, **77** (2000) 2481; TERHAL B. M., quant-ph/0101032.

[16] BENNETT C. H., BRASSARD G., POPESCU S., SCHUMACHER B., SMOLIN J. A., and WOOTTERS W. K., *Phys. Rev. Lett.*, **76** (1996) 722; BENNETT C. H., DIVINCENZO D. P., SMOLIN J. A. and WOOTTERS W. K., *Phys. Rev. A*, **54** (1996) 3824.

[17] GISIN N., *Phys. Lett. A*, **210** (1996) 151.

[18] GIEDKE G., BRIEGEL H. J., CIRAC J. I. and ZOLLER P., *Phys. Rev. A*, **59** (1999) 2641.

[19] See, for example, DIVINCENZO D. P., *Fortschr. Phys.*, **48** (2000) 771.

[20] Special issue of BRAUNSTEIN S. and LO H.-K. (Editors), *Fortschr. Phys.*, **48** (2000) 769.

[21] SHOR P. W., *Phys. Rev. A*, **52** (1995) 2493.

[22] STEANE A. M., *Phys. Rev. Lett.*, **77** (1996) 793.

[23] LAFLAMME R. *et al.*, *Phys. Rev. Lett.*, **77** (1996) 198.

[24] GOTTESMAN D., *Phys. Rev. A*, **54** (1996) 1862.

[25] SHOR P. W., in *37th Symposium on Foundations of Computing* (IEEE Computer Society Press) 1996, pp. 56-65.

[26] BENNETT C. H., BRASSARD G., CREPEAU C., JOZSA R., PERES A. and WOOTTERS W. K., *Phys. Rev. Lett.*, **70** (1993) 1895.

[27] BOUWMEESTER D., PAN J.-W., MATTLE K., EIBL M., WEINFURTER H. and ZEILINGER A., *Nature*, **390** (1997) 575.

[28] BOSCHI D., BRANCA S., DE MARTINI F., HARDY L. and POPESCU S., *Phys. Rev. Lett.*, **80** (1998) 1121.

[29] FURUSAWA A., SORENSEN J. L., BRAUNSTEIN S. L., FUCHS C. A., KIMBLE H. J. and POLZIK E. S., *Science*, **282** (1998) 706.

[30] CIRAC J. I. and ZOLLER P., *Phys. Rev. Lett.*, **74** (1995) 4091.

[31] PELLIZZARI T., GARDINER S. A., CIRAC J. I. and ZOLLER P., *Phys. Rev. Lett.*, **75** (1995) 3788.

[32] CIRAC J. I., ZOLLER P., KIMBLE H. J. and MABUCHI H., *Phys. Rev. Lett.*, **78** (1997) 3221.

[33] SORENSEN A. and MOLMER K., *Phys. Rev. Lett.*, **82** (1999) 1971.

[34] POYATOS J. F., CIRAC J. I. and ZOLLER P., *Phys. Rev. Lett.*, **81** (1998) 1322.

[35] JAKSCH D., BRIEGEL H. J., CIRAC J. I., GARDINER C. W. and ZOLLER P., *Phys. Rev. Lett.*, **82** (1999) 1975.

[36] BRENNEN G., CAVES C., JESSEN P. and DEUTSCH I., *Phys. Rev. Lett.*, **82** (1999) 1060.

[37] JAKSCH D., CIRAC J. I., ZOLLER P., ROLSTON S. L., COTE R. and LUKIN M. D., *Phys. Rev. Lett.*, **85** (2000) 2208.

[38] DUAN L.-M., CIRAC J. I. and ZOLLER P., *Science*, **292** (2001) 1695.

[39] PACHOS J., ZANARDI P. and RASETTI M., *Phys. Rev. A*, **61** (2000) 10305(R).

[40] CIRAC J. I., PARKINS A. S., BLATT R. and ZOLLER P., *Adv. Atom. Mol. Opt. Phys*, **37** (1996) 237.

[41] STEANE A., *Appl. Phys. B*, **64** (1997) 623.

[42] WINELAND D. J., MONROE C., ITANO W. M., LEIBFRIED D., KING B. E. and MEEKHOF D. M., *Res. J. NIST*, **103** (1998) 259.

[43] CIRAC J. I. and ZOLLER P., *Nature*, **404** (2000) 579.

[44] CALARCO T., CIRAC J. I. and ZOLLER P., *Phys. Rev. A*, **63** (2001) 62304.

[45] LAW C. K. and EBERLY J. H., *Phys. Rev. Lett.*, **76** (1996) 1055.

[46] GARDINER S. A., CIRAC J. I. and ZOLLER P., *Phys. Rev. A*, **55** (1997) 1683.

[47] GARDINER C. W. and ZOLLER P., *Quantum Noise* (Springer) 1999.

[48] STENHOLM S., *Rev. Mod. Phys.*, **58** (1986) 699.

[49] CIRAC J. I., BLATT R., ZOLLER P. and PHILLIPS W. D., *Phys. Rev. A*, **46** (1992) 2668.

[50] MARZOLI I., CIRAC J. I., BLATT R. and ZOLLER P., *Phys. Rev. A*, **49** (1994) 2771.

[51] MONROE C., MEEKHOF D. M., KING B. E., ITANO W. M. and WINELAND D. J., *Phys. Rev. Lett.*, **75** (1995) 4714.

[52] KING B. E., WOOD C. S., MYATT C. J., TURCHETTE Q. A., LEIBFRIED D., ITANO W. M., MONROE C. and WINELAND D. J., *Phys. Rev. Lett.*, **7** (1998) 1525.

[53] ROOS C. F., LEIBFRIED D., MUNDT A., SCHMIDT-KALER F., ESCHNER J. and BLATT R., *Phys. Rev. Lett.*, **85** (2000) 5547.

[54] NAGERL H. C., LEIBFRIED D., ROHDE H., THALHAMMER G., ESCHNER J., SCHMIDT-KALER F. and BLATT R., *Phys. Rev. A*, **60** (1999) 145.

[55] TURCHETTE Q. A., WOOD C. S., KING B. E., MYATT C. J., LEIBFRIED D., ITANO W. M., MONROE C. and WINELAND D. J., *Phys. Rev. Lett.*, **81** (1998) 3631.

[56] SACKETT C. A., KIELPINSKI D., KING B. E., LANGER C., MEYER V., MYATT C. J., ROWE M., TURCHETTE Q. A., ITANO W. M. and WINELAND D. J., *Nature*, **404** (2000) 256.

[57] KIELPINSKI D., MEYER V., ROWE M. A., SACKETT C. A., ITANO W. M., MONROE C. and WINELAND D. J., *Science*, **291** (2001) 1013.

[58] ROWE M. A., KIELPINSKI D., MEYER V., SACKETT C. A., ITANO W. M., MONROE C. and WINELAND D. J., *Nature*, **409** (2001) 791.

[59] JONATHAN D., PLENIO M. B. and KNIGHT P. L., *Phys. Rev. A*, **62** (2000) 42307.

[60] ZANARDI P. and RASETTI M., *Phys. Lett. A*, **264** (1999) 94.

[61] PACHOS J. and ZANARDI P., LANL preprint available at `http://xxx.lanl.gov/abs/quant-ph/0007110`.

[62] CIRAC J. I., PELLIZZARI T. and ZOLLER P., *Science*, **273** (1996) 1207.

[63] PELLIZZARI T., *Phys. Rev. Lett.*, **79** (1997) 5242.

[64] TURCHETTE Q. A., HOOD C. J., LANGE W., MABUCHI H. and KIMBLE H. J., *Phys. Rev. Lett.*, **75** (1995) 4710.

[65] MAITRE X., HAGLEY E., NOGUES G., WUNDERLICH C., GOY P., BRUNE M., RAIMOND J. M. and HAROCHE S., *Phys. Rev. Lett.*, **79** (1997) 769.

[66] RAUSCHENBEUTEL A., NOGUES G., OSNAGHI S., BERTET P., BRUNE M., RAIMOND J. M. and HAROCHE S., *Phys. Rev. Lett.*, **83** (1999) 5166.

[67] RAUSCHENBEUTEL A., NOGUES G., OSNAGHI S., BERTET P., BRUNE M., RAIMOND J. M. and HAROCHE S., *Science*, **288** (2000) 2024.

[68] OSNAGHI S., BERTET P., AUFFEVES A., MAIOLI P., BRUNE M., RAIMOND J. M. and HAROCHE S., *Phys. Rev. Lett.*, **87** (2001) 37902.

[69] BRATTKE S., VARCOE B. T. H. and WALTHER H., *Phys. Rev. Lett.*, **86** (2001) 3534.

[70] ZHENG S.-B. and GUO G.-C., *Phys. Rev. Lett.*, **85** (2000) 2392.

[71] GARDINER C. W., *Phys. Rev. Lett.*, **70** (1993) 2269.

[72] CARMICHAEL H. J., *Phys. Rev. Lett.*, **70** (1993) 2273.

[73] BOSE S., KNIGHT P. L., PLENIO M. B. and VEDRAL V., *Phys. Rev. Lett.*, **83** (1999) 5158.

[74] VAN ENK S. J., CIRAC J. I. and ZOLLER P., *Phys. Rev. Lett.*, **78** (1997) 4293.

[75] VAN ENK S. J., CIRAC J. I. and ZOLLER P., *Science*, **279** (1998) 205.

[76] CALARCO T., HINDS E. A., JAKSCH D., SCHMIEDMAYER J., CIRAC J. I. and ZOLLER P., *Phys. Rev. A*, **61** (2000) 22304.

[77] BRENNEN G. K., DEUTSCH I. H. and JESSEN P. S., *Phys. Rev. A*, **61** (2000) 62309.

[78] BRENNEN G. K., DEUTSCH I. H., WILLIAMS C. J., *Phys. Rev. A*, **65** (2002) 022313.

[79] WEINER J., BAGNATO V. S., ZILIO S. and JULIENNE P. S., *Rev. Mod. Phys.*, **71** (1999) 1.

[80] TIESINGA E., WILLIAMS C. J., MIES F. H. and JULIENNE P. S., *Phys. Rev. A*, **61** (2000) 63416.

[81] SCHLOSSER N., REYMOND G., PROTSENKO I. and GRANGIER P., *Nature*, **411** (2001) 1024.

[82] BUCHKREMER F. B. J., DUMKE R., VOLK M., MUETHER T., BIRKL G. and ERTMER W., quant-ph/0110119.

[83] JAKSCH D., BRUDER C., CIRAC J. I., GARDINER C. W. and ZOLLER P., *Phys. Rev. Lett.*, **81** (1998) 3108.

[84] GALLAGHER T. F., *Rydberg Atoms* (Cambridge University Press, New York) 1994.

[85] HALD J., SORENSEN J. L., SCHORI C. and POLZIK E. S., *Phys. Rev. Lett.*, **83** (1999) 1319.

[86] ROCH J.-F., VIGNERON K., GRELU PH., SINATRA A., POIZAT J.-PH. and GRANGIER PH., *Phys. Rev. Lett.*, **78** (1997) 634.

[87] HAU L. V., HARRIS S. E., DUTTON Z. and BEHROOZI C. H., *Nature*, **397** (1999) 594.

[88] LIU C., DUTTON Z., BEHROOZI C. H. and HAU L. V., *Nature*, **409** (2001) 490.

[89] KASH M. M., SAUTENKOV V. A., ZIBROV A. S., HOLLBERG L., WELCH G. R., LUKIN M. D., ROSTOVTSEV Y., FRY E. S. and SCULLY M. O., *Phys. Rev. Lett.*, **82** (1999) 5229.

[90] PHILLIPS D. F., FLEISCHHAUER A., MAIR A., WALSWORTH R. L. and LUKIN M. D., *Phys. Rev. Lett.*, **86** (2001) 783.

[91] JULSGAARD B., KOZHEKIN A. and POLZIK E. S., *Nature*, **413** (2001) 400.

[92] KUZMICH A., BIGELOW N. P. and MANDEL L., *Europhys. Lett. A*, **42** (1998) 481.

[93] LUKIN M. D., YELIN S. F. and FLEISCHHAUER M., *Phys. Rev. Lett.*, **84** (2000) 4232.

[94] DUAN L.-M., CIRAC J. I. and ZOLLER P., talk at the IQEC-CLEO/Europe (Nice, France) 2000.

[95] DUAN L.-M., LUKIN M. D., CIRAC J. I. and ZOLLER P., *Nature*, **414** (2001) 413.

[96] RAYMER M. A. and MOSTOWSKI J., *Phys. Rev. A*, **24** (1981) 1980.

[97] FLEISCHHAUER M. and LUKIN M. D., *Phys. Rev. Lett.*, **84** (2000) 5094.

[98] DUAN L.-M., CIRAC J. I., ZOLLER P. and POLZIK E. S., *Phys. Rev. Lett.*, **85** (2000) 5643.

[99] RAYMER M. G., WALMSLEY I. A., MOSTOWSKI J. and SOBOLEWSKA B., *Phys. Rev. A*, **32** (1985) 332.

[100] KOZHEKIN A. E., MOLMER K. and POLZIK E. S., quant-ph/9912014.

[101] KNILL E., LAFLAMME R. and MILBURN G. J., *Nature*, **409** (2001) 46.

[102] LUKIN M. D., FLEISCHHUAER M., COTE R., DUAN L.-M., JAKSCH D., CIRAC J. I. and ZOLLER P., *Phys. Rev. Lett.*, **87** (2001) 037901.

[103] FLEISCHHAUER M. and LUKIN M. D., quant-ph/0106066.

[104] YE J., VERNOOY D. W. and KIMBLE H. J., *Phys. Rev. Lett.*, **83** (1999) 4987.

[105] HAPPER W., *Rev. Mod. Phys.*, **44** (1972) 169.

[106] Happer W. and Athur B. S., *Phys. Rev. Lett.*, **18** (1967) 577.

[107] Kuzmich A., Mandel L. and Bigelow N. P., *Phys. Rev. Lett.*, **85** (2000) 1594.

[108] Takahashi Y., Honda K., Tanaka N., Toyoda K., Ishikawa K. and Yabuzaki T., *Phys. Rev. A*, **60** (1999) 4974.

[109] Molmer K., *Eur. Phys. J. D*, **5** (1999) 301.

[110] Kuzmich A. and Polzik E. S., *Phys. Rev. Lett.*, **85** (2000) 5639.

[111] Ekert A., *Phys. Rev. Lett.*, **67** (1991) 661.

[112] Knill E., Laflamme R. and Zurek W. H., *Science*, **279** (1998) 342.

[113] Preskill J., *Proc. R. Soc. London, Ser. A*, **454** (1998) 385.

[114] Zukowski M., Zeilinger A., Horne M. A. and Ekert A., *Phys. Rev. Lett.*, **71** (1993) 4287.

[115] Hood C. J., Lynn T. W., Doherty A. C., Parkins A. S. and Kimble H. J., *Science*, **287** (2000) 1447.

[116] Pinkse P. W. H., Fischer T., Maunz T. P. and Rempe G., *Nature*, **404** (2000) 365.

[117] Cabrillo C., Cirac J. I., G-Fernandez P. and Zoller P., *Phys. Rev. A*, **59** (1999) 1025.

[118] Lo H. K. and Chau H. F., *Science*, **283** (1999) 2050.

[119] Shor P. W. and Preskill J., *Phys. Rev. Lett.*, **85** (2000) 441.

[120] Clauser J. F., Horne M. A., Shimony A. and Holt R. A., *Phys. Rev. Lett.*, **23** (1969) 880.

[121] Budker D., Yashuk V. and Zolotorev M., *Phys. Rev. Lett.*, **81** (1998) 5788.

[122] Brassard G., Lütkenhaus N., Mor T. and Sanders B. C., *Phys. Rev. Lett.*, **85** (2000) 1330.

[123] Zeilinger A., *Rev. Mod. Phys.*, **71** (1999) S288.

[124] Braunstein S. L. and Kimble H. J., *Phys. Rev. Lett.*, **80** (1998) 869.

[125] Braunstein S. L., Fuchs C. A. and Kimble H. J., quant-ph/9910030.

[126] Kim J., Benson O., Kan H. and Yamamoto Y., *Nature*, **397** (1999) 500.

[127] Michler P., Kiraz A., Becher C., Schoenfeld W. V., Petroff P. M., Zhang L., Hu E. and Imamoglu A., *Science*, **290** (2000) 2282.

[128] Beveratos A., Brouri R., Gacoin T., Poizat J.-P. and Grangier P., quant-ph/0104028.

[129] Gheri K. M., Saavedra C., Torma P., Cirac J. I. and Zoller P., *Phys. Rev. A*, **58** (1998) R2627.

[130] Braunstein S. L. and Lloyd S., *Phys. Rev. Lett.*, **82** (1999) 1789.

[131] Gottesman D. and Preskill J., *Phys. Rev. A*, **63** (2001) 022309 and references therein.

[132] Duan L.-M., Giedke G., Cirac J. I. and Zoller P., *Phys. Rev. Lett.*, **84** (2000) 4002; *Phys. Rev. A*, **62** (2000) 032304.

[133] Cerf N. J. and Rottenberg A. X., *Phys. Rev. Lett.*, **85** (2000) 1754.

[134] Braunstein S. L., *Nature*, **394** (1998) 47.

[135] Duan L.-M., Giedke G., Cirac J. I. and Zoller P., *Phys. Rev. Lett.*, **84** (2000) 2722.

[136] Imamoglu A., Schmidt H., Woods G. and Deutsch M., *Phys. Rev. Lett.*, **79** (1997) 1467.

[137] Lukin M. D. and Imamoglu A., *Phys. Rev. Lett.*, **84** (2000) 1419.

[138] Gheri K. M., Alge W. and Grangier P., *Phys. Rev. A*, **60** (1999) R2673.

Basic concepts in quantum error correction

C. Macchiavello

Dipartimento di Fisica "A. Volta" & INFM, Unità di Pavia
Via Bassi 6, 27100 Pavia, Italy

1. – Introduction

Error correction arises from the need of protecting information when this is stored, transmitted or processed in the presence of noise. In the classical scenario information consists of classical messages, in the quantum case it is represented by quantum states. In all cases the key idea of error correction is based on redundancy, namely the information is encoded in a system which is larger than the one needed in the absence of noise.

For example, consider the simplest case in a communication scenario, where Alice wants to transmit a message consisting of a classical bit of information to Bob through a noisy channel where the bit is flipped with probability p and with probability $1 - p$ is left unchanged (this kind of channel is called binary symmetric channel and p is the probability of error). The easiest way to make their communication more robust is to encode the bit of information to be transmitted (which we will call the logical bit) into three bits with the same value, so that the logical zero 0_L is encoded into 000 and the logical one 1_L into 111 (this simple method is called repetition code). The three bits are then sent through the channel and at the receiving end Bob performs a decoding procedure to establish whether 0_L or 1_L was sent. This procedure is based on a majority voting, namely the received state of the three bits is decoded as 0_L if the majority of the bits (namely at least two of them) is in state 0, and 1_L otherwise.

In this way when no errors occur (none of the bits is flipped) or one error occurs (one of the bits is flipped) the message is decoded correctly, while in the case of two or more errors the decoded state is wrong. Since the probability that no bits are flipped is $(1-p)^3$ and the probability that only one of the three bits is flipped is $3p(1-p)^2$, the overall probability of error-free communication is therefore given by $P = (1-p)^3 + 3p(1-p)^2$. If we compare this value with the probability $1-p$ of error-free communication, in the absence of error correction, we can see that for all values of $p < 1/2$ the use of this simple repetition code leads to an enhancement of the probability of error-free communication over the channel.

As mentioned above, in the quantum case we want to protect quantum states against the effects of quantum noise. If we want to protect, for example, the state of a two-dimensional quantum system (qubit) $|\psi\rangle = \alpha|0\rangle + \beta|1\rangle$, we can see that it is not possible to apply the classical method of error correction as it is because i) in general the noise affecting quantum states is more complex than in the classical case, where the bit can be only flipped, and ii) at the decoding step it is not possible to simply "read" the received state by measuring it because a quantum measurement in general perturbs the state, and therefore more sophisticated methods must be adopted.

In the quantum case several approaches have been studied: symmetrisation procedures [1, 2], quantum error-correcting codes based on the classical theory of error correction [3, 4], decoherence-free subspaces [5]. In these notes we will briefly review the first method, mainly because of potential applications it may have, and then concentrate on the second approach, starting from a short introduction to the classical theory of error-correcting codes. We will not review the last approach.

2. – Reducing errors via symmetrisation

The first proposed procedure to control quantum noise in quantum computation is based on a symmetrisation method [1]. The basic idea is the following. Suppose we want to protect an initial quantum state $|\psi_i\rangle$, which we want then to simply store or to let evolve according to some unitary transformation $U(t)$ ($U(t)$ can be for example the free evolution operator of the system or a quantum computation we want to perform). The symmetrisation procedure consists in starting with a number N of copies of the initial state $|\psi_i\rangle$ and performing some repeated projections on the symmetric subspace of the N systems (the subspace containing all states which are invariant under any permutation of the subsystems).

Since in the absence of noise the global state of the N systems $(U(t)|\psi_i\rangle)^{\otimes N}$ would always lie in the symmetric subspace, the component which is projected outside the subspace is surely affected by some error. Therefore, upon successful projection part of the error will have been removed. Note however that the projected state is generally not error-free since the symmetric subspace contains states which are not of the simple product form $|\psi\rangle|\psi\rangle \ldots |\psi\rangle$. Nevertheless, it has been shown that the error probability will be suppressed by a factor of $1/N$ [2].

We illustrate here this effect in the simplest case of two qubits. The projection into the symmetric subspace is performed in this case by introducing the symmetrisation operator

$$(1) \qquad\qquad S = \frac{1}{2}(I + P_{21}),$$

where I represents the identity and P_{21} the permutation operator which exchanges the states of the two qubits. The induced map on mixed states of two qubits (including renormalisation) is given by

$$(2) \qquad \rho_1 \otimes \rho_2 \longrightarrow \frac{S(\rho_1 \otimes \rho_2)S^\dagger}{\mathrm{Tr}\left[S(\rho_1 \otimes \rho_2)S^\dagger\right]}.$$

The state of either qubit separately is then obtained by partial trace over the other qubit.

Let us assume that the two copies are initially prepared in pure state $\rho_0 = |\psi\rangle\langle\psi|$ and that they interact with independent environments. After some short period of time δt the state of the two copies $\rho^{(2)}$ will have undergone an evolution

$$(3) \qquad \rho^{(2)}(0) = \rho_0 \otimes \rho_0 \longrightarrow \rho^{(2)}(\delta t) = \rho_1 \otimes \rho_2,$$

where $\rho_i = \rho_0 + \varrho_i$ for some Hermitian traceless ϱ_i. We will retain only terms of first order in the perturbations ϱ_i so that the overall state at time δt is

$$(4) \qquad \rho^{(2)} = \rho_0 \otimes \rho_0 + \varrho_1 \otimes \rho_0 + \rho_0 \otimes \varrho_2 + O(\varrho_1\varrho_2).$$

We can calculate the average purity of the two copies before symmetrisation by calculating the average trace of the squared states:

$$(5) \qquad \frac{1}{2}\sum_{i=1}^{2}\mathrm{Tr}\left((\rho_0 + \varrho_i)^2\right) = 1 + 2\,\mathrm{Tr}(\rho_0\tilde{\varrho}),$$

where $\tilde{\varrho} = (1/2)(\varrho_1 + \varrho_2)$. Note that $\mathrm{Tr}(\rho_0\tilde{\varrho})$ is negative, so that the expression above does not exceed 1. After symmetrisation each qubit is in state

$$(6) \qquad \rho_{\mathrm{s}} = [1 - \mathrm{Tr}(\rho_0\tilde{\varrho})]\rho_0 + \frac{1}{2}\tilde{\varrho} + \frac{1}{2}(\rho_0\tilde{\varrho} + \tilde{\varrho}\rho_0)$$

and has purity

$$(7) \qquad \mathrm{Tr}(\rho_{\mathrm{s}}^2) = 1 + \mathrm{Tr}(\rho_0\tilde{\varrho}).$$

Since $\mathrm{Tr}\,\rho_{\mathrm{s}}^2$ is closer to 1 than (5), the resulting symmetrised system ρ_{s} is left in a purer state.

Let us now see how the fidelity changes by applying the symmetrisation procedure. The average fidelity before symmetrisation is

$$(8) \qquad F_{\mathrm{bs}} = \frac{1}{2}\sum_i \langle\psi|\rho_0 + \varrho_i|\psi\rangle = 1 + \langle\psi|\tilde{\varrho}|\psi\rangle,$$

while after successful symmetrisation it takes the form

$$(9) \qquad F_{\mathrm{as}} = \langle \psi | \rho_{\mathrm{s}} | \psi \rangle = 1 + \frac{1}{2} \langle \psi | \tilde{\varrho} | \psi \rangle.$$

The state after symmetrisation is therefore closer to the initial state ρ_0.

For the generic case of N copies the purity of each qubit after symmetrisation is given by [2]

$$(10) \qquad \mathrm{Tr}\left(\rho_{\mathrm{s}}^2 \right) = 1 + 2\frac{1}{N}\,\mathrm{Tr}(\rho_0 \tilde{\varrho}),$$

where now $\tilde{\varrho} = (1/N) \sum_{i=1}^{N} \varrho_i$, and the fidelity F_N takes the form

$$(11) \qquad F_N = 1 + \frac{1}{N} \langle \psi | \tilde{\varrho} | \psi \rangle.$$

Formulae (10) and (11) must be compared with the corresponding ones before symmetrisation, *i.e.* (5) and (8). As we can see, ρ_{s} approaches the unperturbed state ρ_0 as N tends to infinity. Thus by choosing N sufficiently large and the rate of symmetric projection sufficiently high, the residual error at the end of a computation can, in principle, be controlled to lie within any desired small tolerance.

The symmetrisation method described above is less efficient than the use of quantum error-correcting codes, which were discovered later and will be described in the next sections. Nevertheless, symmetrisation may have potential new applications which go beyond the context of quantum computation and quantum communication. For example, it was proved that the use of repeated symmetrisations is useful to stabilise the evolution of ion-based frequency standards against the effects of decoherence and as a result it leads to an enhancement of their precision [6].

3. – Basics of classical error correction

Let us review some basics of the classical theory of error correction. Suppose Alice wants to send a message m to Bob. $m = (m_1, m_2, \ldots, m_k)$ is a string of binary digits of length k. As mentioned above, in order to protect it from the noise affecting the communication channel, the message is first encoded into a codeword c_m, which is a string of binary digits of length $n > k$. The codeword is then transmitted through the noisy channel, where some errors may occur, and at the receiving end Bob has to identify the most likely codeword sent by Alice.

In order to have an efficient error correction scheme, the encoding procedure performed by Alice and the decoding procedure applied by Bob must be chosen in a suitable way depending on the form of the noise in the communication channel. In particular, correctable errors must be the most likely to be generated. For simplicity, let us consider a binary symmetric channel, as defined in the introduction. In this case the codeword c_m can be transformed into one of the 2^n possible strings of n binary digits. We define the distance between two binary sequences $d(u,v)$, the so-called Hamming distance, as the number of digits in which the two strings differ (for example, for $u = (0101101)$ and $v = (0011100)$, $d(u,v) = 3$). At the receiving end Bob then decodes the received string as the closest codeword, since the most likely codeword sent by Alice is the one with minimum distance from the string received by Bob.

Clearly the larger the distance between codewords the more they are distinguishable in the presence of errors and therefore the more the code is robust against noise. In particular, if $d(c_i, c_j) \geq 2t + 1$ for $i \neq j$, then up to t errors can be corrected [7].

We will now review two bounds, the Hamming bound and the Gilbert-Varshamov bound, that relate the number of codewords in a code to the number of errors which can be corrected and the length of the codewords [7]. The Hamming bound provides an upper bound to the number of codewords in a code able to correct up to t errors as a function of the length of the codewords. The idea is to consider each codeword c_i as the center of a sphere of radius t containing all binary sequences v with Hamming distance $d(c_i, v) \leq t$, $i.e.$ differing from c_i in up to t locations. If the code can correct up to t errors these spheres must be disjoint. Obviously the number of sequences in each sphere times the number of spheres (which is 2^k if we think of encoding messages of length k) must be smaller than the total number of sequences of length n. Since each sphere contains a codeword c plus all the sequences differing from it in $1, 2, \ldots, t$ positions we must have

$$(12) \qquad 2^k \left\{ 1 + n + \binom{n}{2} \cdots \binom{n}{t} \right\} = 2^k \sum_{i=0}^{t} \binom{n}{i} \leq 2^n.$$

In the limit of large n the above bound takes the asymptotic form

$$(13) \qquad \frac{k}{n} \leq 1 - H\left(\frac{t}{n}\right),$$

where H is the entropy function $H(x) = -x \log_2 x - (1 - x) \log_2(1 - x)$.

With a similar reasoning an upper bound on the length of the encoded states for efficient codes can be proved:

$$(14) \qquad 2^k \sum_{i=0}^{2t} \binom{n}{i} \geq 2^n.$$

The above bound is called Gilbert-Varshamov bound. It states that the number of codewords times the number of sequences of length n that can be reached by applying up to $2t$ errors from each codeword must not be smaller than the total number of sequences of length n, because otherwise there would be enough space to accommodate an additional codeword and all the sequences obtained by applying to the codeword up to t errors. The asymptotic form of the Gilbert-Varshamov bound is given by

$$(15) \qquad \frac{k}{n} \geq 1 - H\left(\frac{2t}{n}\right).$$

We will now describe more explicitly how codes work and will consider linear codes. Let us denote by $\mathcal{C}$ the set of codewords. In a linear code

$$(16) \qquad u + v \in \mathcal{C} \quad \forall u, v \in \mathcal{C},$$

where $u + v = (u_1 + v_1, u_2 + v_2, \ldots, u_n + v_n)$ and the addition in each component is performed modulo 2. $\mathcal{C}$ is a linear subspace of the Hamming space containing all the sequences of length n. In these codes the codewords c_i are chosen in such a way as to satisfy a set of linear equations, specified by the parity check matrix M as follows:

$$(17) \qquad M \cdot c_i = 0.$$

The parity check matrix is an $(n - k) \times n$ matrix. For example, in a code with parity check matrix

$$(18) \qquad M = \begin{pmatrix} 1 & 0 & 1 & 0 \\ 0 & 1 & 1 & 1 \end{pmatrix}$$

the codevectors can be chosen as

$$(19) \qquad c_1 = (0000), \qquad c_2 = (0101), \qquad c_3 = (1110), \qquad c_4 = (1011).$$

In the above example $k = 2$ and $n = 4$. Linear codes are usually specified as $[n, k, d]$, where n is the number of bits in the codeword, k is the number of bits of the message to be transmitted and d is the Hamming distance of the code, namely the minimum Hamming distance between codewords. During the transmission the encoded message, which we call c_0, may be affected by errors so that Bob actually receives a string of n bits, which we call z. We can write $z = c_0 + e$, where the error vector e is a string of n binary digits with 1's in the positions where errors occurred. At the receiving end Bob's task is to identify the most likely codeword sent by Alice, given the received string z. To do this Bob applies the parity check matrix to the received string:

$$(20) \qquad Mz = M(c_0 + e) = Mc_0 + Me.$$

Due to the definition of the codewords (17), we then have

$$(21) \qquad Mz = Me = s,$$

where s is an $n - k$ string called error syndrome. The above equation shows an important feature of linear codes: the error syndrome depends only on the error vector, not on the transmitted codeword. Before starting the transmission, a one-to-one correspondence between error syndromes and correctable errors is established. Therefore, Bob, after detecting the received state z and learning the error syndrome by applying the parity check test, can detect the most likely error that occurred and correct it.

As an example, let us consider the [7,4,3] code specified by the parity check matrix

$$(22) \qquad M = \begin{pmatrix} 0 & 0 & 0 & 1 & 1 & 1 & 1 \\ 0 & 1 & 1 & 0 & 0 & 1 & 1 \\ 1 & 0 & 1 & 0 & 1 & 0 & 1 \end{pmatrix}.$$

This code is called Hamming code and it is perfect, *i.e.* it saturates the Hamming bound (12): $2^4(1+7) = 2^8$. Since $d = 3$, it is a one-error-correcting code.

Assume that the codeword $c_0 = (0110011)$ is transmitted and an error occurs on the second bit, so that the string $v = (0010011)$ is received. The receiver applies the parity check matrix

$$(23) \qquad Mv = \begin{pmatrix} 0 \\ 1 \\ 0 \end{pmatrix}.$$

(010) is the error syndrome corresponding to $e = (0100000)$. In this code the error syndrome is the binary representation of the error position. In this way an error on the second bit is detected and the received string v is correctly decoded as $v - e = c_0$.

Let us mention at the end of this section the concept of dual of a code. The dual, denoted by $\mathcal{C}^{\perp}$, is the set of sequences u such that $u \cdot c = 0$ for all sequences $c \in \mathcal{C}$. The inner product $u \cdot v$ is defined by multiplying corresponding components of the two strings and then adding them modulo $2 : u_1 v_1 + u_2 v_2 + \cdots + u_n v_n$. We will see in the following section how the dual is needed to construct quantum error-correcting codes.

4. – Quantum error-correcting codes

As mentioned in the introduction, when we try to extend the error-correcting techniques illustrated above to the quantum scenario we encounter the problems that quantum noise is more complex than a bit flip process and the state at the receiving end cannot be simply measured because this would destroy the quantum superpositions. We will show in the following how these problems can be overcome and how the classical theory of error-correcting codes can be adapted to the quantum case if we use as "codevectors" $|c\rangle$ entangled states of n qubits. The information we want to protect is therefore spread by the entanglement over all the n qubits. Moreover, codevectors can be chosen in such a way that an error will move them into mutually orthogonal subspaces. A measurement of the syndrome will therefore reveal only which subspace the codevector has moved to. In what follows we will assume that each qubit can undergo an error with probability ϵ and that errors on different qubits are independent (the meaning of errors in the quantum scenario will be clarified in the next subsections). With these assumptions the probability of errors on two qubits is of order $O(\epsilon^2)$. For sufficiently small values of ϵ we can reasonably assume that only one error has occurred. The probability of successful computation in the absence of error correction is $(1-\epsilon)$. If we can implement a quantum error correction routine able to correct single errors, the probability of success can be increased to $(1 - O(\epsilon^2))$. In general a code able to correct up to t errors increases the probability of successful computation to $(1 - O(\epsilon^{t+1}))$ [8].

4'1. *The three-qubit code.* – In order to gain some familiarity with quantum error-correcting codes and to illustrate the ideas sketched above, we will start by analyzing the three-qubit code. Suppose that we want to protect the quantum state $\alpha|0\rangle + \beta|1\rangle$ of a qubit and the qubit undergoes the following interaction with the environment:

$$(24) \qquad |0\rangle|E\rangle \longrightarrow |0\rangle|E_0\rangle + |1\rangle|E_1\rangle,$$

$$|1\rangle|E\rangle \longrightarrow |1\rangle|E_1\rangle + |1\rangle|E_0\rangle,$$

where $|E\rangle$ is the initial state of the environment and $|E_{0,1}\rangle$ are the states after the interaction. This kind of noise is also called amplitude noise. As in the classical case, the first step is a suitable encoding procedure. The simplest way to protect the state of the qubit against this kind of noise is to encode it into the state of three qubits, initially prepared in an arbitrary state $|0\rangle$, as follows:

$$(25) \qquad |0\rangle \longrightarrow |000\rangle \qquad |1\rangle \longrightarrow |111\rangle.$$

In this way the initial state of the qubit to be protected is encoded into the state of the three qubits $\alpha|000\rangle + \beta|111\rangle$. The encoded state is then exposed to the action of noise and we assume, as mentioned above, that each qubit interacts independently with the environment, so that the most likely process is the one affecting only one qubit at a time, according to eq. (24). The decoding procedure is performed by projecting the global state of the three qubits into one of four orthogonal subspaces: S_{00}, spanned by the states $|000\rangle$ and $|111\rangle$; S_{01} spanned by $|100\rangle$ and $|011\rangle$; S_{10} spanned by $|010\rangle$ and $|101\rangle$; S_{11} spanned by $|001\rangle$ and $|110\rangle$.

In this way we learn some information on the error that may have affected the encoded state: if, as a result of the projection measurement, the state was projected into S_{00}, the final state is error-free; if the state was projected into S_{01} we say that the error occurred on the first qubit, and so on. After learning the result of the measurement, we perform the correction operation: do nothing if the result is "00", flip the first qubit (applying the σ_x operation) if the result is "01", flip the second if the result is "10" and flip the third if the result is "11". Notice that the encoding and decoding procedures can be represented by simple quantum networks. The encoding operation (25) corresponds to a sequence of two controlled-not (CNOT) gates, where the control qubit is the first in both cases and the targets are the second and the third, respectively. This operation can be written as

$$(26) \qquad |a, b, c\rangle \longrightarrow |a, a + b, a + c\rangle,$$

where a, b, c are binary numbers representing the initial states of the first, the second and the third qubit, respectively, and the addition is performed modulo 2.

The decoding operation can be performed either by a projective measurement, as shown above, or also by a unitary operation consisting of a sequence of two CNOT gates as in the encoding procedure, followed by a Toffoli gate that has the second and the third qubit as control qubits and the first one as the target. The operation of the Toffoli gate is represented in general as

$$(27) \qquad |a, b, c\rangle \longrightarrow |a, b, a \cdot b + c\rangle,$$

where the first two qubits act as control qubits and the third one as target, and $a \cdot b$ is the multiplication of the two binary numbers a and b.

Let us now consider a different kind of noise affecting the qubits, namely decoherence, which can be modelled as

$$(28) \qquad |0\rangle|E\rangle \longrightarrow |0\rangle|E_0\rangle,$$
$$|1\rangle|E\rangle \longrightarrow |1\rangle|E_1\rangle.$$

To protect the state of one qubit against decoherence, we can construct a three-qubit code similarly to the previous case. To do this, we first define a new basis $\{|\bar{0}\rangle, |\bar{1}\rangle\}$ as follows:

$$(29) \qquad |\bar{0}\rangle = H|0\rangle = \frac{1}{\sqrt{2}}(|0\rangle + |1\rangle),$$
$$|\bar{1}\rangle = H|1\rangle = \frac{1}{\sqrt{2}}(|0\rangle - |1\rangle),$$

where the operator H is called Hadamard transform. In the new basis the decoherence process (28) takes the form

$$(30) \qquad |\bar{0}\rangle|E\rangle \longrightarrow \frac{|\bar{0}\rangle(|E_0\rangle + |E_1\rangle)}{2} + \frac{|\bar{1}\rangle(|E_0\rangle - |E_1\rangle)}{2},$$
$$|\bar{1}\rangle|E\rangle \longrightarrow \frac{|\bar{0}\rangle(|E_0\rangle - |E_1\rangle)}{2} + \frac{|\bar{1}\rangle(|E_0\rangle + |E_1\rangle)}{2},$$

namely it has the same form as the amplitude noise (24). Therefore, the code to protect one qubit against decoherence (or equivalently against phase errors) can be constructed in the same way as in the case of amplitude noise, provided that before the encoding step we apply the Hadamard transform (29).

$4{}^{\cdot}2.$ *Conditions for quantum error correction.* – We can now turn our attention to codes able to correct the most general kind of qubit-environment interaction, which is of the form

$$(31) \qquad |0\rangle|E\rangle \longrightarrow |0\rangle|E_{00}\rangle + |1\rangle|E_{01}\rangle,$$
$$|1\rangle|E\rangle \longrightarrow |1\rangle|E_{10}\rangle + |1\rangle|E_{11}\rangle.$$

For a general state of a qubit $|\psi\rangle = \alpha|0\rangle + \beta|1\rangle$ this can be conveniently written as

$$(32) \qquad |\psi\rangle|E\rangle \longrightarrow |\psi\rangle|E_0\rangle + [\sigma_x|\psi\rangle]|E_x\rangle + [\sigma_z|\psi\rangle]|E_z\rangle + [\sigma_y|\psi\rangle]|E_y\rangle,$$

where $|E_0\rangle = (|E_{00}\rangle + |E_{11}\rangle)/2$, $|E_x\rangle = (|E_{01}\rangle + |E_{10}\rangle)/2$, $|E_y\rangle = i(|E_{10}\rangle - |E_{01}\rangle)/2$, $|E_z\rangle = (|E_{00}\rangle - |E_{11}\rangle)/2$, σ_x is the error operator for bit flips, σ_z the error operator for phase flips and $\sigma_y = -i\sigma_z\sigma_x$ is the operator for both errors. As we can see from eq. (32), a general qubit-environment interaction can be expressed as a superposition of the identity operator and the Pauli operators σ_x, σ_y and σ_z acting on the qubit. This means that the qubit state is evolved into a superposition of an error-free component and three erroneous components, with errors of the σ_x, σ_y and σ_z type. Therefore, in order to be able to protect the state of a qubit against any kind of noise it is enough to be able to correct σ_x, σ_y and σ_z types of errors: even if quantum noise is "continuous", we can talk about finite sets of correctable errors (this aspect of quantum error correction is often called discretisation, or digitation, of errors).

The operation of a quantum error correction scheme can be phrased in general as follows. For an initial encoded state $|\phi\rangle$ the interaction with the environment can be expressed as

$$(33) \qquad |E\rangle|\phi\rangle \longrightarrow \sum_s |E_s\rangle Q_s|\phi\rangle,$$

where Q_s are error operators acting on the encoded state and $|E_s\rangle$ are the final states of the environment. If the encoded state consists of an arbitrary number of qubits, the error operators Q_s are tensor products of single-qubit operators $I, \sigma_x, \sigma_y, \sigma_z$. Notice however that the general structure of quantum error correction that we will describe in this subsection does not necessarily refer to two-dimensional systems. After the action of noise the original state is recovered by employing an auxiliary system (called ancilla), initially prepared in an arbitrary state $|0\rangle_a$, and performing the recovery operation, which can be described by a unitary interaction A between the encoded state and the ancilla:

$$(34) \qquad A(|0\rangle_a Q_s|\phi\rangle) = |s\rangle_a Q_s|\phi\rangle,$$

where the states of the ancilla $|s\rangle_a$ are orthonormal for all the values of s or for the ones corresponding to the leading terms Q_s in the noise process (33). In this way the final state of the ancilla $|s\rangle_a$ contains the information on the error that occurred, but not on the encoded state $|\phi\rangle$, so it is the quantum version of the error syndrome.

Starting from the general state (33), the interaction with the ancilla leads to

$$(35) \qquad A(|0\rangle_a \sum_s |E_s\rangle Q_s|\phi\rangle) = \sum_s |s\rangle_a |E_s\rangle Q_s|\phi\rangle.$$

The final step of a quantum error-correcting scheme consists in learning the syndrome and correcting the encoded state. This can be done by performing a measurement of the ancilla state in the basis $\{|s\rangle\}$. Suppose that the result of the measurement is $\bar{s}$: the state of the environment and the encoded state after the measurement is then $|E_{\bar{s}}\rangle Q_{\bar{s}}|\phi\rangle$. Since the result of the measurement is known, the encoded state can then be corrected by applying the operator $Q_{\bar{s}}^{-1}$.

Notice that the recovery operation can also be performed unitarily: after the interaction A of the encoded state and the ancilla, according to eq. (35), we can apply a unitary transformation B acting on the encoded state and the ancilla as follows:

$$(36) \qquad B(|s\rangle_a|\phi\rangle) = |s\rangle_a Q_s^{-1}|\phi\rangle,$$

so that the final state is

$$(37) \qquad |\phi\rangle \sum_s |E_s\rangle|s\rangle_a.$$

We can see from the above equations some of the essential features of quantum error correction: the entanglement between the environment and the encoded state, that arises in the noise process (33), is transformed into entanglement between the environment and the ancilla, as shown in eq. (37). Moreover, the final state of the environment and the ancilla does not depend on the encoded state $|\phi\rangle$. Therefore, the method works for any encoded state, even if we do not have any knowledge of it. In addition, we can see that the procedure also works for any linear combination of the error operators Q_s.

We will now see more explicitly what are the conditions for having a successful error correction scheme. A quantum code is a set of orthonormal quantum states $\{|C_i\rangle\}$, which are also called quantum codewords. Depending on the kind of noise we have to deal with, we identify a set of error operators $S = \{Q_s\}$ which we call set of correctable errors (the correctable errors correspond to the errors that are more likely to occur in the given noise process). The conditions on the quantum codewords and on the correctable errors in order to have successful error correction are given by [9]

$$(38) \qquad \langle C_i|Q_p^\dagger Q_q|C_j\rangle = 0,$$

$$\langle C_i|Q_p^\dagger Q_q|C_i\rangle = \mu_{pq}.$$

Notice that the overlaps μ_{pq} are independent of the codewords. In the particular case where

$$(39) \qquad \langle C_i|Q_p^\dagger Q_q|C_i\rangle = 0$$

the code is called orthogonal, or also nondegenerate. This condition requires that all states which are obtained by applying the error operators to the encoded states are all orthogonal to each other, and therefore distinguishable. This ensures that by performing suitable projections of the encoded state we are able to detect the kind of error which occurred and "undo" it to recover the desired error-free state. The three-qubit code described in subsect. 4'1 belongs to this class. Condition (39), even though more restrictive than (38), is quite useful because it allows to establish bounds on the resources needed in constructing efficient nondegenerate codes, as we will see in the next subsection.

$4'3.$ *Quantum bounds.* – Let us assume that we want to protect an initial state of k qubits by encoding it in a redundant Hilbert space of n qubits. If we want to encode 2^k input basis states and correct up to t errors we must choose the dimension of the encoding Hilbert space 2^n such that all the necessary orthogonal states can be accommodated. According to eq. (39), the total number of orthogonal states that we need in order to be able to correct i errors of the three types σ_x, σ_y and σ_z in an n-qubit state is $3^i \binom{n}{i}$ (this is the number of different ways in which the errors can occur). The argument based on counting orthogonal states then leads to the following condition [10]:

$$(40) \qquad 2^k \sum_{i=0}^{t} 3^i \binom{n}{i} \leq 2^n.$$

Equation (40) is the quantum version of the Hamming bound for classical error-correcting codes (12); given k and t, it provides a lower bound on the dimension of the encoding Hilbert space for nondegenerate codes.

We can also derive the quantum version of the classical Gilbert-Varshamov bound (14), which gives an upper bound on the dimension of the encoding Hilbert space for optimal nondegenerate codes [10]

$$(41) \qquad 2^k \sum_{i=0}^{2t} 3^i \binom{n}{i} \geq 2^n.$$

This expression can be proved from the observation that in the 2^n-dimensional Hilbert space with a maximum number of encoded basis vectors (or codevectors) $|C_i\rangle$ any vector which is orthogonal to $|C_i\rangle$ (for any i) can be reached by applying up to $2t$ error operations of σ_x, σ_y, and σ_z type to any of the 2^k encoded basis vectors. Clearly all vectors which cannot be reached in the $2t$ operations can be added to the encoded basis states $|C_i\rangle$ as all the vectors into which they can be transformed by applying up to t amplitude and/or phase transformations are orthogonal to all the others. This situation cannot happen because we have assumed that the number of codevectors is maximal. Thus the number of orthogonal vectors that can be obtained by performing up to $2t$ transformations on the codevectors must be at least equal to the dimension of the encoding Hilbert space.

It follows from eq. (40) that a one-bit quantum error-correcting code to protect a single qubit ($k = 1$, $t = 1$) requires at least 5 encoding qubits and, according to eq. (41), this can be achieved with less than 10 qubits. The first construction, involving nine qubits, was given in [11]. A seven-qubit code, which we will discuss in the next subsection, was presented in [3], and then explicit constructions for an optimal five-qubit code were first given in [12, 9].

The asymptotic form of the quantum Hamming bound (40) in the limit of large n is given by

$$(42) \qquad \frac{k}{n} \leq 1 - \frac{t}{n} \log_2 3 - H\left(\frac{t}{n}\right).$$

The corresponding asymptotic form for the quantum Gilbert-Varshamov bound (41) is

$$(43) \qquad \frac{k}{n} \geq 1 - \frac{2t}{n} \log_2 3 - H\left(\frac{2t}{n}\right).$$

As we can see from eq. (42), in quantum error correction there is an upper bound on the error rate t/n which a code can tolerate. In fact, differently from the classical case (13), where any arbitrary error rate can be corrected by a suitable code, in the quantum world the ratio t/n cannot be larger than 0.18929 for nondegenerate codes. This bound corresponds to the value of t/n for which the r.h.s. of eq. (42) equals zero. For other bounds regarding quantum error-correcting codes see [13].

$\mathbf{4}\!\cdot\!4.$ *Quasi-classical codes.* – We will briefly explain in this subsection how quantum error-correcting codes can be explicitly constructed starting from the classical theory of error-correcting codes reviewed in sect. **3**. This method leads to the so-called quasi-classical codes, derived by Calderbank and Shor [4], and Steane [3]. We will introduce the method following the approach of Steane [3, 14].

Let us first consider a simplified scenario and assume that only amplitude errors, described by the noise process (24), occur. In this case the use of classical codes can be easily adapted to protect the encoded state of n qubits as follows. The encoding step can be done as in the classical case. If we consider a linear $[n, k, d]$ code, the quantum codewords are quantum systems of n qubits, where the state of the i-th qubit is specified by the i-th bit which characterises the classical codeword. The recovery operation is performed by the use of an auxiliary system, which consists of $n - k$ qubits initially prepared in the state $|0\rangle$. The $n - k$ parity checks specified by the parity check matrix M which defines the code are evaluated on the state of the ancilla by applying an operation $\mathrm{CNOT}^{(M)}$ which is a sequence of CNOT quantum gates such that

$$(44) \qquad \mathrm{CNOT}^{(M)}(|0\rangle_a |C + e\rangle) = |Me\rangle_a |C + e\rangle.$$

The $\mathrm{CNOT}^{(M)}$ operator acts on the ancilla state and on the encoded state as follows: when the matrix element $M_{i,j} = 1$, then a CNOT gate is introduced, whose target is the ancillary qubit i and whose control is qubit j of the codevector. Notice that in eq. (44) the final state of the ancilla depends on the error $|e\rangle$ but not on the codeword. The final step is then the measurement of the ancilla: for example, if only one of the qubits has been affected by the noise process (24), a measurement of the ancilla will project the codevector either in the correct state or in a state with one bit flip. The result of the measurement of the ancilla will tell the location of the qubits in error, which can then be corrected by applying the operator σ_x.

As an example consider the [7,4,3] Hamming code, described by the parity check matrix (22). In this case the first parity check is performed by a sequence of CNOT gates that all have the first qubit of the ancilla as target qubit and the last four qubits of the encoded state as control qubits. The second parity check is performed by a sequence of four CNOT gates where the second, the third and the last two qubits of the encoded system act as control qubits, while the second qubit of the ancilla is the target. In the last parity check the first, the third, the fifth and the seventh qubits of the encoded state are the control qubits of four CNOT gates, whose target is the third qubit of the ancilla.

The technique described above can be easily extended to correct also phase errors, described by the process (28). As shown before, phase flips in the basis $\{|0\rangle, |1\rangle\}$ become bit flips in the Hadamard rotated basis $\{|\bar{0}\rangle, |\bar{1}\rangle\}$. The problem reduces therefore to correct bit flips in the rotated basis. The codevectors are then first transformed by applying a Hadamard transform to each qubit in the encoded state, the recovery operation is performed in the rotated basis as shown above for amplitude noise, and in the end the qubits are transformed back to the original basis by applying again the Hadamard transform.

Noise of general form can be dealt with by combining the above techniques. Notice that, due to the linearity of unitary operators, the transformation (44) applied to quantum superpositions of codevectors takes the form

$$(45) \qquad \mathrm{CNOT}^{(M)}\left(|0\rangle_a \sum_i \alpha_i |C_i + e\rangle\right) = |Me\rangle_a \sum_i \alpha_i |C_i + e\rangle.$$

Therefore, the ancilla still contains information only on the location of the error and not on the encoded state. The quasi-classical codes [4,3] work for general noise by correcting separately amplitude and phase errors (in this way also σ_y-type of errors, which is a sequence of σ_x and σ_z, is corrected). The method is based on the dual code theorem [3]

$$(46) \qquad H \sum_{i \in \mathcal{C}} |i\rangle = \sum_{i \in \mathcal{C}^\perp} |i\rangle,$$

where $\mathcal{C}$ and $\mathcal{C}^\perp$ represent, respectively, the classical code of interest and its dual.

The quantum codewords are then constructed as

$$(47) \qquad |C_u\rangle = \sum_{i \in \mathcal{C}^\perp} |i + u\rangle,$$

where u belongs to $\mathcal{C}$ but not to its dual, and $\mathcal{C}^\perp \in \mathcal{C}$. In this way, from a $[n, k, d]$ code and its dual $[n, n-k, d^\perp]$ we can construct a quantum code $[[n, 2k - n, d]]$. The recovery operation can then be performed by introducing two sets of ancilla states with $n - k$

qubits each, which we denote with a_x and a_z, and by applying the following operation on the encoded state and the ancilla:

$$(48) \qquad (H \, \mathrm{CNOT}_{a_x}^{(M)} \, H) \mathrm{CNOT}_{a_z}^{(M)},$$

where the Hadamard transform is applied to the n qubits of the encoded state.

As an example, a [[7,1,3]] quantum code [3], which protects the quantum state of one qubit against noise of general form affecting at most one qubit in the encoded state, can be constructed starting from the $[7,4,3]$ Hamming code. The quantum codewords are given by

$$
\begin{aligned}
(49) \qquad |C_0\rangle = {}& |0000000\rangle + |1010101\rangle + |0110011\rangle + |1100110\rangle + \\
& + |0001111\rangle + |1011010\rangle + |0111100\rangle + |1101001\rangle, \\
|C_1\rangle = {}& \sigma_{x1}\sigma_{x2}\sigma_{x3}\sigma_{x4}\sigma_{x5}\sigma_{x6}\sigma_{x7}|C_0\rangle,
\end{aligned}
$$

where the normalisation was omitted. Notice that the quantum codewords are entangled states of seven qubits.

Two sets of three ancilla qubits a_x and a_z are introduced at the recovery step. As explained above, a sequence of CNOT gates acting on the first set of ancilla qubits is introduced to correct amplitude errors and then, after applying the Hadamard transform on the encoded state, a second sequence of CNOT gates with the same structure as the first is applied to the encoded state and the ancilla a_z to correct phase errors.

5. – Conclusion

The aim of these notes was to review in a pedagogical way the basic ideas that are needed to understand why quantum error correction can be performed successfully. Of course, there would be much more to say about quantum codes: we have barely scratched the surface of the recent progress in the field, neglecting relevant topics such as group-theoretical ways of constructing good quantum codes [15, 16], concatenated codes [17], quantum fault-tolerant computation [18], the threshold theorem [19] and many others. Excellent more advanced reviews about quantum error correction, which also contain a more exhaustive reference to the original literature in the field, can be found in [14,20-22].

REFERENCES

[1] BERTHIEAUME A., DEUTSCH D. and JOZSA R., in *Proceedings of Workshop on Physics and Computation — PhysComp94* (IEEE Computer Society Press, Dallas, Texas) 1994.
[2] BARENCO A., BERTHIEAUME A., DEUTSCH D., EKERT A., JOZSA R. and MACCHIAVELLO C., *SIAM J. Comput.*, **26** (1997) 1541.
[3] STEANE A., *Phys. Rev. Lett.*, **77** (1996) 793; *Proc. R. Soc. London, Ser. A*, **452** (1996) 2551.
[4] CALDERBANK A. R. and SHOR P. W., *Phys. Rev. A*, **54** (1996) 1098.

[5] PALMA G. M., SUOMINEN K.-A. and EKERT A. K., *Proc. R. Soc. London, Ser. A*, **452** (1996) 567; ZANARDI P. and RASETTI M., *Phys. Rev. Lett.*, **79** (1997) 3306; LIDAR D. A., CHUANG I. L. and WHALEY K. B., *Phys. Rev. Lett.*, **81** (1998) 2594.

[6] HUELGA S. F., KNIGHT P. L., MACCHIAVELLO C., PLENIO M. B. and VEDRAL V., *Appl. Phys. B*, **67** (1998) 723.

[7] MACWILLIAMS F. J. and SLOANE N. J. A., *The Theory of Error-Correcting Codes* (North-Holland, Amsterdam) 1977.

[8] KNILL E. and LAFLAMME R., *Phys. Rev. A*, **55** (1997) 900.

[9] BENNETT C. H., DIVINCENZO D. P., SMOLIN J. A. and WOOTTERS W. K., *Phys. Rev. A*, **54** (1996) 3824.

[10] EKERT A. and MACCHIAVELLO C., *Phys. Rev. Lett.*, **77** (1996) 2585.

[11] SHOR P. W., *Phys. Rev. A*, **52** (1995) R2493.

[12] LAFLAMME R., MIQUEL C., PAZ J. P. and ZUREK W. H., *Phys. Rev. Lett.*, **77** (1996) 198.

[13] CLEVE R., *Phys. Rev. A*, **55** (1997) 4054.

[14] STEANE A., in *Introduction to Quantum Computation and Information*, edited by H.-K. LO, S. POPESCU and T. SPILLER (World Scientific, Singapore) 1998.

[15] GOTTESMAN D., *Phys. Rev. A*, **54** (1996) 1862.

[16] CALDERBANK A. R., RAINS E. M., SHOR P. W. and SLOANE N. J. A., *Phys. Rev. Lett.*, **78** (1997) 405.

[17] KNILL E. and LAFLAMME R., e-print quant-ph/9608012 (1996).

[18] SHOR P. W., *Proceedings of the 37th Annual Symposium on the Foundations of Computer Science*, vol. **56** (IEEE Press, Los Alamitos, CA) 1996, also e-print quant-ph/9605011 (1996); DIVINCENZO D. P. and SHOR P. W., *Phys. Rev. Lett.*, **77** (1996) 3260; for a very good review see PRESKILL J., in *Introduction to Quantum Computation and Information*, edited by H.-K. LO, S. POPESCU and T. SPILLER (World Scientific, Singapore) 1998.

[19] See for example: PRESKILL J., *Proc. R. Soc. London, Ser. A*, **454** (1998) 385; KNILL E., LAFLAMME R. and ZUREK W. H., *Proc. R. Soc. London, Ser. A*, **454** (1998) 365.

[20] NIELSEN M. and CHUANG I., *Quantum Computation and Quantum Information* (Cambridge University Press) 2000.

[21] PRESKILL J., in `http://www.theory.caltech.edu/people/preskill/ph229`.

[22] BOUWMEESTER D., EKERT A. and ZEILINGER A. (Editors), in *The Physics of Quantum Information* (Springer) 2000.

Geometric quantum computation
with Josephson qubits

G. Falci

NEST-INFM e Dipartimento di Metodologie Fisiche e Chimiche (DMFCI)
Università di Catania - viale A.Doria 6, I-95125 Catania, Italy

R. Fazio

NEST-INFM e Scuola Normale Superiore - I-56126 Pisa, Italy

G. M. Palma

Dipartimento di Tecnologie dell'Informazione
Università degli Studi di Milano
via Bramante 65, I-26013 Crema (CR), Italy

1. – Introduction

Quantum computers can be viewed as large-scale programmable quantum interferometers in which a coherent superposition of input states evolves in such a way that the probability amplitudes interfere constructively towards the desired output state [1]. Indeed most of the known quantum algorithms can be phrased in terms of a controlled interference pattern [2, 3], the role of quantum gates being precisely to generate such pattern. In most of the proposed physical implementations of quantum computing devices the relative phase of the probability amplitudes are due to the coherent dynamics of the system and are controlled by switching on and off suitable terms of the many-qubit Hamiltonian. A different class of phases appear when a quantum system undergoes a cyclic evolution. They are known as geometric phases since they do not depend on time but rather on the geometry of the cyclic evolution [4, 5]. Recently there has been a lot of interest in the implementation of quantum gates with geometric means (*geometric quantum computation*) [6-9]. In this lecture we will concentrate our attention to the so-called Berry phases [4], which appear when such cyclic evolution takes place adiabatically. In the next section we will describe the physics and mathematics of geometric phases, we will then proceed to describe geometric quantum gates to conclude with their implementation with Josephson qubits.

2. – Geometric phases

Following ref. [4], let us suppose that a quantum system is described by a Hamiltonian H dependent on a set of parameters $\mathbf{R}$ and assume that $\mathbf{R}$ is varied in time slowly enough to make valid the adiabatic approximation. In this situation if the system is initially prepared in a nondegenerate energy eigenstate it will evolve along the instantaneous eigenstate of corresponding eigenenergy. Of particular interest for us is the situation in which $\mathbf{R}$ varies along a closed loop C in the parameter space, *i.e.* when after a time T we have $\mathbf{R}(0) = \mathbf{R}(T)$. In such case the state of the system will be ($\hbar = 1$)

$$
(1) \qquad |\psi(t)\rangle = \exp\left[-i \int_0^T E_n(\mathbf{R}(t))\,\mathrm{d}t \right] e^{i\gamma_n(C)} |n(\mathbf{R}(T))\rangle,
$$

where $E_n(\mathbf{R})$ and $|n(\mathbf{R})\rangle$ obey the instantaneous eigenvalue equation

$$
(2) \qquad H(\mathbf{R}(t))|n(\mathbf{R}(t))\rangle = E_n(\mathbf{R}(t))|n(\mathbf{R}(t))\rangle.
$$

In (1) $\gamma_n(C)$ is a phase factor, called Berry phase, which appears in addition to the usual dynamical phase. Direct substitution of eq. (1) in the time-dependent Schroedinger equation leads to the following expression:

$$
(3) \qquad \gamma_n(C) = i \oint_C \langle n(\mathbf{R})|\nabla_{\mathbf{R}} n(\mathbf{R})\rangle \cdot \mathrm{d}\mathbf{R},
$$

which clearly depends only on the geometry of the path in the parameter space. It is the line integral of the so-called Berry connection, whose components are defined as

$$
(4) \qquad A_k^n = i\langle n(\mathbf{R})|\frac{\partial}{\partial R_k}|n(\mathbf{R})\rangle.
$$

Note that $\mathbf{A}$ plays a role analogous to the vector potential in the Aharonov-Bohm [10] effect, which is another example of topological phase.

Making use of Stokes theorem the line integral (3) can be conveniently transformed into a surface integral:

$$
(5) \qquad \gamma_n(C) = i \oint_\Sigma \mathbf{F} \cdot \mathrm{d}s,
$$

where Σ is a surface contoured by the closed loop C and

$$\mathbf{F} = \nabla \times \mathbf{A}. \tag{6}$$

Equation (5) makes clear that no Berry phase appears if C encloses no area. When the parameter space is not three-dimensional, the above formulas are generalized as follows:

$$\nabla \times \mathbf{A} \cdot \mathrm{ds} \longrightarrow \sum_{\mu\nu} \left(\partial_\mu A_\nu - \partial_\nu A_\mu \right) \mathrm{d}x_\mu \wedge \mathrm{d}x_\nu = F_{\mu\nu}\, \mathrm{d}x_\mu \wedge \mathrm{d}x_\nu. \tag{7}$$

3. – Geometric quantum computation

We will now specialize our discussion to geometric phases in qubits and on their use for the implementation of quantum gates, in the version proposed by Jones *et al.* [6]. In the following we will describe a single qubit as a spin-1/2 coupled to a magnetic field, whose Hamiltonian can be written as

$$H = -\frac{1}{2} B_0(t)\hat{n}(t) \cdot \vec{\sigma}, \tag{8}$$

where $\vec{\sigma} = (\sigma_x, \sigma_y, \sigma_z)$, σ_i are the Pauli operators and the unit vector is defined as $\hat{n} = (\sin \vartheta \cos \varphi, \sin \vartheta \sin \varphi, \cos \vartheta)$. If $\mathbf{B}(t) = B_0(t)\hat{n}(t)$ is varied slowly in time the instantaneous energy eigenstates are

$$|\uparrow\rangle_n = \cos \frac{\vartheta}{2}|\uparrow\rangle + \sin \frac{\vartheta}{2} e^{i\varphi}|\downarrow\rangle, \tag{9}$$

$$|\downarrow\rangle_n = \cos \frac{\vartheta}{2}|\downarrow\rangle - \sin \frac{\vartheta}{2} e^{-i\varphi}|\uparrow\rangle, \tag{10}$$

where $|\uparrow\rangle, |\downarrow\rangle$, the eigenstates of the σ_z operator, are the $|0\rangle, |1\rangle$ qubit logical states. It is important to note that, while the eigenenergies depend on $B_0(t)$, the eigenstates depend only on $\hat{n}(t)$. As a consequence all Berry phases will depend only on ϑ, φ.

It is straightforward to calculate the components of $\mathbf{A}$:

$$A_\varphi^\uparrow = i\langle\uparrow_n |\frac{\partial}{\partial\varphi}|\uparrow_n\rangle = -\frac{1}{2}(1 - \cos \vartheta), \tag{11}$$

$$A_\vartheta^\uparrow = i\langle\uparrow_n |\frac{\partial}{\partial\vartheta}|\uparrow_n\rangle = 0, \tag{12}$$

with $A_{\vartheta,\varphi}^\downarrow = -A_{\vartheta,\varphi}^\uparrow$, and of $\mathbf{F}$:

$$F_{\vartheta\varphi}^\uparrow = -F_{\vartheta\varphi}^\downarrow = \partial_\varphi A_\vartheta - \partial_\vartheta A_\varphi = -\frac{1}{2}\sin \vartheta. \tag{13}$$

These formulas allow us to easily evaluate $\gamma(C)$ when $\hat{n}$ is varied slowly in time. As an example we will consider the typical case in which ϑ is kept fixed, while φ is slowly rotated by an angle 2π. If (3) is used, we obtain for the state $|\!\uparrow\rangle_n$

$$(14) \qquad \gamma(C) = -\int_0^{2\pi} \frac{1}{2}(1 - \cos\vartheta)\, d\varphi = -\pi(1 - \cos\vartheta),$$

which coincides with the result obtained using (5):

$$(15) \qquad \gamma(C) = -\int_0^{2\pi}\int_0^{\vartheta} \frac{1}{2}\sin\vartheta'\, d\varphi'\, d\vartheta' = -\pi(1 - \cos\vartheta).$$

Of course the state $|\!\downarrow\rangle_n$ will acquire a phase of opposite sign. Furthermore the sign of $\gamma(C)$ depends on the direction of the path in parameter space. An interesting geometrical picture is gained when we note that $\gamma(C)$ is equal to half the solid angle subtended by $\hat{n}$ as it moves along the closed loop. This is a general result, which does not depend on our particular choice of path. Another interesting way to visualize Berry phases is in terms of parallel transport. We will not, however, address such issue here. The interested reader can find a discussion of this topic in [5].

The detection of $\gamma(C)$ is possible by a simple interferometric setup. Here we will describe a general procedure in the spin-$(1/2)$ language. Consider the following sequence of steps:

1. start with state $|\psi(0)\rangle = |\!\downarrow\rangle_z$ with $\hat{n}(0) = \hat{z}$;

2. prepare state $|\psi\rangle = (1/\sqrt{2})(|\!\downarrow\rangle_x + |\!\uparrow\rangle_x)$; this can be done by suddenly switching $\hat{z} \to \hat{x}$;

3. move adiabatically $\hat{n}$ around a nonzero area loop C; at the end of such cyclic evolution the state of the system will be

$$(16) \qquad |\psi\rangle = \frac{1}{\sqrt{2}}\left(e^{i\delta(t)}e^{i\gamma(C)}|\!\downarrow\rangle_x + e^{-i\delta(t)}e^{-i\gamma(C)}|\!\uparrow\rangle_x\right),$$

    where $\delta(t)$ is the dynamical phase; at the end of the loop $\hat{n} = \hat{x}$;

4. switch suddenly $\hat{x} \to \hat{z}$; the state at the end of this step is

$$(17) \qquad |\psi\rangle = \cos(\delta + \gamma)|\!\downarrow\rangle_z + \sin(\delta + \gamma)|\!\uparrow\rangle_z;$$

5. measure the state of the qubit in the $|\!\downarrow\rangle_z, |\!\uparrow\rangle_z$ basis; the probability to detect state $|\!\downarrow\rangle_z$ will be $P(\uparrow_z) = \sin^2(\delta + \gamma)$; note that the role of step 4 is to freeze the spin precession during the measurement; it is possible to dispense with it if the detection takes place instantaneously.

The analogy between the above setup and the single-photon interference in a Mach-Zender interferometer is evident. In such a textbook interference experiment a single photon is "splitted" with equal probability amplitude by a 50/50 beam splitter (corresponding to step 2 above) and propagates along the two arms of the interferometer (step 3). The photon is then recombined again at a second beam splitter (step 4) and then detected (step 5). The probability to detect the photon at each of the two output arms of the second beam splitter is a sinusoidal function of the phase difference between the two arms of the interferometer.

In (17) we can get rid of the dynamical phase isolating the purely geometrical one with a simple spin echo technique. The idea is to apply a NOT gate, *i.e.* a π pulse, to the state at the end of step (3) and then to adiabatically move $\hat{n}$ along the same closed loop but in the reverse direction. In such a way the dynamical phases will cancel while the geometrical one will add. Just before the detection step the state will be

$$(18) \qquad |\psi\rangle = \cos(2\gamma)|\downarrow\rangle_z + \sin(2\gamma)|\uparrow\rangle_z.$$

We now show the possibility to build a geometric two-qubit gate, in particular we will show how it is possible to implement a conditional phase shift. Consider the situation in which two spins $\vec{\sigma}_1, \vec{\sigma}_2$ are coupled to local magnetic fields $\mathbf{B}_1, \mathbf{B}_2$ and to each other by their z component with strength ΔB. The Hamiltonian of such system will be

$$(19) \qquad H = -\frac{1}{2}B_1(t)\hat{n}_1(t)\cdot\vec{\sigma}_1 - \frac{1}{2}B_2(t)\hat{n}_2(t)\cdot\vec{\sigma}_2 + \Delta B\sigma_{z1}\sigma_{z2}.$$

Assume now that $\hat{n}_2$ is kept fixed while $\hat{n}_1$ is again varied around a closed loop. The effective magnetic field acting on spin 1 is $\mathbf{B}_{E_1} = B_1(t)\hat{n}_1(t) \pm \Delta B$, the sign depending on whether the state of spin 2 is $|\uparrow\rangle_{z_2}$ or $|\downarrow\rangle_{z_2}$; therefore the solid angle subtended by $\mathbf{B}_{E_1}$ in its cyclic evolution, and the corresponding geometric phase, will be conditional on the state of spin 2. It is possible, with a suitable choice of cyclic path, to obtain a purely geometrical conditional phase gate. For instance, let us carry $\mathbf{B}_1$ along a 8-shaped loop and travel along the two loops of the 8-shaped path in opposite directions; for example, the upper loop, in the northern hemisphere, clockwise, and the lower loop, in the southern hemisphere, counterclockwise. The net geometric phase $\Delta\gamma = \gamma_n - \gamma_s$ will be therefore proportional to the difference of the areas subtended, which in turn depends on the logical state of qubit 2 as B_{z1} depends on it. To undo the dynamic phase the path is followed twice with a NOT in between.

The corresponding unitary operator will be

$$(20) \qquad U = \begin{pmatrix} e^{-i\Delta\gamma_0} & 0 & 0 & 0 \\ 0 & e^{i\Delta\gamma_0} & 0 & 0 \\ 0 & 0 & e^{-i\Delta\gamma_1} & 0 \\ 0 & 0 & 0 & e^{i\Delta\gamma_1} \end{pmatrix},$$

where $\Delta\gamma_0$ and $\Delta\gamma_1$ are the phase shifts induced when the state of spin 2 is $|\uparrow\rangle_{z_2}$ and $|\downarrow\rangle_{z_2}$, respectively. We could choose the areas of the loops in such a way that $\Delta\gamma_0 = 0$. In this case

$$(21) \qquad U = \begin{pmatrix} 1 & 0 & 0 & 0 \\ 0 & 1 & 0 & 0 \\ 0 & 0 & e^{-i\Delta\gamma} & 0 \\ 0 & 0 & 0 & e^{i\Delta\gamma} \end{pmatrix},$$

which is a controlled phase shift. The important point is that conditional phase shifts are universal two-qubit gates [11]. An experiment along lines very similar to the ones illustrated above has been realized in the framework of NMR quantum computing [6]. Generalizations relaxing the adiabaticity condition (Aharonov-Anandan phase) have recently been proposed as well [12, 13].

Before concluding this section we would like to briefly address two issues. First of all it is clear that it is impossible to implement a universal set of quantum gates simply by means of Berry phases. Indeed while geometrical conditional phase shifts are universal two-qubit gates, dynamical manipulation of a single qubit is needed to obtain single-qubit gates other than trivial phase shifts. The way to obtain purely geometrical single-qubit gates is to make use of degenerate subspaces. Purely geometrical (holonomic) quantum computation has been proposed with nonlinear quantum optical systems in [7], while a realistic scheme with trapped ions has been considered in [9]. If the computational states belong to a degenerate subspace which remains such during the cyclic adiabatic evolution, then such states will not simply acquire a relative phase shift but will in general be unitarily mixed [14]. In our proposed scheme with superconducting charge qubits we will however not make use of single-qubit geometric gates.

A final question is whether geometrical quantum computation is better than the conventional dynamical one. The common belief is that geometric quantum computation is fault tolerant against certain class of errors. However, a quantitative analysis of this issue is still lacking. The most likely answer is that geometrical and dynamical quantum gates are simply two different ways of implementing quantum computation. Which of the two is better or easier to implement will in general depend on the specific hardware used.

4. – Geometric phases in Josephson nanocircuits

At present different proposals have been put forward to use superconducting nanocircuits for the implementation of quantum computation [15]. Depending on the operating regime, they are commonly referred to as charge and flux qubits. As the physics of superconducting qubits is illustrated in detail elsewhere in the reviews by M. Devoret, J. Martinis and G. Castellano in this volume, here we will restrict ourselves to a description of the salient features of such devices which are relevant to geometric computation. The first task is to show how a geometric phase can be generated in a system of small

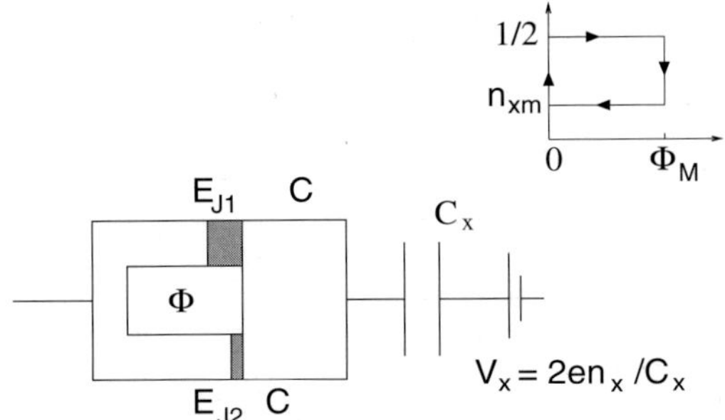

Fig. 1. – The asymmetric SQUID consists of a superconducting loop interrupted by two (small) junctions with different Josephson couplings. An external magnetic generates a flux Φ piercing the SQUID. In addition an applied gate voltage V_x polarizes the Cooper pair box. The offset charge n_x is proportional to V_x. In the inset the loop in the parameter space chosen for the computation of the Berry phase is shown.

coupled Josephson junctions. Then we discuss how it can be used to implement one- and two-qubit gate operations [8].

The required setup consists of a Cooper pair box formed by an asymmetric SQUID, pierced by a magnetic flux and with an applied gate voltage and it is schematically shown in fig. 1. The associated Hamiltonian is

$$(22) \qquad H = E_{\mathrm{C}}(n - n_x)^2 - E_{\mathrm{J}}\cos(\phi - \alpha),$$

where E_{C} is the electrostatic energy of the SQUID and $E_{\mathrm{J}1,2}$ are the Josephson coupling energies. The two control paramters are the flux Φ piercing the SQUID and the external charge n_x proportional to the gate voltage that polarizes the Cooper pair box. The flux controls both the effective Josephson coupling

$$(23) \qquad E_{\mathrm{J}}(\Phi) = \sqrt{(E_{\mathrm{J}1} - E_{\mathrm{J}2})^2 + 4E_{\mathrm{J}1}E_{\mathrm{J}2}\cos^2\left(\pi\frac{\Phi}{\Phi_0}\right)}$$

and the offset phase

$$(24) \qquad \tan\alpha = \frac{E_{\mathrm{J}1} - E_{\mathrm{J}2}}{E_{\mathrm{J}1} + E_{\mathrm{J}2}}\tan\left(\pi\frac{\Phi}{\Phi_0}\right)$$

(Φ_0 is the flux quantum). In the case of a symmetric setup α vanishes. As will be clear in the following, this is a key feature to have geometric phases with superconducting nanocircuits. The electrostatic energy (first term in eq. (22)) is controlled by the external gate voltage. By varying these parameters it is possible to drag the single-qubit part of the Hamiltonian along any direction. The devices operate in the charge regime, *i.e.*

$E_{J1}, E_{J2} \ll E_{ch}$. At temperatures much lower than E_C, if n_x varies around the value $1/2$, only two charge states, $n = 0, 1$, are important. They constitute the basis $\{|0\rangle, |1\rangle\}$ of the computational Hilbert space of the qubit. The effective Hamiltonian is obtained by projecting eq. (22) on the computational Hilbert space, and reads

$$H_B = -\frac{1}{2}\vec{B}\cdot\vec{\sigma},$$

where we have defined the fictitious field

$$B_x = E_J(\Phi)\cos\alpha,$$

$$B_y = -E_J(\Phi)\sin\alpha,$$

$$B_z = E_C(1 - 2n_x).$$

Charging couples the system to B_z whereas the Josephson term determines the projection in the xy-plane. By changing n_x and Φ the qubit Hamiltonian H_B describes a cylindroid in the parameter space $\{\vec{B}\}$.

We now consider the adiabatic time evolution of the qubit Hamiltonian. By changing H_B around the circuit in the parameter space the eigenstates will accumulate a geometric phase. The Berry phase $\gamma_B(\Phi_M, n_{xm})$ accumulated by the two lowest eigenstates along a circuit where n_x is varied from n_{xm} to the degeneracy and the flux from zero to Φ_M $\{\vec{B}\}$, see the inset of fig. 1, is given by

$$\gamma_B(\Phi_M, n_{xm}) = \pm\frac{E_{ch}(1 - 2n_{xm})}{4(E_{J1}E_{J2})^{1/2}}\lambda\mu\Pi\left(\frac{\pi\Phi_M}{\Phi_0}, 1 - \mu^2, \lambda\right),$$

and is plotted in fig. 2. Here $\Pi(x, n, k)$ is the elliptic function of the third kind,

$$\lambda^2 = \frac{4E_{J1}E_{J2}}{(E_{ch}(1 - 2n_{xm}))^2 + (E_{J1} + E_{J2})^2},$$

and

$$\mu = \frac{E_{J1} - E_{J2}}{E_{J1} + E_{J2}}.$$

Notice that nontrivial loops with a controllable Berry phase are possible in our setup thanks to the asymmetry in the SQUID. In the symmetric case $\alpha(\Phi) = 0$ and the evolution takes place in a strip of the plane $B_y = 0$; *i.e.* the Berry phase is zero. The distinct feature of the (geometric) interferometry proposed here is that the geometric phase does depend on *both* gate voltage and flux, in contrast with "classical" interference in mesoscopic junctions, achieved either by changing the magnetic field (in the flux regime) or by a gate voltage modulation (in the charge regime).

                    365

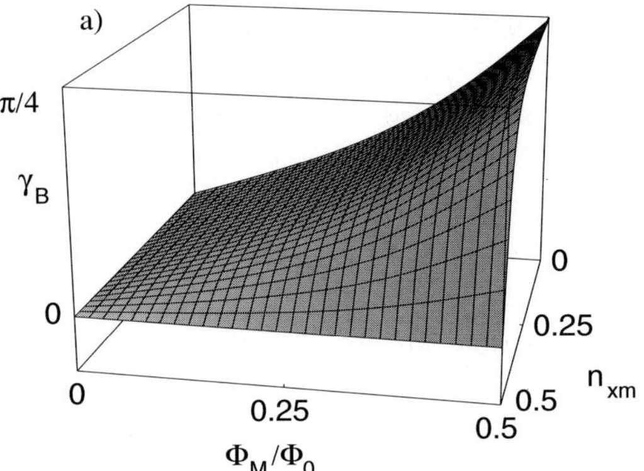

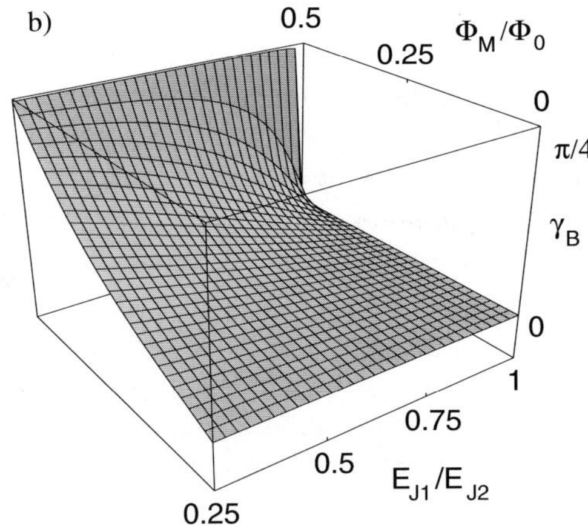

Fig. 2. – a) The Berry phase is calculated for the loop shown in the inset of fig. 1 as a function of the flux and the external charges. The Josephson coupling are $E_{J1} = 0.25E_{J2}$ and $E_{ch} = 5(E_{J1} + E_{J2})$; b) the geometric phase is plotted as a function of the asymmetry parameter E_{J1}/E_{J2} and the flux for the path where $n_{xm} = 0$.

Geometric phases in superconducting nanocircuits emerge also in the opposite (flux) regime, as has recently been shown in ref. [16]. Throughout the paper, we tacitly assumed that the environment played no role during the adiabatic evolution. This is a new aspect which should be addressed when considering geometric phases in mesoscopic systems [17].

In the rest of this contribution we discuss how to perform one- and two-qubit gates with Josephson junctions employing Berry phases. Rather than giving just the prescription for realizing the gates we present a complete scheme to measure single-bit and conditional Berry phases following the steps discussed in sect. **3**.

4`1. *Single-qubit geometric phase.* – The ability of performing single-bit operation is intimately linked to the possibility to detect the geometric phases. A crucial step will be the preparation of a linear superposition of charge states $|0\rangle, |1\rangle$. If we assume that between gate operations $\vec{B} \approx B_z$, this state can be prepared either by suddenly rotating $\vec{B}$ from the north pole to the equator or by rotating $\vec{B}$ adiabatically and then applying a suitable bias voltage pulse. This step is the analogous of the "splitting" the photon wave function at the first beamsplitter of a Mach-Zender interferometer.

A phase difference between the two states can then be introduced by dragging $\vec{B}$ along any closed loop we like back to the initial direction. The phases acquired, however, will have both a geometrical and a dynamical component. To eliminate the latter, it is sufficient to apply a voltage bias pulse to invert the $|0\rangle, |1\rangle$ states, *i.e.* to apply a NOT gate. If the same loop as before is covered by $\vec{B}$, the geometrical component of the phase will add while the dynamical contributions will cancel against each other. The final step, in order to detect the geometric phase, is the measurement of such phase difference. This can be straightforwardly achieved by measuring the charge state while $\vec{B}$ lies on the equator. As the charge states are the eigenstates of σ_z, this measurement automatically mixes the energy eigenstates. The measuring process is performed as reviewed in [15]. This step is therefore equivalent to the detection of the photon after the remix at the second beam splitter of an interferometer. The experimental setup to perform mesurements of this type is already available, as described in [18]. Once more we stress that γ depends on both voltages and flux and therefore it is not simply an Aharonov-Bohm phase.

4`2. *Controlled geometric phase.* – In order to implement two-qubit gates a suitable coupling must be introduced between qubits. In our proposal this is achieved by a capacitive coupling between the SQUIDS. The Hamiltonian of two coupled adjacent qubits, in a pseudo-spin language, reads

$$(25) \qquad\qquad H = \vec{B}_1 \cdot \vec{\sigma}_1 + \vec{B}_2 \cdot \vec{\sigma}_2 + \Delta B_z \sigma_{z1} \sigma_{z2}.$$

The ΔB_z term in eq. (25) describes an additional electrostatic interaction between charges on neighboring islands. Its effect is to modify the effective z component of the $\vec{B}$ acting on one qubit in a way dependent on the state of its neigbouring one. In other words $\vec{B}_{1z} \to \vec{B}_{1z} + \Delta B_z \vec{\sigma}_2$. As shown in the previous section, this can be used to implement a controlled phase shift, *i.e.* to make the geometric phase of one qubit (the so-called target) depend on the state on the neighboring one (the control qubit). From the experimental viewpoint the procedure is analogous to the one described above for the single-qubit gate.

The first step is again the preparation of the initial state. This can be done with a procedure similar to the one described in the one-qubit case: to bring $\vec{B}_2$ to the equatorial plane and to apply an a.c voltage bias pulse (the frequency of this pulse does not depend on the value of the control qubit).

$$* \quad * \quad *$$

We thank J. Siewert and V. Vedral with whom the work reviewed here has been carried on. We also acknowledge A. Ekert, L. Faoro and F. Plastina for useful discussions. This works has been supported by the EU under RTD-IST contracts EQUIP and SQUBIT, by the European Science Foundation PESC program QIT and by the Istituto Nazionale di Fisica della Materia under contract PRA SSQI.

REFERENCES

[1] Ekert A. and Jozsa R., *Rev. Mod. Phys.*, **68** (1996) 733.

[2] Macchiavello C., Palma G. M. and Zeilinger A. (Editors), *Quantum Computation and Quantum Information Theory* (World Scientific, Singapore) 2000.

[3] Cleve R., Ekert A., Henderson L., Macchiavello C. and Mosca M., *On Quantum Algorithms Complexity*, Vol. **4**, No. 1, 1998, pp. 33-42, quant-phys/9903061, reprinted in ref. [2].

[4] Berry M. V., reprinted in *Proc. R. Soc. London, Ser. A*, **392** (1984) 47.

[5] Shapere A. and Wilczek F. (Editors), *Geometric Phases in Physics* (World Scientific, Singapore) 1989.

[6] Jones J., Vedral V., Ekert A. K. and Castagnoli C., *Nature*, **403** (2000) 869.

[7] Zanardi P. and Rasetti M., *Phys. Lett. A*, **264** (1999) 94.

[8] Falci G., Fazio R., Palma G. M., Siewert J. and Vedral V., *Nature*, **403** (2000) 869.

[9] Duan M.-M., Cirac I. and Zoller P., *Science*, **292** (2001) 1695.

[10] Aharonov Y. and Bohm D., *Phys. Rev.*, **115** (1959) 485; Sakurai J. J., *Modern Quantum Mechanics* (Benjamin Cummings) 1985.

[11] Deutsch D., Barenco A. and Ekert A., *Proc. R. Soc. London, Ser. A*, **449** (1995) 669.

[12] Wang X.-B. and Matsumoto K., quant-ph/0105024 and quant-ph/0108111.

[13] Blais A. and Tremblay A.-M. S., quant-ph/0105006.

[14] Wilczek F. and Zee A., *Phys. Rev. Lett.*, **52** (1999) 2111.

[15] Makhlin Y., Schön G. and Shnirman A., *Rev. Mod. Phys.*, **73** (2001) 357, and references therein.

[16] Wilhelm F. K. and Mooij J. E., *Large Berry Phase in Gated Persistent-Current Quantum Bits*, preprint 2001.

[17] Whitney R. S. and Gefen Y., *Proceedings of the XXXVI Rencontres de Moriond* (January 2001), *Electronic Correlations: from Meso- to Nano-physics*; cond-mat/0107359.

[18] Nakamura Y., Pashkin Yu. A. and Tsai J. S., *Nature*, **398** (1999) 786.

Wave packet dynamics and factorization of numbers

H. Mack, M. Bienert, F. Haug, F. S. Straub,
M. Freyberger and W. P. Schleich

Abteilung für Quantenphysik, Universität Ulm - 89069 Ulm, Germany

We connect three phenomena of wave packet dynamics: Talbot images, revivals of a particle in a box and fractional revivals. The physical origin of these effects is deeply rooted in phase factors which are quadratic in the quantum number. We show that the characteristic structures in the time evolution of these systems allow us to factorize large integers.

1. – Talbot effect, revivals and factorization

Although so much has been explained in optical science by the aid of the undulatory hypothesis, yet when any well-marked phænomena occur which present unexpected peculiarities, it may be of importance to describe them, for the sake of comparison with the theory.

This quotation is the opening sentence of a paper [1] by H. F. Talbot entitled *Facts relating to optical sciences*, published in 1836. In this article Talbot reports his experiments on interference of light. We quote again from his seminal paper:

About ten or twenty feet from the radiant point, I placed in the path of the ray an equidistant grating made by Fraunhofer, with its lines vertical. I then viewed the light which had passed through this grating with a lens of considerable magnifying power. The appearance was very curious, being a regular alternation of numerous lines or bands of red and green colour, having their direction parallel to the lines of the grating. On removing the lens a little further from the grating, the bands gradually changed their colours, and became alternately blue and yellow. When the lens was a little more removed,* *the bands again became red and green. And this change continued to take place for an indefinite number of times, as the distance between the lens and grating increased. In all cases the bands exhibited two complementary colours. It was very curious to observe that though the grating was greatly out of the focus of the lens, yet the appearance of the bands was perfectly distinct and well defined.*

** A plate of glass covered with gold-leaf, on which several hundred parallel lines are cut, in order to transmit the light at equal intervals.*

In today's language Talbot has considered the diffraction of light from a grating in the near-field zone. He found that the intensity distribution immediately after the grating repeats itself periodically at multiples of a characteristic distance, later called Talbot length. Moreover, at fractions of the Talbot length the pattern reappears in a rescaled version. Indeed, at the fraction $1/r$ of the Talbot length we find r substructures of the original pattern. The theoretical explanation of this phenomenon was provided in 1881 by Lord Rayleigh [2].

The Talbot effect occurs not only for light waves, but also for matter waves. It has been observed for atoms [3], and most recently for C_{60} molecules [4]. For a comprehensive review of the Talbot effect, we refer to ref. [5].

A phenomenon closely related to the Talbot effect occurs in the time evolution of wave packets [6], such as a Rydberg electron. An initially well-localized packet spreads over its orbit but regains its original shape after a characteristic time, which is much larger than the classical period. Moreover, at fractions of this revival time, the wave packet splits into multiple copies. This effect of fractional revivals appears most clearly in the well-known problem of the particle in a box [7, 8]. Revivals and oscillations can also be used for a new type of interferometer for light in a planar multimode waveguide [9].

Fractional Talbot images as well as the fractional revivals originate from phase factors which are quadratic in the quantum number. They give rise to Gauss sums [10], which have interesting number-theoretical properties.

This immediately points to the newly emerging field of quantum information processing [11], which has received an enormous drive from the discovery of the Shor algorithm [12, 13] to factorize large numbers — a paradigm of number theory. The existence of distinct phenomena at fractions of a characteristic time suggests that the Talbot effect or wave packet physics may have links to the problem of factorizing numbers. Indeed, a recent proposal [14] makes use of a N-slit Young interferometer, where N is the number to be factorized. Such a device is described by the Talbot effect due to a finite grating. The interference structure along the transverse direction at a fixed position after the grating serves as an indicator for a factor. For another approach we refer to ref. [15].

In the present notes we show that the quadratic phase factors inherent in the time evolution of many quantum systems provide a tool to find the prime factors of a large number. Our article is organized as follows: In sect. **2** we briefly summarize the essential ingredients of the Talbot effect in the language of atom optics. We devote sect. **3** to the calculation of the free propagator of an array of wave packets. In sect. 4 we connect this Green's function with the one of the particle in a box. In both cases we arrive at quadratic phase factors. We then show in sect. 5 that similar phases appear in many discrete quantum systems. They manifest themselves in the autocorrelation function. We dedicate sect. 6 to cast the relevant sum into a form which brings out most clearly the phenomenon of fractional revivals. This form allows us in sect. 7 to test our factorization scheme. We suggest a different approach towards factoring in sect. 8 based on properties of Gauss sums. We conclude in sect. 9 by presenting a brief outlook.

2. – Model of Talbot effect

In order to set the stage for the mathematical treatment of the Talbot effect presented in the next section, we now define the principle setup. We consider the diffraction of an atomic wave from a grating. In principle our treatment is also correct for electromagnetic waves in paraxial approximation. Indeed, in this limit the d'Alembert wave equation reduces to the Schrödinger equation. For the influence of the higher-order corrections we refer to ref. [16].

The grating, aligned along the x-axis, could be a mechanical or an optical one. Since we are interested in the subsequent propagation, it suffices to assume that the grating creates a wave function $\phi(x)$ with the period d of the grating. In the case of an infinite array of slits, whose width is much smaller than their separation, the wave function

$$(1) \qquad \phi(x) = \sum_{n=-\infty}^{\infty} \varphi(x - nd)$$

consists of an infinite number of independent initial wave packets $\varphi(x)$ separated by d.

We assume that initially the atomic wave is under normal incidence, that is the wave vector $\vec{k}$, aligned along the z-azis, is orthogonal to the grating. When we consider an atom with an energy much larger than the recoil energy of the grating we can treat its motion along the z-axis classically. In this case time translates into the z-coordinate of the atom [17], that is $z = vt$ where v is its velocity along the z-direction.

We now consider the distribution of atoms along the x-axis for fixed propagation time t. For a defined velocity v of the incoming atoms this time corresponds to a fixed position z behind the grating. At the Talbot time T, that is at the distance $z_T \equiv vT$ the initial wave function $\phi(x)$ repeats itself. At time $T/2$ the interference pattern is identical to the initial one but shifted by half a period. At fractions of the Talbot time the period of the initial wave packet is a fraction of the original one. In the following sections we derive these results and show that they are a consequence of an intricate interference of phases.

3. – Mathematics of the Talbot effect

In the present section we briefly review the mathematical treatment of the Talbot effect. Our key tool is the Poisson summation formula. This analysis provides the foundation for the next section, where we emphasize the close connection between the Talbot effect and the particle in a box.

  3˙1. *Free time evolution of a periodic structure.* – We consider the free motion of the wave packet

$$(2) \qquad \phi(x) = \sum_{n=-\infty}^{\infty} \varphi(x - nd),$$

consisting of an array of infinitely many identical partial waves $\varphi(x)$ separated by a distance d. The propagator [6]

$$(3) \qquad G_{\text{free}}(x, t | y, t = 0) \equiv \mathcal{N}(t)\, e^{i\alpha(t)(x-y)^2}$$

of the free particle of mass M with the scaling factor

$$(4) \qquad \alpha(t) \equiv \frac{M}{2\hbar t}$$

and the normalization

$$(5) \qquad \mathcal{N}(t) \equiv \sqrt{\frac{M}{2\pi i \hbar t}} = \sqrt{\frac{\alpha(t)}{\pi i}}$$

allow us to find the wave function

$$(6) \qquad \psi(x, t) = \int_{-\infty}^{\infty} dy\, G_{\text{free}}(x, t | y, t = 0)\phi(y)$$

at a later time t.

Indeed, when we substitute the initial wave function, eq. (2), and the Green's function, eq. (3), into the propagation equation, eq. (6), we arrive at

$$(7) \quad \psi(x, t) = \int_{-\infty}^{\infty} dy\, \mathcal{N} e^{i\alpha(x-y)^2} \sum_{n=-\infty}^{\infty} \varphi(y - nd) = \int_{-\infty}^{\infty} d\bar{y}\, \mathcal{N} \sum_{n=-\infty}^{\infty} e^{i\alpha(x-\bar{y}-nd)^2} \varphi(\bar{y}),$$

where in the last step we have introduced the integration variable $\bar{y} \equiv y - nd$.

Therefore, eq. (7) takes the form

$$(8) \qquad \psi(x, t) = \int_{-\infty}^{\infty} dy\, G_{\text{Talbot}}(x, t | y, t = 0)\varphi(y),$$

where

$$(9) \qquad G_{\text{Talbot}}(x, t | y, t = 0) \equiv \mathcal{N}(t) \sum_{n=-\infty}^{\infty} e^{i\alpha(t)(x-y-nd)^2}$$

denotes the Talbot propagator.

3'2. *Quadratic phase factors.* – In order to bring out the relation of G_{Talbot} to the propagator G_{box} of the particle in a box, discussed in the next section, we rewrite eq. (9) with the help of the Poisson summation formula

$$(10) \qquad \sum_{n=-\infty}^{\infty} f_n = \sum_{m=-\infty}^{\infty} \int_{-\infty}^{\infty} dn\, f(n) e^{-2\pi i m n},$$

where $f(n)$ is a continuous extension of f_n such that $f(n) = f_n$ at integer values n.

This relation allows us to replace the summation by a sum of Fourier integrals. Hence, the Green's function G_{Talbot}, eq. (9), takes the form

$$(11) \qquad G_{\text{Talbot}}(x, t | y, t = 0) = \sum_{m=-\infty}^{\infty} \mathcal{N} \int_{-\infty}^{\infty} dn\, e^{i\alpha(x-y-nd)^2} e^{-2\pi i m n}$$

$$= \sum_{m=-\infty}^{\infty} \frac{1}{d} e^{-i\kappa_m(x-y)} \sqrt{\frac{\alpha}{i\pi}} \int_{-\infty}^{\infty} d\xi\, e^{i\alpha\xi^2} e^{i\kappa_m\xi},$$

where we have introduced the new integration variable $\xi \equiv x - y - nd$ and the wave number

$$(12) \qquad \kappa_m \equiv m \frac{2\pi}{d}.$$

Moreover, we have recalled the definition, eq. (5), of the normalization $\mathcal{N}(t)$.

When we perform the Gauss integral

$$(13) \qquad \sqrt{\frac{\alpha}{i\pi}} \int_{-\infty}^{\infty} d\xi\, e^{i\alpha\xi^2} e^{i\kappa_m\xi} = \exp\left[-i\frac{\kappa_m^2}{4\alpha}\right],$$

and make use of the definition, eq. (4), of the parameter α, we can identify the phase

$$(14) \qquad \frac{\kappa_m^2}{4\alpha} = \frac{(\hbar\kappa_m)^2}{2M} \frac{t}{\hbar} \equiv E_m \frac{t}{\hbar} \equiv 2\pi m^2 \frac{t}{T}.$$

Here we have introduced the energies

$$(15) \qquad E_m \equiv \frac{(\hbar\kappa_m)^2}{2M}$$

and the Talbot time

$$(16) \qquad T \equiv \frac{M d^2}{\hbar \pi}.$$

Hence, the propagator G_{Talbot} takes the form

$$(17) \qquad G_{\text{Talbot}}(x, t | y, t = 0) = \frac{1}{d} \sum_{m=-\infty}^{\infty} \exp\left[- im \frac{2\pi}{d}(x - y) \right] \exp\left[- 2\pi i m^2 \frac{t}{T} \right].$$

In the time-dependent phase factor we recognize the quadratic dependence on the summation index m.

3‘3. *Integers and half integers of Talbot time.* – According to eq. (17) the propagator for $t = s \cdot T$ with an integer s is identical to the propagator at time $t = 0$. Consequently, at integer multiples of the Talbot time the wave packet regains its initial shape.

Furthermore, at $t = (s + (1/2)) \cdot T$ the relation

$$(18) \qquad \exp\left[- 2\pi i m^2 \left(s + \frac{1}{2} \right) \right] = \exp\left[- \pi i m^2 \right] = (-1)^m = \exp[-im\pi]$$

casts the propagator into the form

$$(19) \qquad G_{\text{Talbot}}\left(x, t = \left(s + \frac{1}{2} \right) T | y, t = 0 \right) = \frac{1}{d} \sum_{m=-\infty}^{\infty} \exp\left[- im \frac{2\pi}{d}\left(x - y + \frac{d}{2} \right) \right].$$

We can identify the right-hand side of this equation with the propagator at time $t = 0$ shifted by half the period d. Hence, the initial wave packet repeats itself at half integer multiples of the Talbot time but is displaced by $d/2$.

4. – Particle in a box

We now turn to the standard problem of a particle of mass M in a box of length L. We first briefly review the essential equations and then derive the corresponding propagator. We conclude by comparing this result to the Talbot propagator.

4‘1. *Propagation.* – The time evolution of an initial wave packet $\psi(x, t = 0) \equiv \varphi(x)$ in a box reads

$$(20) \qquad \psi(x, t) = \sum_{n=1}^{\infty} \psi_n u_n(x) e^{-i E_n t / \hbar},$$

where the expansion coefficients

$$(21) \qquad \psi_n \equiv \int_0^L \mathrm{d}y\, \varphi(y) u_n(y)$$

are in terms of the energy eigenfunctions

$$(22) \qquad u_n(x) \equiv \sqrt{\frac{2}{L}} \sin(k_n x)$$

with wave numbers

$$(23) \qquad k_n \equiv n\frac{\pi}{L}$$

and energies

$$(24) \qquad E_n \equiv \frac{(\hbar k_n)^2}{2M}.$$

When we substitute the expansion coefficients ψ_n, eq. (21), into the expression, eq. (20), for the wave function, we find the propagation equation

$$(25) \qquad \psi(x,t) = \int_0^L dy\, G_{\text{box}}(x,t|y,t=0)\,\varphi(y),$$

in which

$$(26) \qquad G_{\text{box}}(x,t|y,t=0) \equiv \frac{2}{L}\sum_{n=1}^{\infty} \sin(k_n x)\sin(k_n y)e^{-iE_n t/\hbar}$$

is the Green's function of the box.

The quadratic dispersion relation, eq. (24), together with the definition, eq. (23), of the wave number, yields phases

$$(27) \qquad E_n\frac{t}{\hbar} = 2\pi n^2 \frac{t}{T}$$

with characteristic time

$$(28) \qquad T \equiv \frac{4ML^2}{\hbar\pi}.$$

Hence, the Green's function of a particle in a box reads

$$(29) \qquad G_{\text{box}}(x,t|y,t=0) = \frac{2}{L}\sum_{n=1}^{\infty} \sin(k_n x)\sin(k_n y)\exp\left[-2\pi i n^2 \frac{t}{T}\right].$$

In complete analogy to the Talbot effect, we find that for $t=T$ the phases, eq. (27), are integer multiples of 2π. At this time the Green's function is identical to the one at time $t=0$. Consequently, the initial wave function revives, that is $\psi(x,t=T) = \psi(x,t=$

$0) = \varphi(x)$. The similarity of the phases in the Talbot effect and the box problem justifies the use of the same symbol T for the Talbot time and the revival time.

We conclude this section by noting that the quadratic dispersion relation, eq. (24), is also the origin of quantum carpets [8, 18] observed in a box [16, 19] and in many other quantum systems [20, 21].

4‘2. *Relation to the Talbot propagator.* – It is instructive to cast the sum of products of sine functions in the propagator, eq. (29), into a slightly different form. For this purpose we recall the relation

$$(30) \qquad \sin(k_n x) \sin(k_n y) = \frac{1}{4}\left[e^{ik_n(x-y)} + e^{-ik_n(x-y)} - e^{ik_n(x+y)} - e^{-ik_n(x+y)} \right].$$

The terms with negative phase can be combined with the ones having positive phase by extending the summation in eq. (29) to negative values, that is

$$(31) \qquad G_{\text{box}}(x, t|y, t = 0) = \frac{1}{2L} \sum_{n=-\infty}^{\infty} \left[e^{-ik_n(x-y)} - e^{-ik_n(x+y)} \right] \exp\left[-2\pi i n^2 \frac{t}{T} \right].$$

The term $n = 0$ does not contribute since $k_0 = 0$.

When we introduce the Green's function

$$(32) \qquad G_T(x, t|y, t = 0) \equiv \frac{1}{2L} \sum_{n=-\infty}^{\infty} \exp\left[-in\frac{\pi}{L}(x - y) \right] \exp\left[-2\pi i n^2 \frac{t}{T} \right],$$

eq. (31) takes on the form

$$(33) \qquad G_{\text{box}}(x, t|y, t = 0) = G_T(x, t|y, t = 0) - G_T(x, t| - y, t = 0).$$

Hence, the propagator of the box is the difference of two Green's functions with starting points y and $-y$, both leading to x at time t. This difference ensures the boundary condition that the wave function vanishes at the walls.

When we compare the Green's function G_T, eq. (32), to the Talbot propagator G_{Talbot}, eq. (17), we find that they are identical, provided $L = d/2$.

5. – Time evolution and autocorrelation function

So far we have concentrated on the time evolution of two specific quantum systems: a wave periodic in space and a particle confined to a box. Both systems have shown quadratic phase factors and complete revivals. In the present section we generalize this treatment and show that quantum systems with a discrete energy spectrum can display the same phenomena.

5˙1. *Definition*. – We consider a discrete superposition

$$(34) \qquad |\psi(t=0)\rangle = \sum_{n=0}^{\infty} \psi_n |n\rangle$$

of energy eigenstates $|n\rangle$ with energy eigenvalues E_n. Possible quantum systems are the internuclear motion of a diatomic molecule [22], a Rydberg electron [23], the center-of-mass motion of an atom in a standing light wave [24] or a single mode of an electromagnetic field in a cavity [25].

Due to the time evolution the energy eigenstates accumulate phases $E_n t/\hbar$ leading to the state

$$(35) \qquad |\psi(t)\rangle = \sum_{n=0}^{\infty} \psi_n e^{-iE_n t/\hbar} |n\rangle.$$

The autocorrelation function

$$(36) \qquad |S(t)|^2 \equiv \left| \langle \psi(t=0) | \psi(t)\rangle \right|^2$$

with

$$(37) \qquad S(t) \equiv \sum_{n=0}^{\infty} W_n e^{-iE_n t/\hbar}$$

is a measure for the overlap between the initial state $|\psi(t=0)\rangle$ and the state $|\psi(t)\rangle$ at time t. This overlap is a sum over all quantum numbers n of weights $W_n \equiv |\psi_n|^2$ with phases $E_n t/\hbar$.

5˙2. *Quadratic expansion of the energy spectrum*. – When the occupation probability W_n has a dominant maximum around a quantum number $\bar{n}$ and the energy spectrum is only slightly changing with n, we can expand E_n into a Taylor series

$$(38) \qquad E_n \approx E_{\bar{n}} + \left.\frac{\partial E_n}{\partial n}\right|_{n=\bar{n}} (n - \bar{n}) + \frac{1}{2}\left.\frac{\partial^2 E_n}{\partial n^2}\right|_{n=\bar{n}} (n - \bar{n})^2,$$

retaining at most terms quadratic in $n - \bar{n}$. With the definitions

$$(39) \qquad \left.\frac{\partial E_n}{\partial n}\right|_{n=\bar{n}} \equiv \hbar \frac{2\pi}{T_{\mathrm{cl}}}$$

and

$$(40) \qquad \frac{1}{2}\left.\frac{\partial^2 E_n}{\partial n^2}\right|_{n=\bar{n}} \equiv \hbar \frac{2\pi}{T},$$

where T_{cl} and T denote the classical period and the revival time, respectively, we arrive at the approximation

$$(41) \qquad E_n \frac{t}{\hbar} \approx E_{\bar{n}} \frac{t}{\hbar} + \frac{2\pi t}{T_{\mathrm{cl}}}(n - \bar{n}) + \frac{2\pi t}{T}(n - \bar{n})^2$$

of the phases.

We substitute this expression into the definition, eq. (37), of $S(t)$, introduce the new summation index $m \equiv n - \bar{n}$ and arrive at

$$(42) \qquad S(t) \approx e^{-iE_{\bar{n}} t/\hbar}\, \mathcal{S}(t),$$

where

$$(43) \qquad \mathcal{S}(t) \equiv \sum_{m=-\infty}^{\infty} \widetilde{W}_m \exp\left[-2\pi i\left(\frac{m}{T_{\mathrm{cl}}} + \frac{m^2}{T}\right)t\right]$$

with $\widetilde{W}_m = |\psi_{m+\bar{n}}|^2$. Here we have extended the lower bound $-\bar{n}$ of the summation to $-\infty$ since the dominant contributions arise for $m \approx 0$.

In complete analogy to the Green's functions, eqs. (17) and (31), of the Talbot effect or the particle in a box, the autocorrelation function, eq. (36), is determined by a sum where the summation index enters the phase in a quadratic way. However, in contrast to these two examples, we now also have a linear contribution providing two distinct time scales T_{cl} and T.

6. – Fractional revivals

In ref. [26] we have devised a method to rewrite the sum $\mathcal{S}(t)$, eq. (43), in an exact way as to bring out the features of $\mathcal{S}(t)$ typical for the different time domains. We now concentrate on times

$$(44) \qquad t = \ell\, T_{\mathrm{cl}} + \Delta t = \frac{q}{r} T + \varepsilon_{q/r} T_{\mathrm{cl}} + \Delta t$$

that are close to a fraction q/r of T and are close to a large integer multiple ℓ of T_{cl}. The contribution $\varepsilon_{q/r} T_{\mathrm{cl}}$ is a correction term, since in general we have

$$(45) \qquad \ell\, T_{\mathrm{cl}} \neq \frac{q}{r} T.$$

According to ref. [26] we can cast $\mathcal{S}(t)$ into the form

$$(46) \qquad \mathcal{S}\left(t = \frac{q}{r} T + \varepsilon_{q/r} T_{\mathrm{cl}} + \Delta t\right) = \sum_{m=-\infty}^{\infty} \mathcal{W}_m^{(r)} \mathcal{I}_m^{(r)}(\Delta t)$$

with the Gauss sums [10]

$$(47) \qquad \mathcal{W}_m^{(r)} \equiv \frac{1}{r} \sum_{p=0}^{r-1} \exp\left[-2\pi i \left(p^2 \frac{q}{r} + p \frac{m}{r} \right) \right]$$

and the shape functions

$$(48) \qquad \mathcal{I}_m^{(r)}(\Delta t) \equiv \int_{-\infty}^{\infty} d\mu\, \widetilde{W}(\mu) \exp\left[-2\pi i \left[\left(\frac{\Delta t}{T_{\mathrm{cl}}} - \frac{m}{r} \right) \mu + \left(\varepsilon_{q/r} + \frac{\Delta t}{T_{\mathrm{cl}}} \right) \frac{T_{\mathrm{cl}}}{T} \mu^2 \right] \right].$$

Here $\widetilde{W}(\mu)$ denotes the continuous extension of $\widetilde{W}_m$.

The importance of this representation stands out most clearly for the example of a Gaussian weight function

$$(49) \qquad \widetilde{W}(\mu) \equiv \sqrt{\frac{1}{2\pi \Delta n^2}} \exp\left[-\frac{1}{2} \left(\frac{\mu}{\Delta n} \right)^2 \right]$$

of width Δn.

In this case we can perform the integral, eq. (48), and find the explicit expression

$$(50) \qquad \mathcal{I}_m^{(r)}(\Delta t) = \widetilde{\mathcal{N}}(\Delta t) \exp\left[-\frac{(\Delta t - (m/r)T_{\mathrm{cl}})^2}{2\sigma_r^2(\Delta t)} \right] \exp\left[-i \frac{(\Delta t - (m/r)T_{\mathrm{cl}})^2}{2\sigma_i^2(\Delta t)} \right]$$

for the shape function. Here we have introduced the complex amplitude

$$(51) \qquad \widetilde{\mathcal{N}}(\Delta t) \equiv \frac{1}{\sqrt{1 - i4\pi \Delta n^2 (\varepsilon_{q/r} T_{\mathrm{cl}} + \Delta t)/T}}$$

and the widths

$$(52) \qquad \sigma_r^2(\Delta t) \equiv \left[\frac{1}{4\pi^2 \Delta n^2} + 4\Delta n^2 \left(\frac{\varepsilon_{q/r} T_{\mathrm{cl}} + \Delta t}{T} \right)^2 \right] T_{\mathrm{cl}}^2$$

and

$$(53) \qquad \sigma_i^2(\Delta t) \equiv \left[\frac{1}{16\pi^3 \Delta n^2 (\varepsilon_{q/r} T_{\mathrm{cl}} + \Delta t)/T} + \frac{1}{\pi} \Delta n^2 \frac{\varepsilon_{q/r} T_{\mathrm{cl}} + \Delta t}{T} \right] T_{\mathrm{cl}}^2$$

of the real and the imaginary Gaussians.

According to eq. (46) the sum $\mathcal{S}(t)$, determining the autocorrelation function, contains the product $\mathcal{W}_m^{(r)} \mathcal{I}_m^{(r)}(\Delta t)$ of the Gauss sum $\mathcal{W}_m^{(r)}$ and the shape function $\mathcal{I}_m^{(r)}(\Delta t)$. The latter consists of the product of a complex-valued square root, a real and an imaginary Gaussian. The Gaussians only take on non-vanishing values in the neighborhood of $\Delta t = mT_{\mathrm{cl}}/r$. When the separation T_{cl}/r between two neighboring Gaussians is larger than their width σ_r they do not overlap. In this case the sum over m, that is, over the individual Gaussians, separates into a sequence of Gaussians.

Hence, the autocorrelation function in the neigborhood of a fractional revival, that is at a time $t = (q/r)T + \varepsilon_{q/r}T_{\mathrm{cl}} + \Delta t$ consists of a sequence of Gaussians separated by T_{cl}/r provided r is odd or $2T_{\mathrm{cl}}/r$ for r even. This dependence on the odd-even-property of r is a consequence [6] of the Gauss sums $\mathcal{W}_m^{(r)}$, eq. (47). When the Gaussians do not overlap the m-th term in the summation, eq. (46), represents the m-th fractional revival.

When neighboring non-vanishing terms $\mathcal{I}_m^{(r)}(\Delta t)$ and $\mathcal{I}_{m'}^{(r)}(\Delta t)$ overlap, interferences between these terms arise. Then the phases of the complex Gaussian and the square root start to play an important role. Consequently, the sum $\mathcal{S}(t)$ exhibits a more complicated pattern.

7. – Factorization using wavepackets

In the preceding section we have cast the sum $\mathcal{S}(t)$ determining the autocorrelation function into a sequence of complex-valued Gaussians. We now show, that this form suggests a scheme to factorize numbers.

$7\dot{}1$. *General principle*. – For this purpose we return to the non-overlap criterion and recall from eq. (52), that the width σ_r of each Gaussian is different since it depends on time Δt. The minimal width

$$(54) \qquad \sigma_r^{(\mathrm{min})} = \frac{T_{\mathrm{cl}}}{2\pi \Delta n}.$$

occurs for the time Δt_{min} defined by

$$(55) \qquad \varepsilon_{q/r}T_{\mathrm{cl}} + \Delta t_{\mathrm{min}} = 0.$$

If $\varepsilon_{q/r} = 0$ the time Δt_{min} of minimal width vanishes. Since the Gaussian with index $m = 0$ also has its maximum at $\Delta t = 0$, this Gaussian is of minimal width. Moreover, it has no overlap with neighboring ones provided $\sigma_r < (1/r)T_{\mathrm{cl}}$. This condition puts the constraint

$$(56) \qquad \Delta n > \frac{r}{2\pi}$$

on the width Δn of the Gaussian weight function $\widetilde{W}$, eq. (49).

According to eq. (44), a vanishing correction term

$$(57) \qquad \varepsilon_{q/r} = \ell - \frac{q}{r}\frac{T}{T_{\mathrm{cl}}} = 0$$

corresponds to the condition

$$(58) \qquad \ell \cdot r = q \cdot N,$$

where $N = T/T_{\mathrm{cl}}$ is the number to be factorized.

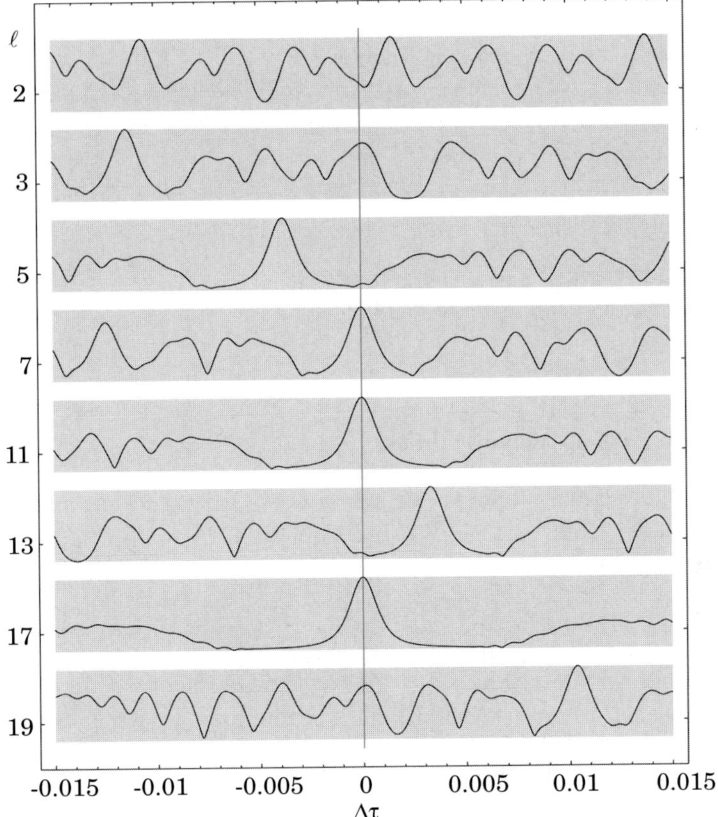

Fig. 1. – Autocorrelation function $|\mathcal{S}_N(\tau = \ell + \Delta\tau)|^2$, defined in eq. (59), as a function of dimensionless time $\Delta\tau$ in the vicinity of various integers $\ell = 2, 3, 5, \ldots$. The goal is to find the factors of $N = 1309 = 7 \cdot 11 \cdot 17$. The autocorrelation function has a maximum at an integer, that is at the origin of each horizontal axis, provided this integer is a factor of N. In the present case we clearly recognize 7, 11 and 17 as factors.

Hence, a well-localized Gaussian at time $\ell \cdot T_{\mathrm{cl}}$ indicates $\varepsilon_{q/r} = 0$ and thus provides the factor ℓ of N.

We conclude this section by briefly discussing the case when ℓ is not a factor of N and therefore $\varepsilon_{q/r}$ is non-zero. In this case the structure at $\Delta t = 0$ is not a Gaussian of minimal width. Indeed, the widths of this Gaussian and its neighbors are so large, that they overlap considerably and the phase factors in $\widetilde{N}$ and in the imaginary Gaussian, eq. (50), lead to complicated interference structures. No clear fractional revival occurs at $\Delta t = 0$.

$7\!\cdot\!2.$ *Simulation.* – In order to test our predictions, we encode the number N in the ratio T/T_{cl} of the two time scales. We then search the autocorrelation function for maxima at integer multiples of T_{cl}.

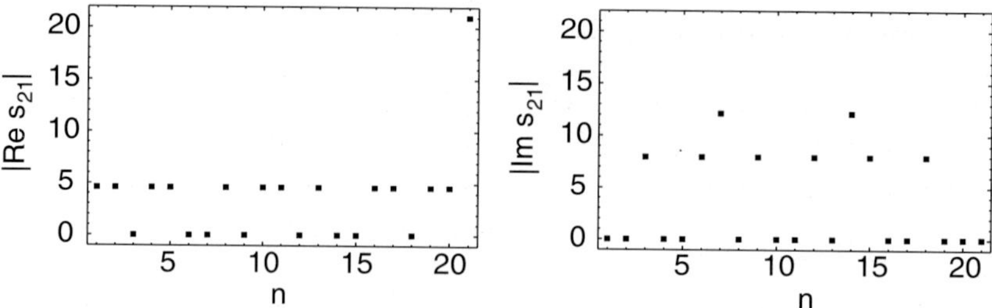

Fig. 2. – Absolute value of real part (left) and imaginary part (right) of the Gauss sum $s_N(n)$, eq. (60), for $N = 21 = 3 \cdot 7$. The factors and their integer multiples appear in the imaginary part of $s_{21}(n)$.

In fig. 1 we show the autocorrelation function

$$(59) \qquad \left| \mathcal{S}_N(\tau) \right|^2 \equiv \left| \sum_{m=-\infty}^{\infty} W(m) \exp\left[-2\pi i \left(m + \frac{m^2}{N} \right) \tau \right] \right|^2$$

as a function of dimensionless time $\tau \equiv t/T_{\mathrm{cl}}$ for the Gaussian weight function, eq. (49). Here we have chosen the width $\Delta n = 250$. The number we want to factorize is $N = 1309 = 7 \cdot 11 \cdot 17$. Since the behavior of $\mathcal{S}_N(\tau)$ in the vicinity of an integer is important, we show $\left| \mathcal{S}_N(\tau) \right|^2$ around various integers. We recognize dominant maxima at $\tau = 7$, $\tau = 11$ and $\tau = 17$, which are indeed the factors of N.

8. – Gauss sums and factorization

The success of our factorization scheme relies on the interference of quadratic phase factors. Why not to concentrate on the bare essentials of the method and eliminate the weight factors and the linear phase term altogether? We therefore consider the sum

$$(60) \qquad s_N(n) \equiv \sum_{m=0}^{N-1} \exp\left[-2\pi i m^2 \frac{n}{N} \right],$$

for fixed N as a function of n.

This sum was also investigated in ref. [27] and gives rise to the so-called curlicues. The emphasis of ref. [27] was on the self-similarity of the emerging structures. However, in the present discussion we focus on the possibility of finding factors by considering the real and imaginary parts of this sum.

In fig. 2 we display real and imaginary parts of $s_{N=21}$. We recognize that the factors 3 and 7 appear as the periods in the imaginary part.

9. – Conclusions and outlook

Interference of quadratic phase factors emerging in the Talbot effect, the particle in a box or in fractional revivals of wave packets has the potential to factorize large numbers. So far our technique relies solely on interference and does not make use of entanglement, a purely quantum-mechanical degree of freedom. In a next step we want to combine the effect of the quadratic phases with the advantages entanglement can offer. We therefore have to consider composite quantum systems, such as highly dimensional spin systems, and develop generalized measurements (POVMs). Besides establishing a novel link to the problem of factorizing large numbers, this approach will povide new insight into the connections between quantum physics and number theory.

$$* * *$$

We thank I. SH. AVERBUKH, M. V. BERRY, H. MAIER and I. MARZOLI for many fruitful discussions. Moreover, two of us (FH and WPS) are grateful to F. DE MARTINI, P. MATALONI and C. MONROE for organizing a most stimulating summer school in the wonderful surroundings of Lake Como. We also thank the editors of these proceedings for patiently awaiting the completion of our manuscript. The work of HM, FSS, MF and WPS was supported by the Deutsche Forschungsgemeinschaft and by the European Commission through the IST network QUBITS.

REFERENCES

[1] TALBOT H. F., *Philos. Mag.*, **9** (1836) 401.
[2] RAYLEIGH L., *Philos. Mag.*, **11** (1881) 196.
[3] CLAUSER J. F. and LI S., *Phys. Rev. A*, **49** (1994) R2213; CHAPMAN M. S., EKSTROM C. R., HAMMOND T. D., SCHMIEDMAYER J., TANNIAN B. E., WEHINGER S. and PRITCHARD D. E., *Phys. Rev. A*, **51** (1995) R14; NOWAK S., KURTSIEFER CH., DAVID C. and PFAU T., *Opt. Lett.*, **22** (1994) 1430.
[4] BREZGER B., HACKERMÜLLER L., UTTENTHALER S., PETSCHINKA J., ARNDT M. and ZEILINGER A., *Phys. Rev. Lett.*, **88** (2002) 100404.
[5] ROHWEDDER B., *Fortschr. Phys.*, **47** (1999) 883.
[6] SCHLEICH W. P., *Quantum Optics in Phase Space* (Wiley-VCH, Berlin) 2001.
[7] STIFTER P., SCHLEICH W. P. and LAMB W. E., *Frontiers of Quantum Optics and Laser Physics*, edited by S. Y. ZHU, M. S. ZUBAIRY and M. O. SCULLY (Springer, Heidelberg) 1997; ARONSTEIN D. L. and STROUD C. R. jr., *Phys. Rev. A*, **55** (1997) 4526.
[8] FRIESCH O., MARZOLI I. and SCHLEICH W. P., *New J. Phys.*, **2** (2000) 4.1.
[9] OVCHINNIKOV Y. B. and PFAU T., *Phys. Rev. Lett.*, **87** (2001) 123901.
[10] LANG S., *Algebraic Number Theory* (Addison-Wesley, New York) 1970.
[11] LO H.-K., SPILLER T. and POPESCU S., *Introduction to Quantum Computation and Information* (World Scientific Publishing, Singapore) 1998; GRUSKA J., *Quantum Computing* (McGraw Hill, London) 1999; BOUWMEESTER D., EKERT A. and ZEILINGER A., *The Physics of Quantum Information* (Springer, Berlin) 2000; NIELSEN M. A. and CHUANG I. L., *Quantum Computation and Quantum Information* (Cambridge University Press, Cambridge) 2000; ALBER G., BETH T., HORODECKI M., HORODECKI P., HORODECKI R., RÖTTELER M., WEINFURTER H., WERNER R. and ZEILINGER A.,

Quantum Information: An Introduction to Basic Theoretical Concepts and Experiments (Springer, Berlin) 2001.

[12] Shor P. W., *Proceedings of the 35th Annual Symposium on the Foundations of Computer Science* (IEEE Computer Society, Los Alamitos) 1994, p. 124 (short version); also in *SIAM J. Sci. Stat. Comp.*, **26** (1997) 1484 or alternatively quant-ph/9508027v2 (1996).

[13] Ekert A. and Josza R., *Rev. Mod. Phys.*, **68** (1996) 733.

[14] Clauser J. F. and Dowling J. P., *Phys. Rev. A*, **53** (1996) 4587.

[15] Harter W. G., *Phys. Rev. A*, **64** (2001) 012312.

[16] Berry M. V. and Klein S., *J. Mod. Opt.*, **43** (1996) 2139.

[17] Kazantsev A. P., Surdutovich G. I. and Yakovlev V. P., *Mechanical Action of Light on Atoms* (World Scientific, Singapore) 1990.

[18] Berry M. V., Marzoli I. and Schleich W. P., *Phys. World*, **14** (2001) 39.

[19] Berry M. V., *J. Phys. A*, **26** (1996) 6617.

[20] Berry M. V. and Bodenschatz E., *J. Mod. Opt.*, **46** (1999) 349.

[21] Kaplan A. E., Marzoli I., Lamb W. E. jr. and Schleich W. P., *Phys. Rev. A*, **61** (2000) 032101.

[22] Vrakking M. J. J., Villeneuve D. M. and Stolow A., *Phys. Rev. A*, **54** (1996) R37.

[23] Yeazell J. A. and Stroud C. R. jr., *Phys. Rev. A*, **43** (1991) 5153.

[24] Raithel G., Phillips W. D. and Rolston S. L., *Phys. Rev. Lett.*, **81** (1998) 3615.

[25] Eberly J. H., Narozhny N. B. and Sanchez-Mondragon J. J., *Phys. Rev. Lett.*, **44** (1980) 1323.

[26] Leichtle C., Averbukh I. S. and Schleich W. P., *Phys. Rev. Lett.*, **77** (1996) 3999; *Phys. Rev. A*, **54** (1996) 5299.

[27] Berry M. V. and Goldberg J., *Nonlinearity*, **1** (1988) 1.

Tomographic methods for universal estimation in quantum optics

G. M. D'Ariano

Quantum Optics & Information Group, Istituto Nazionale di Fisica della Materia
Unità di Pavia e Dipartimento di Fisica "A. Volta" - via Bassi 6, I-27100 Pavia, Italy

"Quantum Tomography" is a general method for estimating arbitrary ensemble averages —including the density matrix itself— of any quantum system, through the measurement of a "quorum" of observables. Recently the method has been extended to the estimation of the matrix form of any "quantum operation" (*i.e.* quantum evolutions and measurements), using only a fixed "entangled state" as the input state, the entangled state playing the role of all possible input states in "quantum parallel". A short review of the theory is presented, with a list of examples of applications for different quantum systems, and with particular focus on quantum optics. Some results from experiments in quantum optics are re-examined. Hints and perspectives on future developments are given at the end of the paper.

1. – Introduction

The possibility of "measuring the quantum state" has puzzled physicists in the last half century, since the earlier theoretical studies of Fano [1]. W. Pauli [2] in a footnote of the *Encyclopedia of Physics* wrote: "The mathematical problem, as to whether for given functions $W(\vec{x})$ and $W(\vec{p})$, the wave function ψ, if such function exists, is always uniquely determined, has still not been investigated in all its generality" (by $W(\vec{x})$ and $W(\vec{p})$ Pauli denoted the probability distributions of position and momentum of a particle, respectively, whereas, for "if such function exists" he meant if $W(\vec{x})$ and $W(\vec{p})$ are compatible). The answer to Pauli's question was clearly negative, since the probabilities $W(\vec{x})$ and $W(\vec{p})$ alone cannot determine the correlation between position and momentum, which could be obtained, for example, from a *joint* measurement of $\vec{x}$ and $\vec{p}$. However, a joint measurement of two conjugated observables would exhibit an additional noise equivalent to an effective quantum efficiency $\eta = 1/2$ [3], and as we will see in subsect. **3·6**, this is exactly the threshold below which the density matrix cannot be measured.

That more than two observables—actually a complete set of them—are needed for a complete determination of the density matrix was clear from the work of Fano [1], and it is explicitly remarked in the book of d'Espagnat [4]. However, since it is difficult to devise concretely measurable observables—other than position, momentum and energy(^{1})— such a fundamental problem—measuring the quantum state!—has remained at the level of mere speculation for many years. The issue finally entered the realm of experiments only less than ten years ago, after the pioneering experiments by Raymer's group [6], in the domain of quantum optics.

What is so special with quantum optics? In quantum optics, differently from quantum mechanics of particles, there is the unique opportunity of measuring all possible linear combinations of position Q and momentum P of a harmonic oscillator, which is represented by a single mode of the electromagnetic field. As explained in subsect. **3·1**, such measurement can be achieved by means of a balanced homodyne detector, which measures the quadrature $X_\phi = (1/2)(a^\dagger e^{i\phi} + a e^{-i\phi})$ of a field mode at any desired phase ϕ with respect to the local oscillator (LO) (as usual a denotes the annihilator of the field mode). The first technique to reconstruct the density matrix from homodyne measurements—so-called *homodyne tomography*—originated from the observation by Vogel and Risken [7] that the collection of probability distributions $\{p(x, \phi)\}$ for $\phi \in [0, \pi)$ is just the Radon transform—*i.e.* the *tomography*—of the Wigner function W. Therefore, by a Radon transform inversion, one can obtain W, and from W the matrix elements of the density operator ρ. This first method, however, was affected by uncontrollable approximations, since, as we will see in subsect. **3·4**, the inversion of the Radon transform needs an *analytic knowledge* of the probability distributions $p(x, \phi)$. In practice, the method works quite well for many photons and *quasi*-classical states, but fails when truly nonclassical states need to be determined experimentally. The main tool, however—*i.e.* using homodyning—still remains good: one only needs to avoid the intermediate step of determining W.

In ref. [8] the first exact technique was given for measuring experimentally the matrix elements of ρ in the photon-number representation, by just averaging functions of homodyne data. After that, the method was further simplified [9], and the feasibility for nonunit quantum efficiency $\eta < 1$ at detectors—above some bounds—was established. Further improvements in the numerical algorithms made the method so simple and fast that it could be implemented easily on small PCs, and the method became quite popular in the laboratories (for the earlier progresses and improvements the reader can see the old review [10]). In the meanwhile there has been an explosion of interest on the subject

---

(^{1}) One can adopt a Schrödinger-picture point of view, and instead of measuring varying operators one can vary the state itself in a controlled way, and eventually measure its energy [5].

of *measuring quantum states*, with hundreds of papers, both theoretical and experimental. The exact homodyne method has been implemented experimentally to measure the photon statistics of a semiconductor laser [11], and the density matrix of a squeezed vacuum [12]. The success of optical homodyne tomography has then stimulated the development of state reconstruction procedures for atomic beams [13], the experimental determination of the vibrational state of a molecule [14], of an ensemble of helium atoms [15], and of a single ion in a Paul trap [16], and different state reconstruction methods have been proposed (for an extensive list of references, see ref. [17]).

In more recent years, the method of quantum tomography has been generalized to the estimation of an arbitrary observable of the field [18], with any number of modes [19], and, finally, to arbitrary quantum systems via group theory [20-22], and with a general method for unbiasing noise [20, 21]. The use of maximum-likelihood strategies [23] has made it possible to reduce dramatically the number of experimental data (by a factor 10^3–10^5!) with negligible bias for most practical cases of interest. Finally, very recently, a method for tomographic estimation of the unknown quantum operation [24] of a quantum device has been presented [25], exploiting the "quantum parallelism" of an entangled input state which plays the role of a "superposition of all possible input states". By another kind of quantum parallelism, one can also estimate the ensemble average of all operators of a quantum system by measuring only one fixed "universal" observable on an extended Hilbert space—the method which I call *Quantum Holography* [26]. Eventually, after the last developments [27], now for the first time we are in a position of getting a first *theory* [28] that is based on the mathematical method of "frames" of operators. This theory will allow to classify all possible "quorums" of observables—*i.e.* those sets of observables that are sufficient to make a tomography of a given quantum system—and generally we will be able to answer to the question: "given a set of available measuring devices and transformation apparatuses, which ensemble averages can be estimated with them? Are they sufficient for a tomographic estimation?"

After briefly giving the general definition of what is quantum tomography in sect. **2**, I will review the method of quantum homodyne tomography in sect. **3**, with a brief introduction to balanced homodyne detection in subsect. **3**˙**1**. The first Radon transform approach is explained in subsect. **3**˙**2**, with the connection to the imaging procedure—which gave the name "tomography" to the method—in subsect. **3**˙**3**. The limitations of the Radon transform approach are explained in subsect. **3**˙**4**, and then, in subsect. **3**˙**5** the exact tomographic approach is given, with the method for unbiasing noise from nonunit quantum efficiency in subsect. **3**˙**6**. A very short account of the multimode homodyne tomography is given in subsect. **3**˙**7**, and improvements based on adaptive techniques and on the maximum-likelihood strategy are rapidly introduced in subsect. **3**˙**8**, with some basic hints on how to estimate the ensemble averages of unbounded operators in subsect. **3**˙**9**. Most of the technical difficulties in homodyne tomography are due to the infinite dimension of the Hilbert space: for finite dimensions everything becomes particularly easy, as in the case of the Pauli tomography given in sect. **4**. With modern words from quantum information theory, we would regard homodyne tomography as the tomography of the so-called *continuous variables*, and the Pauli tomography as the

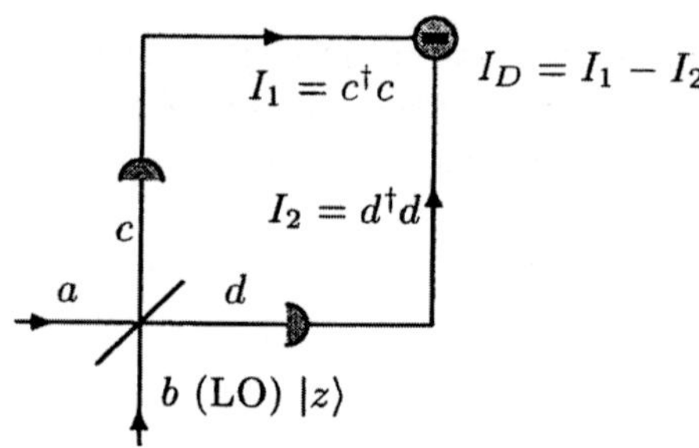

Fig. 1. – Basic scheme of a balanced homodyne detection.

tomography for *qubits*. Some experimental results from ref. [34] on tomography of a twin-beam from parametric downconversion of vacuum are reported in sect. **5**. The general idea of how to perform tomography of a quantum device is explained in sect. **6**. Finally, sect. **7** concludes the paper, with some basic hints and perspectives on a future general theory.

2. – Definition of the problem

Quantum Tomography is a method for estimating the ensemble average $\langle O \rangle$ of arbitrary operator O of a quantum system from measurements of a set—so-called *quorum*—of noncommuting observables. Obviously, the observables of the *quorum* are measured each at a time, with many repeated measurements on an ensemble of equally prepared quantum systems. In the practical situation we want the possibility of unbiasing the estimation from instrumental noise. Moreover, for concrete applications, we need to extend it to the estimation of the matrix of the "quantum operation" that describes the evolution in a device: this will really make quantum tomography a kind of *quantum radiography* of devices.

How can we make it? Let us first analyze the original method that originated from ref. [7], and then see what we can really do better than that.

3. – Homodyne tomography, *i.e.* tomography for *continuous variables*

3'1. *The balanced homodyne detector.* – The balanced homodyne detector is certainly one of the most powerful experimental tools that we have in quantum optics. In fact, by the homodyne detector we can measure all quadratures $X_\phi = (1/2)(a^\dagger e^{i\phi} + a e^{-i\phi})$ of a field mode at any desired phase ϕ with respect to the local oscillator, which is equivalent to measuring all possible linear combinations of position and momentum of the field-mode harmonic oscillator. This is a truly fortunate situation, which does not occur in the quantum mechanics of massive particles.

The basic scheme of the balanced homodyne detector is depicted in fig. 1. The "signal" mode a is combined by means of a 50-50 beam splitter with a "local oscillator" (LO) mode

b operating at the same frequency of a, and prepared in an "intense" coherent state $|z\rangle$. The signal mode a here plays the role of the "system of interest", whereas mode b has to be considered as a part of the apparatus. The field at the output of the beam splitter is described by a "sum" mode $c = (a+b)/\sqrt{2}$ and a "difference" mode $d = (a-b)/\sqrt{2}$. These output modes are detected by two identical photodetectors, and finally the difference of the photocurrents (at zero frequency) is rescaled by $2|z|$. Thus, the output of the detector is given by the following operator:

$$I_{\mathrm{D}} = \frac{c^\dagger c - d^\dagger d}{2|z|} = \frac{a^\dagger b + b^\dagger a}{2|z|}. \tag{1}$$

From eq. (1) we can immediately see that the expectation of the output I_{D} coincides with the expectation of the quadrature X_ϕ, with $\phi = \arg(z)$ being the tunable phase of the LO. Moreover, one can prove rigorously [10] that *in the strong-LO limit $z \to \infty$ the full probability distribution of the output current I_{D} approaches exactly the full probability distribution $p(x,\phi)$ of the quadrature X_ϕ, and this for any state ρ of the signal mode a.* In practice, we need $\langle a^\dagger a\rangle \ll |z|^2$, which is what we actually have in the real experiment. However, what we do not have in the laboratory is a couple of perfect photodetectors for the two modes c and d. On the other hand, since the input currents of the two detectors are both very intense, we can use in practice detectors that behave very linearly for intense inputs, without dark current, and with the only practical limitation that they have nonperfect quantum efficiency, namely that they do not reveal all input photons. The fraction of actually revealed photons is called *quantum efficiency*, and is usually denoted by η. A detector that is sensitive to a very small number of photons—as an avalanche detector—not only saturates soon and exhibits dark current, but it also has very low quantum efficiency. On the contrary, for intense inputs we can easily find detectors that have really good quantum efficiencies, such as $\eta \simeq 0.9$ or better. A theoretical analysis based on the Mandel-Kelley-Kleiner formula [3] shows that in all respects a detector with quantum efficiency $\eta < 1$ is equivalent to an ideal detector preceded by a beam splitter with transmissivity η, and the output becomes a Bernoulli convolution of the response of an ideal detector. In the balanced homodyne detector with two identical detectors with $\eta < 1$, the output photocurrent is reduced by a factor η, and in order to measure the quadrature X_ϕ we now need to rescale the photocurrent by $2|z|\eta$. Then, one can prove rigorously [3] that the probability distribution of the output photocurrent is just the Gaussian convolution of the ideal distribution, with rms $\Delta_\eta \equiv \sqrt{(1-\eta)/(4\eta)}$. In the actual case, further losses in the beam splitter will also reduce the overall effective η, and reasonable values that can be achieved are $\eta = 0.7$–0.8.

3'2. *Homodyne tomography.* – The first hint [7] on how to reconstruct the density matrix of a field mode from homodyne measurements originated from the simple fact that the collection of probability distributions $\{p(x,\phi)\}$ for variable phase $\phi \in [0,\pi)$ is just the Radon transform of the Wigner function $W(\alpha, \overline{\alpha})$, $\alpha \in \mathbb{C}$, namely

$$(2) \qquad p(x, \phi) = \int_{-\infty}^{+\infty} \mathrm{d}y \, W\left((x + iy)e^{i\phi}, (x - iy)e^{-i\phi}\right),$$

where the Wigner function is defined as usual as

$$(3) \qquad W(\alpha, \overline{\alpha}) = \int \frac{\mathrm{d}^2\lambda}{\pi^2} e^{\alpha\overline{\lambda} - \overline{\alpha}\lambda} \, \mathrm{Tr}\left(\rho e^{\lambda a^\dagger - \overline{\lambda} a}\right).$$

Then, from the set of probability distributions $\{p(x, \phi)\}$ one can obtain the Wigner function by inversion of eq. (2), namely

$$(4) \qquad W(\alpha, \overline{\alpha}) = \int_{-\infty}^{+\infty} \frac{\mathrm{d}k|k|}{4} \int_0^\pi \frac{\mathrm{d}\phi}{\pi} \int_{-\infty}^{+\infty} \mathrm{d}x \, p(x, \phi) \exp[ik(x - \alpha_\phi)],$$

where $\alpha_\phi = \Re(\alpha e^{-i\phi})$, and from the knowledge of $W(\alpha, \overline{\alpha})$ one can recover the matrix elements of the density operator ρ via the Fourier transform steps

$$(5) \qquad \langle x + x' | \rho | x - x' \rangle = \int_{-\infty}^{\infty} \mathrm{d}y \, e^{2ix'y} W(x + iy, x - iy),$$

$$(6) \qquad \rho_{nm} = \sqrt{\frac{2^{1-n-m}}{\pi n! m!}} \int_{-\infty}^{\infty} \mathrm{d}x \int_{-\infty}^{\infty} \mathrm{d}x' e^{-(x^2 + x'^2)} H_n\left(\sqrt{2}x\right) H_m\left(\sqrt{2}x'\right) \langle x | \rho | x' \rangle.$$

3`3. *Why the name "tomography"?* – The essential problem of tomographic imaging is to recover a distribution of mass in a 2d slab from a finite collection of one-dimensional projections at different angles ϕ. The situation is schematically sketched in fig. 2, where the distribution of mass describes two circular holes in a uniform background. The tomographic machine—for example, an X-ray equipment—collects many stripe-photos of the sample from various directions ϕ, and then numerically performs a mathematical transform in order to reconstruct the density of mass from its radial profiles at different ϕ. The word "tomography" is customary to denote such imaging procedure starting from radial projections. The situation is strictly analogous to the inversion from the set of quadrature probability distributions $\{p(x, \phi)\}$ to the Wigner function $W(\alpha, \overline{\alpha})$, where now $p(x, \phi)$ plays the role of the radial projection at angle ϕ, and $W(\alpha, \overline{\alpha})$ is the density of mass on $\mathbb{C}$. The collection of all projections $\{p(x, \phi)\}$ at different ϕ's is called *Radon transform*. The reconstruction of the "image" $W(\alpha, \overline{\alpha})$ from its "projections" $p(x, \phi)$—this reconstruction is also called "back-projection"—is given by the inverse Radon transform (4).

3`4. *Limitations of the Radon transform method.* – Suppose that now you want to obtain the Wigner function as an average over ϕ and over homodyne outcomes. This

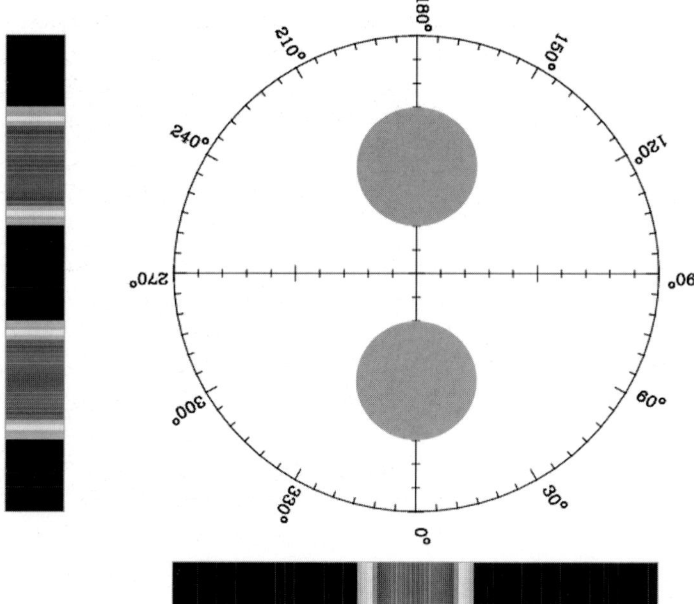

Fig. 2. – Illustration of the tomographic reconstruction of a 2d image (here two holes in a uniform background) from its 1d transmission profiles at different angles ϕ.

means that in eq. (4) the integral over k must be exchanged with those over ϕ and x, obtaining

$$(7) \qquad W(\alpha, \overline{\alpha}) = \int_0^\pi \frac{\mathrm{d}\phi}{\pi} \int_{-\infty}^{+\infty} \mathrm{d}x \, p(x, \phi) \left[-\frac{1}{2} \, \mathsf{P} \frac{1}{(x - \alpha_\phi)^2} \right],$$

$$(8) \qquad \mathsf{P}\frac{1}{z^2} \equiv \lim_{\varepsilon \to 0^+} \Re \frac{1}{(z + i\varepsilon)^2}, \qquad \alpha_\phi = \Re(\alpha e^{-i\phi}),$$

P denoting the Cauchy principal value. Now $W(\alpha, \overline{\alpha})$ is the "expectation" (7) of an unbounded function over data distributed according to $p(x, \phi)$, with random phase ϕ. However, we cannot estimate the expectation as an average over experimental data, since the averaged function is unbounded, and does not satisfy the conditions for the central-limit theorem. Averaging unbounded functions (non square summable) leads to results that never approach any definite value for a large number of data, with rms errors that do not rescale as the inverse square root of the number of data. An example of this behaviour from computer-simulated data is given in fig. 3.

3'5. *The exact method*. – In the Radon transform method the Radon transform inversion is achieved by analytical approximations of the histograms of data and/or introducing a cut-off, for example the parameter ε in eq. (8). The effect on the final matrix elements via eqs. (5) and (6) is a bias whose size depends on ρ and on the number of data. However, since the state ρ is unknown—we want to determine it!—we have now a

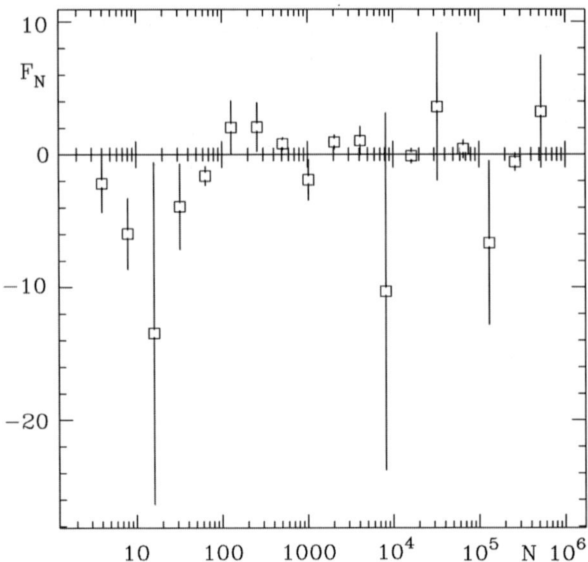

Fig. 3. – Numerical simulation of experiments for estimating the "average" of $f(x) = 1/x$ with uniform probability $p(x) = 1/2$ for $x \in [-1, 1]$.

logical loophole. Worst than that, we do not have any formula to bound the bias, and without any analytical control of inverse Radon transform, we cannot realize that, for example, there are bounds for quantum efficiency, as we shall see in subsect. **3**`6.

The main idea at the basis of the exact method is that the unbounded kernel to be averaged is an artifact of passing through the Wigner function via the Radon transform inversion. Why then not bypass the evaluation of $W(\alpha, \overline{\alpha})$ and go directly to the matrix elements? As you will see soon, in this way not only one gets a method that has no approximation, but also one can recognize the route for generalizing the method to the estimation of any ensemble average for any quantum system. Before deriving the exact method we want to emphasize that for infinite-dimensional Hilbert spaces, as the one of the harmonic oscillator, an experimental knowledge of the density matrix elements ρ_{nm} does not necessarily provide an estimate of the ensemble average of any desired observable. In fact, if we try to obtain the ensemble average $\langle H \rangle$ of the operator H by summing the series $\langle H \rangle = \sum_{nm} \rho_{nm} H_{nm}$ with the matrix elements H_{nm} of H and the measured matrix elements ρ_{nm} of the state, we incur into the situation that even though the series for $\langle H \rangle$ converges in average, it may not converge in error! In fact, as we will see soon, the statistical errors $\varepsilon^2[\rho_{nm}]$ do not vanish for large n, m, and, as a result, for any finite number of data when we increase the series cut-off we ultimately get an unbounded error $\varepsilon^2[H] \simeq \sum_{nm} \varepsilon^2[\rho_{nm}] |H_{nm}|^2$. This is the case, for example, of the photon-number operator $H = a^\dagger a$. Again, we are repeating the same error of the inverse-Radon-transform method: we do not need to estimate the ensemble average of H via the matrix elements ρ_{nm}, but we can bypass the estimation of ρ_{nm} and directly evaluate a function whose average over data and over ϕ gives us the desired expectation value. Let us see now how this can be done.

The displacement operators $D(\alpha) = e^{\alpha a^\dagger - \alpha^* a}$ for $\alpha \in \mathbb{C}$ are an orthonormal basis (in the Dirac sense) for the Hilbert space of Hilbert-Schmidt operators, since $\mathrm{Tr}[D^\dagger(\beta)D(\alpha)] = \pi \delta^{(2)}(\alpha - \beta)$. This means that we have the expansion

$$(9) \qquad H = \int \frac{\mathrm{d}^2\alpha}{\pi} \, \mathrm{Tr}[HD(\alpha)]D^\dagger(\alpha).$$

By changing to polar variables: $\alpha = (i/2)ke^{i\phi}$ and using the symmetry $X_{\phi+\pi} = -X_\phi$, we can rewrite eq. (9) as follows:

$$(10) \qquad H = \int_0^\pi \frac{\mathrm{d}\phi}{\pi} \int_{-\infty}^{+\infty} \frac{\mathrm{d}k|k|}{4} \, \mathrm{Tr}\left[He^{ikX_\phi}\right]e^{-ikX_\phi}.$$

By taking the ensemble average of both sides, and exchanging the integrals with the ensemble average, we can rewrite eq. (10) as the double average of an *estimator* $E_H(X_\phi, \phi)$ over ϕ and over the ensemble

$$(11) \qquad \langle H \rangle = \int_0^\pi \frac{\mathrm{d}\phi}{\pi} \langle E_H(X_\phi, \phi) \rangle, \qquad E_H(x, \phi) \doteq \int_{-\infty}^{+\infty} \frac{\mathrm{d}k|k|}{4} \, \mathrm{Tr}\left[He^{ikX_\phi}\right]e^{-ikx}.$$

Notice that now, thanks to the trace with H, the last integral does not necessarily diverge as the one that gives the Cauchy principal value in eq. (8). In a general theory [28] one classifies the operators H which gives a bounded trace in eq. (11), and such that the integral over k converges. In the present simple form the method will need H at least Hilbert-Schmidt—which is the case of the outer product $|n\rangle\langle m|$ that gives the estimate of the matrix elements ρ_{nm}. However, as we will see in subsect. **3**'9, the method can be extended to a large class of unbounded operators. Anyway, just for the matrix elements ρ_{nm}, bypassing the step of the Wigner function has awarded us with a kernel that is now perfectly bounded, as we will see in the next subsection.

3'6. *Unbiasing noise from nonunit quantum efficiency.* – As we have seen in subsect. **3**'1, the effect of nonunit quantum efficiency of the homodyne detector results in an additional Gaussian noise. This can be conveniently described as a completely positive (CP) map Γ_η, whose effect on the displacement operators is

$$\Gamma_\eta\big(\exp[ikX_\phi]\big) = \exp[ikX_\phi]e^{-(1-\eta)/(8\eta)k^2},$$

Now, we can "unbias" the tomographic estimation by finding a new estimator $E_H^{(\eta)}(x, \phi)$ such that

$$(12) \qquad \langle H \rangle = \int_0^\pi \frac{\mathrm{d}\phi}{\pi} \left\langle E_H^{(\eta)}(X_\phi, \phi) \right\rangle_\eta,$$

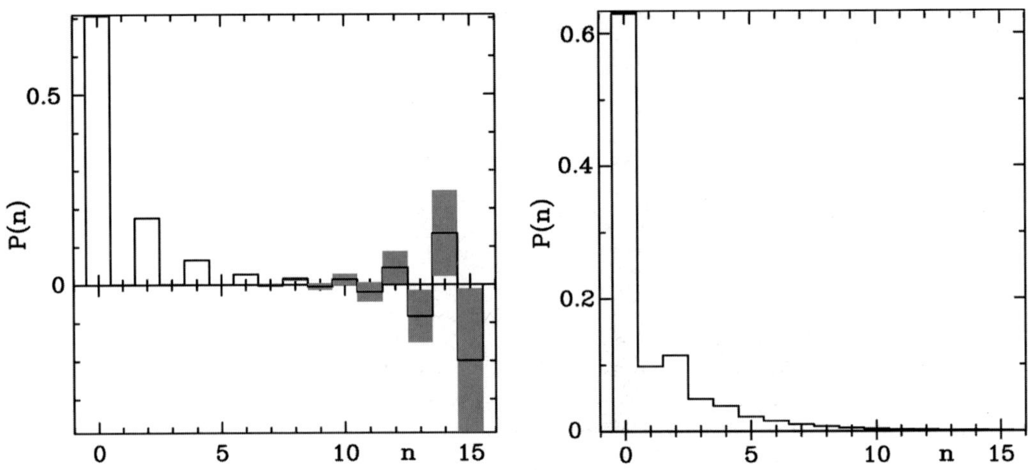

Fig. 4. – Tomographic reconstruction of the photon-number probability $P(n) \equiv \rho_{nn}$ of a squeezed vacuum ($\langle a^\dagger a \rangle = 1$) with detection efficiency $\eta = 0.8$. Homodyne data are computer simulated. (Here we averaged over 27 phases using 200 blocks of 5×10^5 data for each phase.) Experimental errors (confidence intervals) are represented by the gray-shaded thickness of horizontal lines. Left: unbiased reconstruction. Right: reconstruction without unbiasing. From ref. [9].

where $\langle \ldots \rangle_\eta$ denotes the *experimental* ensemble average, *i.e.* with the noisy state $\Gamma_\eta^\tau(\rho)$— Γ_η^τ denoting the same noise map in the Schrödinger picture, *i.e.* the *dual* or *transposed* map. One has

$$(13) \qquad E_H^{(\eta)}(X_\phi, \phi) \doteq \Gamma_\eta^{-1}\{E_H(X_\phi)\} = \int_{-\infty}^{+\infty} \frac{dk|k|}{4} e^{(1-\eta)/(8\eta)k^2} \, \mathrm{Tr}\left[He^{ikX_\phi}\right]e^{-ikX_\phi}.$$

As an example, consider the case of the estimation of the matrix element $\rho_{n+d,n}$, *i.e.* $H = |n\rangle\langle n+d|$. The derivation of the estimator is the following:

$$(14) \qquad E_{|n\rangle\langle n+d|}^{(\eta)}(x, \phi) = \int_{-\infty}^{\infty} \frac{dk|k|}{4} e^{-(2\eta-1)/(8\eta)k^2 - ikx} \langle n+d| : e^{ikX_\phi} : |n\rangle =$$

$$= e^{id(\phi+(\pi/2))} \sqrt{\frac{n!}{(n+d)!}} \int_{-\infty}^{\infty} dk|k| e^{-(2\eta-1)/(2\eta)k^2 - i2kx} k^d L_n^d(k^2),$$

where $: \ldots :$ denotes normal ordering, and $L_n^d(x)$ are generalized Laguerre polynomials. Notice that the estimator is bounded only for $\eta > \eta_b = (1/2)$, and below this bound the method would give unbounded statistical errors. However, as we have seen in subsect. **3**˙1, this bound is well below the values that are reasonably achieved in the laboratory. Here I want to remind that a more efficient algorithm than the estimator in eq. (14) is available, which uses factorization formulas that hold for $\eta = 1$ [29,30], and then un-biases the noise

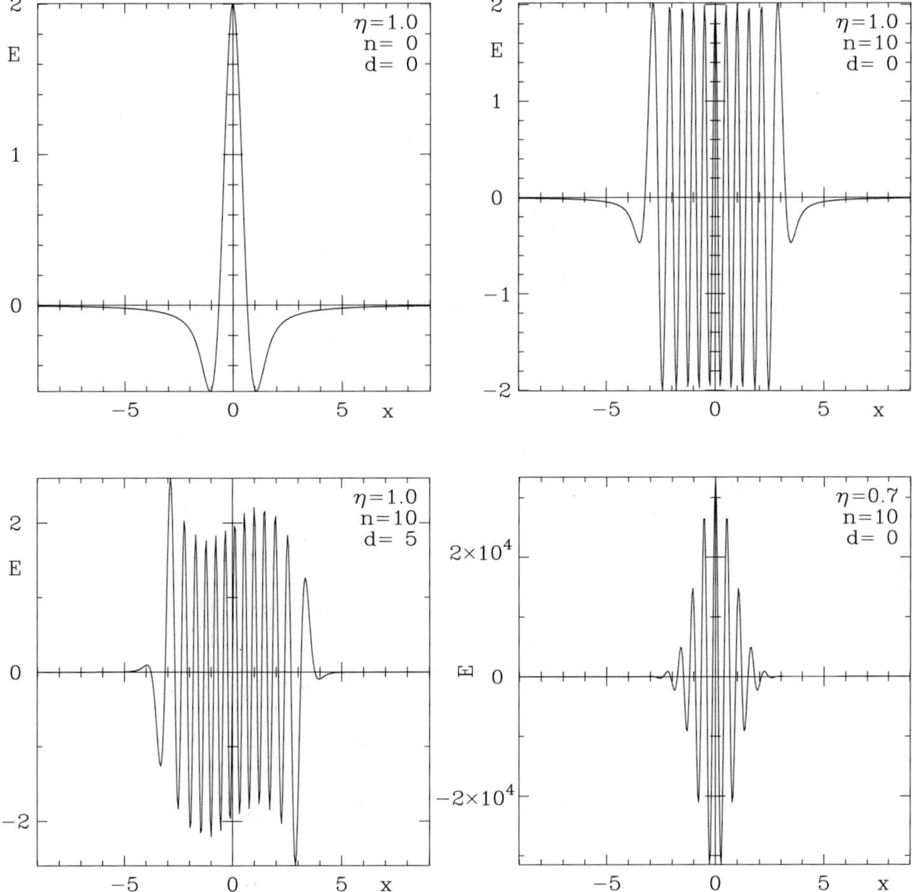

Fig. 5. – Estimator $E^{(\eta)}_{|n\rangle\langle n+d|}(x,\phi)$ of the matrix element $\rho_{n+d,n}$ for $\phi = 0$ and different values of n, d and η.

from quantum efficiency via the inversion of the Bernoulli convolution, which obviously holds above the bound $\eta_b = (1/2)$ (see ref. [10] for a concise review).

An example of application of the estimator (14) to computer-simulated data is given in fig. 4 for the diagonal matrix element ρ_{nn}. There we can see that the price to be paid for the unbiasing procedure is to have statistical errors exponentially growing for increasingly large n. On the contrary, for $\eta = 1$—with no need for unbiasing—the error would remain bonded for $n \to \infty$. In ref. [31] the following asymptotic estimate of the statistical variance has been derived for $n \gg (2\eta - 1)/(1 - \eta)$ and $\eta < 1$:

$$(15) \qquad \sigma^2[\rho_{n,n}] \simeq \frac{\eta^{3/2}}{\sqrt{\pi(1-\eta)}n} \, e^{(1/4n)((2\eta-1)^2/(\eta(1-\eta)))} \left(\frac{1}{2\eta-1}\right)^{2n+1},$$

whereas for $\eta = 1$ one simply has $\sigma^2[\rho_{n,n}] \simeq \sqrt{2}$. The mechanism that develops statistical errors is related to the oscillations of the estimator $E^{(\eta)}_{|n\rangle\langle n+d|}(x, \phi)$. In fig. 5 we report some plots of the estimator $E^{(\eta)}_{|n\rangle\langle n+d|}(x, \phi)$ for different values of n, d and η. One can see that for $d = 0$—along the diagonal of the matrix—the range of the kernel is bounded between -2 and 2, and increases slowly *versus* the distance d from the diagonal. For increasing n and d the kernel oscillates fast, with an increasing number of nodes. Fast oscillations make the average of the kernel—hence the measured value ρ_{nm}—more sensitive to fluctuations of the quadrature outcomes x, producing confidence intervals that increase *versus* n and d. On the other hand, the bounded range makes errors themselves bounded, so they saturate at large n's. For $\eta < 1$ the behavior of the kernel changes dramatically, with its range increasing *versus* n more and more fast as η approaches the lower bound $\eta = 0.5$.

3'7. Multimode homodyne tomography. – For many radiation modes the method is easily generalized by using estimators for tensor product operators which are just the products of their relative estimators, *i.e.* for $M+1$ modes one has $E^{(\eta)}_{\otimes_{n=0}^{M} O_n}(\{x_n\}, \{\phi_n\}) = \prod_{n=0}^{M} E^{(\eta)}_{O_n}(x_n, \phi_n)$. The case of a general operator is then obtained by linearity. However, this method needs a separate measurement—whence a separate LO—for each mode. In ref. [19] it is shown that it is possible to estimate the expectation value of any multimode observable using a single LO, scanning all possible linear combinations of modes on the LO. For the derivation of the method the reader is addressed to ref. [19]. Here we just report the final form of the estimator,

$$(16) \qquad E^{(\eta)}_O(x; \boldsymbol{\theta}, \boldsymbol{\psi}) = \frac{\kappa^{M+1}}{M!} \int_0^\infty dt\, e^{-t+2i\sqrt{\kappa t}x}\, t^M\, \mathrm{Tr}\left\{O : \exp[-2i\sqrt{\kappa t}X(\boldsymbol{\theta}, \boldsymbol{\psi})] : \right\},$$

where $: \ldots :$ denotes normal ordering, $\kappa = (2\eta)/(2\eta - 1)$, and the quadrature operator $X(\boldsymbol{\theta}, \boldsymbol{\psi})$ is the following linear combination of single-mode quadratures:

$$(17) \qquad X(\boldsymbol{\theta}, \boldsymbol{\psi}) = \frac{1}{2}\left[A^\dagger(\boldsymbol{\theta}, \boldsymbol{\psi}) + A(\boldsymbol{\theta}, \boldsymbol{\psi})\right], \qquad A(\boldsymbol{\theta}, \boldsymbol{\psi}) = \sum_{l=0}^{M} e^{-i\psi_l} u_l(\boldsymbol{\theta}) a_l,$$

a_l and $a_l^\dagger$ $(l = 0, \ldots, M)$ being the annihilation and creation operators of the $M + 1$ independent modes with $[a_l, a_{l'}^\dagger] = \delta_{ll'}$, $\boldsymbol{\theta} = (\theta_0, \ldots, \theta_M)$ and $\boldsymbol{\psi} = (\psi_0, \ldots, \psi_M)$ denoting hyper-polar angles with ranges $\psi_l \in [0, 2\pi]$ and $\theta_l \in [0, \pi/2]$, whereas $u_l(\boldsymbol{\theta})$ are hyper-spherical coordinates, such that $\sum_{l=0}^{M} u_l^2(\boldsymbol{\theta}) = 1$, with $u_0(\boldsymbol{\theta}) = \cos\theta_1$, $u_1(\boldsymbol{\theta}) = \sin\theta_1 \cos\theta_2$, $u_2(\boldsymbol{\theta}) = \sin\theta_1 \sin\theta_2 \cos\theta_3, \ldots$, $u_{M-1}(\boldsymbol{\theta}) = \sin\theta_1 \sin\theta_2 \ldots \sin\theta_{M-1} \cos\theta_M$, $u_M(\boldsymbol{\theta}) = \sin\theta_1 \sin\theta_2 \ldots \sin\theta_{M-1} \sin\theta_M$. The ensemble average $\langle O \rangle$ is obtained by averaging the estimator (16) as follows:

$$(18) \qquad \langle O \rangle = \int d\mu[\boldsymbol{\psi}] \int d\mu[\boldsymbol{\theta}]\, p(x, \boldsymbol{\theta}, \boldsymbol{\psi})\, E^{(\eta)}_O(x, \boldsymbol{\theta}, \boldsymbol{\psi}),$$

where

$$\int \mathrm{d}\mu[\boldsymbol{\psi}] \doteq \prod_{l=0}^{M} \int_{0}^{2\pi} (d\psi_l)/(2\pi)$$

and

$$\int \mathrm{d}\mu[\boldsymbol{\theta}] \doteq 2^{M} M! \prod_{l=1}^{M} \int_{0}^{\pi/2} \mathrm{d}\theta_l \sin^{2(M-l)+1} \theta_l \cos\theta_l.$$

In particular, one can estimate the matrix element $\langle\{n_l\}|R|\{m_l\}\rangle$ of the joint density matrix of modes. This will be obtained by averaging the following estimator:

$$(19) \qquad E_{|\{m_l\}\rangle\langle\{n_l\}|}^{(\eta)}(x, \boldsymbol{\theta}, \boldsymbol{\psi}) = e^{-i\sum_{l=0}^{M}(n_l-m_l)\psi_l} \frac{\kappa^{M+1}}{M!} \prod_{l=0}^{M} \left\{ [-i\sqrt{\kappa}u_l(\boldsymbol{\theta})]^{\mu_l-\nu_l} \sqrt{\frac{\nu_l!}{\mu_l!}} \right\} \times$$

$$\times \int_{0}^{\infty} \mathrm{d}t\, e^{-t+2i\sqrt{\kappa}tx}\, t^{M+(1/2)\sum_{l=0}^{M}(\mu_l-\nu_l)} \times$$

$$\times \prod_{l=0}^{M} L_{\nu_l}^{\mu_l-\nu_l}[\kappa u_l^2(\boldsymbol{\theta})t],$$

where $\mu_l = \max(m_l, n_l)$, and $\nu_l = \min(m_l, n_l)$. In fig. 6 we report some computer simulations from ref. [19].

3˙8. *Improving statistical errors*. – One of the major limitations of the unbiased tomographic methods are the quite large statistical errors. These, however, can be improved in several ways. One way of doing it is to exploit the nonunicity of the estimators. As a matter of fact, there are "null estimators", which have zero expectation for arbitrary probability $p_\eta(x, \phi)$, *i.e.* for arbitrary state ρ. Null estimators are obtained as linear combinations of the following functions [32]:

$$(20) \qquad N_{k,n}(X_\phi) = X_\phi^k e^{\pm i(k+2+2n)\phi}, \qquad k, n \geq 0.$$

The reader can easily check that they have zero average over ϕ, independently of ρ. Hence, for every operator O one actually has an equivalence class of infinitely many unbiased estimators, which differ by a linear combination of functions $N_{k,n}$. It is then possible to minimize the rms error in the equivalence class by the least-squares method, obtaining in this way an optimal estimator that is *adapted* to the particular set of experimental data. For details on this *adaptive techniques*, the reader is addressed to ref. [32].

Another relevant strategy, the maximum-likelihood method, can be used for measuring unknown parameters of a unitary transformation on a given state, or for measuring

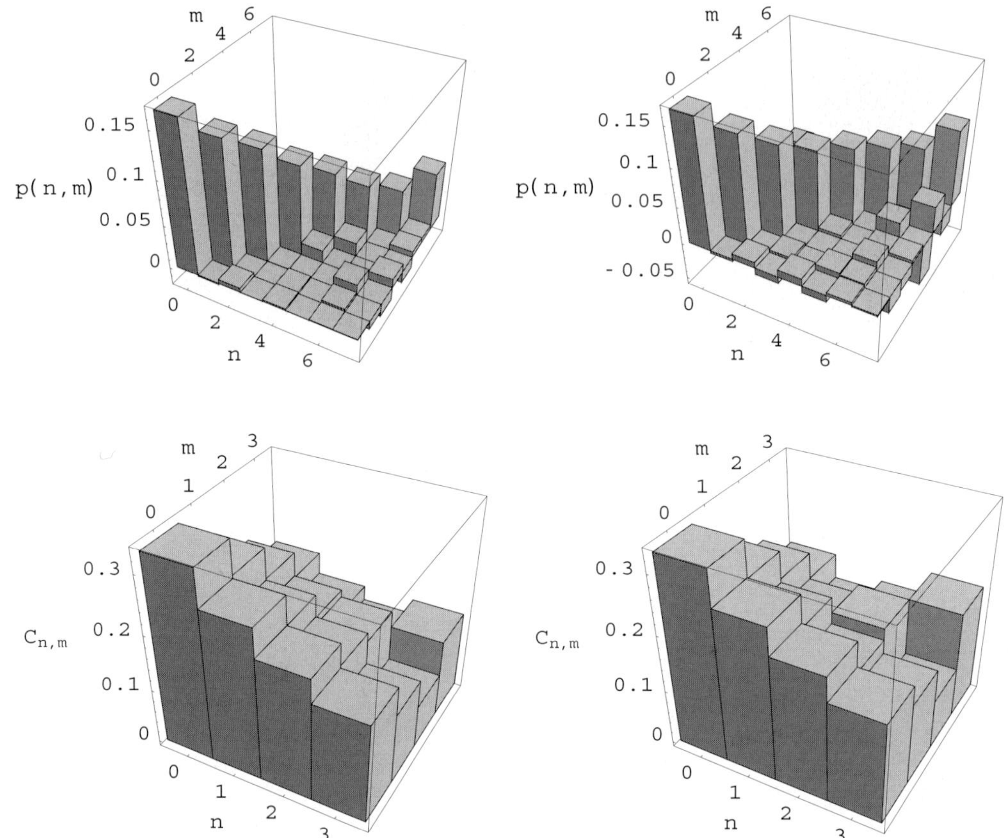

Fig. 6. – The two top figures show the two-mode photon-number probability $p(n, m)$ of the twin-beam state of parametric fluorescence for average number of photons per beam $\bar{n} = |\xi|^2/(1 - |\xi|^2) = 5$ obtained by a Monte Carlo simulation with random parameters $\cos 2\theta$, ψ_1, and ψ_2. The first figure is for quantum efficiency $\eta = 1$, and a sample of 10^6 data. The second figure is for quantum efficiency $\eta = 0.9$, and a sample of 5×10^6 data. The two bottom figures show the tomographic reconstruction of the matrix elements $C_{n,m} \equiv {}_a\langle m|_b\langle m|\Psi\rangle\langle\Psi|n\rangle_a|n\rangle_b$ of the twin-beam state $|\Psi\rangle$ from parametric fluorescence for average number of photons per beam $\bar{n} = 2$. The first figure if for quantum efficiency $\eta = 0.9$, and a sample of 10^6 data, the second figure is for quantum efficiency $\eta = 0.8$, and a sample of 3×10^6 data. From ref. [19].

the matrix elements of the density operator itself [23]. For the full joint density matrix R of many modes, for example, the likelihood function would be

$$(21) \qquad \mathcal{L} = \sum_i \log\left(\sum_{kn} |\langle\{n_i\}|TW_k^\dagger|\vec{x}_i\rangle_{\lambda_i}|^2\right) - \beta \operatorname{Tr}(T^\dagger T),$$

where β is a Lagrange multiplier, T is an upper triangular matrix in the Cholesky decomposition $R = T^\dagger T$ of the density matrix R, W_k—with $\sum_k W_k^\dagger W_k = I$—are a Kraus

   399

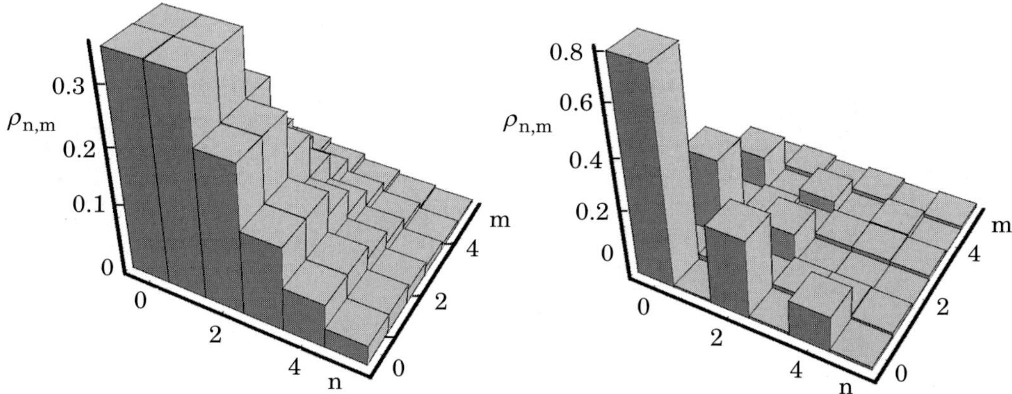

Fig. 7. – From ref. [32]. Monte Carlo simulation of the tomographic reconstruction of the density matrix using the maximum-likelihood technique. Left: density matrix for a coherent state with $\langle a^\dagger a \rangle = 1$; right: squeezed vacuum with $\langle a^\dagger a \rangle = 0.5$. Both: 100 phases with 5000 data each. Hilbert space truncation set to $N_{\mathrm{H}} = 5$; quantum efficiency $\eta = 0.8$.

decomposition $\Gamma[A] = \sum_k W_k^\dagger A W_k$ of the noise Γ (*e.g.*, quantum efficiency), $|\{n_j\}\rangle$ denotes the photon-number basis, $|\vec{x}_i\rangle$ is the joint eigenvector of quadratures, and, finally, the sum runs over the label of the i-th measurement. Notice that, since this method needs a finite parametrization of the density matrix, we need to truncate the Hilbert space dimension. However, much smaller statistical errors are obtained, as compared to the averaging procedure of subsect. **3**`6. An example of computer-simulated experiment is given in fig. 7. We want to emphasize that the maximum-likelihood method is not always the optimal solution of the tomographic problem, since it suffers from some major limitations. Besides being biased due to the Hilbert space truncation—even though the bias can be very small if, from other methods, we know where to truncate—moreover it cannot be generalized to the estimation of any ensemble average, but just of a set of parameters from which ρ depends. In addition, for increasing number of modes the method has exponential complexity, since it requires a search of a maximum over all possible matrix elements. This should be compared with the procedure of subsect. **3**`7, which has polynomial (linear) complexity, and where we can estimate just one matrix element at the time. Therefore, the maximum-likelihood strategy has to be regarded more as a complementary method of the unbiased averaging method. Finally, we want to notice that both the adaptive and the maximum-likelihood improvements can be implemented for the general tomographic approach to arbitrary quantum system [21]: especially for the adaptive technique the general mathematical theory that uses operator frames [28] is particularly advantageous.

3`9. *Estimating ensemble averages of unbounded operators.* – The estimator (11), which comes from the expansion (9) with displacement operators, cannot be used for estimating the ensemble average of operators which are not traceclass. However, one can easily recognize that, for example, an estimator for the field operator a is given by $E_a^{(\eta)}(X_\phi, \phi) = 2e^{i\phi}X_\phi$, and for the number operator easily one gets $E_{a^\dagger a}^{(\eta)}(X_\phi, \phi) = 2X_\phi^2 - (1/2\eta)$. By analytic methods Richter [33] has found the following estimators for normal-ordered monomials in the field operators (then extended to $\eta < 1$ and s-ordering in ref. [20]):

$$(22) \qquad E_{:a^{\dagger m}a^n:_s}^{(\eta)}(X_\phi, \phi) = e^{i(n-m)\phi} \frac{H_{m+n}(\sqrt{2/s_\eta}\, X_\phi)}{\binom{n+m}{n}(\sqrt{2/s_\eta})^{n+m}}, \qquad s_\eta \doteq s - 1 + \eta^{-1},$$

where s is the usual ordering parameter ($s = 1$ for normal ordering, $s = -1$ for antinormal ordering, $s = 0$ for symmetrical ordering), and $H_n(x)$ denote Hermite polynomials. From eq. (22) one can recover a general expansion for a large class of unbounded operators, that here we report just for $\eta = 1$:

$$(23) \qquad H = \int_0^\pi \frac{\mathrm{d}\phi}{\pi} \int_{-\infty}^{\infty} \mathrm{d}t \, \mathrm{Tr}\left[HG_{t,\phi}^\dagger\right]F_{t,\phi}, \qquad F_{t,\phi} = \frac{1}{\sqrt{2\pi}} \exp\left[2\left(X_\phi + \frac{it}{2}\right)^2\right],$$

$$G_{t,\phi}^\dagger = \frac{\mathrm{d}}{\mathrm{d}t}t \int_0^1 \mathrm{d}\theta \exp\left[\theta(1-\theta)t^2\right] e^{-ie^{i\phi}\theta t a^\dagger} |0\rangle\langle 0| \, e^{-ie^{-i\phi}(1-\theta)ta}.$$

A systematic method for finding estimators for unbounded operators is provided by the general theory based on operator frames [28], and is based on the idea of changing the definition of the scalar product in the operator expansion (see sect. **7**).

4. – Pauli tomography, *i.e.* tomography for *qubits*

Most of the technical difficulties in homodyne tomography are due to the infinite dimension of the Hilbert space. For finite dimensions everything becomes particularly easy, since we can expand operators without concerns about convergence of the expansion. The easiest case is the two-state system—the *qubit*—for which the Pauli matrices with the identity $\{I, \sigma_x, \sigma_y, \sigma_z\}$ play the role of an orthonormal basis of observables. One has the following simple expansion:

$$(24) \qquad H = \frac{1}{2}\{\vec{\sigma} \cdot \mathrm{Tr}[\vec{\sigma}H] + I\,\mathrm{Tr}[H]\},$$

which gives the tomographic estimation

$$(25) \qquad \langle H \rangle = \frac{1}{3}\sum_{\alpha=x,y,z}\langle E_H(\sigma_\alpha; \alpha)\rangle, \qquad E_H(\sigma_\alpha; \alpha) = \frac{3}{2}\mathrm{Tr}[H\sigma_\alpha]\sigma_\alpha + \frac{1}{2}\mathrm{Tr}[H].$$

It is also very simple to unbias the noise, by just inverting its CP map. For example, for a Pauli channel with $0 \leq p \leq 1$,

$$(26) \qquad \Gamma_p(H) = (1-p)H + \frac{p}{2}\mathrm{Tr}[H],$$

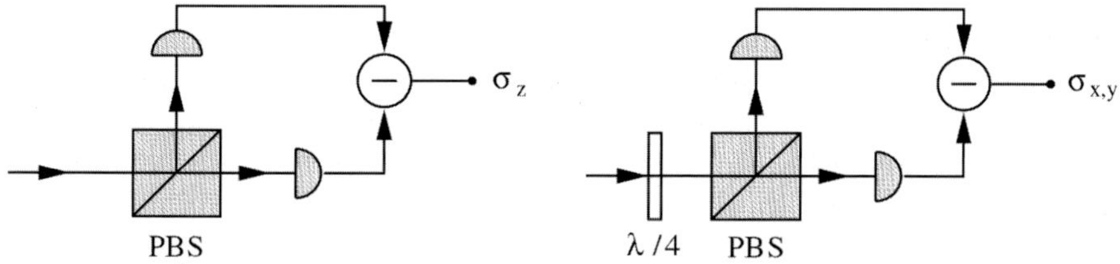

Fig. 8. – Pauli-matrix detectors for photon-polarization qubits (PBS = polarizing beam splitter).

one has the unbiased estimator

$$(27) \qquad E_H^{(p)}(\sigma_\alpha; \alpha) = \frac{3}{2(1-p)} \operatorname{Tr}[H\sigma_\alpha]\sigma_\alpha + \frac{1}{2}\operatorname{Tr}[H].$$

In quantum optics it is easy to implement a qubit using single-photon polarized states, such as $|\uparrow\rangle \equiv |1\rangle_h|0\rangle_v$ and $|\downarrow\rangle \equiv |0\rangle_h|1\rangle_v$, where h and v denote horizontal and vertical polarization, respectively. The Pauli matrix σ_z is then given by the difference $\sigma_z = h^\dagger h - v^\dagger v$ of the horizontally and the vertically polarized photon-numbers, and can be measured by means of a polarizing beam splitter as in fig. 8. Similarly, it is easy to recognize that the other two Pauli matrices can be measured by the same scheme, preceded by a suitably oriented $\lambda/4$ plate, which transforms vertical-horizontal polarization to left-right circular polarization for σ_y, and to orthogonal linear diagonal polarizations for σ_x (see fig. 8). This easily follows from the identities $\sigma_y = e^{i(\pi/4)\sigma_x}\sigma_z e^{-i(\pi/4)\sigma_x}$, which gives $e^{-i(\pi/4)\sigma_x}|1\rangle_h|0\rangle_v = (1/\sqrt{2})[|1\rangle_h|0\rangle_v - i|0\rangle_h|1\rangle_v] \equiv |1\rangle_l|0\rangle_r$, and $\sigma_x = e^{-i(\pi/4)\sigma_y}\sigma_z e^{i(\pi/4)\sigma_y}$, which gives $e^{i(\pi/4)\sigma_y}|1\rangle_h|0\rangle_v = (1/\sqrt{2})[|1\rangle_h|0\rangle_v - |0\rangle_h|1\rangle_v] \equiv |1\rangle_\nearrow|0\rangle_\searrow$.

5. – Some experimental results

We will not attempt to review the large experimental literature in quantum tomography: some references have been given in the introduction of this paper. Here we want to report some results from the experiment of ref. [34]—in which the present author has been involved—since the same apparatus can be used for a homodyne tomography of a quantum device, as explained in sect. **6**. The experiment of ref. [34] is actually the first measurement of the joint photon-number probability distribution for a two-mode quantum state created by a nondegenerate optical parametric amplifier. In this experiment, the two twin beams are detected separately by two balanced-homodyne detectors. A schematic of the experimental setup is reported in fig. 9, and some experimental results are reported in fig. 10. As expected for parametric fluorescence, the experiment has shown a measured joint photon-number probability distribution that exhibited up to 1.9 dB of quantum correlation between the two modes, with thermal marginal distributions.

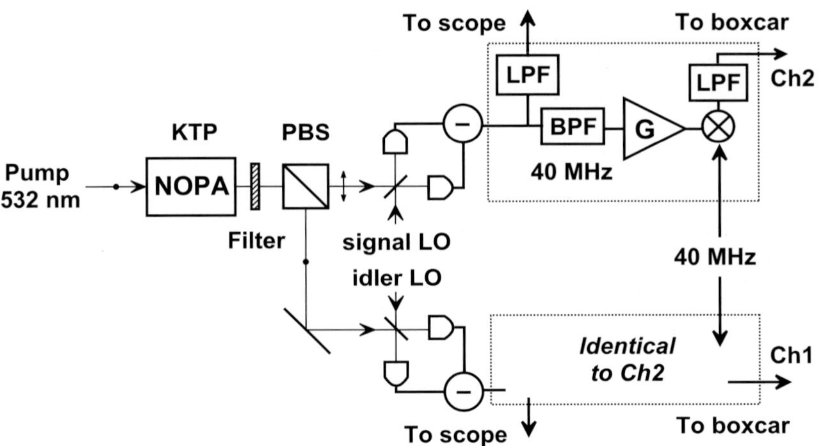

Fig. 9. – A schematic of the experimental setup. NOPA: nondegenerate optical parametric amplifier; LOs: local oscillators; PBS: polarizing beam splitter; LPFs: low-pass filters; BPF: band-pass filter; G: electronic amplifier. Electronics in the two channels are identical. From ref. [34].

6. – Tomography of a quantum device

What does performing the quantum tomography of a device mean? In quantum mechanics, the most general input-output evolution of a device—such as an amplifier, a measuring apparatus, etc.—is described by the state transformation

$$(28) \qquad \rho \longrightarrow \frac{\mathrm{E}(\rho)}{\mathrm{Tr}\left(\mathrm{E}(\rho)\right)},$$

which occurs with probability $p = \mathrm{Tr}[\mathrm{E}(\rho)] \leq 1$. The *quantum operation* E is a linear, trace-decreasing CP map.

Suppose now that we have a quantum device that performs an unknown quantum operation E, and we want to determine it experimentally. How can we do? We can exploit the one-to-one correspondence $\mathrm{E} \leftrightarrow R_\mathrm{E}$ between quantum operations and positive operators R_E on two copies of the Hilbert space $\mathsf{H} \otimes \mathsf{H}$, which is given by

$$(29) \qquad R_\mathrm{E} = \mathrm{E} \otimes \mathrm{I}_\mathsf{H}(|I\rangle\!\rangle\langle\!\langle I|), \qquad \mathrm{E}(\rho) = \mathrm{Tr}_2[I \otimes \rho^\tau R_\mathrm{E}],$$

where for an operator $O = \sum_{nm} O_{nm}|n\rangle\langle m|$ the notation O^τ means the transposed operator $O^\tau = \sum_{nm} O_{mn}|n\rangle\langle m|$ with respect to some pre-chosen orthonormal basis $\{|n\rangle\}$. In the following we will use the notation $|A\rangle\!\rangle \doteq \sum_{nm} A_{nm}|n\rangle\otimes|m\rangle \equiv A\otimes I|I\rangle\!\rangle \equiv I\otimes A^\tau|I\rangle\!\rangle$ that exploits the isomorphism between the Hilbert space of the Hilbert-Schmidt operators $A, B \in \mathsf{HS}(\mathsf{H})$ with scalar product $\langle A, B\rangle \doteq \mathrm{Tr}[A^\dagger B]$ and the Hilbert space of bipartite vectors $|A\rangle\!\rangle, |B\rangle\!\rangle \in \mathsf{H} \otimes \mathsf{H}$, where one has $\langle\!\langle A|B\rangle\!\rangle \equiv \langle A, B\rangle$. If we consider an entangled

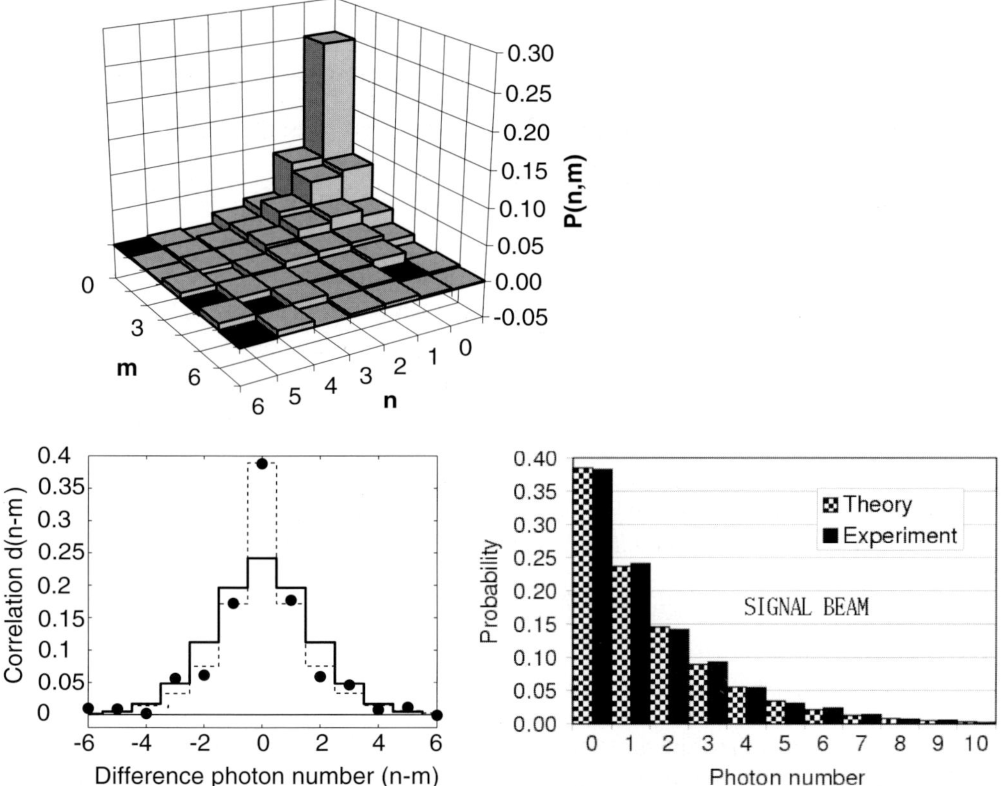

Fig. 10. – Top: Measured joint photon-number probability distributions for the twin-beam state. Bottom left: Difference photon-number distributions corresponding to the left graphs. Filled circles: experimental data; solid lines: theoretical predictions; dashed lines: difference photon-number distributions for two independent coherent states with the same total mean number of photons and $\overline{n} = \overline{m}$. 400000 samples, $\overline{n} = \overline{m} = 1.5$, $N = 10$. Bottom right: Marginal distributions for the signal beam for the same data. Theoretical distributions for the same mean photon numbers are also shown. Very similar results are obtained for the idler beam. From ref. [34].

input state $|\psi\rangle\rangle$ and operate with E only on one party of $|\psi\rangle\rangle$, as in fig. 11, the output state is the joint density matrix

$$(30) \qquad |\psi\rangle\rangle\langle\langle\psi| \longrightarrow R(\psi) \equiv \mathrm{E} \otimes \mathrm{I}(|\psi\rangle\rangle\langle\langle\psi|).$$

But now the quantum operation E is in correspondence with $R_{\mathrm{E}} \equiv R(\psi)$ for $\psi = I$, and for invertible ψ the two matrices $R(I)$ and $R(\psi)$ are connected as follows:

$$(31) \qquad R(I) = \left(I \otimes \psi^{-1^{\tau}}\right) R(\psi) \left(I \otimes \psi^{-1*}\right),$$

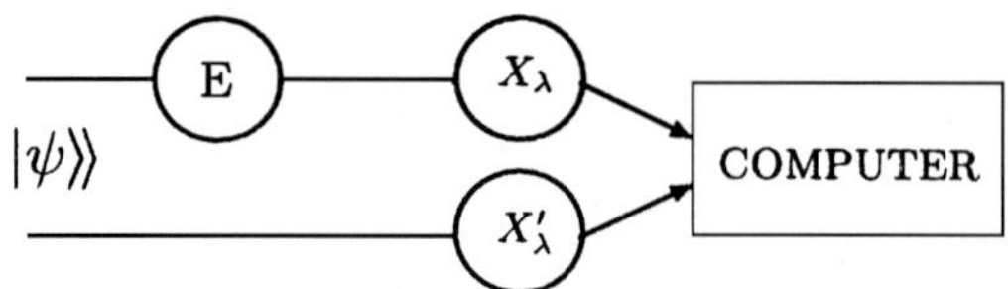

Fig. 11. – General experimental scheme of the method for the tomographic estimation of a quantum operation. Two identical quantum systems are prepared in an entangled state $|\psi\rangle\!\rangle$. One of the two systems undergoes the quantum operation E, whereas the other is left untouched. At the output one makes a quantum tomographic estimation, by measuring jointly two observables X_λ and X'_λ from two quorums $\{X_\lambda\}$ and $\{X'_\lambda\}$ of observables for the two Hilbert spaces, such as two different quadratures of the two field modes in a two-mode homodyne tomography.

where $O^* = \sum_{nm} O^*_{nm}|n\rangle\langle m|$ denotes the conjugate operator of $O^* = \sum_{nm} O_{nm}|n\rangle\langle m|$. Hence, the quantum operation (four-index) matrix R_{E} can be obtained by estimating via a joint double quantum tomography the following output ensemble averages:

$$(32) \qquad \langle\!\langle i,j|R(I)|l,k\rangle\!\rangle = \left\langle |l\rangle\langle i| \otimes |\psi^{-1*}(k)\rangle\langle\psi^{-1*}(j)| \right\rangle,$$

where $|v^*\rangle$ denotes the conjugate vector $|v^*\rangle = \sum_n v^*_n|n\rangle$ of $|v\rangle = \sum_n v_n|n\rangle$.

In fig. 12 the results from a homodyne tomography of an optical displacement of one of the two twin beams from parametric downconversion of the vacuum are reported from ref. [25] for a simulated experiment, for displacement parameter $z = 1$, and for some typical values of the quantum efficiency η at homodyne detectors and of the mean thermal photon number $\bar{n}$ of the twin beams. As one can see, a meaningful reconstruction of the matrix can be achieved in the given range with 10^6–10^7 data, but this number can be decreased of a factor 100–1000 using the tomographic maximum-likelihood techniques of subsect. **3**`8, however, at the expense of the complexity of the algorithm. Homodyne quantum efficiencies and amplifier gains (for the twin beams) typical of the experimental setup of ref. [34] are considered. Improving quantum efficiency and increasing the amplifier gain (toward a maximally entangled state) have the effect of making statistical errors smaller and more uniform *versus* the photon labels n and m of the matrix A_{nm}. Meaningful reconstructions can be achieved with as few as $\bar{n} \sim 1$ thermal photons, and with quantum efficiency as low as $\eta = 0.7$.

7. – Conclusions and future perspectives

What is quantum tomography in general, in simple words? As a method to estimate the ensemble average $\langle H \rangle$ of any arbitrary operator H on a Hilbert space H by using only measurement outcomes of a *quorum* of observables $\{O(l)\}$, it must be essentially a technique for expanding operators over a set of functions $f_n(O(l))$ of the observables $\{O(l)\}$. What makes a general theory nontrivial is the fact that the set of observables

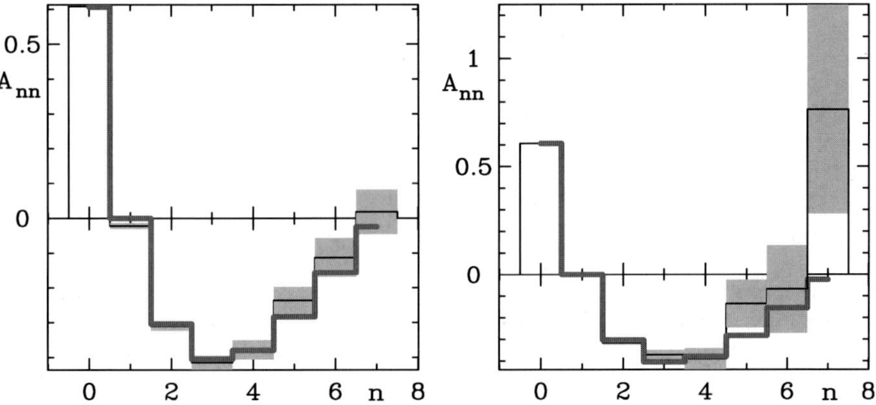

Fig. 12. – From ref. [25]. Homodyne tomography of the quantum operation corresponding to the unitary displacement of one mode of the radiation field. Diagonal elements D_{nn} of the displacement operator (shown by thin solid line on an extended abscissa range), with their respective error bars in gray shade, compared to the theoretical probability (thick solid line). Similar results are obtained for all upper and lower diagonals of the quantum operation matrix A. The reconstruction has been achieved using an entangled state $|\psi\rangle\rangle$ at the input corresponding to parametric downconversion of vacuum with mean thermal photon $\bar{n}$ and quantum efficiency at homodyne detectors η. Left: $z = 1$, $\bar{n} = 5$, $\eta = 0.9$, and 150 blocks of 10^4 data have been used. Right: $z = 1$, $\bar{n} = 3$, $\eta = 0.7$, and 300 blocks of $2 \cdot 10^5$ data have been used. The plot on the right corresponds to the same parameters of the experiment in ref. [34].

$\{O(l)\}$ generally does not span the operator space, whereas it is the set of operators $f_n(O(l))$—*nonlinear* in $\{O(l)\}$—that is complete for varying n and l. Let us denote by $P(j)$ such complete set of operators. Once you have the $P(j)$, then the problem is reduced to the *linear* problem of expanding an operator as $H = \sum_j \langle Q^\dagger(j), H \rangle P(j)$, for a suitable "dual" set of operators $\{Q(j)\}$. Now the point is to have a scalar product $\langle A, B \rangle$ in the expansion that is not just the Hilbert-Schmidt $\langle A, B \rangle \equiv \mathrm{Tr}[A^\dagger B]$, since we want to expand also operators that are not Hilbert-Schmidt. This can be done by changing the definition of the scalar product. The mathematical *theory of frames* [35, 36] is the perfect tool for establishing completeness and for finding dual sets $\{Q(j)\}$. Notice that in most practical situations, the nonorthogonal set $\{P(j)\}$ is over-complete, and there are many alternate dual sets $\{Q(j)\}$: such nonunicity is the basis for general adaptive techniques. Finally, I want to emphasize that a general theory should also classify the operators H which have bounded scalar product with $\{Q(l)\}$ and bounded expansion $H = \sum_j \langle Q^\dagger(j), H \rangle P(j)$, and also assess how these classes are restricted when a noise is unbiased. Only in this way one would be able to say which ensemble averages can be actually estimated and which noise can be unbiased. For the reader that is interested in these future developments, I suggest to look on the Los Alamos ArXive for the forthcoming manuscript [28].

REFERENCES

[1] Fano U., *Rev. Mod. Phys.*, **29** (1957) 74.

[2] Pauli W., *Encyclopedia of Physics*, Vol. V (Springer, Berlin) 1958, p. 17.

[3] D'Ariano G. M., *Quantum Estimation Theory and Optical Detection*, in *Concepts and Advances in Quantum Optics and Spectroscopy of Solids*, edited by T. Hakioğlu and A. S. Shumovsky (Kluwer Academic Publishers, Amsterdam) 1997, pp. 139-174.

[4] d'Espagnat B., *Conceptual Foundations of Quantum Mechanics* (W. A. Benjamin, Mass.) 1976.

[5] Royer A., *Found. Phys.*, **19** (1989) 3.

[6] Smithey D. T., Beck M., Raymer M. G. and Faridani A., *Phys. Rev. Lett.*, **70** (1993) 1244.

[7] Vogel K. and Risken H., *Phys. Rev. A*, **40** (1989) 2847.

[8] D'Ariano G. M., Macchiavello C. and Paris M. G. A., *Phys. Rev. A*, **50** (1994) 4298.

[9] D'Ariano G. M., Leonhardt U. and Paul H., *Phys. Rev. A*, **52** (1995) R1801.

[10] D'Ariano G. M., *Measuring Quantum States*, in the same book of ref. [3], pp. 175-202.

[11] Munroe M., Boggavarapu D., Anderson M. E. and Raymer M. G., *Phys. Rev. A*, **52** (1995) R924.

[12] Schiller S., Breitenbach G., Pereira S. F., Müller T. and Mlynek J., *Phys. Rev. Lett.*, **77** (1996) 2933; Breitenbach G., Schiller S. and Mlynek J., *Nature*, **387** (1997) 471.

[13] Janicke U. and Wilkens M., *J. Mod. Opt.*, **42** (1995) 2183; Wallentowitz S. and Vogel W., *Phys. Rev. Lett.*, **75** (1995) 2932; Kienle S. H., Freiberger M., Schleich W. P. and Raymer M. G., in *Experimental Metaphysics: Quantum Mechanical Studies for Abner Shimony*, edited by S. Cohen *et al.* (Kluwer, Lancaster) 1997, p. 121.

[14] Dunn T. J., Walmsley I. A. and Mukamel S., *Phys. Rev. Lett.*, **74** (1995) 884.

[15] Kurtsiefer C., Pfau T. and Mlynek J., *Nature*, **386** (1997) 150.

[16] Leibfried D., Meekhof D. M., King B. E., Monroe C., Itano W. M. and Wineland D. J., *Phys. Rev. Lett.*, **77** (1996) 4281.

[17] D'Ariano G. M., Vasilyev M. and Kumar P., *Phys. Rev. A*, **58** (1998) 636.

[18] D'Ariano G. M., *Homodyning as universal detection*, in *Quantum Communication, Computing, and Measurement*, edited by O. Hirota, A. S. Holevo and C. M. Caves (Plenum Publishing, New York and London) 1997, p. 253.

[19] D'Ariano G. M., Kumar P. and Sacchi M., *Phys. Rev. A*, **61** (2000) 13806.

[20] D'Ariano G. M., *Latest developments in quantum tomography*, in *Quantum Communication, Computing, and Measurement*, edited by P. Kumar, G. M. D'Ariano and O. Hirota (Kluwer Academic/Plenum Publishers, New York and London) 2000, p. 137.

[21] D'Ariano G. M., *Phys. Lett. A*, **268** (2000) 151.

[22] Cassinelli G., D'Ariano G. M., De Vito E. and Levrero A., *J. Math. Phys.*, **41** (2000) 7940.

[23] Banaszek K., D'Ariano G. M., Paris M. G. A. and Sacchi M., *Phys. Rev. A*, **61** (2000) 010304.

[24] Kraus K., *States, Effects, and Operations* (Springer-Verlag, Berlin) 1983.

[25] D'Ariano G. M. and Lo Presti P., *Phys. Rev. Lett.*, **86** (2001) 4195.

[26] D'Ariano G. M., *Phys. Lett. A*, **300** (2002) 1.

[27] D'Ariano G. M., Maccone L. and Paris M. G. A., *J. Phys. A*, **34** (2001) 34; *Phys. Lett. A*, **276** (2000) 25.

[28] D'Ariano G. M., to appear on LANL arXive quant-ph/.

[29] RICHTER T., *Phys. Lett. A*, **221** (1996) 327.
[30] LEONHARDT U., MUNROE M., KISS T., RAYMER M. G. and RICHTER T., *Opt. Commun.*, **127** (1996) 144.
[31] D'ARIANO G. M., MACCHIAVELLO C. and STERPI N. A., *Quantum Semiclass. Opt.*, **9** (1997) 929.
[32] D'ARIANO G. M. and PARIS M. G. A., *Phys. Rev. A*, **60** (1999) 518.
[33] RICHTER T., *Phys. Rev. A*, **53** (1996) 1197.
[34] VASILYEV M., CHOI S.-K., KUMAR P. and D'ARIANO G. M., *Phys. Rev. Lett.*, **84** (2000) 2354.
[35] CASAZZA P., *Taiwanese J. Math.*, **4** (2000) 129.
[36] HAN D. and LARSON D. R., *Mem. Amer. Math. Soc.*, vol. **147** (AMS, Providence, Rhode Island) 2000, n. 697 0-94.

NMR, SQUID, STOPPING OF LIGHT

Introduction to NMR quantum information processing

R. Laflamme

University of Waterloo - Waterloo, ON N2L 3G1, USA

E. Knill

LANL, MS B265 - Los Alamos, NM 87545, USA

C. Negrevergne

LANL, MS B288 - Los Alamos, NM 87545, USA

R. Martinez

LANL, MS E529 - Los Alamos, NM 87545, USA

S. Sinha

NW14-2311 - 77 Massachusetts Avenue, Cambridge, MA 02139, USA

D. G. Cory

NW14-2217 - 77 Massachusetts Avenue, Cambridge, MA 02139, USA

The theory of quantum information is changing the foundations of, and how we think about, information and computation. Using quantum physics to represent and to manipulate information makes possible surprising improvements in the efficiency with which some problems can be solved. But can these improvements be realized experimentally? If we consider the history of the implementation of theoretical ideas about classical information and computation, we find that initially, small numbers of simple devices were used to explore the advantages and the difficulties of information processing. For example, in 1933 Atanasoff and his colleagues at the Iowa State College were able to implement digital calculations using around 300 vacuum tubes. Although the device was never practical because its error rate was too large, it was probably the first instance of a programmable computer using vacuum tubes and it opened the way for more stable and reliable devices. Progress toward implementing quantum information processors is also initially confined to limited capacity and error-prone devices.

There are numerous proposals for implementing quantum information processing (QIP) prototypes. To date (2001) only three of them have been used to successfully manipulate more than one qubit: cavity quantum electrodynamics (cavity QED), ion traps and nuclear magnetic resonance with molecules in a liquid (liquid state NMR). One explanation for the difficulty in realizing QIP devices is that there is an intrinsic conflict between two of the most important requirements: On the one hand, it is necessary for the device to be well isolated from and therefore interact only weakly with its environment. Otherwise, the crucial quantum correlations on which the advantages of QIP are based are destroyed. On the other hand, it is necessary for the different parts of the device to interact strongly with each other and for some of them to be coupled strongly with the measuring device, which is needed to read out "answers". That few physical systems have these properties naturally is apparent from the absence of obvious quantum effects in the macroscopic world.

One system whose properties constitute a reasonable compromise between the two requirements is the one that consists of the nuclear spins in a molecule dissolved in a liquid. The spins, particularly those with spin half, provide a natural representation of quantum bits. They interact weakly but reliably with each other and the effects of the environment are often small enough. They can be controlled by using radio-frequency (RF) pulses and observed collectively by measuring the magnetic fields generated by the spins. Liquid state NMR has now been used to demonstrate control of up to seven physical qubits.

It is important to remember that the idea of QIP is less than two decades old, and, with the notable exception of quantum cryptography, experimental proposals and efforts aimed at realizing modern QIP only began in the last five years of the 20th century. Increasingly advanced experiments are being implemented, but from an information processing point of view, we are a long way from using quantum technology to solve an independently posed problem not solvable on a standard personal computer. To get to the point where such problems can be solved using QIP, current experimental efforts are devoted toward understanding the behavior of, and how to control, the various quantum systems, and how to overcome their limitations. The work on NMR QIP has focused on the control of quantum systems by algorithmically implementing various unitary transformations as precisely as possible. Within the limitations of the device this has been surprisingly successful, thanks to the many scientists and engineers who have perfected NMR spectrometers over the last 50 years.

After a general introduction to NMR, we give the basics of implementing quantum algorithms. We describe how qubits are realized and controlled using RF pulses, the internal interactions and gradient fields. A peculiarity of NMR is that the internal interactions (given by the internal Hamiltonian) are always on. We discuss how it can be effectively turned off using a standard NMR method called "refocusing". Liquid state NMR experiments are done at room temperature, leading to an extremely mixed (that

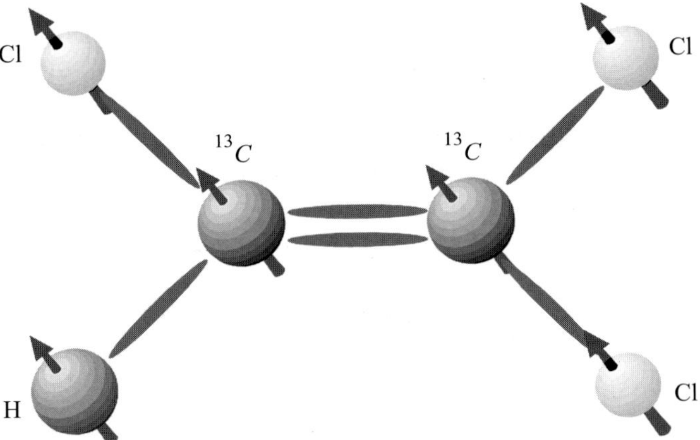

Fig. 1. – Trichloroethylene. Schematic of a typical molecule used for quantum information processing. There are three useful nuclei for realizing qubits. They are the proton (H), and the two carbons (^{13}C). The molecule is "labeled", which means that the nuclei are carefully chosen isotopes. In this case, the normally predominant isotope of carbon, ^{12}C (a spin-zero nucleus), is replaced by ^{13}C (which has spin half).

is, nearly random) initial state. Despite this high degree of randomness, it is possible to investigate QIP because the relaxation time (the time scale over which useful signal from a computation is lost) is sufficiently long. We explain how this leads to the crucial ability of simulating a pure (non-random) state using "pseudo-pure" states. We discuss how the "answer" provided by a computation is obtained by measurement and how this differs from the ideal, von Neumann projective measurement of QIP. We then give implementations of some simple quantum algorithms with a typical experimental result. We conclude with a discussion of what we have learned from NMR QIP so far and what the prospects for future NMR QIP experiments are. For an elementary, device-independent introduction to quantum information and definitions of the states and operators used here, see [1].

1. – Liquid state NMR

1`1. *NMR basics*. – Many atomic nuclei possess a magnetic moment, which means that like small bar magnets, they respond to electromagnetic fields and can be detected by their effect on these fields. Although single nuclei are impossible to detect directly by these means using today's technology, if sufficiently many are available so that their effects add, they can be observed as an ensemble. In liquid state NMR, the nuclei belong to atoms forming a molecule, a very large number of which are dissolved in a liquid. An example is ^{13}C trichloroethylene (TCE) (fig. 1). The hydrogen nucleus (that is the proton) of each TCE molecule has a relatively strong magnetic moment. When the sample is placed in a powerful external magnetic field, each proton's spin prefers to align itself with the field. It is possible to induce the spin direction to "tip" off-axis by means of RF pulses, at which point the effect of the static field is to induce a rapid precession of the proton spins. The precession frequency is called the Larmor frequency. The field generated by the protons now oscillates rapidly and a coil judiciously placed around the sample can pick up this oscillation, making it possible to observe the entire ensemble of protons by "magnetic induction". This is the fundamental idea of NMR. The device that applies the static magnetic field, radio-frequency control pulses and detects the magnetic induction is called an NMR spectrometer.

Magnetic induction of nuclear spins was observed for the first time in 1946 by the groups of Purcell [2] and Bloch [3]. This opened a new field of research leading to many important applications such as molecular structure determination, dynamics studies both in the liquid and solid state [4], and magnetic resonance imaging [5]. The application of NMR to QIP is related to that of structure determination. Many of the same techniques are used, but instead of using uncharacterized molecules, the ones that are used are well understood. In this setting, one can manipulate the nuclear spins as quantum information so that it becomes possible to experimentally demonstrate the fundamental ideas of QIP.

Perhaps the clearest example of early connections of NMR to information theory is the spin echo [6]. When the static field is not "homogeneous" (that is, it is not constant across the sample), the spins precess at different frequencies depending on their location in the sample. As a result, the magnetic induction signal rapidly vanishes. The spin echo is used to "refocus" this effect by inverting the spins, which effectively reverses their precession until they are all aligned. Based on spin echos, the idea of using nuclear spins for (classical) information storage was suggested and patented by Anderson and Hahn as early as 1955 [7, 8].

NMR would not be possible if it were not for relatively long "relaxation" times. Relaxation is the process that tends to re-orient the nuclear spins and randomize their phases, an effect that leads to complete loss of the information represented in such a spin. In liquid state, relaxation times of the order of seconds are common and attributed to the weakness of nuclear interactions and a fast averaging effect associated with the rapid, tumbling motions of molecules in the liquid state.

Today, "off-the-shelf" NMR spectrometers are robust and straightforward to use. The requisite control is to a large extent computerized, so most NMR experiments involve few custom device adjustments after the sample has been obtained. Given that the underlying nature of the nuclear spins is intrinsically quantum mechanical, it is not surprising that after Shor's discovery of the quantum factoring algorithm, NMR was soon studied as a potentially useful device for QIP.

1`2. *A brief survey of NMR quantum information.* – Concrete and workable proposals for using liquid state NMR for quantum information were first given in 1996/7 by two groups: Cory, Fahmy and Havel [9] and Gershenfeld and Chuang [10]. Three difficulties had to be overcome for NMR QIP to become possible. The first was that the standard definitions of quantum information and computation require that quantum information is stored in a single physical system. In NMR, the obvious such system consists of some of the nuclei in a single molecule. But it is not possible to detect single molecules with available NMR technology. The solution that makes NMR QIP possible can be used in general: Consider the large collection of available molecules as an ensemble of identical systems. As long as they all perform the same task, the desired answers can be read off collectively. The second difficulty was that the standard definitions require that read-out takes place by means of a projective quantum measurement where the answer for measuring a qubit is either "0" or "1" and the state is collapsed to the corresponding outcome. The measurement in NMR is much too weak for that. But due to the additive effects of the ensemble, one can observe a (noisy) signal that represents the expectation of the probabilistic answer obtained by the ideal measurement. It turns out that this suffices for realizing most quantum algorithms. The final and most severe difficulty was that even though in equilibrium there is a tendency for the spins to align with the magnetic field, the energy associated with this tendency is very small compared to room temperature. This means that the equilibrium states of the molecules' nuclei are nearly random, with only a small fraction pointing in the right direction. This difficulty was overcome by devising a method (anticipated in 1977 [11]) for singling out the small fraction of the observable signal that represents the desired initial state.

Soon after these difficulties were shown to be overcome or circumventable, two groups were able to implement short quantum algorithms using NMR with small molecules [12, 13]. Although there have been claims that these algorithms beat their classical counterparts, when compared to an equivalent ensemble of probabilistic classical computers, no real advantage has been demonstrated. The main achievement of experiments in liquid state NMR QIP is the demonstration that we can control the unitary evolution of physical qubits sufficiently well to implement simple quantum information processing tasks. The control methods borrowed from NMR and developed for the more complex experiments in NMR QIP are applicable to other device technologies, enabling better control in general.

2. – Principles of liquid state NMR QIP

In order to physically realize quantum information it is necessary to find a way of representing qubits, ways of manipulating them, ways of coupling them to implement non-trivial quantum gates, a means for preparing a useful initial state and the ability to read out the answer. The next subsections show how to accomplish these tasks in liquid state NMR.

$2 \cdot 1$. *Realizing qubits*. – The first step for implementing QIP is to have a physical system that will carry the quantum information, for example by defining qubits in terms of the physical system's state space. There are different ways in which qubits can be instantiated in a physical system. An approach based on subsystems is in [14]. For nuclear spins with spin half, the realization of qubits is direct, thus providing an example for which the physical particles are naturally equivalent to qubits.

A spin-half nucleus is a quantum-mechanical two-state system. Once we fix the direction along the strong external magnetic field, its state space consists of the superpositions of "up" and "down" states. That is, we can imagine that the nucleus is somewhat like a small magnet, with a definite axis, which can point either "up" (logical state $|0\rangle$) or "down" (logical state $|1\rangle$). By the superposition principle, every quantum state of the form $|\psi_0\rangle = \alpha|0\rangle + \beta|1\rangle$ with $|\alpha|^2 + |\beta|^2$ is a possible (pure) state for the nucleus. In the external magnetic field, the two basis states have different energies. This means that the time evolution of $|\psi\rangle$ is given by

$$(1) \qquad |\psi_t\rangle = e^{i\omega\sigma_z t/2}|\psi_0\rangle$$

$$(2) \qquad = e^{i\omega t/2}\alpha|0\rangle + e^{-i\omega t/2}\beta|1\rangle.$$

Here, σ_z is the standard "z" Pauli matrix (see [1]). The number ω is the angular frequency (in radians per second, if t is in seconds) associated with the nucleus in this magnetic field. This frequency is called the Larmor frequency and is observed (see below) as a precession of the spin axis around the z-axis. It is proportional to the energy difference between the "up" and "down" states, which implies that for nuclei with different magnetic moments, the precession frequency can vary substantially. For example, at 11.7 T, the precession frequency for protons is 500 MHz and for ^{13}C it is 125 MHz. This is exploited in measurement and control to distinguish between the different types of nuclei. Nuclei in different chemical environments vary in how they experience external magnetic fields, which has a small effect on the precession frequency and can be used to distinguish, for example, between the two ^{13}C in TCE: The frequency difference (called the "chemical shift") is 600–900 Hz at 11.7 T, depending on the solvent, the temperature and the concentration.

The idea that a single nuclear spin half has a spin direction (as would be expected for a tiny magnet) can be made explicit by means of the Bloch sphere interpretation of a state of the nucleus. See fig. 2. The Bloch sphere is a useful tool for thinking about one- and sometimes about two-qubit processes. Probabilistic mixtures of pure states are represented by vectors of length less than one, inside the Bloch sphere.

2`2. *One-qubit gates.* – The second step for realizing QIP is to give a means for controlling the qubits so that quantum algorithms can be implemented. There are many different ways in which sufficient control can be achieved. The simplest one is to demonstrate that enough basic unitary control steps can be realized to achieve "universal" control. This means that, in principle, every unitary operator can be implemented. It is usually accomplished by techniques for implementing a small set of one- and two-qubit gates [15-17].

One-qubit rotations can be implemented in NMR by applying electromagnetic pulses. A magnetic field applies a torque to a nuclear spin that induces a rotation around the orientation of the magnetic field. The frequency of the rotation is proportional to the strength of the field and to the value of the spin's magnetic moment. Because of the strong magnetic field along the z-axis, the nuclear spin rapidly precesses around the z-axis. A pulse consists of turning on an additional, relatively weak magnetic field with orientation in the (x, y)-plane. Suppose we wish to tip the spin from the z-axis into the

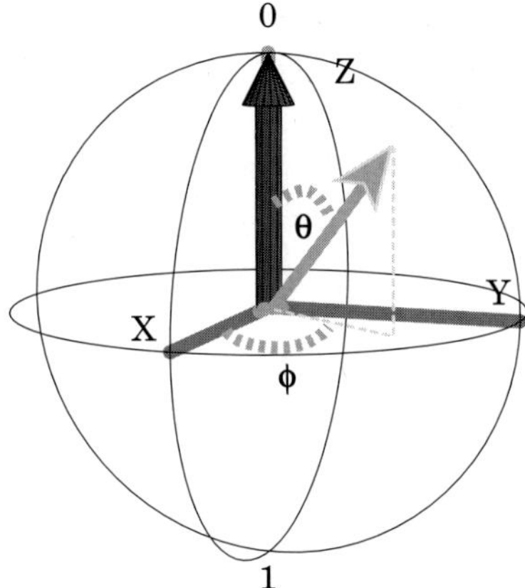

Fig. 2. – Bloch sphere representation of a qubit state. The light grey arrow represents $|\psi\rangle = \cos(\theta/2)|0\rangle + e^{i\phi}\sin(\theta/2)|1\rangle$. The cylindrical arrow along the z-axis indicates the orientation of the magnetic field and the vector for $|0\rangle$. If we write the state in operator form as an expansion in terms of Pauli matrices, $\rho = |\psi\rangle\langle\psi| = (1/2)(\mathbb{1} + \sin(\theta)\cos(\phi)\sigma_x + \sin(\theta)\sin(\phi)\sigma_y + \cos(\theta)\sigma_z)$, then the coefficients of the Pauli matrices form the desired vector. For a pure state this vector is on the surface of the unit sphere and for a mixed state it is inside.

plane. The pulse field will cause such a tipping, but in order to avoid the effect's being reversed by the rapid precession due to the main field, the pulse field needs to rotate at the same frequency. This ensures that the pulse field remains perpendicular to the spin as it tips toward the plane. Therefore a typical NMR control pulse consists of a radio-frequency (RF) electromagnetic field whose frequency is the precession frequency of the spin. This is called a "resonant" pulse. The integrated strength of the pulse determines how far the spin tips. To visualize the process, it is convenient to use the "rotating frame". In this frame, we imagine rotating at the same frequency as the precession frequency of the nucleus. The pulse's magnetic field now looks like a constant magnetic field applied, for example, along the x-axis. The nucleus responds to the resulting torque by tipping perpendicular to the field as expected. If it starts along the z-axis, it tips toward $-y$, then goes to $-z$, to y and finally back to the z-axis. See fig. 3. Any oscillating component of the applied RF field along the z-axis averages to zero on a time scale very short compared to the tipping processes and therefore can be ignored. Thus applied resonant RF fields can usually be thought of as fields in the (x, y)-plane in the rotating frame.

The applied field's direction in the plane of the rotating frame is called the "phase" of the pulse. For a spin half, the Bloch sphere picture of the spin and the effect of applied fields directly corresponds to the unitary gate implemented by a pulse. For example, a pulse applied along the x-axis that causes a rotation around the x-axis by θ acts as $e^{-i\sigma_x\theta/2}$. It is a fact that all rotations of the Bloch sphere can be decomposed into rotations around axes in the plane. For rotations around the z-axis, an easier technique is possible. In choosing the rotating frame, we can arbitrarily set the phase of the x-axis

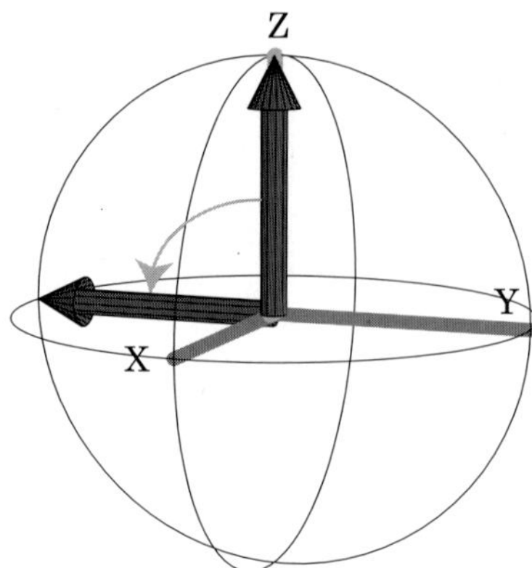

Fig. 3. – Single bit rotation around the x-axis (an "x-rotation"), bringing the nuclear spin direction from the equilibrium position along the z-axis to the (minus) y-axis.

to determine the "logical" x-direction. Once it is set, it rotates with the precession of the nucleus. A rotation by a known angle around the z-axis can be implemented by changing the logical x-direction by the negative of this angle. In this sense, z-pulses can be implemented exactly. In practice it means that the absolute phases of future pulses are shifted by the same amount. Happily, phase control in modern equipment is extremely reliable so that phase errors are negligible compared to other sources of errors.

So far we considered just one nucleus in a molecule. But the RF fields are experienced by the other nuclei also. Nuclei of different composition, such as those of other species of atoms, usually have precession frequencies that differ from the target nucleus by many MHz. In the rotating frame of such a nucleus, the RF field is not constant, but rotates very rapidly compared to its strength. As a result, its effect averages to zero, leading to virtually no tipping of its spin. This is not the case for nuclei of the same composition. Although the chemical environment means that they have frequencies often differing by many kHz, a "hard" RF pulse is strong enough to be short in comparison, requiring only 10's of μs (only 1/100 of a period of a 1 kHz frequency rotation). This means that every nucleus of that species experiences a similar tipping. To target a specific nucleus or nuclei within a narrow frequency band, one can use longer "soft" pulses instead. In summary: To rotate all the nuclei of a given species (like the two ^{13}C of TCE) by a desired angle, we apply a "hard" RF pulse of short duration. To target just one having a distinct precession frequency, we apply "soft" RF pulses. The latter are usually carefully shaped to minimize "cross-talk" (a term that describes unintended effects on other spins, nuclei or qubits). One of the issues an NMR programmer has to deal with is compensating for all the side effects coming from cross-talk and variations in precession frequencies.

2˙3. *Two-qubit gates.* – Two nuclear spins in a molecule in liquid state interact, as you would expect of two magnets. The details of the interactions are more complicated as they are mediated by the electrons. The resulting interaction is called the J-coupling. When

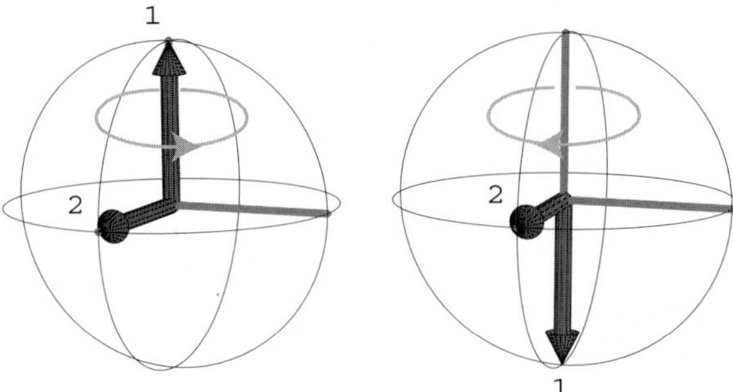

Fig. 4. – Effect of the J-coupling: In the weak-coupling regime with a positive coupling constant, the coupling between two spins can be interpreted as a slightly larger precession frequency of the second spin when the first spin is "up" and a smaller frequency when the first spin is "down". The diagram on the left has spin 1 pointing along the z-axis. In the rotating frame of the second spin, it precesses from x to y. The diagram on the right has spin 1 pointing down instead, causing a precession in the opposite direction in the second spin.

the difference of the precession frequencies between the coupled nuclear spins is large compared to the strength of the coupling, it is a good approximation to write the coupling Hamiltonian as a product of the z Pauli operator for each spin: $H_J = C\sigma_z^{(1)}\sigma_z^{(2)}$. This is the "weak-coupling" regime. Under this Hamiltonian, an initial state $|\psi_0\rangle$ of two nuclear spin qubits evolves as $|\psi_t\rangle = e^{-iC\sigma_z^{(1)}\sigma_z^{(2)}t}|\psi_0\rangle$ (using rotating frames for both nuclei). Since the Hamiltonian is diagonal in the logical basis, the effect of the coupling can be understood as an increase of the (signed) precession frequency of the first spin if the second one is up and a decrease if the second one is down. Since the Hamiltonian is symmetric under exchange of the spins, one can also think of it the other way around (fig. 4). The changes in precession frequency for adjacent nuclei in organic molecules are typically in the range of 20–200 Hz. They are usually much smaller for non-adjacent nuclei. The strength of the coupling is called the "coupling constant" and is usually given as the change in the precession frequency. In terms of the constant C used above, the coupling constant is given by $J = 2C/\pi$ in Hz. For example, the coupling constants in TCE are close to 100 Hz between the two carbons, 200 Hz between the proton and the adjacent carbon, and 9 Hz between the proton and the far carbon.

The J-coupling and the one-qubit pulses suffice for realizing the controlled-not operation usually taken as the starting point for QIP. However, this requires being able to turn the coupling on only when needed, while in our situation the coupling is constant. How to effectively turn it off is the subject of the next subsection. A pulse sequence for implementing the controlled-not in terms of the J-coupling constitutes the first quantum algorithm of sect. **3**.

$2\dot{}4$. *Turning off the coupling*. – The coupling between the nuclear spins in a molecule is always "on", that is, we cannot physically turn it off. But for QIP, we need to be able

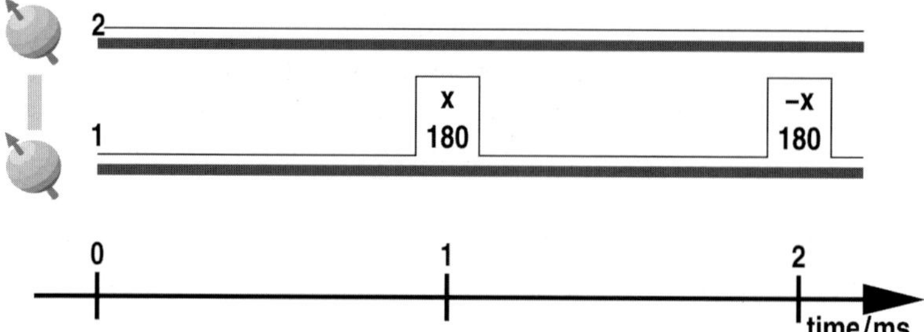

Fig. 5. – Pulse sequence for refocusing the coupling. The sequence of events is shown with time running from left to right. The two spins' lifelines are shown in grey with the RF power targeted at each spin indicated by the black line above. Pulses are applied to spin 1 only, as indicated by the rectangular rises in RF power at 1 ms and 2 ms. The phase for each pulse is given with the pulse. The angle is determined by the area under the pulse and is also given explicitly. Ideally for pulses of this type, the pulse times (width of the rectangles) should be zero. In practice, for hard pulses, they can be as small as ≈ 0.01 ms. Any $\sigma_z^{(1)}\sigma_z^{(2)}$ coupling's effect is refocused by the sequence shown, so that the final state of the two spins is the same as the initial state.

to maintain a state in memory and to couple qubits selectively. This is a problem that NMR spectroscopists solved well before the development of modern quantum information concepts. The idea is to use the control of single spins to cancel the interaction's effect over a given time period. This technique is called refocusing and requires applying a 180° pulse to one of two coupled spins at the midpoint of the desired time period. To understand how this works, consider again the visualization of fig. 4. A general state is in a superposition of the four logical states of the two spins. By the linear superposition rules, it suffices to consider the evolution with spin 1 being in one of its two logical states, up or down along the z-axis. Suppose we wish to remove the effects of the coupling over a period of 2 ms. To do so, wait 1 ms. The effect on spin 2 in its rotating frame is to precess counter-clockwise if spin 1 is up, and clockwise for the same angle if it is down. Now apply a pulse that rotates spin 1 by 180° around the x-axis. This is called an "inversion", or in the current context, a "refocusing" pulse. It exchanges the up and down states. For the next 1 ms, the effect of the coupling on spin 2 is to undo the earlier rotation. At the end of the period, we can apply another 180° pulse to reverse the earlier inversion. Everything is back to where it started. The pulse sequence is depicted in fig. 5.

Turning off couplings between more than two nuclei can be quite complicated, unless one takes advantage of the fact that non-adjacent nuclei tend to be relatively weakly coupled. Methods that scale polynomially with the number of nuclei and that can be used to selectively couple pairs of nuclei can be found in [18, 19]. These techniques can be used in other physical systems where couplings exist that are difficult to turn off directly. An example is qubits represented by the state of one or more electrons in tightly packed quantum dots.

2˙5. *Measurement.* – To determine the "answer" of a quantum computation, it is necessary to make a measurement. As noted earlier, the technology for making a projective measurement of individual nuclei does not yet exist. In liquid state NMR, instead of using just one molecule to define a single quantum register, we use a large ensemble of molecules in a test tube. Ideally, their nuclei are all placed in the same initial state, and the subsequent RF pulses affect each molecule in the same way. As a result, weak signals from (say) the proton nuclei in TCE can add up to make a detectable signal. The signal we are interested in is the magnetization in the (x, y)-plane, to be picked up by coils placed transversely to the main field. Because the interaction of any given nucleus with the coil is very weak, we can think of the situation (almost) classically. That is, each nucleus behaves like a tiny bar magnet and the fields from the nuclei add up, creating a "bulk magnetization". As the nuclei precess, so does the bulk magnetization. This is detected by the coil (if tuned near the precession frequency). By observing the induced oscillating signal coming from the coil, we can infer both the magnetization in the plane and the phase of the magnetization with respect to the current rotating frame.

Consider the TCE molecule with three spin-half nuclei used for information processing. The proton magnetization can be distinguished from the carbon magnetization by the precession frequency (requiring two coils tuned to 500 MHz and 125 MHz in an 11.7 T field, respectively). For simplicity, we restrict our attention to the two carbons. (It is possible to actively remove all effects of the proton by using a technique called "decoupling".) Assume that the state of the carbon nuclei in each molecule is the same, described by the two spin-half (or qubit) density matrix $\rho(t)$, evolving according to the internal Hamiltonian. Choosing one rotating frame for both carbons at the precession frequency of carbon 1, the Hamiltonian is approximately given by

$$(3) \qquad H = \pi 900\,\mathrm{Hz}\,\sigma_z{}^{(2)} + \pi 50\,\mathrm{Hz}\,\sigma_z{}^{(1)}\sigma_z{}^{(2)}.$$

The magnetization detected in the x-direction at time t is proportional to $M_x(t) = \mathrm{tr}(\rho(t)(\sigma_x{}^{(1)} + \sigma_x{}^{(2)}))$, where $\mathrm{tr}(\sigma)$ denotes the trace, that is, the sum of the diagonal elements of the matrix σ. This links the magnetization signal to the Bloch sphere representation for single qubits. The constant of proportionality depends on the size of the ensemble and the magnetic moments of the nuclei. From the point of view of NMR, it corresponds to an arbitrary scale. What matters is how strong this signal is compared to the noise in the system. We can also detect the magnetization $M_y(t)$ in the y-direction and form the complex number representing the planar magnetization

$$(4) \qquad M(t) = M_x(t) + iM_y(t)$$

$$(5) \qquad = \mathrm{tr}\left(\rho(t)\left(\sigma_+{}^{(1)} + \sigma_+{}^{(2)}\right)\right),$$

where we defined $\sigma_+ = \sigma_x + i\sigma_y = \left(\begin{smallmatrix} 0 & 2 \\ 0 & 0 \end{smallmatrix}\right)$. What can we infer about $\rho(0)$ from observing $M(t)$ for some time? For starters, neglect the coupling Hamiltonian. Under the chemical

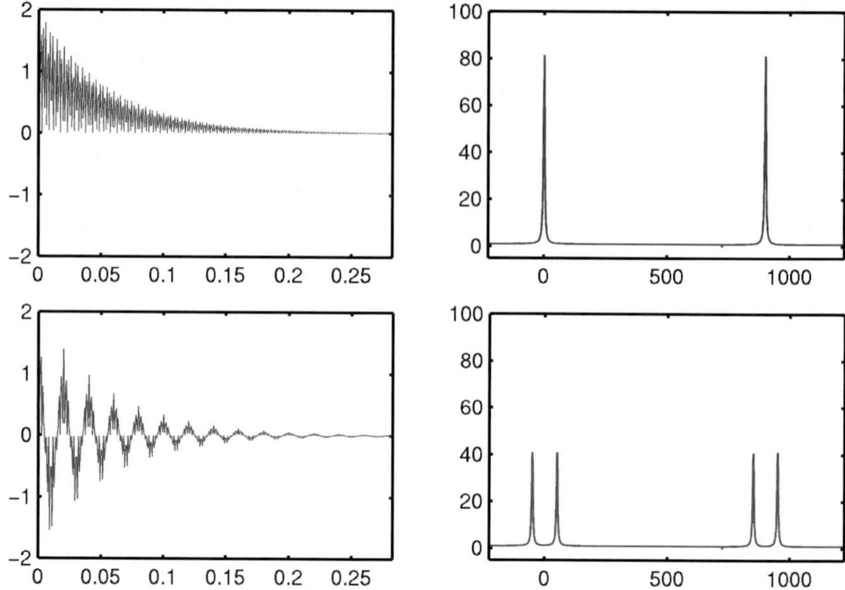

Fig. 6. – Simulated signals (left) and spectra (right). Top left: The x-magnetization signal as a function of time acquired for a pair of uncoupled spins with a relative chemical shift of 900 Hz. The signal (called the "free induction decay") decays with a half-time of 0.0385 s due to simulated relaxation processes. Typically, the half-times are much longer. A short one was chosen to broaden the peaks for visual effect. Top right: The spectrum, that is, the Fourier transform of the combined x- and y-magnetization. It shows peaks at frequencies of 0 Hz and 900 Hz due to the independently precessing pair of spins. Bottom left: The x-magnetization signal when the two spins are coupled as described in the text. Bottom right: The corresponding spectrum. Each peak from the previous spectrum "splits" into two. The left peak of each pair is associated with the other spin being in the $|1\rangle$ state and the right one with the $|0\rangle$ state. The x-axis units are relative intensity with the same constant of proportionality for the two spectra.

shift Hamiltonian $H_{\mathrm{CS}} = \pi 900\,\mathrm{Hz}\,\sigma_z{}^{(2)}$, $M(t)$ evolves as

$$
\begin{aligned}
(6) \qquad M(t) &= \mathrm{tr}\left(e^{-iH_{\mathrm{CS}}t}\rho(0)e^{iH_{\mathrm{CS}}\,t}\left(\sigma_+{}^{(1)} + \sigma_+{}^{(2)}\right)\right) \\[2mm]
&= \mathrm{tr}\left(\rho(0)e^{iH_{\mathrm{CS}}\,t}\left(\sigma_+{}^{(1)} + \sigma_+{}^{(2)}\right)e^{-iH_{\mathrm{CS}}\,t}\right) \\[2mm]
&= \mathrm{tr}\left(\rho(0)\left(\sigma_+{}^{(1)} + e^{iH_{\mathrm{CS}}\,t}\sigma_+{}^{(2)}e^{-iH_{\mathrm{CS}}\,t}\right)\right) \\[2mm]
&= \mathrm{tr}\left(\rho(0)\left(\sigma_+{}^{(1)} + e^{i2\pi 900\,\mathrm{Hz}\,t}\sigma_+{}^{(2)}\right)\right) \\[2mm]
&= \mathrm{tr}\left(\rho(0)\sigma_+{}^{(1)}\right) + \mathrm{tr}\left(e^{i2\pi 900\,\mathrm{Hz}\,t}\sigma_+{}^{(2)}\right),
\end{aligned}
$$

where we used the fact that $\mathrm{tr}(AB) = \mathrm{tr}(BA)$ in the first step. Thus the signal is a combination of a constant signal given by the magnetization in the plane of the first spin, and a signal oscillating with a frequency of 900 Hz with amplitude given by the planar magnetization of the second spin. The two contributions can be separated by Fourier transforming $M(t)$, which results in two distinct peaks, one at 0 Hz and a second at 900 Hz. See fig. 6.

To see what happens when we add the coupling, we recall that the up/down states are invariant under the Hamiltonian, and the only observables seen directly are single qubit operators. So

$$(7) \qquad M(t) = \mathrm{tr}\left(\rho(t)\sigma_+{}^{(1)}\right) + \mathrm{tr}\left(\rho(t)\sigma_+{}^{(2)}\right)$$

$$= \mathrm{tr}\left(\rho(t)\sigma_+{}^{(1)}\mathbb{1}\right) + \mathrm{tr}\left(\rho(t)\sigma_+{}^{(2)}\mathbb{1}\right)$$

$$= \mathrm{tr}\left(\rho(t)\sigma_+{}^{(1)}\left(e_\uparrow{}^{(2)} + e_\downarrow{}^{(2)}\right)\right) + \mathrm{tr}\left(\rho(t)\left(e_\uparrow{}^{(1)} + e_\downarrow{}^{(1)}\right)\sigma_+{}^{(2)}\right),$$

where $e_\uparrow = \left(\begin{smallmatrix} 1 & 0 \\ 0 & 0 \end{smallmatrix}\right)$ and $e_\downarrow = \left(\begin{smallmatrix} 0 & 0 \\ 0 & 1 \end{smallmatrix}\right)$. Using a similar calculation to the one leading to eq. (6), the first term can be written as

$$(8) \qquad M_1(t) = \mathrm{tr}\left(e^{-iH\,t}\rho(0)e^{iHt}\sigma_+{}^{(1)}\left(e_\uparrow{}^{(2)} + e_\downarrow{}^{(2)}\right)\right)$$

$$(9) \qquad = e^{i2\pi 50\,\mathrm{Hz}\,t}\,\mathrm{tr}\left(\rho(0)\left(\sigma_+{}^{(1)}e_\uparrow{}^{(2)}\right)\right) + e^{-i2\pi 50\,\mathrm{Hz}\,t}\,\mathrm{tr}\left(\rho(0)\sigma_+{}^{(1)}e_\downarrow{}^{(2)}\right),$$

and similarly for the second term, but with an offset frequency of 900 Hz due to the chemical shift. It can be seen that the zero-frequency signal splits into two signals, one with frequency -50 Hz, the other with 50 Hz. The difference between the two frequencies is the coupling constant. The amplitudes of the different frequency signals can be used to infer the expectations (up to an overall scale) of operators like $\sigma_+{}^{(1)}e_\uparrow{}^{(2)}$, given by $\mathrm{tr}(\rho(0)\sigma_+{}^{(1)}e_\uparrow{}^{(2)})$. For k spin-half nuclei where every pair is measurably coupled, the peaks associated with each nucleus split into 2^{k-1} peaks, each associated with operators like $\sigma_+ e_\uparrow e_\downarrow \dots$. From these, expectations of single-spin operators can be obtained by summing over the amplitudes of the peaks associated with a given nucleus. Product operators with only one σ_x or σ_y, like $\sigma_x\sigma_z\mathbb{1}\sigma_z$, can usually be inferred by taking different linear combinations of the peaks in a group.

In addition to the unitary evolution due to the internal Hamiltonian, relaxation processes tend to decay $\rho(t)$ toward the equilibrium state. In liquid state, equilibrium is close to $\rho(t) = \mathbb{1}/N$, where N is the total dimension of the state space. The equilibrium deviation from this density matrix has only magnetization along the z-axis (see subsect. 2`6). Thus, the signal in the plane that is observed decays to zero. To a good approximation we can write

$$(10) \qquad \rho(t) = \frac{1}{N}\mathbb{1} + e^{-\lambda t}\rho'(t),$$

where $\rho'(t)$ has trace zero and evolves unitarily under the Hamiltonian. The effect of the relaxation is that $M(t)$ has an exponentially decaying envelope, explaining the conventional name for $M(t)$: the "free induction decay" (FID). Typical half-times for the decay are 0.1–2 s for nuclei used for QIP.

For QIP, we wish to measure the probability p that a given (say the first) qubit is in the state $|1\rangle$. We have $2p - 1 = \mathrm{tr}(\rho \sigma_z^{(1)})$. In principle, σ_z can be measured using $M(t)$ by first giving a pulse which rotates z to x (say) in the Bloch sphere representation. A measurement of $\mathrm{tr}(\rho \sigma_x^{(1)})$ then gives the desired number. However, there is a problem in that the states ρ that are available are highly mixed (close to $\mathbb{1}/N$). The next subsection discusses how to compensate for this problem.

2˙6. *The initial state*. – Because the energy difference between the nuclear spins' up and down states is so small compared to room temperature, the equilibrium distribution of states is nearly random. In the liquid samples that we use, equilibrium is established after 10–40 s if no RF fields are being applied. As a result, all computations performed start with the sample in equilibrium. One way to think of this initial state is that every nucleus in each molecule begins in the highly mixed state $(1 - \epsilon)\mathbb{1}/2 + \epsilon|0\rangle\langle0|$, where ϵ is a small number (of the order of 10^{-5}). This is a nearly random state with a small excess of the state $|0\rangle$. The expression for the initial state is obtained by using the fact that the equilibrium state ρ_{thermal} is proportional to $e^{-H/kT}$, where H is the internal Hamiltonian of the nuclei in a molecule (in energy units), T is temperature and k is the Boltzmann constant. In our case, H/kT is very small and the coupling terms are negligible. Therefore

$$\tag{11} e^{-H/kT} \approx e^{-\epsilon_1 \sigma_z^{(1)}/kT} e^{-\epsilon_2 \sigma_z^{(2)}/kT} \ldots$$

$$\tag{12} e^{-\epsilon_1 \sigma_z^{(1)}/kT} \approx \frac{\mathbb{1} - \epsilon_1 \sigma_z^{(1)}}{kT}$$

$$\tag{13} e^{-H/kT} \approx \frac{\mathbb{1} - \epsilon_1 \sigma_z^{(1)}}{kT} - \frac{\epsilon_2 \sigma_z^{(2)}}{kT} - \ldots,$$

where ϵ_i is half of the energy difference between the up and down states of nucleus i.

Clearly the initial state available is very far from what is needed for standard QIP. However, it can still be used to perform interesting computations. To see how, first note that the observables available in NMR (which are sufficient for obtaining the answers of quantum algorithms) have trace zero, which means that the contribution to the state proportional to $\mathbb{1}$ is not visible, contributing nothing to the measured expectations. The only component of the state ρ that is seen is the so-called "deviation" density matrix $\delta = \rho - \mathbb{1}/2^n$, where n is the number of spin-half nuclei used. In general, a deviation for a state is an operator obtained from the density matrix by subtracting some multiple of $\mathbb{1}$. Thus $\epsilon|0\rangle\langle0|$ is a deviation for the initial equilibrium state of a single nucleus. The second observation is that all the unitary operations used, as well as the non-unitary ones to be discussed below, preserve the completely mixed state $\mathbb{1}/2^n$ ([1]). As a result our observations depend only on the initial deviation. Since the scales are relative, the most important issue is that the signals are strong enough to be observable over the noise, not the absolute scale of the deviation.

The simplest example consists of one qubit/spin-half. If U is the total unitary operator associated with a computation, then the initial deviation $\epsilon|0\rangle\langle 0|$ is transformed to $\epsilon U|0\rangle\langle 0|U^\dagger$. For QIP purposes, the goal is to determine what the final probability of measuring $|0\rangle$ is, given that $|0\rangle$ is the initial state. This can be determined by measuring the expectations a_i of σ_z for the initial state, $a_0 = \epsilon$, and for the final state, $a_1 = \epsilon\,\mathrm{tr}(U|0\rangle\langle 0|U^\dagger\sigma_z)$. The desired answer is $(1 - (a_1/a_0))/2$. This method generalizes to many qubits if a preparable initial state is

$$(14) \qquad \rho_0 = \frac{1-\epsilon}{2^n}\mathbb{1} + \epsilon|00\ldots\rangle\langle 00\ldots|.$$

This state is called a "pseudo-pure" state, because the deviation is up to scale equivalent to that of a pure state. Among the most important enabling techniques in NMR QIP are the methods that can be used to transform the initial thermal equilibrium state to a standard pseudo-pure state. An example of how that can be done will be given as the second algorithm in sect. **3**. The basic principle for each method is to create, directly or indirectly by summing over multiple experiments, a new initial state as a sum $\rho_0 = \sum_i U_i\rho_{\text{thermal}}U_i^\dagger$, where the U_i are carefully and sometimes randomly chosen [9,10,20,21]. One of the most useful tools for realizing such sums is gradient fields.

2'7. *Gradient fields*. – Modern NMR spectrometers are equipped with the capability of applying, for a chosen, brief amount of time, a magnetic-field gradient in any direction. If the direction is along the z-axis, this means that while the gradient is on, the field in the z-direction varies as $B(z) = B_0 + \gamma z B_1$, where B_0 is the strong, main field. The effect of the gradient is that nuclei at different positions along z have precession frequencies varying with $B(z)$. One of the most important applications of these techniques is NMR imaging, as gradients provide one method for distinguishing different parts of the sample along the z-axis. Different positions along the x- or the y-axis can be distinguished by gradients in those directions.

The effect of applying the gradient can be visualized for the example where there is only one observable nucleus per molecule. Suppose that before the gradient pulse, the deviation density matrix is (proportional to) σ_x and the gradient is applied for a time t. Then the new state for a nucleus at position z along the z-coordinate is given by $\cos(\mu zt)\sigma_x + \sin(\mu zt)\sigma_y$, where the constant μ depends linearly on the strength of the gradient and the magnetic moment of the nucleus. See fig. 7. Only the total magnetization is measured by the observation coil. Thus an idealized observation adds

---

([1]) The intrinsic relaxation process does not satisfy this. But its contribution is either negligible over the time scale of typical experiments or can be removed by using subtractive phase cycling.

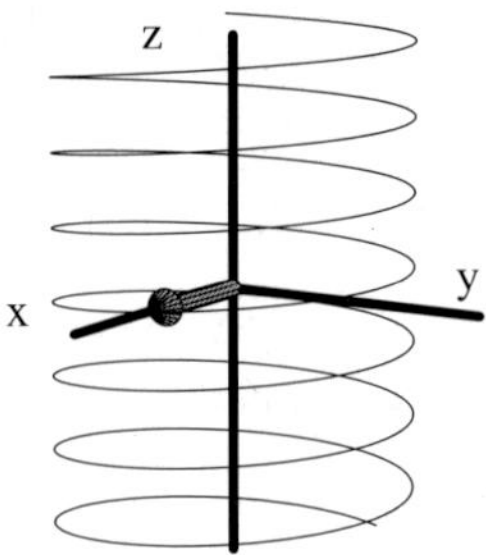

Fig. 7. – Effect of a pulsed gradient field along the z-axis on initial x-magnetization: A spin at $z = 0$ is not affected but the ones above and below are rotated by an amount proportional to z. As a result, the local magnetization follows a spiral curve.

the magnetization from each nucleus over a fixed interval along the z-axis, so that the observed x magnetization is

$$(15) \qquad M_x = \int_{-a}^{a} dz \, \mathrm{tr} \left(\sigma_x (\cos(\mu z t)\sigma_x + \sin(\mu z t)\sigma_y) \right)$$

$$(16) \qquad = 2 \int_{-a}^{a} dz \, \cos(\mu z t).$$

For large a, this is small compared to the magnetization $2 \int_{-a}^{a} dz = 4a$ seen before the gradient was applied. From the point of view of our observations, the effect of the gradient is the same as the averaging map:

$$(17) \qquad \rho \longrightarrow \frac{1}{2a} \int_{-a}^{a} dz \, e^{-i\sigma_z \mu t/2} \rho \, e^{i\sigma_z \mu t/2},$$

just the kind of sum (or integral) useful for making states like pseudo-pure states. In the case of more than one nucleus in a molecule used for QIP, one can use selective refocusing between gradient pulses to restrict the effect to a single nucleus.

An interesting property of the gradient field is that, in principle, its effect can be reversed by applying an opposite gradient for the same amount of time. This only works if it is reversed sufficiently soon. Diffusion moves the molecules around and will randomize the phase before the echo, leading to an effective irreversibility that can be fine tuned by changing the delay between or the strength of the two gradient pulses. This turns out to be a very useful tool for studies of decoherence, believed to be one of the most significant problems for QIP.

For multiple nuclei, a gradient pulse has the effect of labeling "transitions" $|a\rangle\langle b|$ in the density matrix by a z-dependent phase that also depends on which nuclei "flip" between a and b. After subsequent processing, the labeling can be undone by using a gradient pulse on one nucleus (for example) of a length chosen so that only the desired transitions contribute to the observed signal. This effect is used for the pseudo-pure state preparation discussed in subsect. **3**`2.

3. – Examples of quantum algorithms for NMR

We give three examples of quantum algorithms with their implementations. The first two are fundamental to QIP with NMR. Realizations of the controlled-not operation are needed to translate quantum algorithms into the language of NMR. Explicit procedures for making pseudo-pure states are necessary before any of these algorithms are tried. The last example shows how these elements can be combined to confirm the behavior of simple error-correction procedures.

3`1. *The controlled-not*. – One of the standard gates used in most quantum algorithms is the controlled-not. An important problem is to determine how to implement the controlled-not using basic NMR control methods. This is the subject of our first example. The controlled-not gate acts on two qubits. Its action can be described by "if the first qubit is $|0\rangle$, then flip the second qubit". As a unitary operator, it is defined by the action on logical states

$$(18) \qquad\qquad |00\rangle \longrightarrow |00\rangle,$$

$$|01\rangle \longrightarrow |01\rangle,$$

$$|10\rangle \longrightarrow |11\rangle,$$

$$|11\rangle \longrightarrow |10\rangle,$$

or as the operator

$$(19) \qquad |0\rangle_1^1\langle 0| + |1\rangle_1^1\langle 1|\sigma_x^{(2)} = \frac{\left(\mathbb{1} + \sigma_z^{(1)}\right) + \left(\mathbb{1} - \sigma_z^{(1)}\right)\sigma_x^{(2)}}{2}.$$

The unitary operations that are implemented by simple NMR manipulations are rotations around an axis (or Pauli matrix, $e^{-i\sigma_u\theta/2}$), or around a product of two Pauli matrices (the coupling, $e^{-i\sigma_z^{(i)}\sigma_z^{(j)}\theta/2}$). Such rotations and their effects on deviation density matrices are readily understood by a generalization of the Bloch sphere picture called the "product operator formalism" introduced by Sörensen *et al.* [22]. This underlies the implementation of the controlled-not below.

To implement the controlled-not using NMR techniques one can decompose it into a sequence of 90° pulses around the main axes on each of the two qubits, and a 90° rotation around $\sigma_z^{(1)}\sigma_z^{(2)}$, which can be realized by using the coupling. One way to find a decomposition is to first realize that $e^{-i\sigma_z^{(1)}\sigma_z^{(2)}\pi/4}$ can be viewed as a combination of two conditional gates. The first applies $e^{-i\sigma_z\pi/4}$ to qubit 2 conditional on qubit 1's state being $|0\rangle$. The second applies $e^{i\sigma_z\pi/4}$ conditional on the state being $|1\rangle$. By following this with a $-90°$ rotation around the z-axis $(e^{i\sigma_z\pi/4})$ on qubit 2, we obtain an operation

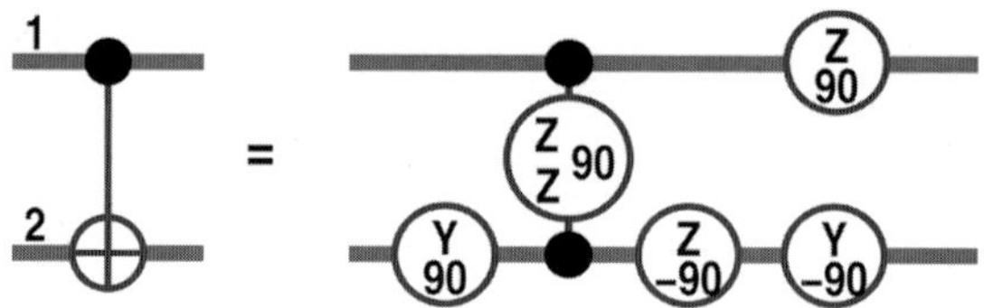

Fig. 8. – Quantum network for implementing the controlled-not using operations available in NMR. The conventions are as explained in [1]. The two one-qubit z-rotations can be implemented by a change of rotating frame reference phase without applying any RF pulses.

that, conditional on qubit 1 being $|1\rangle$, applies the $-180°$ rotation $U = i\sigma_z$. If we precede this sequence with $e^{-i\sigma_y\pi/4}$ and follow it by $e^{i\sigma_y\pi/4}$ (this is called "conjugating" the gate by a $-90°$ y-rotation), the total effect is a conditional $-i\sigma_x$ operation. Note how the conjugation rotated the operation's axis according to the Bloch sphere rules. The controlled-not is obtained by eliminating the $-i$ using a $90°$ z-rotation on qubit 1: The total effect is $e^{-i\pi/4}|0\rangle_1^1\langle 0| + e^{-i\pi/4}|1\rangle_2^2\langle 1|\sigma_x^{(2)}$, which up to a global phase is the controlled-not. A quantum network for this sequence of gates is shown in fig. 8. The pulse sequence implementation is in fig. 9.

The effect of the NMR pulse sequence for the controlled-not can be visualized for specific logical initial states using the Bloch-sphere picture of the states. This is shown for two initial states in fig. 10.

The ability to depict the evolution of the NMR pulse sequence implementing the controlled-not using the Bloch sphere is due partly to the fact that it is implemented using

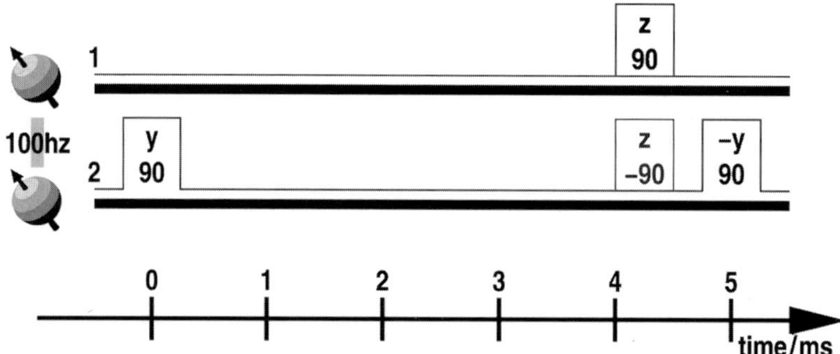

Fig. 9. – Pulse sequence for realizing the controlled-not. The control bit is spin 1 and the target is spin 2. The pulses are shown using the representation discussed in fig. 5. The z-pulses are "virtual", requiring only a change of reference frame. Their timing between the RF pulses is immaterial, because they commute with the coupling that evolves in between. The delay between the two RF pulses is $1/(2J)$ (5 ms if $J = 100\,\mathrm{Hz}$), which realizes the desired coupling term by internal evolution. Note that an ideal $90°$ pulse with phase $-x$ has the same effect as a $-90°$ pulse with phase x, except for an overall constant. The pulse widths are exaggerated and should be as short as possible to avoid errors due to coupling evolution during the RF pulses. Alternatively, there are techniques to compensate for some of these errors [23].

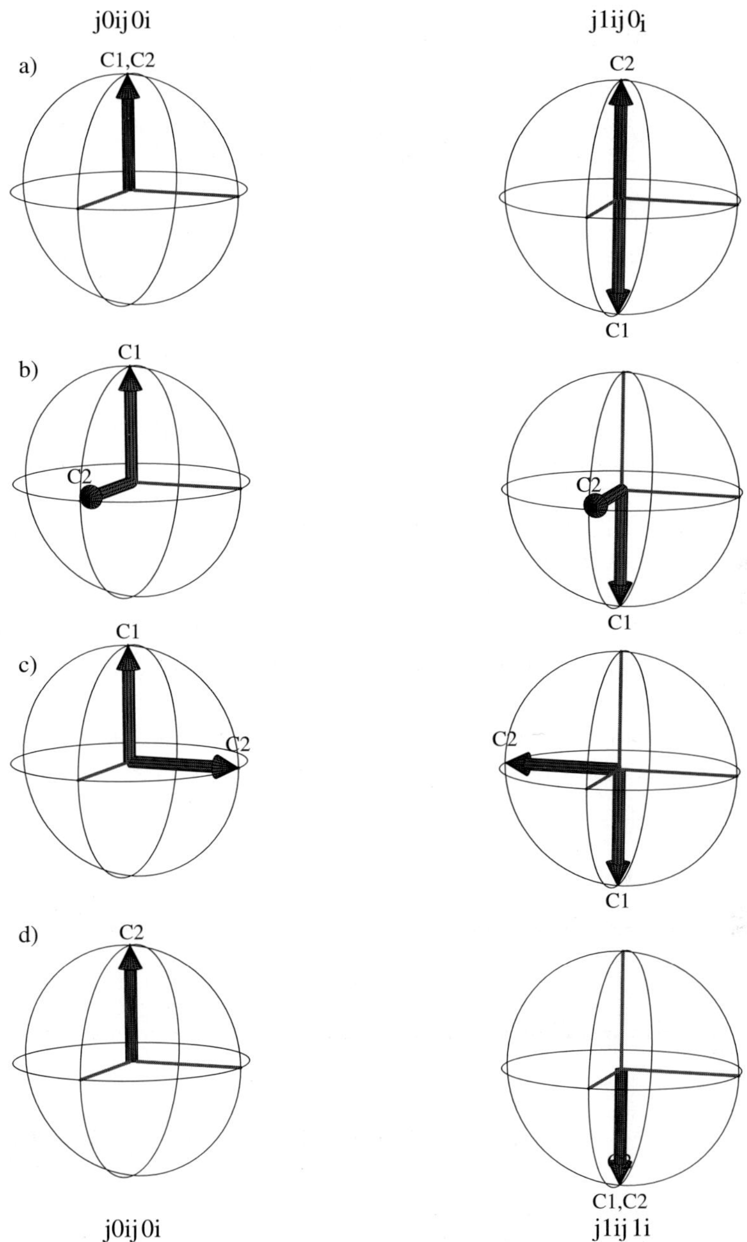

Fig. 10. – Sequence of states for the controlled-not pulse sequence. The first column has both spins initially in the logical $|0\rangle$ state, represented by two arrows pointing up. The second has the first spin initially in the $|1\rangle$ state, indicated by its arrow pointing down. The shown configurations (a-d) are achieved initially, after the $90°$ y-rotation, after the J-coupling, and at the end. The conditional effect is realized by the second spin's pointing down at the end of the second column. The effect of the J-coupling causing the evolution from (b) to (c) is best understood as a conditional rotation around the z-axis (forward by $90°$ if the first spin is up, backward if it is down).

only 90° rotations, and partly because the controlled-not preserves the set of logical basis states. The latter means that it can be interpreted as a classical reversible operation. Things are no longer so simple if the initial state of the spins is $(1/\sqrt{2})(|0\rangle + |1\rangle)|0\rangle$, for example. This is representable as spin 1's arrow pointing along the x-axis, but the J-coupling leads to a superposition of states (a maximally entangled state) no longer representable by a simple combination of arrows in the Bloch sphere.

3\`2. *Creating a labeled pseudo-pure state.* – The goal of making a pseudo-pure initial state is to remove all detectable signals from initial states different from the desired $|0\ldots0\rangle\langle0\ldots0|$. There are several different ways of accomplishing this. For example, one can perform multiple experiments where signal from other states averages to zero (temporal averaging); or one can use gradients to remove the unwanted states in one experiment (spatial averaging). Instead of creating the standard pseudo-pure initial state $(1 - \epsilon/2^n)\mathbb{1} + \epsilon|0\ldots\rangle\langle0\ldots|$, one can create the labeled pseudo-pure state $(1/2^n)\mathbb{1} + \delta\sigma_x^{(1)}|0\ldots\rangle\langle0\ldots|$. This state is easily recognizable by making an NMR observation of the first spin. Only one of the usual set of peaks arising from couplings to the other spins should be seen. The labeled pseudo-pure state can be used as a standard pseudo-pure state on one less qubit. Observation of the final answer of a computation is possible by observing spin 1, provided that the coupling to the answer-containing qubit is sufficiently strong. This is done by comparing the signal in the set of peaks corresponding to the answer qubit being in $|0\rangle$, and in $|1\rangle$, respectively. Alternatively it can be used to investigate a process applied to one qubit with a generic input state. Examples include experimental verification of one-qubit error-correcting codes as explained in subsect. **3\`3**.

Here we describe one version of the spatial averaging technique for obtaining a labeled two-qubit pseudo-pure state. The technique can be extended to any number of qubits and, at the same time, serves as one of the simplest benchmarks to establish the quality of the quantum control used.

For concreteness, consider the two carbon nuclei in labeled TCE with the proton decoupled so that its effect can be ignored. A "transition" in the density matrix for this system is an element of the density matrix of the form $|ab\rangle\langle cd|$. Let $\Delta(ab,cd) = (a - c) + (b - d)$. Applying a pulsed gradient along the z-axis evolves the transitions according to: $|ab\rangle\langle cd| \rightarrow e^{-i\Delta(ab,cd)\mu z}|ab\rangle\langle cd|$, where μ is determined by the gradient strength and pulse time and z is the position along the z-coordinate of the molecule. For example, $|01\rangle\langle10|$ has $\Delta = 0$ and is not affected, while $|00\rangle\langle11|$ acquires a phase of $e^{i2\mu z}$. There are only two transitions (the "two-coherences") whose acquired phase has a rate of ±2 along the z-axis, these are $|00\rangle\langle11|$ and $|11\rangle\langle00|$. The idea is to first recognize that these transitions can be used to define a labeled pseudo-pure "cat" state, then to exploit their unique behavior under the gradient to extract that pseudo-pure state, and finally to "decode" to a standard labeled pseudo-pure state. Note that the property that two-coherences' phases evolve at twice the basic rate is a uniquely quantum phenomenon for two spins. No such effect is observed for a pair of classical spins.

The standard two-qubit labeled pseudo-pure state's deviation from the identity is proportional to $\rho_{\mathrm{std}_x} = \sigma_x^{(1)}(1/2)(\mathbb{1} + \sigma_z^{(2)})$. We can consider other deviations like this, with the two (products of) Pauli operators occurring in the expression commuting. An example is

$$(20) \qquad \rho_{\mathrm{cat}_x} = \left(\sigma_x^{(1)}\sigma_x^{(2)}\right)\frac{1}{2}\left(\mathbb{1} + \sigma_z^{(1)}\sigma_z^{(2)}\right),$$

where we replaced $\sigma_x^{(1)}$ by $\sigma_x^{(1)}\sigma_x^{(2)}$ and $\sigma_z^{(2)}$ by $\sigma_z^{(1)}\sigma_z^{(2)}$. Observe that the two Pauli products commute. We will show that there is a simple sequence of $90°$ rotations whose effect is to map $\sigma_x^{(1)}\sigma_x^{(2)} \to \sigma_x^{(1)}$ and $\sigma_z^{(1)}\sigma_z^{(2)} \to \sigma_z^{(2)}$, thus converting the state ρ_{cat_x} to ρ_{std}. It is straightforward to check that

$$(21) \qquad \rho_{\mathrm{cat}_x} = |00\rangle\langle 11| + |11\rangle\langle 00|,$$

so that it consists only of two-coherences. Another such state is

$$(22) \qquad \rho_{\mathrm{cat}_y} = \left(\sigma_x^{(1)}\sigma_y^{(2)}\right)\frac{1}{2}\left(\mathbb{1} + \sigma_z^{(1)}\sigma_z^{(2)}\right)$$

$$(23) \qquad = -i|00\rangle\langle 11| + i|11\rangle\langle 00|.$$

Suppose we can create a state that has a deviation of the form $\rho = \alpha\rho_{\mathrm{cat}_x} + \beta\rho_{\mathrm{rest}}$ such that ρ_{rest} contains no two-coherences or zero-coherences. After applying a gradient pulse, the state becomes

$$(24) \qquad \alpha\left(\cos(2\mu z)\rho_{\mathrm{cat}_x} + \sin(2\mu z)\rho_{\mathrm{cat}_y}\right) + \beta\rho_r(z),$$

where $\rho_r(z)$ has only periodicities along z with rates μ, not 2μ or 0. We can then decode this to the state

$$(25) \qquad \varrho = \alpha\left(\cos(2\mu z)\rho_{\mathrm{std}_x} + \sin(2\mu z)\rho_{\mathrm{std}_y}\right) + \beta\rho_r'(z)$$

$$= \alpha\left(\cos(2\mu z)\sigma_x^{(1)} + \sin(2\mu z)\sigma_y^{(1)}\right)\frac{1}{2}\left(\mathbb{1} + \sigma_z^{(1)}\right) + \beta\rho_r'(z).$$

If we now apply a gradient pulse of twice the total strength and opposite orientation, the first term is restored to $\alpha\rho_{\mathrm{std}_x}$, but the second term retains non-zero periodicities along z. Thus, if we no longer use any operations to distinguish different molecules along the z-axis, or if we let diffusion erase the memory of the position along z, then the second term is eliminated from observability by averaging to zero. The desired labeled pseudo-pure state is obtained. Zero-coherences during the initial gradient pulse are acceptable provided that the decoding transfers them to coherences different from zero or two during the final pulse to ensure that they also average to zero. A pulse sequence that realizes a version of the above procedure starting with an initial deviation of $\sigma_z^{(1)}$ is shown in fig. 11.

We can follow what happens to an initial deviation density matrix of $\sigma_z^{(1)}$ as the network of fig. 11 is executed. We use product operators with the abbreviations $I = \mathbb{1}$,

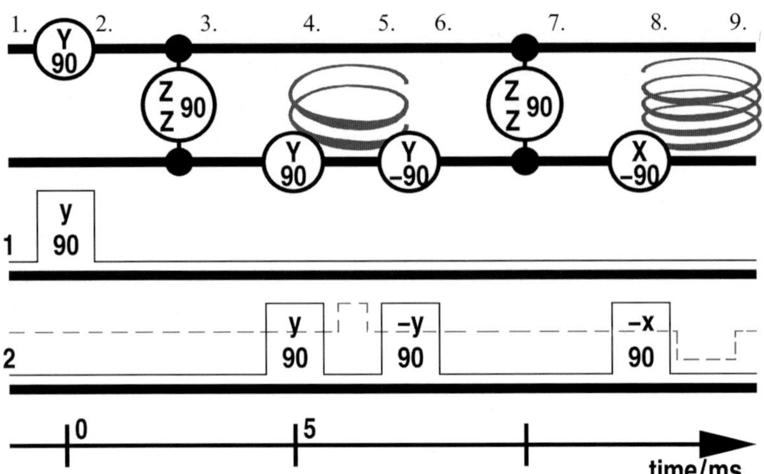

Fig. 11. – Quantum network and pulse sequence to realize a two-qubit labeled pseudo-pure state. The network is shown above the pulse sequence realizing it. A coupling constant of 100 Hz is assumed. Gradients are indicated by spirals in the network. The gradient strength is given as the dashed line in the pulse sequence. The doubling of the integrated gradient strength required to achieve the desired "echo" is indicated by a doubling of the gradient pulse time. The numbers above the quantum networks are checkpoints used in the discussion below. The input state's deviation has to be $\sigma_z^{(1)}$. This can be obtained from the thermal equilibrium state either by applying a 90° rotation to spin 2 followed by a gradient pulse along another axis to remove $\sigma_z^{(2)}$; or by using phase cycling, which involves performing two experiments, the second with the sign of the phase in the first y pulse changed, and then subtracting the measured signals.

$X = \sigma_x$, $Y = \sigma_y$, $Z = \sigma_z$, and, for example $XY = \sigma_x^{(1)}\sigma_y^{(2)}$. For the checkpoints indicated in the figure:

(26)

   1. ZI

   2. XI

   3. YZ

   4. $YX \propto YX + XY$ $\qquad\qquad\qquad\qquad\qquad +YX - XY$

   5. $\cos(2\mu z)(YX + XY) + \sin(2\mu z)(YY - XX)$ $\quad +YX - XY$

   6. $\cos(2\mu z)(YZ + XY) + \sin(2\mu z)(YY - XZ)$ $\quad +YZ - XY$

   7. $\cos(2\mu z)(-XI + XY) + \sin(2\mu z)(YY - YI)$ $\quad \pm XI - XY$

   8. $\cos(2\mu z)(-XI - XZ) + \sin(2\mu z)(-YZ - YI)$ $\quad \pm XI + XZ$

   9. $-X(I + Z)$ $\qquad\qquad\qquad\qquad\qquad \pm\,(\cos(-2\mu z)X + \sin(-2\mu z)Y)(I - Z).$

Except for a sign, the desired state is obtained. The right-most term is eliminated after integrating over the sample, or after diffusion erases memory of z.

This method for making a two-qubit labeled pseudo-pure state can be generalized to arbitrarily many (n) qubits by exploiting the two n-coherences instead of the two-coherences. By observing the intensity of the final state with a measurement of the size of the single peak observed on spin 1 and comparing to the size of the corresponding peak in a spectrum obtained for the initial state $\sigma_x^{(1)}$, one can make a measurement of how well the requisite quantum control was implemented. We performed this experiment on a seven-spin system and measured that the output was $73 \pm 2\%$ of the input, indicating that an error rate of about 2% per two-qubit gate is achievable for nuclei in this setting [23].

3˙3. *Quantum error correction for phase errors.* – Currently envisaged scalable quantum computers require the use of quantum error correction to enable relatively error-free computation on a platform of physical systems that are inherently error-prone. Thus some of the most commonly used "subroutines" will be associated with controlling encoded information. This motivates experimental realizations of quantum error correction to determine whether adequate control can be achieved to implement these subroutines and to see in a practical setting that error correction has the desired effects. Experiments to date have realized a version of the three-qubit repetition code [24] and the five-qubit one-error-correcting code (the shortest possible such code) [25]. In this subsection we discuss the experimental implementation of the former.

In NMR one of the primary sources of error is phase decoherence due to both systematic and random fluctuations in the field along the z-axis experienced by the nuclei. At the same time, using gradient pulses and diffusion, phase decoherence is readily induced artificially and in a controlled way. The three-bit quantum repetition code can be adapted to protect against phase errors to first order. Define $|+\rangle = (1/\sqrt{2})(|\mathbf{0}\rangle + |\mathbf{1}\rangle)$ and $|-\rangle = (1/\sqrt{2})(|\mathbf{0}\rangle - |\mathbf{1}\rangle)$. The code we want is defined by the logical states

$$(27) \qquad |\mathbf{0}\rangle_\mathsf{L} = |+\rangle|+\rangle|+\rangle,$$

$$(28) \qquad |\mathbf{1}\rangle_\mathsf{L} = |-\rangle|-\rangle|-\rangle.$$

It is readily seen that the three one-qubit phase errors, $\sigma_z^{(1)}, \sigma_z^{(2)}, \sigma_z^{(3)}$, and "no error" ($\mathbb{1}$) map the code to orthogonal subspaces. As a result this set of errors is correctable. See the introduction to quantum error correction [26]. The simplest way to use this code is to encode one qubit's state into it, wait for some errors to happen and then decode to an output qubit. Success is indicated by the output qubit's state being significantly closer to the input state with error correction compared to not using error correction at all. The error in the output state compared to the input state without any errors happening in between measures how well the encoding and decoding operations were implemented.

The phase-correcting repetition code can be obtained from the standard repetition code by applying Hadamard transforms or $90°$ y-rotations to each qubit. The quantum network shown in fig. 12 was obtained in this fashion from the network given in [26].

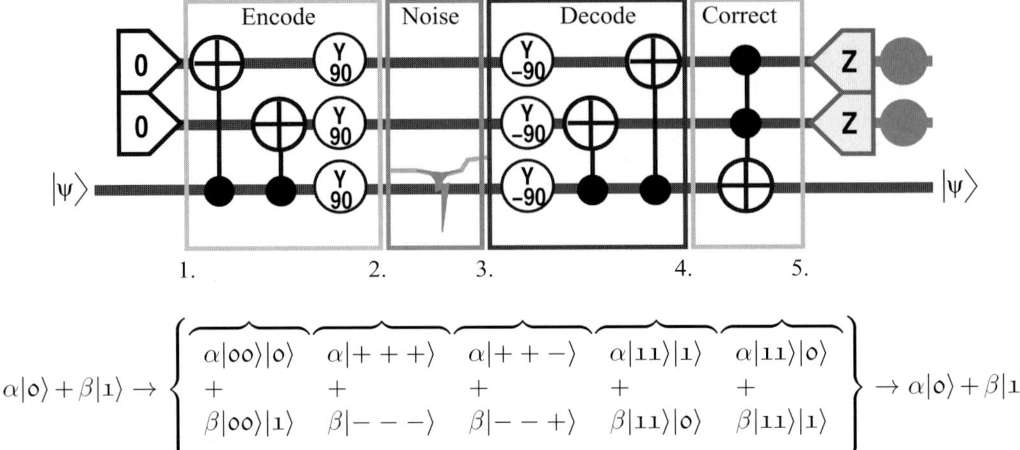

$$\alpha|0\rangle + \beta|1\rangle \rightarrow \left\{ \begin{array}{ccccc} \overbrace{\alpha|00\rangle|0\rangle} & \overbrace{\alpha|++\,+\rangle} & \overbrace{\alpha|++\,-\rangle} & \overbrace{\alpha|11\rangle|1\rangle} & \overbrace{\alpha|11\rangle|0\rangle} \\ + & + & + & + & + \\ \beta|00\rangle|1\rangle & \beta|---\rangle & \beta|--\,+\rangle & \beta|11\rangle|0\rangle & \beta|11\rangle|1\rangle \end{array} \right\} \rightarrow \alpha|0\rangle + \beta|1\rangle$$

Fig. 12. – Quantum network for the three-qubit phase-error-correcting repetition code. The bottom qubit is encoded with two controlled-nots and three y-rotations. In the experiment, either physical or controlled noise is allowed to act, and the encoded information is decoded using the inverse of the encoding circuit. The last step is to use the error information in the syndrome qubits (the top two) to restore the encoded information. This requires a Toffoli gate, which conditionally on the syndrome qubits' state being $|11\rangle$ flips the output qubit. The Toffoli gate can be realized with NMR pulses and delays by using more sophisticated versions of the implementation of the controlled-not. The syndrome qubits can be "dumped" at the end of the procedure. The behavior of the network is shown for a generic state in the case where the bottom qubit experiences a σ_z error. See also [26].

To determine the behavior and the quality of the implementation for various σ_z-error models in an actual NMR realization, one can use as initial states labeled pseudo-pure states with deviations $\sigma_u|00\rangle\langle00|$ for $u = x, y, z$. Without error, the total output signal on spin 1 along σ_u for each u should be the same as the input signal. The measured relative signals for the different inputs are used to calculate a "fidelity" (technically, the entanglement fidelity, see [25] for how it is obtained for a situation like this). One way of evaluating the success of the experiment is by how close to one the fidelity is. Some of the data reported in [24] is shown in fig. 13.

Work on benchmarking error control methods using liquid state NMR is continuing. Other experiments include the implementation of a two-qubit code with an application to phase errors [28] and the verification of the shortest non-trivial noiseless subsystem on three qubits [14]. The latter demonstrates that for some physically realistic noise models, it is possible to store quantum information in such a way that it is completely unaffected by the noise. See the introduction to quantum error correction [26] to learn about the basic principles of subsystems.

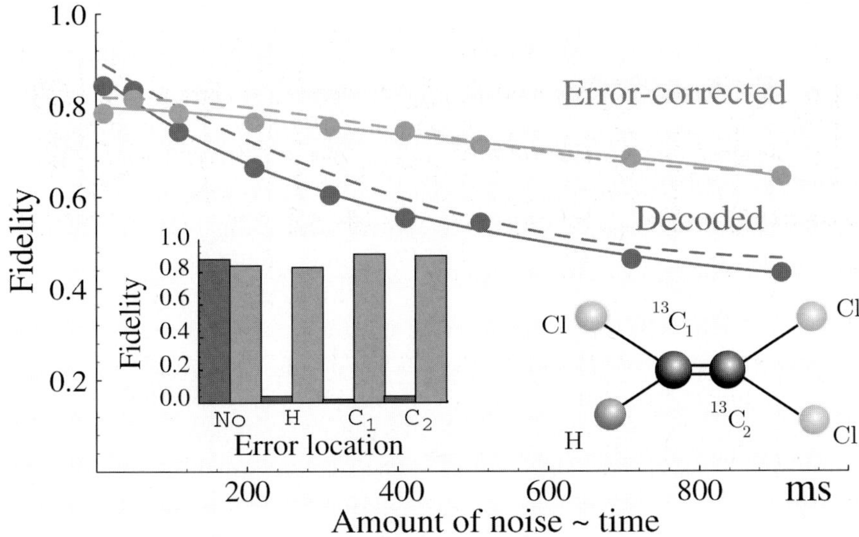

Fig. 13. – Experimentally obtained fidelities for the error correction experiment. The inset bar graph shows fidelities for explicitly applied errors. The reduction from 1 of the light grey bars (showing fidelity for the full procedure) is due to errors in our implementation of the pulses and from relaxation processes. The dark grey bars are the fidelity for the output before the last error correction step and contain the effects of the errors. The main graph shows the fidelities for the physical relaxation process. Here, the evolution consisted of a delay varying up to 1000 ms. The effect of error correction can be seen by a significant flattening of the curve, since first-order correction of phase errors implies that residual errors are quadratic in time. The light grey curve starts lower than the dark grey one due to additional errors incurred by the implementation of the last error correction step (the Toffoli gate). The dashed curves are obtained by simulation using estimated phase relaxation rates with half-times of 2 s (proton), 0.76 s (first carbon) and 0.42 s (second carbon). Errors in the data points are approximately 0.05. For a more thorough implementation and analysis of a three-qubit phase-error-correcting code, see [27].

4. – Discussion

4`1. *Overview of contributions to QIP*. – An important issue in current experimental efforts toward realizing QIP is how to achieve the necessary quantum control and whether sufficiently low error rates are possible. Liquid state NMR is the only extant system (as of 2001) with the ability to realize relatively universal manipulations on more than two qubits (restricted control has been demonstrated in four ions [29]). For this reason, NMR serves as a useful platform for developing and experimentally verifying techniques for QIP and for establishing simple procedures for benchmarking information processing tasks. The "cat-state" and the various error correction benchmarks [23, 25] consist of a set of quantum control steps and measurement procedures that can be used with any general-purpose QIP system to determine, in a device-independent way, the degree of control achieved. The demonstration of error rates in the few percent per non-trivial operation is encouraging. For existing and proposed experimental systems other than NMR, achieving such error rates is still a great challenge.

Prior research in NMR independent of quantum information has proven to be a rich source of basic quantum control techniques useful for physically realizing quantum information in other settings. We mention four examples. 1) The development of sophisticated shaped pulse techniques that can selectively control transitions or spins while being robust against typical errors. These techniques are finding applications to quantum control involving laser rather than RF pulses [30] and are likely to be very useful when using coherent light to accurately control transitions in atoms or quantum dots, for example. 2) The recognition that there are simple ways in which imperfect pulses can be combined to eliminate systematic errors like those associated with miscalibration of power or side effects on off-resonant nuclei. Although many of these techniques were originally developed for problems like accurate inversion of spins, they are readily generalized to other quantum gates [31,32]. 3) Decoupling to reduce unwanted external interactions. For example, a common problem in NMR is to eliminate the interactions between protons and labeled carbon nuclei to observe "decoupled" carbon spins. In this case, the protons constitute an external system with an unwanted interaction. The interaction is eliminated by frequently applying inversions ("flips") to the protons. Sophisticated techniques for ensuring that the interactions are effectively turned off independent of pulse errors have been developed (see, for example, [4]). These techniques have been greatly generalized and shown to be useful for actively creating protected qubit subsystems in any situation where the interaction has relatively long correlation times [33-35]. 4) Refocusing to undo unwanted internal interactions. The technique for "turning off" the coupling between spins that is so important for realizing QIP in liquid state NMR is a special case of much more general methods of turning off or refocusing Hamiltonians. For example, a famous technique in solid state NMR is to reverse the dipolar coupling Hamiltonian using a clever sequence of 180° pulses at different phases (see, for example, [4], p. 48). Many other proposed QIP systems suffer from such internal interactions while having similar control opportunities.

The contributions of NMR QIP research extend beyond those directly applicable to experimental QIP systems. It is due to NMR that the idea of ensemble quantum computation with weak measurement was introduced and recognized as being, for pure initial states, as powerful for solving algorithmic problems as the standard model of quantum computation. (It cannot be used in settings involving quantum communication.) One implication is that to a large extent, the usual assumption of projective measurement can be replaced by any measurement that can statistically distinguish between the two states of a qubit. Scalability still requires that qubits can be "reset" during the computation. Another interesting concept emerging from NMR QIP is that of "computational cooling" [36], which can be used to efficiently extract initialized qubits from a large number of noisy qubits in initial states that are only partially biased toward $|0\rangle$. This is a very useful tool for better exploiting otherwise noisy physical systems.

The last example of interesting ideas arising from NMR studies is the "one-qubit" model of quantum computation [37]. This is a useful abstraction of the capabilities of liquid state NMR. In this model, it is assumed that initially, one qubit is in the state $|0\rangle$ and all the others are in random states. Standard quantum gates can be applied and the final measurement is destructive. Without loss of generality, one can assume that all qubits are re-initialized after the measurement. This model can perform interesting physics simulations with no known efficient classical algorithms. On the other hand, with respect to oracles, it is strictly weaker than quantum computation. It is also known that it cannot "faithfully" simulate quantum computers [38].

4`2. *Capabilities of liquid state NMR.* – One of the main issues in liquid state NMR QIP is the highly mixed initial state. The methods of extracting pseudo-pure states are not practical for more than 10 (or so) nuclear spins. The problem is that for these methods, the pseudo-pure state signal decreases exponentially with the number of qubits prepared while the signal to noise ratio is constant. This limits the ability to explore and benchmark standard quantum algorithms even in the absence of noise. There are in fact ways in which liquid state NMR can be usefully applied to many more qubits. The first and less practical is to use computational cooling for a (unrealistically) large number of spins to obtain less mixed initial states. Versions of this technique have been studied and used in NMR to increase signal to noise [39].

The second is to use the one-qubit model of quantum computation instead of trying to realize pseudo-pure states. For this purpose, liquid state NMR is limited only by relaxation noise and pulse control errors, not by the number of qubits. Noise still limits the number of useful operations, but non-trivial physics simulations are believed to be possible with less than 100 qubits [40]. Remarkably, a one-qubit quantum computer matches the efficiency of standard quantum computation for obtaining spectral information about a generic Hamiltonian [37, 41, 42]. Consequently, although QIP with molecules in liquid state cannot realistically be used to realize standard quantum computations involving more than about 10 qubits, its capabilities have the potential of exceeding the resource limitations of available classical computers for some applications.

4`3. *Prospects for NMR QIP.* – There are many more algorithms and benchmarks that can be usefully explored using the liquid state NMR platform. We hope to soon have a molecule with 10 or more useful spins and good properties for QIP. Initially this can be used to extend and verify the behavior of existing scalable benchmarks. Later, experiments testing basic ideas in physics simulation or more sophisticated noise-control methods are likely.

Liquid state NMR QIP is one of many ways in which NMR can be used for quantum information. One of the most promising proposals for quantum computation is based on phosphorus embedded in silicon [43] and involves controlling phosphorus nuclei using NMR methods. In this proposal, couplings and frequencies are controlled using locally applied voltages. RF pulses can be used to implement universal control. It is also possible to scale up NMR QIP without leaving the basic paradigms of liquid state NMR while adding features like high polarization, the ability to dynamically reset qubits (required for scalability) and much faster two-qubit gates. One proposal for doing this is to use dilute molecules in a solid state matrix instead of molecules in liquid [44]. This proposal may be able to realize pure-state quantum computation for significantly more than 10 qubits.

REFERENCES

[1] Knill E. and Laflamme R., *Introduction to quantum information processing*, Technical Report LAUR-01-4761, Los Alamos National Laboratory, 2001. To appear in *LA Science*.

[2] Purcell E. M., Torrey H. C. and Pound R. V., *Phys. Rev.*, **69** (1946) 37.

[3] Bloch F., *Phys. Rev.*, **70** (1946) 460.

[4] Ernst R. R., Bodenhausen G. and Wokaun A., *Principles of Nuclear Magnetic Resonance in One and Two Dimensions* (Oxford University Press, Oxford) 1994.

[5] Mansfield P. and Morris P., *Adv. Mag. Res.*, **S2** (1982) 1.

[6] Hahn E. L., *Phys. Rev.*, **80** (1950) 580.

[7] Anderson A. G., Garwin R., Hahn E. L., Horton J. W. and Tucker G. L., *J. App. Phys.*, **26** (1955) 1324.

[8] Anderson A. G. and Hahn E. L., *Spin echo storage technique*, US Patent # 2714714, 1955.

[9] Cory D. G., Fahmy A. F. and Havel T. F., *Proc. Natl. Acad. Sci. USA*, **94** (1997) 1634.

[10] Gershenfeld N. A. and Chuang I. L., *Science*, **275** (1997) 350.

[11] Stoll M. E., Vega A. J. and Vaughan R .W., *Phys. Rev. A*, **16** (1977) 1521.

[12] Chuang I. L., Vandersypen L. M. K., Zhou X., Leung D. W. and Lloyd S., *Nature*, **393** (1998) 143.

[13] Jones J. A., Mosca M. and Hansen R. H., *Nature*, **392** (1998) 344.

[14] Viola L., Knill E. and Laflamme R., *J. Phys. A*, **34** (2001) 7067 (LAUR-00-5877).

[15] Barenco A., Bennett C. H., Cleve R., DiVincenzo D. P., Margolus N., Shor P., Sleator T., Smolin J. and Weinfurter H., *Phys. Rev. A*, **52** (1995) 3457.

[16] DiVincenzo D. P., *Phys. Rev. A*, **51** (1995) 1015.

[17] Lloyd S., *Phys. Rev. Lett.*, **75** (1995) 346.

[18] Leung D. W., Chuang I. L., Yamaguchi F. and Yamamoto Y., *Efficient implementation of selective recoupling in heteronuclear spin systems using hadamard matrices*, quant-ph/9904100, 1999.

[19] Jones J. A. and Knill E., *J. Mag. Res.*, **141** (1999) 322.

[20] Knill E., Chuang I. and Laflamme R., *Phys. Rev. A*, **57** (1998) 3348.

[21] Sharf Y., Havel T. F. and Cory D. G., *Phys. Rev. A*, **6205** (2000) 052314/1.

[22] Sörensen O. W., Eich G. W., Levitt M. H., Bodenhausen G. and Ernst R. R., *Prog. Nucl. Mag. Res. Spectrosc.*, **16** (1983) 163.

[23] Knill E., Laflamme R., Martinez R. and Tseng C.-H., *Nature*, **404** (2000) 368.

[24] Cory D. G., Maas W., Price M., Knill E., Laflamme R., Zurek W. H., Havel T. F. and Somaroo S. S., *Phys. Rev. Lett.*, **81** (1998) 2152.

[25] Knill E., Laflamme R., Martinez R. and Negrevergne C., *Phys. Rev. Lett.*, **86** (2001) 5811.

[26] Knill E. and Laflamme R., *Introduction to quantum error correction*, Technical report, Los Alamos National Laboratory 2001. To appear in *LA Science*.

[27] Sharf Y., Cory D. G., Somaroo S. S., Knill E., Laflamme R., Zurek W. H. and Havel T. F., *Mol. Phys.*, **98** (2000) 1347.

[28] LEUNG D., VANDERSYPEN L., ZHOU X. L., SHERWOOD M., YANNONI C., KUBINEC M. and CHUANG I., *Phys. Rev. A*, **60** (1999) 1924.

[29] SACKETT C. A., KIELPINSKI D., KING B. E., LANGER C., MEYER V., MYATT C. J., ROWE M., TURCHETTE Q. A., ITANO W. M. and WINELAND D. J., *Nature*, **404** (2000) 256.

[30] WARREN W. S., RABITZ H. and DAHLEH M., *Science*, **59** (1993) 1581.

[31] LEVITT M. H., *J. Mag. Res.*, **48** (1982) 234.

[32] CUMMINS H. K. and JONES J. A., *Use of composite rotations to correct systematic errors in NMR quantum computation*, quant-ph/9911072, 1999.

[33] VIOLA L. and LLOYD S., *Phys. Rev. A*, **58** (1998) 2733.

[34] VIOLA L., KNILL E. and LLOYD S., *Phys. Rev. Lett.*, **82** (1999) 2417.

[35] VIOLA L., LLOYD S. and KNILL E., *Phys. Rev. Lett.*, **83** (1999) 4888.

[36] SCHULMAN L. J. and VAZIRANI U., *Scalable NMR quantum computation*, in *Proceedings of the 31st Annual ACM Symposium on the Theory of Computation (STOC), El Paso, Texas* (ACM Press, New York) 1998, pp. 322-329.

[37] KNILL E. and LAFLAMME R., *Phys. Rev. Lett.*, **81** (1998) 5672.

[38] AMBAINIS A., SCHULMAN L. J. and VAZIRANI U., *Computing with highly mixed states*, in *Proceedings of the 32nd Annual ACM Symposium on the Theory of Computation (STOC), New York* (ACM Press, New York) 2000, pp. 697-704.

[39] GLASER S. J., SCHULTE-HERBRÜGGEN T., SIEVEKING M., SCHEDLETZKY O., NIELSEN N. C., SÖRENSEN O. W. and GRIISIGNER C., *Science*, **280** (1998) 421.

[40] LLOYD S., *Science*, **273** (1996) 1073.

[41] SOMMA R., ORTIZ G., GUBERNATIS J. E., LAFLAMME R. and KNILL E., *Simulating physical phenonmena by quantum networks*, manuscript in preparation, 2001.

[42] MIQUEL C., PAZ J. P., SARACENO M., KNILL E., LAFLAMME R. and NEGREVERGNE C., *State tomography and spectroscopy as quantum computations*, quant-ph/0109072, 2001.

[43] KANE B. E., *Nature*, **393** (1998) 133.

[44] CORY D. G., LAFLAMME R., KNILL E., VIOLA L., HAVEL T. F., BOULANT N., BOUTIS G., FORTUNATO E., LLOYD S., MARTINEZ R., NEGREVERGNE C., PRAVIA M., SHARF Y., TEKLEMARIAM G., WEINSTEIN Y. S. and ZUREK W. H., *Fort. Phys.*, **48** (2000) 875.

Solid-state crystal lattice NMR quantum computation

Y. Yamamoto, T. D. Ladd, J. R. Goldman and F. Yamaguchi

Quantum Entanglement Project, ICORP, JST and NTT Basic Research Laboratories
E. L. Ginzton Laboratory, Stanford University - Stanford, CA 94305, USA

1. – Introduction

A quantum computer is a machine that starts with an initial quantum state, performs unitary logic operations on that state, and measures a resulting final state. The superposition principle of quantum mechanics and the quantum interference observed by a projective measurement bring about "quantum parallelism", by which certain problems can be computed more efficiently than with any classical computers [1-7].

Several physical systems are known to provide the coherent superposition states that carry quantum bits ("qubits") of information. Cold trapped ions [8-10], atomic cavity quantum electrodynamics [11,12], and optical lattices [13] require immense experimental exertion, but the computation and final readout can be performed on individual qubits in these systems. Single-photon gates [14,15], which are compatible with quantum communication links [8], are made possible with photon trapping techniques using atomic ensembles [16] or quantum teleportation techniques [17]. Systems of nuclear spins in molecules in solution [18,19] have the ease of use of chemically synthesized structures and well-established pulsed nuclear magnetic resonance (NMR) techniques. Despite the large ensemble of molecules needed to detect the output signals, these NMR systems are the most successful experimental realizations of multi-qubit, multi-gate quantum computers to date. Solid-state systems, including nuclear spins of implanted phosphorus ions in silicon [20], Josephson junctions [21,22], and quantum dots [23], hold promise for scalability to large numbers of distinct qubits.

We explore several systems which attempt to bridge the ensemble techniques of solution NMR with the initialization and sensitive measurement techniques available in the solid state. As in solution NMR, qubits are ensembles of spin-(1/2) nuclei. We imagine that these nuclei are spatially well-ordered due to their natural placement in a *crystal lattice*; this is in contrast to most other solid-state nuclear spin proposals which require atomic scale placement of impurity spins [20,24,25]. Regular arrays of qubits require the use of elements with 100% abundance of nuclear spin 1/2. Only five nuclei, ^{19}F, ^{31}P, ^{89}Y, ^{103}Rh and ^{169}Tm, carry this abundance naturally; other nuclei such as ^{1}H, ^{13}C, and ^{29}Si require isotope engineering.

Initialization of the nuclear spins by simple cooling of the lattice requires millikelvin temperatures with fields of 20 T or higher; such extremes are difficult to achieve by themselves, and they are nearly impossible when coupled with the high-power RF pulses employed for quantum control. We presume, therefore, that initialization of the quantum computer will require a polarization transfer from cold electron spins to the nuclear spins. Electron spins, however, tend to cause thermal relaxation and decoherence of the nuclear spins; these detrimental effects can be avoided by using a material in which magnetic fluctuations due to electrons are somehow suppressed. An insulating crystal with only a diffuse population of isolated electron impurities may be considered. At the opposite extreme, a strong ferromagnet or antiferromagnet maintains electron spin order due to the very large exchange couplings between electrons. Finally, a semiconducting crystal such as silicon holds great promise, since a bath of cold electrons can be optically excited for initialization, but their quick recombination can remove them as a decoherence source during the computation. In this article, we will consider all of the above-mentioned systems for crystal lattice quantum computation.

Measurement of the nuclear spins is another challenge. We imagine that for control and measurement, the nuclei in the crystal lattice are distinguished by a large magnetic-field gradient. Thus, nuclei in different atomic positions will have distinguishable Larmor frequencies. To create an ensemble, however, the gradient must allow an atomic plane of nuclei with the same Larmor frequency. The need to maintain such homogeneity limits the size of the ensemble. The ensemble sizes in the schemes to be proposed below are too small for the usual inductive measurement techniques of NMR. A more sensitive means of measurement is required; fortunately, in the solid state, more sensitive means are available. In particular, we have focused on the ultra-sensitive spin measurement techniques of magnetic resonance force microscopy (MRFM) [26,27].

We begin by considering the general issue of decoherence timescales in different materials. We then focus on fluorapatite (an insulating crystal), CeP (an antiferromagnet), and silicon (a semiconductor). In these discussions we will assume that the nuclei are well-distinguished by the large magnetic-field gradient and that their spin states may be controlled and measured as an ensemble average; details of how to generate such a large magnetic-field gradient, how to apply quantum logic, and how to perform the force measurement will be discussed immediately after. Finally, we will discuss the scalability of crystal lattice quantum computation, which requires consideration of initialization, measurement, logic-gate design, and decoherence.

2. – Nuclear interactions and decoherence

We begin by establishing notation, indicating the relevant interaction Hamiltonians, and defining the important timescales.

2‵1. *Nuclear interactions.* – We consider as a qubit an ensemble of spin-$(1/2)$ nuclei. The nuclear spin operators are notated $I_{ij}^{\{x,y,z\}}$, where the first subscript indicates to which qubit the spin operator applies, and the second subscript indicates the ensemble member. The i-th qubit has a unique Larmor frequency ω_i; qubits are distinguished by a large magnetic-field gradient which leads to a frequency separation of $\omega_i - \omega_{i'} = \Delta\omega_{ii'}$. For adjacent qubits, we notate $\Delta\omega_{i+1,i} = \Delta\omega$.

The Hamiltonian for the nuclear spins in a solid with negligible covalency may be summarized as [28]

$$\mathcal{H} = \mathcal{H}_{\mathrm{nZ}} + \mathcal{H}_{\mathrm{RF}} + \mathcal{H}_{\mathrm{D}}, \tag{1}$$

where $\mathcal{H}_{\mathrm{nZ}}$ is the nuclear Zéeman Hamiltonian,

$$\mathcal{H}_{\mathrm{nZ}} = -\hbar \sum_{ij} \omega_i I_{ij}^z, \tag{2}$$

and $\mathcal{H}_{\mathrm{RF}}$ is the RF control Hamiltonian, generated by considering the Zéeman energy of spins coupled to an oscillating field in the y-direction,

$$\mathcal{H}_{\mathrm{RF}} = -\hbar \sum_{ij} 2\Omega_i(t) I_{ij}^y \cos[\omega_i t + \phi_i(t)]. \tag{3}$$

The time dependence of the Rabi frequency Ω_i and the phase ϕ_i near each Larmor frequency ω_i indicates that these parameters are under arbitrary control. The dipole Hamiltonian $\mathcal{H}_{\mathrm{D}}$ is

$$\mathcal{H}_{\mathrm{D}} = \frac{\mu_0}{4\pi} \gamma^2 \hbar^2 \frac{1}{2} \sum_{\{i,j\}\neq\{k,l\}} \frac{\mathbf{I}_{ij} \cdot \mathbf{I}_{kl} - 3(\mathbf{I}_{ij} \cdot \hat{\mathbf{r}}_{ij,kl})(\mathbf{I}_{kl} \cdot \hat{\mathbf{r}}_{ij,kl})}{r_{ij,kl}^3}, \tag{4}$$

where γ is the gyromagnetic ratio of the nuclei and $\mathbf{r}_{ij,kl}$ is the vector between nucleus ij and nucleus kl.

We now rewrite these Hamiltonians in "the rotating frame" and neglect terms which oscillate at the Larmor frequency. The motivation for this approximation is to notice that the speed of $\mathcal{H}_{\mathrm{nZ}}$ is on the order of 100 MHz, whereas the other terms are several kilohertz; it is therefore sufficient to consider first-order perturbations due to these smaller terms. In the reference frame of the applied RF, assuming $\phi(t)$ changes sufficiently slowly in comparison to ω_i such that we may write $\phi(t) = \phi_0 + t\partial\phi/\partial t$, the truncated Hamiltonian involves the terms

$$\mathcal{H}_{\mathrm{RF}}^* = -\hbar \sum_{ij} \left[\frac{\partial\phi}{\partial t} I_{ij}^z + \Omega_i \cos(\phi_0) I_{ij}^y + \Omega_i \sin(\phi_0) I_{ij}^x \right], \tag{5}$$

and

$$(6) \qquad \mathcal{H}_{\mathrm{D}}^* = \mathcal{H}_{\mathrm{D,qq}}^* + \mathcal{H}_{\mathrm{D,cc}}^* + \mathcal{H}_{\mathrm{D,qc}}^*,$$

where

$$(7) \qquad \mathcal{H}_{\mathrm{D,qq}}^* = \sum_{k>i}\sum_j \frac{1 - 3\cos^2\theta_{ij,kj}}{r_{ij,kj}^3}\left[I_{ij}^z I_{kj}^z - \frac{1}{4}\left(I_{ij}^+ I_{kj}^- e^{-i\Delta\omega_{ik}} + I_{ij}^- I_{kj}^+ e^{i\Delta\omega_{ik}}\right)\right],$$

$$(8) \qquad \mathcal{H}_{\mathrm{D,cc}}^* = K\sum_i\sum_{l>j} \frac{1 - 3\cos^2\theta_{ij,il}}{2r_{ij,il}^3}\left[3I_{ij}^z I_{il}^z - \mathbf{I}_{ij}\cdot\mathbf{I}_{il}\right],$$

$$(9) \qquad \mathcal{H}_{\mathrm{D,qc}}^* = K\sum_{k>i}\sum_{l>j} \frac{1 - 3\cos^2\theta_{ij,kl}}{r_{ij,kl}^3}\left[I_{ij}^z I_{kj}^z - \frac{1}{4}\left(I_{ij}^+ I_{kj}^- e^{-i\Delta\omega_{ik}} + I_{ij}^- I_{kj}^+ e^{i\Delta\omega_{ik}}\right)\right].$$

Here, $\theta_{ij,kl}$ is the polar angle of $\mathbf{r}_{ij,kl}$ and $K = (\mu_0/4\pi)\gamma^2\hbar^2$.

In many solid-state NMR experiments, these terms are sufficient to describe all of the decoherence and control of nuclear spins. In our case, we control our qubits using the Hamiltonian $\mathcal{H}_{\mathrm{RF}}^*$ and the "qubit-qubit" dipole Hamiltonian $\mathcal{H}_{\mathrm{D,qq}}^*$; the "copy-copy" term $\mathcal{H}_{\mathrm{D,cc}}^*$ and the "qubit-copy" term $\mathcal{H}_{\mathrm{D,qc}}^*$ must be suppressed with pulse-sequence and materials engineering in order to maintain coherence in the ensemble, as will be discussed. If such suppression is sufficient, then the interaction of nuclei with the magnetic fields of any nearby electrons become important for decoherence; also, such interactions cause thermal relaxation and allow initialization. The interactions of the nuclei with a set of electrons with spin operators $\mathbf{S}_k$ and orbital angular momenta $\mathbf{L}_k$ may be summarized by the Hamiltonian [28]

$$(10) \quad \mathcal{H}_{\mathrm{en}} = -\hbar\gamma\sum_{ij}\mathbf{I}_{ij}\cdot\left[\frac{\mu_0}{4\pi}g\mu_{\mathrm{B}}\sum_k\left(\frac{\mathbf{L}_k}{r_{ij,k}^3} - \frac{\mathbf{S}_k - 3\mathbf{r}_{ij,k}(\mathbf{S}_k\cdot\mathbf{r}_{ij,k})}{r_{ij,k}^3} + \frac{8\pi}{3}\mathbf{S}_k\delta(\mathbf{r}_{ij,k})\right)\right]$$

$$\equiv -\hbar\gamma\sum_{ij}\mathbf{I}_{ij}\cdot\mathbf{B}_{\mathrm{e}},$$

which describes both dipolar interactions (first two terms) and contact hyperfine interactions (third term).

2˙2. Decoherence. – Generally, solid-state NMR linewidths are dominated by dephasing processes, in which an ensemble of rotating nuclei maintain their energy but go out of phase due to the varying magnetic fields caused by other nuclei. Such processes are in principle reversible, as decoupling pulse sequences can be used to refocus the dephased nuclei. Of more severe consequence are intrinsic decoherence processes, in which a nucleus exchanges energy or loses phase coherence to other, uncontrollable degrees of freedom in the lattice. The advantage of solid-state NMR for quantum computation is that spin-(1/2) nuclei can remain relatively uncoupled from their environment. Such isolation is spoiled only by internal magnetic fields due to electrons and extra nuclei.

The definition of the longitudinal relaxation time T_1 is straightforward. The probability for a nucleus to give up its Zéeman energy to the lattice is nearly always independent of time, leading to an exponential decay of a non-equilibrium polarization of spins. The time scale for this decay is T_1, and it represents the maximum time available for a quantum computation. Fermi's Golden Rule leads to the formula [28]

$$(11) \qquad \frac{1}{T_1} = \frac{\gamma^2}{2} \int_{-\infty}^{\infty} \left(\langle \delta B^x(t) \delta B^x(0) \rangle + \langle \delta B^y(t) \delta B^y(0) \rangle \right) \cos(\omega_i t) \mathrm{d}t,$$

where $\delta \mathbf{B}$ is the randomly fluctuating local field at the nucleus. Typically, the relevant fluctuating field will be $\mathbf{B}_{\mathrm{e}}$. The order of magnitude of T_1 will thus depend heavily on the electron content of the material. In pure, solid insulators or low-temperature semiconductors, T_1 can be 100's of hours or more.

The transverse relaxation or decoherence time T_2 is more complicated. It may be empirically defined as $T_2 = 1/\pi \Delta \nu$, where $\Delta \nu$ is the full-width half-maximum of the NMR spectral line. However, this definition is not convenient for a theoretical analysis. In fact, we use two theoretical definitions of T_2, depending on the process being considered. If the decoherence results from phase kicks from a randomly fluctuating, "memoryless" external bath, a T_2^{ext} process may be calculated in much the same way as a T_1 process; one finds that the time-independent transition rate between two transverse spin states is given by

$$(12) \qquad \frac{1}{T_2^{\mathrm{ext}}} = \frac{1}{2T_1} + \frac{\gamma^2}{2} \int_{-\infty}^{\infty} \langle \delta B^z(t) \delta B^z(0) \rangle \mathrm{d}t.$$

As in the case of T_1, this contribution to T_2 will depend highly on the density and character of nearby electrons; hence we will discuss this contribution in the context of specific materials in ensuing sections. The T_2 due to coupling between ensemble nuclei, which typically dominates decoherence in solid-state NMR, may not be calculated this way due to the high degree of correlation, or "memory", of the many-body system. Instead, the evolution of the many-body system can be calculated for short times and the decay of coherence may be presumed Gaussian, so that knowledge of the second moment is sufficient to estimate T_2.

As an example, consider the dephasing due to the coupling between identical qubits, governed by $\mathcal{H}_{\mathrm{D,cc}}^*$. The second moment is given by the Van Vleck formula [28]

$$(13) \qquad M_2^{\mathrm{cc}} = -\frac{\mathrm{tr}\{[\mathcal{H}_{\mathrm{D,cc}}^*, I_{il}^x]^2\}}{\mathrm{tr}\{(I_{il}^x)^2\}} = \frac{9}{16} \left(\frac{\mu_0}{4\pi} \right)^2 \gamma^4 \hbar^2 \sum_{j \neq l} \frac{(1 - 3 \cos^2 \theta_{ij,il})^2}{r_{ij,il}^6},$$

and T_2^{cc} is approximately $(M_2^{\mathrm{cc}})^{-1/2}$. For a typical, non-magnetic insulating crystal such as CaF_2, this timescale is on the order of 10^{-5} seconds, which would seem prohibitively short for successful quantum computation. It is important to realize, however, that the dephasing caused by $\mathcal{H}_{\mathrm{D,cc}}^*$ is, in principle, completely reversible. In practice, pulse sequences such as WAHUHA and its offspring have been able to lengthen this T_2^{cc} by more than three orders of magnitude [29, 30].

Also, the $1/r^3$ dependence of the dipolar coupling allows us to lengthen T_2^{cc} by spatially separating ensemble members while maintaining a small qubit-qubit distance. A geometry by which this is possible is for each quantum computer to be an atomic chain, and for the chains comprising the ensemble to be well separated. Thus, different qubits are relatively densely spaced in a one-dimensional array, whereas the ensemble corresponding to a single qubit is a large plane of relatively diffuse sites. Physical examples of such "one-dimensional" materials will be discussed in following sections. Let us focus on the geometry where such straight atomic chains are parallel to the applied magnetic field and to the field gradient.

Consider the form of the qubit-qubit coupling $\mathcal{H}_{\mathrm{D,qq}}^*$. We see in eq. (7) that the flip-flop terms carry a time dependence in the rotating frame. This is because the field gradient imposes an energy cost $\sim \hbar\Delta\omega$ to such a flip-flop process. For such terms to be important, this energy must have a source or destination. The lengthy T_1 times which are typical of solid-state NMR indicate that processes in which this energy is exchanged with the lattice are extremely improbable. It is known that this energy can be absorbed into the dipolar bath [31]; the rates for such processes scale as $(T_2^{cc}\Delta\omega)^{-2}$. Let us assume for now that our crystal is sufficiently one-dimensional and our field gradient sufficiently large to render such terms negligible; we will return to the $T_2^{cc}\Delta\omega$ figure of merit when we consider specific experimental designs. Hence we neglect the time-dependent terms in $\mathcal{H}_{\mathrm{D,qq}}^*$.

To simplify notation, we write the nearest-neighbor qubit-qubit distance as $r_{ij,(i+1)j} = a$. Equation (7) may then be rewritten as

$$(14) \qquad \mathcal{H}_{\mathrm{D,qq}}^* = \frac{\mu_0}{4\pi}\gamma^2\hbar^2 \sum_{i,j}\sum_{m>0} \frac{1-3\cos^2(0)}{(am)^3} I_{ij}^z I_{i+m,j}^z \equiv -\hbar \sum_{i,j}\sum_{m>0} \frac{\delta\omega}{m^3} I_{ij}^z I_{i+m,j}^z.$$

This is the interaction we will use for two-qubit logic gates. The sufficiency of this interaction for universal quantum logic has been demonstrated [19]; we discuss how in sect. **5**. A problem arises in a solid crystal, however. If we consider the coupling between a qubit and its m-th nearest neighbor, we find that the couplings described by $\mathcal{H}_{\mathrm{D,qc}}^*$ become important, especially for higher m. To quantify this statement, let us rewrite eq. (9), truncating the time-dependent terms as justified above, noting the distance between chains as $r_{ij,i(j+k)} = a\lambda_k$, and considering only the coupling between the i-th qubit and another qubit m planes away:

$$(15) \qquad \mathcal{H}_{\mathrm{D,qc}}^{*(i,i+m)} = \hbar\delta\omega \sum_{j,k} \frac{\lambda_k^2/m^2 - 2}{2(\lambda_k^2/m^2 + 1)^{5/2}} I_{ij}^z I_{i+m,j+k}^z.$$

These unwanted terms lead to an effective decoherence, whose T_2 may again be estimated by the method of moments. We find

$$(16) \qquad \left(\frac{1}{T_2^{\text{qc}}(m)}\right)^2 = \frac{1}{16}\left(\frac{\delta\omega}{m^3}\right)^2 \sum_i \frac{(\lambda_i^2/m^2 - 2)^2}{(\lambda_i^2/m^2 + 1)^5} \equiv \left(\frac{\delta\omega}{m^3}\Gamma(m)\right)^2.$$

The function $\Gamma(m)$ characterizes the amount of error present due to parasitic couplings between ensemble members when attempting to couple two qubits m planes away from each other. For a very one-dimensional crystal, $\Gamma(m)$ can be quite small until m exceeds λ_1, the ratio of the nearest-neighbor chain-chain distance to the nearest-neighbor qubit-qubit distance. An important parameter for a crystal lattice quantum computer is therefore λ_1; if this number is large then quantum gates between nearest-neighbor qubits can be performed with acceptable error.

In this section, we have considered several figures of merit related to timescales and decoherence for a solid-state NMR quantum computer. We now consider specific materials for the computer.

3. – Materials for crystal lattice quantum computation

$3^\cdot1$. *Fluorapatite* [32]. – A natural crystal which approximates the description of "one-dimensional" is fluorapatite, $Ca_5F(PO_4)_3$, whose ^{19}F structure is shown in the inset of fig. 1. Quantitatively, we find that $\lambda_1 = 2.74$ for this crystal, so $\Gamma(m)$ is only small for $m \sim 1$. However, if during a computation we *always* suppress all but nearest-neighbor couplings, the errors in quantum gates will be on the order of $\Gamma(1) = 1/58$. Although not perfect, this is nearly an order of magnitude improvement over simpler crystal structures such as the cubic structure of CaF_2. The one-dimensional nuclear structure of fluorapatite has been recognized since early NMR experiments [33]. It has provided a one-dimensional testing ground for lineshape analysis investigations [34] and, more recently, experiments studying multiple quantum coherence [35], a topic intimately related to quantum computation.

Fluorapatite, like the well-studied CaF_2, is an ionic insulator with a high Debye temperature. In such a material, interactions of the nuclei with the intrinsic electrons can be completely ignored; the effect of phonon-induced displacements leads to nuclear relaxation estimates that are extremely long. Instead, it is observed that nuclear relaxation in fluorapatite and similar crystals is entirely caused by paramagnetic impurities [36]. In very pure crystals at low temperature, T_1 can range from minutes to weeks.

While a long T_1 removes the worry of thermal relaxation causing errors in a quantum computation, the lack of electron spins leaves very few practical avenues for nuclear polarization. One possible method of nuclear polarization in fluorapatite is to employ the "solid effect" with deliberately introduced paramagnetic impurities [28]. This approach requires a well-resolved electron spin resonance (ESR) frequency for the impurities and a short electron spin-lattice relaxation time T_{1e}. Such an approach is complicated by the presence of the magnetic-field gradient. Moreover, introduction of paramagnetic impurities has severe consequences for decoherence. If we assume a spin-$(1/2)$ electron

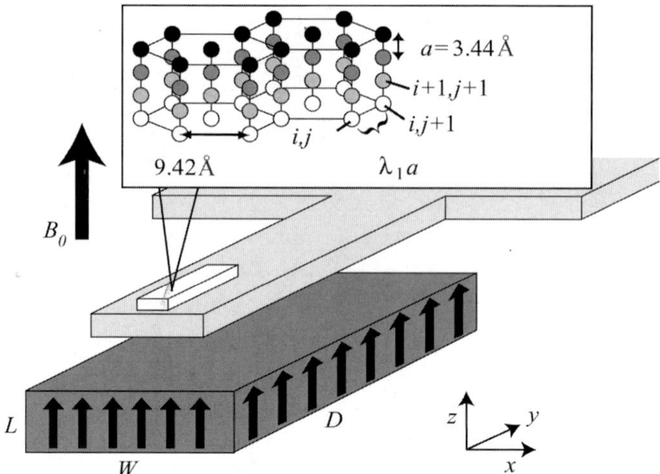

Fig. 1. – A schematic for a fluorapatite crystal lattice quantum computer. The fluorapatite crystal is shown mounted on a silicon cantilever, whose oscillations provide MRFM readout. The cantilever and crystal are aligned with the micromagnet which generates the field gradient. The inset shows the "one-dimensional" structure of ^{19}F nuclei in fluorapatite.

impurity which obeys the Bloch equations and we combine the dominant terms of eq. (10) with eq. (12), we find that a nearby nucleus experiences a T_2^{ext} of

$$(17) \qquad \frac{1}{T_2^{\text{ext}}} = \left(\frac{\mu_0}{4\pi}\right)^2 \left(\frac{\gamma g \mu_{\text{B}}}{2}\right)^2 \left(\frac{1 - 3\cos^2\theta}{r^3}\right)^2 T_{1\text{e}}(1 - P_{\text{e}}^2),$$

where r is the length and θ the polar angle of the vector connecting the nucleus to the impurity and P_{e} is the electron polarization, given by $\tanh(g\mu_{\text{B}}B_0/2k_{\text{B}}T)$ for isolated impurities. If r is in units of nanometers and $T_{1\text{e}}$ in units of seconds, this equation estimates that $T_2^{\text{ext}} \approx 10\,\text{ps} \times r^6/T_{1\text{e}}(1 - P_{\text{e}}^2)$. At very low temperature, P_{e} may be high but $T_{1\text{e}}$ is long; for example, Cr^{5+} impurities in fluorapatite show a $T_{1\text{e}}$ of $0.1\,\text{s}$ at $4\,\text{K}$ [37]. There does not seem to be a reasonable combination of temperature, $T_{1\text{e}}$, and impurity concentration for effective use of the "solid effect" in a fluorapatite quantum computer.

A better approach for initialization is to epitaxially grow the fluorapatite crystal on a semiconductor substrate, optically excite spin-polarized electrons in that semiconductor, and allow cross-polarization across the semiconductor/fluorapatite surface [38]. Such an approach remains highly speculative; we will return to such optical pumping techniques when we consider using silicon for the quantum computer.

Another approach to the initialization problem is to use electrons for which $P_{1\text{e}}$ is very large at reasonably low temperatures due to strong electron-electron interactions— in other words, to use a magnetic material such as CeP, to which we now turn.

3˙2. CeP [39]. – Rare-earth elements form compounds with the group Vb elements (N, P, As, Sb, Bi) known as "rare-earth salts". Among them, cerium monophosphide

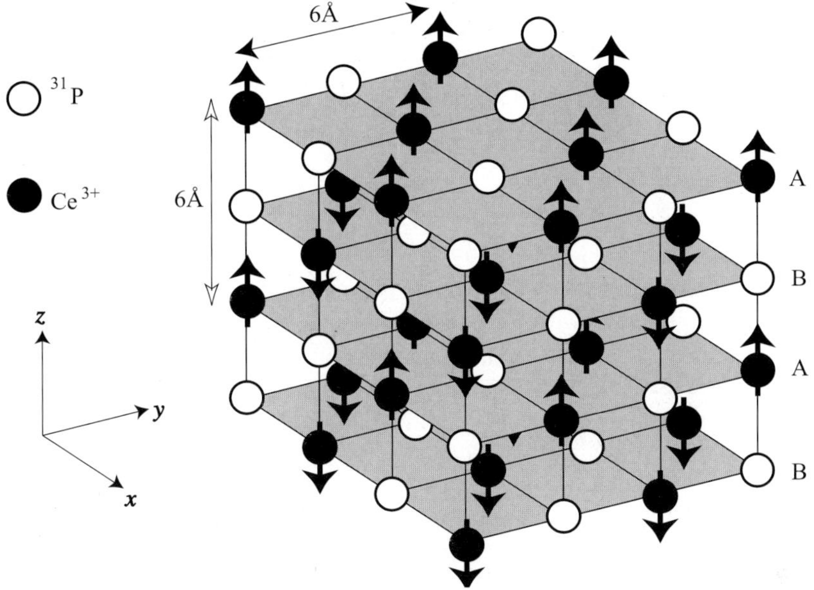

Fig. 2. – Crystal structure of cerium phosphide compound (NaCl structure with a lattice constant $\sim 6\,\text{Å}$) and type-I antiferromagnetic (AF) order of the $4f$ electron spins of cerium ions under an external magnetic field along the z-axis. In A layers, electrons have effective spins $1/2$ ($|+\rangle$ state), and in B layers electrons have effective spins $-1/2$ ($|-\rangle$ state).

(CeP) would seem to be the simplest crystal in terms of nuclear and electron spins, since a cerium atom has zero nuclear spin (all the other rare-earth elements have nonzero nuclear spins), ^{31}P is 100% spin $1/2$, and there is only one unpaired $4f$ electron per Ce atom outside the xenon core. CeP crystallizes in the NaCl structure with a lattice constant of $6\,\text{Å}$. High-purity single crystals of CeP are now obtainable with modern growth techniques [40, 41]. CeP is categorized as a semimetal with a free carrier number ~ 0.01 per cerium atom. Below the Néel temperature $T_\text{N} \sim 8.5\,\text{K}$ [42], it shows type-I antiferromagnetic (AF) order under an applied magnetic field $\leq 0.2\,\text{T}$ [43, 44]. This ordered state has the $4f$ electron spins of cerium ions aligned in each layer, but the direction of alignment of adjacent layers are opposite, as shown in fig. 2.

In fact, the spin-related physics of CeP is far from simple. The low concentration of conduction electrons leads to RKKY-type interactions, Kondo effects, and related phenomena. Moreover, the conduction electron and hole states are known to mix with the localized $4f$ states, which lie particularly close to the Fermi energy in CeP. These effects lead to anomalies in nuclear relaxation and to complicated magnetic phases at low temperature [45]. However, for sufficiently low magnetic fields ($< 0.2\,\text{T}$) and low temperatures where the AF order is simply type-I, it seems a reasonable approximation to ignore $4f$ mixing phenomena and treat the electron system as a Heisenberg antiferromagnet, as we show in the following discussion. More detail is available in refs. [39, 46].

The Hamiltonian that we consider here for $4f$ electrons of Ce^{3+} ions consists of the spin-orbit coupling, $\mathcal{H}_{SO}$; the crystalline field with cubic symmetry, $\mathcal{H}_{crystal}$, created by neighboring anions (phosphorus ions); the electron Zéeman energy, $\mathcal{H}_{eZ}$; and the exchange interaction among electrons, $\mathcal{H}_{exch}$. The spin-orbit coupling,

$$(18) \qquad \mathcal{H}_{SO} = \lambda_{SO}\mathbf{L} \cdot \mathbf{S} = \frac{\lambda_{SO}}{2}(J^2 - L^2 - S^2),$$

gives rise to the splitting of states into the $^2F_{5/2}$ multiplet ($J = 5/2$) and $^2F_{7/2}$ multiplet ($J = 7/2$), which are energetically separated by $(7/2)\lambda_{SO} \sim 280\,\text{meV}$ [47]. In the case of rare-earth compounds, where the spin-orbit coupling is relatively large compared to the crystalline field, the $^2F_{7/2}$ multiplet can be neglected at low temperature. The crystalline field Hamiltonian then splits the $^2F_{5/2}$ multiplet into a ground-state Γ_7 doublet and excited-state Γ_8 quartet, which are energetically separated by $\sim 15\,\text{meV}$ [42]. At low temperature ($T \ll 170\,\text{K}$), only the ground-state Γ_7 doublet is populated. In a weak magnetic field, its two states may be written as

$$(19) \qquad |\pm\rangle \equiv \frac{1}{\sqrt{6}}\left|J^z = \pm\frac{5}{2}\right\rangle - \sqrt{\frac{5}{6}}\left|J^z = \mp\frac{3}{2}\right\rangle.$$

This doublet is represented by a "spin-Hamiltonian" with an effective spin $\tilde{S} = 1/2$. The states $|\pm\rangle$ are identified by $\tilde{S}_z = \pm 1/2$, respectively. The electron Zéeman term,

$$(20) \qquad \mathcal{H}_{eZ} = \mu_B\mathbf{B} \cdot [2\mathbf{S} + \mathbf{L}] = g_J\mu_B\mathbf{B} \cdot \mathbf{J},$$

where g_J is the Landé g-factor,

$$(21) \qquad g_J = \frac{3}{2} + \frac{S(S+1) - L(L+1)}{2J(J+1)} = \frac{6}{7},$$

is then described within the Γ_7 doublet with the effective spin $\tilde{S}$ and a g-factor of the form

$$(22) \qquad \mathcal{H}_{eZ} = g\mu_B\mathbf{B} \cdot \tilde{\mathbf{S}},$$

where

$$(23) \qquad g = 2g_J\langle+|J_z|+\rangle = -\frac{5}{3}g_J = -\frac{10}{7} < 0.$$

The exchange Hamiltonian is difficult to write down *a priori*. It is governed by some combination of traditional Heisenberg exchange and RKKY couplings; it may be anisotropic and involve long-range terms. Empirically, however, the observed type-I order may mean that a Heisenberg exchange interaction with isotropic nearest-neighbor and next-nearest-neighbor terms is a sufficient first approximation [48]. Hence, we have

$$(24) \qquad \mathcal{H}_{\text{exch}} = J_1 \sum_{\langle l,l' \rangle} \tilde{\mathbf{S}}_l \cdot \tilde{\mathbf{S}}_{l'} + J_2 \sum_{\langle\langle l,l' \rangle\rangle} \tilde{\mathbf{S}}_l \cdot \tilde{\mathbf{S}}_{l'},$$

where the exchange parameters may be related to empirically measured temperatures via mean-field theory [49] as

$$(25) \qquad J_1 = \frac{3}{8} \frac{T_{\text{N}} - \theta}{S(S+1)}, \qquad J_2 = -\frac{1}{4} \frac{3T_{\text{N}} + \theta}{S(S+1)}.$$

Here, θ is the Curie-Weiss temperature, which has been measured as roughly $3.6\,\text{K}$ [50], yielding $J_1/k_{\text{B}} \sim 3\,\text{K}$ and $J_2/k_{\text{B}} \sim -11\,\text{K}$.

The coupling of the ^{31}P nuclear spins to the effective spin-$(1/2)$ electrons given by eq. (10) is dominated by the hyperfine term, so it is sufficient to write

$$(26) \qquad \mathcal{H}_{\text{en}} = \hbar \mathcal{A}_{\text{hf}} \sum_{\langle ij,k \rangle} \mathbf{I}_{ij} \cdot \tilde{\mathbf{S}}_k,$$

where the magnitude of the isotropic coupling is estimated as $\mathcal{A}_{\text{hf}}/2\pi \sim 3\,\text{MHz}$ [48, 50].

To summarize this section, we have reduced the Hamiltonian of the $4f$ electrons to the simple case of fixed spin-$(1/2)$ electrons with the Zéeman Hamiltonian of eq. (22) and the exchange interaction eq. (24). We would now like to consider how this simplified electronic system may be useful for initialization, and also its effect on the decoherence of the qubit nuclei.

Our goal for initialization is to transfer the perfect electron polarization to the nuclei in a scheme similar to the "solid effect". In the following, we will make the approximation that an electron spin in a B layer (see fig. 2) sees a very large effective field $\mathbf{B}_{\text{exch}} \sim 30\,\text{T}$ due to its exchange couplings with neighboring spins, and that this field is unaffected by RF manipulation of the spin in question. This "mean-field" approach is a crude approximation to a description of the physics of such a process. A more precise approach would be to consider the excitation of electron spin waves to achieve the polarization.

As shown in fig. 2, a nuclear spin in an A layer at position (x, y, z), whose spin operator is here denoted by simplified subscripting as $\mathbf{I}_{\text{A}}(x, y, z)$, is surrounded by six electron spins, four of which are in the same xy-plane with $\langle \tilde{S}^z_{\text{A}} \rangle \sim +1/2$ and two of which are located above and below the nuclear spin in the z-direction $(x, y, z \pm a)$ with $\langle \tilde{S}^z_{\text{B}} \rangle \sim -1/2$. If one chooses a frequency of an alternating magnetic field along the x-axis that corresponds to an electron resonance of $\tilde{S}_{\text{B}}(x, y, z \pm a)$, the other four electron spins $\tilde{S}_{\text{A}}(x \pm a, y, z)$, $\tilde{S}_{\text{A}}(x, y \pm a, z)$ are not affected by this field since they have different resonant frequencies primarily due to the different exchange field. Thus, it is sufficient to consider double resonance of the nuclear spin $I_{\text{A}}(x, y, z)$ and one of the two identical spins $\tilde{S}_{\text{B}}(x, y, z \pm a)$, say $\tilde{S}_{\text{B}}(x, y, z + a)$. The full Hamiltonian for the system of $I^{\text{A}}(x, y, z)$

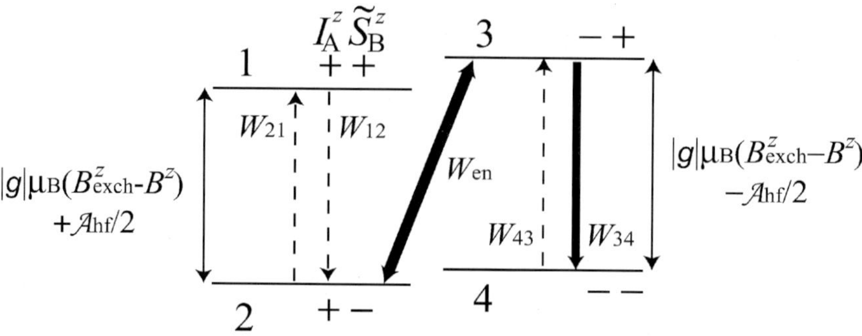

Fig. 3. – The energy diagram of a system consisting of the nuclear spin $I_A(x, y, z)$ and the electron spin $\tilde{S}_B(x, y, z + a)$. The first signs correspond to the eigenvalues of $I_A^z(x, y, z)$ and the second signs are those of $\tilde{S}_B^z(x, y, z + a)$. Transitions $1 \leftrightarrow 2$ and $3 \leftrightarrow 4$ are electron transitions, and transitions $1 \leftrightarrow 3$ and $2 \leftrightarrow 4$ (neglected) are nuclear transitions. Transition $2 \leftrightarrow 3$ is a combined nucleus-electron spin flip. If the transition $2 \leftrightarrow 3$ is saturated by an alternating field at a frequency $[|g|\mu_B(B_{exch}^z - B^z)]/\hbar + \gamma B_z - 3\mathcal{A}_{hf}/2$, the population of the state 4 becomes unity.

and $\tilde{S}_B(x, y, z + a)$, including effective mean-field Zéeman and hyperfine terms, is thus taken as

$$(27) \quad \mathcal{H} = g\mu_B[B^z(x, y) - B_{exch}^z]\tilde{S}_B^z(x, y, z + a) - \hbar\left[\gamma B^z(x, y) - \frac{3}{2}\mathcal{A}_{hf}\right]I_A^z(x, y, z) +$$

$$+ \hbar\mathcal{A}_{hf}I_A^z(x, y, z)\tilde{S}_B^z(x, y, z + a) + \frac{1}{2}\hbar\mathcal{A}_{hf}\left\{I_A^+(x, y, z)\tilde{S}_B^-(x, y, z + a) +\right.$$

$$\left. + I_A^-(x, y, z)\tilde{S}_B^+(x, y, z + a)\right\},$$

and the energy diagram of this reduced system is shown in fig. 3. Allowed transitions are electron transitions ($1 \leftrightarrow 2$, $3 \leftrightarrow 4$) and a combined nucleus-electron spin flip ($2 \leftrightarrow 3$). Here nuclear transitions ($1 \leftrightarrow 3$, $2 \leftrightarrow 4$) are neglected because we now consider the initialization of nuclear spins within a time scale much shorter than the nuclear spin relaxation time, T_1.

In thermal equilibrium, populations of the energy levels $i = 1, 2, 3$ and 4 (as labeled in fig. 3) with the energy eigenvalues E_i are given by the Boltzmann factors. Introducing energy differences,

$$(28) \quad E_1 - E_2 \sim E_3 - E_4 \sim |g|\mu_B B_{exch}^z \equiv \Delta,$$

$$E_3 - E_1 \sim E_4 - E_2 \sim \gamma\hbar B_0 \equiv \delta,$$

we obtain the probabilities of occupation of the states,

$$(29) \qquad p_2 = \frac{1}{(1 + e^{-\Delta/kT})(1 + e^{-\delta/kT})} \approx \frac{1}{2}, \qquad p_4 = e^{-\delta/kT} p_2 \approx \frac{1}{2},$$

$$p_1 = e^{-\Delta/kT} p_2 \approx 0, \qquad\qquad p_3 = e^{-(\Delta+\delta)/kT} p_2 \approx 0,$$

where the approximations assume $\delta \ll k_{\mathrm{B}} T \ll \Delta$.

If we saturate the transition between states 2 and 3, their populations are clamped (*i.e.*, $p_2 = p_3$). This transition is induced by an applied alternating magnetic field perpendicular to the z-axis at a frequency

$$(30) \qquad \omega_{23} = \frac{|g|\mu_{\mathrm{B}}}{\hbar}\left[B^z_{\mathrm{exch}} - B^z(x,y)\right] + \gamma B^z(x,y) - \frac{3}{2}\mathcal{A}_{\mathrm{hf}}.$$

The probabilities of occupation of the states, in the presence of the transition induced by this alternating field at a rate W_{en}, obey

$$(31) \qquad \frac{\mathrm{d}p_1}{\mathrm{d}t} = p_2 W_{21} - p_1 W_{12},$$

$$(32) \qquad \frac{\mathrm{d}p_2}{\mathrm{d}t} = p_1 W_{12} - p_2 W_{21} + p_3 W_{32} - p_2 W_{23} + (p_3 - p_2)W_{\mathrm{en}},$$

$$(33) \qquad \frac{\mathrm{d}p_3}{\mathrm{d}t} = p_2 W_{23} - p_3 W_{32} + p_4 W_{43} - p_3 W_{34} + (p_2 - p_3)W_{\mathrm{en}},$$

$$(34) \qquad \frac{\mathrm{d}p_4}{\mathrm{d}t} = p_3 W_{34} - p_4 W_{43},$$

where W_{ij} is the thermally induced relaxation rate from state i to state j and satisfies $W_{ij}e^{-E_i/kT} = W_{ji}e^{-E_j/kT}$ (detailed balance). The probabilities of occupation add to unity, $p_1 + p_2 + p_3 + p_4 = 1$. In the steady state, we have

$$(35) \qquad p_4 = \frac{1}{1 + 2e^{-\Delta/kT} + e^{-2\Delta/kT}}, \qquad p_2 = p_3 = e^{-\Delta/kT} p_4,$$

$$p_1 = e^{-2\Delta/kT} p_4,$$

on the assumption that W_{en} is sufficiently large to produce complete saturation, $W_{\mathrm{en}} \gg W_{23}, W_{32}$. At low temperatures, p_4 deviates from unity by approximately $2e^{-\Delta/kT}$, which is about 10^{-8} at $1\,\mathrm{K}$, and the other occupation probabilities are approximately zero.

We see that following the saturation of the combined nucleus-electron spin flip transition, the induced state where the electron is excited decays rapidly due to the large exchange interaction with neighboring electron spins. Thereby, nuclear spins are efficiently polarized, *i.e.*, nuclear spins in A layers are prepared in $I_z^{\mathrm{A}} = -1/2$ state with probability ~ 1.

We must consider that the antiferromagnetic electrons that allow the initialization scheme above will also lead to decoherence. The decoherence time T_2 has two new contributions in an antiferromagnet.

First there is the T_2^{ext} due to small fluctuations of the coupled electron spin system. These fluctuations are described by spin waves. In ref. [46], we derive theoretically a 4-sublattice spin wave spectrum for the type-I antiferromagnetic order seen in CeP. This spectrum is then used with eq. (12) to deduce an approximate T_2^{ext}. The result is

$$(36) \qquad \frac{1}{T_2^{\text{ext}}} \approx 0.3 \, T^2 \ln\left(\frac{k_{\text{B}}T}{g\mu_{\text{B}}B_0}\right) + 10^{-4}T^3 + 10^{-5}T^5,$$

where the temperature T is measured in kelvin. At $4\,\text{K}$, this formula estimates the T_2 as $20\,\text{ms}$ due to electron fluctuations alone.

Even worse is the second contribution to T_2, which is provided by a new nuclear-nuclear interaction. In second-order perturbation theory, the hyperfine coupling $\mathcal{H}_{\text{en}}$ allows the possibility for a nucleus at one site to create a spin wave which may then be absorbed by a nucleus at another site. These processes lead to the effective coupling Hamiltonian

$$(37) \qquad \mathcal{H}_{\text{eff}}^* = -\sum_i \sum_{k>l} D_{ikl}\left(I_{ik}^+ I_{il}^- + I_{ik}^+ I_{il}^-\right),$$

where D_{ikl} is a complicated, anisotropic coefficient. The form of D_{ikl} has been calculated using the same spin wave physics in ref. [46]. This effective coupling, known as the Suhl-Nakamura interaction [51], may not be decoupled by WAHUHA-like sequences. Its effect on the T_2 may be estimated using the method of moments. The resulting T_2 is of order $\hbar J_2^2/\mathcal{A}_{\text{hf}}^2 J_1 \sim 1\,\text{ms}$. This is of the same order as the time for a single, nearest-neighbor, two qubit logic gate, $2\pi/\delta\omega$, thus prohibiting quantum computation in this material.

A CeP quantum computer has another difficulty still: the face-centered-cubic crystal structure of the phosphor qubits is not one-dimensional. In the geometry imagined in ref. [39], $\lambda_1 = 1/\sqrt{2} < 1$. Thus, even nearest-neighbor interactions have a large error due to parasitic dipolar couplings between ensemble members.

CeP was historically the first crystal proposed for crystal lattice quantum computation. Since then, however, the theoretical considerations presented above would seem to favor a material in which electrons used for initialization can be removed during a computation. A proposal which combines this idea with a one-dimensional structure similar to fluorapatite is to use isotopically engineered silicon, as described in the next section.

3˙3. *Silicon* [52]. – The use of optical pumping in a semiconductor for initialization is an important reason for considering a silicon implementation. In addition, from an engineering perspective, there are many other advantages. Perhaps the most important motivation is that the crystal growth and processing technologies for silicon are highly matured. Although methods exist for the growth of small single crystals of fluorapatite and CeP, for example, the technology to incorporate these materials in micrometer-scale devices has not yet developed. With its wealth of fabrication technologies and its favorable material properties, pure silicon has proven to be an excellent material for micrometer-scale mechanical oscillators, which will be important when we discuss force detection in sect. **6**.

Another advantage of silicon is that, unlike III-V semiconductors, the family of stable nuclear isotopes is quite simple: 95.33% of natural silicon is ^{28}Si or ^{30}Si, which are both spin-0, and 4.67% is ^{29}Si, which is spin 1/2, perfect for the qubit. Thus, silicon is well suited for nuclear-spin isotope engineering [53].

A crucial element of the all-silicon proposal we present here is that, following optical polarization, there is no need for electrons for the control or measurement of nuclear spins, unlike a number of other silicon-based proposals which require bound donor electrons and (usually) nanometer-scale metallic gates [20, 24, 25]. The absence of any spurious spins, nuclear or electronic, should leave the ^{29}Si nuclei well decoupled from their environment.

Of course, the diamond crystal structure of silicon is not "one-dimensional", as in the case of fluorapatite; especially if one considers the randomly placed ^{29}Si nuclei in a sample with natural abundance. To make an effective crystal lattice quantum computer, we must somehow fabricate long, straight atomic chains of the active ^{29}Si isotopes in a single-crystal matrix of the other, spin-0 isotopes. We emphasize that, considering the wide range of silicon processing methods available, there may be more than one way to fabricate such a system; in the following we consider one particular method. We note that this procedure is similar to one already proposed for a Kane-like implementation [54].

We start with a nearly isotopically pure ^{28}Si(111) silicon-on-oxide (SOI) wafer with the surface miscut by an angle of about $1°$ towards $(\bar{1}\bar{1}2)$. The buried oxide layer is needed to release the silicon structure to make a vibrating bridge, as will be discussed in sect. **6**. It has been demonstrated [55, 56] that such a miscut, followed by a simple multi-step annealing sequence, leads to atomically straight terrace-steps on the vicinal Si(111)7 × 7 surface. These steps run along the $(1\bar{1}0)$ direction for up to 2×10^4 lattice sites, with an average terrace width of $\sim 15\,\mathrm{nm}$. The lack of kinks in these steps is due to the high energy cost of interrupting or misplacing the 7×7 domains.

This miscut wafer is processed in a multi-chamber molecular beam epitaxy (MBE) machine equipped with a scanning tunneling microscope (STM) and a pre-heating stage. The STM is used for verification of the step edges; afterwards, the wafer is transferred to the growth chamber to form atomic chains of ^{29}Si using the step-flow mode at temperatures $T \sim 850\,°\mathrm{C}$. Arrival of ^{29}Si isotopes at the substrate surface induces a surface transition from 7×7 to 1×1 [57], so that ^{29}Si isotopes travel to the step edges and form atomically straight lines. We terminate the evaporation of ^{29}Si when the atomic chains are one atom wide and run completely along the step. A ^{28}Si capping layer of $\sim 15\,\mathrm{nm}$ is grown on top of the ^{29}Si chains and we repeat the same sequence of multi-step annealing, ^{29}Si chain growth, and capping to produce replicas of parallel ^{29}Si chains. The last step in the growth process is the capping of ensembles of ^{29}Si chains by a thick ^{28}Si layer.

The atomic chains resulting from this procedure lead to a highly one-dimensional structure; on average, we find $\lambda_1 \approx 15\,\mathrm{nm}/2\,\text{Å} = 75$. In particular, $\Gamma(1) \sim 10^{-6}$, so the

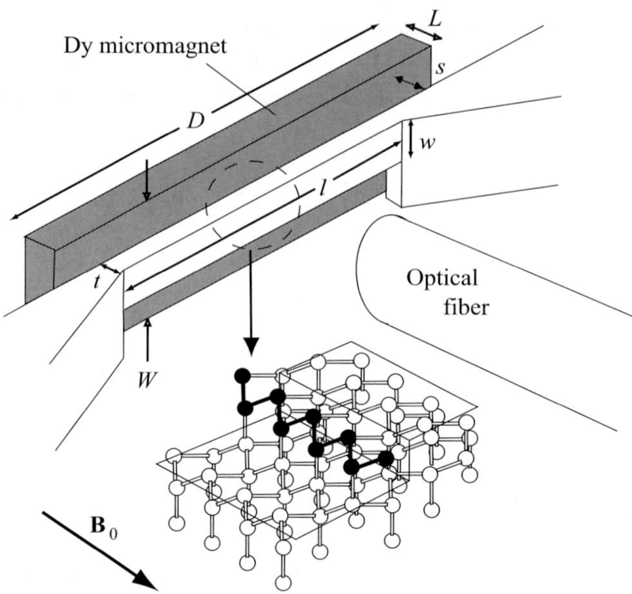

Fig. 4. – A schematic for an all silicon quantum computer. The figure shows the integrated micromagnet and bridge structure needed for distinguishing qubits and for readout. The bridge has length $l = 300\,\mu$m, width $w = 4\,\mu$m, and thickness $t = 0.25\,\mu$m. The micromagnet has $D = 400\,\mu$m, $L = 4\,\mu$m, and $W = 10\,\mu$m, and produces a field gradient of $\partial B^z/\partial z = 1.4\,T/\mu$m, uniform over a $100\,\mu$m by $0.2\,\mu$m region inside the bridge. The insert shows the structure of the silicon matrix and the terrace edge. The darkened spheres represent the ^{29}Si nuclei, which preferentially bind at the edge of the Si step.

error in the nearest-neighbor gate operation is quite tolerable. We will return to $\Gamma(m)$ for the silicon implementation, as well as other decoherence-related issues, in sect. **7**, when we discuss scalability. We also note that the atomic chains, while very regular, are not straight rows of atoms. As shown in fig. 4, the atoms zig-zag with polar angle satisfying $\cos^2\theta_{ij,(i+1)j} = 2/3$, leaving $\delta\omega = 2\pi \times 0.4\,$kHz. The $1/m^3$-dependence in eq. (14) is not precisely correct for small qubit displacements m, however we will ignore this small discrepancy when we consider scalability calculations.

The ability to perform optical pumping in silicon [58] aids the initialization problem, though such techniques are unlikely to yield 100% polarization. Fortunately, the techniques of algorithmic cooling [59, 60] and the use of pseudo-pure states [61], both originally developed for solution-NMR quantum computers, may be employed. These techniques introduce scaling limitations which are only restrictive if the physical polarization provided by optical pumping is too low.

The premise of optical pumping is that nuclei exchange Zéeman energy with a bath of electrons which have been preferentially excited into a single spin state by circularly polarized light. The nuclei thereby relax thermally to an effective spin temperature corresponding to the non-equilibrium electron-spin polarization. Once those electrons recombine, the nuclei retain their spin polarization for the "dark" T_1 time, which is extremely long (200 hours in ref. [58]). A solution of the rate equations for this system yields the steady-state nuclear polarization in the strong-pumping limit as

$$(38) \qquad \langle I^z \rangle \approx \frac{T_{1e}}{\tau + T_{1e}} \frac{G_+ - G_-}{2},$$

where τ is the lifetime of the photoexcited electrons and $G_\pm$ is the transition probability for generating a spin $m^z = \pm 1/2$ with circularly polarized light.

The indirect bandgap of silicon limits the nuclear polarization for two reasons. First, the two transition probabilities $G_\pm$ are affected by the band structure; the indirect bandgap leads to a reduced value of $(G_+ - G_-)$ relative to direct bandgap semiconductors. Second, the ratio of the conduction electron spin-lattice relaxation time T_{1e} to its recombination time τ is small since the indirect processes necessary for electron recombination lead to a long τ. As a result, low-field ($\sim 1\,\mathrm{G}$) experiments at $77\,\mathrm{K}$ [58,62] have seen nuclear polarizations that have not exceeded 0.1%. Improved nuclear polarization in silicon may be observable in higher magnetic fields ($\sim 10\,\mathrm{T}$) and lower temperatures ($\sim 1\,\mathrm{K}$); in these regimes, the transition probabilities $G_\pm$ may be more favorable, and T_{1e} is substantially longer [63]. A small sample will also help, since rapid recombination of electrons via surface states can reduce τ [64].

4. – Gradient design

We now discuss the design and operation of the control and measurement aspects of the crystal lattice quantum computer. We begin by describing the method for producing the magnetic-field gradient.

4‘1. *Design considerations.* – A requirement for the gradient is that the adjacent-nucleus frequency separation be large compared to the nuclear dipole coupling; thus we require a ratio $\Delta\omega/\delta\omega$ greater than unity. The reason for this criterion will be discussed in sect. **5** in the context of quantum logic. We have already discussed in subsect. **2**‘2 that the unitless parameter $\Delta\omega T_2^{\mathrm{cc}}$ must be much greater than one, in order to neglect non-secular terms of $\mathcal{H}_{\mathrm{qq}}^*$. These criteria require a minimum field gradient of more than $1\,\mathrm{T}/\mu\mathrm{m}$, depending on the material.

We have discussed so far the one-dimensional quantum computer, which has a field gradient in one dimension with the orthogonal dimensions having minimal field variation. Maxwell's equations for a static system that does not enclose any currents state that $\nabla \cdot \mathbf{B} = 0$ and $\nabla \times \mathbf{B} = 0$. Thus, a gradient or spatial variation of one component of the magnetic field implies a change of a different component in an orthogonal dimension. However, application of a large external field introduces an asymmetry so that nuclei along the direction in which there is no intended field gradient have negligible Zéeman energy shifts. The major design constraint is to maximize the field gradient in one dimension while maintaining the homogeneity in the orthogonal direction over a sufficiently large area to enclose a detectable number of nuclear spins with identical Larmor precession frequencies.

A spatial variation in the magnetic field can also be introduced in two dimensions. This can be used to create a two-dimensional quantum computer with ensemble copies existing along the third dimension. A two-dimensional quantum computer requires an anisotropic field gradient with no spectral overlap along both gradient directions. The anisotropy must be sufficient so that nuclei along the dimension with the larger field gradient are separated in frequency by more than the maximal frequency separation of qubits in the direction with the smaller field gradient. If pulse techniques permit two-bit operations between nearest neighbors, then a one-dimensional quantum computer allows for up to two neighboring nuclei to be involved in logic operations, whereas a two-dimensional computer permits at least 4. This geometry will allow more neighboring nuclei to carry out logic operations, thus reducing bit-swapping.

Current-carrying wires allow one to construct an electromagnet with arbitrary geometries, and the induced magnetic field can be directly controlled through changes in the applied current. Unfortunately, the field gradients with such systems are limited by the current-carrying capacity of the wires. Even an optimal metal (for example Au) arranged in a Helmholtz configuration operating at a current density of 10^{12} A/m^2 yields a maximal field gradient on the order of 0.01 to 0.1 T/μm, which extends for less than 1 μm. Theoretical proposals and reported results with current-carrying wires for micromagnetic atom traps on the order of 0.001 T/μm are presented in [65-67] and up to 0.1 T/μm field gradients are described in [68].

Larger gradients have been realized with ferromagnetic materials [69-71] such as Tb, Fe, and Dy. Dysprosium (Dy) has one of the largest values for saturated magnetic polarization (3.5 to 3.7 T) [71, 72]. Polarization is achieved by the application of an external field of a few tesla. Provided that Dy is placed in a sufficiently high field (greater than 1 T) it will exhibit ferromagnetism below 180 K and paramagnetism above this temperature [73, 74].

4$^.$2. *Magnetic-field calculation.* – The calculations begin with the equation describing the field due to a single magnetic dipole:

$$(39) \qquad \mathbf{B}_{\mathrm{dip}}(r, \theta) = \frac{\mu_0}{4\pi} \frac{m}{r^3} (2 \cos\theta \hat{\mathbf{r}} + \sin\theta \hat{\theta}),$$

where m is the magnetic moment of the electron, and r and θ are the spherical coordinates shown in fig. 5. To calculate the magnetic field due to a slab of magnetic material, we integrate this expression over the slab's dimensions and replace the prefactor with the saturated magnetic polarization, J_s, assuming full saturation is reached. The total

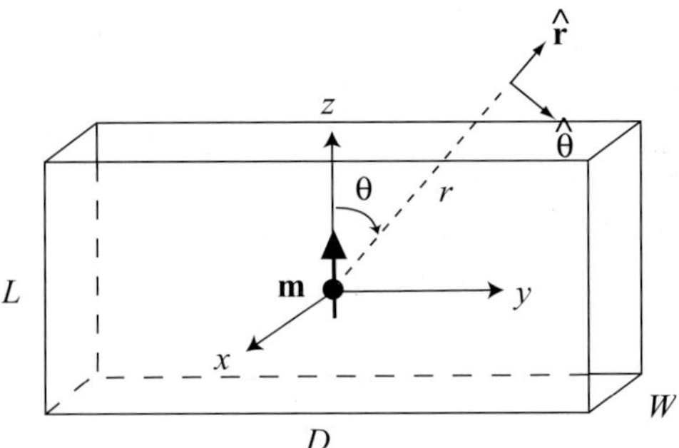

Fig. 5. – A single dipole is shown to be aligned along the z-axis. The standard spherical coordinates r and θ are used. To calculate the field due to a rectangular slab of material one integrates this expression along the 3 dimensions, as shown.

magnetic field is then given by

$$(40) \quad \mathbf{B}(x,y,z) = J_s \int_{-L/2}^{L/2} \int_{-D/2}^{D/2} \int_{-W/2}^{W/2} [3(x-x')(z-z')\hat{\mathbf{x}} + 3(y-y')(z-z')\hat{\mathbf{y}} +$$

$$+ 2(z-z')^2 - (x-x')^2 - (y-y')^2 \hat{\mathbf{z}}] \frac{\mathrm{d}x'\mathrm{d}y'\mathrm{d}z'}{[(x-x')^2 + (y-y')^2 + (z-z')^2]^{5/2}},$$

where W, D, and L correspond to the dimensions of the magnet along x, y, and z, respectively, and the origin of the coordinate system is located at the center of the magnet (fig. 5). This expression is only valid for $x > W/2$, $y > D/2$, and $z > L/2$, i.e., outside of the bar magnet. The integration of eq. (40) yields

$$(41) \quad \mathbf{B}(x,y,z) = J_s \sum_{i,j,k=0}^{1} (-1)^{i+j+k} \mathbf{G}\left[x,y,z,(-1)^i\frac{W}{2},(-1)^j\frac{D}{2},(-1)^k\frac{L}{2}\right],$$

$$G^x(x,y,z,x',y',z') = -\sinh^{-1}\left[\frac{(y-y')}{\sqrt{(x-x')^2+(z-z')^2}}\right],$$

$$G^y(x,y,z,x',y',z') = -\sinh^{-1}\left[\frac{(x-x')}{\sqrt{(y-y')^2+(z-z')^2}}\right],$$

$$G^z(x,y,z,x',y',z') = \frac{1}{2}\tan^{-1}\left[\frac{2(z-z')(y-y')(x-x')r}{(z-z')^2r^2-(x-x')^2(y-y')^2}\right],$$

where $r = \sqrt{(x-x')^2 + (y-y')^2 + (z-z')^2}$.

$4\cdot3.$ *One-dimensional quantum computer.* – The objective of this design is to achieve a large one-dimensional field gradient while suppressing inhomogeneity in the transverse directions. In this way, a plane of nuclei will have nearly identical Zéeman energies.

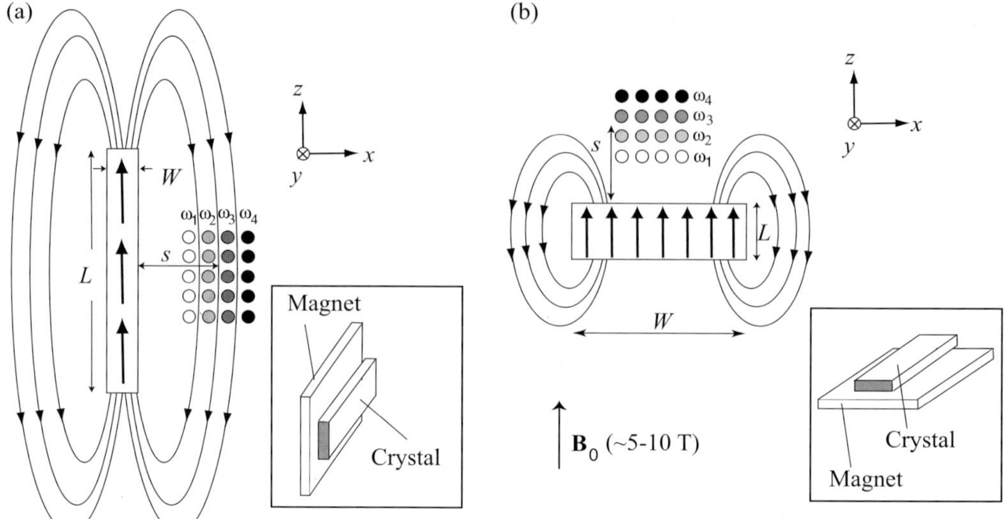

Fig. 6. – Two configurations for a one-dimensional quantum computer. (a) The sample is placed at $z \approx 0$ and the gradient is along the x-axis. In this configuration the yz-plane houses ensemble copies of qubits. (b) The sample is placed directly above the polarized magnet such that $\partial B^z / \partial z$ corresponds to the large field gradient, and the xy-plane contains ensemble copies of qubits.

Figure 6 presents two possible configurations using the magnetized material discussed in the previous subsection. At a properly chosen distance s from the magnet in the x-direction as shown in fig. 6(a) and near the line $z = 0$, the total field is independent of z. However, $\partial B^z / \partial x$ can be large and may be used as the field gradient to separate the Zéeman energy splitting of the spins. Alternatively, the homogeneous magnetic field can lie in the xy-plane and a large field gradient can be realized along the z-direction as depicted in fig. 6(b). The magnet extends considerably in y for both designs, so there are a large number of ensemble copies of qubits in this direction.

The inhomogeneity in the yz- or xy-planes for the two designs depicted in fig. 6(a) and (b), respectively, must be considered. Here we consider the case of the design depicted in fig. 6(b). As the value of x moves away from $x = 0$, the field varies, and the number of ensemble qubits permitted in this dimension is limited by the allowed inhomogeneity. When the crystal is placed in a large external field of strength B_0 aligned along the z-direction, the shift in the magnetic field is

$$(42) \qquad \triangle B(x, z_0) = \sqrt{[B^x(x, z_0)]^2 + [B_0^z + B^z(x, z_0)]^2} - [B_0^z + B^z(x = 0, z_0)],$$

where z_0 is the location of a plane of a single qubit nuclear ensemble. Note that the y-component of the magnetic field is ignored, since the magnet is very long in the y-direction.

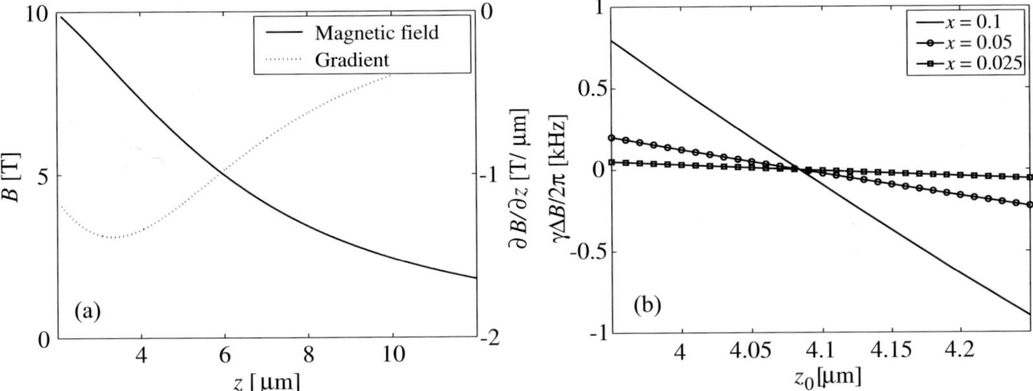

Fig. 7. – For the one-dimensional design of fig. 6(b), with $L = 4\,\mu$m, and $W = 10\,\mu$m, one can achieve a field gradient of $\sim 1\,$T$/\mu$m over a few μm distance, as shown in (a). Plot (b) shows the magnetic-field shift as a function of distance z_0 in the z-direction for several displacements in the x-direction.

Figure 7 shows the magnetic field and gradient $\partial B^z/\partial z$ with respect to z for $L = 4\,\mu$m and $W = 10\,\mu$m. With these parameters, the maximum field gradient achieved is larger than $1\,$T$/\mu$m. By increasing the value of L or decreasing W, larger field gradients can be achieved, but the trade-off with inhomogeneity reduces the number of available ensemble copies of qubits. The gradient in this configuration persists for the range of a few micrometers along the z-direction, so the number of the distinct qubits can be on the order of a few thousand. The permitted inhomogeneity will determine the number of ensemble copies. Assuming a required homogeneity better than $10\,$ppm and an external field of $10\,$T, we can allow for approximately 10^3 ensemble copies of qubits in x. Since y could be made to be 10 to 100's of micrometers, the number of ensemble copies of qubits could be on the order of 10^5 along this direction, leading to an ensemble of 10^8 copies. This number depends on the lattice spacing and the interchain spacing.

For the particular case of the all-silicon proposal detailed in subsect. **3**$\cdot$3, the design shown in fig. 6(b) is chosen with $D = 400\,\mu$m, $L = 4\,\mu$m, and $W = 10\,\mu$m, as shown in fig. 4. The field gradient reaches $\partial B^z/\partial z = 1.4\,T/\mu$m in the center of the bridge microstructure, and the homogeneous area containing the active nuclei is $100\,\mu$m by $0.2\,\mu$m. The spacing of ^{29}Si nuclei in the z-direction is $0.19\,$nm, so the gradient leads to a qubit-qubit frequency difference of $\Delta\omega = a\gamma\partial B^z/\partial z = 2\pi \times 2\,$kHz. The homogeneity for the active region is shown in fig. 7(b), where it can be seen that the ensemble copies of qubits in an xy-plane all lie within a bandwidth of $0.6\,$kHz $< \Delta\omega/2\pi$. The parameter $\Delta\omega/\delta\omega = 5$. For an interchain spacing of $\sim 15\,$nm, the order of magnitude for T_2^{cc} is $4\pi(15\,\mathrm{nm})^3/\gamma^2\hbar\mu_0 \sim 100\,$s, yielding $\Delta\omega T_2^{\mathrm{cc}} \sim 10^6$, well-justifying the secular approximation.

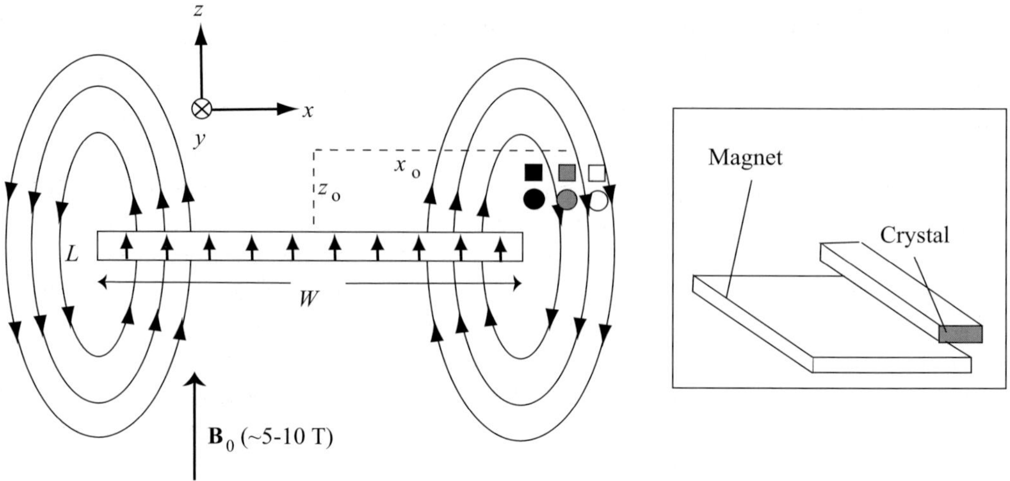

Fig. 8. – A configuration for a two-dimensional quantum computer. The different shapes and shading are meant to show an anisotropy in the field gradient.

4˙4. *Two-dimensional quantum computer*. – One possible configuration for a two-dimensional quantum computer is shown in fig. 8. By using the edge of the magnet one can achieve a large field variation in an additional dimension, as plotted in fig. 9. This is possible due to the considerable curvature of the field lines near the edge.

This configuration has only one dimension of ensemble copies of qubits, so it loses about 10^3 spins compared with the designs shown in fig. 6. As for the actual size of the computer, the increased dimensionality compensates for the decreased extent over which these larger field gradients extend, so thousands of distinct qubits are still available. More details and more designs may be found in ref. [75].

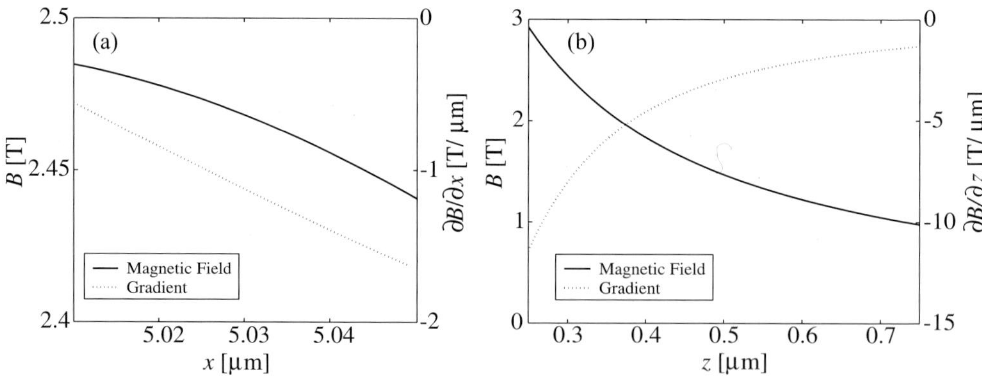

Fig. 9. – These plots show the field variation in the (a) x- and (b) z-directions for a magnet in the configuration shown in fig. 8, with dimension $W = 10\,\mu$m, and $L = 0.1\,\mu$m. $x_0 = 5.35$, $z_0 = 0.55\,\mu$m.

5. – Decoupling and recoupling pulse sequences

Single-qubit operations are performed by applying pulsed radio frequency (RF) magnetic fields according to $\mathcal{H}_{\mathrm{RF}}$ (eq. (3)). The Rabi frequency Ω_i of an on-resonant pulse is given by γB_1^y, where B_1 is the strength of the RF field. The RF power used is limited by the fact that, in order to control only a single qubit, the duration τ of a pulse should be longer than $1/\Delta\omega$. For a pulse achieving a constant angle of rotation, a long pulse requires a small power. A crucial assumption we will make in this section, however, is that Ω_i is substantially larger than dipolar couplings or inhomogeneous frequency offsets so that coupled evolution can be ignored during the pulse; this requirement thus recalls the criterion $\Delta\omega/\delta\omega \gg 1$ mentioned in sect. **4**. In this limit and in the rotating frame, the unitary evolution generated by each pulse of duration τ, frequency ω_i, and phase ϕ, according to eq. (5), is $\exp[i\theta \sum_j (I_{ij}^x \sin\phi + I_{ij}^y \cos\phi)]$, where the angle of rotation θ is $\Omega_i \tau$. We denote such a rotation as $R_i^\alpha(\theta)$, where $\alpha = Y$ for $\phi = 0$ and $\alpha = X$ for $\phi = \pi/2$. The Lie group of all single qubit operations can be generated by these two rotations.

Coupled operations such as controlled-phase-shift or controlled-NOT acting on the i-th and the k-th spins can be performed given the primitive,

$$(43) \qquad ZZ_{ik} = e^{-i\pi \sum_j I_{ij}^z I_{kj}^z},$$

available from $\mathcal{H}_{\mathrm{D,qq}}^*$ (eq. (7)) for duration $\tau = |k - i|^3 \pi/\delta\omega$. For instance, a controlled-NOT from the i-th spin to the k-th spin can be implemented by compositing the gates $R_i^Y(-\pi/2)R_i^X(\pi/2)R_k^Y(\pi/2)R_j^X(-\pi/2)R_k^Y(\pi/2)ZZ_{ik}R_k^Y(-\pi/2)$. The existence of this gate and all single qubit rotations assures universal quantum logic [76].

The difficulty is that, while we may activate or remove $\mathcal{H}_{\mathrm{RF}}^*$ at will, the coupling Hamiltonian $\mathcal{H}_{\mathrm{D,qq}}^*$ is always "on" for all spin pairs. In the following, we develop a scheme where $\mathcal{H}_{\mathrm{D,qq}}^*$ may be turned altogether "off" (heteronuclear decoupling) and then reintroduced only between two arbitrary qubits (selective recoupling).

5‘1. *Heteronuclear decoupling concepts.* – To motivate the general construction, we analyze the simplest example of decoupling two spins. Consider the evolution operator for an arbitrary duration t given by $U(t) = e^{-i\delta\omega t I_1^z I_2^z}$. We define X_i to be the gate $\sigma_i^x = 2I_i^x$. In the notation defined above, X is achieved by $R_i^X(\pi)$, up to an irrelevant overall phase. The important observation is

$$(44) \qquad X_2 U(t) X_2 = e^{-i\delta\omega t I_1^z X_2 I_2^z X_2} = e^{-i\delta\omega t I_1^z(-I_2^z)} = U(-t),$$

where the first equality is obtained using a Taylor series expansion of the matrix exponents and the identity $(X_2)^2 = 1$. This observation implies that adding the gate X_2 before and after the evolution $U(t)$ results in $U(-t)$, so that the sequence of events $X_2 U(t) X_2 U(t) = 1$ has no net coupling although the spins are actually coupled all the time. This is called *refocussing*.

We now extract the essential features of the above decoupling scheme by rewriting the sequence $X_2U(t)X_2U(t)$ as

$$(45) \qquad e^{-i\delta\omega t(+I_1^z)(-I_2^z)} \times e^{-i\delta\omega t(+I_1^z)(+I_2^z)},$$

and referring to $U(t)$ and $X_2U(t)X_2$ as time intervals. We note the following facts:

1) Since the matrix exponents commute, negating the coupling for exactly half of the total time is sufficient to cancel out the coupling.

2) Since the coupling is bilinear in I_1^z and I_2^z, it is unchanged (negated) when the signs of I_1^z and I_2^z agree (disagree).

3) The sign of I_i^z is $(+)$ or $(-)$ depending on whether or not X_i gates are applied before and after the interval. In other words, the sign of I^z for each spin in each time interval is controlled by inserting X gates for that spin before and after that interval.

Following these observations, we generalize the result to n qubits. We consider schemes which concatenate a certain number of equal-time intervals and use X_i gates to control the signs of I_i^z for each spin. The essential information on the signs can be represented by a "sign matrix" defined as follows. The sign matrix of a pulse scheme for n spins with m time intervals is the $n \times m$ matrix with the (i, a) entry being the *sign* of I_i^z in the a-th time interval. We denote any sign matrix for n spins by S_n. For example, the sequence in eq. (45) can be represented by the sign matrix

$$(46) \qquad S_2 = \begin{bmatrix} + & + \\ + & - \end{bmatrix}.$$

Each column represents a time interval; each row thus represents a sequence of m intervals for a particular spin. An entry "$-$" in the i-th row represents an interval that is preceded and followed by X_i gates. Therefore, each sign matrix corresponds to a sequence of events for the whole system. Decoupling is achieved whenever any two rows in the sign matrix disagree in exactly half of the entries (all couplings are negated for exactly half of the time). The general construction of the decoupling scheme is now reduced to finding sign matrices satisfying the above criteria.

As an illustration, we construct a decoupling scheme for four spins. We first find a correct sign matrix and then derive the corresponding pulse sequence. A possible sign matrix is given by

$$(47) \qquad S_4 = \begin{bmatrix} + & + & + & + \\ + & + & - & - \\ + & - & - & + \\ + & - & + & - \end{bmatrix},$$

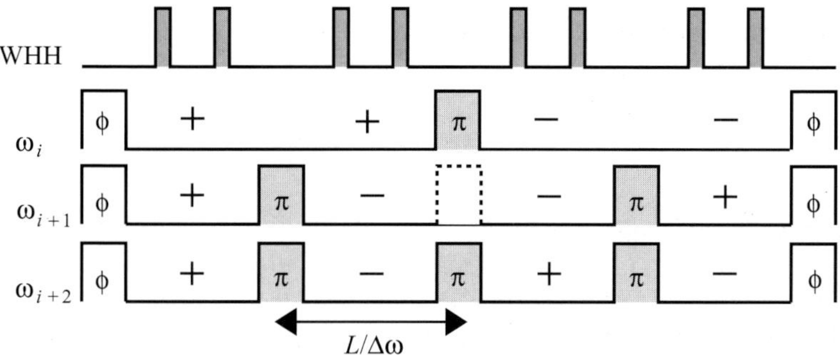

Fig. 10. – (a) Pulse sequence corresponding to eq. (49). This is constructed directly from S_4 by translating each change of sign in the i-th row to a $R_i^X(\pi)$ pulse, labeled π, at ω_i. The first, all-positive row of S_4 has been removed to eliminate inhomogeneous broadening. Single-qubit rotations, as indicated by ϕ, must occur between full cycles of the sequence. Recoupling between qubits $i+1$ and $i+2$ may be achieved by inserting a π pulse where indicated by the dotted line. The top row represents a broadband WAHUHA sequence for homonuclear decoupling, as discussed in subsect. **5·4**. The time duration between π pulses is $L/\Delta\omega$, as discussed in subsect. **7·2**.

where we note that any two rows disagree in exactly two entries. The sequence corresponding to S_4 can be obtained by converting each column to a time interval before and after which X pulses are applied to spins (rows) given by $-$'s. No pulses are applied to spins (rows) with $+$'s. The resulting sequence,

$$(48) \qquad U(t)(X_3 X_4 U(t) X_3 X_4)(X_2 X_3 U(t) X_2 X_3)(X_2 X_4 U(t) X_2 X_4),$$

is the identity by construction. Note that $U(t)$ is generated by a Hamiltonian containing a sum of six possible coupling terms for four spins. Note also that eq. (48) is written in such a way that it corresponds visually to the sign matrix, though the evolutions are actually in reverse time order relative to S_4. However, such ordering is irrelevant for commuting evolutions. Since $X_i X_i = 1$, eq. (48) can be simplified to

$$(49) \qquad U(t)(X_3 X_4 U(t) X_4)(X_2 U(t) X_3)(X_4 U(t) X_2 X_4).$$

This simplified pulse sequence can also be obtained directly from eq. (47) by converting columns to time intervals and inserting X_i between intervals whenever the i-th row changes sign. This sequence is illustrated in fig. 10.

The above scheme can be generalized to decouple n spins with m time intervals as follows: Construct the $n \times m$ sign matrix S_n, with entries $+$ or $-$, such that *any* two rows disagree in exactly half of the entries. For each $-$ sign in the i-th row and the a-th column, apply X_i before and after the a-th time interval.

5·2. *Hadamard decoupling and recoupling scheme.* – Hadamard matrices have applications in many areas such as the construction of combinatorial designs, error-correcting codes, and Hadamard transformations [77-79].

A Hadamard matrix of order n, denoted by $H(n)$, is an $n \times n$ matrix with entries ± 1, such that

$$(50) \qquad H(n)H(n)^T = nI.$$

The rows are pairwise orthogonal; therefore any two rows agree in exactly half of the entries. Likewise columns are pairwise orthogonal. It is immediately clear that each $H(n)$ is a valid sign matrix giving a decoupling scheme for n spins using only n time intervals.

Whenever $H(n)$ exists, there is a decoupling scheme for n spins concatenating only n time intervals. However, $H(n)$ may or may not exist for a given n. For an arbitrary integer n, let $\overline{n}$ be the smallest integer that satisfies $n \leq \overline{n}$ with *known* $H(\overline{n})$. To construct a decoupling scheme for n spins when $H(n)$ does not necessarily exist, we start with $H(\overline{n})$ and take S_n to be any $n \times \overline{n}$ submatrix of $H(\overline{n})$. In other words, S_n is formed by choosing n rows from $H(\overline{n})$, which still achieves decoupling because subsets of rows of $H(\overline{n})$ are still pairwise orthogonal. The resulting decoupling scheme for n spins requires $\overline{n}$ time intervals.

This decoupling scheme may be used to "turn off" $\mathcal{H}^*_{\mathrm{D,qq}}$ and $\mathcal{H}^*_{\mathrm{D,qc}}$. It may also remove any frequency offsets caused by inhomogeneous broadening (*i.e.*, it may incorporate the spin-echo technique). To see how, note that such frequency offsets correspond to terms in the Hamiltonian linear in I_i^z, so negating I_i^z for half of the time results in no net evolution for these terms. Therefore, such evolution for all spins can be removed if the sign matrix has identically zero row sum. We note that complete negations of rows or columns of Hadamard matrices leave their essential properties intact; thus we may always perform such negations to generate a Hadamard matrix which has only $+$'s in the first row and column. The sign matrix with zero row sum can be made by then removing this first, all-positive row. All other rows have zero row sums by orthogonality. Such construction is possible unless $n = \overline{n}$, in which case construction should start with $H(\overline{n+1})$.

To implement selective recoupling between the i-th and the k-th spins, the sign matrix should have equal i-th and k-th rows but any other two rows should be orthogonal. The coupling term $I_i^z I_k^z$ never changes sign and that coupling is implemented selectively, while all other couplings are removed. The needed sign matrix can be obtained from $H(\overline{n})$ by replacing the k-th row with a copy of the i-th row. This scheme also removes frequency offsets and requires no more than $\overline{n}$ time intervals. To implement ZZ_{ik}, the duration of each interval t is chosen to satisfy $\delta\omega \overline{n} t / |i - k|^3 = \pi$. Note that the total time used to implement ZZ_{ij} is the shortest possible, since the coupling is always "on".

5'3. *Efficiency*. – The decoupling and recoupling schemes require $\overline{n}$ time intervals. They require at most $n\overline{n}$ pulses, since $X_i X_i = I$ and the X pulses are only used in pairs. The remaining question is: how does $\overline{n}$ depend on n? Some Hadamard matrices are missing, either because no construction methods are known or they simply cannot exist; therefore, $\overline{n} = cn$, where $c \geq 1$. $H(n)$ can only exist for $n = 1$, $n = 2$ or $n \equiv 0 \bmod 4$. Hadamard conjectured [80] that $H(n)$ exists for *every* $n \equiv 0 \bmod 4$. This famous conjecture is verified for all $n < 428$. Therefore, $\overline{n} - n \leq 3$ for all $n < 428$. We argue for *arbitrary* n that the schemes are still very efficient. First, we prove that $c < 2$. For each n, there exists r such that $2^{r-1} \leq n < 2^r$. Since $H(2^r)$ exists by Sylvester's construction [81], $cn = \overline{n} \leq 2^r < 2n$. We now show that c is close to the ideal value 1 in most cases, due to the existence of Hadamard matrices of orders other than powers of 2. This is why the full connection to Hadamard matrices is useful. We note that $\overline{n} - n \leq 31$ for all $n \leq 10000$. Within this technologically relevant range of n, c deviates significantly from 1 only for a few exceptional values of n *when n is small*. For completeness, we present arguments for $c \approx 1$ for *arbitrarily large* n in ref. [82]. This is based on Paley's construction [83] and the prime number theorem. Finally, if Hadamard's conjecture is proven, $\overline{n} - n \leq 3$ for all n.

$5 \cdot 4$. *Homonuclear decoupling*. – In subsect. $2 \cdot 2$, we discussed the possibility of reducing T_2^{cc} by introducing such "homonuclear decoupling" pulse sequences as WAHUHA to remove the coupling between ensemble members with equal frequency. The physics behind these pulse sequences have been presented in several textbooks [29, 30]. Such sequences must be performed simultaneously with the Hadamard construction described above.

The pulses for homonuclear decoupling have no need to be narrow band. The lower limit for the duration of the needed $\pi/2$ pulse is set by the available RF power, which is constrained only by the technical issues of RF amplification, resonant tuning, and heat dissipation. Ignoring such technical issues, the most efficient pulse sequence uses extremely short $\pi/2$ pulses which affect all qubits. This requires a bandwidth of several hundred kilohertz, or pulses in the μs regime, which is possible with commercial solid-state NMR systems. Full cycles or appropriate subcycles of the $\pi/2$ homonuclear decoupling pulses may appear during the "windows" of the π pulse scheme. Since such cycles leave nuclei in the state in which they started, the homonuclear decoupling pulse sequences do not affect the Hadamard decoupling scheme.

Treating all pulses as δ-functions in time, as we have done in the previous subsection, we see that the effect of the π pulses is to negate I_{ij}^z for the entire ensemble of i-th qubits. Such negation does not affect the homonuclear couplings $\mathcal{H}_{\mathrm{D,cc}}^*$, since $(-I_{ij}^z)(-I_{il}^z) = I_{ij}^z I_{il}^z$. In this respect, the Hadamard decoupling does not affect the homonuclear decoupling sequence. However, when we consider the evolution which occurs *during* the π pulses, in which a spin rotates over a time $1/\Delta\omega$, we see that this evolution generates error terms in the WAHUHA-type pulse sequences. Given the substantial amount of progress over the past 35 years in the design of pulse sequences to eliminate similar error terms, it seems likely that such errors can be avoided with an improved design.

6. – MRFM readout

We now discuss one proposed method of readout: magnetic resonance force microscopy (MRFM). In this discussion, we will focus on the proposed silicon implementation, although these techniques may also be employed with other materials by attaching the small crystal to the silicon oscillator.

The presence of a large magnetic-field gradient provides a natural means for performing MRFM on a magnetization M^z, since this technique is sensitive to the gradient force given by $F^z = M^z \partial B^z / \partial z$. If M^z is made to oscillate at the resonant frequency of a mechanical oscillator, this force can be deduced by measuring the displacement of that oscillator.

Precise alignment of the atomic planes of the silicon quantum computer with the dysprosium micromagnet motivates the use of a double-clamped structure; a single-clamped cantilever can more easily become warped and misaligned. A schematic of this structure is shown in fig. 4. This device may be fabricated as follows: after the formation of the ^{29}Si wire block discussed in subsect. **3**`3, the narrow bridge with length $300\,\mu$m and thickness $0.25\,\mu$m is created with electron-beam lithography. A plasma etcher removes unprotected silicon $3\,\mu$m down to the buried oxide layer. An HF vapor etch or acid solution removes the exposed oxide to release the structure; it is then dried in a critical-point dryer. After etching $3\,\mu$m further into the substrate and through the oxide layer, the dysprosium (Dy) micromagnet can be evaporatively deposited $s = 2.1\,\mu$m from the bridge and defined as a parallelopiped using a shadow mask. In this way, the magnet and silicon are aligned during fabrication.

The bridge is now free to oscillate in the z-direction. A lumped harmonic-oscillator model for this bridge yields an estimated spring constant of $k \approx 0.0042\,\mathrm{N/m}$ and a resonance frequency $\omega_{\mathrm{c}}/2\pi \approx 23\,\mathrm{kHz}$.

The readout is performed in high vacuum ($< 10^{-5}\,\mathrm{Torr}$) and at low temperatures (4 K). The coil used to generate the RF pulses for logic operations and decoupling sequences also generates continuous-wave (CW) radiation for readout. An optical-fiber–based displacement sensor is used to monitor deflection of the bridge using interferometry. Sub-Ångstrom oscillations can be detected; larger oscillations can be damped with active feedback which avoids additional broadening while maintaining high sensitivity [84].

Readout is performed using cyclic adiabatic inversion [27, 28], which modulates the magnetization of a plane of nuclei at a frequency near or on resonance with the bridge. Suppose the result of the quantum algorithm is encoded in the spin state of the i-th qubit; we seek a method to readout whether that spin and its ensemble copies are up or down. We begin by ramping the power of the CW, RF field tuned to frequency $\omega_i + \Delta$, where Δ is the frequency excursion chosen to be less than $\Delta\omega$. From eq. (5), we see that for $\Omega_i \ll \Delta$, in the rotating frame, the nuclei see an effective field Δ/γ. A spin-up or spin-down nucleus is thus in an eigenstate of this control Hamiltonian. If we increase Ω_i sufficiently slowly, the spins remain in an eigenstate by the adiabatic theorem. Once the power is ramped up, the field is frequency modulated, so $B^y(t) = 2B_1^y \cos\{\omega_i t - (\Delta/\omega_m)\cos(\omega_m t)\}$, where ω_m is the modulation frequency chosen to be near the resonance of bridge oscillations. Adiabaticity is maintained since $\omega_m \ll \omega_i$; in this regime the nuclei "follow" the effective magnetic field. In the rotating frame, we see from eq. (5) that the effective field seen by the nuclei has z-component $(\Delta/\gamma)\sin(\omega_m t)$. Hence the z-component of the spin magnetization M^z, and also the mechanical force, oscillates at the cantilever resonance. The z-component of the i-th plane's magnetization is deduced from the phase of the resulting bridge oscillation. Simultaneous detection of signals from multiple planes is possible if the different planes to be measured are driven at distinct modulation frequencies ω_m.

The force resolution for MRFM is limited by thermal fluctuations of the mechanical oscillator [85]. Force resolutions of $5.6 \times 10^{-18}\,\mathrm{N}/\sqrt{\mathrm{Hz}}$ have been reported for single crystal silicon cantilevers at $4\,\mathrm{K}$ [86]. The thermal noise threshold of the bridge structure in fig. 4 is estimated to be $\sim 1.2 \times 10^{-17}\,\mathrm{N}/\sqrt{\mathrm{Hz}}$ assuming a modest quality factor Q of 10^4 [87]. This minimum force resolution will be an important factor for the scalability of crystal lattice quantum computation, which we discuss in the next section.

The presence of such a sensitive measurement device usually implies the existence of a decoherence source as well. Indeed, nuclei will receive random phase kicks from the oscillation of the bridge through the large magnetic gradient. A calculation using the thermal noise statistics of a high-sensitivity mechanical oscillator [85] with eq. (12) estimates the T_2 timescale due to the bridge's thermal drift as $T_2^{\mathrm{c}} = k\omega_{\mathrm{c}}Qa^2/\Delta\omega^2 k_{\mathrm{B}}T \approx 25\,\mathrm{s}$. Active feedback stabilization is expected to increase this timescale by as many as four orders of magnitude, bringing it close to the lengthy T_1 timescale.

7. – Scalability

In this section we consider two aspects of scalability for crystal lattice quantum computation: the number of qubits available, and the number of logic gates available. The number of qubits depends on the device geometry and the available polarization; we discuss these factors in view of the all-silicon proposal. The number of logic gates available within the decoherence time will depend on the external decoherence timescale T_2^{ext}, the copy-copy coupling timescale T_2^{cc}, and on errors due to parasitic couplings $\Gamma(m)$; again we consider these parameters for the silicon proposal. We note that the polarization and readout procedures in this proposal take much longer than the coherence time, so efficient quantum error correction is difficult without expending many qubits. Therefore, we consider the scaling limits before error correcting codes are employed.

7·1. *Number of qubits.* – Optical pumping, as discussed in subsect. **3·3**, cannot achieve a pure initial state. Nonetheless, solution-NMR implementations [18, 19, 61] have shown how quantum computation is still possible with a highly mixed state. "Pseudo-pure" states will still be useful for crystal lattice quantum computation, but the initial polarization should be sufficiently high to avoid the exponential downscaling seen in solution NMR.

Additionally, the polarization due to physical cooling may be increased by an algorithmic cooling technique introduced by Schulman and Vazirani [59, 60]. This technique redistributes the entropy among a register of qubits to a subregister that is then discarded (decoupled and ignored). This method is expensive; it takes time to perform the (classical) logic operations and, worse, it sacrifices many qubits. Entropy conservation for the closed system requires that if n_0 initial qubits each start with entropy H and we end up with n perfectly polarized final qubits (*i.e.*, with 0 entropy), we must have $n = (1 - H)n_0$. For small initial polarization p_0, $n \to n_0 p_0^2/2\ln 2$ if the procedure is

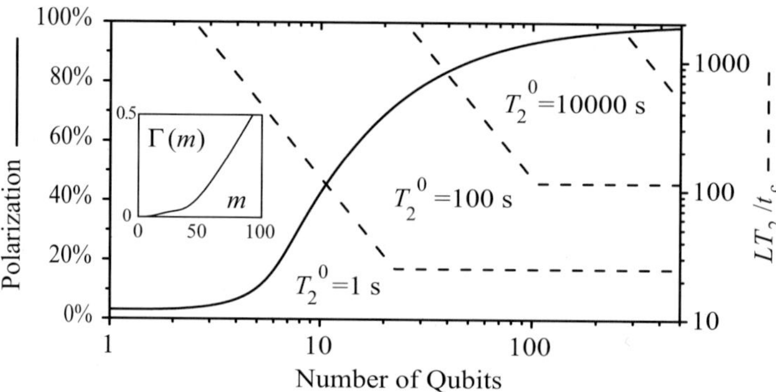

Fig. 11. – A plot showing the scalability of the all-silicon scheme. The solid curve, corresponding to the left axis, shows the polarization p needed in order for a number of qubits n to be measurable. The dashed lines, corresponding to the right axis, plot the number of logic gates (decoherence time T_2 divided by the pulse sequence cycle time t_c) times the length of a "block" of the decoupling sequence L against n for several values of T_2^0. The inset, showing $\Gamma(m)$, is explained in the text.

taken to the entropy limit. However, the very long T_1 in solid-state NMR affords ample time for the procedure, and the number of available qubits in our configuration is limited only by the width of the bridge structure. A bridge $0.25\,\mu$m wide accommodates $n_0 = 0.25\,\mu\text{m}/1.9\,\text{Å} = 1300$ initial qubits. Moreover, the algorithm need not be taken to the entropy limit, since large but still incomplete polarizations can be handled with pseudo-pure state techniques. Finally, we note that the entropy limit may be escaped if the hot subregister can be selectively recooled in a time much less than T_1; this possibility is discussed in ref. [88]. This procedure may be possible with optical pumping if the polarized light couples only to a few surface layers of nuclei, perhaps with a total internal reflection geometry.

Ultimately, the available polarization p following optical pumping and algorithmic cooling will be best determined experimentally. This p is crucial for scalability, since the force from the subensemble magnetization corresponding to a pseudo-pure state is estimated [89] as

$$
(51) \qquad F^z = \frac{\hbar\Delta\omega}{2a} N\left[\left(\frac{1+p}{2}\right)^n - \left(\frac{1-p}{2}\right)^n\right].
$$

This equation was derived assuming a "best-case" scenario for a pseudo-pure state, but we find that in the high-n, high-p case it is nearly equivalent to similar formulae specific to actual pseudo-pure state techniques, such as that given for exhaustive averaging in ref. [88]. The number of qubits available in the silicon-based quantum computer may be found by maximizing n in eq. (51) such that the force exceeds the thermal noise threshold. The results of such maximization are shown in fig. 11. At low p, exponential improvements in p are needed to increase the number of measurable qubits n. Once

p exceeds about 60%, however, we find $n \sim (1+p)/(1-p)$, escaping the exponential downscaling which plagues solution NMR.

7'2. *Number of logic gates.* – To estimate the number of logic gates available, we divide the total decoherence time T_2 by the pulse sequence cycle time, t_c. This cycle time results from the fact that the pulse sequences developed in sect. **5** must undergo a full cycle in order for the nuclei to be completely refocussed, and only refocussed nuclei may undergo low-error single-qubit logic gates. How long is t_c? We showed in subsect. **5'3** that the Hadamard-matrix based decoupling/recoupling sequence requires cn^2 time intervals for n qubits, where $c \approx 1$. (At the level of approximation present for this argument, we may take $c = 1$.) One time interval must be sufficient for the π pulses of duration greater than $1/\Delta\omega$ and a homonuclear decoupling sequence (or subsequence); this length of time will depend on several technical factors, but $1/\Delta\omega$ is the only fundamentally limiting timescale. Hence, let us denote the length of one interval by $L/\Delta\omega$, where L is a design detail between about 1 and 10. These arguments lead us to $t_c = Ln^2/\Delta\omega$.

In this article, we have discussed several possible sources for T_2. Let us group all of the uncontrollable decoherence sources (for example, residual T_2^{cc} after homonuclear decoupling and cantilever drift) into one rate denoted by T_2^0. We then find that the number of logic gates is $T_2^0 \Delta\omega / Ln^2$. In view of this scaling, we consider the following alteration to the decoupling pulse sequence. As n increases, distant qubits begin to become uncoupled due to the $1/m^3$ dependence of eq. (14). We may thus consider leaving these weak couplings in place, instead only decoupling qubits in sets of $l < n$. The weak couplings that remain following pulse sequence truncation therefore contribute to a new T_2 decoherence source. This T_2 may be estimated from the second moment using both $\mathcal{H}_{D,qq}^*$ and $\mathcal{H}_{D,qc}^*$ as

$$(52) \qquad T_{2l}^t = \frac{l^3}{\delta\omega\sqrt{1+\Gamma^2(l)}}.$$

Given a large number of qubits, the number of gates available including this decoherence source is $(1/T_2^0 + 1/T_{2l}^t)^{-1}\Delta\omega/Ll^2$. For long T_2^0 and $l \ll \lambda_1$, this function at first increases as l. Once $l \geq \lambda_1$, so that $\Gamma(l)$ begins to scale roughly like l, the function flattens out and starts to decrease for finite T_2^0. In between, the function has a maximum; the l^* which yields this maximum is the optimum number of qubits to decouple, as decoupling fewer qubits than l^* leads to excessive decoherence while decoupling more than l^* leads to excessively long pulse sequences. The $1/n^2$ scaling for the number of gates therefore becomes flat once $n > l^*$. Figure 11 shows this scaling behavior and the $\Gamma(m)$ function used to calculate it.

8. – Conclusion

We have discussed a complete scheme for implementing a quantum computer using solid-state NMR in a crystal lattice, including methods for initialization, quantum control, and readout. We emphasize that this scheme was constructed without the need for exotic electron-mediated control and measurement strategies; in fact most of the technologies discussed have already been experimentally demonstrated. This scheme seems to have more room for qubits than the solution NMR systems that have preceded it. We believe that solid-state NMR in crystal lattices will provide important steps towards the final goal of a large, scalable quantum computer.

* * *

The authors wish to thank D. Leung, A. Verhulst, A. Dâna, C. Master, K. Itoh, E. Abe, T. Shimizu, H. Tsubaki, and I. Fisher for their needful discussions.

REFERENCES

[1] Deutsch D., *Proc. R. Soc. London, Ser. A*, **400** (1985) 97.
[2] Shor P. W., *Proceedings of the 35th Annual Symposium on Foundations of Computer Science* (IEEE Computer Society Press, Los Alamitos, CA) 1994, pp. 124-134.
[3] Shor P. W., *SIAM J. Comput.*, **26** (1997) 1484.
[4] Simon D. R., *Proceedings of the 35th Annual Symposium on Foundations of Computer Science* (IEEE Computer Society Press, Los Alamitos, CA) 1994, pp. 116-123.
[5] Simon D. R., *SIAM J. Comput.*, **26** (1997) 1474.
[6] Grover L. K., *Phys. Rev. Lett.*, **80** (1998) 4329.
[7] Kitaev A. Yu, Los Alamos National Laboratory E-print quant-ph/9511026 (1995).
[8] Zoller P. *et al.*, this volume, p. 263.
[9] Wineland D., this volume, p. 165.
[10] Blatt R. *et al.*, this volume, p. 197.
[11] Haroche S. *et al.*, this volume, p. 3.
[12] Rempe G. *et al.*, this volume, p. 37.
[13] Rolston S. L., this volume, p. 101.
[14] Yamamoto Y., Kitagawa M. and Igeta K., *Proceedings of the 3rd Asia Pacific Physics Conference*, edited by Y. Chan *et al.* (World Scientific, New Jersey) 1988, pp. 779-799.
[15] Milburn G., *Phys. Rev. Lett.*, **62** (1989) 2124.
[16] Fleischhauer M. *et al.*, this volume, p. 511.
[17] De Martini F. *et al.*, this volume, pp. 233, 241.
[18] Cory D. G., Fahmy A. F. and Havel T. F., *Proceedings of the Fourth Workshop on Physics and Computation*, in *Proc. Natl. Acad. Sci. USA* (New England Complex Systems Institute, Boston, MA) 1997, p. 1634.
[19] Gershenfeld N. A. and Chuang I. L., *Science*, **275** (1997) 350.
[20] Kane B. E., *Nature*, **393** (1998) 133.
[21] Devoret M. H. *et al.*, this volume, p. 475.
[22] Martinis J., oral presentation.
[23] DiVincenzo D. P., this volume, p. 251.
[24] Mozyrsky D., Privman V. and Glasser M. L., *Phys. Rev. Lett.*, **86** (2001) 5112.
[25] Berman G. P. *et al.*, *Phys. Rev. Lett.*, **86** (2001) 2894.
[26] Sidles J. A., *Appl. Phys. Lett.*, **58** (1991) 2854.
[27] Rugar D., Yannoni C. S. and Sidles J. A., *Nature*, **360** (1992) 563.
[28] Abragam A., *The Principles of Nuclear Magnetism* (Oxford University Press, Oxford) 1961.
[29] Haeberlin U., *High Resolution NMR in Solids: Selective Averaging* (Academic Press, New York) 1976.

[30] MEHRING M., *High Resolution NMR in Solids* (Springer-Verlag, Berlin) 1983.

[31] GENACK A. Z. and REDFIELD A. G., *Phys. Rev. Lett.*, **31** (1973) 1204.

[32] LADD T. D. *et al.*, Los Alamos National Laboratory E-print quant-ph/0009122 (2000).

[33] VAN DER LUGT W. and CASPERS W. J., *Physica*, **30** (1964) 1658.

[34] ENGELSBERG M., LOWE I. J. and CAROLAN J. L., *Phys. Rev. B*, **7** (1973) 924.

[35] CHO G. and YESINOWSKI J. P., *J. Phys. Chem.*, **100** (1996) 15716.

[36] BLOEMBERGEN N., *Physica*, **15** (1949) 386.

[37] KURKIN I. N. and TSVETKOV E. A., *Sov. Phys. Solid State*, **20** (1978) 870.

[38] TYCKO R., *Solid State Nucl. Magn. Reson.*, **11** (1998) 1.

[39] YAMAGUCHI F. and YAMAMOTO Y., *Appl. Phys. A*, **68** (1999) 1.

[40] KWON Y. S. *et al.*, *Physica B*, **171** (1991) 324.

[41] HAGA Y. *et al.*, *Physica B*, **206 & 207** (1995) 792.

[42] HEER H. *et al.*, *J. Phys. C*, **12** (1979) 5207.

[43] TAKEUCHI T. *et al.*, *J. Phys. Soc. Jpn.*, **67** (1998) 2094.

[44] KOHGI M. *et al.*, *Physica B*, **213 & 214** (1995) 110.

[45] TERASHIMA T. *et al.*, *Phys. Rev. B*, **55** (1996) 4197.

[46] LADD T. D. *et al.*, *Appl. Phys. A*, **71** (2000) 27.

[47] LANG R. J, *Canad. J. Res.*, **14** (1936) 127.

[48] MYERS S. M. and NARATH A., *Phys. Rev. B*, **9** (1974) 227.

[49] SMART J. S., *Effective Field Theories of Magnetism* (W. B. Saunders) 1966.

[50] JONES E. D., *Phys. Rev.*, **180** (1968) 455.

[51] NAKAMURA T., *Prog. Theor. Phys.*, **20** (1958) 542.

[52] LADD T. D. *et al.*, Los Alamos National Laboratory E-print quant-ph/0109039 (2001).

[53] ITOH K. M. and HALLER E. E., *Physica E*, **10** (2001) 463.

[54] SHLIMAK I., SAFAROV V. I. and VAGNER I. D., *J. Phys. Condens. Matter*, **13** (2001) 6059.

[55] VIERNOW J. *et al.*, *Appl. Phys. Lett.*, **72** (1998) 948.

[56] LIN J.-L. *et al.*, *J. Appl. Phys.*, **84** (1998) 255.

[57] LATYSHEV A. V. *et al.*, *Phys. Status Solidi A*, **113** (1989) 421.

[58] LAMPEL G., *Phys. Rev. Lett.*, **20** (1968) 491.

[59] SCHULMAN L. J. and VAZIRANI U. V., *Proceedings of the 31st ACM Symposium on Theory of Computing* (ACM Press, New York) 1999, p. 322.

[60] CHANG D. E., VANDERSYPEN L. M. K. and STEFFAN M., *Chem. Phys. Lett.*, **338** (2001) 337.

[61] KNILL E., CHUANG I. and LAFLAMME R., *Phys. Rev. A*, **57** (1998) 3348.

[62] BAGRAEV N. T., VLASENKO L. S. and ZHITNIKOV R. A., *JETP Lett.*, **25** (1977) 190, and references therein.

[63] FEHER G. and GERE E. A., *Phys. Rev.*, **114** (1958) 1245.

[64] DELERUE C., ALLAN G. and LANNOO M., *Light Emission in Silicon*, edited by D. J. LOCKWOOD (Academic, New York) 1998, Chapt. 7.

[65] WEINSTEIN J. D. and LIBBRECHT K. G., *Phys. Rev. A*, **52** (1995) 4004.

[66] VULETIC V. *et al.*, *Phys. Rev. Lett.*, **80** (1998) 1634.

[67] REICHEL J. *et al.*, *Phys. Rev. Lett.*, **83** (1999) 3398.

[68] DRNDIĆ M. *et al.*, *Appl. Phys. Lett.*, **72** (1998) 2906.

[69] MATULIS A. and PEETERS F. M., *Phys. Rev. Lett.*, **72** (1994) 1518.

[70] IBRAHIM I. S. and PEETERS F. M., *Phys. Rev.*, **52** (1995) B17321.

[71] TSUBAKI K., *Physica B*, **456-458** (1998) 392.

[72] STEPANKIN V., *Physica B*, **211** (1995) 345.

[73] ANDRIANOV A. V. *et al.*, *Magn. Mater.*, **97** (1991) 246.

[74] ALKHAFAJI M. T. and ALI N., *J. Alloys Comp.*, **250** (1997) 659.

[75] Goldman J. R. *et al.*, *Appl. Phys. A*, **71** (2000) 11.
[76] Barenco A. *et al.*, *Phys. Rev. A*, **52** (1995) 3457.
[77] Colbourn C. and Dinitz J. (Editors), *The CRC Handbook of Combinatorial Designs* (CRC Press, Boca Raton) 1996.
[78] van Lint J. and Wilson R., *A Course in Combinatorics* (Cambridge University Press, Cambridge) 1992.
[79] MacWilliams F. and Sloane N., *The Theory of Error-Correcting Codes* (North Holland, Amsterdam) 1977.
[80] Hadamard J., *Bull. Sciences Math.*, **17** (1893) 240.
[81] Sylvester J., *Philos. Mag.*, **34** (1867) 461.
[82] Leung D. *et al.*, *Phys. Rev. A*, **61** (2000) 42310.
[83] Paley R. E. A. C., *J. Math. Phys.*, **12** (1933) 311.
[84] Dürig U., Steinauer H. R. and Blanc N., *J. Appl. Phys.*, **8** (1997) 3641.
[85] Gabrielson T. B., *IEEE Trans. Electron Devices*, **40** (1993) 903.
[86] Stowe T. D. *et al.*, *Appl. Phys. Lett.*, **71** (1997) 288.
[87] Blom F. R., *J. Vac. Sci. Technol. B*, **10** (1992) 19.
[88] Boykin P. O. *et al.*, Los Alamos National Laboratory E-print quant-ph/0106093 (2001).
[89] Warren W. S., *Science*, **277** (1997) 1688.

Superconducting quantum bit based on the Cooper pair box

D. Vion, A. Aassime, A. Cottet, P. Joyez, H. Pothier,
C. Urbina, D. Esteve and M. H. Devoret

Quantronics Group, Service de Physique de l'Etat Condensé, CEA-Saclay
F-91191 Gif-sur-Yvette cedex, France

1. – Introduction: why superconducting tunnel junction circuits?

The more advanced proposals so far for the implementation of qubits and quantum gates are based on ions or atoms in vacuum (see articles by Wineland and Haroche in this volume). These systems have been manipulated individually in a controlled fashion for about 20 years and techniques have reached a high level of sophistication. However, it is not clear yet if these proposals can be extended to the fabrication of a quantum processor which would be "scalable", a jargon term meaning that fabrication costs would scale sufficiently "gently" with the number of quantum bits and gates that quantum computation remains advantageous compared with its classical counterpart.

This is why there is currently interest in various solid-state implementations of qubits [1-10] which could be fabricated in parallel by lithography methods, hence benefiting of the technological advances that have made classical processors so powerful today.

In this article we will focus on one particular solid-state implementation for which a lot of experimental data on feasibility has recently been gathered, namely qubits based on superconducting tunnel circuits.

Ion systems have remarkable level structures. The ground state is split by the hyperfine interaction, thus yielding basis states for the qubit, while well-resolved excited states hover well above the ground state at energies corresponding to visible or infra-red photons, thus providing a very good isolation between the degrees of freedom representing the qubits and perturbing ones. At the same time, the ground-to-excited states transitions can be exploited to read out the value of the qubit by fluorescence photons which can be easily detected.

A similar, although not as ideal, situation is encountered with a superconducting junction, *i.e.* two superconducting electrodes separated by an oxide layer acting as a tunnel barrier. Here we suppose that the electrodes are sufficiently small that the superconducting order parameter is homogeneous inside the electrode. In order to describe the degrees of freedom of this so-called Josephson tunnel junction, it is useful to first begin with only one isolated BCS-type superconducting electrode with an even number of electrons. This system has a unique many-body ground state. The excited states of the electrode have a minimum energy of 2Δ, where Δ is the superconducting quasiparticle gap, an energy of the order of a few kelvin which we suppose large in comparison with all the environment energies like temperature fluctuations energies. The energy 2Δ, needed to break a Cooper pair into independent quasiparticles, is equivalent to the energy difference between ground and excited state of the ion. If we now imagine having two such isolated electrodes come together until they become separated only by a tunnel barrier, we will have a combined system possessing a set of quasi-degenerate ground states corresponding to all the possible number N of pairs of electrons which can be transferred from one electrode to the other. The degree of freedom corresponding to N has to be understood as an operator and will thus be noted $\hat{N}$. We will use the usual Dirac notation for the eigenstates and eigenvalues of $\hat{N}$:

$$\hat{N}|N\rangle = N|N\rangle.$$

The degeneracy between the eigenstates of $\hat{N}$, the so-called charge states, is lifted by two energy terms which play the role in the junction of the hyperfine interaction in ions.

The first term is the Coulomb energy, whose origin is the electric field in the tunnel oxide due to the charge on the two electrodes. With the previously introduced notation the Coulomb Hamiltonian can be written

$$H_{\mathrm{C}} = 4E_{\mathrm{C}} \sum_{p} \left(\hat{N} - q\right)^{2}|N\rangle\langle N|,$$

where the Coulomb energy,

$$E_{\mathrm{C}} = \frac{e^2}{2C_{\mathrm{j}}},$$

corresponds to the electrostatic energy of one charge quantum on the capacitance $C_{\mathrm{j}} = \epsilon A/d$ of the tunnel junction, which we suppose here to be well described by a parallel-plate capacitor with area A, dielectric constant ϵ and dielectric thickness d. The charge q, which in general is not an integer, corresponds to the offset charge on the tunnel junction and reflects the fact that even when two electrodes are at the same electrochemical potential, there can be a non-zero electric field between them due to the difference in their work function. We suppose that the temperature T and the superconducting gap Δ satisfy $k_{\mathrm{B}}T \ll \Delta/\ln \mathsf{N}$ and $E_{\mathrm{C}} < \Delta$, where N is the total number of paired electrons in the island. Under these conditions, there are no unpaired electrons in the island [11, 12].

The second term is the Josephson energy. It correspond to the delocalization energy of Cooper pairs through the tunnel barrier.

The Josephson Hamiltonian has the form

$$H_{\mathrm{J}} = \frac{E_{\mathrm{J}}}{2} \sum_p |N+1\rangle\langle N| + \text{h.c.}$$

The Josephson energy,

$$E_{\mathrm{J}} = \frac{1}{8}\frac{G_t}{e^2/h}\Delta,$$

is simply proportional, via the tunnel conductance of the barrier

$$G_t = \frac{e^2}{h}M\mathcal{T},$$

to the transparency $\mathcal{T}$ of the barrier and the number of independent electronic M modes passing through it. The constant e^2/h appearing here is the conductance quantum. The number of modes is proportional to the area A of the junction and inversely proportional to the Fermi wavelength λ_{F}:

$$M = \frac{4A}{\lambda_{\mathrm{F}}^2}.$$

It is worth noting that the operator $\hat{N}$ is a macroscopic quantum variable: its dynamics is governed by quantum mechanics, but the parameters appearing in the Hamiltonian are not God-given constants like the Rydberg constant or the Bohr magneton. They correspond to macroscopic parameters like the junction area which can be engineered by nanolithography. Also, the various N states correspond to a complete rearrangement of the charge on the electrode, and not to the filling of single-electron wave functions. The $\hat{N}$ operator is a collective variable like the angular momentum of a top.

It is useful at this point to introduce the operator $\hat{\theta}$ canonically conjugate to $\hat{N}$. Since $\hat{N}$ has integer eigenvalues, some care is needed in the definition of $\hat{\theta}$:

$$e^{i\alpha\hat{\theta}}e^{-i\alpha\hat{\theta}} = \hat{N} - \alpha.$$

The operator $\hat{\theta}$ is called the phase difference between the two electrodes. At this level, the name "phase" can be misleading. Its meaning is that of a position variable in a cyclic system, like the center of mass of a ball constrained to roll in a circle-shaped groove, not the phase of a single-particle wave function. In the $\hat{\theta}$ representation, the Josephson Hamiltonian is diagonal and takes a very simple form,

$$H_{\mathrm{J}} = -E_{\mathrm{J}} \cos \hat{\theta}.$$

The pair number $\hat{N}$ and the phase difference $\hat{\theta}$ bear some resemblance with the photon number and the phase operators of a cavity mode in quantum optics. However, one must realize that the pair number eigenvalues can take both positive and negative values, whereas the photon number is always non-negative. This difference is fundamental. The topology of the Hilbert space spanned by $\hat{N}$ and $\hat{\theta}$ is that of a cylinder and is intrinsically curved, while the harmonic oscillator Hilbert space is flat.

The Josephson tunnel junction is the unique quantum non-linear electric system. Although many other non-linear electrical system exist, their internal degrees of freedom all get frozen by the time one reaches the temperatures at which quantum effects implying collective variables like charge can be observed.

If we express the total tunnel junction Hamiltonian in terms of the "macrovariable" operators,

$$H_0 = 4E_{\mathrm{C}}\left(\hat{N} - q\right)^2 - E_{\mathrm{J}} \cos \hat{\theta},$$

we see that the Josephson term plays the role of a potential term while the Coulomb energy corresponds to a kinetic energy term.

If $E_{\mathrm{C}} \ll E_{\mathrm{J}}$, we can forget about the discreetness of $\hat{N}$, associated with the cyclic property of $\hat{\theta}$. If furthermore $q = 0$, the Hamiltonian H_0 becomes the Hamiltonian of a anharmonic oscillator, the quadratic potential term being replaced by a sinusoidal one. We have then a Hamiltonian analogous to that of a pendulum making an angle $\hat{\theta}$ with the vertical:

$$H_0 = 8E_{\mathrm{C}} \frac{\hat{N}^2}{2} + E_{\mathrm{J}} \frac{\left(2\sin(\hat{\theta}/2)\right)^2}{2}.$$

The small oscillation frequency is thus given by

$$\omega_{\mathrm{p}} = \frac{\sqrt{8E_{\mathrm{C}}E_{\mathrm{J}}}}{\hbar}.$$

This quantity is called the plasma frequency of the junction. The junction surface scales out of this parameter which is essentially only determined by the oxide thickness and dielectric constant (the Fermi wavelength does not vary much from one metal to another). In practice its value is in the 5–50 GHz range.

In the limit $E_{\mathrm{C}} \ll E_{\mathrm{J}}$ the mean square fluctuations of $\hat{\theta}$ in the ground state is given by

$$\langle \hat{\theta}^2 \rangle_0 = \sqrt{\frac{2E_{\mathrm{C}}}{E_{\mathrm{J}}}}.$$

We can thus see the ratio E_C/E_J as a sort of "quanticity" parameter. If $E_C \ll E_J$, the junction behaves as a semiclassical anharmonic oscillator and θ is a good quantum number. If $E_C \gg E_J$, it behaves as a free particle on a one-dimensional ring and its momentum N is a good quantum number. Neither the charge nor the phase are good quantum numbers when $E_C/E_J \sim 1$.

The Josephson junction is thus a macroscopic one-dimensional atom.

We will see in the next section that we can manipulate the states of this macroatom with electric signals traveling along "wires" instead of light.

We have so far a "bare" junction with certain macrostates. How can we address them? We need fields that couples to the junction macrovariables. There are three types of fields we can impose, each one of them corresponding to a different type of electrical bias. We can impose a current through the junction and get the so-called current-biased Josephson junction [13]. We can also link the two sides of the junction by a superconducting loop and impose an external magnetic flux through the loop: this forms the so-called RF-SQUID. Finally, the simplest form of bias has been proposed theoretically by Büttiker in 1989 [14], well after the initial developments of the Josephson effect. It was realized experimentally several years later by our group [15]. The associated device is called the single Cooper pair box.

2. – The single Cooper pair box

The idea of Büttiker was to subject the Josephson junction superconducting electrodes to a minimally invasive influence: the junction is placed between two capacitor plates connected to a voltage source. By electrostatic influence, this applied voltage induces surface charge on the junction electrodes and modify the offset charge:

$$(1) \qquad H_U = 4E_C \left(\hat{N} - C_g \frac{U}{2e} \right)^2 - E_J \cos \hat{\delta}.$$

Here, we have lumped the junction intrinsic offset charge into the voltage U. The value $U = 0$ corresponds here to a nulling out of the offset charge, not necessarily to zero voltage applied to the capacitor plates. The parameter C_g is called the gate capacitance. It is half the value of the capacitance between plate and junction electrode, if they influence each other completely. In practice, we do not need two capacitors on each side of the junction. In the Cooper pair box, the voltage U is applied between one junction electrode and the gate capacitor (see fig. 1). Note that for the box the value of the Coulomb energy now takes the form

$$E_C = \frac{e^2}{2C_\Sigma},$$

where $C_\Sigma = C_j + C_g$.

In fig. 2 we show the eigenvalues of the Hamiltonian (1) as a function of the reduced gate charge $n_g = C_g U/e$ for different values of the ratio E_J/E_C. When $E_C \gg E_J$

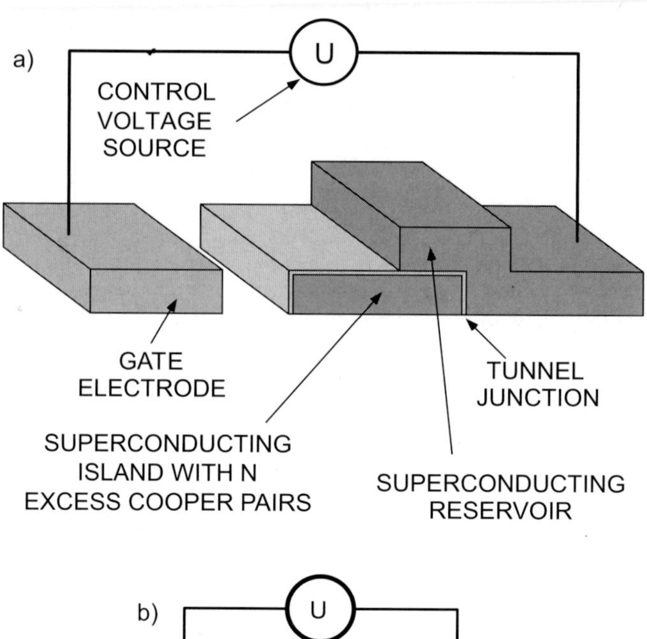

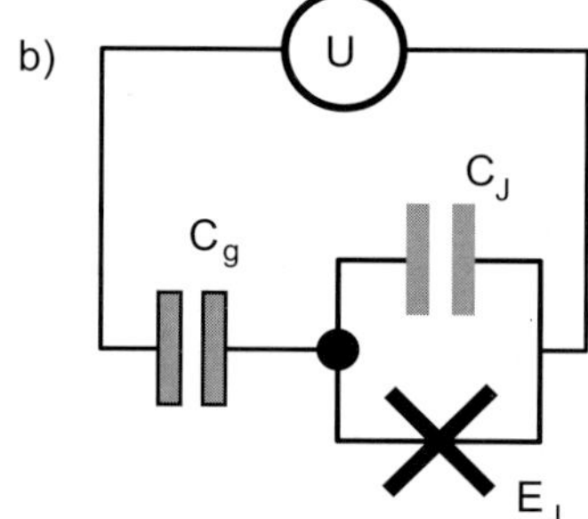

Fig. 1. – a) Schematic representation of the single Cooper pair box. b) Equivalent circuit diagram for the single Cooper pair box. The capacitances C_g and C_j correspond to the gate and junction, respectively. E_J is the Josephson energy of the tunnel junction.

the Cooper pair number is almost everywhere a good quantum number, except at the degeneracy points of the electrostatic part of the Hamiltonian, *i.e.* for odd values of n_g. In the vicinity of these degeneracy points, we can limit our analysis to two neighboring charge states $|N_0\rangle$ and $|N_0 + 1\rangle$, which are coupled by the Josephson energy and which play the role of the 2 basis states of an effective spin 1/2. Since the Hamiltonian is invariant under the transformation $\hat{N} \to \hat{N} - N_0$, we can limit ourselves to the reduced Hilbert space spanned by $|N = 0\rangle$ and $|N = 1\rangle$.

In this reduced Hilbert space, we have a spin Hamiltonian

$$H = -\vec{h} \cdot \vec{s},$$

where the spin $\vec{s}$ is given by

$$\vec{s} = \frac{1}{2}\left(\sigma_X \vec{X} + \sigma_Y \vec{Y} + \sigma_Z \vec{Z}\right),$$

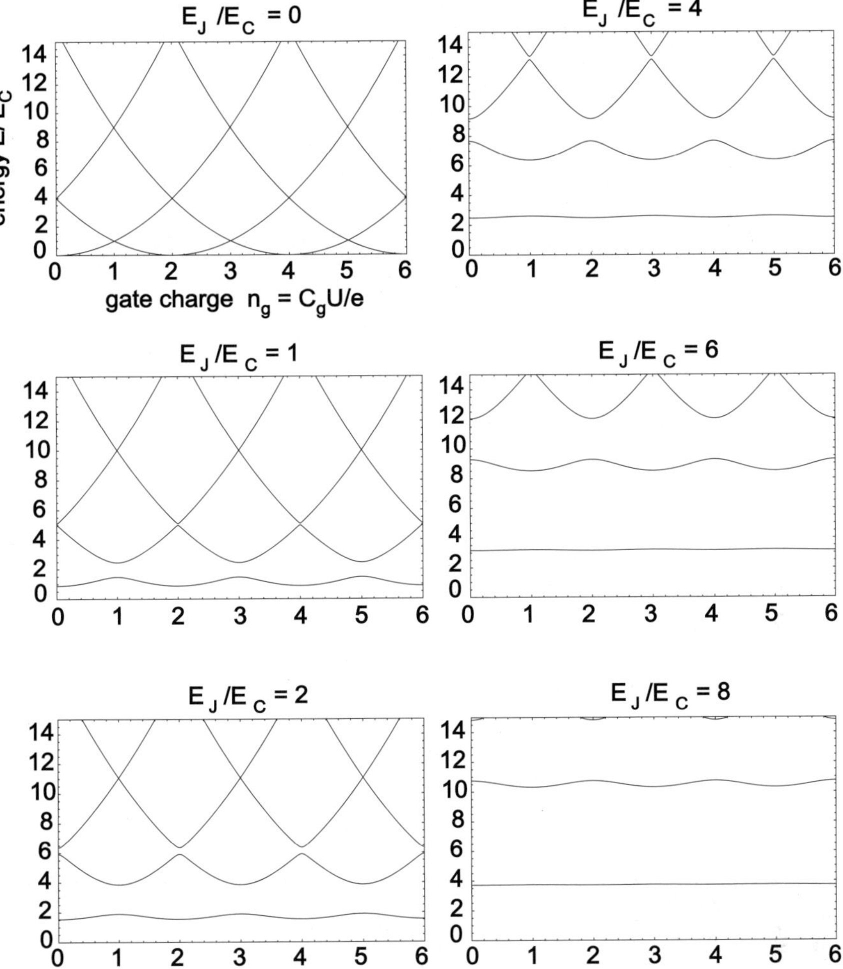

Fig. 2. – Energy spectrum of the single Cooper pair box as a function of the gate charge $n_\mathrm{g} = C_\mathrm{g}U/e$, for several values of the ratio $E_\mathrm{J}/E_\mathrm{C}$.

with the Pauli spin matrices $\sigma_{X,Y,Z}$ given by

$$\sigma_X = \begin{bmatrix} 0 & 1 \\ 1 & 0 \end{bmatrix},$$

$$\sigma_Y = \begin{bmatrix} 0 & i \\ -i & 0 \end{bmatrix},$$

$$\sigma_Z = \begin{bmatrix} 1 & 0 \\ 0 & -1 \end{bmatrix}$$

and where the effective field $\vec{h}$ is given

$$\vec{h} = 4E_{\mathrm{C}}\left(n_{\mathrm{g}} - 1\right)\vec{X} + E_{\mathrm{J}}\vec{Z}.$$

The Josephson energy thus plays the role of the Zeeman energy, while the applied voltage plays the role of the transverse field.

At $T = 0$, the effective spin points in the direction of the effective field. This latter has a direction $\vec{\zeta}$ whose angle θ with $\vec{Z}$ is given by

$$\tan\theta = \frac{4E_{\mathrm{C}}\left(n_{\mathrm{g}} - 1\right)}{E_{\mathrm{J}}}$$

and an intensity

$$h_\zeta = \sqrt{E_{\mathrm{J}}^2 + \left[4E_{\mathrm{C}}\left(n_{\mathrm{g}} - 1\right)\right]^2}.$$

The spin density matrix can thus be written as

$$\rho = \frac{1}{2}\left[1 + \vec{m}\cdot\vec{\sigma}\right]$$

with

$$\vec{m} = \vec{\zeta}\tanh\frac{h_\zeta}{2k_{\mathrm{B}}T}.$$

The average Cooper pair number is directly proportional to the polarization of the spin along the X-axis:

$$\langle\hat{p}\rangle = \frac{m_X + 1}{2}$$

with

$$(2) \qquad m_X = \tanh\left(\frac{h_\zeta}{2k_{\mathrm{B}}T}\right)\cdot\sin\theta.$$

As we vary U, the average Cooper pair number varies in a stepwise fashion, with the shape of the step being described by expression (2). At $T = 0$, the width of the step is broadened by the quantum fluctuations of the charge governed by the Josephson energy. This behavior has been measured experimentally [15].

Rabi oscillations between two charge states of the box have subsequently been measured [5], thus further demonstrating the macrospin character of the system and its promising features for the implementation of a qubit.

3. – Cooper pair box with orthogonal write and read ports

Unlike the electric dipoles of isolated atoms or ions, the state variables of a circuit like voltages and currents usually undergo rapid quantum decoherence [16] because, in general, they are coupled to an environment with a large number of uncontrolled degrees of freedom [4]. Tunnel junction circuits [5-8, 17] have displayed so far coherence quality factors Q_φ of several hundreds [10]. Here, the coherence quality factor for a pair of quantum levels $|a\rangle$ and $|b\rangle$ is defined by $Q_\varphi = \pi \nu_{ab} T_\varphi$, where ν_{ab} is the transition frequency and T_φ is the inverse of decoherence rate of quantum superpositions of these two states. In the following, we describe an experiment performed by our group which has shown that coherence quality factors of more than 10^4 can experimentally be achieved in a new circuit built around the Cooper pair box. Two orthogonal access ports are used for preparing and measuring the quantum state [9, 18]. This leads us to an improvement of decoherence time by a factor of more than 10^2 compared to the best previously published results [5].

In our experiment $E_{\rm J} \simeq 5 E_{\rm C}$ (see upper right panel in fig. 2) and neither $\hat{N}$ nor $\hat{\theta}$ are good quantum numbers. However, we can still restrict ourselves to the Hilbert space spanned by the ground $|0\rangle$ and first excited energy eigenstate $|1\rangle$ since the system is sufficiently non-harmonic. The effective spin $\vec{s}$ has a Zeeman energy $h\nu_{01}$ which goes to the minimal value $E_{\rm J}$ when $N_{\rm g} = 1/2$. The spin-up ($s_z = 1/2$, state $|0\rangle$) and -down ($s_z = -1/2$, state $|1\rangle$) states are there approximately $(|N = 0\rangle \pm |N = 1\rangle)/2$. Both states have the same average charge $\langle \hat{N} \rangle = 1/2$, and consequently the system is immune to first-order fluctuations of the gate charge. With appropriate NMR-like microwave pulses on the gate $u(t) = U_{\mu w}(t) \sin 2\pi \nu t$, where $\nu \simeq \nu_{01}$, any superposition $|\Psi\rangle = \alpha|0\rangle + \beta|1\rangle$ can be prepared [19].

For readout, instead of measuring the charge $\langle \hat{N} \rangle$ which requires moving away from $N_{\rm g} = 1/2$ [5, 20], we entangle $\vec{s}$ with a new degree of freedom more advantageous to measure. For this purpose, the single junction of the basic Cooper pair box has been split into two nominally identical junctions inserted into a superconducting loop (see fig. 3). The new degree of freedom is the phase difference $\hat{\delta}$ across the two junctions, and the Josephson energy $E_{\rm J}$ in eq. (1) becomes $E_{\rm J} \cos(\hat{\delta}/2)$ [21]. The information about $\vec{s}$ is transferred onto $\hat{\delta}$ through the supercurrent in the loop $\hat{I} = (2e/\hbar)\partial\hat{H}/\partial\hat{\delta} \simeq i_0 s_z \sin(\hat{\delta}/2)$, where $i_0 = eE_{\rm J}/\hbar$. The currents $\langle 0|\hat{I}|0\rangle$ and $\langle 1|\hat{I}|1\rangle$ in the two states grow with opposite signs when $\delta = \langle \hat{\delta} \rangle$ moves away from zero.

Our readout strategy is to exploit the two corresponding evolutions of δ generated by its entanglement with $\vec{s}$. It has been implemented by inserting in the loop a large Josephson junction biased with a current $I_{\rm b}$ [22]. This junction has a Josephson energy $E_{\rm J0}$ which is about 20 times larger than $E_{\rm J}$. Its phase difference is denoted by $\hat{\gamma}$. We have designed the loop dimensions to be as small as possible, thus making the loop inductance much smaller than the effective inductance of all junctions. This condition imposes the constraint $\hat{\delta} = \phi + \hat{\gamma}$, where $\phi = 2e\Phi/\hbar$, Φ being the externally imposed flux through the loop. By placing a large capacitance C in parallel with the large junction so that $(2e)^2/2C \ll E_{\rm J0}$, the phase $\hat{\gamma}$, and consequently $\hat{\delta}$, can be treated in first approximation

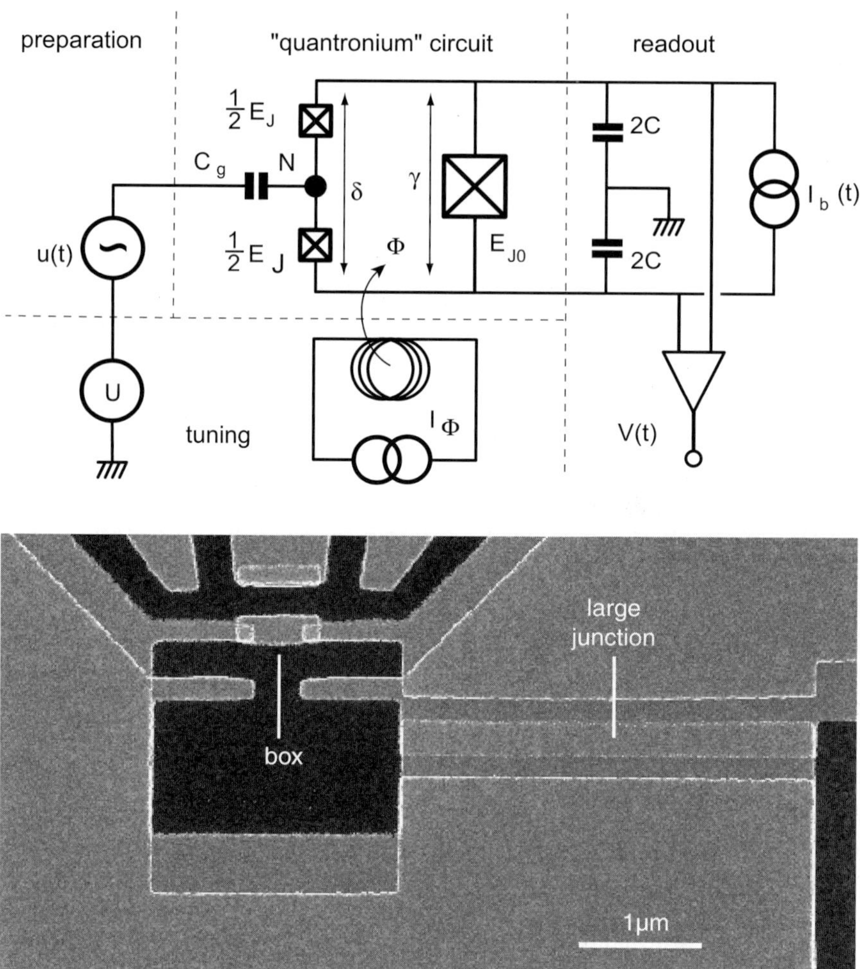

Fig. 3. – Top: schematic diagram of our quantum coherent circuit with its tuning, preparation and readout blocks. The coherent circuit, nicknamed "quantronium", consists of a Cooper pair box island (black node) delimited by two small Josephson junctions (cross symbols) in a superconducting loop. The loop also includes a third, much larger Josephson junction shunted by a capacitance C. The Josephson energy of the box and the large junction are E_J and E_{J0}. The Cooper pair number N and the phases δ and γ are the degrees of freedom of the circuit. A dc voltage U applied to the gate capacitance C_g and a dc current I_ϕ applied to a coil producing a flux Φ in the circuit loop tune the quantum energy levels. Microwave pulses $u(t)$ applied to the gate prepare arbitrary quantum states of the circuit. The states are read out by applying a current pulse $I_b(t)$ to the large junction and by monitoring the voltage $V(t)$. Filters in the microwave and dc lines have not been represented for simplicity. Bottom: scanning electron micrograph of a quantronium sample.

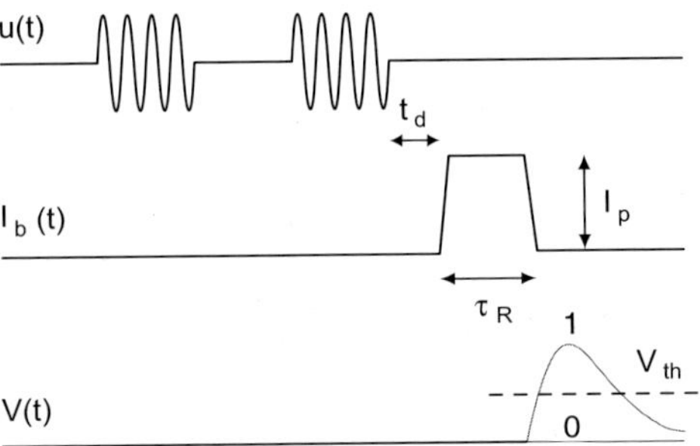

Fig. 4. – Signals involved in quantum state manipulations and measurement of the "quantron-ium". Top: microwave voltage pulses are applied to the gate for state manipulation. Middle: a readout current pulse $I_b(t)$ with amplitude I_p is applied to the large junction t_d after the last microwave pulse. Bottom: voltage $V(t)$ across the junction. The occurrence of a pulse depends on the occupation probabilities of the energy eigenstates. A discriminator with threshold V_{th} converts $V(t)$ into a Boolean 0/1 output for statistical analysis.

as classical variables. Under these conditions, $\delta \simeq \phi + \arcsin(I_b/I_0)$, neglecting terms of order $i_0/2I_0 \simeq 0.01$, where $I_0 = 2eE_{J0}/\hbar$. Just as the system is immune to charge noise at $N_g = 1/2$, it is immune to flux and current noise at $\phi = 0$ and $I_b = 0$, where $\hat{I} = 0$. The preparation of the quantum state and its manipulation are therefore performed at this working point in charge-flux-current space. Readout is then achieved by displacing the system adiabatically along the I_b axis to make the loop supercurrent non-zero.

In practice, a trapezoidal pulse of current with amplitude slightly below I_0 is applied to the large junction (see fig. 4). Depending on whether the effective spin $\vec{s}$ is up or down, a value of order $i_0/2$ will be subtracted or added to I_b in the large junction. The junction is extremely sensitive to the total current running through it when a value close to $\gamma = \pi/2$ is reached at the top of the pulse. A small excess current induces the switching of the junction from the zero-voltage state to the finite-voltage state. With a precise adjustment of the amplitude and duration of the $I_b(t)$ pulse, the large junction switches to the voltage state with a large probability if $s_z = -1/2$ and with a small probability if $s_z = +1/2$ [9]. For the parameters given above, the efficiency of this projective measurement should be $\eta = p_1 - p_0 = 0.95$, where p_1 and p_0 are the switching probabilities in the excited and ground states, respectively, for optimum readout conditions.

The readout is also designed so as to minimize the relaxation of $\vec{s}$ using a Wheatstone-bridge–like symmetry. The large ratios E_{J0}/E_J and C/C_j provide further protection from the environment.

It is worthwhile to develop analogies between our experiment on an electrical cir-cuit and experiments on atoms. Our circuit can be viewed as an artificial two-level atom, which we have nicknamed "quantronium". The gate voltage and the magnetic flux through the loop, the two bias parameters used to tune the transition frequency, play a role similar to that of the static electric and magnetic fields for atoms. As in many atomic physics experiments, transitions are induced by microwave pulses. Finally, the

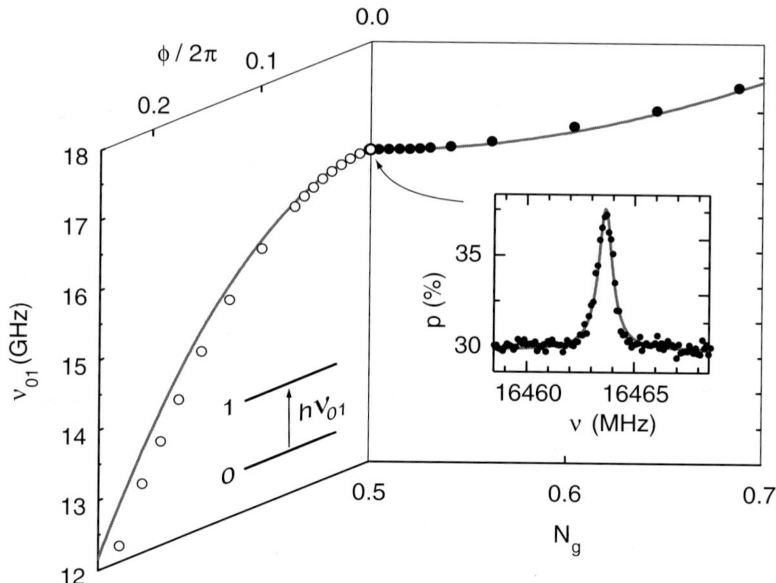

Fig. 5. – Measured center frequency (symbols) of the resonance as a function of reduced gate charge N_g for reduced flux $\phi = 0$ (right panel) and as a function of ϕ for $N_g = 0.5$ (left panel), at 15 mK. Spectroscopy is performed by measuring the switching probability p (10^5 events) when a continuous microwave irradiation of variable frequency is applied to the gate before readout ($t_d < 100$ ns). Continuous line: theoretical best fit (see text). Inset: minimum width lineshape measured at the optimal working point $\phi = 0$ and $N_g = 0.5$ (dots). Lorentzian fit with a FWHM $\Delta\nu_{01} = 0.8$ MHz and a center frequency $\nu_{01} = 16463.5$ MHz (curve).

readout scheme can be compared with a Stern and Gerlach experiment: $\vec{s}$ is the analog of the spin of the Ag atom, while the applied bias current is the analog of the magnetic-field gradient. The loop current response is to be compared with the transverse acceleration of the Ag atom, the phase γ being the analog of the transverse position of the atom. The presence or absence of the voltage pulse corresponds to the impact of the Ag atom in the upper or lower spot of the "screen".

The actual sample on which measurements have been performed is shown in fig. 3(bottom). Tunnel junctions have been fabricated using the standard technique of Al evaporation through a shadow-mask obtained by e-beam lithography [23]. With the external microwave circuit capacitor $C = 1$ pF, the plasma frequency of the large junction with $I_0 = 0.77\,\mu$A is $\omega_p/2\pi \simeq 8$ GHz. The sample and last filtering stage were anchored to the mixing chamber of a dilution refrigerator with 15 mK base temperature. The switching of the large junction to the voltage state is detected by measuring the voltage across it with a room temperature preamplifier followed by a discriminator with a threshold voltage V_{th} well above the noise level (fig. 4). By repeating the experiment, we can determine the switching probability, and hence, the occupation probabilities $|\alpha|^2$ and $|\beta|^2$.

We have first tested the readout part of the circuit by measuring at thermal equilib-

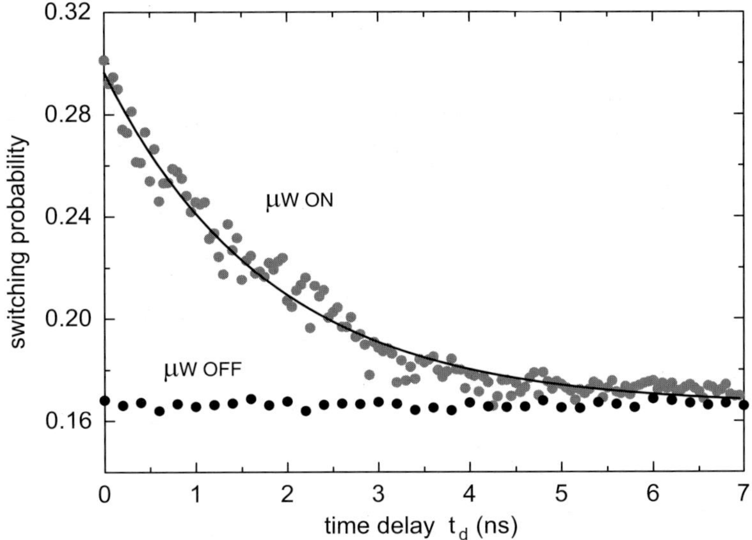

Fig. 6. – Decay of the switching probability as a function of the delay time after a continuous excitation at the center frequency of the line shown in the inset of fig. 3 (data points labeled "μW ON"). A control experiment has been performed by doing the same measurement as a function of the delay time, but without excitation (data points labeled "μW OFF").

rium, for a current pulse duration of $\tau_{\rm r} = 100\,\text{ns}$, the switching probability p as a function of the pulse height $I_{\rm p}$. We have found that the discrimination between the currents corresponding to the $|0\rangle$ and $|1\rangle$ states had an efficiency of $\eta = 0.6$, lower than the expected $\eta = 0.95$. Measurements of the switching probability as a function of temperature and repetition rate indicate that the discrepancy between the theoretical and experimental readout efficiency could be due to an incomplete thermalization of our last filtering stage in the bias current line.

We have then performed spectroscopic measurements of ν_{01} by applying to the gate a weak continuous microwave irradiation suppressed just before the readout current pulse. The variations of the switching probability as a function of the irradiation frequency display a resonance whose center frequency evolves as a function of the dc gate voltage and flux as the Hamiltonian (1) predicts, reaching $\nu_{01} \simeq 16.5\,\text{GHz}$ at the optimal working point (see fig. 5). We attribute the small discrepancy between theory and experiment to a residual flux penetration in the small junctions not taken into account in the model. We have used these spectroscopic data to determine precisely the relevant circuit parameters and found $i_0 = 18.1\,\text{nA}$ and $E_{\rm J}/E_{\rm C} = 5.08$. At the optimal working point, the linewidth was found to be minimal with a 0.8 MHz full width at half-maximum. When varying the delay between the end of the irradiation and the measurement pulse, the peak height decays with a time constant $T_1 = 1.8\,\mu\text{s}$ (see fig. 6). Supposing that the energy relaxation of the system is only due to the bias circuitry, a calculation along the lines of ref. [24]

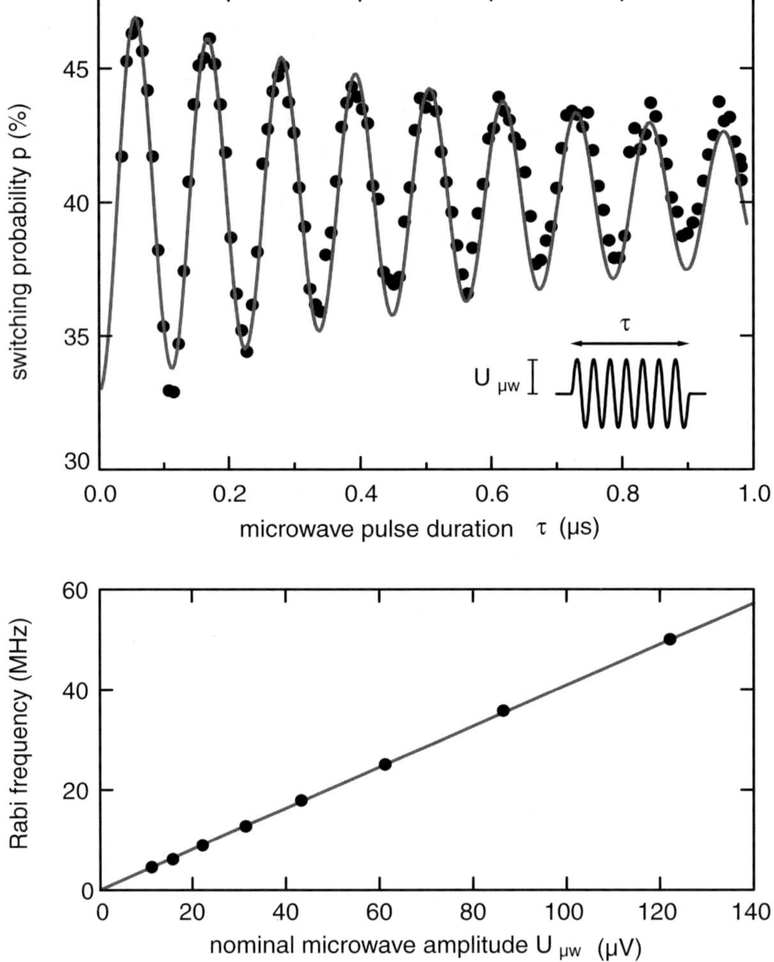

Fig. 7. – Top: Rabi oscillations of the switching probability p (5×10^4 events) measured just after a resonant microwave pulse of duration τ. Data taken at $15\,\mathrm{mK}$ for a nominal pulse amplitude $U_{\mu\mathrm{w}} = 10\,\mu\mathrm{V}$ (dots). The Rabi frequency is extracted from an exponentially damped sinusoidal fit (continuous line). Bottom: the measured Rabi frequency (dots) varies linearly with $U_{\mu\mathrm{w}}$, as expected.

predicts that $T_1 \sim 10\,\mu$s for a crude discrete element model. This result shows that no detrimental sources of dissipation have been seriously overlooked in our circuit design.

We have then addressed the fidelity of controlled rotations of $\vec{s}$ around an axis x perpendicular to the quantization axis z. Prior to readout, a single pulse at the transition frequency with variable amplitude $U_{\mu\mathrm{w}}$ and duration τ was applied. The resulting change in switching probability is an oscillatory function of the product $U_{\mu\mathrm{w}}\tau$ (see fig. 7), in agreement with the theory of Rabi oscillations [19, 25]. It provides direct

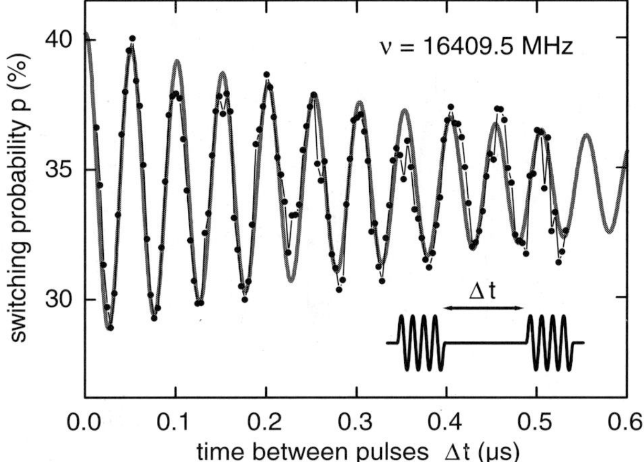

Fig. 8. – Ramsey fringes of the switching probability p (5×10^4 events) after two phase-coherent microwave pulses separated by Δt. Dots: data at 15 mK; the total acquisition time was 5 mn. Continuous line: fit by exponentially damped sinusoid with time constant $T_\varphi = 500 \pm 50$ ns. The oscillation corresponds to the "beating" of the free evolution of the spin with the external microwave field. Its period indeed coincides with the inverse of the detuning frequency (here $\nu - \nu_{01} = 20.6$ MHz).

evidence that the resonance indeed corresponds to an effective spin rather than to a spurious harmonic-oscillator resonance in the circuit. The proportionality ratio between the Rabi period and $U_{\mu w}\tau$ was used to calibrate microwave pulses for the application of controlled rotations of $\vec{s}$.

The main result of this paper, the coherence time of $\vec{s}$ during free evolution, was obtained by performing the classic Ramsey fringes experiment [26] on which atomic clocks are based. One applies on the gate two phase-coherent microwave pulses corresponding each to a $\pi/2$ rotation around x [27] and separated by a delay Δt during which the spin precesses freely around z. For a given detuning of the pulse center frequency, we have observed decaying oscillations of the switching probability as a function of Δt (see fig. 8), which correspond to the "beating" of the spin precession with the external microwave field. The oscillation period agrees exactly with the inverse of the detuning, allowing a measurement of the transition frequency with an accuracy of 6×10^{-6} [28]. The envelope of the oscillations yields the decoherence time $T_\varphi \simeq 0.5\,\mu s$. Given the transition period $1/\nu_{01} \simeq 60$ ps, this means that $\vec{s}$ can perform on average 8000 coherent free precession turns.

In all our time domain experiments, the oscillation period of the switching probability closely agrees with theory, meaning a precise control of the preparation of $\vec{s}$ and of its evolution. However, the amplitude of the oscillations is smaller than expected by a factor of three to four. This loss of contrast is likely to be due to a relaxation of the level population during the measurement itself. In principle the current pulse, whose rise time is 50 ns, is sufficiently adiabatic not to induce transitions directly between the two levels. Nevertheless, it is possible that the readout or even the preparation pulses excite resonances in the bias circuitry which in turn could induce transitions in our two-level manifold. Experiments using better shaped readout pulses and a bias circuitry with better controlled high-frequency impedance are needed to clarify this point.

In order to understand what limits the coherence time of the circuit, we have performed measurements of the linewidth $\Delta\nu_{01}$ of the resonant peak as a function of U and Φ. The linewidth varies linearly when departing from the optimal point ($N_{\mathrm{g}} = 1/2$, $\phi = 0$, $I_{\mathrm{b}} = 0$), the proportionality coefficients being $\partial\Delta\nu_{01}/\partial N_{\mathrm{g}} \simeq 250\,\mathrm{MHz}$ and $\partial\Delta\nu_{01}/\partial(\phi/2\pi) \simeq 430\,\mathrm{MHz}$. These values can be translated into RMS deviations $\Delta N_{\mathrm{g}} = 0.004$ and $\Delta(\phi/2\pi) = 0.002$ of the transition frequency during the time needed to record the resonance. The residual linewidth at the optimal working point is well explained by the second-order contribution of these noises and is not therefore limited by any fundamental factor or unknown noise sources. The amplitude of the charge noise is in agreement with measurements of $1/f$ charge noise [29], and its effect could be minimized by increasing the $E_{\mathrm{J}}/E_{\mathrm{C}}$ ratio. By contrast, the amplitude of the flux noise is unusually large [30], and we think that implementation of magnetic shielding and better Al layer configuration will significantly reduce it. An improvement of Q_φ by an order of magnitude seems thus possible [31].

$$* \quad * \quad *$$

The indispensable technical work of Pief Orfila is gratefully acknowledged. This work has greatly benefited from direct inputs from J. M. Martinis and Y. Nakamura. The authors acknowledge discussions with P. Delsing, G. Falci, D. Haviland, H. Mooij, R. Schoelkopf, G. Schön and G. Wendin. This work is partly supported by the European Union through contract IST-10673 SQUBIT.

REFERENCES

[1] Bocko M. F., Herr A. M. and Feldman M. F., *IEEE Trans. Appl. Supercond.*, **7** (1997) 3638.

[2] Kane B. E., *Nature*, **393** (1998) 133.

[3] Loss D. and DiVincenzo D. P., *Phys. Rev. A*, **57** (1998) 120.

[4] Makhlin Yu., Schoen G. and Shnirman A., *Nature*, **398** (1999) 786; Makhlin Y., Schön G. and Shnirman A., *Rev. Mod. Phys.*, **73** (2001) 357.

[5] Nakamura Y., Pashkin Yu. A. and Tsai J. S., *Nature*, **398** (1999) 786; *Phys. Rev. Lett.*, **87** (2001) 246601; to be published (cond-mat/0111402).

[6] Mooij J. E. *et al.*, *Science*, **285** (1999) 1036; Caspar H., van der Wal *et al.*, *Science*, **290** (2000) 773.

[7] Siyuan-Han, Rouse R. and Lukens J. E., *Phys. Rev. Lett.*, **84** (2000) 1300.

[8] Siyuan-Han, Yang-Yu, Xi-Chu, Shih-I-Chu and Zhen-Wang, *Science*, **293** (2001) 1457.

[9] Cottet A., Vion D., Joyez P., Aassime A., Esteve D. and Devoret M. H., to be published in *Physica C*.

[10] Martinis J. M., Sae Woo Nan, Aumentado J. and Urbina C. (unpublished) have recently obtained Q_φ's of the order of 500 for a current-biased Josephson junction.

[11] TUOMINEN M. T., HERGENROTHER J. M., TIGHE T. S. and TINKHAM M., *Phys. Rev. Lett.*, **69** (1992) 1997.

[12] LAFARGE P., JOYEZ P., ESTEVE D., URBINA C. and DEVORET M. H., *Nature*, **365** (1993) 422.

[13] LIKHAREV K. K., *Dynamics of Josephson Junctions and Circuits* (Gordon and Breach, New York) 1986.

[14] BÜTTIKER M., *Phys. Rev. B*, **36** (1987) 3548.

[15] BOUCHIAT V., VION D., JOYEZ P., ESTEVE D. and DEVORET M. H., *Phys. Scr.*, **T76** (1998) 165.

[16] ZUREK W. H., *Phys. Today*, **44** (1991) 36; ZUREK W. H. and PAZ J. P., in *Coherent Atomic Matter Waves*, edited by R. KAISER, C. WESTBROOK and F. DAVID (Springer-Verlag, Heidelberg) 2000 (quant-ph/0010011).

[17] MARTINIS J. M., DEVORET M. H. and CLARKE J., *Phys. Rev. B*, **35** (1987) 4682; DEVORET M. H., ESTEVE D., URBINA C., MARTINIS J. M., CLELAND A. N. and CLARKE J., in *Quantum Tunneling in Condensed Media*, edited by Y. KAGAN and A. J. LEGGETT (Elsevier Science Publishers) 1992.

[18] Another two-port design has been proposed by ZORIN A. B., cond-mat/0112351; to be published in *Physica C*.

[19] ABRAGAM A., *The Principles of Nuclear Magnetism* (Oxford University Press) 1961.

[20] AASSIME A., JOHANSSON G., WENDIN G., SCHOELKOPF R. J. and DELSING P., *Phys. Rev. Lett.*, **86** (2001) 3376.

[21] AVERIN D. V. and LIKHAREV K. K., in *Mesoscopic Phenomena in Solids*, edited by B. L. ALTSHULER, P. A. LEE and R. A. WEBB (Elsevier, Amsterdam) 1991.

[22] A different Cooper pair box readout scheme using a large Josephson junction is discussed by HEKKING F. W. J., BUISSON O., BALESTRO F. and VERGNIORY M. G., in *Electronic Correlations: from Meso- to Nanophysics*, edited by T. MARTIN, G. MONTAMBAUX and J. THANH VÂN TRÂN (EDPSciences, Les Ulis) 2001, p. 515.

[23] DOLAN G. J. and DUNSMUIR J. H., *Physica B (Amsterdam)*, **152** (1988) 7.

[24] COTTET A., STEINBACH A. H., JOYEZ P., VION D., POTHIER H., ESTEVE D. and HUBER M. E., in *Macroscopic Quantum Coherence and Quantum Computing*, edited by D. V. AVERIN, B. RUGGIERO and P. SILVESTRINI (Kluwer Academic, Plenum Publishers, New York) 2001, pp. 111-125.

[25] RABI I. I., *Phys. Rev.*, **51** (1937) 652.

[26] RAMSEY N. F., *Phys. Rev.*, **78** (1950) 695.

[27] In practice, the rotation axis does not need to be x, but the rotation angle of the two pulses is always adjusted so as to bring a spin initially along z into a plane perpendicular to z.

[28] At fixed Δt, the switching probability displays a decaying oscillation as a function of detuning, the maximum corresponding to zero detuning.

[29] WOLF H., AHLERS F.-J., NIEMEYER J., SCHERER H., WEIMANN TH., ZORIN A. B., KRUPENIN V. A., LOTKHOV S. V. and PRESNOV D. E., *IEEE Trans. Instrum. Meas.*, **46** (1997) 303.

[30] WELLSTOOD F. C., URBINA C. and CLARKE J., *Appl. Phys. Lett.*, **50** (1987) 772.

[31] Critical current noise (SAVO B., WELLSTOOD F. C. and CLARKE J., *Appl. Phys. Lett.*, **50** (1987) 1758) seems to be of lesser concern since none of our results forces us to invoke it.

Superconducting devices for quantum logic gates using magnetic flux states

M. G. Castellano

Istituto di Elettronica dello Stato Solido del CNR
Via Cineto Romano 42, 00156 Roma, Italy
and INFN, Sezione di Roma 1 - Roma, Italy

F. Chiarello

INFM, Università dell'Aquila - Monteluco di Roio, 67040 L'Aquila, Italy

1. – Introduction

Superconducting devices, namely Josephson junctions and SQUIDs (Superconducting Quantum Interference Devices), have been recognized since the '80s as excellent systems to perform tests on the quantum behavior of macroscopic systems [1]. Being described by macroscopic variables (the superconducting phase in the case of a junction, the magnetic flux in the case of a SQUID) governed by truly quantum dynamics, these devices have been used to demonstrate experimental evidence of quantum tunnelling, resonant tunnelling [2-6] and most recently even quantum coherence on a macroscopic scale [7]. Besides providing the source of the quantum state, superconducting devices can also provide the instruments necessary for its probing, especially as regards flux states, which can be read out by SQUID magnetometers with a sensitivity approaching the quantum limit.

Josephson devices have also been proposed for the physical implementation of bits, gates and registers for quantum computing [8,9]; in this respect, they have the advantage of being easily scaled and engineered, using standard lithography and thin-film techniques. A physical system suitable for quantum computation must satisfy several requirements [10]. The qubit candidate must be well approximated by a two-level system that can be prepared in a specific initial state and has a phase coherence time long enough to allow for many manipulations. It must be possible to perform unitary operations (quantum gates) on a single qubit. Quantum registers, that is arrays of qubits, must be constructed, by using interacting single qubits that form entangled many-qubit states; a good control of the inter-qubit interaction is required, preferably with the option of switching on or off. Finally, the qubit status must be read out by a suitable detector. From an experimental point of view, the main steps of the program are: finding a suitable two-state system; observing the coherent superposition of the two states; manipulating the qubits and preparing them in specific superposition of states; then, two qubits must be put together to originate entangled states.

Josephson junctions with low capacitance, showing charging effects, have been proposed [11] and examined from the point of view of construction of single qubits, qubit registers and quantum state engineering. The Josephson effect allows to have a mixing of different charge states at the degeneracy point, thus forming the two-state system. A recent experiment performed on such a system [12] has shown the presence of coherent time oscillations of the charge states and has allowed an experimental evaluation of the decoherence times, enforcing the expectations for a solid-state qubit implementation. Another possibility, which we discuss in this paper, is using the magnetic flux states of superconducting devices exploiting the Josephson effect [13]. We will illustrate the working principle, the physical implementation and how to construct the universal gates for such quantum bits. We will refer to a thin-film integrated system of SQUIDs [14] that we designed and fabricated and we are presently testing in order to observe quantum-coherent time oscillations of the flux states: this system should be feasible for quantum computing implementation.

2. – The building blocks: superconducting phase, magnetic flux quantization, Josephson effects

Superconductivity is a well-established property of many metals, alloys and ceramics that, below some critical temperature T_c, exhibit a transition to a state of null resistance (at zero frequency). At the same time the magnetic flux density **B** inside the superconductor vanishes (Meissner effect). As a result, the current density inside a superconductor must be zero; shielding currents, which are confined at the surface of the superconducting body, are not damped and can go on circulating permanently.

The microscopic theory of superconductivity [15] states that a superconductor is a Bose-Einstein condensate of Cooper pairs, with electrons coupled two by two. In a Cooper pair, an attractive interaction due to virtual phonon exchange prevails over the repulsive force due to the screened Coulomb interaction. The total charge of a pair is $2e$, twice the charge of the electron, and the coupling energy is $2\Delta(T)$, related to the critical temperature of the material: $2\Delta(0) = 3.52 k_B T_c$ at zero temperature. The coupling energy produces a gap in the excitation spectrum of the superconductor, across the Fermi energy. The two electrons are paired in a singlet state and all the pairs carry the same total momentum; the distance between the two electrons of a pair is larger than the mean distance between pairs.

2˙1. *Superconducting phase*. – All the pairs behave as if they had condensed into a single state, described by a wave function of the kind

$$(1) \qquad \psi(\mathbf{r}) = \sqrt{\rho(\mathbf{r})}e^{i\theta(\mathbf{r})},$$

where $\rho(\mathbf{r})$ and $\theta(\mathbf{r})$ are real functions of position $\mathbf{r}$. $\rho(\mathbf{r}) = \psi^*(\mathbf{r}) \cdot \psi(\mathbf{r})$ represents the density of pairs and $\theta(\mathbf{r})$ is the phase of the macroscopic wave function. The phase of the superconducting wave function has a particular importance: we will see that the gradient of the superconducting phase is associated to a physically measurable quantity, breaking the symmetry of gauge invariance, which, in electrodynamics, leaves the electromagnetic potentials unchanged if the four-dimensional gradient of a vector is added. The second-order superconducting transition is related to this symmetry breaking.

From quantum mechanics, the generalized expression for the current density, in the presence of a magnetic field derived from a vector potential $\mathbf{A}$, is

$$(2) \qquad \mathbf{J} = \frac{i\hbar 2e}{2m}(\psi^*\nabla\psi - \psi\nabla\psi^*) + \frac{(2e)^2}{2m}|\psi|^2\mathbf{A}.$$

By inserting expression (1) for the wave function, there follows a relationship linking the phase gradient, the vector potential and the current density:

$$(3) \qquad \nabla\theta = \frac{m}{2e\hbar\rho}\mathbf{J} + \frac{2e}{\hbar}\mathbf{A}.$$

The phase gradient is then associated to measurable quantities.

2˙2. *Magnetic flux quantization*. – For a superconducting ring, eq. (3) can be integrated on a path along the loop. Since the wave function must be single-valued, the phase change after a complete turn must be $2\pi n$; on the right side, the integral of $\mathbf{J}$ is zero (the current does not flow inside the superconductor), while the integral of the vector potential gives $2e/\hbar$ times the magnetic flux Φ through the ring. We get then

$$(4) \qquad \Phi = n\Phi_0,$$

where n is an integer and the quantity $\Phi_0 = h/2e = 2.07 \cdot 10^{-15}$ Wb is the flux quantum; hence, the magnetic flux inside a superconducting ring is quantized in units of Φ_0. The screening current flows on the ring surface, allowing to satisfy the quantization condition.

2˙3. *Josephson effects*. – A Josephson junction is made by two superconducting electrodes, separated by a thin oxide barrier: by tunnel effect, the wave functions on each side extend across the barrier, so that there is a finite superposition. As a consequence, the two phases are not free to change independently; the phase difference δ across the tunnel junction is governed by the two Josephson equations

$$(5) \qquad I = I_0 \sin \delta,$$

$$(6) \qquad \frac{\mathrm{d}\delta}{\mathrm{d}t} = \frac{2e}{\hbar} V.$$

The first Josephson equation (dc Josephson effect) states that at zero voltage a pair current can flow in the junction, up to a maximum value I_0 (critical current); the phases on the two sides adjust to satisfy the Josephson condition. Above the critical value, a voltage V appears across the junction, related to a constant rate of change of the phase difference (eq. (6), ac Josephson effect). The ratio between frequency and voltage, $\nu/V = 483.594\,\mathrm{MHz}/\mu\mathrm{V}$, is at the basis of many Josephson devices.

3. – Basic Josephson devices

By arranging a superconducting loop and one or more Josephson junctions, it is possible to build several devices where macroscopic quantum phenomena can be observed. The device dynamics can be usually reduced to that of a particle, described by a proper macroscopic variable (the phase or the magnetic flux) and subjected to a potential with several maxima and minima, corresponding to metastable states. At low enough temperatures and dissipation, macroscopic quantum phenomena, such as tunnelling through the potential barriers, can be observed. We will briefly recall the main devices, in each case stressing the appropriate variable describing the dynamics and the corresponding equivalent potential.

$3^{\cdot}1$. *Josephson junctions.* – A Josephson junction can be conveniently modelled by a lumped element circuit (RCSJ model): a non-linear Josephson element (graphically represented by a cross) in parallel with a capacitance (the physical capacitance of the junction sandwich) and with a resistor (junction intrinsic resistance or external load). By expressing the voltage in terms of the time derivative of the phase difference δ by using the Josephson relations, the equation of motion becomes

$$(7) \qquad \frac{\hbar C}{2e}\frac{\mathrm{d}^2\delta}{\mathrm{d}t^2} + \frac{\hbar}{2eR}\frac{\mathrm{d}\delta}{\mathrm{d}t} + I_0 \sin \delta = I + I_\mathrm{n}(t).$$

Neglecting the noise term I_n, this is the equation of a damped nonlinear oscillator, the velocity of which is the phase derivative $\dot\delta$, proportional to the voltage across the device. The dynamics of the Josephson junction can be understood by using a mechanical analog, where a particle of mass $C(\Phi_0/2\pi)^2$ with dissipation $(\Phi_0/2\pi)^2(1/R)$, described by the mechanical degree of freedom δ (the phase difference across the junction), is subjected

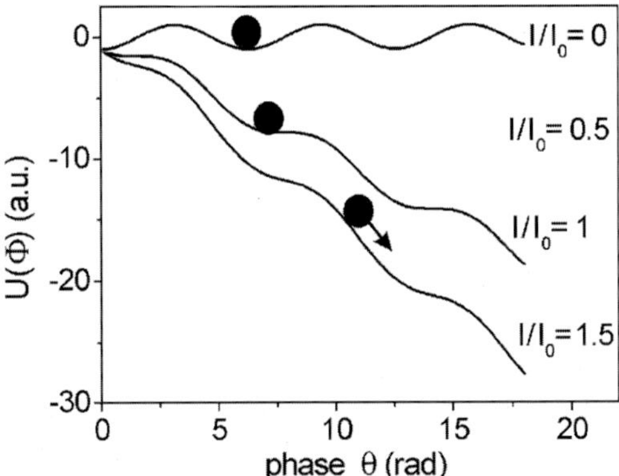

Fig. 1. – The tilted-washboard potential for three different values of the bias current. When the bias current is below the critical current, the particle, which represents the Josephson junction, is sitting inside one of the metastable minima: the current is carried by Cooper pairs and no voltage is developed across the junction. If the critical current value is exceeded, the particle starts rolling down the potential, θ changes with time and a voltage appear across the junction.

to a tilted washboard potential U (fig. 1):

$$(8) \qquad U(\delta) = -I_0 \frac{\Phi_0}{2\pi} \left(\delta \frac{I}{I_0} + \cos\delta \right).$$

Here I_0 is the junction critical current and I is the bias current.

The bias current tilts the corrugated curve. If the critical current is not exceeded, the particle remains trapped in a potential metastable well, oscillating at the plasma frequency; since the phase is not varying, there is no voltage across the junction. When the potential slope is increased at the critical value, the minima become flex points and the particle starts rolling down the potential (running state, δ is varying, hence eq. (6) gives $V \neq 0$). If the temperature is not zero, the thermal fluctuations allow escape from the metastable states even for bias values lower than the critical one, when the barrier between adjacent wells has not disappeared yet. The activation over the residual barrier produces a spreading of the switching current, which is now related to a stochastic process. If the dissipation and the temperature are low enough, the switching may occur because of quantum tunnelling through the potential barrier: this is one of the first macroscopic quantum phenomena that have been observed in Josephson devices [3].

From an experimental point of view, two main different technological processes are used to fabricate Josephson junctions suitable for our purposes, based on aluminum or niobium electrodes. Al/AlO$_x$/Al junctions with side less than 100 nm are used for the charge states qubits and also for the so-called superconducting persistent-current qubit [16], which relies on magnetic flux states; if larger junctions can be used (from 1 μm on), the choice of election is the Nb/AlO$_x$/Nb trilayer technique, which produces robust and very high-quality junctions: actually, this is the case for our devices. A schematic

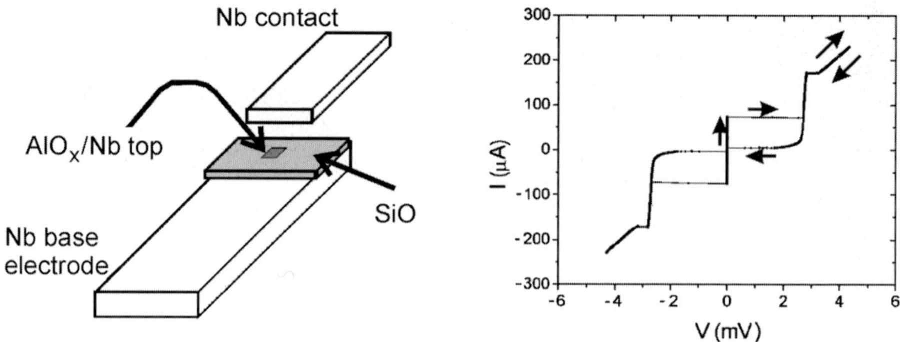

Fig. 2. – Schematic of a Josephson junction and current-voltage characteristics.

of a thin-film $\mathrm{Nb/AlO}_x/\mathrm{Nb}$ tunnel junction is shown in fig. 2. The junction is fabricated starting with a $\mathrm{Nb/AlO}_x/\mathrm{Nb}$ trilayer, deposited in a single run over the whole substrate to form a reliable, uniform, thin oxide barrier. The area of the junction (a square window, with side ranging from units to hundreds of micrometers) is then protected by resist (through standard optical lithography) and the top Nb layer is removed around this zone by Reactive Ion Etching: this leaves a pillar, whose cross-section at the AlO_x level defines the junction area. Next, a Nb contact wire must be connected to the junction top electrode; this requires a careful isolation around the pillar, which is achieved by depositing a silicon oxide layer just after the etching. In alternative, after the junction area definition, it is possible to use liquid anodization to turn the surrounding Nb into Nb-oxide, acting as an insulator. The current-voltage characteristic of the junction is shown on the right side of fig. 2. A constant current can pass through the junction at zero voltage until the switching value is reached; at this point the junction switches to the gap voltage (about 2.7 mV for niobium junctions) and approaches the slope of the normal tunnel resistance R_n. By decreasing the bias current, the representative point follows the lower branch of the curve (quasiparticle branch) as indicated by the arrows, until retrapping in the zero-voltage state occurs. The characteristic is hysteretic whenever McCumber's parameter, $\beta_\mathrm{c} = 2\pi I_0 R^2 C/\Phi_0$ (where R is the total resistance across the junction, any external shunt included), is larger than 1.

3`2. *The superconducting interferometer*. – When two Josephson junctions are placed in a superconducting loop, the phase across each junction is coherently related to the phase of the other junction by the flux quantization condition eq. (3), so that quantum interference effects between the currents carried by each junction are produced [17, 18]. Across the (identical) junctions 1 and 2, the phase difference is, respectively, δ_1 and δ_2. The device is biased by a constant current I, while a circulating current i can flow in the ring, so a current $I/2 + i$ flows in one junction, while $I/2 - i$ flows in the other. Each junction can be described by the lumped element model discussed above, with the conditions imposed by the Josephson equations and that imposed by the flux

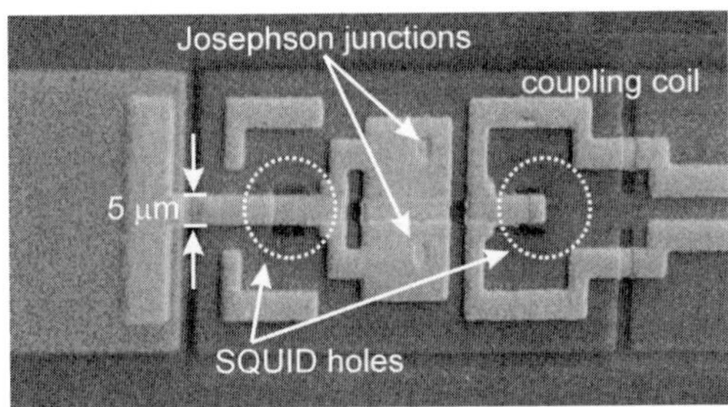

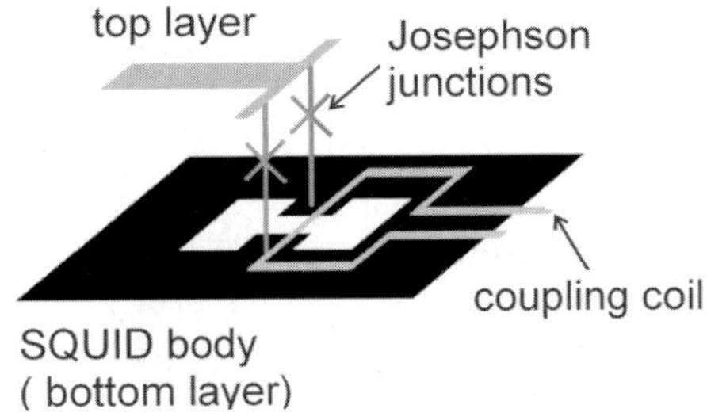

Fig. 3. – A SEM micrograph of a superconducting interferometer. The geometry of the device is sketched in the drawing at the right.

quantization, which relates the magnetic flux to the phase difference along the loop.

The equivalent potential, in this case, is a surface with the shape of an egg-carton, with several wells where the particle can be trapped:

$$(9) \qquad U(\chi,\psi) = -\frac{2I_0\Phi_0}{2\pi}\left(\frac{I}{2I_0}\psi + \cos\chi\cos\psi\right) + \left(\frac{\Phi_0}{2\pi}\right)^2\frac{1}{L}\left(\chi - \frac{\pi\Phi_x}{\Phi_0}\right)^2,$$

where $\chi = (\delta_1 + \delta_2)/2$ is the average phase difference across the two junctions, $\psi = (\delta_1 - \delta_2)/2$ is a quantity proportional to the magnetic flux Φ_0 in the ring, L is the loop inductance. An important parameter in the description of the device behavior is $\beta_L = 2\pi I_0 L/\Phi_0$. If $\beta_L \ll 1$, this device works as a single Josephson junction with a critical current that can be modulated periodically (with period of one flux quantum) by the magnetic flux according to the expression

$$(10) \qquad I_c(\Phi_0) = 2I_0\left|\cos\left(\pi\frac{\Phi}{\Phi_0}\right)\right|.$$

If β_L is not much smaller than one, the current is modulated between a minimum and a maximum value, which are a function of β_L [19]. When the dc-SQUID has to be used as a practical device, for instance for detecting magnetic fields or currents, the hysteresis in the I-V characteristics is removed by shunting the junctions with low-value resistors: in this way, the change in critical current determined by an applied magnetic flux is converted into a voltage signal that is a periodic function of the flux itself. The details of the motion are determined by the value of β_L, bias current and external magnetic flux.

A picture of our hysteretic dc-SQUID is shown in fig. 3, together with a sketch of the device scheme. As usual in thin-film superconducting devices, the superconducting loop is realized by a square hole in a superconducting layer. The inductance value can be numerically calculated as follows, neglecting the effect of the film thickness:

$$(11) \qquad L = \mu_0 d\frac{2}{\pi}\left[1.96 + 1.15\cdot\ln\left(\frac{x+0.096}{x}\right)\right],$$

where d is the internal side of the hole, w is the width of the film around it and $x = w/d$. In the limit of large x the formula reduces to $L = 1.25\mu_0 d$ [20]. The superconducting ring is interrupted by a slit, across which the two junctions are placed; the slit adds a stray inductance that can be reduced by covering the slit with a superconducting patch, electrically insulated from the SQUID body.

The device shown in the picture has a gradiometer configuration, that is, the loop is made by two counterwound loops ($10\,\mu$m inner side), with the junctions placed across them. In this way, a uniform field like the one produced by ambient disturbances produces opposite shielding currents in each loop, which cancel out in the SQUID. An input coil surrounds only one of the two holes, so that a current flowing in it produces a net flux in the SQUID. The measured SQUID inductance, given by the parallel of the two loops plus stray terms, is around $10\,$pH. The coupling coil has a mutual inductance of $5\,$pH.

In the design of our setup, we used two samples of this underdamped dc-SQUID for two different functions: inside the rf-SQUID (qubit) and as readout for the qubit (see the next subsections).

3`3. *The rf-SQUID*. – The rf-SQUID is made of a superconducting loop of inductance $L_{\rm rf}$ interrupted by one Josephson junction. In this case, the variable suitable for a semiclassical treatment of the system dynamics is the magnetic flux linked to the ring Φ, which is related to the phase difference; the external bias is given by the external magnetic flux Φ_x. The corresponding potential is

$$(12) \qquad U(\Phi) = -\frac{I_0\Phi_0}{2\pi}\cos\left(\frac{2\pi\Phi}{\Phi_0}\right) + \frac{(\Phi - \Phi_x)^2}{2L_{\rm rf}}.$$

If the parameter β_L (defined as before) is greater than unity, this curve describes a corrugated parabola, as if the tilted washboard had now been rolled into a parabola; smaller β_L makes the corrugation (*i.e.* the barriers between wells) vanish. By varying the external flux Φ_x, the parabola sweeps and rolls at the same time on one side, so that the lowest-energy state changes from corrugation to corrugation. If the value of the dimensionless factor β_L is between 1 and $5\pi/2$ and at an external bias of exactly $\Phi_0/2$, the bottom of the parabola is shared by two symmetric wells and the whole system dynamics can be reduced to the well-known problem of a two-state system. It is precisely this region that can be used to build a qubit out of the magnetic flux states. Figure 4 shows how the characteristics (internal magnetic flux Φ *vs.* applied magnetic flux Φ_x) of a rf-SQUID vary with the parameter β_L; in this case, the rf-SQUID is actually replaced by a double SQUID (see below), which acts as a rf-SQUID with tunable β_L. The experimental curves show that, by increasing the applied flux, the device reacts with a shielding current to keep the total flux constant; at some point, it becomes energetically convenient for the device to change the flux state by one quantum and let the current circulate in the opposite direction. Hysteresis corresponds to $\beta_L > 1$.

3`4. *The double SQUID*. – In order to have an additional "knob" to control the qubit, the single junction of the rf-SQUID can be replaced by a hysteretic dc-SQUID, acting as

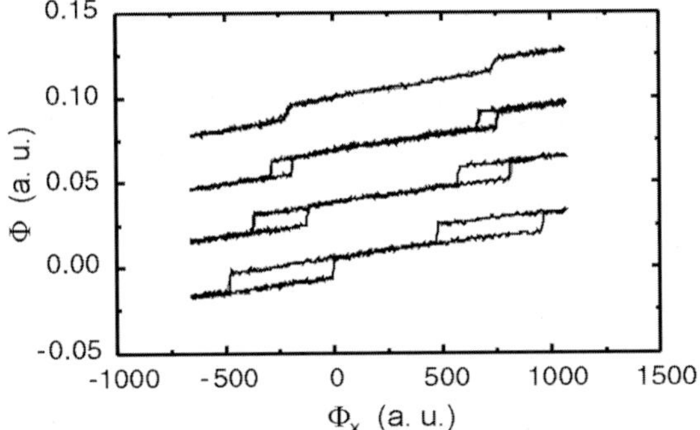

Fig. 4. – Flux characteristics of a double SQUID, acting as a rf-SQUID with variable β_L. The rf-SQUID flux Φ is measured by a non-hysteretic dc-SQUID, magnetically coupled to one of the two holes of the device, while the input flux Φ_x is applied by using the other SQUID hole. β_L is varied from $\simeq 1$ (non-hysteretic characteristic) to $\beta_L > 1$ (increasingly hysteretic characteristics) by sending a dc current in a coil coupled to the inner dc-SQUID.

a Josephson junction whose critical current can be tuned through a magnetic flux [21]. In order to be in this limit, the inner dc-SQUID must have a self-inductance l much smaller than the rf-SQUID. The degrees of freedom that describe the system are the magnetic flux in the dc-SQUID Φ_{dc} and in the larger loop Φ, while the external bias is given by external fluxes Φ_{xdc} and Φ_x. Figure 5 shows our implementation, along with a schematic. The rf-SQUID has once again a gradiometric configuration, with a total inductance of 85 pH; the two loops are coupled to two different coils: one serves for the biasing or excitation flux (mutual inductance with the SQUID), the other couples the

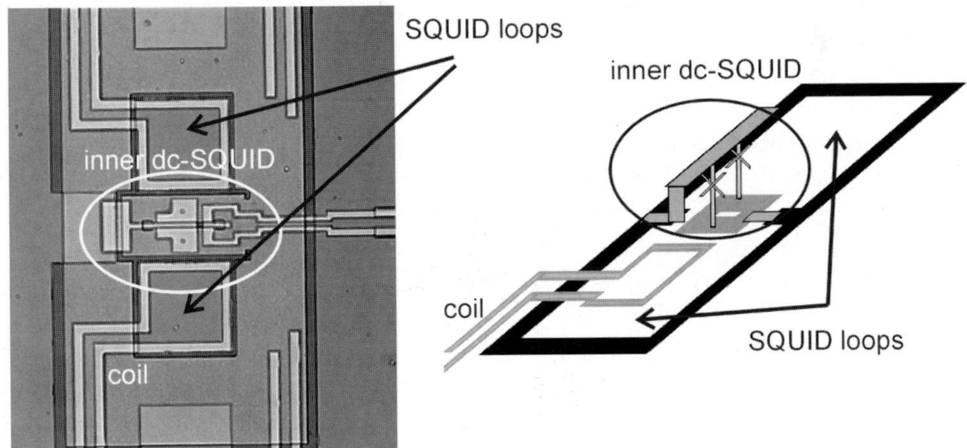

Fig. 5. – SEM micrograph of a double SQUID; the corresponding 3-dim schematic (with only one coupling coil for clarity) is shown on the right.

     M. G. Castellano and F. Chiarello

Table I. – *Summary: coordinates, parameters and equivalent potential for Josephson devices.*

Device	Coordinate	Bias	Parameters	Potential
JJ	δ	I	I_0	$U(\delta) = -\frac{I_0\Phi_0}{2\pi}\cdot\left(\delta\frac{I}{I_0}+\cos\delta\right)$
rf-SQUID	Φ	Φ_x	I_0	$U(\Phi) = -\frac{I_0\Phi_0}{2\pi}\cos\frac{2\pi\Phi}{\Phi_0}+\frac{(\Phi-\Phi_x)^2}{2L}$
dc-SQUID	$\chi = \frac{\delta_1-\delta_2}{2}\propto\frac{\Phi}{\Phi_0},$ $\psi = \frac{\delta_1-\delta_2}{2}$	I,Φ_x	I_0, L	$U(\chi,\psi)=-\frac{2I_0\Phi_0}{2\pi}\left(\frac{I}{2I_0}\psi+\cos\chi\cos\psi\right)$ $+\left(\frac{\Phi_0}{2\pi}\right)^2\frac{1}{L}\left(\chi-\frac{\pi\Phi_x}{\Phi_0}\right)^2$
double SQUID	Φ,Φ_{dc}	$\Phi_x,\Phi_{\mathrm{xdc}}$	I_0, L, l	$U(\Phi,\Phi_{\mathrm{dc}})=-\frac{2I_0\Phi_0}{2\pi}\cos\frac{\pi\Phi_{\mathrm{dc}}}{\Phi_0}\cos\frac{2\pi\Phi}{\Phi_0}+$ $+\frac{(\Phi-\Phi_x)^2}{2L}+\frac{(\Phi_{\mathrm{dc}}-\Phi_{\mathrm{xdc}})^2}{2l}$

rf-SQUID response to a readout device. A third, smaller coil is coupled to the inner dc-SQUID to vary its critical current.

Table I summarizes some relevant quantities for these Josephson devices.

A picture of the integrated system that we realized for the detection of macroscopic quantum coherence, and suitable to be transformed into a system for quantum computing, is shown in fig. 6, together with a schematic of the circuit.

4. – Qubit implementation

In order to build qubits, one must have a suitable quantum two-level system. The rf-SQUID, when $\beta_L = 1$–$5\pi/2$ and the external flux is $\Phi_0/2$, provides this system. The potential is a symmetric double well, where the barrier between wells ΔU depends on the critical current of the Josephson junction (or, in the case of the double SQUID, of

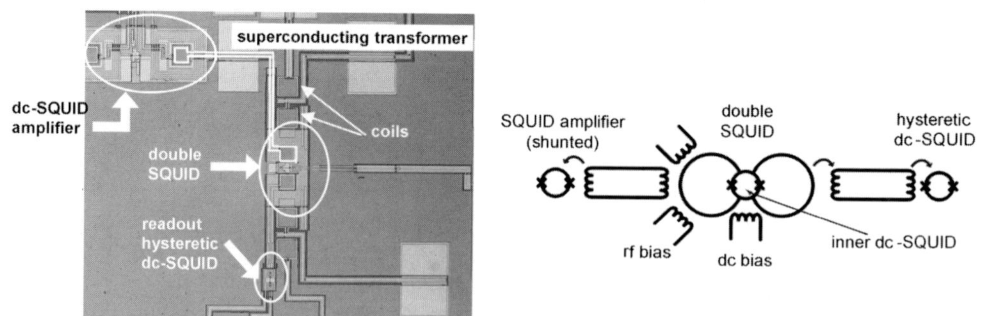

Fig. 6. – Picture of an integrated system formed by a double SQUID, read out by a hysteretic dc-SQUID or a dc-SQUID amplifier.

the inner dc-SQUID); an approximate expression, valid in the limit $0 < \beta_L - 1 \ll 1$, is

$$(13) \qquad \Delta U \simeq \frac{3}{2} \frac{\Phi_0^2}{(2\pi)^2 L} \left(\frac{\beta_L - 1}{\beta_L} \right)^2.$$

In the energy spectrum, the levels in the well are arranged in doublets, with close levels in the same doublet and a greater spacing between different doublets. Supposing that only the lowest doublet is populated, the flux eigenstates are two distinct sharp peaks, with fixed position (the bottom of the two wells), corresponding to shielding current circulating either clockwise or counterclockwise: we will call them $|L\rangle$ (for "left") and $|R\rangle$ (for "right"). They are not energy eigenstates; as a fact, in terms of the energy levels $|0\rangle$ and $|1\rangle$, they are given by a linear combination:

$$(14) \qquad |L\rangle = \frac{1}{\sqrt{2}}(|0\rangle + |1\rangle)$$

$$(15) \qquad |R\rangle = \frac{1}{\sqrt{2}}(-|0\rangle + |1\rangle).$$

Quantum coherence of flux states should manifest itself as coherent oscillations of the magnetic flux between the two potential wells; this is a direct consequence of the tunnelling through the barrier between them. This coherent oscillation has not yet been observed, but there is a spectroscopic proof of the state superposition [7], although not for the ground-state levels. Of course, only if the coherent oscillation occurs and the decoherence time does not destroy it too fast, it is possible to build a qubit with such a system.

In the absence of decoherence, a state initially in, say, $|L\rangle$ will evolve according to

$$(16) \qquad |\psi(t)\rangle = \cos(\Omega t)|L\rangle - i\sin(\Omega t)|R\rangle,$$

where the frequency Ω (frequency of Rabi's oscillations) is approximately

$$(17) \qquad \Omega \simeq 6\frac{\Delta U}{\hbar} \exp\left[-\frac{8\Delta U}{\hbar\omega} \right]$$

with $\omega \simeq 1/\sqrt{LC} \cdot \sqrt{2(\beta_L - 1)}$ (frequency of small oscillations in the wells). The exponential dependence of the tunnelling frequency on the barrier height (hence on the critical current of the junction—or of the dc-SQUID which replaces it) makes the latter a crucial parameter to be controlled.

5. – Qubit readout

In order to read out the flux state of the qubit, it is natural to use a SQUID, the most sensitive magnetometer existing; it has the advantage of being easily integrable with the qubit, since it can be fabricated with the same technology as the qubit. It is

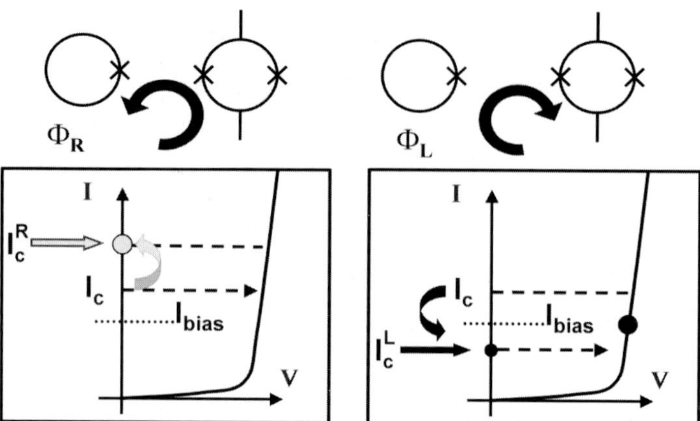

Fig. 7. – Schematic showing the working principle of the threshold detector. The hysteretic dc-SQUID is biased at a current I_{bias} lower than the critical current I_c. Under the action of the additional flux Φ_R, the critical current increases to I_c^R and the working point remains at $V = 0$. On the contrary, the flux state Φ_L makes the critical current decrease at $I_c^L < I_{bias}$, so that the working point switches on the branch with $V \neq 0$.

possible to use a non-hysteretic dc-SQUID or a hysteretic dc-SQUID (used as a threshold detector). The non-hysteretic dc-SQUID (commercially available as magnetometer or current amplifier) has a resistive shunt across the two junctions, whose task is to remove the junction hysteresis and convert the critical current modulation into a flux-modulated voltage output, periodic with period $1\Phi_0$. With a proper control electronics, which linearizes the output voltage, one gets an output signal proportional to the input flux. Since the flux noise of such a dc-SQUID can be as low as $1\mu\Phi_0/\sqrt{\text{Hz}}$, this method provides a very accurate readout of the qubit flux state. However, this device is also intrinsically dissipative, due to the low value of the shunting resistor, and this dissipation leads to decoherence of the qubit. The alternative is to use an underdamped (hysteretic) dc-SQUID as a flux-activated threshold detector to probe one flux eigenstate of the rf-SQUID qubit. Figure 7 shows the working principle. The device is inductively coupled to the rf-SQUID; the bias (both current and flux) must be such that the underdamped SQUID is by itself in the zero-voltage state but it overcomes the threshold when reading one of the two flux states of the qubit, while it remains under the threshold for the other flux state. In this way, only part of the information about the qubit state is retrieved, namely whether the shielding current in the qubit is circulating clockwise or counterclockwise, by observing if the underdamped SQUID has switched to the voltage state or not. In other terms, the underdamped SQUID tells which of the two bottom wells is occupied, without performing a quantitative measure of the flux. A drawback of this scheme is due to the fact that the switching of the underdamped dc-SQUID to its dissipative voltage state is in itself a random process, activated by thermal processes or, at low temperature, by quantum fluctuations; so it must be ascertained that the resulting spread in the switching values is not larger than the difference in switching currents corresponding to the two qubit states.

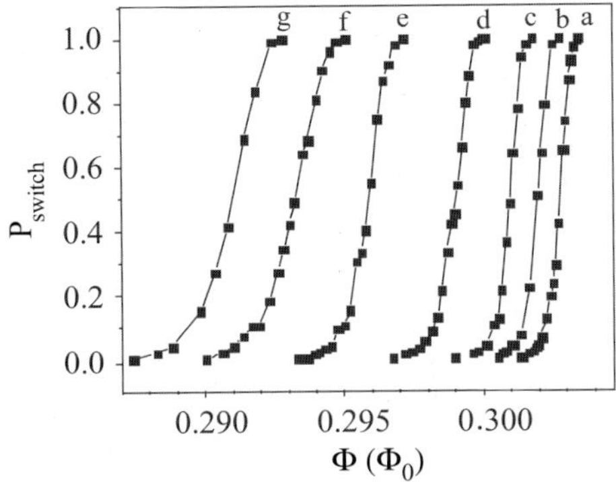

Fig. 8. – Switching probabilities for our underdamped dc-SQUID as a function of the bias magnetic flux, for different temperatures: (a) $T = 18\,\text{mK}$, (b) $T = 73\,\text{mK}$, (c) $T = 150\,\text{mK}$, (d) $T = 283\,\text{mK}$, (e) $T = 373\,\text{mK}$, (f) $T = 535\,\text{mK}$, (g) $T = 626\,\text{mK}$. At low temperature, the spread due to the stochastic nature of the switching process is determined by quantum tunnelling and is about $1m\Phi_0$, a value lower than the additional flux coming from the qubit.

In our experiment, we used two superconducting transformers to couple the double SQUID with both a shunted dc-SQUID and an underdamped dc-SQUID (fig. 6), which can be turned on or off independently: the first one is used in all the preliminary tests, since it provides a complete information on the flux state of the double SQUID and allows a better characterization of the device.

We measured the spread in the switching flux for our underdamped dc-SQUID; the results are shown in fig. 8, where it can be noted how the spread of the switching distribution is reduced by lowering the temperature, until, at low temperature ($T < 300\,\text{mK}$), quantum tunnelling occurs and no further reduction is found. The limit spread (defined as the difference between the points for each curve where the switching probability is 10% and 90%) is $1m\Phi_0$. This spread is smaller than the signal produced by the rf-SQUID in the detector, estimated to be (considering the transforming ratio) a few $m\Phi_0$: so in our setup we should be able to observe the single switching event, not obscured by the stochastic behavior of the readout system.

6. – Logic quantum gates

Qubits must be prepared in a well-defined state, manipulated and coupled each other; the logic operations (gates) on the qubit are represented by unitary manipulation of the quantum state. We now will describe how to perform these operations on the SQUID system.

In order to prepare the initial state of the qubit in one of the two bottom wells (left or right flux state), one can proceed in two ways: it is possible to produce a collapse of the wave function through a flux measurement, or follow the scheme of fig. 9a: the potential shape is altered, so as to unbalance the two wells; the system will go into the

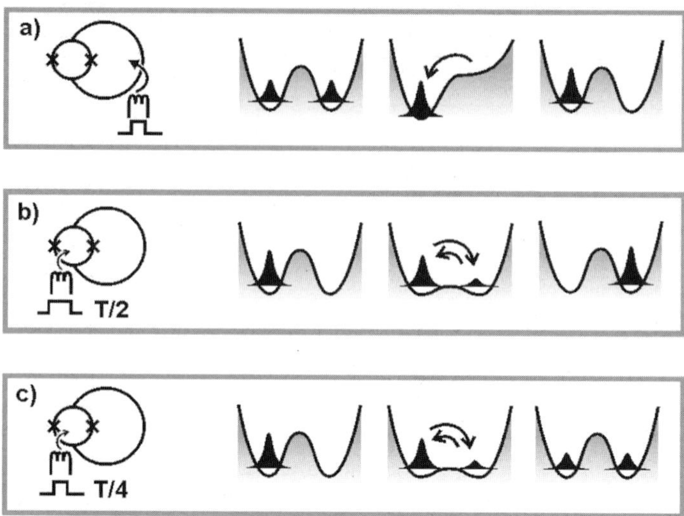

Fig. 9. – (a) State preparation: the potential well is made asymmetric, so that the wave function collapses in the lower well; by adiabatically balancing the potential, this configuration is maintained. (b) Qubit manipulation: from the left state to the right state, by lowering the barrier, letting the system evolve freely and coherently for half a period, then raising the barrier again to freeze the configuration. (c) Qubit manipulation: from a left state to a superposition of states, again lowering the barrier and waiting for a time of one-fourth of period.

new ground state, corresponding to the flux state of the lowest well. At this point, the balancing is nonadiabatically restored.

If the rf-SQUID is replaced by a double SQUID, two parameters are available for controlling the qubit: one is the biasing flux, which mainly modifies the balance of the two potential wells, the other is the flux in the inner dc-SQUID, which raises or lowers the potential barrier between the wells. By increasing the barrier, the tunnelling between wells is exponentially suppressed and the quantum state of the qubit is frozen; on the contrary, a lower barrier allows tunnelling, which will occur with a precise oscillation frequency, determined by the barrier height (eq. (17)). In this framework, the simplest operation on a single qubit, a NOT-like gate that inverts the (unknown) qubit state, is obtained by lowering the barrier to a precise level and letting the state evolve for exactly half a period. The new state is frozen by raising again the barrier. In the same way, for instance, the Hadamard gate that transforms one basis vector into a superposition, is obtained by letting the initial state evolve for one fourth of period. All these operations require a precise knowledge of the tunnelling frequency corresponding to any barrier height.

Building the controlled-NOT (C-NOT) gate is the next task, since it is known [22] that any other gate can be obtained starting from the C-NOT gate together with the single-bit operations (they form a universal set). The C-NOT consists of the conditional negation of a target qubit, controlled by the status of a first qubit. A scheme for a C-NOT

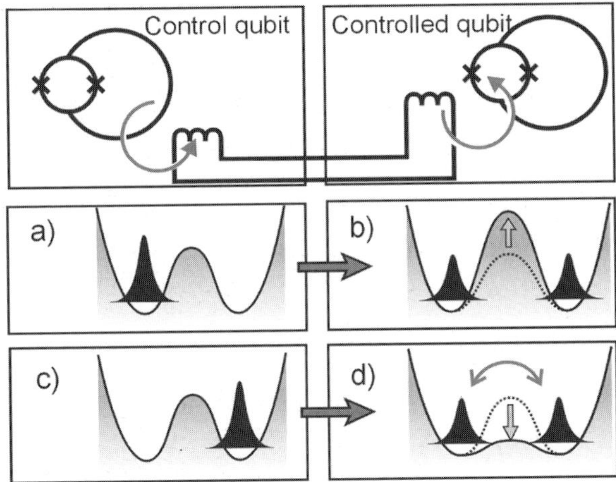

Fig. 10. – Scheme of interaction between two qubits. Through a superconducting transformer, the flux state of the control qubit is coupled to the inner dc-SQUID of the target qubit. One of the states (left, in this example) enhances the target barrier and freezes the state in the controlled qubit; the right state, instead, lowers the barrier so that the target qubit can evolve freely and coherently.

gate is shown in fig. 10. Here, the larger loop of a control qubit is inductively coupled, through a superconducting transformer, to the inner loop of a target qubit. The coupling can be turned on or off by using a suitable switch to break the superconducting circuit. In the example of fig. 10a, b, the left flux state of the control qubit enhances the barrier height of the target qubit: the state of the latter remains frozen. The right state fig. 10c, d, instead, produces a lowering of the barrier in the target qubit, which is then brought into an evolving state, with a known oscillation frequency: after half a period, the natural evolution has exchanged the flux state of the target (NOT operation), whatever its initial state. This coupling scheme can be extended, in principle, to many qubits.

7. – Decoherence

All the above operations rely on the possibility of keeping the qubit in a coherent superposition of different states for a time long enough to allow manipulations. However, if the system is coupled to an external environment, and so it is subjected to dissipation and to random fluctuations, the coherence is progressively lost, and the evolution tends to a classical statistical mixture. This problem has been studied for different situations, in particular for a two-level system coupled to a bath of harmonic oscillators [23,24]. The statistical and coherence properties of a quantum system are described by the density matrix operator, defined as

$$(18) \qquad \hat{\rho}(t) = \sum_{\alpha} p_{\alpha} |\psi_{\alpha}(t)\rangle \langle \psi_{\alpha}(t)|,$$

where $|\psi_\alpha(t)\rangle$ is one of the possible quantum states, and p_α is the probability to have this state.

We now consider the two-level system used to describe the dynamics of a rf-SQUID under symmetric bias. In the absence of coupling with the environment (no decoherence), the density matrix operator in the flux base $|L\rangle$, $|R\rangle$ is given by

$$\hat{\rho}(t) = \frac{1}{2}[1 + \cos(\Omega t + \varphi)]|L\rangle\langle L| + \frac{1}{2}[1 - \cos(\Omega t + \varphi)]|L\rangle\langle L| -$$

$$- \frac{i}{2}\sin(\Omega t + \varphi)|L\rangle\langle R| + \frac{i}{2}\sin(\Omega t + \varphi)|R\rangle\langle L|,$$

where Ω is Rabi's oscillation frequency and φ is a phase that contains the effect of the initial preparation (for example for an initial $|L\rangle$ state it is $\varphi = 0$). The diagonal terms are the probabilities to find the system in a well-defined flux state, while the off-diagonal terms are different from zero for a coherent superpositions, but must be zero for a classical mixture of states.

To describe the coupling with an external ohmic bath at temperature T one can introduce an effective resistance R in parallel to the junction. This simple model is generally sufficient to describe the system considered in this paper. In our case the strength of the coupling to the environment, given by the adimensional Kondo parameter k, is

$$(19) \qquad\qquad k = \frac{\Delta\Phi^2}{2\pi\hbar R},$$

where $\Delta\Phi$ is the distance between the two magnetic flux states. In the presence of a weak coupling to the environment (for $k \ll 1$), the evolution of the density matrix is given by

$$\hat{\rho}(t) = \frac{1}{2}[1 + e^{-\gamma t}\cos(\Omega t + \varphi)]|L\rangle\langle L| + \frac{1}{2}[1 - e^{-\gamma t}\cos(\Omega t + \varphi)]|L\rangle\langle L| -$$

$$- \frac{i}{2}e^{-\gamma t}\sin(\Omega t + \varphi)|L\rangle\langle R| + \frac{i}{2}e^{-\gamma t}\sin(\Omega t + \varphi)|R\rangle\langle L|,$$

where the decoherence rate γ is

$$(20) \qquad\qquad \gamma = \pi k \frac{k_{\mathrm{B}} T}{\hbar}.$$

As expected, the off-diagonal terms tend to zero for $t \to \infty$ (decoherence), while the probabilities tend to $1/2$.

The effective resistance can be estimated by means of measurements of the system relaxation times in quantum regime [6].

In a system of SQUIDs similar to that necessary to realize a qubit, we have measured the escape rate out of one flux state for a rf-SQUID, observing the effect of the quantized energy levels in the potential well [6]; this measurement allowed us to give an estimate for the system dissipation at $35\,\text{mK}$, $R \sim 1\,\text{M}\Omega$, corresponding to a Kondo parameter $k \sim 10^{-4}$ under typical conditions. Probably this value can be reduced by means of more technological efforts. At $T = 10\,\text{mK}$ this value corresponds to a decoherence rate $\gamma \sim 10^6\,\text{s}^{-1}$. Since the typical internal times of a SQUID are $\sim 10^{11}\,\text{s}^{-1}$, one can suppose to perform operations on these quantum devices at a clock $\sim 10^9\,\text{s}^{-1}$. In this case, one has $\sim 10^3$ operation performed before decoherence occurs, that is enough for quantum computations. Work is in progress to make further tests on the system and finally observe the coherent time oscillations, which would represent a starting point for quantum computing with SQUIDs.

$$* * *$$

This work has been supported by INFN under the MQC project. We thank C. COSMELLI and P. CARELLI for reading and discussing the manuscript, and all our colleagues of the MQC experiment.

REFERENCES

[1] LEGGETT A. J., *Prog. Theor. Phys. Suppl.*, **69** (1980) 80.
[2] VOSS R. F. and WEBB R. A., *Phys. Rev. Lett.*, **47** (1981) 265.
[3] MARTINIS J. M., DEVORET M. H. and CLARKE J., *Phys. Rev. B*, **35** (1987) 4682.
[4] ROUSE R., HAN S. and LUKENS J. S., *Phys. Rev. Lett.*, **75** (1995) 514.
[5] SILVESTRINI P., PALMIERI V. G., RUGGIERO B. and RUSSO M., *Phys. Rev. Lett.*, **79** (1997) 3046.
[6] COSMELLI C., CARELLI P., CASTELLANO M. G., CHIARELLO F., DIAMBRINI PALAZZI G., LEONI R. and TORRIOLI G., *Phys. Rev. Lett.*, **82** (1999) 5357.
[7] FRIEDMAN J. R., PATEL V., CHEN W., TOLPYGO S. K. and LUKENS J. E., *Nature*, **406** (2000) 43.
[8] MAKHLIN Y., SCHÖN G. and SHNIRMAN A., *Rev. Mod. Phys.*, **73** (2001) 357.
[9] AVERIN D. V., *Fortschr. Phys.*, **48** (2000) 1055.
[10] DIVINCENZO D., *Fortschr. Phys.*, **48** (2000) 771.
[11] SHNIRMAN A., SCHÖN G. and HERMON Z., *Phys. Rev. Lett.*, **79** (1997) 2371.
[12] NAKAMURA Y., PASHKIN YU. and TSAI J. S., *Nature*, **398** (1999) 786.
[13] CHIARELLO F., *Phys. Lett. A*, **277** (2000) 189.
[14] CARELLI P., CASTELLANO M. G., CHIARELLO F., COSMELLI C., LEONI R. and TORRIOLI G., *IEEE Trans. Appl. Supercond.*, **11** (2001) 210.
[15] TINKHAM M., *Introduction to Superconductivity* (McGraw-Hill, Singapore) 1996.
[16] ORLANDO T. P., MOOIJ J. E., TIAN L., VAN DER WAAL C. H., LEVITOV L. S., LLOYD S. and MAZO J. J., *Phys. Rev. B*, **60** (1999) 15398.
[17] JAKLEVIC R. C., LAMBE J., SILVER A. H. and MERCEREAU J. E., *Phys. Rev. Lett.*, **12** (1964) 159.
[18] JAKLEVIC R. C., LAMBE J., MERCEREAU J. E. and SILVER A. H., *Phys. Rev. A*, **140** (1965) 1628.
[19] CASTELLANO M. G., INTELISANO A., LEONI R., MILANESE N., TORRIOLI G., COSMELLI C., CARELLI P. and CHIARELLO F., *Int. J. Mod. Phys. B*, **14** (2001) 3056.
[20] KETCHEN M. B., *IEEE Trans. Magn.*, **MAG-23** (1987) 1650.

[21] HAN S., LAPOINTE J. and LUKENS J. E., *Phys. Rev. Lett.*, **63** (1989) 1712.
[22] BARENCO A., BENNET C., CLEVE R., DIVINCENZO D., MARGOLUS N., SHOR P., SLEATOR T., SMOLIN J. and WEINFURTER H., *Phys. Rev. A*, **52** (1995) 3457.
[23] CALDEIRA A. O. and LEGGETT A. J., *Phys. Rev. Lett.*, **46** (1981) 211.
[24] WEISS U., *Quantum Dissipative Systems* (World Scientific, Singapore) 1999.

"Stopping" of light and quantum memories for photons

M. FLEISCHHAUER and C. MEWES

Fachbereich Physik, Universität Kaiserslautern
E.-Schrödinger Str. 46, D-67663 Kaiserslautern, Germany

1. – Introduction

Among the many challenges for the implementation of quantum information process-ing (QIP) is the transport of unknown quantum states between separated locations as well as the realization of quantum memories with short access times [1]. Quantum opti-cal systems are very attractive for this, since photons provide ideal carriers of quantum information and nuclear-spin or hyperfine states of atoms are ideal storage systems. Iso-lation from environmental interactions on the one hand and controllable coupling to the photon field on the other can most easily be realized through Raman transitions with a classical laser providing the Stokes field. Combining Raman coupling with adiabatic following provides control over the coherent absorption or emission of photons.

The application of Stimulated Raman adiabatic passage (STIRAP) [2] to single-atom cavity systems has led to a number of important proposals for qubit transfer between atoms and photons as well as for quantum-logic gates [3]. However, due to the small absorption cross-section of an isolated atom, it is necessary to employ strongly coupling resonators. The realization of the strong-coupling regime for optical frequencies remains a technically very challenging task, despite the enormous experimental progress in this field [4]. Furthermore a single-atom system is by construction highly susceptible to the loss of atoms and requires a high degree of control over atomic positions. For these reasons the prospects of cavity-QED systems in qubit transfer and storage may be limited.

On the other hand a photon is absorbed with certainty, if a sufficiently large number of atoms are present. Normally such absorption is accompanied by *dissipation* and only a partial mapping of quantum properties of light to atomic ensembles can be achieved [5]. If dissipation is involved, it is in general not possible to store the quantum state of photons on the level of *individual* quanta (single qubits) in a reversible manner. Rather a quasi-stationary source is required, whose output can be considered as a train of wave packets in identical quantum states. Similar limitations apply to techniques of classical optical-data storage in the time domain based on spin- and photon echo or Raman photon echo [6].

Recently we have proposed a method that combines the enhancement of the absorption cross-section in many-atom systems with dissipation-free adiabatic passage [7]. It is based on Raman adiabatic transfer of the quantum state of photons to *collective atomic excitations* using electromagnetically induced transparency (EIT) [8].

In EIT a strong coherent Stokes field renders the otherwise optically thick medium transparent for a resonant pump. Associated with the transparency is a large linear dispersion, which has been demonstrated to lead to a substantial reduction of the group velocity of light [9]. When a pump pulse propagates in such a medium, its front end gets coherently "absorbed" and the corresponding excitation is transferred to the Stokes field as well as into a spin excitation. At the back end the process is exactly reversed and all excitation returned to the field, *i.e.* there is no net transfer from photons to atoms or vice versa. From the point of view of the atoms slow light in an EIT systems corresponds to a complete adiabatic return. Nevertheless part of the photonic excitation is temporarily stored in the atoms. During this time interval photons and atoms form quasi-particles, called dark-state polaritons [7] that propagate with a velocity determined by the ratio of the electromagnetic to the matter component. Unfortunately EIT systems have only limited capabilities as a temporary memory, since the achievable ratio of storage time to pulse length is rather limited [10].

The limitations of EIT can be overcome, however, when the group velocity is changed *in time* and is adiabatically reduced to zero by dynamically decreasing the strength of the classical Stokes field. The reduction of the Stokes field in time breaks the symmetry of the process and net flow of excitation from photons to atoms or vice versa can be achieved. By adiabatically reducing the group velocity the light pulse can eventually be brought to a full stop. In this process its excitation as well as its quantum state are completely transferred to the atomic spins. The process can be reversed and the light pulse regenerated. Recent experiments [11] have already demonstrated some basic principles of this technique—the dynamic group velocity reduction to a full stop and adiabatic following in dark-state polaritons.

2. – Single-atom cavity QED

To introduce the basic idea of a controlled and reversible quantum state transfer via Raman adiabatic passage, let us first consider the case of an individual 3-level atom coupled to a single quantized mode as pump and a classical field as Stokes field. This is illustrated in fig. 1. Both fields are assumed to be resonant with the corresponding transitions, and decay from the excited state out of the system is taken into account. The

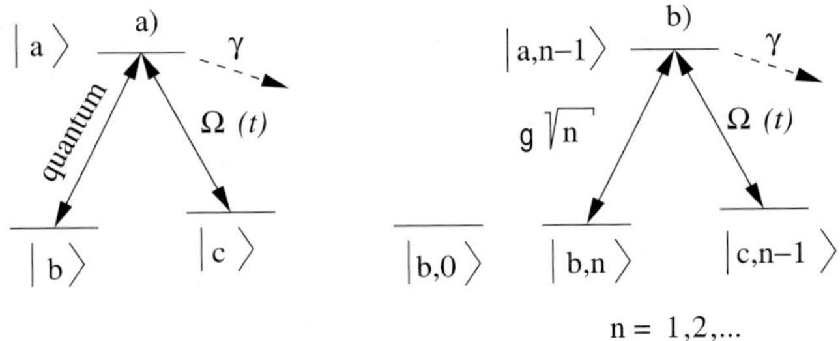

Fig. 1. – a) Three-level atoms coupled to single quantized mode and classical control field of (real) Rabi frequency $\Omega(t)$. b) Coupling of relevant bare eigenstates; g-vacuum Rabi frequency.

(complex) interaction "Hamiltonian" in rotating-wave approximation reads in a rotating frame

$$(1) \qquad H = \hbar g a\, \sigma_{ab} + \hbar \Omega(t) \sigma_{ac} + \text{h.c.} - i\hbar\gamma\sigma_{aa},$$

where $a, a^\dagger$ are the annihilation and creation operators of the quantized pump mode and $\Omega(t)$ is the (in general time-dependent) Rabi frequency of the classical Stokes field. $\sigma_{ij} \equiv |i\rangle\langle j|$ is the atomic spin-flip operator from state j to state i and $g = \wp\sqrt{\omega/\hbar\epsilon_0 V}$ characterizes the vacuum Rabi frequency of the quantized mode, $\wp$ being the dipole moment and V the mode volume. The imaginary part of the Hamiltonian effectively describes spontaneous emission from the excited state with rate γ. The interaction couples only triplets of bare eigenstates of the combined atom-field system, viz. $|b, n + 1\rangle \leftrightarrow |a, n\rangle \leftrightarrow |c, n\rangle$, where $n = 0, 1, \ldots$ denotes the number of photons in the mode. In addition there is the total ground state $|b, 0\rangle$ which is completely decoupled. In the basis of these states the Hamiltonian separates into 3×3 block-matrices of the form

$$(2) \qquad \mathsf{H}_n = \hbar \begin{bmatrix} -i\gamma & g\sqrt{n} & \Omega(t) \\ g\sqrt{n} & 0 & 0 \\ \Omega(t) & 0 & 0 \end{bmatrix}.$$

Each of these matrices has one instantaneous (adiabatic) eigenstate $|\phi_0^{(n)}\rangle$ with eigenvalue $\epsilon_0^{(n)} = 0$ and two eigenstates $|\phi_\pm^{(n)}\rangle$ with eigenvalues $\epsilon_\pm^{(n)} = \pm\sqrt{\Omega^2(t) + g^2 n}$. The zero eigenstates read

$$(3) \qquad \left|\phi_0^{(n)}\right\rangle = \cos\theta_n(t)\, |b, n + 1\rangle - \sin\theta_n(t)\, |c, n\rangle, \quad n = 0, 1, 2, \ldots,$$

where the mixing angles $\theta_n(t)$ are defined as $\tan\theta_n(t) \equiv g\sqrt{n}/\Omega(t)$. The important feature of the zero eigenstates is that they do not contain the excited state and are thus immune to spontaneous emission. For this reason these states are called dark states [12]. Furthermore by changing the strength of the classical Stokes field, $i.e.$ $\Omega(t)$, the mixing angles θ_n can be rotated from $\theta_n = 0$, where $|\phi_0^{(n)}\rangle = |b, n+1\rangle$, to $\theta_n = \pi/2$, where $|\phi_0^{(n)}\rangle = -|c, n\rangle$. If the atom-field system is initially prepared in the dark state and if $\Omega(t)$ changes sufficiently slowly, the state vector will follow the rotation. Thus by stimulated Raman adiabatic passage [2] a controlled and reversible transfer of excitation from the quantized radiation mode to the atom is possible:

$$(4) \qquad |b, n+1\rangle \longleftrightarrow -|c, n\rangle.$$

This mechanism is the basis of several proposals for engineering of quantum states of the radiation field in resonators, for the transfer of quantum states between atoms through a resonator mode and for the transfer of quantum states between different cavities [3].

Adiabatic following requires that the rate of change of the mixing angles should be sufficiently slow. In particular the characteristic time of transfer T should obey the condition

$$(5) \qquad \frac{g^2 n}{\gamma} T \gg 1.$$

The transfer time T is usually limited by the finite decoherence time of the field. In a resonator set-up the lifetime is determined, for example, by mirror losses and coupling to the outside. If κ denotes the characteristic rate of photon loss, the decoherence time of a Fock state $|n\rangle$ is $1/(n\kappa)$. Thus adiabaticity requires

$$(6) \qquad g^2 \gg \kappa \gamma.$$

This condition is referred to as strong-coupling regime. Condition (6) is technically very difficult to satisfy. The physical origin of this strong condition is the small absorption cross-section σ of atoms in the optical frequency domain. In fact as $g^2 \sim \sigma/A \sim \lambda^2/A$, where A is the cross-section of the light field at the position of the atoms, tight focusing is required. As a consequence cavity-QED techniques with individual atoms in the strong-coupling regime are very sensitive to an exact positioning of the atom. Resonator systems have the further disadvantage that communication between them requires a careful timing of the control fields to avoid losses due to impedance mismatch at the cavity interfaces.

3. – Temporary storage of photons in many-atom systems: electromagnetically induced transparency and slow light

An obvious way to overcome the limitations of single-atom systems caused by the small optical cross-section is to use optically thick ensembles of atoms. This also allows to abandon the use of resonators and thus to avoid impedance matching problems. In order to identify appropriate transfer schemes in ensembles of 3-level atoms, let us first discuss pulse propagation in such media.

One of the most important phenomena associated with pulse propagation in these systems is called electromagnetically induced transparency [8]. If, as indicated in fig. 2,

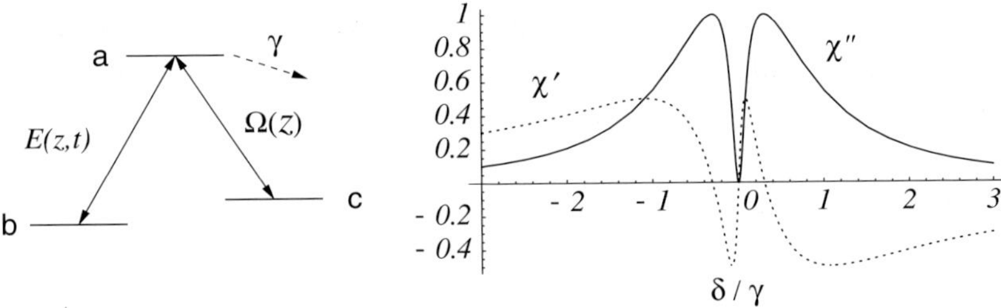

Fig. 2. – Left: 3-level Λ-type medium resonantly coupled to a coherent Stokes field with Rabi frequency Ω and a (quantum) probe field $E(z,t)$. Right: Typical susceptibility spectrum for probe field E as a function of normalized detuning for resonant and constant drive field. The real part χ' describes the refractive index contribution and imaginary part χ'' absorption.

a sufficiently strong coherent field of constant Rabi frequency Ω couples the excited state $|a\rangle$ to a metastable state $|c\rangle$, the absorption of the probe field is exactly cancelled at two-photon resonance. In fig. 2 the imaginary part of the susceptibility

$$(7) \qquad \chi = \eta \, \frac{\gamma\delta}{|\Omega|^2 - \delta^2 - i\gamma\delta} \approx \eta \left[\frac{\gamma\delta}{|\Omega|^2} + i\left(\frac{\gamma\delta}{|\Omega|^2} \right)^2 + \mathcal{O}(\delta^3) \right]$$

is shown as a function of the detuning from resonance. Here $\eta = (3/8\pi^2)\rho\lambda^3$ is a density parameter that gives the number of atoms per cubic wavelength, δ is the probe detuning and γ the excited-state decay rate.

At the same time the real part χ', which is responsible for the refractive index, shows a large normal linear dispersion. Associated with the linear dispersion is a reduction of the group velocity

$$(8) \qquad v_{\mathrm{gr}} = \frac{c}{1 + n_{\mathrm{g}}(z)}, \quad \text{with } n_{\mathrm{g}}(z) = \omega \frac{\mathrm{d}\chi}{\mathrm{d}\omega} = \eta \, \frac{kc\gamma}{|\Omega|^2}.$$

Since the medium is non-absorbing, high densities can be used and rather small group velocities can be achieved. In fact values as low as few meters per second have been observed [9]. In the region of linear dispersion the propagation equation for the slowly varying complex field amplitude reads

$$(9) \qquad \left(\frac{\partial}{\partial t} + v_{\mathrm{gr}}(z) \frac{\partial}{\partial z} \right) E(z,t) = 0,$$

where we have taken into account a possible space dependence of the group velocity. The solution of this equation,

$$(10) \qquad E(z,t) = E\left(0, t - \int_0^z \mathrm{d}z' \, \frac{1}{v_{\mathrm{gr}}(z')} \right),$$

describes slow propagation with an *invariant temporal pulse shape*. Note that the spatial shape is not conserved and depends on the spatial profile of $v_{\mathrm{gr}}(z)$. The slow-down is a lossless linear process and hence all properties of the slowed pulse are conserved. In particular also its quantum state. The slow-down of the group velocity is thus interesting from the point of view of quantum memories as it represents a temporary storage.

If a light pulse propagates in an absorption-free medium with linear dispersion, its total number of photons is reduced by the ratio of the group velocity v_{gr} to the vacuum speed of light $n/n_0 = v_{\mathrm{gr}}/c$. However the time-integrated number of photons crossing a plane perpendicular to the propagation direction is constant. Thus photons must be temporarily stored in the combined system of atoms and control field.

The transfer mechanism between probe field and atomic system can be identified as STIRAP. To see this, let us consider a coherent probe pulse with all atoms initially in the ground state $|b\rangle$. This is a natural assumption since optical pumping will prepare the atoms in this state. At the initial time the ground state $|b\rangle$ is then identical to the dark state

$$(11) \qquad |d(z)\rangle = \cos\vartheta(z,t)\,|b\rangle - \sin\vartheta(z,t)\,|c\rangle \longrightarrow |b\rangle,$$

where the mixing angle ϑ is now given by $\tan\vartheta(z,t) = \Omega_{\mathrm{p}}(z,t)/\Omega$ with $\Omega_{\mathrm{p}}(z,t)$ being the Rabi frequency of the probe pulse. When the front end of the probe pulse arrives at an atom, the mixing angle makes a small excursion from zero. In this process a fraction of the state amplitude rotates from $|b\rangle$ to $|c\rangle$ and energy is taken out of the probe pulse. When the probe pulse reaches its maximum, the excursion of the mixing angle stops and ϑ returns to zero. The state vector of the atoms follows and rotates back to the ground state. Thus the entire energy is put back into the probe pulse at its back end. The slow-down of light in EIT can be understood as a Raman adiabatic return of the atoms. In order for the process to be entirely adiabatic, the condition

$$(12) \qquad \frac{\Omega^2}{\gamma}T_{\mathrm{p}} \gg \sqrt{\eta k L}$$

must be fulfilled, where T_{p} is the pulse duration, L the length of the medium, and it was assumed that $|\Omega| \gg |\Omega_{\mathrm{p}}|$. The presence of the factor $\sqrt{\eta k L}$ instead of unity in (12), where $\alpha = \eta k L$ is the opacity of the medium in the absence of EIT, is due to the fact that propagation effects need to be taken into account [2, 13, 14].

The inequality (12) has a very simple physical meaning. It states that the spectrum of the pulse should be much less than the spectral window of transparency in EIT given by

$$(13) \qquad \Delta\omega_{\mathrm{p}} \ll \Delta\omega_{\mathrm{tr}} = \frac{|\Omega|^2}{\gamma}\frac{1}{\sqrt{\eta k l}}.$$

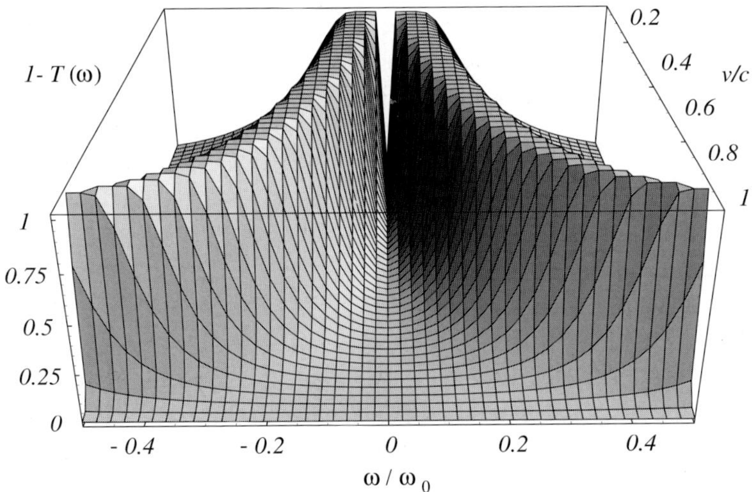

Fig. 3. – Transmission spectrum of EIT in units of $\omega_0 = \eta k c$ as a function of the group velocity. When the transmission spectrum becomes narrower than the (constant) pulse spectrum, strong absorption sets in. Parameters are $\alpha = \eta k L = 20$, $\eta k c / \gamma = 10$.

It is instructive to express this condition in terms of the group velocity. This yields

$$(14) \qquad \Delta\omega_\mathrm{p} \ll \frac{v_\mathrm{gr}}{l} \sqrt{\eta k l}.$$

One immediately recognizes that a vanishing group velocity implies a zero spectral width of transparency. This is illustrated in fig. 3 which shows the absorption of a homogeneously broadened EIT medium as a function of the detuning from resonance and the group velocity.

Since EIT does not change the spectrum of the probe pulse, in this way it is not possible to bring a photon wave packet to a complete stop. This is illustrated in fig. 4, where the propagation of a pulse in an EIT medium with spatially decreasing group velocity is depicted. The left curve shows the effect of the linear dispersive part alone (χ'). Since the front end of the pulse always sees a smaller group velocity than the back end, the wave packet becomes spatially compressed during the deceleration according to $\Delta l(z)/\Delta l(0) = v_\mathrm{gr}(z)/v_\mathrm{gr}(0)$. The right curve shows for comparison a numerical solution of the 1D propagation equations. One clearly recognizes that at a particular position, where the pulse spectral width becomes comparable to the decreasing transparency width, dissipation and partial back-reflection sets in.

Although EIT does not allow to bring a pulse to a complete stop, the question arises whether it can be used as a *temporary* memory. Expressing the transparency width in terms of the pulse delay time $\tau_\mathrm{d} = n_\mathrm{g} L / c$ for the medium yields

$$(15) \qquad \Delta\omega_\mathrm{tr} = \sqrt{\eta k L} \, \frac{1}{\tau_\mathrm{d}}.$$

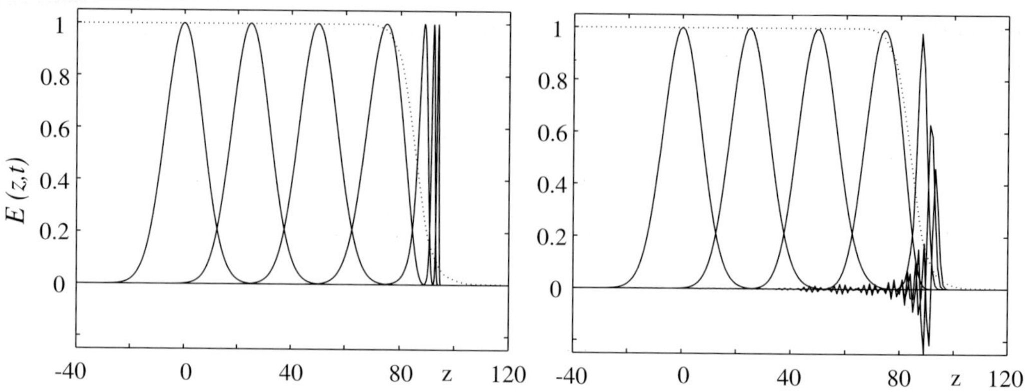

Fig. 4. – Probe pulse as it propagates into a soft "road-block", *i.e.* a region where $v_{\mathrm{gr}}(z)/c \to 0$, for times $t = 0, 25, 50, 75, 100, 125, 150$; left: if only linear dispersion is taken into account, right: full numerical solution of probe pulse propagation. Axes are in arbitrary units with $c = 1$. The dotted line shows v_{gr}/c.

Thus large delay times imply a narrow transparency window, which in turn requires a long pulse time. Hence there is an upper bound for the ratio of achievable delay (storage) time to the initial pulse duration of a photon wave packet

$$(16) \qquad \frac{\tau_{\mathrm{d}}}{\tau_{\mathrm{p}}} \leq \sqrt{\eta k L}.$$

$\tau_{\mathrm{d}}/\tau_{\mathrm{p}}$ is the figure of merit for any memory device. In practice, the achievable opacity $\alpha = \eta k L$ of atomic vapor systems is limited to values below 10^4 resulting in upper bounds for the ratio of time delay to pulse length of the order of 100. Thus EIT media with ultra-small group velocity are only of limited use as temporary storage devices.

The delay-time limitations are a common feature of any EIT-based scheme which does not affect the spectrum of the probe pulse and sets strong limitations, *e.g.*, to freezing of light in moving media [15] and so-called optical black holes based on EIT [16]. For instance a light pulse propagating in an EIT medium with a vortex matter flow, as suggested in [16], will be absorbed or reflected before it reaches the analog of the Schwarzschild radius.

4. – Light stopping by adiabatic rotation of dark-state polaritons

4`1. *Time-dependent group velocity*. – In the last section we have seen that the essential limitation of EIT for a temporary memory and light stopping is the proportionality between spectral transmission width and group velocity. Below a certain value of v_{gr} the transmission window becomes narrower than the pulse spectrum and the pulse is absorbed unless its spectrum narrows as well. Changing the spectrum of a pulse in a linear medium is only possible if the medium properties change in time. Let us thus consider a hypothetical medium with a time-dependent group velocity. The corresponding wave equation reads

$$(17) \qquad \left(\frac{\partial}{\partial t} + v_{\mathrm{gr}}(t)\frac{\partial}{\partial z}\right) E(z,t) = 0.$$

It has the solution

$$(18) \qquad E(z,t) = E\left(z - \int_0^t \mathrm{d}\tau v_{\mathrm{gr}}(\tau), 0\right),$$

which in contrast to the case of $v_{\mathrm{gr}} = v_{\mathrm{gr}}(z)$ describes a *spatially invariant* propagation. At the same time the temporal profile is however not conserved and the spectrum of the probe field changes during propagation. Assuming that v_{gr} changes only slowly compared to the field amplitude, one finds that the spectral width narrows (broadens) according to

$$(19) \qquad \Delta\omega_{\mathrm{p}}(t) \approx \Delta\omega_{\mathrm{p}}(0)\,\frac{v_{\mathrm{gr}}(t)}{v_{\mathrm{gr}}(0)}.$$

In this way the bandwidth limitations of EIT can be overcome. The spectrum of the probe pulse narrows in the same way as the transparency width when the group velocity is reduced. It is thus worthwhile to consider the propagation of a probe pulse in an EIT medium with an explicitly time-dependent control field.

$4^{\cdot}2.$ *Polariton picture of EIT and "stopping" of light.* – In order to describe the propagation of a weak probe pulse in an explicitly time-dependent EIT medium, let us consider a one-dimensional model. Two fields, a strong and undepleted Stokes field of Rabi frequency $\Omega(t)$ and a weak probe pulse with dimensionless field amplitude $\mathcal{E}(z,t)$, couple resonantly the transitions in a 3-level medium as shown in fig. 2. The propagation equation can be written in the form

$$(20) \qquad \left(\frac{\partial}{\partial t} + c\frac{\partial}{\partial z}\right)\mathcal{E} = igN\,\rho_{ab}(z,t),$$

where g is the vacuum Rabi frequency in an interaction volume V, with $g^2 N = \eta kc\gamma$, N being the number of atoms in V. $\rho_{ab}(z;t)$ is the density matrix element of the atoms between the excited and ground states at position z. Without the probe pulse, the atoms are assumed to be in state $|b\rangle$, *e.g.* as a result of optical pumping by the Stokes field, *i.e.* in zeroth-order of the probe field one has $\rho_{bb} = 1$ and all other elements vanish. In first order of $\mathcal{E}$ the following relevant density-matrix equations are obtained:

$$(21) \qquad \dot{\rho}_{ab} = -\gamma\rho_{ab} + ig\mathcal{E} + i\Omega\rho_{cb},$$

$$(22) \qquad \dot{\rho}_{cb} = i\Omega\,\rho_{ab}.$$

It is now convenient to introduce two new dimensionless field variables corresponding to quasi-particles, called dark- and bright-state polaritons [17]:

$$(23) \qquad \Psi(z,t) = \cos\theta(t)\,\mathcal{E}(z,t) - \sin\theta(t)\,\sqrt{N}\,\rho_{cb}(z;t),$$

$$(24) \qquad \Phi(z,t) = \sin\theta(t)\,\mathcal{E}(z,t) + \cos\theta(t)\,\sqrt{N}\,\rho_{cb}(z;t),$$

where $\tan^2\theta = g^2 N/\Omega^2 = n_{\mathrm{g}}$. The group velocity of slow-light propagation is thus given by $v_{\mathrm{gr}} = c\cos^2\theta$. One can transform the equations of motion for the electric field and the atomic variables into the new variables. This yields

$$(25) \qquad \left[\frac{\partial}{\partial t} + c\cos^2\theta\frac{\partial}{\partial z}\right]\Psi = -\dot\theta\Phi - \sin\theta\cos\theta\,c\frac{\partial}{\partial z}\Phi,$$

and

$$(26) \qquad \Phi = \frac{\sin\theta}{g^2 N}\left(\frac{\partial}{\partial t} + \gamma\right)\left(\tan\theta\frac{\partial}{\partial t}\right)(\sin\theta\,\Psi - \cos\theta\,\Phi).$$

Introducing an adiabaticity parameter $\varepsilon \equiv (g\sqrt{N}T)^{-1}$ with T being a characteristic time, one can expand the equations of motion in a power series in ε. In lowest order, *i.e.* in the adiabatic limit, one finds $\Phi \approx 0$ and thus the electric-field amplitude as well as the spin coherence are entirely determined by the amplitude of the dark-state polariton

$$(27) \qquad \mathcal{E} = \cos\theta\,\Psi, \qquad \text{and} \qquad \sqrt{N}\rho_{cb} = -\sin\theta\,\Psi.$$

Furthermore all terms on the r.h.s. of eq. (25) vanish and one finds

$$(28) \qquad \left[\frac{\partial}{\partial t} + c\cos^2\theta(t)\frac{\partial}{\partial z}\right]\Psi = 0.$$

4'3. *"Stopping" of light.* – Equation (28) describes a shape-preserving propagation with instantaneous velocity $v = v_{\mathrm{gr}}(t) = c\cos^2\theta(t)$:

$$(29) \qquad \Psi(z,t) = \Psi\left(z - c\int_0^t \mathrm{d}\tau\cos^2\theta(\tau), 0\right).$$

For $\theta \to 0$, *i.e.* for a strong external drive field $\Omega^2 \gg g^2 N$, the polariton has purely photonic character $\Psi = \mathcal{E}$ and the propagation velocity is that of the vacuum speed of light. In the opposite limit of a weak drive field $\Omega^2 \ll g^2 N$ such that $\theta \to \pi/2$, the polariton becomes spin-wave–like, $\Psi = -\sqrt{N}\rho_{cb}$, and its propagation velocity approaches zero. Thus the following mapping can be realized:

$$(30) \qquad \mathcal{E}(z) \longleftrightarrow \rho_{cb}(z')$$

with $z' = z + z_0 = z + \int_0^\infty \mathrm{d}\tau c\cos^2\theta(\tau)$.

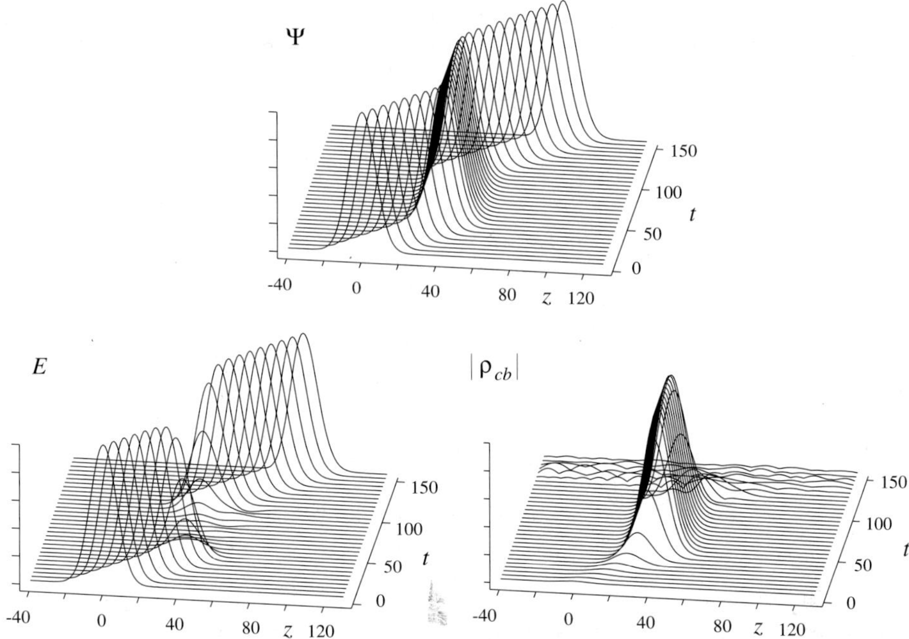

Fig. 5. – Numerical simulation of dark-state polariton propagation with envelope $\exp[-(z/10)^2]$. The mixing angle is rotated from 0 to $\pi/2$ and back according to $\cot\theta(t) = 100(1-0.5\tanh[0.1(t-15)] + 0.5\tanh[0.1(t-125)])$. Top: polariton amplitude; bottom-left: electric-field amplitude; bottom-right: atomic spin coherence; all in arbitrary units. Axes are in arbitrary units with $c = 1$.

Equation (30) is the essence of the transfer technique of quantum states from photon wave packets propagating at the speed of light to stationary atomic excitations (stationary spin waves). Adiabatically rotating the mixing angle from $\theta = 0$ to $\theta = \pi/2$ decelerates the polariton to a full stop, changing its character from purely electromagnetic to purely atomic. Due to the linearity of eq. (28), the quantum state of the polariton is not changed during this process.

Likewise the polariton can be re-accelerated to the vacuum speed of light; in this process the stored quantum state is transferred back to the field. This is illustrated in fig. 5, where we have shown the coherent amplitude of a dark-state polariton which results from an initial light pulse as well as the corresponding electromagnetic and matter components obtained from a numerical solution of the 1D Maxwell-Bloch equations.

As noted above, the spectrum of the polariton and thus of the probe pulse narrows in the same way as the EIT transmission window when the group velocity is reduced as a function of time. The adiabatic slow-down of a dark-state polariton thus avoids the bandwidth limitation of EIT and a light pulse can indeed be brought to a full stop. This is the essential difference of the present scheme from the light-freezing proposal of ref. [15]. In the adiabatic transfer process all information carried by the pulse, *i.e.* also

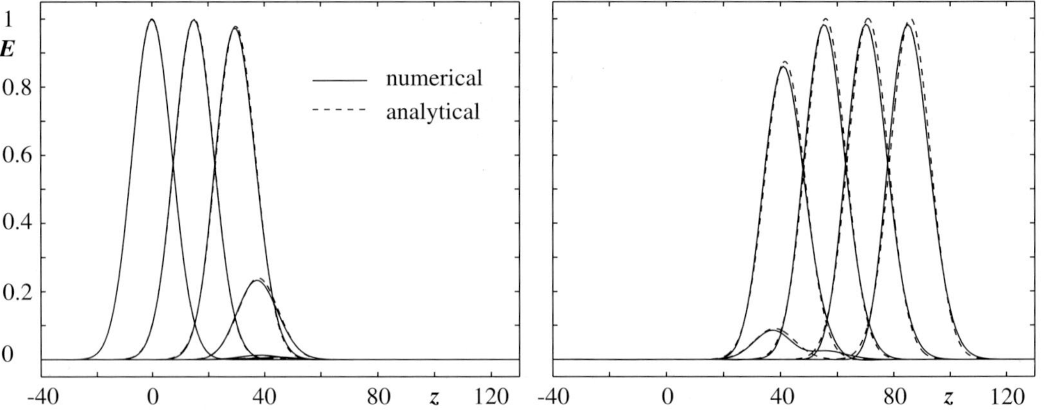

Fig. 6. – Comparison of numerical (full line) and approximate analytical results (dashed line) following from eq. (29) for the example of fig. 5. Left: electric-field amplitude in arbitrary units for $t = 0, 15, 30, 45, 60$. Right: the same for $t = 90, 105, 120, 135, 150$.

its quantum state, is transferred to a collective Raman or spin excitation of atoms.

In order to prove the validity of the simple analytic solution (29), we have compared in fig. 6 exact numerical and analytical results for the situation of fig. 5. As can be recognized there is an excellent agreement.

4'4. *Adiabatic condition and non-adiabatic corrections.* – The analysis of the conditions for Raman adiabatic passage in single-atom cavity QED in sect. **2** led to the experimentally challenging strong-coupling requirement. Let us now investigate the corresponding requirements for the present scheme. To this end, we take into account the first-order correction to the adiabatic solution (29). Making use of the zeroth-order relation $\dot{\Psi} = -c\cos^2\theta\,\partial\Psi/\partial z$ to eliminate time derivatives in favor of spatial derivatives we find

$$(31) \qquad \left[\frac{\partial}{\partial t} + c\cos^2\theta\,\frac{\partial}{\partial z}\right]\Psi = -A(t)\Psi + B(t)c\frac{\partial}{\partial z}\Psi + C(t)c^2\frac{\partial^2}{\partial z^2}\Psi - D(t)c^3\frac{\partial^3}{\partial z^3}\Psi,$$

where

$$(32) \qquad A(t) = \left(\gamma + \frac{1}{2}\frac{\partial}{\partial t}\right)\left(\frac{\dot{\theta}^2\sin^2\theta}{g^2 N}\right), \qquad B(t) = \frac{\sin\theta}{3g^2 N}\frac{\partial^2}{\partial t^2}\sin^3\theta,$$

$$(33) \qquad C(t) = \left(\gamma + \frac{1}{2}\frac{\partial}{\partial t}\right)\frac{\sin^4\theta\cos^2\theta}{g^2 N}, \qquad D(t) = \frac{\sin^4\theta\cos^4\theta}{g^2 N}.$$

$A(t)$ describes homogeneous losses due to non-adiabatic transitions followed by spontaneous emission. $B(t)$ gives rise to a correction of the polariton propagation velocity, $C(t)$ results in a pulse spreading by dissipation of high-spatial-frequency components and

$D(t)$ leads to a deformation of the polariton. Since all coefficients depend only on time, eq. (31) can be solved by spatial Fourier transform and simple integration in time. In order to neglect dissipation, the terms containing $A(t)$ and $C(t)$ must be small, which results in the two conditions

$$(34) \qquad \frac{\gamma c^2}{L_{\mathrm{p}}^2} \int_0^\infty dt\, \frac{\cos^2 \theta(t)}{g^2 N} = \frac{\gamma c L}{g^2 N L_{\mathrm{p}}^2} \ll 1,$$

and

$$(35) \qquad \gamma \int_0^\infty dt\, \frac{\dot\theta^2 \sin^2 \theta}{g^2 N} = \gamma \int_0^\infty dt\, \frac{\dot\theta^2}{g^2 N + \Omega^2(t)} \ll 1.$$

To simplify the first expression we have used here the approximation $\sin \theta \approx 1$. L_{p} denotes the pulse length in the medium at the initial time.

The first condition should be compared and contrasted to the strong-coupling condition of sect. **2**. Using $T_{\mathrm{p}} = L_{\mathrm{p}}/c$ one finds

$$(36) \qquad \frac{g^2 N'}{\gamma} T_{\mathrm{p}} \gg 1 \longleftrightarrow \frac{g^2}{\gamma} T_{\mathrm{p}} \gg 1,$$

where $N' \equiv NL/L_{\mathrm{p}}$ is the number of atoms in a volume of length L_{p} (rather than L). The large factor N' is a signature of the collective interaction and strongly alleviates the requirements in the many-atom system as compared to the single-atom one. Condition (34) has again a simple physical meaning. It states that the *initial* pulse spectral width has to be much less than the *initial* transparency width, *i.e.* at the time when the pulse enters the medium:

$$(37) \qquad \Delta\omega_{\mathrm{p}}(0) \ll \Delta\omega_{\mathrm{tr}}(0).$$

This is easily satisfied, if the probe pulse is not too short. Practically it does prevent, however, light storage of pulses shorter than a few nanoseconds.

The second condition is well known from adiabatic passage [13, 18] and sets a limit to the rotation velocity $\dot\theta$ of the mixing angle and hence to the deceleration/acceleration of the polariton. In order to stay in the adiabatic regime at all times the characteristic time scale T has to fulfill the condition

$$(38) \qquad T \gg \frac{l_{abs}}{c} \frac{v_{\mathrm{gr}}^0}{c},$$

where $l_{abs} = c\gamma/g^2 N$ is the absorption length in the absence of EIT. We note that for realistic experimental parameters the quantity on the right-hand side is extremely small on all relevant time scales. It should be noted that a sudden and sharp switch-off of the drive field leads to a complete loss of the entire electromagnetic component of the polariton. This is illustrated by numerical simulations in fig. 7. The left figure shows

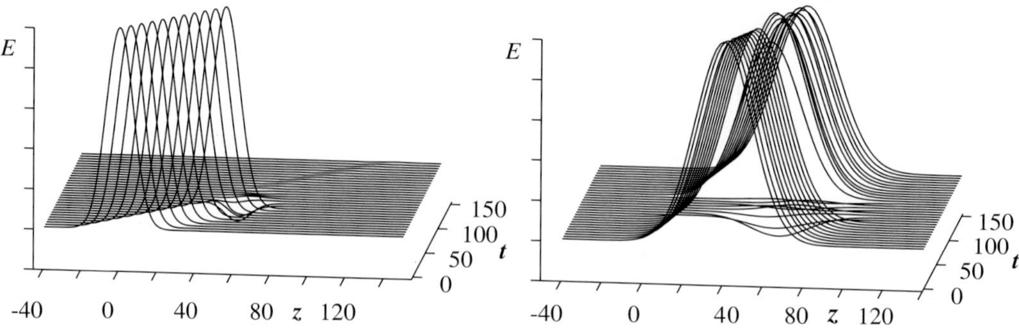

Fig. 7. – Propagation of light pulse in EIT medium with sudden switching of $\cos\theta$. Left: $\cos\theta$ is switched from 1 to zero at $t = 50$ and back at $t = 90$, right: same switch, but with initial and final values of $\cos^2\theta(\pm\infty) = 0.1$, only the electromagnetic component of the polariton (here 10%) gets lost; $\gamma = 1$.

the propagation of a light pulse in an EIT medium, where the cosine of the mixing angle is switched from 1 to 0 at $t = 50$ and back to 1 at $t = 90$. Apart from some small and rapidly decaying Rabi oscillations, the light field is immediately absorbed and cannot be regenerated. On the other hand the losses are small if the cosine of the mixing angle is initially significantly less than unity. This is shown in the right picture, where $v_{\mathrm{gr}}/c = \cos^2\theta(\pm\infty) = 0.1$ and hence the output amplitude is only reduced by 10%.

5. – Quantum memory and decoherence

An important question when discussing the application of the many-atom system to a quantum memory is the sensitivity to errors. Thus in the following the influence of decoherence processes on the fidelity of the state storage will be discussed.

The most essential properties of a quantum memory based on collective excitations can be understood looking at the case of a single mode of the radiation field as realized, *e.g.*, in a single-mode optical cavity. Consider a collection of N 3-level atoms as shown in fig. 1. The dynamics of this system is described by the complex Hamiltonian ($E_{\mathrm{b}} = \hbar\omega_{\mathrm{b}} = 0$)

$$(39) \qquad H = \hbar\omega a^\dagger a + \hbar(\omega_a - i\gamma)\sum_{j=1}^{N}\sigma_{aa}^j + \hbar\omega_c\sum_{j=1}^{N}\sigma_{cc}^j +$$

$$+ \hbar g\sum_{i=1}^{N} a\sigma_{ab}^i + \hbar\Omega(t)\mathrm{e}^{-i\nu t}\sum_{i=1}^{N}\sigma_{ac}^i + \mathrm{h.c.}$$

Here $\sigma_{\mu\nu}^i = |\mu\rangle_{ii}\langle\nu|$ is the flip operator of the i-th atom and the vacuum Rabi frequency is assumed to be equal for all atoms. We have also introduced an imaginary part to the Hamiltonian to take into account losses from the excited state.

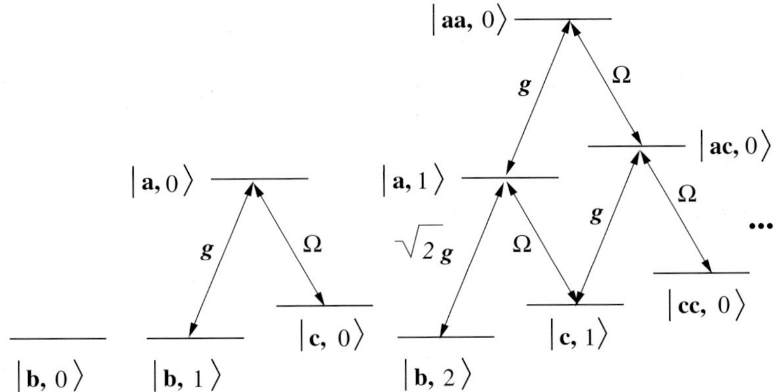

Fig. 8. – Coupling of bare eigenstates of the atom + cavity system for at most two photons.

When all atoms are prepared initially in level $|b\rangle$ the only states coupled by the interaction are the totally symmetric Dicke states [19]

$$\text{(40)} \qquad |\mathbf{b}\rangle_N = |b_1, b_2, \ldots, b_N\rangle,$$

$$\text{(41)} \qquad |\mathbf{a}\rangle_N = \frac{1}{\sqrt{N}} \sum_{j=1}^{N} |b_1, \ldots, a_j, \ldots, b_N\rangle,$$

$$\text{(42)} \qquad |\mathbf{c}\rangle_N = \frac{1}{\sqrt{N}} \sum_{j=1}^{N} |b_1, \ldots, c_j, \ldots, b_N\rangle,$$

$$\text{(43)} \qquad |\mathbf{aa}\rangle_N = \frac{1}{\sqrt{2N(N-1)}} \sum_{i \neq j = 1}^{N} |b_1, \ldots, a_i, \ldots, a_j, \ldots, b_N\rangle, \quad \text{etc.}$$

The couplings within the subsystems corresponding to a single and a double excitation are shown in fig. 8.

The interaction of the N-atom system with the quantized radiation mode has now a family of dark states

$$\text{(44)} \qquad |D, n\rangle_N = \sum_{k=0}^{n} \sqrt{\frac{n!}{k!(n-k)!}} (-\sin\theta)^k (\cos\theta)^{n-k} |\mathbf{c}^k, n-k\rangle_N,$$

$$\tan\theta(t) \equiv \frac{g\sqrt{N}}{\Omega(t)}.$$

It should be noted that although the dark states $|D, n\rangle_N$ are degenerate they belong to exactly decoupled subsystems. This means there is no transition between them even if non-adiabatic corrections are taken into account. Adiabatically rotating the mixing angle θ from 0 to $\pi/2$ leads to a complete and reversible transfer of all photonic states to a collective atomic excitation if the maximum number of photons n is less than the number of atoms N. If the initial quantum state of the single-mode light field is in a mixed state described by the density matrix $\hat{\rho}_f = \sum_{n,m} \rho_{nm} |n\rangle\langle m|$, the transfer process generates a quantum state of collective excitations according to

$$(45) \qquad \sum_{n,m} \rho_{nm} |n\rangle\langle m| \otimes |\mathbf{b}\rangle_N \,_N\langle\mathbf{b}| \longleftrightarrow |0\rangle\langle 0| \otimes \sum_{n,m} \rho_{nm} |\mathbf{c}^n\rangle_N \,_N\langle\mathbf{c}^m|.$$

Thus the storage state corresponding to a photon Fock state $|n\rangle$ is an entangled many-particle state $|\mathbf{c}^n\rangle_N$. These states are known to be highly susceptible to decoherence. For example, if anyone of the N atoms undergoes a spin flip, corresponding to $|b\rangle_j \leftrightarrow |c\rangle_j$, an orthogonal state is created. When the probability of a single-atom error is p, the total probability that an orthogonal state is created is $P = 1 - (1 - p)^N \sim Np$. Thus one might conclude that collective quantum memories display a substantially enhanced rate of decoherence, which is a substantial drawback that outweights the advantage of ensembles for the coupling of light and matter. Although the following discussion is far from complete, some reasons why this is indeed not the case will be presented now [20].

For this the system Hamiltonian will first be brought into a more appropriate form. Separating the oscillatory factor $e^{-i\nu t}$ by a canonical transformation, assuming two-photon resonance, *i.e.* $\omega = \omega_c + \nu$, and adiabatically eliminating the excited state $|a\rangle$ yields

$$(46) \qquad H = \hbar\omega\left(\Psi^\dagger\Psi + \sum_l \Phi_l^\dagger\Phi_l\right) - i\hbar\frac{\Omega^2(t)}{\gamma}\sum_l \Phi_l^\dagger\Phi_l.$$

Here we have assumed adiabatic conditions and introduced the dark-state polariton operator

$$(47) \qquad \Psi = \cos\theta(t)\, a - \sin\theta(t)\, \frac{1}{\sqrt{N}}\sum_{j=1}^N \sigma_{bc}^j,$$

as well as the N bright-state polariton operators

$$(48) \qquad \Phi_0 = \sin\theta(t)\, a + \cos\theta(t)\, \frac{1}{\sqrt{N}}\sum_{j=1}^N \sigma_{bc}^j,$$

$$(49) \qquad \Phi_l = \frac{1}{\sqrt{N}}\sum_{j=1}^N \sigma_{bc}^j \exp\left[2\pi i\frac{lj}{N}\right] \quad \text{for } l \in \{1, 2, \ldots, N - 1\}.$$

Expression (46) shows that under adiabatic conditions all excitations of the dark-state polariton $(\Psi^\dagger)^n |\mathbf{b}, 0\rangle_N$ are conserved, while all bright-polariton excitations decay via optical pumping into the excited state and subsequent spontaneous emission.

It can also be seen from eqs. (46)-(49) that in the read-out process of the memory, where $\theta : \pi/2 \to 0$, only dark-polariton excitations will be transferred to field excitations. Thus from the point of view of photon storage all collective states with the same number of dark-polariton excitations are equivalent. For example the symmetric dark state

$$(50) \qquad |D, n\rangle_N = \frac{1}{\sqrt{n!}} \left(\Psi^\dagger\right)^n |\mathbf{b}, 0\rangle_N,$$

which is a storage state of a radiation-mode Fock state with n photons, is equivalent to all states with arbitrary number of bright-polariton excitations as long as there are exact n excitations of Ψ:

$$(51) \qquad |D, n\rangle_N \hat{=} \left(\Phi_1^{(\dagger)}\right)^k \ldots \left(\Phi_2^{(\dagger)}\right)^l \left(\Psi^\dagger\right)^n |\mathbf{b}, 0\rangle_N.$$

The existence of these *equivalence classes* of storage states has an important consequence. All transitions within an equivalence class of states caused by environmental interactions leave the fidelity of the quantum memory unaffected. This leads to an exact compensation of the enhanced probability of a single-atom error to occur.

To illustrate this, let us consider the case of a Fock state of the radiation field $|n\rangle$ stored in the symmetric dark state $|D, n\rangle_N$. A random spin flip of atom j from state $|b\rangle$ to $|c\rangle$ can be described by the operation $|D, 1\rangle_N \to \sigma_{cb}^j |D, 1\rangle_N$. Assuming $\theta = \pi/2$ and making use of

$$(52) \qquad \sigma_{cb}^j = \frac{1}{\sqrt{N}} \left(\sum_{l=1}^{N-1} \exp\left[-2\pi i \frac{lj}{N} \right] \Phi_l^\dagger + \Psi^\dagger \right),$$

one finds

$$(53) \qquad |D, n\rangle \longrightarrow \sum_{j=l}^{N-1} \frac{\exp\left[-2\pi i (lj/N) \right]}{\sqrt{N}} \Phi_l^\dagger |D, n\rangle + \frac{1}{\sqrt{N}} \Psi^\dagger |D, n\rangle.$$

Only the last component belongs to a state that is not in the same equivalence class as $|D, n\rangle_N$. Thus the probability to leave the equivalence class is only $1/N$. This factor exactly compensates the factor N of the total probability of a spin flip error to occur in any atom.

One can show that similar arguments hold for all possible non-cooperative decoherence mechanisms. Hence the sensitivity of the collective quantum memory does not depend on the number of particles involved. The sensitivity of many-particle quantum memories to decoherence is not enhanced as compared to the single-atom counterparts, given that there are no cooperative decay processes.

6. – Summary

A technique for a controlled and coherent transfer of the quantum state of photons to and from collective spin excitations of ensemble of atoms has been discussed. The underlying mechanism is an adiabatic rotation of a dark-state polariton between a purely photonic and a purely atomic excitation. Due to the collectively enhanced interaction cross-section, the strong-coupling requirement of single-atom cavity QED is eliminated. The mapping technique, although superficially similar looking, is quite different from conventional methods of optical data storage based on optical pumping. In the present set-up no spontaneous emission is present, making a reversible storage on the level on individual light quanta possible. It was shown that, despite the entangled character of the storage states, the collective quantum memory is not more sensitive to decoherence than a single-particle system. This is due to the existence of equivalence classes of many-particle excitations representing the same stored quantum state of light.

While many-particle excitations may have potential advantages over single-atom ones with respect to storage purposes, processing of the information in collective and thus delocalized qubits is more difficult. Recently a method was suggested, however, to implement quantum logic operations between collective excitations [21]. The underlying physical mechanism is here a blockade effect caused by resonant dipole-dipole interactions in Rydberg states which is not sensitive to the actual separation between two atoms.

$$* \quad * \quad *$$

The authors would like to thank M. D. Lukin, L. M. Duang, J. I. Cirac and P. Zoller for stimulating discussions. The financial support of the Deutsche Forschungsgemeinschaft under contract no. FL210/10 is gratefully acknowledged.

REFERENCES

[1] DiVincenzo D. P., *Fortschr. Phys.*, **48** (2000) 771.

[2] Bergmann K., Theuer H. and Shore B. W., *Rev. Mod. Phys.*, **70** (1998) 1003; Vitanov N. V., Fleischhauer M., Shore B. W. and Bergmann K., *Adv. Atom. Mol. Opt. Phys.*, Vol. **46** (Academic Press) 2001, pp. 57-190.

[3] Parkins A. S., Marte P., Zoller P. and Kimble H. J., *Phys. Rev. Lett.*, **71** (1993) 3095; Pelizzari T., Gardiner S. A., Cirac J. I. and Zoller P., *Phys. Rev. Lett.*, **75** (1995) 3788; Cirac J. I., Zoller P., Mabuchi H. and Kimble H. J., *Phys. Rev. Lett.*, **78** (1997) 3221.

[4] Kimble H. J., *Phys. Scr.*, **76** (1998) 127.

[5] Kuzmich A., Mølmer K. and Polzik E. S., *Phys. Rev. Lett.*, **79** (1997) 4782; Hald J., Sørensen J. L., Schori C. and Polzik E. S., *Phys. Rev. Lett.*, **83** (1999) 1319; Kozhekin A. E., Mølmer K. and Polzik E. S., *Phys. Rev. A*, **62** (2000) 033809.

[6] Hahn E. L., *Phys. Rev.*, **80** (1950) 580; Abella J. D., Kurnit N. A. and Hartmann S. R., *Phys. Rev.*, **141** (1966) 391; Hemmer P. R., Cheng K. Z., Kierstead J., Shariar M. S. and Kim M. K., *Opt. Lett.*, **19** (1994) 296; Ham B. S., Shariar M. S., Kim M. K. and Hemmer P. R., *Opt. Lett.*, **22** (1997) 1849.

[7] Fleischhauer M. and Lukin M. D., *Phys. Rev. Lett.*, **84** (2000) 5094;

[8] Harris S. E., *Phys. Today*, **50** (1997) 36.

[9] Hau L. V., Harris S. E., Dutton Z. and Behroozi C. H., *Nature*, **397** (1999) 594; Kash M. *et al.*, *Phys. Rev. Lett.*, **82** (1999) 5229; Budker D. *et al.*, *Phys. Rev. Lett.*, **83** (1999) 1767.

[10] Harris S. E. and Hau L. V., *Phys. Rev. Lett.*, **82** (1999) 4611.

[11] Liu C., Dutton Z., Behroozi C. H. and Hau L. V., *Nature*, **409** (2001) 490; Phillips D. F., Fleischhauer A., Mair A., Walsworth R. L. and Lukin M. D., *Phys. Rev. Lett.*, **86** (2001) 783.

[12] See, *e.g.*, Arimondo E., *Prog. Opt.*, **35** (1996) 259.

[13] Fleischhauer M. and Manka A. S., *Phys. Rev. A*, **54** (1996) 794.

[14] Lukin M. D., Fleischhauer M., Zibrov A. S., Robinson H. G., Velichansky V. L., Hollberg L. and Scully M. O., *Phys. Rev. Lett.*, **79** (1997) 2959.

[15] Kocharovskaya O., Rostovtsev Y. and Scully M. O., *Phys. Rev. Lett.*, **86** (2001) 628.

[16] Leonhardt U. and Piwnicki P., *Phys. Rev. Lett.*, **84** (2000) 822.

[17] Fleischhauer M. and Lukin M. D., preprint quant-ph/0106066.

[18] Vitanov N. V. and Stenholm S., *Opt. Commun.*, **127** (1996) 215; Vitanov N. V. and Stenholm S., *Phys. Rev. A*, **56** (1997) 1463.

[19] Dicke R. H., *Phys. Rev.*, **93** (1954) 99.

[20] For a more detailed discussion see: Fleischhauer M., Lukin M. D. and Mewes C., to be published.

[21] Lukin M. D., Fleischhauer M., Cote R., Duan L. M., Jaksch D., Cirac J. I. and Zoller P., *Phys. Rev. Lett.*, **87** (2001) 037901.

International School of Physics "Enrico Fermi"

Villa Monastero, Varenna

Course CXLVIII

17 – 27 July 2001

"Experimental Quantum Computation and Information"

Directors

Francesco DE MARTINI
Dipartimento di Fisica
Università "La Sapienza"
P.le Aldo Moro 2
00185 ROMA
Italy
Tel.: ++39 06 49913518
Fax: ++39 06 4454778
demartini@caspur.it

Christopher MONROE
Department of Physics
University of Michigan
2477 Randall Laboratory
ANN ARBOR, MI 48109-1120
USA
Tel.: ++1 734 615 9625
Fax: ++1 734 764 5153
crmonroe@umich.edu

Scientific Secretary

Paolo MATALONI
Dipartimento di Fisica
Università "La Sapienza"
P.le Aldo Moro 2
00185 ROMA
Italy
Tel.: ++39 06 49913478
Fax: ++39 06 4463158
p.mataloni@caspur.it

Lecturers

Phillip BUCKSBAUM
Department of Physics
University of Michigan
2477 Randall Laboratory
ANN ARBOR, MI 48109-1120
USA
Tel.: ++1 734 7644348
Fax: ++1 734 7645153
phb@umich.edu

Eric CORNELL
JILA
University of Colorado
440 UCB
BOULDER, CO 80309-0440
USA
Tel.: ++1 303 4926281
Fax: ++1 303 4925235
ecornell@jilau1.Colorado.edu

David CORY
M.I.T.
Department of Nuclear Engineering
77 Massachusetts Ave.
CAMBRIDGE, MA 02139
USA
Tel.: ++1 617 2533806
Fax: ++1 617 2535405
dcory@mit.edu

Michel DEVORET
Groupe Quantronique SPEC
CEA/Saclay
91191 GIF-SUR-YVETTE Cedex
France
Tel.: ++33 1 69087341
Fax: ++33 1 69087342
devoret@drecam.saclay.cea.fr

David DI VINCENZO
IBM J. Watson Research Center
P.O. Box 218
YORKTOWN HEIGHTS, NY 10598
USA
Tel.: ++1 914 9453076
Fax: ++1 914 9454421
divince@watson.ibm.com

Serge HAROCHE
Ecole Normale Supérieure
Rue Lhomond 24
75231 PARIS
France
Tel. ++33 1 44323420
Fax: ++33 1 45350076
serge.haroche@lkb.ens.fr

John MARTINIS
Electromagnetic Technology Division (814.00)
National Institute of Standards
and Technology
325 Broadway
BOULDER, CO 80305-3328
USA
Tel.: ++1 303 4973597
martinis@boulder.nist.gov

Gerhard REMPE
Max-Planck-Institut für Quantenoptik
Hans-Kopfermann-Str. 1
D-85748 GARCHING
Germany
Tel.: ++49 89 32905 711
gerhard.rempe@mpq.mpg.de

Yoshihisa YAMAMOTO
Edward L. Ginzton Laboratory
Stanford University
STANFORD, CA 94305
USA
Tel.: ++1 650 7239723
Fax: ++1 650 7235320
yamamoto@loki.stanford.edu

Hugo ZBINDEN
Group of Applied Physics
Université de Geneve
CH-1211 GENEVE 4
Switzerland
Tel.: ++41 22 7026883
Fax: ++41 22 7810980
hugo.zbinden@physics.unige.ch

Peter ZOLLER
Institute of Theoretical Physics
Universität Innsbruck
Technikerstraße 25
6020 INNSBRUCK
Austria
Tel.: ++43 512 5076203
Fax: ++43 512 5072919
peter.zoller@uibk.ac.at

David WINELAND
NIST Time and Frequency Division
325 Broadway
BOULDER, CO 80303
USA
Tel.: ++ 1 303 4975286
Fax: ++ 1 303 4977375
djw@central.bldrdoc.gov

Seminar Speakers

Ennio ARIMONDO
Dipartimento di Fisica dell'Università
Via Buonarroti 2
56126 PISA
Italy
Tel. ++39 050 844515
Fax: ++39 050 844333
arimondo@mail.df.unipi.it

Rainer BLATT
Institut für Experimentalphysik
Universität Innsbruck
Technikerstraße 25
6020 INNSBRUCK
Austria
Tel.: ++43 512 5076350
Fax: ++43 512 5072952
rainer.blatt@uibk.ac.at

Gabriella CASTELLANO
I.E.S.S. Laboratories
Via Tiburtina 770
00185 ROMA
Italy
Tel.: ++39 06 415221
mgcastellano@iess.rm.cnr.it

Ignacio CIRAC
Institute of Theoretical Physics
Universität Innsbruck
Technikerstraße 25
6020 INNSBRUCK
Austria
Tel.: ++43 512 5076230
Fax: ++43 512 5072919
ignacio.cirac@uibk.ac.at

G. Mauro D'ARIANO
Dipartimento di Fisica
Via Bassi 6
27100 PAVIA
Italy
Tel.: ++39 0382 507484
dariano@pv.infn.it

Michael FLEISCHAUER
Fachbereich Physik, Uni Kaiserslautern
Erwin-Schroedinger Str.
D-67663 KAISERSLAUTERN
Germany
Tel.: ++49 631 205 3206
mfleisch@physik.uni-kl.de

Massimo INGUSCIO
Dipertimento di Fisica e L.E.N.S
Università di Firenze
Largo E. Fermi 2
50125 FIRENZE
Italy
Tel.: ++39 055 2307610
Fax: ++39 055 224072
inguscio@lens.unifi.it

Chiara MACCHIAVELLO
Dipartimento di Fisica
Via Bassi 6
27100 PAVIA
Italy
Tel.: ++39 0382 507674
macchiavello@pv.infn.it

Pierpaolo MALINVERNI
European Commission
Directorate General Information Society
Office BU 33 3/9, Rue de la Loi 200
B-1049 BRUXELLES
Belgium
Tel.: ++32 2 2994282
Fax: ++32 2 2968390
pierpaolo.malinverni@cec.eu.int

Massimo PALMA
Dipartimento di Tecnologia
dell'Informazione
Università di Milano
Via Bramante 65
26013 CREMA
Italy
Tel.: ++39 02 50330052

Steven ROLSTON
NIST
100 Bureau Drive
Stop 8424
GAITHERSBURG, MD 20899-8424
USA
Tel.: ++1 301 975-6581
Fax: ++1 301 990-1350
steven.rolston@nist.gov

Wolfgang SCHLEICH
Abteilung für Quantenphysik
Universität Ulm
Albert-Einstein-Allee 11
D-89069 ULM
Germany
Tel.: ++49 731 5023080
Fax: ++49 731 5023086
wolfgang.schleich@physik.uni-ulm.de

Herbert WALTHER
Sektion Physik der Universitat Munchen
Am Coulombwall 1
D-85748 GARCHING
Germany
Tel.: ++ 49 89 289 14142
Fax: ++ 49 89 289 14141
herbert.walther@physik.uni-muenchen.de

Students

Axel ANDRE
Physics Department
Harvard University
CAMBRIDGE, MA 02138
USA
andre@physics.harvard.edu

José AUMENTADO
NIST MS 814.03
National Institute of Standards
and Technology
325 Broadway
BOULDER, CO 80305-3328
USA
Tel.: ++1 303 4974137
Fax: ++1 303 4973042
jose.aumentado@boulder.nist.gov

Martin AUSTWICK
Oxford University
Materials Science Department
Parks Road
OXFORD OX1 3PH
UK
Tel.: ++44 1865 273742
Fax: ++44 1865 273789
martin.austwick@new.ox.ac.uk

Andrea BERTOLDI
Università di Trento
Dipartimento di fisica
Via Sommarive 14
38050 TRENTO
Italy
Tel.: ++39 0461 881642
Fax: ++39 0461 881696
abertold@science.unitn.it

Boris BLINOV
University of Michigan
Department of Physics
2477 Randall Laboratory
ANN ARBOR, MI 48109-1120
USA
Tel.: ++1 734 2224628
Fax: ++1 734 7645153
bblinov@umich.edu

Antoine BROWAEYS
Laser Cooling and trapping Group
NIST, Mail stop 8424, Bureau Drive 100
GAITHERSBURG, MD 20877
USA
Tel.: ++1 301 9752353
Fax: ++1 301 9758252
antoine.browaeys@nist-gov

Andrea CAMPOSEO
Dipartimento di Fisica
Università di Pisa
Via F. Buonarroti, 2
56127 PISA
Italy
Tel.: ++39 050 844293
Fax: ++39 050 844333
camposeo@df.unipi.it

Gianluca CAPITANI
Via Don Minzoni 431
47023 CESENA
Italy

Angelo CAROLLO
Dipartimento di Scienze Fisiche
ed Astronomiche
Università di Palermo
Via Archirafi 36
90123 PALERMO
Italy
Tel.: ++39 091 6234237
Fax: ++39 091 6234282
angrollo@yahoo.com

Fabio CHIARELLO
Dipartimento di Fisica
Università di Roma "La Sapienza"
P.le Aldo Moro 2
00185 ROMA
Italy
Tel.: ++39 06 49914279
Fax: ++39 06 4957697
fabio.chiarello@roma1.infn.it

Richard CONROY
CUA, Physics Department
Harvard University
CAMBRIDGE
MA 02138
USA
Tel.: ++1 617 4954483
Fax: ++1 617 4950416
RConroy@fas.harvard.edu

Valentina CORATO
Dipartimento di Fisica
Università "La Sapienza"
P.le Aldo Moro 2
00185 ROMA
Italy
vale.corato@libero.it

Ruiner DUMKE
Institute of Quantum Optics
University of Hannover
Welfengurtenl
30167 HANNOVER
Germany
Tel.: ++49 511 7625769
Fax: ++49 511 7622211
dumke@190.uni-hannover.de

Lara FAORO
ISI
Institute for Scientific Interchange
Viale Settimio Severo 65
10133 TORINO
Italy
Tel.: ++39 011 6603090
Fax: ++39 011 6600049
faoro@isiosf.isi.it

Matthew FENSELAU
Sloane Physics Laboratory
Yale University
NEW HAVEN, CT06570
USA
Tel.: ++1 203 4364997
Fax: ++1 203 4329710
Matthew.Fenselau@Yale.edu

Gianni DI DOMENICO
Observatoire Cantonal
58, Rue de l'Observatoire
CH-2000 NEUCHATEL 1
Switzerland
Tel.: ++41 32 8896870
Fax: ++41 32 8896281
Gianni.DiDomenic@ne.ch

Marco GRAMEGNA
Istituto Elettrotecnico Nazionale
"Galileo Ferraris"
Strada delle Cacce 91
10135 TORINO
Italy
Tel.: ++39 011 3919234
Fax: ++39 011 3919259
gramegna@tf.ien.it

David HARBER
University of Colorado
BOULDER, CO 80309-0440
USA
Tel.: ++1 303 4928442
Fax: ++1 303 4925235
david.harber@colorado.edu

Hayk HAROUTYUNYAN
Leiden University
Huygens Laboratory
Niels Bohrweg 2
2333 CA LEIDEN
The Netherlands
Tel.: ++31 71 5275928
Fax: ++31 71 5275819
hayk@molphys.leidenuniv.nl

S. Abbas HOSSEINI
Tata Institute of Fundamental Research
B.124, Atomic & Molecular group
Homi Bhabha Road, Colaba
MUMBAI 400005
India
Tel.: ++91 22 2152971 Ext.2687
Fax: ++91 22 2152110
abbas@tifr.res.in

Hermann KAMPERMANN
Nilchplatz 5
47495 RHEINBERG
Germany
hkampermann@gmx.de

David KIELPINSKI
National Institute of Standards
and Technology
MS 847, 325 Broadway
BOULDER, CO 80301
USA
Tel.: ++1 303 4977470
Fax: ++1 303 4977375
davidk@boulder.nist.gov

Brian E. KING
NIST, 100 Bureau Drive, Mail stop 8424
GAITHERSBURG, MD 20899-8424
USA
Tel.: ++1 301 9758571
Fax: ++1 301 9758272
brian.king@nist.gov

Patricia LEE
University of Michigan
Department of Physics
2477 Randall Laboratory
ANN ARBOR, MI 48109-1120
USA
Tel.: ++1 734 6475465
Fax: ++1 734 7645153
juitl@umich.edu

Michael LIM
NIST Mail stop 8424
100 Bureau Drive
GAITHERSBURG, MD 20899-8424
USA
Tel.: ++1 301 9758569
Fax: ++1 301 9758272
mlim@nist.gov

Brendon LOVETT
Department of Materials
Oxford University - Parks Road
OXFORD, OXI 3PH
UK
Tel.: ++44 077 511 66201
Fax: ++44 01865 272400
brendon.lovett@materials.ox.ac.uk

Filippo MAIMONE
Dipartimento di Fisica
Università di Salerno
Via S. Allende
84081 BARONISSI SA
Italy
filippo278@interfree.it

Jason McKEEVER
Caltech (Quantum Optics Group)
Caltech 12-33
PASADENA, CA 91125
USA
Tel.: ++1 626 3958343
Fax: ++1 626 7939506
jasonmck@caltech.edu

Claudia K. A. MEWES
Department of Physics
University of Kaiserslautern
Erwin-Schroedinger-Street
67663 KAISERSLAUTEN
Germany
Tel.: ++49 631 2052271
Fax: ++49 631 2053907
cmewes@rhrk.uni-kl.de

Rosanna MIGLIORE
Dipartimento di Scienze Fisiche
ed Astronomiche
Università di Palermo
Via Archirafi 36
90123 PALERMO
Italy
Tel.: ++39 091 6234243
Fax: ++39 091 6234243
migliorer@yahoo.com

Anders MORTENSEN
Institute of Physics and Astronomy
University of Aarhus
Bygn. 520, Ny Munkegade
DK 8000 AARHUS C,
Denmark
Tel.: ++45 89 424826
Fax: ++45 86 120740
andersvm@ifa.au.dk

Ahsan NAZIR
Department of Materials
Oxford University
Parks Road,
OXFORD, OXI 3PH
UK
Tel.: ++44 01865 273742
Fax: ++44 01865 273789
ahsan.nazir@materials.ox.ac.uk

Sumant S. R. OEMRAWSINGH
Leiden University
Huygens Laboratory
Niels Bohrweg 2
2333 CA LEIDEN
The Netherlands
Tel.: ++31 71 5275929
Fax: ++31 71 5275819
soemraws@molphys.leidenuniv.nl

Stefano OSNAGHI
Laboratoire Kastler Brossel E.N.S.
24, Rue Lhomond
75231 PARIS Cedex 05
France
Tel.: ++33 1 44323299
Fax: ++33 1 44323434
stefano.osnaghi@lkb.ens.fr

Elisabetta PALADINO
Università di Catania
Dipartimento di Metodologie Fisiche
e Chimiche per l'Ingegneria
Viale A. Doria 6, ed. 10
95128 CATANIA
Italy
Tel.: ++39 095 7382813/7382806
Fax: ++39 095 333231
elisa@meso.if.ing.unict.it

Nikolay PANEV
Solid State Physics
Lund University
Box118
SE-22100 LUND
Sweden
Tel.: ++46 46 2220929
Fax: ++46 46 2223637
Nikolay.Panev@ftf.lth.se

Francesco PLASTINA
Dipartimento di Scienze Fisiche
ed Astronomiche
Università di Palermo
Via Archirafi 36
90123 PALERMO
Italy
Tel.: ++39 091 6234248
Fax: ++39 091 6234281
plastina@fis.unical.it

Jacob SAUER
Georgia Institute of Technology
837 State Street,
ATLANTA, GA 30332
USA
Tel.: ++1 404 3850294
Fax: ++1 404 3850044
jakesauer@mindspring.com

Sergio SEVERINI
INFM - Dipartimento di Energetica
Via Scarpa 16
00161 ROMA
Italy
Tel.: ++39 339 7461499
Fax: ++39 06 44240183
s_severini@hotmail.com

Stefan RINNER
Institut für Theoretische Physik/Lehrstuhl
Universitatstr. 31
D-93053 REGENSBURG
Germany
Tel.: ++49 941 9466768
Fax: ++49 941 9431734
stefan.rinner@physik.uni-regensburg.de

Giacomo ROATI
LENS
Largo Fermi 2
50125 FIRENZE
Italy
Tel.: ++39 055 2307809
Fax: ++39 055 224072
roati@lens.unifi.it

Riccardo SAPIENZA
Leiden University
Huygens Laboratory
Niels Bohrweg 2,
2333 CA LEIDEN
The Netherlands
Tel.: ++39 049 680873
Fax: ++39 049 680873
r.icarus@tin.it

Fabio SCIARRINO
INFM - Dipartimento di Energetica
Via Scarpa 16
00161 ROMA
Italy
fabio.sciarrino@uniroma1.it

Thomas STACE
University of Cambridge
Madingley Road
CAMBRIDGE CB3 OHE
UK
Tel.: ++44 1223 337478
Fax: ++44 1223 337271
tms29@cam.ac.uk

Ari TUCHMAN
Yale University
217 Prospect St.
YALE, USA
Tel.: ++1 203 4364997
Fax: ++1 203 4329710
ari.tuchman@yale.edu

Chikako UCHIYAMA
Yamanashi University
4-3-11, Takeda, Kofu
YAMANASHI, 400
Japan
Tel.: ++81 55 2208578
Fax: ++81 55 2208578
uchiyama@es.yamanashi.ac.jp

Dries VAN OOSTEN
University of Utrecht
Postbus 80000
3508 TA UTRECHT
The Netherlands
Tel.: ++31 302 532372
Fax: ++31 302 537468
dvoosten@phys.uu.nl

Timur ZAINIEV
University of Cambridge
TCM group, Cavendish Lab.
CAMBRIDGE CB3 OHE
UK
Tel.: ++44 1223 337460
Fax: ++44 1223 337356
tz209@phy.cam.ac.uk

Alessandro ZAVATTA
Istituto Nazionale di Ottica Applicata
Largo E. Fermi 6
50125 FIRENZE
Italy
Tel.: ++39 055 23081
Fax: ++39 055 2337755
azavatta@ino.it

Observers

Maurizio ARTONI
LENS
Largo E. Fermi 2
50125 FIRENZE
Italy
Tel.: ++39 055 2307829
Fax: ++39 055 224072
artoni@lens.unifi.it

Francesco Saverio CATALIOTTI
LENS
Largo E. Fermi 2
50125 FIRENZE
Italy
Tel.: ++39 055 2307821
Fax: ++39 055 224072
fsc@lens.unifi.it

Matteo CHERCHI
Pirelli Cavi e Sistemi S.p.A.
Viale Sarca 222, c.2119
Fabbricato 11 ingr. E
20126 MILANO
Italy
Tel.: ++39 02 64427789
Fax: ++39 02 64426699
matteo.cherchi@pirelli.com

Gabriele FERRARI
LENS
Largo E. Fermi 2
50125 FIRENZE
Italy
Tel.: ++39 055 2307809
Fax: ++39 055 224072
ferrari@lens.unifi.it

Karl-Lugwig KOMPA
Max-Plank Institut fuer Quantenoptik
Hans-Kopfermann-Str.1
D-85748 GARCHING
Germany
Tel.: ++49 89 32905-703
Fax: ++49 89 32905-313
Karl.Kompa@mpg.mpg.de

Reuben SHUKER
Department of Physics
Ben Gurion University
P.O. Box 653
BEER SHEVA 84105
Israel
Tel.: ++972 8 6461169
shuker@bgumail.bgu.ac.il

Giuseppe MAINO
ENEA Physics Division
Via Fiammelli 2
40129 BOLOGNA
Italy
Tel.: ++39 051 6098206
Fax: ++39 051 6098359
maino@bologna.enea.it

Saewoo NAM
NIST MS 814.03
National Institute of Standards
and Technology
325 Broadway
BOULDER, CO 80305-3328
USA
Tel.: ++1 303 4973148
Fax: ++1 303 4973042
nams@boulder.nist.gov

Andrei SHELANKOV
Theoretical Physics Department
Umea University
90187 UMEA
Sweden
Tel.: ++46 90 7869986
Fax: ++46 90 789556
shelankov@tp.umu.se

PROCEEDINGS OF THE INTERNATIONAL SCHOOL OF PHYSICS «ENRICO FERMI»

Course I (1953)
Questioni relative alla rivelazione delle particelle elementari, con particolare riguardo alla radiazione cosmica
edited by G. PUPPI

Course II (1954)
Questioni relative alla rivelazione delle particelle elementari, e alle loro interazioni con particolare riguardo alle particelle artificialmente prodotte ed accelerate
edited by G. PUPPI

Course III (1955)
Questioni di struttura nucleare e dei processi nucleari alle basse energie
edited by C. SALVETTI

Course IV (1956)
Proprietà magnetiche della materia
edited by L. GIULOTTO

Course V (1957)
Fisica dello stato solido
edited by F. FUMI

Course VI (1958)
Fisica del plasma e relative applicazioni astrofisiche
edited by G. RIGHINI

Course VII (1958)
Teoria della informazione
edited by E. R. CAIANIELLO

Course VIII (1958)
Problemi matematici della teoria quantistica delle particelle e dei campi
edited by A. BORSELLINO

Course IX (1958)
Fisica dei pioni
edited by B. TOUSCHEK

Course X (1959)
Thermodynamics of Irreversible Processes
edited by S. R. DE GROOT

Course XI (1959)
Weak Interactions
edited by L. A. RADICATI

Course XII (1959)
Solar Radioastronomy
edited by G. RIGHINI

Course XIII (1959)
Physics of Plasma: Experiments and Techniques
edited by H. ALFVÉN

Course XIV (1960)
Ergodic Theories
edited by P. CALDIROLA

Course XV (1960)
Nuclear Spectroscopy
edited by G. RACAH

Course XVI (1960)
Physicomathematical Aspects of Biology
edited by N. RASHEVSKY

Course XVII (1960)
Topics of Radiofrequency Spectroscopy
edited by A. GOZZINI

Course XVIII (1960)
Physics of Solids (Radiation Damage in Solids)
edited by D. S. BILLINGTON

Course XIX (1961)
Cosmic Rays, Solar Particles and Space Research
edited by B. PETERS

Course XX (1961)
Evidence for Gravitational Theories
edited by C. MØLLER

Course XXI (1961)
Liquid Helium
edited by G. CARERI

Course XXII (1961)
Semiconductors
edited by R. A. SMITH

Course XXIII (1961)
Nuclear Physics
edited by V. F. WEISSKOPF

Course XXIV (1962)
Space Exploration and the Solar System
edited by B. ROSSI

Course XXV (1962)
Advanced Plasma Theory
edited by M. N. ROSENBLUTH

Course XXVI (1962)
Selected Topics on Elementary Particle Physics
edited by M. CONVERSI

Course XXVII (1962)
Dispersion and Absorption of Sound by Molecular Processes
edited by D. SETTE

Course XXVIII (1962)
Star Evolution
edited by L. GRATTON

Course XXIX (1963)
Dispersion Relations and their Connection with Causality
edited by E. P. WIGNER

Course XXX (1963)
Radiation Dosimetry
edited by F. W. SPIERS and G. W. REED

Course XXXI (1963)
Quantum Electronics and Coherent Light
edited by C. H. TOWNES and P. A. MILES

Course XXXII (1964)
Weak Interactions and High-Energy Neutrino Physics
edited by T. D. LEE

Course XXXIII (1964)
Strong Interactions
edited by L. W. ALVAREZ

Course XXXIV (1965)
The Optical Properties of Solids
edited by J. TAUC

Course XXXV (1965)
High-Energy Astrophysics
edited by L. GRATTON

Course XXXVI (1965)
Many-Body Description of Nuclear Structure and Reactions
edited by C. BLOCH

Course XXXVII (1966)
Theory of Magnetism in Transition Metals
edited by W. MARSHALL

Course XXXVIII (1966)
Interaction of High-Energy Particles with Nuclei
edited by T. E. O. ERICSON

Course XXXIX (1966)
Plasma Astrophysics
edited by P. A. STURROCK

Course XL (1967)
Nuclear Structure and Nuclear Reactions
edited by M. JEAN and R. A. RICCI

Course XLI (1967)
Selected Topics in Particle Physics
edited by J. STEINBERGER

Course XLII (1967)
Quantum Optics
edited by R. J. GLAUBER

Course XLIII (1968)
Processing of Optical Data by Organisms and by Machines
edited by W. REICHARDT

Course XLIV (1968)
Molecular Beams and Reaction Kinetics
edited by CH. SCHLIER

Course XLV (1968)
Local Quantum Theory
edited by R. JOST

Course XLVI (1969)
Physics with Intersecting Storage Rings
edited by B. TOUSCHEK

Course XLVII (1969)
General Relativity and Cosmology
edited by R. K. SACHS

Course XLVIII (1969)
Physics of High Energy Density
edited by P. CALDIROLA and H. KNOEPFEL

Course IL (1970)
Foundations of Quantum Mechanics
edited by B. D'ESPAGNAT

Course L (1970)
Mantle and Core in Planetary Physics
edited by J. COULOMB and M. CAPUTO

Course LI (1970)
Critical Phenomena
edited by M. S. GREEN

Course LII (1971)
Atomic Structure and Properties of Solids
edited by E. BURSTEIN

Course LIII (1971)
Developments and Borderlines of Nuclear Physics
edited by H. MORINAGA

Course LIV (1971)
Developments in High-Energy Physics
edited by R. R. GATTO

Course LV (1972)
Lattice Dynamics and Intermolecular Forces
edited by S. CALIFANO

Course LVI (1972)
Experimental Gravitation
edited by B. BERTOTTI

Course LVII (1972)
History of 20th Century Physics
edited by C. WEINER

Course LVIII (1973)
Dynamics Aspects of Surface Physics
edited by F. O. GOODMAN

Course LIX (1973)
Local Properties at Phase Transitions
edited by K. A. MÜLLER and A. RIGAMONTI

Course LX (1973)
C-Algebras and their Applications to Statistical Mechanics and Quantum Field Theory*
edited by D. KASTLER

Course LXI (1974)
Atomic Structure and Mechanical Properties of Metals
edited by G. CAGLIOTI

Course LXII (1974)
Nuclear Spectroscopy and Nuclear Reactions with Heavy Ions
edited by H. FARAGGI and R. A. RICCI

Course LXIII (1974)
New Directions in Physical Acoustics
edited by D. SETTE

Course LXIV (1975)
Nonlinear Spectroscopy
edited by N. BLOEMBERGEN

Course LXV (1975)
Physics and Astrophysics of Neutron Stars and Black Holes
edited by R. GIACCONI and R. RUFFINI

Course LXVI (1975)
Health and Medical Physics
edited by J. BAARLI

Course LXVII (1976)
Isolated Gravitating Systems in General Relativity
edited by J. EHLERS

Course LXVIII (1976)
Metrology and Fundamental Constants
edited by A. FERRO MILONE, P. GIACOMO and S. LESCHIUTTA

Course LXIX (1976)
Elementary Modes of Excitation in Nuclei
edited by A. BOHR and R. A. BROGLIA

Course LXX (1977)
Physics of Magnetic Garnets
edited by A. PAOLETTI

Course LXXI (1977)
Weak Interactions
edited by M. BALDO CEOLIN

Course LXXII (1977)
Problems in the Foundations of Physics
edited by G. TORALDO DI FRANCIA

Course LXXIII (1978)
Early Solar System Processes and the Present Solar System
edited by D. LAL

Course LXXIV (1978)
Development of High-Power Lasers and their Applications
edited by C. PELLEGRINI

Course LXXV (1978)
Intermolecular Spectroscopy and Dynamical Properties of Dense Systems
edited by J. VAN KRANENDONK

Course LXXVI (1979)
Medical Physics
edited by J. R. GREENING

Course LXXVII (1979)
Nuclear Structure and Heavy-Ion Collisions
edited by R. A. BROGLIA, R. A. RICCI and C. H. DASSO

Course LXXVIII (1979)
Physics of the Earth's Interior
edited by A. M. DZIEWONSKI and E. BOSCHI

Course LXXIX (1980)
From Nuclei to Particles
edited by A. MOLINARI

Course LXXX (1980)
Topics in Ocean Physics
edited by A. R. OSBORNE and P. MALANOTTE RIZZOLI

Course LXXXI (1980)
Theory of Fundamental Interactions
edited by G. COSTA and R. R. GATTO

Course LXXXII (1981)
Mechanical and Thermal Behaviour of Metallic Materials
edited by G. CAGLIOTI and A. FERRO MILONE

Course LXXXIII (1981)
Positrons in Solids
edited by W. BRANDT and A. DUPASQUIER

Course LXXXIV (1981)
Data Acquisition in High-Energy Physics
edited by G. BOLOGNA and M. VINCELLI

Course LXXXV (1982)
Earhquakes: Observation, Theory and Interpretation
edited by H. KANAMORI and E. BOSCHI

Course LXXXVI (1982)
Gamow Cosmology
edited by F. MELCHIORRI and R. RUFFINI

Course LXXXVII (1982)
Nuclear Structure and Heavy-Ion Dynamics
edited by L. MORETTO and R. A. RICCI

Course LXXXVIII (1983)
Turbulence and Predictability in Geophysical Fluid Dynamics and Climate Dynamics
edited by M. GHIL, R. BENZI and G. PARISI

Course LXXXIX (1983)
Highlights of Condensed-Matter Theory
edited by F. BASSANI, F. FUMI and M. P. TOSI

Course XC (1983)
Physics of Amphiphiles: Micelles, Vesicles and Microemulsions
edited by V. DEGIORGIO and M. CORTI

Course XCI (1984)
From Nuclei to Stars
edited by A. MOLINARI and R. A. RICCI

Course XCII (1984)
Elementary Particles
edited by N. CABIBBO

Course XCIII (1984)
Frontiers in Physical Acoustics
edited by D. SETTE

Course XCIV (1984)
Theory of Reliability
edited by A. SERRA and R. E. BARLOW

Course XCV (1985)
Solar-Terrestrial Relationships and the Earth Environment in the Last Millennia
edited by G. CINI CASTAGNOLI

Course XCVI (1985)
Excited-State Spectroscopy in Solids
edited by U. M. GRASSANO and N. TERZI

Course XCVII (1985)
Molecular-Dynamics Simulations of Statistical-Mechanical Systems
edited by G. CICCOTTI and W. G. HOOVER

Course XCVIII (1985)
The Evolution of Small Bodies in the Solar System
edited by M. FULCHIGNONI and Ľ. KRESÀK

Course XCIX (1986)
Synergetics and Dynamic Instabilities
edited by G. CAGLIOTI and H. HAKEN

Course C (1986)
The Physics of NMR Spectroscopy in Biology and Medicine
edited by B. MARAVIGLIA

Course CI (1986)
Evolution of Interstellar Dust and Related Topics
edited by A. BONETTI and J. M. GREENBERG

Course CII (1986)
Accelerated Life Testing and Experts Opinions in Reliability
edited by C. A. CLAROTTI

Course CIII (1987)
Trends in Nuclear Physics
edited by P. KIENLE, R. A. RICCI and A. RUBBINO

Course CIV (1987)
Frontiers and Borderlines in Many-Particle Physics
edited by R. A. BROGLIA and J. R. SCHRIEFFER

Course CV (1987)
Confrontation between Theories and Observations in Cosmology: Present Status and Future Programmes
edited by J. AUDOUZE and F. MELCHIORRI

Course CVI (1988)
Current Trends in the Physics of Materials
edited by G. F. CHIAROTTI, F. FUMI and M. TOSI

Course CVII (1988)
The Chemical Physics of Atomic and Molecular Clusters
edited by G. SCOLES

Course CVIII (1988)
Photoemission and Absorption Spectroscopy of Solids and Interfaces with Synchrotron Radiation
edited by M. CAMPAGNA and R. ROSEI

Course CIX (1988)
Nonlinear Topics in Ocean Physics
edited by A. R. OSBORNE

Course CX (1989)
Metrology at the Frontiers of Physics and Technology
edited by L. CROVINI and T. J. QUINN

Course CXI (1989)
Solid-State Astrophysics
edited by E. BUSSOLETTI and G. STRAZZULLA

Course CXII (1989)
Nuclear Collisions from the Mean-Field into the Fragmentation Regime
edited by C. Detraz and P. Kienle

Course CXIII (1989)
High-Pressure Equation of State: Theory and Applications
edited by S. Eliezer and R. A. Ricci

Course CXIV (1990)
Industrial and Technological Applications of Neutrons
edited by M. Fontana and F. Rustichelli

Course CXV (1990)
The Use of EOS for Studies of Atmospheric Physics
edited by J. C. Gille and G. Visconti

Course CXVI (1990)
Status and Perspectives of Nuclear Energy: Fission and Fusion
edited by R. A. Ricci, C. Salvetti and E. Sindoni

Course CXVII (1991)
Semiconductor Superlattices and Interfaces
edited by A. Stella

Course CXVIII (1991)
Laser Manipolation of Atoms and Ions
edited by E. Arimondo, W. D. Phillips and F. Strumia

Course CXIX (1991)
Quantum Chaos
edited by G. Casati, I. Guarneri and U. Smilansky

Course CXX (1992)
Frontiers in Laser Spectroscopy
edited by T. W. Hänsch and M. Inguscio

Course CXXI (1992)
Perspectives in Many-Particle Physics
edited by R. A. Broglia, J. R. Schrieffer and P. F. Bortignon

Course CXXII (1992)
Galaxy Formation
edited by J. Silk and N. Vittorio

Course CXXIII (1992)
Nuclear Magnetic Double Resonance
edited by B. Maraviglia

Course CXXIV (1993)
Diagnostic Tools in Atmospheric Physics
edited by G. Fiocco and G. Visconti

Course CXXV (1993)
Positron Spectroscopy of Solids
edited by A. Dupasquier and A. P. Mills jr.

Course CXXVI (1993)
Nonlinear Optical Materials: Principles and Applications
edited by V. Degiorgio and C. Flytzanis

Course CXXVII (1994)
Quantum Groups and their Applications in Physics
edited by L. Castellani and J. Wess

Course CXXVIII (1994)
Biomedical Applications of Synchrotron Radiation
edited by E. Burattini and A. Balerna

Course CXXIX (*) (1994)
Observation, Prediction and Simulation of Phase Transitions in Complex Fluids
edited by M. Baus, L. F. Rull and J.-P. Ryckaert

Course CXXX (1995)
Selected Topics in Nonperturbative QCD
edited by A. Di Giacomo and D. Diakonov

Course CXXXI (1995)
Coherent and Collective Interactions of Particles and Radiation Beams
edited by A. Aspect, W. Barletta and R. Bonifacio

Course CXXXII (1995)
Dark matter in the Universe
edited by S. Bonometto and J. Primack

Course CXXXIII (1996)
Past and Present Variability of the Solar-Terrestrial System: Measurement, Data Analysis and Theoretical Models
edited by G. Cini Castagnoli and A. Provenzale

Course CXXXIV (1996)
The Physics of Complex Systems
edited by F. Mallamace and H. E. Stanley

Course CXXXV (1996)
The Physics of Diamond
edited by A. Paoletti and A. Tucciarone

Course CXXXVI (1997)
Models and Phenomenology for Conventional and High-Temperature Superconductivity
edited by G. Iadonisi, J. R. Schrieffer and M. L. Chiofalo

(*) This course belongs to the *NATO ASI Series C*, Vol. 460 (Kluwer Academic Publishers).

Course CXXXVII (1997)
Heavy Flavour Physics: a Probe of Nature's Grand Design
edited by I. BIGI and L. MORONI

Course CXXXVIII (1997)
Unfolding the Matter of Nuclei
edited by A. MOLINARI and R. A. RICCI

Course CXXXIX (1998)
Magnetic Resonance and Brain Function: Approaches from Physics
edited by B. MARAVIGLIA

Course CXL (1998)
Bose-Einstein Condensation in Atomic Gases
edited by M. INGUSCIO, S. STRINGARI and C. E. WIEMAN

Course CXLI (1998)
Silicon-Based Microphotonics: From Basics to Applications
edited by O. BISI, S. U. CAMPISANO, L. PAVESI and F. PRIOLO

Course CXLII (1999)
Plasma Astrophysics
edited by B. COPPI, A. FERRARI and E. SINDONI

Course CXLIII (1999)
New Directions in Quantum Chaos
edited by G. CASATI, I. GUARNERI and U. SMILANSKY

Course CXLIV (2000)
Nanometer Scale Science and Technology
edited by M. ALLEGRINI, N. GARCÍA and O. MARTI

Course CXLV (2000)
Protein Folding, Evolution and Design
edited by R. A. BROGLIA, E. I. SHAKHNOVICH and G. TIANA

Course CXLVI (2000)
Recent Advances in Metrology and Fundamental Constants
edited by T. J. QUINN, S. LESCHIUTTA and P. TAVELLA

Course CXLVII (2001)
High Pressure Phenomena
edited by R. J. HEMLEY, G. L. CHIAROTTI, M. BERNASCONI and L. ULIVI